# Brewing Yeast and Fermentation

## Chris Boulton
### *and*
## David Quain

**Blackwell**
Science

Editorial offices:
Blackwell Science Ltd, 9600 Garsington Road, Oxford OX4 2DQ, UK
Tel: +44 (0) 1865 776868
Blackwell Publishing Professional, 2121 State Avenue, Ames, Iowa 50014-8300, USA
Tel: +1 515 292 0140
Blackwell Science Asia Pty Ltd, 550 Swanston Street, Carlton, Victoria 3053, Australia
Tel: +61 (0)3 8359 1011

Hardback edition published 2001 by Blackwell Science Ltd
Reissued in paperback 2006
2   2008

ISBN: 978-1-4051-5268-6

Library of Congress Cataloging-in-Publication Data (from the hardback edition)
Boulton, Chris.
Brewing yeast and fermentation / Chris Boulton and David Quain.
p. cm.
Includes bibliographical references and index.
ISBN 1-4051-5268-0
1. Yeast. 2. Brewing. I. Quain, David. II. Title.
TP580 .B68 2001
663'.42—dc21
2001043598

A catalogue record for this title is available from the British Library

Set in 10/12 pt Times by DP Photosetting, Aylesbury, Bucks

For further information on Blackwell Publishing, visit our website:
www.blackwellpublishing.com

# Contents

# Preface

As is perhaps suggested by the title, this book was written with the intention of filling two perceived gaps in the current literature. In the first instance the authors hope that they have provided a comprehensive description of brewery fermentation, both traditional and modern. Secondly, to underpin this, a detailed review is presented of current understanding of those aspects of the biochemistry and genetics of yeast, which impact on brewery fermentation.

Of course, both of these related topics have received considerable attention elsewhere. For example, the multi-volume texts, *Brewing Science* (Pollock, 1979) and *Malting and Brewing Science* (Hough *et al.*, 1981; Briggs *et al.*, 1981), provide detailed descriptions of the brewing process in its entirety. Both of these are effectively standard works of reference and are highly regarded. However, both now have a vintage of some twenty years. During the period since their publication much has happened to the brewing industry as it has perforce adapted to changes in the global market place. Economic and social forces coupled with developments in science and technology have resulted in profound changes in the practice of fermentation and the range of beers produced. Hopefully these developments, together with a full account of the still much used traditional processes, are captured in this book.

The scientific literature devoted to yeast belonging to the genus *Saccharomyces* is extensive and diverse. This is fitting, since strains of this organism are of paramount importance in the biotechnological processes of brewing, wine making and baking. In addition, the type species, *Saccharomyces cerevisiae*, has been much employed worldwide in research laboratories as *the* model eukaryotic cell. Indeed, the publication of the sequence of the *S. cerevisiae* genome in April 1996 was a milestone in late twentieth-century biology. The plethora of primary literature concerning the *Saccharomyces* and other yeast genera, is reviewed in depth elsewhere. Most notably in the consistently excellent multi-authored series, *The Yeasts* (Rose & Harrison, 1987 *et seq.*; Rose *et al.*, 1995). However, it is perhaps true that the mainstream of yeast research has dwelt little on the elucidation of the biochemical activities of yeast in the complex systems associated with growth on poorly characterised media such as brewers' wort, under conditions of transient aerobiosis. Indeed in some instances it has perhaps been assumed that yeast metabolism under the conditions of a brewery fermentation are more similar to those which pertain to aerobic growth on a defined medium than is, in fact, the case. These issues, together with a review of the misunderstood but steadfastly important subject of brewing microbiology, are addressed in this present work.

The foregoing discussion raises the question as to who is the target audience for this volume? The answer is everyone who has an interest, either as a seasoned professional or student, of the noble art of brewing. In addition, we hope that the wider academic community of zymologists will also find items of interest. Much of the content of the book is devoted to fermentation as practised by large scale commercial brewers, however, we have endeavoured to include material of interest to those who operate the time-honoured traditional processes.

The authors are associated with the UK brewing industry. We have provided an international perspective by highlighting aspects of the fermentation of beers which are peculiar to individual countries. Of course, the promulgation of developments in brewing science, in common with any other academic discipline, has no regard for national boundaries and we humbly dedicate this book to the spirit of that tradition.

# Acknowledgements

Chris Boulton would like to thank many friends within the brewing industry and the academic community who have contributed both directly and indirectly to this project. In particular, I acknowledge the help of colleagues, past and present, with whom I have had the privilege of working with at Bass Brewers. Without their collective wisdom, this book would not have been possible. To name individuals is always invidious, however, I acknowledge Dr Peter Large and Professor Colin Ratledge of Hull University for teaching me to appreciate the delights of biochemistry and yeast physiology. Thanks to Professor Charlie Bamforth of Davis University, California who was responsible for diverting my attention from obese yeasts to the alcoholic varieties. I thank Dr Vincent Clutterbuck, Dr Sean Durnin, Dr Stuart Molzahn, David Lummis, Preston Besford and particularly Wendy Box for numerous fruitful discussions and collaborations. Special thanks to my co-author, David Quain, for perseverance and inspiration. Finally, I am indebted to my wife, Wendy, for patience and understanding above and beyond the call of duty; without her it would not have happened.

David Quain would like to thank the many people who in various ways provided support in what turned out to be a mammoth adventure. Firstly, Sue and our children – Ben, Rosie and Sophie – who endured five years of evenings, weekends and holidays restricted by the looming presence of 'the book'. This is for you all! I also thank my co-author and friend – Chris Boulton – for his patience and support that (for the most part!) survived what life threw at us. Numerous people at Bass Brewers contributed both directly and indirectly. Although too many to mention all by name it would be inappropriate not to acknowledge Wendy Box, Alisdair Hamilton, Ian Whysall, Steph Valente and, generally, the various incarnations of the 'Research Team'. Finally, thanks to Poppy the Labrador who, via long weekend walks, provided welcome time to think!

I would also like to formally acknowledge those who made a difference to my career in a variety of ways. These include my Mother and lifelong friends, Dave Fenton and Alistair Stewart. Looking back over the last 30 years, a number of people 'made a difference' and provided direction and support. These include Dr John Hudson, Professor Anna MacLeod, Ted Hickman, Dr Pat Thurston, the late (and sadly missed) Dr Andrew Goodey, Professor Graham Stewart, Dr Philip Meaden, Dr Jim Haslam, Dr Tony Portno, Dr Fintan Walton, Roger Putman, Dr Harry White, Professor Charlie Bamforth, and especially Dr Stuart Molzahn.

We thank the Directors of Bass Brewers for permission to publish this work.

# 1 Beer and brewing

## 1.1 Introduction

In order to delineate the boundaries of brewery fermentation it is necessary to define the product of the process, beer. In particular, how is beer distinguished from other beverages, which are produced by fermentation. The etymology of the word gives little clue other than inferring a certain universality, it being derived from the Latin *bibere*, to drink. The *Shorter Oxford Dictionary* describes beer as 'alcoholic liquor obtained by the fermentation of malt (or other saccharin substance) flavoured with hops or other bitters'. For the purposes of this book this definition is too proscriptive. Beer is now used as a generic term for a multitude of beverages of widely differing appearance and flavour and produced from a variety of raw materials. Whilst it is true that in most cases malted barley provides the major source of fermentable sugars, many other sources of extract may be used. Beers may or may not contain hops or other bittering substances. A whole range of other flavourings may be added including fruits, spices and various plant extracts. In some cases beers may not even contain alcohol.

Another way of defining beer is to consult the relevant legislation. In fact. this is also relatively unhelpful in that it has usually been framed with the specific intent of establishing standards for alcohol content, setting up a framework for collecting taxes and as a means of controlling the activities of manufacturers and consumers. Frequently the product itself is relatively undefined. Thus, in the United Kingdom, the Alcoholic Liquor Duties Act (Her Majesty's Stationery Office, 1979) describes beer as:

> 'including ale, porter, stout and any other description of beer, and any liquor which is made or sold as a description of beer or as a substitute for beer and which on analysis of a sample thereof at any time is found to be of a strength exceeding 2° of proof [1.2% ethanol v/v].'

The same Act has specific categories for other alcoholic beverages such as wine, cider/perry and spirits. These are more obviously distinctive in that manufacture involves distillation, or fermentation of a medium largely obtained from a single specific raw material not associated with brewing such as grape or apple juice. However, the Act does include another catch-all category, 'Made Wine', which is defined as

> 'any liquor obtained from the alcoholic fermentation of any substance, or by mixing a liquor so obtained or derived from a liquor so obtained with any other liquor or substance but does not include wine, beer, black beer, spirits or cider'.

Even allowing for the somewhat impenetrable prose it is apparent that legal distinctions are made in the United Kingdom between beer and other beverages, which may at the very least, be open to interpretation elsewhere.

1

Most other countries have a legal definition of beer, which is similar to and equally as vague as that of the United Kingdom. A notable exception is Germany where a very distinct view is taken as to what is and what is not a beer. Germany has very stringent laws, which define precisely what may be labelled as beer and the nature of the materials that may be used during its manufacture. The *Reinheitsgebot* or beer purity laws are based on legislation introduced into Bavaria, in 1493, by Duke Albrecht IV. The original law stipulated that beer could be made only from malted barley, hops and water; yeast was not included since at that time it was undiscovered (see Section 1.2). Subsequent manifestations of the laws included this vital ingredient when it was observed that more consistent fermentations were obtained when inoculated with the crop harvested from the previous fermentation. The purity laws are still in operation and prohibit the use, in Germany, of many ingredients and process aids which are legally acceptable elsewhere (Narziss, 1984).

For the purposes of this book and in order to maintain a proper international perspective, beer is simply defined as a beverage produced by fermentation of an aqueous medium which contains sugars derived mainly from cereals. The fermentation step is catalysed principally by yeast. The latter may be a pure monoculture, or a mixture of yeast strains, with or without the involvement of bacteria.

Having defined the product, it is necessary to consider the scope of the fermentation process used in its production. The range of fermentation practices that may be encountered differ mainly in terms of scale and sophistication. At its simplest the process may be little more than a domestic operation producing beer in volumes sufficient for family consumption. Such undertakings tend to use procedures which are relatively undefined and have been developed by empirical observation and practice. At the other end of the scale, modern industrial breweries may have fermentation capacities of several million litres and use a process which approaches pharmaceutical standards in terms of control, hygiene and use of defined microbial cultures. Intermediate to these extremes is a continuum that ranges from purely traditional breweries, which may produce beer in considerable volume but also use empirically derived practices and equipment which has changed little over many centuries, through to the 'modernised' traditional brewery. In the latter case the basic process is unchanged and perhaps still relatively undefined but new plant has been installed, usually with greater capacity and improved hygienic properties (see Chapter 5).

Irrespective of the scale of operation, brewing has long been – and with few notable exceptions – continues to be practised as a batch process. It consists of three principal stages. First, wort production in which raw materials are treated to produce an aqueous medium rich in sugars and various other yeast nutrients. Second, fermentation, the stage in which certain components of the wort are assimilated by yeast cells to produce ethanol, carbon dioxide and a multitude of other metabolic products, which collectively form beer. Third, post-fermentation processing in which the immature beer is rendered into a form in which it is considered suitable for drinking.

Although this book concentrates on the fermentation stage of brewing, it is necessary to include some discussion of other parts of the process (see Chapter 2). With regard to the initial stages, fermentation performance is much influenced by wort composition. Since the latter is affected by the methods and materials used in its production, it is necessary to provide here a description of the processes involved. After primary fermentation is completed, various procedures may be carried out which require further activity by yeast (or other micro-organisms). These processes are grouped under the heading of secondary fermentation and obviously merit discussion (see Section 6.9).

Whilst it is true that the vast majority of brewers produce beer by batch fermentation there is at present a renaissance of interest in continuous processes. This follows an earlier foray, which occurred during the 1960s and 1970s. At that time continuous fermentation was seen by many of the larger companies as a potential cost-effective method for satisfying the need for highly efficient and high-productivity brewing. During this period several brewers invested in large-scale continuous fermentation plant. Unfortunately, practical experience failed to live up to the theoretical promise, and, with a few exceptions, application at production scale was largely discontinued by the 1980s. In the last decade, however, there has been a resurgence of interest in continuous brewing fermentation largely because of the development of new systems which employ immobilised yeast (see Sections 5.6 and 5.7).

Parallel to a discussion of the practice of brewery fermentation is a consideration of the biochemical changes, that encompass the conversion of wort to beer (see Chapter 3). These are as fascinating as they are complex. Thus, the growth and metabolism of the yeast, *Saccharomyces cerevisiae*, has been subject to intensive scrutiny over a number of years and it is now one of the best characterised of all eukaryotic cells. Indeed many of the groundbreaking discoveries on which modern biochemistry is based were founded on research performed with brewers' or bakers' yeast. However, much of this work has perforce studied metabolism under carefully controlled conditions; usually aerobiosis, frequently non-growth, growth in continuous culture on defined media, or batch growth on defined media. Of all of these batch growth, even on a defined medium, is the most difficult to characterise since conditions are in a constant state of flux and a steady state is never achieved.

This difficulty applies to a batch brewery fermentation; however, superimposed on this are several other layers of complexity. The growth medium, wort, is a natural product that contains a multitude of components, many yet to be identified and fluctuating in concentration from batch to batch. Several of the constituents, in isolation, are known to exert positive and negative influences on yeast metabolism; however, in combination they may interact to produce new and totally unexpected synergistic and antagonistic effects. This situation is further complicated by the role of oxygen in brewery fermentations. It is often assumed by the uninitiated that a beer fermentation is an entirely anaerobic process but, in fact, this is not so. For prolonged growth under anaerobic conditions, brewing yeast must have sources of unsaturated fatty acids and sterols. Some of this requirement may be satisfied by assimilation from wort but most has to be synthesised, *de novo*, and this requires molecular oxygen (see Section 3.5). Thus, at the beginning of fermentation, wort is provided with a single dose of oxygen, which is utilised by yeast during the first few hours of fermentation producing a transition from aerobiosis to anaerobiosis. This unsteady state, coupled with the other changes associated with growth and metabolism, has far-reaching effects on yeast physiology, many of which remain to be fully characterised.

The biochemistry of brewery fermentation is further complicated by virtue of the fact that vessels are not usually provided with forced agitation and neither is there any standardisation with respect to vessel configuration. Mixing of the fermenting wort to encourage adequate mass transfer of metabolites and dispersion of yeast cells is reliant upon natural convection currents brought about by the combined effects of escaping gaseous carbon dioxide and localised differences in fluid density due to differential cooling. During primary fermentation this natural agitation may be vigorous and probably sufficient to ensure homogeneity of the contents of the fermenter. However, when the yeast is relatively quiescent, at the outset and in the later stages, considerable stratification may be evident. This heterogeneity results in individual cells within the yeast population being exposed to widely differing environments depending on their particular location within the vessel. In this respect, physical parameters such as hydrostatic pressure and possibly temperature are important, particularly where very deep vessels are employed (see Section 5.1).

Intimately related to the biochemistry of fermentation is a consideration of the yeast used in the process. This topic may be divided into two areas of discussion. First, a description of the general biology and classification of brewing yeast. This includes a description of cellular and colonial morphology, patterns of growth, nutritional requirements, technological properties and life cycle (see Chapter 4). Second, how brewing yeast is handled on a large scale. As with the fermentation process as practised in individual breweries, there is also a spectrum of sophistication in the manner with which the yeast is handled. In rare cases there is no need for yeast husbandry as such since fermentation results from spontaneous contamination of wort by the microflora of the fermenting hall. In this case the fermentation involves a mixture of several yeast types and bacteria. However, the majority of breweries use their own strains of yeast and these are jealously guarded. Historically these were selected from a wild population, by empirical means, on the basis of their possession of desired fermentation properties and ability to produce beer with a particular flavour. In many traditional breweries the same yeast may have been used continuously for very many years, inoculated into and recovered from successive fermentations and so on *ad infinitum*. In such cases the yeast is commonly a mixture of several strains, present in proportions which have been selected by the conditions of the particular fermentation and method of yeast handling.

In modern breweries it is also usual to inoculate a batch of wort with yeast derived from the crop of a previous fermentation. However, in order to ensure strain purity a newly propagated culture is periodically introduced (see Section 7.2). This does not preclude the use of mixtures of strains, although it is perhaps more difficult to control the relative proportions of each. In any case, in such modern breweries, for simplicity of handling there has been, for some years, a trend towards the use of single strains.

The practice of serial fermentation and periodic propagation has several consequences. It produces a requirement for recovering yeast from a fermenting vessel and then storing it under conditions that are both hygienic and capable of maintaining the yeast in a physiological condition appropriate for use in the next fermentation. Introduction of a new yeast culture into the brewery requires suitable plant in both the laboratory and brewery. This must be capable of producing a pure yeast culture in sufficient volume to achieve a desired inoculation rate in the first generation fermentation. The trend, which has occurred over recent years, towards larger capacity breweries possibly using several separate yeast strains to produce a multiplicity of different beers has produced a need for yeast handing facilities whose sophistication matches the complexity of the operation. This has been accomplished by a combination of greater understanding of the physiological requirements of yeast handling, and, in response, introduction of brewery plant capable of maintaining the yeast under appropriate conditions.

A pre-requisite of propagation is a method for maintenance of laboratory master cultures, coupled with a system of strain identification. These aspects of yeast husbandry are particularly important where several yeast strains are used within a particular brewery and both have seen significant advances in recent years. Thus, there are now several options for maintaining yeast culture collections, ranging from the traditional agar slope approach through to storage in liquid nitrogen (see Section 7.1). Identification of yeast strains may use conventional microbiological methods based on patterns of growth and morphology. However, relatively new and much more precise approaches have now been introduced; for example, the use of genetic fingerprinting (see Section 4.1 *et seq.*).

The use of pure yeast cultures in brewing has been well established for many years. In tandem with this there have also been efforts to develop new strains with enhanced properties. Classical genetic improvement techniques have been applied, although the aneuploid/polyploid nature of most brewing yeast strains coupled with a lack of any sexual stage to the life cycle renders most of these ineffective. However, advances in the understanding of the physiology of *S. cerevisiae* continue to be made and there has now has been a complete characterisation of the genome (see Section 4.2.1.1). Clearly this offers the possibility of targeted strain improvement using recombinant DNA techniques. During the 1980s this approach was pursued with some vigour and several recombinant strains with selected improvements to their brewing characteristics were developed. Notwithstanding this scientific success, perceived consumer resistance to the use of such strains has mitigated against their application for commercial brewing.

## 1.2 Historical perspective

The predilection of mankind for consuming alcohol is common to all civilisations and undoubtedly pre-dates recorded history. Examples of alcoholic beverages may be found on every continent. They may have been consumed simply as a part of everyday diet but because of the physiological affect of alcohol and the apparent magic of the fermentation process they have also often been associated with religious or ritual ceremony. Beverages which may be classified as beers, particularly those produced by primitive societies, are legion. Thus, Forget (1988) in his *Dictionary of Beer and Brewing* lists more than 60 of what may be described as native beers. These have in common being made from a source of fermentable sugar, usually a cereal, to provide the alcohol and with additional flavouring materials. The latter may be from a variety of sources, as shown in the examples in Table 1.1.

Presumably these early fermentations were unwitting and serendipitous. Thus, wherever any natural source of sugar is to be found it will be accompanied by yeast

**Table 1.1**  Examples of native beers (from Forget, 1988).

| Beer | Country of origin | Description |
| --- | --- | --- |
| Aca | Peru | Maize beer |
| Algorobo | Central & S. America | Beer flavoured with carob beans and mesquite |
| Basi | Philippines | Fermented extract of sugar cane with herb flavouring |
| Bilbil | Ancient Egypt | Sorghum beer |
| Bi-se-bar | Ancient Sumeria | Barley beer |
| Boza | Ancient Egypt & Babylon | Millet beer |
| Chi | India | Millet beer |
| Chicha | Peru (Inca) | Maize beer flavoured with fruit juices |
| Chiu | China (Han Dynasty) | Wheat beer |
| Dolo | Africa | Millet beer flavoured with various bitter herbs including sisal, castor oil, cassia, pimento and tobacco |
| Kaffir beer | Africa (Bantu) | Sorghum beer (made to various recipes and is brewed commercially in Zimbabwe and S. Africa) |
| Korma | Ancient Egypt | Barley beer flavoured with ginger |
| Kurunnu | Ancient Babylon | Wheat beer |
| Manioc beer | Amazon Indians | Fermented extract of Manihot plant |
| Okelehao | Hawaii | Fermented extract of root of Ti plant |
| Pachwai | India | Rice beer with added extract of *Cannabis sativa* |
| Pombe | Africa | Sorghum beer flavoured with banana juice and various herbs |
| Sango-lo | China | Rice beer |
| Shimeyane | South Africa | Fermented extract of malted corn, brown bread and brown sugar |
| Tiswin | North American India | Fermented extract of corn and wheat flavoured with Jimson weed |

contamination and therefore, providing a supply of water, ethanolic fermentation will ensue. The phenomenon of carbon catabolite repression (see Section 3.4.1) ensures that ethanol is still the major product of carbon dissimilation, even under aerobic conditions. The serendipity effect has some other surprising twists. Thus, it is known that the preparation of some native beers that used cereals as a source of extract involved a step where the grains where masticated by the brewer. In so doing the addition of saliva, which contains the amylase, ptyalin, would partially degrade the starch content of the grain and thereby increase the fermentability of the wort. It is interesting to conjecture as to the train of empiricism that culminated in this process!

The beneficial effects of fermentation extend to the bacteriocidal qualities of the product. Ethanol itself inhibits the growth of many micro-organisms and this property is reinforced by the lowering of pH caused by other by-products of yeast metabolism. If the beverage contains viable yeast cells these will ensure that anaerobiosis is maintained and so inhibit the growth of aerobic contaminants. Further antiseptic qualities are introduced by many of the supplementary flavouring agents, for example, hops. In addition, the preparation of many beers includes a boiling step which alone would have a sterilising effect. In historical times, therefore, beer was a useful source of dietary calories, minerals and vitamins but could also be viewed as sanitised water. In medieval times this property was of no small significance when one considers the number of potentially fatal diseases which could be contracted after imbibing polluted water. This is illustrated by the story of Saint Arnold, the patron saint of Belgian brewers, who reportedly saved the inhabitants of a village gripped by a cholera epidemic, by blessing the local brewery and advising them to eschew water and from then on drink only beer (Bell, 1995).

The origins of the discovery of alcoholic fermentation are lost in pre-history. However, archaeological records indicate that brewing has been an organised community activity for at least 5000 years, hence the often quoted comment that production of beer, together with baking, represent the world's oldest technologies. Thus, it is reported that in ancient Mesopotamia some 40% of cereal crops were used for brewing (Corran, 1975). It appears that in antiquity the skills of malting cereals, to release fermentable sugars from starch, were also discovered (Samuel & Bolt, 1995; Samuel, 1996). It must be assumed that this also was an accident, although perhaps one that is more difficult to explain. Corran (1975) speculates that it may have been based on the observation that cereal grains used for brewing which had been stored under wet conditions would potentially give an increased yield of alcohol. It seems likely that such experience would have provided a powerful stimulus for further experimentation. However, the same author also contends that primitive malting may also have arisen as part of ordinary cooking, where it would have rendered the grain more nutritious and digestible. Whatever the truth, it is certainly the case that the discovery of malting marked the beginnings of brewing recognisably modern beers and that this probably occurred in the Middle East sometime after the birth of agriculture in 6000 BC.

The classical civilisations (Greece and Rome) have no history of brewing, since the preferred drink was wine, presumably another skill acquired by happy accident, and dependent on a ready supply of grapes. The rise of modern brewing occurred in the more temperate climates of northern Europe. Possibly some of the skills were acquired from the Middle East, although independent discovery may also have occurred. By medieval times brewing was an everyday and universal domestic occupation usually performed by women, the alewives of yore. Larger scale operations occurred wherever the needs of a sizeable community required to be satisfied, for example, religious institutions or large households. Thus, in Burton-upon-Trent in the United Kingdom, the abbey founded in the eleventh century had a brewery whose product formed the reputation of this town as a centre of brewing excellence. Similarly the high regard paid to Belgian beers owes much to the skills of the medieval abbey brewers. Fortunately a few of these monastic breweries have survived and still produce the specialist bottle-conditioned Trappist beers.

Beer production on large country estates could be prodigious, reflecting the fact that it formed the staple drink of all classes. Thus, Sambrook (1996) records that ale consumption in single medieval noble households were usually in the range of 750–1500 hl per annum. In such households, the daily allowances for each servant was one gallon (3.8 litres)! The origins of the inn or tavern are unrecorded but they were certainly in existence in the United Kingdom during Roman times. However, the widespread use of beer, as opposed to wine, was probably heavily influenced by later Saxon and Danish invasions (Hackwood, 1985).

The benefits of using hops in brewing were known in antiquity and records exist detailing their cultivation in ancient Babylon (Corran, 1975). This knowledge was passed to Europe; however, initially hops were used in conjunction with other herbs such as rosemary, bog myrtle, sweet gale, coriander, caraway, nutmeg, cinnamon, ginger, milfoil and yarrow. Mixtures of such herbs were called 'gruit' and with local variations were incorporated into beers to add flavour and improve keeping qualities (Forget, 1988). Widespread use of cultivated hops, as opposed to gruit, began in Germany, probably in the tenth century, and from there spread to the rest of Europe. Hops were introduced to Kent in England probably by Flemish weavers in the fifteenth century (Lawrence, 1990). Apart from altering the flavour of beer, the preservative qualities of hops allowed weaker beers with longer shelf-lives to be brewed since it was no longer necessary to rely entirely on the anti-microbial qualities of ethanol.

The arrival of hops in the United Kingdom marked a need to distinguish between ale and beer. The latter was taken to refer only to a hopped fermented malt beverage; however, as Sambrook (1996) discusses, other meanings were also used. Thus, when commercial and domestic brewing were parallel operations, the product of the town brewery was often termed beer and the home brewed material as ale. In another sense, ale was used to describe the product made from the first strong worts, whereas the term beer derived from subsequent weaker worts, hence the expression 'small-beer'. There was resistance to the use of hops in most countries of Europe, primarily because of vested interests. Thus, purveyors of ales and gruit all had good commercial reasons to discourage production of hopped beers. Ultimately, these sanctions were unsuccessful and in response to public demand the use of hops became the norm.

The mineral composition of local water supplies was influential in the development of centres of brewing excellence in particular areas. For example, the soft waters of Pilsen and Munich are required for brewing lager beers and the water at Burton-upon-Trent with high levels of gypsum is optimal for the production of pale ales by top fermentation (Section 2.3.3). Unlike Burton-upon-Trent, the natural well waters of Scotland are soft and high-quality lager beers have been produced in Glasgow since the late 1880s (Donnachie, 1979). Again, as with much of the history of brewing, it must be assumed that the significance of water composition was based on empirical observations; however, as discussed later, although the causes were unknown it was the major reason for the establishment of brewing in particular areas.

The origins of the use of bottom fermentation to produce lager beers are also unknown but Corran (1975) considered that it probably began in Bavarian monastic breweries. It must be assumed that by chance a bottom-fermenting strain of yeast was selected and was retained because of the superior nature of the product. A natural supply of soft water provided another vital ingredient, although temperature was also important. Lager brewing requires low temperatures and unlike the United Kingdom, with its heritage of top fermentation, this was to be found in continental Europe, at least during the winter. Before the advent of refrigeration the requirement for low temperatures made bottom fermentation a seasonal activity. Thus, in Bavaria, bottom fermentation was discontinued during the summer months and top-fermented beers using higher temperatures were produced instead (Corran, 1975). From a another standpoint, the preponderance of ale production in the United Kingdom can be attributed directly to the effects of the Gulf Stream. According to Miller (1990) it was a Bavarian monk who in 1842 smuggled the first bottom-fermenting yeast into Bohemia and thereby laid the foundation for the production of Pilsener lager beer in the Czech Republic. Soon after this the same style of brewing spread from Munich into the rest of Germany and from there to the remainder of Europe and Scandinavia.

The development of industrial brewing and the decline in domestic production was due to many factors. Beer and ale ceased to be the universal drink of all classes when other beverages became available and consequently there was less need to provide a constant domestic supply. Tea and coffee were introduced to Britain during the middle of the seventeenth century, providing an alternative to beer. Initially both tea and coffee were affordable only by the wealthy; however, in Britain, the cost of tea fell dramatically in the mid-nineteenth century and this period coincided with the greatest decline in domestic brewing. The rise of the public house, the reduction in the number of servants and the discontinuation of the use of beer as part of wages also contributed.

In the United Kingdom, the development of the commercial brewery was part of the Industrial Revolution of the eighteenth and nineteenth centuries. The resultant urbanisation provided a mass market for the product and the introduction of methods of large scale production, together with a ready supply of labour, provided the means for satisfying the need. According to Gourvish and Wilson (1994), in 1830 domestic brewing accounted for 20% of total output in the United Kingdom. In the same year single outlet breweries had 45% of the beer market. By 1900 both of these segments of UK brewing were negligible but between the same years the annual production of commercial brewers rose from less than 13 million hl to nearly 50 million hl.

Initially commercial brewers provided for the needs of local communities. However, the development of efficient methods of freight transport allowed further expansion. Naturally this occurred in recognised centres of brewing. The experiences of Burton-upon-Trent provide an example of these developments as described by Owens (1987). From monastic beginnings brewing in Burton passed into the hands of a local noble family, the Pagets, after the dissolution of the abbey in 1540. The Pagets fostered the brewing tradition, and, by 1604, 46 Burton brewers served the local community as well as meeting the demand of London consumers, via road carriers, for the then-fashionable Burton beer. In the eighteenth century the development of navigable waterways between Burton and Hull, on the east coast of Yorkshire, provided a means of exporting beers to satisfy a growing market in the Baltic. In addition, the links provided by other canals fuelled an ever-increasing domestic trade.

By 1793 the Baltic trade had ceased due to the Napoleonic wars and other markets were required. This was achieved by export of pale ale to India. By 1832, two Burton brewers, Bass and Allsopp, were shipping nearly 10 000 hl of pale ale to Calcutta. A consequence of this was that the qualities of the beer became widely valued within Britain and this led to a much increased domestic market.

The importance of the national home sale trade was facilitated by the growth of the Victorian railway system. In the period 1831 to 1868, the capacity of Bass increased from just under 19 000 hl to nearly 880 000 hl per annum. The realisation of the significance of the local water supply in Burton to the brewing of pale ales led many London brewers to commence operations in Burton. By 1888 Owens (1987) records that 32 brewers were in operation within the town, employing a work force of more than 8000 and producing an annual output of nearly 5 million hl.

The spread of European brewing techniques to the rest of the world was fuelled by emigration and colonisation. The type of beers produced would have been those associated with the countries of origin of the settlers. Thus, because of the empire building activities of the British, the brewing of top-fermented beers became common in many parts of the world, only to be superseded at a later date by lager beers. A result of this expansion was that in many countries the brewing of European-type beers by settlers coexisted with the production of the native beers made by the original inhabitants. The supplies of European beers could also be augmented by imported products.

For example, brewing of lager beers was introduced to Japan and other Asian countries in the late 1800s, when barriers to foreign trade were dropped. Initially the breweries were managed by western in-comers as a separate industry to the indigenous production of Japanese rice beer, *saké*. By 1901, over 100 Japanese breweries were in operation. In subsequent years the number of breweries fell, and, as in other countries, the industry became dominated by fewer larger companies and in this case, managed by native Japanese brewers (Inoue *et al.*, 1992a).

In America, brewing was introduced into Virginia in 1587 by Walter Raleigh. The first commercial brewery opened in Charlestown, Massachusetts in 1637 (Corran, 1975). At first the majority of American brewing was concentrated on the east coast and was based on British procedures. Later, waves of German and Dutch immigration, from 1850 onwards, initiated lager beer production and this style of beer eventually became predominant. The change from top fermentation to bottom-fermented lager was accelerated when commercial refrigeration plant became available. Interestingly, the rise of Milwaukee as a centre of lager brewing was in part a consequence of the availability of large quantities of ice from the adjacent Lake Michigan.

The development of brewing, both domestic and commercial, was obviously based largely on empirical experience and as such a successful outcome owed much to the art of the brewer. However, what of the elucidation of the underlying science? With respect to fermentation, three discoveries were crucial to the development of the modern industrial process. First, an appreciation of the true role of yeast; second, identification of the causes of spoilage; third, invention of apparatus to allow accurate monitoring and control of the fermentation process.

The introduction of methods of improved monitoring of fermentation was an essential adjunct to the increases in scale associated with industrialisation. In particular there was an obvious need to be able to monitor the progress of the various processes involved. Two innovations of note are the introduction of the thermometer by Combrune in 1768 and the saccharometer by Richardson in 1784.

The story of the establishment of the biological nature of ethanolic fermentation in effect marks the end of the old alchemical theories and the beginnings of modern microbiology and biochemistry. Thus, this discovery was of profound significance to the development of all of the biological sciences. In the early stages of this history, as befits the fundamental nature of the subject, the major protagonists were in most part what would now be described as main-stream academics. Since this same group produced most of the written records, they are usually credited with the discovery of the vital nature of yeast and its role in the biochemistry of fermentation. However, as will be seen, a few brewers have left records that indicate that everyday observations and experience had already suggested to some that yeast was more than just an inanimate by-product of fermentation. It is interesting to note that an early English name for yeast was Godisgood (Keller *et al.*, 1982) which perhaps supports the view that common bakers and brewers had suspicions that it played more than a passing role in fermentation. In later years when the biological nature of fermentation was known and fully accepted the major brewing companies themselves employed specialist scientists and it can be shown that for a brief period of time this group of people were very much in the vanguard of fundamental research.

The discovery of the nature of fermentation is described by Florkin (1972) in his excellent *A History of Biochemistry* and the subject is also well documented by Anderson (1989, 1991, 1993), Curtis (1971) and Stewart and Russell, (1986).

The word fermentation derives from the Latin *fevere*, to boil. According to alchemical theory fermentation was an active separation process and thus, in the case of alcoholic fermentation, the ethanol was already present but did not become evident until impurities, in the form of yeast and carbon dioxide, became separated. It was recognised, however, that sugar was required. Subsequently the idea was born that fermentation did involve the formation of new products but these changes were taken to be purely inanimate and involved a chemical simplification. For example, in 1697, Georg Ernst Stahl, the founder of the phlogiston theory of combustion, concluded that the violent activity and heat generation associated with fermentation 'loosened' the particles present in the medium and allowed the formation of new products. Later, towards the end of the eighteenth century, the chemical nature of fermentation was given added credence by Lavoisier who, on the basis of elemental analysis, calculated a stoichiometric relationship between sugar and the products of fermentation (as he believed) ethanol, carbon dioxide and acetic acid.

In 1810, Gay-Lussac proposed that oxygen was the active ingredient that initiated fermentation. Thus, he considered that the oxygen reacted with the sugary liquid to produce the ferment. This conclusion was based on experiments in which he observed that foodstuffs sealed in bottles and heated in boiling water spoiled, or fermented, only when air was admitted. He also performed the key experiment showing that fermentation could be stopped, even after the addition of air, by boiling. Unfortunately this was interpreted as an effect in which the heat altered the ferment produced by the action of oxygen on the liquid into an inactive form. Yeast was discounted as being part of the process on the basis of insolubility. Gay-Lussac did establish that during fermentation one molecule of glucose gives two molecules each of carbon dioxide and ethanol and this equation still bears his name.

The true role of yeast in fermentation was established, apparently independently during the early nineteenth century, by a Frenchman, Charles Cagniard-Latour, and two Germans, Theodor Schwann and Friedrich Traugott Kützing. However, before mention of their work it should be noted that the microscopic nature of yeast cells were of course first recorded by Antonie van Leeuwenhoek, using one of his own microscopes, to examine fermenting beer. His observations were published in a letter to the Royal Society in 1680 (given in translation by Chapman, 1931). Chapman makes clear, however, that although van Leeuwenhoek undoubtedly saw yeast cells in the beer he had no appreciation of their role. Cagniard-Latour also made direct microscopic observations of growing yeast cells but with the benefit of the more advanced instruments available in the 1830s. He noted that the cells proliferated by the formation of buds and concluded that this was evidence of life and went on to show that the presence of yeast cells was necessary for fermentation to occur.

Schwann examined alcoholic fermentation as part of a larger investigation to disprove the theory of spontaneous generation. He had already shown that putrefaction of infusions of meat did not occur when it was heated and air was excluded. Furthermore, the onset of spoilage, which occurred when air was admitted, was prevented if the air was first heated. His interest in alcoholic fermentation was aroused because of Gay-Lussac's contention that oxygen was the causative agent. In fact, Schwann was able to show that contrary to Gay-Lussac's assertion, heated air was essentially unchanged since it remained able to support life, in the form of a frog. He also described the morphology and growth of yeast cells, which he termed 'Zuckerpilz' and noted that their fermentative activity was destroyed by heating.

Kützing also used microscopic observation as the basis of his conclusion that yeast was the agent responsible for alcoholic fermentation. Thus, he wrote 'it is obvious that chemists must now strike yeast off the role of chemical compounds, since it is not a compound but an organised body, an organism' (Anderson, 1989).

Despite the apparently strong evidence of Cagniard-Latour, Schwann and Kützing for the role of yeast in fermentation, eminent members of the pro-chemical lobby remained in their entrenched positions. In particular, Jöns Jacob Berzelius, a Swedish chemist and leading opinion former (he was largely responsible for the universal adoption of the atomic theory and established the current system of chemical notation), ridiculed the suggestion that yeast was animate. This polarisation of views was exacerbated by a scurrilous piece of satire published anonymously in the prestigious *Annalen der Pharmacie* but actually written by one of its editors, Justus Liebig, and his close friend Friedrich Wöhler, a former pupil of Berzelius. This article depicted yeast cells in the form of eggs which on contact with sugar, hatched to form microscopic creatures. These ate the sugar and then expelled alcohol from the digestive tract and excreted carbon dioxide from the bladder, which took the shape of a champagne bottle. Liebig believed in the existence of a vital force which held the atoms and molecules of living organisms together. After death these components were held together only by weak forces which could be overcome by another stronger force such as contact with oxygen or heat. Fermentation or putrefaction of sugary solutions was brought about by the oxygen acting on unspecified nitrogenous substances, which were mysteriously transformed into inducers of fermentation.

As Anderson (1993) discusses, whilst this controversy was raging during the middle of the nineteenth century, a few scientists associated with the brewing industry accepted the role of yeast in fermentation. However, it required Pasteur to convince the academic mainstream. Undoubtedly this was made easier by his then international reputation. Anderson (1995) has also made an excellent chronicle of this part of the story. Pasteur came to consider fermentation whilst occupying the Chair of Chemistry at Lille. One of his interests was a study of the optical activity of molecules, an attribute he associated with living organisms. At Lille, he detected optically active amyl alcohol in the product of beet juice fermentations and, therefore, he concluded that the process must be in some way animate.

Using the famous swan-neck flasks Pasteur augmented the earlier findings of Schwann and proved conclusively that putrefaction occurred as a result of airborne microbial contamination and not via spontaneous generation. His initial studies with wine and latterly with beer confirmed that yeast cells were the causative agents of fermentation. Crucial to the development of controlled and hygienic brewing and oenology, he extended this conclusion to show that infections by other specific micro-organisms were the cause of 'diseased' fermentations. In his *Etudes sur la bière*, published in 1876, he produced designs for hygienically designed industrial scale fermentation vessels, which included plant for aerating wort.

Pasteur is generally credited with introducing practical scientific methods to brewing. For example, introducing the microscope for use as a tool for routine quality control. Although there is truth in this, as Anderson (1995) points out, the microscope had been in common use in some United Kingdom breweries for some time prior to Pasteur's visitations. Nevertheless, it was the work of Pasteur that finally laid to rest the non-animate theories of fermentation. Whatever the precise chronology, it is certainly true that the late Victorian age brought a bloom of scientific enterprise to brewing. Many of these scientists were of the first rank academically and had international reputations to match.

In Burton-upon-Trent, in the United Kingdom, Horace Tabberer Brown, his half-brother, Adrian John Brown, Cornelius O'Sullivan and Johann Peter Griess all contributed greatly to understanding the scientific basis of fermentation and other aspects of brewing. These four, together with four other locals, formed an informal scientific discussion group, the Bacterium Club in 1876. Later in 1886 this became the Laboratory Club, and, as the membership increased, the Institute of Brewing in 1890. It is noteworthy that, due to the influence of scientific discovery, brewing in Burton was transformed from a seasonal October to April activity into an all-year-long process.

Although the effects of the Industrial Revolution were felt most in the United Kingdom, the advance of science in brewing was also evident in other parts of Europe during the late nineteenth century. In Germany, educational courses for brewers were founded at Weinstephan in 1865. These evolved into an Academy for Agriculture and Brewing in 1895. This was incorporated into the Technical University of Munich in 1930. In 1883, in Berlin, the Versuchs-und Lehranstalt für Brauerei (VLB) was established as a centre for practical and scientific research in brewing. In Copenhagen in 1875, J. C. Jacobsen, the founder of the Carlsberg brewery, founded the laboratory of the same name. The following year this became the Carlsberg Foundation under the control of the Royal Danish Academy of Sciences. Of the many eminent alumni of this institute, two are of particular relevance to yeast and fermentation. In 1935, Øjvind Winge demonstrated sexual reproduction in yeast, and, later, in 1937, Mendelian segregation, thereby opening the possibility of developing improved yeast strains using classical genetic breeding techniques. Before this, in 1883, E. C. Hansen introduced pure yeast cultures to brewing (Curtis, 1971).

Emil Hansen applied to fermenting worts the relatively new methods of the medical bacteriologist Robert Koch of isolating pure cultures using solid media. He noted that a number of distinct yeast strains could be isolated from turbid (or infected) beer. Of these only one, when cultivated and inoculated into fresh sterile wort, produced satisfactory beer. Hansen had, therefore, extended the observations of Pasteur that not all yeasts could be classed as beneficial and indeed some strains were responsible for the so-called diseases of beer. Hansen's brewing yeast strain was christened Carlsberg Yeast Number 1 and was used successfully for commercial brewing. This was accomplished using specially designed pure yeast propagation plant. Pure yeast cultures were soon exported to several other European centres of lager brewing where they were used with much success. Thus, the first propagation plant was installed into the Copenhagen Carlsberg brewery in 1884 and similar plant the following year at the Heineken Brewery in Rotterdam. The latter helped promulgate pure yeast brewing since they supplied yeast to several German brewers.

In the United Kingdom Hansen's yeast propagation plant was first installed at the Worthington Brewery in Burton-upon-Trent by G. H. Morris, a colleague of Horace Brown. It was not a success and it was mistakenly claimed that British top-fermenting ale yeasts were unsuited to this approach. The problem was that beers in cask failed to develop sufficient condition. In fact, it seems likely that much of the secondary fermentation in the cask was brought about by contamination with wild yeast, post-fermenter. In addition, since many, if not the majority, of British brewers used mixed cultures it was considered that it would not be practicable to duplicate this with a pure yeast culture plant. The British prejudice not to adopt Hansen's principles persisted for many years and it was not until the mid-1950s that modern yeast culture plant came into common use.

The burgeoning of scientific research in brewing exemplified by the Victorian era ended with World War I and the subsequent depression. Prohibition in the United States and Finland, together with slumps in trade in Germany and much of Europe, heralded the beginning of a period of austerity both economically and scientifically which lasted through to the beginning of World War II. The post-war period marked the start of a new era of modern science. During this time many brewing companies opened or enlarged their own research and development facilities. However, because of the applied nature of the research, these have not offered the opportunities for in-house scientists to establish reputations of the same magnitude as the Victorian 'greats'. No doubt this was also due to a great deal of the work being shrouded in the cloak of commercial secrecy.

On the positive side the post-war period has seen a move towards greater international co-operation. Thus, the Brewing Industry Research Foundation was founded in 1951. In the last decade this organisation has added 'International' to its name and is currently known as Brewing Research International (BRi). Other research associations such as the American Society of Brewing Chemists (ASBC), the Master Brewers Association of America (MBAA) and the Japanese Association of Brewing Chemists (JASBC) were formed at a similar time. To further international co-operation in brewing science, the European Brewing Convention (EBC) was formed in 1947 by Philippe Kreiss. This has established links with the brewing industry in most of the countries of Europe as well as the United States and Japan. In addition to promulgating advances in brewing science, the EBC and the other sister organisations have established standard methods for measurement and analysis.

## 1.3 Current developments

In the post-war years a global market for European-style beers has developed. The bulk of the beer is of the lager variety and is produced by bottom fermentation. Of course, many other beer types are still produced and many may have important domestic markets but in terms of world sales they may be regarded as specialities. In consequence, in most countries the brewing industry has tended to become polarised between a few companies which produce the bulk of the national output and numerous small breweries each servicing a small local market. The major companies frequently own many individual breweries, sometimes in several countries, and have a national or international market.

In Europe, in particular, the major brewing companies have usually arisen via merger and acquisition. This process has occurred over many years but was particularly prevalent during the 1960s (Gourvish & Wilson, 1994). In consequence, the total number of brewing companies in any particular country has tended to decline. For example, in the United Kingdom in 1870 there were 133 840 licensed brewers. By 1906 this had declined to 1418, in 1946 453 and a mere 111 in 1967. By 1997, 80% of the total beer output is produced by just four companies. Indeed, at the time of writing further consolidation seems likely! A similar situation exists in the rest of the world. With the exception of Germany, which has retained a greater number of smaller brewing companies, the bulk of the world brewing volume in developed countries is carried out by a few large companies (Table 1.2).

In parallel, there is now a trend towards the establishment of new small craft or 'pub' breweries. In the United Kingdom, for example, the Brewers' Society *Statistical Handbook* (Thurman & Witheridge, 1995) lists 177 brewing companies which have been launched since 1 January 1971. Most of these are small, frequently servicing a

**Table 1.2** Proportion of total beer output produced by large brewing companies in some leading countries during 1993 (from Thurman & Witheridge, 1995).

| Country | Total annual production (millions of hl) | Output by major brewers (millions of hl) | Companies |
|---|---|---|---|
| Australia | 18.1 | 16.0 | 2 |
| Brazil | 57.0 | 54.9 | 4 |
| Belgium | 14.7 | 13.0 | 3 |
| Canada | 22.1 | 20.7 | 2 |
| China | 122.5 | 18.5 | 11 |
| Denmark | 9.4 | 8.0 | 2 |
| France | 20.8 | 19.6 | 5 |
| Germany | 115.4 | 77.2 | 23 |
| Japan | 69.0 | 68.3 | 4 |
| Mexico | 40.9 | 40.6 | 2 |
| Netherlands | 20.4 | 13.6 | 4 |
| New Zealand | 3.6 | 3.4 | 2 |
| Nigeria | 6.7 | 3.2 | 1 |
| South Africa | 22.6 | 22.2 | 1 |
| United Kingdom | 56.7 | 45.0 | 4 |
| United States | 237.3 | 177 | 3 |

single trade outlet, and using very traditional approaches to brewing. This movement, which in many ways reflects a reaction against the perceived increased blandness and sameness of the products of the major brewers, is common to the entire developed world.

Patterns of beer consumption and other alcoholic beverages have also been influential in shaping the modern industry. For example, in the United Kingdom during the years 1899 to 1993, total beer production has declined from 22.76 to 21.58 million hl per annum (Table 1.3). In the same period the strength of the beer has fallen from an average gravity of 1.055 to 1.037. Although the fall in total beer production may seem modest it should be noted that it was accompanied by a total population increase of 43% from just under 40 million to 58 million. Perhaps most dramatic has been the increase in competition from other alcoholic beverages. Thus, in the United Kingdom, annual consumption of all wines in 1899 was 0.78 million hl. This volume remained relatively unchanged throughout the first five decades of the twentieth century. However, from 1950 to 1993 wine consumption increased 14-fold from 0.53 to 7.7 million hl per annum.

**Table 1.3**  Beer production compared to wine consumption in the United Kingdom (from Thurman & Witheridge, 1995).

| Year | Total population (millions) | Beer production (million hl) | Average gravity | Wine consumption (thousand hl) |
|------|------|------|------|------|
| 1899 | 40.77 | 22.76 | 1.055 | 779.5 |
| 1910 | – | 21.43 | 1.053 | 520.3 |
| 1920 | 43.65 | 21.17 | 1.043 | 560.5 |
| 1930 | – | 14.66 | 1.043 | 828.3 |
| 1940 | – | 16.08 | 1.039 | 965.1 |
| 1950 | – | 15.27 | 1.037 | 533.9 |
| 1960 | 52.37 | 16.63 | 1.037 | 1064.9 |
| 1970 | 55.63 | 21.08 | 1.037 | 1938.4 |
| 1980 | 56.33 | 23.92 | 1.037 | 4828.7 |
| 1990 | 57.56 | 22.93 | 1.038 | 7253.6 |
| 1993 | 58.19 | 21.58 | 1.037 | 7739.3 |

With regard to the large brewing companies the trends described above have provided the stimulus to transform brewery fermentation from a small-volume unsophisticated process to a more regulated, efficient, hygienic and large batch size operation. Several individual causes of the need for greater control in fermentation can be recognised. The international brewing market is fiercely competitive and therefore every effort must be made to ensure that high process efficiencies are maintained. This requires that fermentation is controlled adequately so as to maximise the conversion of sugar to alcohol and minimise the proportion of wort sugars used to fuel yeast growth. Fermentation process efficiency, in terms of both revenue and capital costs, may be increased by economies of scale and thus individual batch sizes have also increased. With ever larger batch sizes the consequences of non-standard behaviour become commensurately more severe, therefore providing another compelling reason to ensure that the process is controlled properly. Many major brewing companies operate several individual breweries. Multi-site breweries may be used to produce beer brands with an international market. Clearly it is important to ensure that fermentations are controlled precisely so that a consistent product is generated at all locations.

## 1.4 Legislation

The universal popularity of alcoholic beverages has had the result that all governments have viewed them as a convenient and lucrative method of raising revenue. In addition, the effect of ethanol on human behaviour has had the consequence that the production and sale of such beverages has usually been subject to stringent regulation. In many countries and at various times in history, the supply of crucial raw materials has been controlled by the state as a means of regulating consumption by controlling production. Occasionally this method has also been used to maintain a state monopoly.

With regard to brewing, the weight of legislation has influenced the materials which are used, the conduct of the process as a whole and fermentation in particular. With respect to raw materials, there may be restrictions on what may be used. Thus, as described in Section 1.1, the German beer purity laws allow only malt, water, yeast and hops to be used in brewing. This precludes the use of alternative sources of fermentable sugar and many so-called 'process aids', such as fining agents to assist with beer clarification after fermentation is complete. Even where the legislation allows greater latitude with respect to raw materials, the precise system of raising excise revenue influences the composition of the wort, in particular the sugar concentration. Thus, because of the limits of accuracy and reproducibility of analytical methods it is usual to assess liability to excise payment in discrete bands of alcohol concentration in the beer. The width of these bands varies from country to country but the result is that worts have to be prepared that contain fermentable sugar at a concentration that will produce a desired alcohol concentration. Of course beers at the top and bottom of any particular alcohol concentration excise band will attract the same excise payment and therefore it is incumbent on the prudent brewer to steer an appropriate course between legality and profitability.

The modern practice of high-gravity brewing (see Section 2.5), where a concentrated wort is fermented, followed by dilution to a desired alcohol concentration, places less onus on the need to achieve an accurate initial wort sugar concentration. However, at the other end of the process after fermentation is complete, a means of precise blending of concentrated beer and water is needed.

The liability for excise payment is usually gauged by self-assessment. Since brewing is still predominantly a batch process, this system has produced a requirement for keeping scrupulous records. Of equal importance, as alluded to already, it has produced a need for accurate measurement of volume, weight and concentration. Whilst many methods may be used in a brewery to measure, for example, ethanol concentration, only those which have been given government approval may be employed for the purposes of assessing excise duty payments.

In the majority of countries, excise payments are now based on the alcohol content of the beer, termed end-product duty systems. This was not always the case and in some countries, for example, Belgium and the United Kingdom, excise was levied on the basis of sugar usage, expressed in terms of the specific gravity of the wort. Such systems arose since in historical times when the precision with which some measurements could be made left much to be desired it was simpler to police a levy based on quantities of raw materials used. In order to manage this system it was necessary to develop the concept of original gravity. For reasons of micro-biological safety, it is not generally considered good practice to store uninoculated (unpitched) wort for any length of time. This, coupled with the possibility of fur-ther sugar additions being made after the primary fermentation had started made it necessary to have a method of computing initial specific gravity from samples taken during or at the end of fermentation. These methods allowed correction for the relative proportions of wort sugars converted to ethanol and that used for new yeast biomass formation.

In the United Kingdom, this measure, the original gravity, was based on data gathered by measurements made in ten British breweries during the years 1909–10 and published as the 'mean brewery table' by Thorpe and Brown (1914). In essence, a representative sample of beer of known volume at a controlled tempera-ture is taken and distilled. The residue and distillate are both diluted to the same initial volume and the specific gravity of each is measured. The difference in degrees of specific gravity between the two, termed the spirit indication, is calcu-lated. This figure is used in the mean brewery table to obtain a value for the pro-portion of the specific gravity of the original wort that must have been used in fermentation. This added to the specific gravity of the residue gives the original gravity of the wort.

The data used to formulate the mean brewery tables were obtained from British ale fermentations. In the case of bottom-fermented lager the proportions of sugar used for yeast biomass formation are altered and in continental Europe an alternative system, the Balling or Plato tables, are used, as in the following relationship:

$$p = \frac{(2.0665 \times A + n) \times 100}{100 + 1.0665\,A}$$

where $p$ is the original gravity in degrees Plato; $A$ is the alcohol content in % by weight; $n$ is the residual gravity.

Wort-based excise assessment has fallen out of favour and, for example, in the United Kingdom a system of end-product duty payment was introduced in 1993. Thus, the wort-based system was unwieldy and the method of arriving at a figure for original gravity was subject to serious inaccuracy. This worked in favour of more efficient breweries since it proved possible by careful control of fermentation to convert a greater proportion of the fermentable extract into ethanol than that assumed in the excise calculation. Hence, there was a significant but legal under-payment of duty and consequent loss to the Exchequer. The development of routine methods allowing precise measurement of ethanol concentration made the intro-duction of end product assessment inevitable and a lucrative loophole has been closed.

# 2 The brewing process

## 2.1 Overview

It is beyond the scope of this book to give a comprehensive description of the brewing process. For this the interested reader should consult the many authoritative texts which are already available (Lloyd Hind, 1940; de Clerck, 1954; Briggs *et al.*, 1981; Hardwick, 1995; Bamforth, 1998). It is important, however, to provide a basic outline of the brewing process in order to be able to place the fermentation stage into its correct context. Of particular note is the subject of wort composition and how this may be influenced by the raw materials used and the practices employed in its preparation. The bulk of this chapter is devoted to this critical aspect of brewing.

It has already been intimated in the previous chapter that beer as a whole comprises a diverse range of products. These are made from an enormous range of raw materials and using plant which varies in sophistication from that associated with a purely domestic undertaking through to the ultramodern brewing 'factory'. Again it is beyond the scope of this book to provide full detail of the manufacture of the many different beers which are encountered throughout the world. Consequently, much of the description provided here applies to the process as practised in a commercial brewery producing a fermented beverage from wort made largely from malted cereals and hops. In this respect the beer most widely brewed and consumed throughout the world is a lager product of the Pilsener type. Hopefully, therefore, it will be understandable that disproportionate coverage is given to the manufacture of such a product. However, for completeness, a brief description is provided of the manufacture of some of the more notable related beverages, particularly those produced at commercial scale.

## 2.2 Beer types

Beers may be classified into types on the basis of the nature of the raw materials, the microbial agents used in the fermentation step(s), and the processes used to convert the starting materials into a beverage ready for consumption. Although there is much overlap, it is possible to use these criteria to identify the unique points of difference of the major beer types. As will become evident from the discussion, some of these distinctions are becoming increasingly tenuous in the modern industry. Examples of representative beer styles are shown in Table 2.1.

With regard to nomenclature, the names applied to some beers have changed their meanings. For example, 'lager' derives from the German *lagern*, to store or rest. This refers to the lengthy low-temperature maturation phase that is a characteristic of the manufacture of traditional lager beers. As will be discussed in later chapters, a major

**Table 2.1** Classic beer types.

| Beer | Origin | Fermentation system | Characteristics |
|------|--------|--------------------|-----------------|
| Pilsener (archetype – Pilsener Urquell) | Bohemia, Czech Republic | Bottom fermented | Pale highly hopped lager brewed with very soft water (5% abv) |
| Dortmunder (Dort) (archetype – Dortmunder Aktien Brauerei) | Westphalia, Germany | Bottom fermented | Golden coloured lager, less hop than Pilsener (5.2% abv) |
| Münchener dunkel | Munich, Bavaria, Germany | Bottom fermented | Dark, malty, lightly hopped lager (4.4–5.0 abv) |
| Münchener helles (archetype – Paulaner) | Munich, Bavaria, Germany | Bottom fermented | Paler version of dunkel |
| Altbier (Düsseldorfer) | Mainly Düsseldorf, some other towns in Westphalia, Germany | Top fermented | Dark colour, well hopped fruity flavoured beer (4.4–5.0% abv) |
| Bockbier | Originally from Einbeck, Saxony, Germany | Originally top fermented | Strong dark brown |
| Doppelbock | Bavaria, Germany | Bottom fermented | Full, estery strong lager (7.5–13.0% abv) |
| Eisbock (ice beer) | Bavaria, Germany | Bottom fermented, very-low-temperature lagering to increase alcohol content by removal of ice | Very strong full bodied lager (13.2% abv) |
| Bock beer (USA) | USA and Canada | Bottom fermented | Dark colour, lightly hopped and estery |
| Vienna (Märzen) bier | Originally Austrian beer but now Bavaria, Germany | Bottom fermented | Pale, well-hopped lager stronger than Helles (5–6% abv) |
| Bitter ale | UK | Originally top fermented | Well bittered pale to copper ale. Sold for draught dispense, originally cask-conditioned (3.5–5.5% abv) |
| Brown ale | UK | Top fermented | Sweetish, full-bodied, lightly hopped bottled ale. 3.5–4.5% abv |
| Burton ale | Burton-upon-Trent, UK | Top fermented | Pale, well-bittered top-fermented ale brewed using gypsum-rich well waters of Burton-upon-Trent |
| India pale ale (IPA) | UK | Top fermented | Highly bittered strong bottled pale ale originally brewed for export to British troops in India |
| Kölschbier | Köln (Cologne) Germany | Top fermented | Pale golden highly hopped ale with acidic/lactic taste (4.6% abv) |
| Mild ale | North and Midlands, England | Top fermented | Lightly-hopped, sweet, dark brown ale (3.5–4.0% abv) |

**Table 2.1**  *Contd*

| Beer | Origin | Fermentation system | Characteristics |
|---|---|---|---|
| Weizenbier | Bavaria, Germany | Top fermented | Lightly hopped, highly carbonated wheat beer with estery, phenolic taste. Usually bottle conditioned (5–5.5% abv) |
| Blanche de Hoegaarden | Flanders, Belgium | Top fermentation | Wheat beer from 50% wheat and 50% malt |
| Lambic | Brussels area, Belgium | Spontaneous top fermentation | Wheat/barley beer with sour or estery flavour depending on age |
| Gueze | Brussels area, Belgium | Bottle fermented | Blend of young and old lambic, bottle fermented and aged for at least a year. Dry, fruity and highly carbonated (5–5.5% abv) |
| Porter | London, UK | Top fermented | Deep ruby colour from brown malts and made with London water, high in bicarbonate (6–7% abv) |
| Provisie | Oudenaarde, Belgium | Top fermented | Sweet dark ale aged for a minimum of 2 years and up to 25 years (5.5–6.5% abv) |
| Rauchbier (smoked beer) | Bamberg, Bavaria, Germany | Bottom fermented | Dark beer with smoked taste due to practice of drying malts over open beech wood fires |
| Saison | Walloon region of Belgium | Top fermented | Estery amber-coloured bottle conditioned ale (5.5–6.0% abv) |
| Scotch ale | Scotland, UK | Bottom fermented, low fermentation temperature | Dark copper brown, strong sweet ale with creamy taste (7–10% abv) |
| Steam beer | California, USA | Top fermentation using bottom yeast at relatively high temperature | Well-hopped amber coloured highly carbonated beer (sound of gas escaping from broached casks gives name). (4.5–5.0% abv) |
| Stout | Sweet stout (UK) Dry stout (Eire) | Top fermentation | Very dark, highly hopped made with up to 10% roasted unmalted barley and possibly caramel malt. Dry version (Guinness) more alcoholic (4.0–5.0% abv) |
| Trappist beers | Five brewing abbeys of Belgium (Chimay, Orval, Rochefort, St Sixtus, Westmalle) and one remaining in Holland (Schaapskooi) | Mainly top fermentation; the Dutch brewery produces a bottom fermented Pilsener-type lager as well | Brown or amber colour, bottle conditioned. Usually several beers of different strengths produced (5–12% abv) |

goal of fermentation technologists has been to identify more rapid maturation procedures. This has been successful to the point that most lager beer is now produced by a process which takes days as opposed to months to complete. The only certain way of distinguishing lagers made by a traditional process from their counterparts made by foreshortened modern processes is by knowledge of specific brands made at breweries where the plant and process are known.

This highlights another aspect of beer classification, particularly those with a long heritage. This is the practice of naming beer styles after the region of their origin. There are numerous examples including 'London porter', 'Pilsener' (from Plzeň, Bohemia in the Czech Republic), 'Münchener' (Munich, Germany) and 'Dortmunder' (Dortmund, Germany). These beers have distinctive characters and the original versions are produced using traditional processes. They owe their geographical origins to several factors, especially local availability of a particular combination of raw materials that lend themselves to the type of beer being produced. In addition, the mineral composition of the local water supply was usually a crucial influence.

Derivative versions of these classic beer styles are now brewed worldwide, arguably not always with great success. The example *par excellence* of an exported beer style is the pale Pilsener type lager which is now the most commonly brewed international beer style. Beers of this variety frequently have the word 'Pils' attached to the brand name. They are now brewed using a large range of techniques ranging from the traditional low-temperature fermentation and long maturation phase through to relatively rapid high-temperature fermentations and maturation in a few hours using immobilised yeast reactors. Various sources of fermentable sugars, other than the original malted barley, are used and the mineral content of the brewing water is adjusted so that any local supply is usable. Even zero or low alcohol variants are available.

Ales of the types brewed in the United Kingdom were originally distinguished from beers in that, although both were made from malted barley, ales were not bittered by the addition of hops, whereas beers were. In another sense, ales were considered to be country beers, whereas beer had urban associations. Thus, hops were introduced into the UK probably from Holland and Belgium in the fifteenth century and naturally their spread from the cities of importation to the provinces took longer. By the eighteenth century most ales were hopped but the term was reserved for paler country beers, which were distinct from dark town-brewed beers (Sambrook, 1996). In the modern sense ales have been distinguished from lagers by the use of fermentations employing top-cropping and bottom-cropping yeast strains, respectively. In addition, ales tend to be fermented at higher temperatures than lager beers.

In fact, the march of progress has rendered obsolete many of these points of distinction. Thus, in modern breweries there has been a tendency to use increasingly higher fermentation temperatures for lagers in order to shorten vessel residence times and this distinction between ales and lagers is becoming blurred in many breweries. The production of traditional ales is associated with the use of open shallow fermenting vessels, which facilitate removal of the yeast top crop. However, in many large modern breweries, deep cylindroconical vessels are now used for both ale and lager fermentations. In these cases manipulation of the conditions in the fermenter allow ale yeast strains to form bottom crops in the same manner as lager yeast strains.

It may be appreciated from the foregoing discussion that there are now almost two parallel brewing industries. The first produces a traditional beer using a process which has been almost frozen at a particular point in time. The products devolving from these processes constitute the parental classic beer types. The second industry is derived from the first and is typified by the modern often large-capacity brewing factory. Here the same beer types may be produced but using traditional processes, which have been adapted to afford greater efficiency, or by the introduction of totally new methods for achieving the same end products.

By a process of elimination it may be supposed that the unique points of difference between beer types on the world stage are the raw materials used to make the wort and the nature of the micro-organisms used to catalyse the conversion to beer. The vast majority of beers are made by fermentation of an aqueous extract of malted barley, possibly supplemented with other sources of fermentable sugars. The types and blend of malts used are characteristic for each beer type. In particular, the malt supplies much of the colour of the beer as well as flavour. Thus, Pilsener lager beers are made from pale malts whereas black and chocolate malts are roasted to supply colour and flavour to dark beers such as stouts. Similarly, the varieties, rate and point of addition of hops to beers are characteristic. True Pilsener lager beers are well hopped and traditionally use the renowned Bohemian Saaz variety of hop.

UK cask-conditioned ales use hops in the copper for bittering purposes, possibly supplemented with other aroma hops added to the cask. Traditional varieties include Fuggles and Goldings.

As with process conditions, both sources of fermentable sugars and hop flavourings have been subject to continuous developments. Brewing in Germany is still subject to the *Reinheitsgebot*, or beer purity laws, which preclude the use of ingredients other than water, malt, hops and yeast. Other countries have less proscriptive legislation and many other sources of fermentable extract are now used. These additional sugars, termed brewing adjuncts, may be obtained from cereals, either malted or unmalted, other crops such as rice or from semi-refined sugar syrups. Similarly, many new varieties of hops are now available and frequently they are used in forms in which the active ingredients have been isolated and rendered into forms which make them easier to dose into beer, or render them into a form in which they have new more desirable properties. It is no longer possible, therefore, to provide a single definitive recipe from which a particular type of beer can be brewed and this represents another way in which specific beer styles have drifted from the original parental type.

Wheat beers are made partially or entirely from malted wheat. Typically they are produced by top fermentation at relatively high temperature. As a group they are lightly hopped and many have a characteristic clove or phenolic flavour. This feature demonstrates the critical influence of the strain of yeast used for fermentation. The clove-like flavour is due to the formation of a group of yeast metabolites, particularly 4-vinyl guaiacol and 4-vinyl phenol via the decarboxylation of $p$-coumaric and ferulic acids, respectively, during fermentation. The production of these compounds from precursors, which derive from malted cereals, is under the control of a gene, termed 'POF' for 'phenolic off-flavour' (Section 8.1.3.1). The presence of the POF gene in the yeasts used to make wheat beers is considered beneficial and essential. As the name implies, the formation of these compounds in other beers would be considered a grave flavour defect. Occasionally they may arise by contamination with a so-called wild yeast strain, which carries the POF gene. In this sense the brewer's use of the term wild yeast is similar to that of the gardener who describes a weed as a flower in the wrong place.

This serves to demonstrate that almost without exception the strain or mixture of strains of yeast used to produce a particular type of beer is the single essential component, which cannot be substituted. Undoubtedly yeast strains were selected by accident as being the most suitable for the particular style of beer. Properties of note would include flocculence characteristics, top or bottom cropping, optimum growth temperature and ethanol tolerance. In addition, and perhaps of greatest importance, is the crucial influence of the yeast strain on the development of beer flavour and aroma constituents. Even this aspect of brewing has been subject to change. It has always been common practice to select new strains of yeast from existing populations, which have properties in some way more suitable than the parental type. In recent years, developments in genetic engineering have afforded new opportunities for more directed manipulation of yeast properties, although such altered yeasts have not yet been used at commercial scale.

Not all beers use yeast strains as the sole fermenting agents. In this respect the Belgian lambic beers, produced in a single area close to Brussels, are unique. The wort is made typically from 40% unmalted wheat and 60% malted barley and contains high concentrations of dextrins. High dosing rates with hops that have already been used for brewing and consequently contain little bittering substances but retain antiseptic properties, are included in the process of wort manufacture. The unique aspect of these beers is that the fermentation is spontaneous. The wort is allowed to cool in large shallow open vessels, which are located in rooms where good ventilation provides the maximum opportunity for contamination with the microbial flora of the room. After holding overnight in the cooling vessel, the wort is transferred to wooden casks where it receives a further natural inoculum. The fermentation takes place over a period of several months in the casks and is typified by successive blooms of several distinct microbial populations.

First, yeasts of the genus *Kloeckera* and various enteric bacteria grow and produce some acetic acid (Martens *et al.*, 1991). After 2–3 weeks *Saccharomyces* yeasts overgrow the initial microbial population resulting in the formation of ethanol and some esters. After some 3–4 months, bacteria belonging to the genera *Lactobacillus* and *Pediococcus* become predominant, resulting in the formation of much higher concentrations of lactic acid. Finally, after 5–8 months, a further yeast population develops, in this case species of the genus *Brettanomyces*, typically *B. bruxellensis* and *B. lambicus*, producing additional flavour and aroma components (see Section 8.1.3.2).

The lambic fermentation may last for up to two years. Young lambic beer may be sold directly from the cask as *lambic doux*, or after three years in cask it may be sold as old *lambic vieux*. Another variant, *faro*, is lambic sweetened by the addition of sugar. More commonly, lambic of various ages is used as a base to produce other beer types. Thus, lambic beers of various ages are blended to obtain what is considered a product with a desired balance of flavours, lightly filtered and bottled. The resultant *gueuze* is subjected to a lengthy secondary fermentation, frequently of two or more years, in bottle to produce carbonation and allow the flavour to mature. Another variation is that in which young lambic is blended with whole macerated fruit prior to bottling for secondary fermentation. Such beers are named after the fruit used, for example, *kriek* (cherries), *framboise* (raspberry), *cassis* (blackcurrant), *pêche* (peach) and *muscat* (grape).

As with other countries, the techniques applied to brewing lambic beers have undergone some development and modifications to the basic process have been introduced both to improve process efficiency and in response to consumer demand. In this case gueuze is made from a blend of traditional lambic and a base beer produced by wort fermentation using a cultured mixture of *Brettanomyces* yeast, *Lactobacillus* and *Acetobacter*.

Some new styles of beers have arisen in recent years. Concerns over drinking and health have provided a market for low- or zero-alcohol beers. Low-alcohol beers are not new, for example, the German *Malzbier* containing 0.5–1.0% abv has a long heritage. Similar beverages, often called 'near beers', are produced in many countries, usually with an alcohol content of less than 0.6% abv. However, these tend to be specific highly flavoured malty beverages sold as tonics for their high nutritive contents. More recently, a market has developed for low or zero alcohol versions of existing beers, in particular, Pilsener-type lagers. Usually there is a legal definition of the alcohol content, typically less than 1.0–2.0% abv for low alcohol and less than 0.05% abv for zero alcohol beers (Davies, 1990).

Several processes have been developed for production of these beers (Muller, 1990; Regan, 1990). These are of two types: first, removal of ethanol from a fully fermented beer; second, manipulation of fermentation conditions to produce a beer with low or zero alcohol content. Both approaches have advantages and disadvantages.

Removal of alcohol from a fully fermented beer may be achieved by vacuum distillation or evaporation, dialysis or reverse osmosis. Distillation and evaporation make use of the high volatility of ethanol compared to other beer components. The use of a vacuum is favoured since it allows comparatively low temperatures (40°C) to be used, thereby, minimising potential flavour perturbations due to thermal damage. A typical single-stage vacuum evaporation system is shown in Fig. 2.1. The beer in-feed is preheated and evaporated and the resultant mixture of beer and vapour fed into a separating column. Here the alcohol vapour is taken off from the top and condensed prior to collection. The de-alcoholised beer is taken from the base of the separator and cooled. With single units the beer is recirculated in order to achieve the desired alcohol content. Multiple units allow a single pass operation, which will reduce ethanol concentrations to less than 0.05% abv, the maximum limit for alcohol-free beers.

Evaporative or distillation processes have the disadvantage that some stripping of flavour volatiles is inevitable. Reverse osmosis and dialysis approaches to de-alcoholisation rely on forcing ethanol across a semi-permeable membrane by generating a pressure differential or by establishing a concentration gradient, respectively. These plants are operated at low temperatures, thereby reducing the risks of heat damage to the beer but they are not practicable methods for achieving very low ethanol concentrations. These methods may also remove desirable flavour components with ethanol.

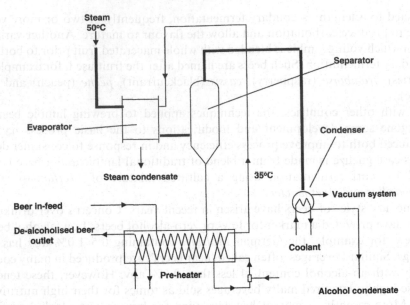

**Fig. 2.1** Schematic of plant for production of de-alcoholised beer using vacuum evaporation (redrawn from Regan, 1990).

Manufacture of low or zero alcohol beers by de-alcoholisation allows the brewery to produce a normal strength beer stream from which a desired proportion may be diverted and treated. However, this flexibility is an expensive option in terms of installation and running costs of the de-alcoholisation plant. The other approach is to manage fermentation so as not to produce alcohol in the first place. This may be achieved in several ways but the general guiding principle is that it is necessary to produce beer flavour and aroma components without ethanol formation and to ensure that undesirable wort flavour components are eliminated. Capital cost associated with these methods is low because little if any modification to existing plant is needed.

Muller (1990) discussed a number of approaches that appear superficially feasible, but, as the author points out, lack practicality. For example, fermentation of a very-low-gravity wort either produced specifically for the purpose, or obtained by dilution of normal-gravity wort. These methods fail since insufficient flavour compounds are generated during fermentation. Production of wort using temperatures below 60°C results in very low fermentability due to lack of gelatinisation and hydrolysis of starches. However, malt flavour components may still be extracted. In practice, run-off rates are unacceptably low because of the presence of starch. Similarly, arresting a standard gravity fermentation at a point where ethanol concentration is low is also unsatisfactory because undesirable wort carbonyls are still present.

A more promising method is dilution of very-high-gravity (1080) beers. In this instance the increased formation of flavour volatiles associated with very-high-gravity brewing (see Section 2.5) is corrected for by post-fermentation dilution to a low ethanol concentration. Pilot scale production of low alcohol beers using this method was described by Muller (1990). A significant concentration of esters and higher alcohols was achieved in a beer with an alcohol content of less than 2% abv. However, foam performance was very poor, an effect attributed to over-dilution of the available total soluble nitrogen.

Another way of producing wort with low fermentability is to perform the mashing step at a high temperature. Here the premise is that degradation of starches by α–amylases produces small quantities of fermentable sugars together with high concentrations of non-fermentable dextrins. Conversely, β-amylases produce mostly fermentable sugars in the form of maltose. β-amylases are less heat stable than α-amylases and thus, use of a high mashing temperature (80–85°C) produces a wort with sufficiently low fermentability to give a low alcohol product. This method has been used successfully at commercial scale. It requires very accurate control of mashing temperature since if this is too low fermentability increases, if too high the wort will have an elevated starch content.

The most often used limited fermentation approach is the cold contact method in which wort is exposed to yeast for a brief period at low temperature and under anaerobic conditions. This treatment precludes yeast growth and ethanol formation but permits removal of wort carbonyls. This method is particularly suited to a continuous process using immobilised yeast reactors and it is in this form that it has found successful commercial application (see Section 5.7.2.1).

Another comparatively modern development in beer styles is the so-called diet, low-carbohydrate or light ('lite') beers. These are fermented so that there is little or no residual gravity. This is achieved by producing worts in which dextrin contents have been reduced by addition of amyloglucosidase. Alternatively, super-attenuating yeast strains may be used which possess amyloglucosidase activity. Several authors have described construction of such yeast strains using genetic engineering techniques although none has been used for commercial brewing (Lancashire et al., 1989; Vakeria & Hinchliffe, 1989; Hansen et al., 1990).

These beers are suitable for diabetics because of their low carbohydrate content and it was to meet this need for which they owe their German origin, where they are known as Diätbier or Diät Pils. Occasionally and incorrectly, they may be referred to as low-calorie beers. Obviously conversion of dextrins to fermentable sugars enhances ethanol formation and this is not inconsiderably calorific. In order to be a truly low-calorie beer it is necessary to reduce both carbohydrates and ethanol.

### 2.2.1 Beverages related to beer

As discussed in Chapter 1 there are many alcoholic beverages that may be categorised as beers in that the principal source of fermentable sugar is a grain but which are obviously distinct from the European beer tradition. There are numerous examples of such beers as described in Table 1.1. Two members of this group are worthy of further discussion since they are of considerable commercial importance: saké and sorghum beer.

2.2.1.1 *Saké.* Excellent reviews of the nature of saké and full details of its production may be found in Kodama (1970, 1993) and Inoue et al. (1992a). It is a beer-type beverage produced by fermentation of sugars derived from rice and typically contains 14–17% alcohol by volume. It is clear, pale yellow, has both sweet and acidic flavour notes and an estery aroma. Saké is the national drink of Japan, although it probably originated in China. It is still made in the latter country although the Japanese and Chinese processes have now diverged. Japanese saké is fermented using specific yeast strains and the mould *Aspergillus oryzae*. Chinese saké is prepared using a microbial flora which develops spontaneously and contains yeasts as well as various moulds belonging to the genera *Rhizopus, Mucor, Penicillium, Absidia* and *Monascus*, in addition to *A. oryzae* (Kodama, 1993).

As with conventional beer brewing, saké can be made using a traditional process, which is highly labour-intensive. However, modern automated alternatives to many of the steps have now been introduced. The mould, *A. oryzae* provides amylases and proteases to release sugars and amino nitrogen from the rice grains.

Large grain rice varieties are used and these are first pre-treated using a mechanical milling treatment to remove the bran. This treatment removes much of the rice protein and lipids, both of which are considered undesirable. The proportion of bran removed, on a weight basis, compared to the remaining grain is termed the polishing ratio. After the polishing treatment the rice is washed then steeped in water. Steeped rice is then steamed to gelatinise starch and denature proteins. After cooling the steeped rice is inoculated with a culture of *A. oryzae* in the form of a dried mass of mould, rich in spores and rice, known as *tane-koji*. The mould proliferates on the wet rice covering it in a mycelium, which penetrates into each grain. This mixture is known as rice-*koji*.

For the main fermentation a starter culture is prepared, known as *moto* ('mother of saké'). In the traditional process a mixture of steamed rice, water and rice-*koji* is prepared and held for a period for saccharification to occur. During this time a complex bloom of bacteria and yeast develops then dies off. This is accompanied by the formation of lactic acid. At this point the infusion is inoculated with a pure culture of saké yeast (*Saccharomyces cerevisiae* var. *saké*). This yeast has the important characteristic of being highly ethanol tolerant. The mixture is held for a number of days during which rapid yeast growth takes place. The *moto* is used as the starter culture for the main fermentation, or *moromi*.

The *moromi* fermentation is conducted in large open vessels in which sequential batches of steamed rice, water and rice-*koji* are added to the *moto* starter. The batch addition process maintains a high yeast concentration and reduces the risk of contamination. The fermentation takes roughly two weeks, during which time the ethanol concentration increases to approximately 17–20% by volume. At this point flavour adjustments may be made by addition of lactic and succinic acids, glucose and sodium glutamate. Ethanol may also be added to achieve a final concentration of 20–22%. The process is completed by allowing solids to sediment and clarifying the liquid by filtration. Following a sometimes lengthy maturation storage phase, the saké is pasteurised and bottled.

2.2.1.2 *Sorghum beer.* Sorghum beer is the traditional native alcoholic beverage of South Africa. Similar beers are associated with several African countries and many names may be used, for example, *joala* (Sesotho), *utshwala* (Zulu) and *utywala*

(Eastern Cape tribes) (van Heerden, 1989). The modern version is called sorghum, Bantu or opaque beer. It has a long heritage and in its many guises a variety of sources of fermentable sugars other than sorghum may be used, for example, maize, millet, cassava and malted barley. Three scales of production are currently in operation: within a tribal context, as an urban domestic undertaking and at commercial brewery scale (Novellie & de Schaepdrijver, 1986).

Sorghum beer is described as being opaque, slightly viscous, usually with a pink tinge and sour tasting. When sold in unpasteurised form sorghum beer may be still slowly fermenting when consumed. The industrial product is now made in a commercial brewery using stainless steel plant. Batch sizes are modest, typically 10 000–20 000 litres.

The beer is made from malted sorghum and an adjunct which is usually maize but can be other cereals, including unmalted sorghum. In the first stage a sorghum malt flour is prepared which is suspended in water at a temperature of 48–50°C. This is inoculated with a culture of *Lactobacillus leichmanii* and a souring process takes place in which lactic acid is formed. The relatively high temperature helps prevent growth of other micro-organisms. Furthermore, little sugar is formed since the acid conditions inhibit sorghum amylases. When sufficient lactic acid is judged to have been generated (approximately pH 3.2) more water is added, together with maize adjunct and the mixture heated to boiling point to gelatinise starch and soften the maize.

After cooling to 60°C, further sorghum malt is added and saccharification takes place due to the amylases present in the sorghum malt. The wort so formed is then clarified by passing the process stream through in- line centrifugal decanters. This part of the process is termed straining, a reference to an earlier method of accomplishing this separation. The fermentation is started by the addition of a dried yeast culture (see Section 7.1.1.3), is conducted at relatively high temperature and is rapid.

In modern plants the beer may be pasteurised and bottled with added sugar and yeast to allow for secondary fermentation, mainly for carbonation. More usually it is sold still fermenting in draught form or in plastic-coated cartons of various capacities and which are fitted with vented closures.

## 2.3 The brewing process

The major stages that together comprise the brewing process for a European-style beer made by a traditional batch process are shown in Fig. 2.2. The process consists of three phases: wort manufacture, fermentation and post-fermentation processing. Each of these phases contains several distinct steps. The precise detail of each step depends, to some extent, on the nature of the beer being made and the plant used. However, there are common themes and these are described in this section.

The raw materials for wort production are water, malt, hops and, possibly, adjuncts. In addition, other materials may be used which assist in the process of wort preparation. The essential features of these ingredients are discussed in following sections. These are termed process aids and are distinguished from raw materials in that they may not persist into the finished beverage.

**Fig. 2.2** Major steps in the brewing process.

### 2.3.1 *Malting*

Several cereals may be malted; however, the description given here is the process applied to barley. In the malting process the barley grain is encouraged to undergo controlled germination (Briggs, 1987). This is initiated by wetting the grains, known as steeping. In the subsequent germination phase enzyme systems are activated as the barley grain begins to mobilise starch reserves to provide carbon and energy for development of the embryo. At an appropriate point the germination process is arrested by application of heat, termed kilning. This stabilises the grain such that in malt the relevant enzymes and reserve materials are available for subsequent extraction and further degradation to release fermentable sugars during wort production.

Particular varieties of barley (*Hordeum distichon* and *H. vulgare*) are cultivated specifically for malting. Key properties are an ability to undergo even germination within a given time period, possession of disease resistance and, most importantly, formation of a large grain containing an appropriate balance of starch and nitrogen. In general, a low nitrogen content is preferred.

A cross-sectional diagram of a barley grain is shown in Fig. 2.3. The endosperm consists of a protein mesh in which starch grains, both large and small, are embedded. Starch accounts for 55–65% of the total grain weight. Some 75–80% of the starch is in the form of the branched polymer, amylopectin (D-glucose, $\alpha$-$(1\rightarrow4)$ and $\alpha$-$(1\rightarrow6)$ linkages) and 20–25% amylose (D-glucose, $\alpha$-$(1\rightarrow4)$ linkages only). Important protein components of the barley grain include globulins, albumins, hordein and glutelin. The first two of these are mainly enzyme proteins, the others are structural and degraded during malting. The relative proportions of starch and protein influence the appearance of the endosperm. Thus, it is described as 'steely' or 'floury' where protein

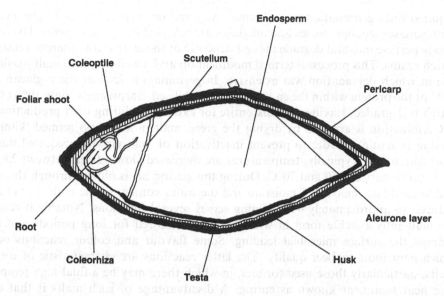

**Fig. 2.3** Diagrammatic representation of a cross-section through a barley grain.

or starch levels are high, respectively. Utilisation of the starchy and nitrogenous components of the endosperm is catalysed by amylases and proteases, which are located in the aleurone layer of the grain.

The barley grain contains many other components that contribute to wort composition including sucrose, vitamins, minerals, polyphenols, nucleotides and lipids. Some components have the potential to produce problems during wort production. For example, malting barley endosperm contains approximately 4% β-glucans. These must be degraded during malting, since barley β-glucanases are heat labile and are inactivated during wort mashing. Failure to degrade β-glucans results in high viscosity worts, which create problems in run-off. Barley varieties containing large quantities of β-glucans are unsuitable for brewing.

Steeping is managed in a variety of ways depending on the sophistication of the maltings. It is common practice to soak the barley grains in water with periods of exposure to air. The steep water may contain a biocide (see Section 8.2.1.1) to minimise surface microbial growth. During steeping the water content of the barley grains increases to 42–45% and this initiates germination, termed chitting. Steep water, which is removed from the barley grains, contains some tannins, which can impart bitter taints in beers. Exposure of the embryo to moisture stimulates the formation of the plant hormone gibberellic acid. This is transported to the aleurone layer, where it stimulates activation of endosperm-degrading enzymes. Occasionally exogenous gibberellic acid may be added to the grains to encourage more rapid germination, although this is now relatively uncommon.

During the germination stage the growing barley is held as a bed which is slowly turned over to keep each grain separate and facilitate even germination. In traditional floor maltings this is a manual process, whereas in modern facilities it is automated. A high humidity is maintained and a temperature between 13 and 16°C. Cooling is required since germination is exothermic. As germination proceeds endo-glucanases, pentosanases, endoproteases and amylases are released from the aleurone layer and slowly perfuse into and degrade the cell structure of the endosperm, thereby releasing starch grains. This process is termed modification and a well-modified malt would be one in which degradation was extensive. In germination 75% of the β-glucan and 40% of the protein within the endosperm is solubilised. Surprisingly, only 10% of the starch is degraded, leaving the bulk entire for extraction during wort production.

Germination is arrested by drying the green malt in a process termed 'kilning'. Drying is gentle in order to prevent inactivation of malt enzymes required during wort formation. Typically, temperatures are increased slowly from between 25 and 30°C up to between 60 and 70°C. During this period air is blown through the malt bed to facilitate removal of moisture and the water content of the malt is gradually reduced to approximately 4%. Kilning serves several functions. Notably it renders the malt into a stable form in which it may be stored for long periods and also reduces the surface microbial loading. Some flavour and colour reactions occur which contribute to beer quality. The latter reactions are characteristic of certain malts, particularly those used for ales, in which there may be a final high temperature heat treatment known as curing. A disadvantage of such malts is that considerable enzyme inactivation occurs and there is now a trend towards using paler malts supplemented with other sources of colour and flavour added later in the wort manufacturing process.

Malt is stored in 'silos' (Fig. 2.4), both in the maltings and at the brewery. On malt intake, malt dust is distributed into the air and may well be a prime source of brewery specific micro-organisms (see Section 8.1.4.2).

**Fig. 2.4** Malt silo (kindly provided by Richard Webster, Bass Brewers).

## 2.3.2 Adjuncts

Adjuncts, or secondary brewing agents, are defined as any source of fermentable sugar that is used in the brewing process and is not derived directly from malted barley. They are widely used in all countries, although the German beer purity law, the *Reinheitsgebot*, prohibits their use in beers for domestic consumption. Indeed in Germany, adjuncts are taken to mean any addition to the brewing process other than water, malt, hops and yeast (Stewart, 1995).

Two groups of adjunct are used, solid and liquid. Solid adjuncts require some processing and are usually incorporated into the mash. Liquid adjuncts can be used with no treatment and are added to the copper. In addition, liquid adjuncts are used post-primary fermentation in the form of 'primings' to provide fermentable sugar for secondary fermentations and as a means of adjusting beer sweetness.

Adjuncts may be used simply as sources of fermentable sugars, which are less expensive than malt, although this is not usually the sole motive. Thus, they are used for their contribution to beer colour and flavour and they may be added to improve beer foam performance. Liquid adjuncts added to the copper serve as a valuable means of improving the productivity of the brewhouse. This is of particular importance in high-gravity brewing (see Section 2.5) where the wort sugar concentration may be conveniently increased by addition of liquid syrups. Perhaps most importantly, they provide the brewer with a positive method of controlling the gross composition of wort. For example, influencing the ratio of carbon to nitrogen as a means of producing beers with particular flavours.

Although individual adjuncts may be less expensive alternatives to malt, other costs are associated with their handling. For example, liquid syrup adjuncts require storage tanks, which may need to be heated to maintain a manageable viscosity. In addition, storage must be hygienic to prevent microbial spoilage. Although many solid adjuncts by-pass the malting stage of brewing, they may require specific plant for extracting fermentable sugars.

Commonly used solid adjuncts are barley, wheat, maize, rice and triticale. These come in many forms to facilitate extraction of sugars. For example, in the form of dried rolled flakes, as a milled grain, in micronised or torrefied form (product of rapid heating of cereal grains), as a flour (brewer's wheat flour) and as 'grits'. The latter category is the commonly used adjunct forms of rice and maize, particularly popular in the United States. Grits are by-products of processing of the cereals and consist mainly of endosperm. As discussed subsequently, a key part of the mashing process is to subject malt and other cereal sources of extract to a heat-step during which starch grains are gelatinised and therefore made accessible to enzymic degradation. In a typical mashing regime this is achieved within a temperature range of 63 to 65°C. However, the starch from maize and rice grits requires a higher temperature to gelatinise starch (65 to 75°C) and where these adjuncts are used it is necessary to use a separate cooker to provide a high temperature pre-treatment prior to addition to the main mash.

Liquid adjuncts come in the form of various syrups, which are characterised by their concentration, purity, colour and sugar spectrum. Thus, they may range from an almost pure glucose syrup, which is entirely fermentable, through to a crude hydrolysate of a cereal starch, which has been subject to treatment with various amylases. In between are a whole range of syrups, which may contain a mixture of fermentable sugars and dextrins. Less pure cereal extracts, particularly those derived from barley and wheat, may contain significant concentrations of nitrogen. Clearly the use of these will influence yeast nutrition and by implication the formation of ethanol and flavour metabolites during fermentation, in a totally different way to pure sugar adjuncts. The choice depends upon cost and the desired application. Caramels are a specific group of liquid adjuncts that are prepared by subjecting sugar solutions to a high temperature heating process. Caramelisation is promoted by aluminium ions which act as catalysts. These are used mainly for colour and flavour adjustment and now frequently in preference to dark roasted malts. The latter group of speciality malts, which are used principally as a source of colour and flavour, as opposed to extract, may also be considered as being adjuncts.

### 2.3.3  Brewing water

The ionic composition of water used for wort preparation exerts a crucial influence on the successful outcome of the brewing process. For optimum process performance and in order to maintain the highest product quality, brewing of different beer styles requires water (liquor) with particular ionic spectra. This fact has long been recognised by empirical observation even though the underlying science was not understood. Thus, availability of appropriate local water supplies accounts for the rise to prominence of the more famous centres of brewing excellence. For example, the reputations of towns and cities such as Pilsen, Dublin, Burton-upon-Trent, Dortmund and Munich are largely based on the qualities of the local water. Pilsen has very soft water with less than 10 ppm each of calcium, magnesium and sulphate and 10–20 ppm of bicarbonate (note: ppm or parts per million is equivalent to $mg\,l^{-1}$). Burton-upon-Trent well water is high in permanent hardness. It contains 250–300 ppm calcium and bicarbonate, more than 600 ppm sulphate and 60–70 ppm magnesium. Munich water is intermediate with 70–80 ppm calcium, 10–20 ppm sulphate and magnesium together with 150 ppm bicarbonate.

For many breweries, particularly those using a traditional process the composition of local water supplies continues to be of importance as it is used without modification to ionic composition. However, in many modern plants this has become an irrelevance since it is usual to de- mineralise all brewing liquor and then add back salts to achieve a desired ionic composition. This allows a single brewery to produce water appropriate for any beer style. It follows that it is advantageous if the local supply of water is soft since de-ionisation requirements are small. This is especially so since in any case most water is used for non-brewing purposes and has to be softened for most applications to avoid corrosion of plant.

The ions in brewing water have vital roles in the various stages involved in wort formation; they contribute to yeast nutrition, they influence yeast technological properties and have an impact on beer flavour. Key ions must be present in sufficient concentrations to exert positive effects but not be too high to cause inhibition or impart undesirable flavours. More importantly key ions must be present in balanced quantities. The sum of the effects on the brewing process and beer quality of the ionic composition of brewing liquor are complex since many synergistic and antagonistic interactions are possible. Some of the more notable effects are summarised in Table 2.2. With regard to wort production the most significant requirement is regulation of pH, especially during mashing and to a lesser extent during the copper boil. It is essential to maintain a low pH for efficient starch breakdown and proteolysis during mashing. Wort buffering capacity is provided by phosphates derived from malt. The pH is reduced by interactions between calcium ions and phosphates and with proteins and polypeptides which contain glutamate and aspartate in reactions which liberate protons. Taylor (1990) argued that interactions between calcium and amino groups was likely to be of greatest importance to control of wort pH since phosphate is a relatively poor buffer at wort pH values (pH 5.0–5.3).

**Table 2.2** Effects of the ionic composition of water on the brewing process.

| Ion | Effect |
| --- | --- |
| Ammonium | Indicative of contamination with decomposed organic material. |
| Calcium | One of the most significant ions with multiple effects. Interacts with phosphate and proteins to reduce pH of mash and promote formation of a bright wort with good run-off. Precipitates oxalate from wort, which can cause hazes and gushing and hazes in beer. Activates α-amylase and proteases in mash. Promotes flocculation of yeast at the end of fermentation. Inhibits extraction of hop resins at high concentrations. Has a bitter astringent taste. |
| Copper | Toxic to brewing yeast at high concentrations. Eliminates $H_2S$ from beer as insoluble sulphide. |
| Iron | Toxic to yeast. Can produce hazes and adverse colour changes in beer. |
| Magnesium | Reduces wort pH by interaction with phosphates but less important than calcium. Important co-factor for many enzymes especially those catalysing dissimilation of pyruvate during fermentation. Essential component of many enzymes involving ATP. |
| Manganese | Co-factor of many yeast and malt enzymes. |
| Potassium | Imparts saline taste to beer. |
| Sodium | In combination with chloride contributes to beer sweetness. |
| Zinc | Inhibitory to yeast growth at high concentration (> 1 ppm) but stimulates fermentation at lower concentrations (0.1–0.3 ppm). |
| Bicarbonate | At high concentrations (> 100 ppm) causes increase of mash pH and concomitant reduction in extract formation. |
| Chloride | Inhibits fermentation at high concentrations (> 600 ppm). Contributes to fullness at low concentration but imparts saline taste at concentrations above 400 ppm. |
| Nitrate | Precursor to nitrosamine formation by *Obesumbacterium proteus* |
| Phosphates | Interacts with calcium and magnesium to reduce wort pH during mashing. Essential nutrient for yeast growth. |
| Sulphate | Precursor for sulphur-containing amino acid synthesis by yeast in worts with low amino acid content. Precursor for sulphite formation by yeast which improves beer flavour stability. Precursor for sulphide formation by yeast. |

The beneficial effects of calcium ions may be defrayed by carbonate. The formation of bicarbonate from carbonate, under acidic conditions in the mash, removes protons and increases wort pH. Two protons are removed for each carbonate ion so a relatively small concentration has the potential to produce a large effect on pH. It is essential, therefore, that temporary hardness is removed from brewing water.

Brewing water must meet standards of purity with respect to contamination, both microbial and chemical. Where municipal water supplies are used, standards are guaranteed by legislation. Since the supplier reserves the right to change the source of water without notice, considerable and abrupt changes in ionic composition are possible. Where brewery borehole water is employed, much less variation in ionic composition would be expected; however, pollution of ground waters can be a problem. To avoid this, where necessary, brewing water should be treated to remove organic contaminants and occasionally treated to reduce bacterial loading (see Section 8.2.1.1).

### 2.3.4  Hops

The hop plant (*Humulus lupulus L.*) is a member of the family *Cannabinaceae* that grows in temperate regions of the world. It is used in brewing to impart both bitterness and floral character. The flavour-active components of hops are resins and essential oils, which are produced in the lupulin glands found at the base seed-bearing bractioles of cones of the female plant. The cones are harvested and it is these that are referred to as hops in the brewing industry. In the traditional process the harvested and separated hop cones were dried in oast houses. Now more usually they are dried in a kiln to a moisture content of no more than 12%. The dried hop cones are packaged in bales, or in the United Kingdom in elongated sacks, termed pockets, for delivery to the brewery.

The bitter character of hops is due to alpha acids, also known as humulones. The three most prevalent alpha acids are humulone, cohumulone and adlupulone (Fig. 2.5). The alpha acid content is a characteristic of particular hop varieties and varies

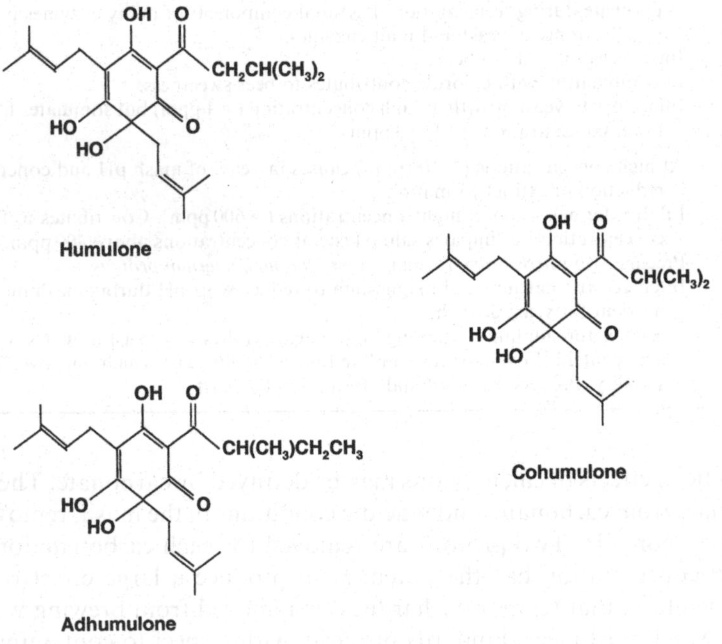

**Fig. 2.5**  Structures of hop alpha acids (kindly provided by Richard Webster, Bass Brewers).

between 2 and 15% of the weight of the hop cone. During the copper boil the alpha acids undergo isomerisation to form the cis and trans forms of the humulones. It is these iso-alpha acids that impart the bitter character to beer. In addition, iso-alpha acids have antiseptic properties and bittered beers have much better keeping properties than the earlier unbittered ales.

The hop oil fraction accounts for 0.05–2.0% of the weight of the cone. They comprise a complex mixture of more than 250 components, which have subtle floral, spicy, citrus and estery aromas and tastes. Varieties of hops which are rich in hop oils are often referred to as aroma hops. These varieties cannot be used throughout the copper boil since the oils are volatile. Thus, they are added towards the end of the boil or possibly post-fermentation, known as dry hopping.

In addition to the dried cones, several other hop-derived products are available for use in commercial brewing. The use of whole hops is now comparatively rare. The simplest processed hop products are powdered and pelleted dried cones. The hop cones are milled and then pressed into pellets and sealed into foil packets from which air has been removed. Removal of oxygen produces a more stable product. Stabilisation may be further promoted by the addition of magnesium hydroxide to form the magnesium salts of the alpha acids. Further heat processing is possible by heat treating the pellet to produce a hop preparation in which the alpha acids are pre-isomerised.

Extracted hop resins are also available. Early commercial processes used organic solvents to extract hop resins, now liquid carbon dioxide is more often employed. The latter avoids the possibility of organic solvent residues being introduced to beers with the hop extracts and by manipulation of the conditions of extraction it is possible to obtain a fraction rich in alpha acids but without unwanted impurities. The same technique can be applied to aroma varieties to obtain hop oil extracts. These extracts may be blended to produce bespoke flavours and aromas for individual customers.

Use of pre-isomerised resins provides a reproducible means of bittering beers with a product which generates a minimum of waste material and is stable. These purified products have the further advantage that they may be modified chemically to introduce desirable properties and remove undesirable characters. Thus, hop derived products are now in use which have enhanced bittering qualities, promote good beer head formation and a reduced potential to generate adverse flavours due to interactions with light (Kember, 1990; Bradley, 1997).

### 2.3.5 Production of sweet wort

The first stage of the brewery process proper is the preparation of 'sweet wort'. The steps involved are (1) milling, in which the malt and any other solid source of extract are converted into a coarse flour known as the grist, (2) 'mashing', in which the grist is suspended in water and heated to prepare an aqueous extract from the grist and finally (3) a separation stage, in which the spent grains are removed to leave partially clarified sweet wort.

The charge of material delivered to the mill (Fig. 2.6) is calculated to produce an appropriately sized batch of wort with a desired sugar concentration. In addition to malt, other solid adjuncts may be included; alternatively, solid adjuncts not requiring

**Fig. 2.6**  Malt mill (kindly provided by Harry White, Bass Brewers).

milling may be added to the grist after milling. Milling is designed to release the contents of the malt endosperm from the husk and to reduce the mixture to a particle size which provides optimum operation in the mashing stage. If the particle size is too large enzymic degradation is inefficient. If the particle size is too small wort separation is impeded. Milling may be wet or dry depending on the composition of the grist and the preferences of the particular brewery.

During mashing the grist is suspended in water and the mixture heated. Many of the components of wort arise at this time by simple solution. Others are formed as a result of reactions between malt-derived enzymes and their substrates, which become possible because of the provision of an aqueous environment and the presence of an appropriate combination of pH, temperature and ions. The elevated temperature gelatinises starch granules, in other words disrupts their crystalline structure, so that they are susceptible to attack by amylases. Both α– and β-amylases from malt are active during mashing, although the latter is more heat labile and its activity does not persist for long in high temperature mashes. Similarly, limit dextrinase is heat labile and is rapidly denatured during mashing.

α-amylase requires calcium ions for activity and if necessary it is usual to add calcium sulphate to the mash liquor to provide this ion. In addition, calcium takes part in reactions with other ions in water and helps maintain a low pH. The optimum pH for amylase activity is approximately pH 5.3. The concerted action of the amylases produces mainly maltose and but leaves significant concentrations of undegraded dextrins. The action of amylases reduces wort viscosity due to starch granules. Elevated viscosity can be caused by malt β-glucans and these carbohydrates may persist since the malt β-glucanase is heat labile. To avoid this problem heat stable commercial enzyme preparations may be added.

Some 35–40% of malt proteins are solubilised during mashing, depending on the temperature and pH. However, malt proteases are heat labile and thus more protein solubilisation occurs during malting than mashing. Lower mash temperatures favour greater concentrations of wort total soluble nitrogen. The lower the mash temperature, the greater the proportion of total soluble nitrogen arising as free α-amino nitrogen. Thus, mashing temperatures of 64–68°C favour rapid starch conversion and high extract formation, whereas, 50–55°C favour high α-amino nitrogen.

Different mashing systems are used depending on the type of beer being brewed. For ale fermentations a single temperature mash, usually 65°C, is employed in a process referred to as infusion mashing. Such a method suits the more highly modified worts, which tend to be used for production of ale worts. In this case the relatively high and constant temperature is appropriate for the formation of both adequate fermentable sugar and total soluble nitrogen.

Lager beers are traditionally brewed using a less well-modified wort. In this case it is necessary to use a relatively low initial temperature to promote proteolysis followed by a higher temperature for starch gelatinisation and amylolytic attack. In a traditional process this is achieved by a process termed decoction mashing in which successive proportions of the mash are removed and boiled then returned so as to increase gradually the temperature of the mash. A disadvantage of this method is that the boiling step denatures a significant proportion of malt enzymes. An alternative and now widely adopted method is programmed temperature mashing in which the temperature is gradually increased in a series of steps so as to allow progressive enzymic degradation of proteins and carbohydrates.

Infusion and decoction mashing require different brewery plant. Infusion mashing systems employ a mash tun for the conversion step (mash conversion stand), often with a Steel's mash mixer to take the milled grist and ensure it is mixed with hot liquor. Care must be taken to ensure that the mash reaches the desired temperature. This is achieved by calculating what temperature of mash liquor is required, termed the striking heat, which corrects for the reduction in temperature due to mixing hot water with the colder grist. This may be assisted by heating the grist with steam prior to mixing with hot liquor. Attemperation in the mash tun may be accomplished by direct injection of steam or by introducing hot liquor into the base of the vessel, termed 'underletting'. In modern installations heating jackets are provided.

The mash tun (Fig. 2.7) is a circular closed stainless steel vessel of hygienic design. Raised in the base of the vessel is a false bottom plate consisting of a mesh of slotted holes. In the headspace of the vessel there are two rotating arms, which spray liquor

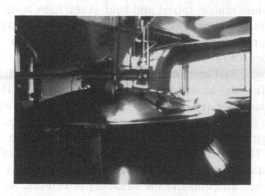

**Fig. 2.7**  Mash tun (kindly provided by Ian Dobbs, Bass Brewers).

onto the surface of the mash, known as 'sparging'. Usually there is a facility for moving the mash using a series of rotating rakes. These may be used during the mashing process itself but mainly serve to remove spent grains by moving them towards a discharge point in the base of the vessel.

The mash is allowed to stand in the vessel for a desired period after which time the grains settle to form a bed through which the wort is filtered. The filtration process is known as the run-off and must be managed carefully so as to avoid compressing the bed. The mash tun is usually fitted with some device to ensure that the wort is removed in an even manner across the whole bed. In the first phase of run-off, the wort is recycled back into the mash tun to allow a stable bed to form and ensure that bright wort is obtained. Once the bulk of the wort has been removed the bed is sparged with hot liquor to remove liquid entrained in the bed. This is allowed to proceed until the concentration of the wort in the run-off falls to a value at which further treatment is uneconomic. The sweet wort may be stored in a holding vessel prior to delivery to the copper.

In traditional decoction mashing systems a number of vessels are used. The simplest arrangement is a three-vessel system using a mash mixing vessel, in which grist and liquor are mixed and where conversion takes place, a copper for heating the proportion of mash removed from the mash mixer, and a lauter tun for separating wort and spent grains. More complex systems double up some of the vessels. In some modern systems a double mashing system is used. This consists of a main mash mixing vessel in which the bulk conversion of the malt grist takes place and a secondary cereal cooker for processing solid adjuncts. The mash mixer is a hygienic stainless steel vessel fitted with a large bottom-located agitator to ensure good mixing. Bottom and side jackets provide controlled heating. Mashing involves a ramped temperature profile in which the first low-temperature stand is provided for maximum activity of proteases and $\beta$-glucanases. This is followed by a second higher-temperature stand for starch gelatinisation and amylolysis. A final even higher-temperature short stand may be incorporated, designed to denature enzymes which may cause problems further down stream.

The cereal cooker is particularly popular in the United States where a large proportion of adjuncts such as rice or corn grits are used to produce pale lager beers with blander flavours than European style lagers. In this vessel the adjunct and a small charge of high amylase malt is mixed and held at the high temperature required to gelatinise adjunct starch grains. Operation of mash mixer and cereal adjunct cooker are arranged so that both processes are completed at a similar time which allows mixing of each prior to wort separation in the 'lauter tun'.

The lauter tun (Fig. 2.8) (German for *refine* or *clarify*) uses a shallower bed than the mash tun and therefore has a considerably larger diameter. Like the mash tun it has a false bottom with slots for retaining the grains and allowing wort to drain away. Sparge arms are provided and a series of rakes. The blades on the rakes can usually be rotated to either agitate the bed in one orientation or push the grains to the outlet main in the other. The provision of raking allows for the use of a finer grind grist which is characteristic of lauter brewing. In addition, pumps may be used to induce a pressure across the bed and improve the efficiency of run-off. As with the mash tun, the bed functions as a filter through which the wort percolates. Initial recycling allows

**Fig. 2.8** Lauter tun (kindly provided by Harry White, Bass Brewers).

the bed to form and sparge liquor is used to extract the maximum volume of wort. Modern lauter tuns are designed to minimise oxygen pick-up which affects beer keeping qualities (Geering, 1996).

Sweet wort and spent grains may be separated using a mash filter (Fig. 2.9). This is a simple plate and frame filter and has the advantage that separation is rapid, efficient and produces a very dry spent grain. However, it has the disadvantage that it is

**Fig. 2.9** Mash filter (kindly provided by Meura Ltd).

complex and does not easily produce a bright wort. A modern version of the mash press overcomes these problems (Hermia & Rahier, 1992). This uses polypropylene filter sheets with a fine pore size such that a very fine grind grist can be used. However, this requires the use of a hammer mill as opposed to the more conventional roller types associated with lauter tun breweries. The fine grind allows use of thin beds since the surface area is concomitantly large. Thus, bright worts are formed at relatively high flow rates.

### 2.3.6    Wort boiling

Clarified sweet wort is delivered to the copper (kettle) where it is subjected to a heat treatment. In beers subject to the *Reinheitsgebot* laws, only hops are added to an all-malt sweet wort and boiled. In other countries additional sources of fermentable sugar in the form of liquid sugar syrup adjuncts may be added with the sweet wort.

The copper boil serves many functions. It sterilises the wort and inactivates all malt enzymes before it is cooled and added to the fermenter. Hop alpha acids are isomerised during the copper boil. This aspect of copper management is dependent on the type of hops used. Varieties which require extensive heat treatment for isomerisation are added early on, whereas those which are added for aromatic hop oil content are added towards the end. Inevitably some compromise is necessary between the need for isomerisation and the potential for loss of aromatic components. In addition, prolonged boiling can result in conversion of alpha acids to humulinic acids, which lack bitterness. Removal of most of the essential oil component of hops in the steam exhaust from the copper is essential for balanced flavour in the final beer. Apart from these hop components other volatiles derived from malts, many of which impart undesirable 'vegetable' odours, are also lost. For example, S-methylmethionine derived from malt decomposes to dimethyl sulphide during mashing and in the copper boil. The latter is an essential contributor to lager character (see Section 3.7.5).

Wort boiling assists with clarification and removes substances which may cause problems in downstage processes. Thus, polyphenols from malt and hops react with some proteins to form insoluble precipitates. In addition, some of the other proteins coagulate. The resultant mixture is termed 'hot break' or 'trub'. Most of this material is separated from the wort during transfer from copper to fermenter. Oxalate, which can form beer haze, is precipitated as the insoluble calcium salt. The high temperatures in the copper promote many other chemical reactions. Maillard reactions between reducing sugars and amines form melanoidins, which contribute to beer colour. These melanoidins may also displace aldehydes from sulphite adducts in packaged beers, thereby contributing to staling. The boiling process provides an opportunity to concentrate the wort. This allows correction for dilution in the mashing stage due to sparging operations.

Several copper designs are in use (Fig. 2.10). The name suggests the metal originally used for their construction; however, modern vessels are invariably made from stainless steel. Early versions were simply open vessels with a flue for escape of steam and heat provided by an underbuilt fire. Modern coppers are usually heated by steam, either with internal heat exchangers or by circulating the wort through an external

**Fig. 2.10** Copper (kindly provided by Richard Webster, Bass Brewers).

loop fitted with a tube and shell heat exchanger known as a calandria. Such systems use convective forces to drive the boiling wort around the loop.

After the boil is completed, solids in the form of 'trub' (see Section 2.4.4) and any hop material have to be separated from the hot wort before it is cooled and delivered to the fermenting vessel. Various types of plant may be used to accomplish this. Where whole hops are used plant is required to separate the spent cones. Traditionally this is achieved in ale breweries using a hop-back, which is similar in design to a mash tun, in that it has a false-slotted base and sparge arms. Like the mash tun the principle of operation is that the spent hops form a bed through which the hot wort passes and in so doing filters out trub. Sparging provides a means of recovering some of the wort entrained in the hop bed. Traditional lager brewers employ a similar device called a hop-jack or montejus. This is a closed tank fitted with a mechanical agitator. Hot wort is fed in at the top of the vessel and is discharged from the base. Solid materials are retained by an internal mesh, which forms a cage inside the vessel.

Where pelleted hops or hop extracts are used there is insufficient material to make hop-backs or hop-jacks practicable. In this case, which applies to the majority of modern breweries, a whirlpool is used to clarify worts. Several designs are used which are claimed to offer various advantages but all use the same basic principle of operation (Andrews, 1988). The whirlpool consists of a cylindrical insulated vessel into which the hot wort is pumped via a tangentially mounted entry main. This induces a hydrocylone effect in the vessel so that as the liquid circulates, the solid material drops out and forms a compact mound in the centre of the base. The clarified wort can then be run off leaving the solid matter in the whirlpool for subsequent removal.

### 2.3.7 *Fermentation and post-fermentation processes*

Clarified wort is cooled using a paraflow or heat exchanger (Fig. 2.11) and pumped to the fermenting vessel where oxygenation and yeast pitching allow the fermentation to

**Fig. 2.11**   Paraflow (kindly provided by Harry White, Bass Brewers).

commence. The various processes and types of plant used in the fermentation stage of brewing form the subject of Chapters 5–7 and are not described further here.

When fermentation is complete the beer must be rendered into a form suitable for consumption. Many options are possible depending on the type of beer. Beers subject to a secondary fermentation may simply be removed from the primary fermentation vessel after the bulk of the yeast has been removed and packaged, either in cask or bottle (see Section 6.9). Such types of beers are comparatively rare. Most beer is subject to post-fermentation processing to produce a packaged product, which is stable, both from microbiological, physical and flavour standpoints. The downstream processes involved include 'conditioning' (or 'maturation', 'ageing'), filtration and pasteurisation/sterile filtration.

In the case of traditional beers, conditioning involves a period of storage in which flavour maturation occurs and this requires the presence of viable yeast cells. The processes involved are described in Section 6.5. For most beers the primary function of this stage of brewing is to produce beer with a colloidal stability which is appropriate for its projected shelf-life. Some positive flavour adjustments may be made. For example, pre-isomerised hop extract additions may be made to conditioning vessels as opposed to the copper since obviously no heating is required. Recovered beer may also be added back at this stage. Small volumes of recovered beer are generated in many stages of the brewing process, usually where there is a separation stage of liquid from solid and the bulk of a batch of beer has been moved forward to the next stage of processing leaving behind a solid–liquid mixture. A typical example is beer recovered from yeast cropped from a fermenter. The recovered product must be of sufficient quality not to compromise the beer with which it is to be blended. This may involve some intermediate treatment, for example, flash pasteurisation to avoid micro-biological contamination.

Colloidal instability of beers may result in the formation of two types of haze, termed chill hazes and permanent hazes. The first of these is a haze that is formed at low temperature but redissolves when the beer is returned to room temperature. Permanent hazes, by definition, once formed persist in beer under all conditions. Hazes are due to interactions between proteins, tannins (polyphenols), polysaccharides and metal ions by mechanisms that are not fully characterised. In the case of chill hazes the interactions are not stabilised, whereas permanent hazes involve the formation of covalent bonds, possibly involving oxygen.

Several chill-proofing treatments are possible. Proteases such as papain reduce the concentration of proteins available for haze formation. However, care must be taken with this approach as it can have a detrimental effect on beer foaming potential. Various adsorbents are used which are themselves insoluble, and therefore are removed with the potential haze-forming material during filtration. Silica gel and bentonite adsorb proteins; Nylon 66 and polyvinylpolypyrrolidone (PVPP) adsorb polyphenols. Another strategy is to add tannic acid, which forms complexes with the proteins that have the potential to produce hazes.

Conditioning is terminated by filtration to produce bright beer. It is advantageous to present beer to filters with as low a solids concentration as possible. To facilitate this beer in conditioning tanks is commonly dosed with isinglass finings to promote sedimentation of suspended solids. Of course, whilst this has the beneficial effect of reducing the solids content in the beer that is to be filtered, there is a concomitant increase in the quantity of solids in the tank bottoms and associated recovered beer. Inevitably, a compromise must be made that suits the individual brewery. Many types of filter are used, depending on the type of operation. Usually two treatments are given, a pre-filtration to remove the bulk of solids followed by a polishing filter to produce brilliantly clear beer. In all cases it is essential to exclude oxygen to prevent adverse staling effects. This is particularly so when all yeast cells have been removed since these have the ability to assimilate low concentrations of oxygen and prevent harmful oxidations.

For beer filtration, plate and frame or leaf filters are most commonly used (Fig. 2.12) although some breweries use a candle filter (Fig. 2.13). In these cases kieselguhr or perlite is used as a filter aid. These filters use a porous membrane on which the filter aid is deposited, termed pre-coating. More filter aid is mixed with the beer as it is fed into the filter, termed body feed. Selection of a suitable grade of filter aid allows operation as rough or polishing filter. The filtered bright beer is held in refrigerated tanks for final adjustments prior to packaging. It is possible at this stage to adjust colour; however, the most important task is to ensure that carbonation is correct.

Beer from bright beer tanks feeds the packaging lines, where the finished product is placed into keg, bottle or can. These processes are clean but not usually sterile and therefore the process culminates with pasteurisation. In the case of keg beers, product is flash pasteurised, in-line, before packaging into steam-sterilised and clean kegs. Cans and bottles are filled, sealed and then tunnel pasteurised.

Severe pasteurisation regimes undoubtedly produce some deterioration in flavour and to overcome this some brewers package canned and bottled beers under sterile conditions. It should be noted that in most countries it is not permitted to use artificial preservatives. Sterile packaging is expensive since it requires that all operations after

**Fig. 2.12** Plate and frame filter (kindly provided by Harry White, Bass Brewers).

**Fig. 2.13** Candle filter (kindly provided by Steve Freckleton, Bass Brewers).

the sterilising step have to be performed under aseptic conditions. In the case of a high-speed canning line this is technologically challenging. Sterilisation of beer is achieved by passage through a membrane or candle filter with a pore size of 0.1–0.4 microns, which is sufficiently small to remove micro-organisms. Further discussion on pasteurisation and sterile filtration can be found in Section 8.2.1.3.

## 2.4 Wort composition

The nature of the materials and procedures used to make wort make it inevitable that its composition is complex and, in fact, not well characterised. Thus, Hudson (1973) described the single condition that wort composition must meet as that which produces a beer of desired character by intention rather than by chance. It is perhaps implicit in this definition that it is not necessary to know the precise composition as long as the materials and methods used in its preparation are themselves constant. Many of the components are of little relevance to the processes which occur during fermentation in that they take no active role. However, they may contribute to beer flavour, aroma and colour.

Wort composition is influenced by the nature of the raw materials and management of the processes used in its preparation. With regard to the raw materials, the nature of the brewing water, the blend and addition rate of hops, the blend and types of malt and the nature of any adjunct are all influential. The latter is of special relevance in that whereas all-malt wort will tend to produce wort with an appropriate balance between fermentable sugar and amino nitrogen, use of high concentrations of sugar adjuncts will have an obvious impact on carbon to nitrogen ratios.

During wort preparation, the conditions of mashing and boiling are of particular significance. Low mashing temperatures favour continued activity of heat-labile malt proteases and therefore elevated concentrations of amino nitrogen are formed. Conversely, higher mashing temperatures favour amylolytic activity at the expense of proteolysis and in consequence the ratio of fermentable sugar to amino nitrogen increases. In addition, the relative activities at different mashing temperatures of $\alpha$– and $\beta$-amylases and limit dextrinase will modulate the sugar spectrum of wort. Furthermore, the method of separation of sweet wort from spent grains and conditions of the boil all influence the precise composition of wort in fermenter.

Wort is a clear liquid, which usually contains a small proportion of suspended trub. The requirement for wort clarity is somewhat controversial. This parameter is often used as a measure of wort quality and the performance of the brewhouse may be judged on the basis of its ability to generate bright wort with a very low trub content. However, others (Olsen, 1981; Schisler et al., 1982; Lentini et al., 1994) argue that some trub is needed during fermentation as a source of lipids (especially unsaturated fatty acids) which are essential to yeast nutrition. Second it has been suggested that trub can provide nucleation sites to assist carbon dioxide bubble formation and breakout (Siebert et al., 1986). The common sense answer is that the individual brewer operates the brewhouse in such a way as to produce wort that gives satisfactory fermentation performance. It is a matter of fact that solids must be removed at some stage in the process between wort production and beer packaging. It is perhaps less important where this is actually achieved and there are obviously advantages and disadvantages whichever choice is made. It may be of practical relevance, however, where new plant is installed which significantly changes wort clarity. In this case changes in fermentation performance may be observed which require corrective action such as a change in yeast pitching rate and/or wort oxygenation regimes.

Worts of specific gravity within the range 1.03–1.06 (7.5–15°Plato) contain approximately 7.5–15% dissolved solids. Wort pH is typically in the range pH 5.0–5.3. The colour is highly dependent on the materials used for production and varies between a very pale brown through to almost black. Wort viscosity is dependent on the total solids content and the concentration of components such as $\beta$-glucans derived from malt. Typical values for worts of 7.5–25°Plato are 1.6–5.0 cP (Hudson, 1970). Wort dissolved solids comprise approximately 90–92% carbohydrate and 4–

5% nitrogenous components (MacWilliam, 1968). The remainder consists of a wide variety of organic compounds which encompass all the major groups of biochemicals which would predictably be extracted from a living cell and withstand the conditions which pertain during wort preparation.

### 2.4.1   Carbohydrates

Maltose is the most abundant sugar, usually accounting for 60–65% of the total fermentable sugar. The remainder consists mainly of glucose, fructose, sucrose, maltotriose, maltotetrose and higher dextrins (MacWilliam, 1968; Enevoldsen, 1974; Hoekstra, 1974; Kieninger & Rottger, 1974; Taylor, 1974). Carbohydrate spectra of some typical worts are shown in Table 2.3. Some of the variations in sugar spectra due to different wort production procedures are summarised in Table 2.4.

**Table 2.3**   Carbohydrate composition of worts (g 100 ml$^{-1}$) (from MacWilliam, 1968).

| Origin | Danish | Canadian | Canadian | Canadian | German | UK | UK |
|---|---|---|---|---|---|---|---|
| Wort type | Lager | Lager | Lager | Lager | Lager | Pale ale | Pale ale |
| Wort concentration (°Plato) | 10.7 | 12.32 | 11.9 | 11.55 | 12.0 | 10.0 | 10.0 |
| Fructose | 0.21 | 0.15 | 0.13 | 0.10 | 0.39 | 0.33 | 0.97 |
| Glucose | 0.91 | 1.03 | 0.87 | 0.50 | 1.47 | 1.00 | – |
| Sucrose | 0.23 | 0.42 | 0.35 | 0.10 | 0.46 | 0.53 | 0.60 |
| Maltose | 5.24 | 6.04 | 5.57 | 5.50 | 5.78 | 3.89 | 3.91 |
| Maltotriose | 1.28 | 1.77 | 1.66 | 1.30 | 1.46 | 1.14 | 1.39 |
| Maltotetrose | 0.26 | 0.72 | 0.54 | 1.27 | – | 0.20 | 0.53 |
| Total dextrins | 2.39 | 3.40 | 3.06 | 4.21 | – | 2.52 | 2.48 |
| Total fermentable sugars | 7.87 | 9.41 | 8.58 | 7.50 | 9.56 | 6.89 | 6.78 |
| Total sugars | 10.26 | 12.81 | 11.64 | 11.71 | – | 9.41 | 9.26 |
| Fermentability (%) | 76.7 | 73.7 | 73.7 | 64.1 | – | 73.3 | 73.2 |

The unfermentable fraction of extract accounts for approximately 25% of the total carbohydrate in worts. This is usually referred to as dextrin, but, as Enevoldsen (1974) points out, it contains small quantities of monosaccharides such as arabinose, xylose and ribose, the disaccharide isomaltose, trisaccharides such as panose and iosopanose and β-glucans. Thus, true dextrins, containing four or more glucose units, comprise approximately 90% of the residual carbohydrate in beer. Some 40–50% of the dextrins are oligosaccharides containing 4–9 glucose units, the remaining 50–60% are higher dextrins with 10 or more glucose units. Enevoldsen and Schmidt (1974) demonstrated that the concentration and spectrum of dextrins in worts and beers were identical. Furthermore, they also observed that the concentrations of individual dextrins for all worts and beers examined showed a characteristic pattern of peaks and troughs (Table 2.5 and Fig. 2.14). Dextrins containing four glucose units are always the most abundant dextrin, and regular peaks occur separated by approximately 4–5 glucose units.

Dextrins are derived from starch of the malt endosperm and therefore they may be linear glucose polymers joined by α-(1→4) linkages or branched molecules containing

**Table 2.4** Fermentable sugar contents of worts produced by the procedures indicated (data abstracted from Hoekstra, 1974).

| Origin of brewery (fermentation) | UK (top) | UK (top) | Denmark (bottom) | Canada (bottom) | Canada (bottom) | Germany (bottom) | Holland (bottom) | Holland (bottom) |
|---|---|---|---|---|---|---|---|---|
| Mashing procedure | Infusion | Infusion | Decoction | Infusion | Infusion | Not given | Decoction | Decoction |
| Adjunct (%) | Sucrose (amount not given) | Some (amount not given) | Not given | 15% | 15% | None | 20% | 20% |
| Total fermentable sugar (g 100 ml$^{-1}$ wort at 12°P) | 7.25 | 8.96 | 8.88 | 9.16 | 8.86 | 8.34 | 10.07 | 9.38 |
| Fructose (%) | 3.5 | 1.8 | 2.7 | 1.6 | 1.6 | 4.8 | 1.4 | 1.4 |
| Glucose (%) | 13.7 | 9.9 | 11.6 | 10.9 | 11.9 | 14.6 | 13.3 | 12.4 |
| Sucrose (%) | 20.1 | 5.5 | 2.9 | 4.5 | 3.2 | 7.7 | 5.5 | 5.9 |
| Maltose (%) | 44.4 | 56.9 | 66.6 | 64.2 | 64.8 | 56.5 | 61.3 | 61.7 |
| Maltotriose (%) | 18.3 | 25.9 | 16.2 | 18.8 | 18.5 | 16.5 | 18.5 | 18.6 |

**Table 2.5** Dextrin spectrum of a 12°Plato lager wort (from Enevoldsen, 1974).

| Composition (degree of polymerisation) | g 100 ml$^{-1}$ wort |
|---|---|
| 1 [non-fermentable sugars] | 0.03 |
| 2 [non-fermentable sugars[ | 0.07 |
| 3 [non-fermentable sugars] | 0.12 |
| 4 | 0.39 |
| 5 | 0.12 |
| 6 | 0.15 |
| 7 | 0.14 |
| 8 | 0.17 |
| 9 | 0.18 |
| 10 | 0.07 |
| 11 | 0.06 |
| 12 | 0.08 |
| 13 | 0.12 |
| 14 | 0.11 |
| 15 | 0.08 |
| 16 | 0.05 |
| 17–21 | 0.29 |
| 22–27 | 0.19 |
| 28–34 | 0.12 |
| 35+ | 0.46 |
| Dextrins (4 or more glucose units) [% total wort extract) | 22.1 |
| Non-fermentable sugars [% total extract] | 2.0 |
| Fermentable sugars [% total extract] | 66.9 |
| Non-carbohydrate [% total extract] | 9.0 |

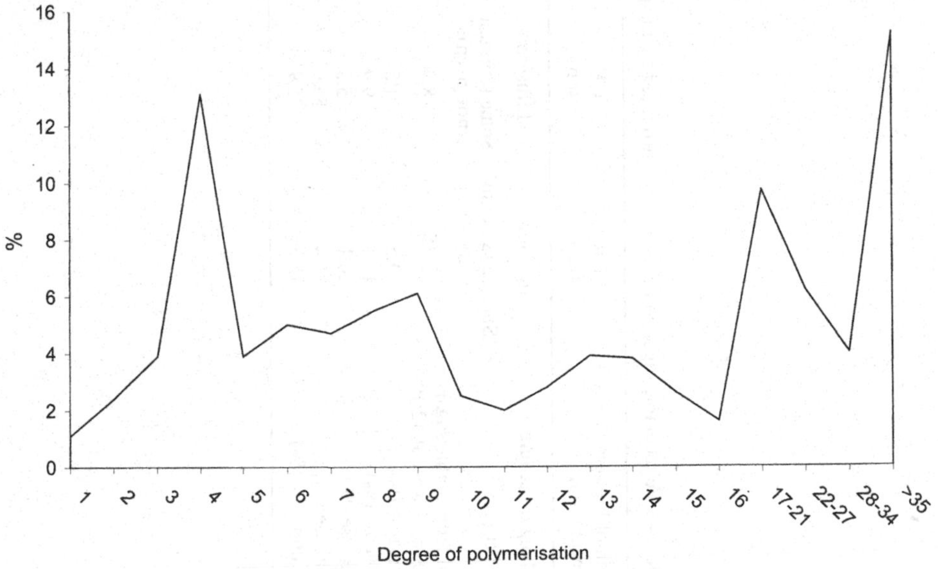


**Fig. 2.14** Distribution of dextrins in malt wort (drawn from data in Table 2.5).

α-(1→6) linkages. Obviously linear dextrins can have one structure only, whereas in the case of larger dextrins several branched structures are possible. Predictably, the proportion of branched dextrins increases with the number of glucose units such that those with four glucose units contain 30% branched molecules; 70% in the case of those with seven glucose units and 100% at ten glucose units and above (Enevoldsen and Schmidt, 1973; Enevoldsen, 1974). The same authors concluded that this was evidence that the majority of α-(1→6) linkages of malt amylopectin survive wort production intact. Furthermore, the relative abundance and patterns of branching in wort dextrins was a reflection of the structure of the starch of malt.

Concentrations of β-glucans [(1→3)(1→4)-β-D-glucans] in worts are dependent on the type of malt used and the mashing regime. Aastrup & Erdal (1987) reported levels of 560–620 mg l$^{-1}$ in sweet worts of 15.7°Plato made with brewing malt. Using the same wort production regimes non-brewing malt produced sweet worts containing 1000–1300 mg l$^{-1}$ β-glucan. β-glucans are derived from malt endosperm cell walls where they exist as polymers with molecular weights between 50 000 and 200 000 (Schur et al., 1974). Complete degradation yields glucose, although during malting endo-b-(1→3)(1→4) glucanases yield saccharides of varying chain-lengths and the disaccharides cellobiose and laminoaribiose. The proportions of degradation products depend on the mashing temperature, since malt β-glucanases are heat labile, or whether or not β-glucanases are added during wort production as process aids.

β-glucans, together with pentosans, contribute to wort viscosity and their occurrence at high concentrations in certain barley varieties is one of the factors which makes them unsuitable for malting for brewing purposes (Viëtor et al., 1993). In malts some 73% of pentosans are in the form of arabinoxylans which consist of a polymer of β-(1→4) linked xylose units with side chains of single β-(1→3) linked arabinose molecules. These polymers are associated with the aleurone layer cell walls of barley grains. During mashing little degradation occurs and the polysaccharides persist in worts at concentrations of the order of 2.0 g per litre of wort (Viëtor et al., 1991).

### 2.4.2 Nitrogenous components

Nitrogenous components of wort account for 4–5% of the total dissolved solids. The bulk (85–90%) of the total is in the form of amino acids, small peptides and proteins (Enari, 1974; Hudson, 1974; Jones, 1974; Lie et al., 1974; Moll et al., 1978). The relative proportions of each of these groups of nitrogenous components depend on the composition of the grist and the conditions of wort production. Thus, controlling factors are the nitrogen content of the malt, dosage rates and type of adjunct, the extent of proteolysis during mashing and precipitation and removal during the wort boil and clarification steps.

Worts can contain more than 1000 mg l$^{-1}$ total nitrogen, depending on the materials used and the process conditions, although 700–800 mg l$^{-1}$ is more typical (Jones, 1974). Clapperton (1971) fractionated the soluble nitrogen of wort and reported that 20% was protein, 22% polypeptides and 58% peptides and free amino acids. Free amino acid concentrations are within the range 150–230 mg l$^{-1}$ in a wort of 10.5°Plato (Lie et al., 1974). Recommended free amino nitrogen concentrations of wort is of the order of 150–200 mg l$^{-1}$ where oxygen is the limiting substrate. A third to a half of the amino acids in wort arise from the action of proteases (mainly car-boxypeptidases) during mashing, the remainder being derived directly from malt and formed during malting. Malt carboxypeptidases have maximal activity at temperatures between 40 and 60°C and are inactivated at 70°C (Jones, 1974). It may be appreciated, therefore, that the temperature at which mashing is conducted has a crucial impact on the free amino nitrogen content of worts. All free amino acids can be assimilated by yeast during fermentation, under appropriate conditions, other than proline, which requires oxygen (see Section 3.3.2). Some 40% of the small peptides are also utilised (Clapperton, 1971).

The free amino acid spectra of worts are relatively constant (Table 2.6). Total concentrations arising in worts are influenced by raw materials and wort production techniques. Jones (1974) reported α-amino nitrogen contents of barley as varying between 592 and 946 µg per grain. Abrading and gibberellic acid treatment during malting both increase α-amino nitrogen contents of malts, whereas there is a negative correlation between nitrogen level and curing temperature. The temperature of mashing has a most significant impact on α-amino nitrogen concentration (Tables 2.7a and b). The data presented here, taken from Chen et al. (1973), shows that a mashing temperature close to 60°C produced a greater yield of α– amino nitrogen than 40 or 70°C. Use of low nitrogen adjuncts, in addition to malt, predictably reduces wort nitrogen content (Table 2.8).

The protein and polypeptide fraction of worts are a diverse group of molecules with molecular weights within the range 5000–100 000 (Enari, 1974). The various protein fractions have importance with respect to their contribution to head retention of beers and ability to interact with polyphenols and form beer hazes.

**Table 2.6** Amino acid content of worts (mg 100 ml$^{-1}$ wort of 10°Plato) (from Otter and Taylor, 1976).

| Amino acid | All malt wort | Brewery wort | Malt + 25% barley adjunct | Malt +25% wheat adjunct | Acidified mash wort |
|---|---|---|---|---|---|
| Alanine | 15.1 | 9.9 | 15.5 | 16.5 | 39.4 |
| Valine | 13.7 | 9.2 | 14.0 | 14.7 | 26.0 |
| Glycine | 7.9 | 5.2 | 9.1 | 10.3 | 19.0 |
| Iso-leucine | 7.5 | 5.3 | 9.4 | 10.7 | 18.6 |
| Leucine | 18.8 | 10.9 | 18.3 | 19.2 | 39.7 |
| Proline | 44.5 | 38.9 | 47.9 | 47.3 | 73.0 |
| Threonine | 17.2 | 2.5 | 11.2 | 23.9 | 1.7 |
| Serine | 7.4 | 4.8 | 7.5 | 9.3 | 28.8 |
| Cysteine | None | None | None | None | None |
| Methionine | None | 2.3 | 4.7 | 4.9 | 12.8 |
| Hydroxyproline | None | None | None | None | None |
| Phenylalanine | 16.2 | 11.4 | 15.8 | 17.3 | 34.4 |
| Aspartate | 18.9 | 17.9 | 23.3 | 24.8 | 29.0 |
| Glutamate | 24.8 | 8.0 | 31.0 | 40.9 | 59.6 |
| Tyrosine | 8.9 | 0.8 | 8.3 | 8.1 | 11.5 |
| Ornithine | 0.9 | None | None | None | 13.3 |
| Lysine | 10.1 | 5.6 | 11.3 | 8.9 | 19.2 |
| Tryprophan | Trace | None | None | None | None |
| Histidine | Trace | 0.3 | None | None | 0.4 |
| Arginine | 15.5 | 10.8 | 17.6 | 13.8 | 18.3 |
| Cystine | Trace | None | None | None | None |

**Table 2.7a** Amino acid contents of worts (mg 100 ml$^{-1}$) produced at various mashing temperatures and green beers from ale fermentations performed at 17–22°C (from Chen *et al.*, 1973).

| Mashing temperature | 40°C | | 60°C | | 70°C | |
|---|---|---|---|---|---|---|
| | Wort | Beer | Wort | Beer | Wort | Beer |
| Glycine | 12.4 | 7.3 | 13.5 | 4.3 | 12.1 | 5.3 |
| Alanine | 22.0 | 7.8 | 30.2 | 10.4 | 22.3 | 9.3 |
| Valine | 27.4 | 17.5 | 47.2 | 22.0 | 22.8 | 14.9 |
| Leucine | 14.2 | 8.3 | 32.4 | 11.5 | 8.2 | 3.5 |
| Isoleucine | 8.9 | 5.9 | 10.4 | 5.5 | 5.3 | 3.6 |
| Phenylalanine | 20.1 | 6.8 | 20.7 | 12.0 | 8.2 | 6.4 |
| Tyrosine | 12.5 | 5.2 | 11.5 | 5.9 | 7.9 | 5.2 |
| Tryptophan | 11.9 | 5.8 | 9.0 | 6.3 | 2.2 | 2.1 |
| Serine | 9.9 | 5.7 | 10.7 | 4.5 | 7.8 | 3.3 |
| Threonine | 17.2 | 10.7 | 20.4 | 7.9 | 12.4 | 6.9 |
| Aspartic acid | 12.9 | 5.8 | 16.0 | 7.4 | 8.1 | 3.5 |
| Glutamic acid | 16.4 | 7.9 | 27.7 | 15.8 | 16.6 | 9.5 |
| Lysine | 12.4 | 8.8 | 21.4 | 9.7 | 21.3 | 10.3 |
| Arginine | 23.4 | 17.3 | 32.9 | 25.3 | 34.3 | 24.0 |
| Histidine | 2.4 | 1.6 | 2.3 | 1.4 | 4.6 | 4.1 |
| Methionine | 7.2 | 3.2 | 9.7 | 4.5 | 6.8 | 3.9 |
| Cystine | Not detected | Not detected | Not detected | Not detected | Not detected | Not detected |
| Cysteine | 13.9 | 6.5 | 6.2 | 5.0 | 12.3 | 7.6 |
| Proline | 48.7 | 45.7 | 61.3 | 58.4 | 49.2 | 45.7 |
| Hydroxyproline | 3.7 | 2.8 | 4.9 | 4.5 | 3.5 | 2.6 |
| Ornithine | Trace | 3.9 | Not detected | 4.9 | 1.0 | 6.4 |
| 4-aminobutyric acid | 15.5 | 13.6 | 21.4 | 16.5 | 11.3 | 9.1 |
| Total | 313.0 | 198.1 | 409.7 | 243.7 | 278.2 | 187.2 |

Wort contains various nucleic acids and their degradation products. MacWilliam (1968) reported that 5.7–6.4% of the total wort nitrogen consisted of purines and of this 33% was free adenine and guanine, the remainder being adenosine and guanosine. During mashing nucleases are active, liberating ribosides and phosphate. Higher temperature mashes are associated with reduced concentrations of free purines. This author concluded that there were no reports of pyrimidines occurring in worts. However, Chen *et al.* (1973) produced data indicating the presence of both. Total concentrations of nucleic acids in worts were given as being of the order of 280–330 mg l$^{-1}$. More nucleic acids arose when the mashing temperature was 60°C, as opposed to 40 or 70°C. Of this approximately 10–20% disappeared during fermentation, although there was considerable variation between individual components (Table 2.9).

Worts of 10.5°Plato contain 25–30 mg l$^{-1}$ ammonia (Lie *et al.*, 1974). Traces (usually less than 10 mg l$^{-1}$) of various amines may also be detected, including methylamine, dimethylamine, ethylamine, butylamine, amylamine, tyramine, hordenine and choline. Certain amines have been implicated in the hygiene-related issue of 'biogenic amines' (see Section 8.1.5.2). Reportedly, only choline is utilised to any great extent during fermentation (MacWilliam, 1968).

**Table 2.7b**  Amino acid contents of worts (mg $100\,ml^{-1}$) produced at various mashing temperatures and green beers from lager fermentations performed at 11–16°C (from Chen *et al.*, 1973).

| Mashing temperature | 40°C Wort | 40°C Beer | 60°C Wort | 60°C Beer | 70°C Wort | 70°C Beer |
|---|---|---|---|---|---|---|
| Glycine | 13.9 | 6.6 | 12.9 | 5.9 | 10.6 | 5.6 |
| Alanine | 26.2 | 19.8 | 31.0 | 24.4 | 20.7 | 23.0 |
| Valine | 29.6 | 18.1 | 39.7 | 33.2 | 33.9 | 26.6 |
| Leucine | 24.1 | 10.5 | 28.6 | 14.6 | 15.0 | 5.1 |
| Isoleucine | 9.5 | 6.7 | 9.3 | 7.0 | 7.9 | 5.8 |
| Phenylalanine | 25.0 | 8.7 | 24.2 | 12.6 | 25.2 | 8.5 |
| Tyrosine | 13.9 | 9.3 | 12.5 | 7.0 | 11.6 | 8.8 |
| Tryptophan | 6.7 | 4.2 | 6.6 | 3.8 | 2.5 | Trace |
| Serine | 12.0 | 5.8 | 16.1 | 6.5 | 11.4 | 3.5 |
| Threonine | 21.9 | 8.6 | 21.6 | 8.3 | 15.7 | 8.0 |
| Aspartic acid | 15.3 | 7.2 | 20.9 | 13.7 | 13.9 | 5.1 |
| Glutamic acid | 16.9 | 8.3 | 25.2 | 15.4 | 22.0 | 16.1 |
| Lysine | 14.5 | 8.0 | 14.5 | 7.4 | 14.3 | 5.1 |
| Arginine | 25.6 | 19.2 | 29.9 | 19.3 | 25.2 | 13.5 |
| Histidine | 3.3 | 2.1 | 4.7 | 1.9 | 9.8 | 5.3 |
| Methionine | 7.9 | 3.9 | 8.9 | 7.3 | 6.4 | 3.4 |
| Cystine | Not detected | Not detected | Not detected | Not detected | Not detected | Not detected |
| Cysteine | 14.0 | 9.4 | 6.5 | 4.5 | 7.6 | 5.1 |
| Proline | 49.6 | 45.9 | 48.8 | 45.3 | 48.7 | 44.9 |
| Hydroxyproline | 2.8 | 3.1 | 6.2 | 4.3 | 5.1 | 4.2 |
| Ornithine | 1.0 | 5.1 | Trace | 8.3 | Trace | 1.8 |
| 4-aminobutyric acid | 15.8 | 11.8 | 15.4 | 13.5 | 10.7 | 9.7 |
| Total | 349.5 | 222.3 | 383.5 | 264.2 | 318.2 | 209.1 |

**Table 2.8**  Analysis of worts prepared from malt and varying proportions of maize grit adjunct (from Klopper, 1974).

| Maize grit adjunct (%) | 0 | 20 | 30 | 40 |
|---|---|---|---|---|
| Wort nitrogen (mg $l^{-1}$) | 1089 | 870 | 809 | 700 |
| α-amino nitrogen (mg $l^{-1}$) | 233 | 211 | 184 | 142 |
| pH | 5.4 | 5.4 | 5.3 | 5.4 |
| Viscosity (Cp) | 1.79 | 1.66 | 1.65 | 1.65 |
| Total extract (g $100\,g^{-1}$) | 12.0 | 11.9 | 11.8 | 12.0 |

### 2.4.3  *Polyphenols*

Polyphenolic constituents of worts arise from both malt and hops. The significance of those derived from malts resides in their ability to form complexes with proteins and form hazes in beer. Formation of such complexes, in the form of insoluble precipitates, is encouraged at several stages in the brewing process such that they may be removed in order to produce finished beer with no propensity to form hazes. For example, formation of hot and cold break during wort boiling and cooling and chill-proofing during conditioning and filtration (described in previous sections of this chapter).

Polyphenols, or tannins, are complex chemically and several chemical species,

**Table 2.9** Nucleic acid components of worts produced at various mashing temperatures, together with concentrations measured in the respective beers (mg l⁻¹). Ale fermentations were performed at 15–22°C and lager fermentations at 11–16°C (from Chen et al, 1973).

| Mashing temperature | 40°C Wort | 40°C Ale | 60°C Wort | 60°C Ale | 70°C Wort | 70°C Ale | 40°C Wort | 40°C Lager | 60°C Wort | 60°C Lager | 70°C Wort | 70°C Lager |
|---|---|---|---|---|---|---|---|---|---|---|---|---|
| Adenine + Adenosine | 15.0 | 7.2 | 32.0 | 5.5 | 13.1 | 2.0 | 15.0 | 5.0 | 32.0 | 10.4 | 13.1 | 9.6 |
| Uracil | 22.0 | 8.0 | 34.6 | 24.4 | 32.1 | 6.8 | 22.0 | 19.3 | 34.6 | 26.6 | 32.1 | 21.7 |
| Uridine + Cytosine + Cytidine | 149.3 | 68.5 | 154.8 | 112.4 | 134.6 | 117.9 | 149.3 | 98.3 | 154.8 | 80.4 | 134.6 | 125.6 |
| Guanine | 33.5 | 34.0 | 8.0 | 38.5 | 34.8 | 22.5 | 33.5 | 28.0 | 8.0 | 31.7 | 34.8 | 23.3 |
| Guanosine | 42.7 | 48.5 | 48.3 | 41.3 | 42.8 | 39.1 | 42.7 | 43.3 | 48.3 | 43.6 | 42.8 | 44.4 |
| Inosine | 1.9 | — | — | 1.0 | 2.1 | — | 1.9 | 0.8 | — | 3.1 | 2.1 | — |
| Xanthine | — | 13.3 | — | 9.2 | — | 5.0 | — | 22.1 | — | 17.0 | — | 5.0 |
| Hypoxanthine | 1.8 | 4.4 | 8.7 | 1.4 | 3.2 | 1.3 | 1.8 | 8.3 | 8.7 | 11.5 | 3.2 | 6.2 |
| Thymine + Thymidine | 12.8 | 19.6 | 8.7 | 22.4 | 6.6 | 21.2 | 12.8 | 14.4 | 8.7 | 15.1 | 6.6 | 15.4 |
| 5'-AMP | 4.7 | 0.9 | 13.7 | 3.7 | 3.2 | 0.7 | 4.7 | 3.2 | 13.7 | 5.4 | 3.2 | 5.5 |
| 2'-AMP | 2.1 | 0.9 | 6.0 | 1.6 | 0.8 | 1.1 | 2.1 | 2.6 | 6.0 | 2.6 | 0.8 | 3.4 |
| 5'-UMP | 0.1 | 0.5 | 1.0 | 0.8 | 0.1 | 0.1 | 0.1 | 0.3 | 1.0 | 0.1 | 0.1 | — |
| 5'-XMP | 0.2 | 0.3 | 0.1 | 0.1 | 1.1 | 0.1 | 0.2 | 0.1 | 0.1 | 0.2 | 1.1 | 0.5 |
| 5'-GMP | 0.4 | 0.6 | 0.1 | 0.4 | 0.1 | 0.4 | 0.4 | 0.1 | 0.1 | 0.1 | 0.1 | — |
| 3'-AMP | 3.8 | 0.9 | 5.1 | 9.5 | 3.9 | 1.1 | 3.8 | 7.8 | 5.1 | 11.4 | 3.9 | 10.4 |
| 2'-, 3'-CMP + 2'-, 3'-UMP | 0.1 | — | — | — | 2.1 | 0.7 | 0.1 | 0.5 | — | — | 2.1 | — |
| 5'-IMP | 1.2 | — | 3.1 | 1.1 | 0.7 | 0.1 | 1.2 | 0.3 | 3.1 | 0.4 | 0.7 | 1.6 |
| Methyl uracil | — | — | — | — | — | 0.3 | — | — | — | — | — | 0.4 |
| Methyl cytosine | 1.1 | — | — | 1.7 | — | — | 1.1 | 2.3 | — | 1.9 | — | 2.5 |
| Total | 292.7 | 207.6 | 324.2 | 275.0 | 281.3 | 220.4 | 292.7 | 256.7 | 324.2 | 260.5 | 281.3 | 275.5 |

monomers, oligomers and polymers, are represented in worts and these are extracted during the wort boil. Three classes of simple polyphenols occur (Hough et al., 1982). The first class are simple phenolic derivatives of hydroxybenzoic and hydroxycinnamic acids. These include among others, p-hydroxybenzoic, vanillic, syringic, p-coumaric and ferulic acids. They have no impact on beer colloidal stability but have roles in beer flavour and aroma. The second class are the flavonols, which are principally extracted from hops. These consist of quercitin and kaempferol and their glycosides. These survive into beer but also have no impact on colloidal stability. The third class are the monomeric and oligiomeric flavanols. Two of the common monomeric flavanols are (+)-catechin and (−)-epicatechin. Dimers such as prodelphinidin B3 and Procyanidin B3 and trimeric forms can be detected in malt. The oligomeric flavanoids consist of short chains of polyhydroxy-flavan-3,4-diol-monomers attached to (+)-catechin or (−)-epicatechin. The most abundant polymers of this type are the proanthocyanidins (formerly anthocyanogens) so called because they yield anthocyanidins when exposed to acid conditions in the presence of oxygen. These compounds have the ability to cross-link and form large molecular weight polymeric flavanols and also to form complexes with proteins and polypeptides, the complex flavanols.

Changes in the concentrations of polyphenols, derived from both malt and hops during wort production, are complex. The most significant stages are the formation of 'hot break' during the copper boil and 'cold break' after wort cooling together with oxidation which accompanies wort oxygenation (McMurrough & Delcour, 1994). McMurrough et al. (1983) reported that a hopped wort contained 46.2 mg l$^{-1}$ total flavanols. Of these 39% were simple types, 18% polymeric and the remaining 43%, complexed.

### 2.4.4   Lipids

Wort lipids may potentially be derived from malt, adjuncts and hops. Up to 4% of the dry weight of barley is lipid and 3.4% in the case of malt. Commonly used adjuncts contain up to 4% lipid although there is some variation. For example, wheat flour contains only 1%, whereas the lipid content of flaked maize is 3.7% (Anness, 1984). Triacylglycerols form the most abundant lipid group in barley and malts. Up to 70% of the total fatty acids are esterified in this form. There is considerable modification to lipid composition and concentration during malting and wort production. In particular, Anness (1984) reported that 30% of the lipid content of barley grains disappeared during germination, an effect ascribed to metabolism of triacylglycerols.

The lipid composition of a typical commercial grist is shown in Table 2.10. Little of this lipid persists into the finished wort; however, that which does is available for yeast nutrition during fermentation and later in the process to be involved in reactions leading to the formation of deleterious flavours. The method of separation of sweet wort from spent grains has the most significant effect on wort lipid content. Anness and Reed (1985) reported that 4.5% of malt lipids were released into wort by a mash press, only 1% with a lauter tun and just 0.3% in the case of a mash tun. Further lipid is lost in the form of trub separated out after the copper boil. Thus, in the same report Anness and Reed (1985) indicated that 91% of wort lipids in the copper were

**Table 2.10** Lipid analysis of a commercial grist (from Anness, 1984).

| Lipid class | Fatty acid (mg g dry wt$^{-1}$) | % |
|---|---|---|
| Phospholipids + Glycolipids | 6.8 | 21.5 |
| Monacylglycerols | 0.4 | 1.3 |
| Diacylglycerols | 0.8 | 2.5 |
| Triacyglycerols | 19.9 | 63.0 |
| Free fatty acids | 2.7 | 8.5 |
| Steryl esters | 1.0 | 3.2 |
| Total | 31.6 | – |

deposited with the trub in the whirlpool in one brewery. Supplementation of wort lipids may occur by extraction from hops. Hop pellets are rich in 18:3 fatty acids and this may be the major source of this material in wort. The major fatty acids found at various stages in wort production using a lauter tun are shown in Table 2.11.

**Table 2.11** Fatty acid composition (%) of components during wort production using a lauter tun (from Anness and Reed, 1985).

| Fatty acid | 16:0 | 18:0 | 18:1 | 18:2 | 18:3 |
|---|---|---|---|---|---|
| Grist | 20.8 | 1.0 | 11.3 | 57.9 | 8.9 |
| Spent grains | 25.2 | 0.8 | 10.5 | 57.5 | 6.0 |
| Sweet wort | 41.4 | 3.4 | 5.4 | 45.3 | 4.4 |
| Hop pellets | 11.4 | 1.3 | 6.5 | 47.4 | 33.4 |
| Whirlpool trub | 24.0 | 1.1 | 6.9 | 49.3 | 18.6 |
| Sugar syrup adjunct | 45.3 | 19.5 | 5.5 | 24.7 | 5.0 |
| Hopped pitching wort | 59.8 | 6.2 | 3.9 | 24.9 | 5.8 |

It may be appreciated from the foregoing discussion that the lipid content of wort is greatly dependent on the methods used in its preparation. Letters (1992) presented data showing the total lipid contents of worts from several breweries using lauter tuns, conventional mash presses and the membrane mash press. There was much variability even where the same wort separation method was used. Thus, lipid contents were given as 10–80 mg l$^{-1}$ (lauter tun), 10–25 mg l$^{-1}$ (membrane filter press) and 70–140 mg l$^{-1}$ (conventional mash press). Free fatty acids are the most abundant lipid, whereas triacylglycerols form the largest esterified fraction (Table 2.12).

### 2.4.5 Sulphur compounds

Wort contains both organic and inorganic sulphur- containing constituents. In wort of 12% solids, Mandl (1974) reported the presence of an average of 90 mg l$^{-1}$ sulphur of which 60% occurred in organic molecules and 40% in inorganic form. Inorganic sulphur compounds include sulphate derived from the brewing water, hydrogen sulphide and sulphur dioxide. The latter two gases, which may be formed from amino acid decomposition during malting and mashing, are reduced in concentration during the copper boil. MacWilliam (1968) reported the presence of 5.9–8.8 mg l$^{-1}$ hydrogen

**Table 2.12** Lipid content of wort (12°Plato) (from MacWilliam, 1968).

| Component | mg $l^{-1}$ |
|---|---|
| Free fatty acids (C$_4$–C$_{10}$) | 0–1 |
| Free fatty acids (C$_{12}$–C$_{18}$) | 18–26 |
| Triacylglycerols | 5–8 |
| Diacylglycerols | 0.2–0.5 |
| Monoacylglycerols | 1.6–1.8 |
| Sterol esters | 0.1–0.2 |
| Free sterols | 0.2–0.4 |
| Phospholipid + Glycolipid | 3.0 |
| Hydrocarbons and waxes | 0.7 |
| Fatty acid esters | 1.2–1.3 |

sulphide in wort. The higher concentrations were reported to be associated with elevated levels of copper. This is surprising in that addition of copper ions to beer was used as a method of reducing hydrogen sulphide concentration in beer by encouraging the formation of insoluble copper sulphide.

Organic constituents of worts include the sulphur-containing amino acids, methionine, cystine and cysteine. In addition, there are other sulphur-containing biochemicals such as glutathione, thiamine and free and esterified coenzyme A. Cooled wort contains dimethyl sulphide, its precursor, S-methylmethionine and the related dimethylsulphoxide. The proportion of each depends on the variety of malt used, the duration of the copper boil and the conditions employed in wort cooling. Anness (1981) suggested that pitching worts (11°Plato) usually contain 200–700 µg $l^{-1}$ dimethyl sulphoxide and 30-50 µg $l^{-1}$ each of S-methylmethionine and dimethylsulphide. However, much variation is possible depending on the combination of effects as discussed above.

Of the many compounds identified in hop oils, about 30 contain sulphur (Seaton et al., 1981). These include dimethyl sulphide, other polymethyl sulphides and several thiolesters, for example, methylthioacetate and ethylthioacetate. In addition, treatment of hops with elemental sulphur can lead to reactions with hop oil constituents to form episulphides such as 1,2-epithiohumelene, 4,5-epithiohumulene, 4,5-epithiocaryophyllene and myrcene disulphide. Many of these sulphur-containing hop components have unpleasant flavours and aromas (Soltoft, 1988). The relative concentrations of these, which persist in pitching wort, are dependent on the varieties of hops, or derived products, which are used and the conditions of wort production.

### 2.4.6  *Minerals*

Minerals in hopped wort derive mainly from brewing water although significant concentrations also arise from the other raw materials. More concentrated worts will inevitably contain a higher proportion of minerals derived from malt and the other sources of extract used. Beers made with hard water such as UK pale ales have worts which contain higher levels of sulphate and other ions compared to

Pilsener-type lagers which use very soft water for wort production. In the case of the latter beers, the bulk of the minerals in worts are derived from malt and hops. The increasingly common practice of using brewing water which has been completely de-mineralised and then dosed with a cocktail of salts considered to be appropriate for a particular beer style will tend towards a simpler ionic composition compared to worts made with relatively untreated bore-hole waters. Thus, there may be much variation in the mineral composition of individual worts. Mandl (1974) gave average mineral values for 12°Plato wort. These are reproduced in Table 2.13.

**Table 2.13**  Mineral composition of a 12% solids hopped wort (from Mandl, 1974).

| Component | Concentration (mg $l^{-1}$) |
|---|---|
| Potassium | 550 |
| Sodium | 30 |
| Calcium | 35 |
| Magnesium | 100 |
| Copper | 0.1 |
| Iron | 0.1 |
| Manganese | 0.15 |
| Zinc | 0.15 |
| Sulphur | 90 |
| Phosphate | 575 |
| Chloride | 45 |

Several wort components can chelate metal ions. Should the metal chelates be insoluble there is an opportunity for the ions to be lost during wort production and thereby produce deficiencies in fermenter. For example, zinc ions may bind to amino acids and calcium ions to polyphenols (Jacobsen & Lie, 1977). Since a proportion of wort amino acids (as polypeptides or protein) and polyphenols are removed as hot and cold break during wort boiling and cooling, a fraction of the bound metal ions may also be lost. This possibility has been confirmed in the case of zinc (Daveloose, 1987) and many brewers add this metal to wort in fermenter before pitching.

Several authors have reported that worts may be deficient in magnesium ions, for example, Dombek and Ingram (1986b); Walker et al. (1996); Rees and Stewart (1997). The latter reports concluded that the relative concentrations of calcium and magnesium could be important and that very high ratios of calcium and magnesium could exert detrimental effects during fermentation by reducing the availability of magnesium to yeast. These observations appear not be of universal applicability, presumably indicating considerable variation in metal contents of worts, or use of yeast strains with differing metal ion requirements. Thus, Bromberg et al. (1997) found no effect on any fermentation parameters when a 16°Plato wort was supplemented with manganese (five-fold), magnesium or calcium (two-fold) or by varying the magnesium–calcium ratio from one to four. Only addition of zinc (0.1–0.15 mg $l^{-1}$) was found to stimulate fermentation rate, but not yeast growth.

### 2.4.7  *Miscellaneous*

A few of the other myriad components of wort are worthy of brief mention. Beer contains considerable quantities of simple organic acids. Most of these arise during fermentation as the result of yeast metabolism; however, small concentrations are present in unpitched wort. Mandl (1974) reported that in a 12°Plato wort, citrate was the most abundant organic acid at 170 mg l$^{-1}$, followed by gluconate and malate at 50 and 60 mg l$^{-1}$, respectively. Pyruvate, D-lactate and L-lactate were present at concentrations of less than 10 mg l$^{-1}$. In addition, MacWilliam (1968) reported the presence of small quantities (approximately 10 mg l$^{-1}$) succinate, fumarate, oxalate and α-ketoglutarate.

Vitamin contents of worts are (µg l$^{-1}$); thiamine, 150–750; pyridoxin, 150–200; *p*-aminobenzoic acids, 20–50; nicotinic acid, 1500–2500; inositol, 40 000–45 000; pantothenate, 150–250; biotin, 5–10; riboflavin, 300–500 and folic acid, 50–100 (MacWilliam, 1968; Graham *et al.*, 1970; Silhankova, 1985).

## 2.5  High-gravity brewing

High-gravity brewing is the practice of producing concentrated wort followed, at some stage, by dilution to produce finished beer of a desired alcohol content. In essence, therefore, it is a strategy that involves brewing using a concentrate, which thereby reduces the requirement for handling water, the major component of wort and beer. In terms of plant efficiency, it follows that it is prudent to delay the dilution step until the last possible process stage. Usually this will be in bright beer tanks, prior to packaging. However, plant permitting, it can be before or during fermentation, although the latter two options fail to provide many of the available benefits (Schaus, 1971; Pfisterer & Stewart, 1976).

High-gravity brewing was originally introduced to the United States in the 1950s. It has since gained widespread popularity throughout the brewing world and is now very common especially for the production of Pilsener-type lagers. In the vast majority of cases a process is operated in which the concentrated wort is fermented to produce a high-gravity beer which is diluted just before packaging. The principal advantage is that fermentation capacity is increased with no need for capital expenditure. Allowing for the fact that there may be some increase in fermentation cycle times, Hackstaff (1978) considered that by high-gravity brewing the capacity of existing plant was increased by 20–30%.

Several other advantages may be recognised. There are reductions in both energy usage and labour costs per unit volume of beer produced. The relatively high ethanol concentrations formed during fermentation promote increased precipitation of polyphenol protein material and, therefore, high-gravity beers have better colloidal stability than standard gravity fermented products (Whitear & Crabb, 1977). Relatively high ethanol yields impart greater microbiological stability to beers. Yeast growth extent increases with increase in wort concentration; however, the relation is not *pro rata* and high gravity fermentations are more efficient and produce a commensurately greater yield of ethanol. Nevertheless, it is necessary to increase the yeast pitching rate in proportion with wort gravity and crop viabilities can be reduced when very concentrated worts are used (Fernandez *et al.*, 1985). After dilution the beers from high-gravity fermentations are generally considered to have a smoother palate than sales gravity beers, probably due to the loss of polyphenols (Hackstaff, 1978).

There are some constraints that limit the benefits of high-gravity brewing. Although yields from individual fermentations are improved, there will be a concomitant increase in the fermentation cycle time, assuming all other control parameters stay the same. In this case the total number of fermentations which can be performed per unit time per vessel will decrease. For this reason, some of the gains in productivity are lost. A major aim of high-gravity fermentation management has been to identify regimes that avoid the penalty of protracted cycle times, yet produce beer, which after dilution is indistinguishable from the same product fermented at sales gravity. This theme is explored in greater detail in various sections of Chapters 3, 5 and 6.

A fundamental requirement of high-gravity brewing is that the brewhouse must be capable of producing concentrated worts. In the case of new breweries designed for the application of this technique, the brewhouse may be sized appropriately to accommodate the wort batch size and concentration that is required. In many cases, however, the process has been introduced to an existing site where the brewhouse was designed to service fermenters requiring less concentrated worts. In this case it has been necessary to adapt the process of wort production to meet the needs for increased capacity. Obviously this may be partially achieved by using a lower grist to liquor ratio. However, this is of limited value since the efficiencies of plant such as lauter tuns will allow only a certain margin of turn-up in their rated capacities.

Schaus (1971) proposed the use of a recycling wort production method, in which the run-off from an initial lautering was directed towards a mash mixer, added to a further charge of grist and used to produce a second concentrated wort. However, this approach has not seen widespread adoption and it would be suspected that wort viscosity would be problematic. In the case of plant designed specifically for high-gravity brewing, the new generation of membrane mash presses seem ideally suited to producing concentrated worts (Hermia & Rahier, 1992; Bühler, 1995). Copper boil of high-gravity worts reduces the efficiency of extraction of hop bitterness and it is necessary to counteract this by increasing dosage rates (Hackstaff, 1978). This is now less of a problem since bitter hop extracts can be added at later stages in the process.

The most commonly used method of increasing wort concentration without modifying the brewhouse is the use of liquid syrup adjuncts which may be added directly to the copper and therefore circumvent the earlier steps of wort production. This approach is superficially very attractive; however, it should be treated with caution since the use of pure sugar syrups dilutes the available nitrogen concentration of the finished wort. Dramatic changes in wort carbon to nitrogen ratios result in large shifts in the concentrations of flavour-active metabolites produced during fermentation, in particular elevated levels of acetate esters (Whitworth, 1978) (Section 3.7.3). The result is that it is difficult to match the flavour of existing beers fermented at low gravity with a high-gravity fermented product.

In terms of wort concentration, high-gravity brewing is taken to mean fermentation of worts of approximately 16–18°Plato (1.064–1.072) which are subsequently diluted to sales gravity beers of 10–12°Plato (1.040–1.048). This upper limit is based upon constraints, which relate to issues both economic and of beer quality. Whitworth (1978) considered that a breakpoint of about 15°Plato (1.060) for UK brewed beers was optimal. More concentrated worts attracted unacceptable penalties in terms of brewhouse efficiency and yeast wetting losses in fermenter. However, of greater significance was the observation that fermentation of worts of greater than 15°Plato was associated with an exponential increase in beer ester levels. Furthermore, use of more concentrated worts may result in sluggish or sticking fermentations and formation of low-viability yeast crops. These effects have been assumed to be a consequence of sensitivity of yeast to ethanol toxicity and high osmotic pressure.

It is now suggested that the adverse effects on yeast and fermentation, which result from the use of very concentrated worts, reflect nutrient deficiencies. These deficiencies may be exacerbated by dilution of malt constituents with syrup adjuncts. Casey and Ingledew (1986) concluded that a normal 12°Plato wort required a minimum free amino nitrogen (FAN) concentration of 160 mg l⁻¹. High-gravity worts (18°Plato) could be used to produce beers with normal levels of esters after dilution and fermentation performance was satisfactory providing at least 280 mg l⁻¹ FAN was provided. Provision of semi-aerobic conditions also produced similar effects (Casey et al., 1984, 1985). The same authors and others, for example, McCaig et al., (1992) have shown that worts of 24–30°Plato may be fermented satisfactorily providing that yeast pitching rate, wort oxygen concentration and FAN levels were increased pro rata. In the case of very-high-gravity brewing (more than 24°Plato) it may be necessary to supplement worts with yeast foods, which contain assimilable nitrogen, metal ions (especially magnesium) and unsaturated fatty acids.

Provided these nutritional deficiencies are remedied it may be demonstrated that brewing yeast strains are not less osmotolerant or ethanol tolerant than strains, such as saké yeasts, which are generally considered to have enhanced abilities in these respects (Mogens & Piper, 1989). Nevertheless, even discounting considerations of brewery plant efficiency, beers diluted from very-high-gravity fermentations are different in character to those brewed at lower gravities. Thus, McCaig et al. (1992) reported that very-high-gravity beers, after dilution, were well received but considered to be lighter, smoother, less sweet and more winey and estery than their normal-gravity counterparts. In addition, it was noted that yeast became less flocculent, a change considered to be a response to the high osmotic pressure, and thus, less yeast was cropped such that the high suspended cell count in green beer would significantly increase solids loading on centrifuges. Younis and Stewart (1998, 1999) concluded that the carbohydrate spectrum of very-high-gravity worts exerted an influence on the extent of volatile synthesis during fermentation. It was demonstrated that volatile levels were reduced when maltose was the predominant carbohydrate, as compared with glucose or fructose. Thus, 20°Plato worts containing 30% high maltose adjunct produced lower levels of higher alcohols and esters compared with similar concentrated wort made from all malt.

The most significant on-cost of high-gravity brewing is the requirement for blending plant. If, as is usually the case, this is performed with bright, beer the process requires careful control. The blending must be accurate to avoid product waste and to meet the requirements of legislation. The cutting water must be of high standard with regard to microbiological purity and it must be free from flavour and colour taints. Most importantly, it must be deaerated and have a mineral composition and pH that will not perturb the colloidal stability of the beer after dilution. An essential part of the blending process is to adjust the level of carbonation.

Dilution water must be very pure and totally free of particulate matter. In modern plants fully demineralised or reverse osmosis purified water is used. Deaeration is usually achieved by gas stripping, via passage through a column against a counter-current flow of carbon dioxide. A typical specification for oxygen is 0.05 ppm maximum dissolved oxygen concentration. Frequently, the purification and deaeration treatment may be followed by pasteurisation or treatment with ultraviolet light to reduce microbiological loading.

The simplest blending system is that in which a calculated volume of deaerated water is added to a batch of high-gravity beer in a tank and the mixture stirred to ensure homogeneity. Much more elegant automatic systems are now in use. These may combine blending and carbonation in one operation. Several approaches are used which differ in sophistication. All demand plant of excellent hygienic design and an operation which prevents oxygen ingress. In the simpler systems, the high gravity feed beer is analysed and the required blend rate calculated. This is then achieved by microprocessor control of flow rates of the feeds of water and high gravity beer. Carbonation is then adjusted in a separate process.

More sophisticated blending systems use in-line monitors to measure alcohol and/or specific gravity. Some of these sensors are described in Section 6.3.2.2. In these systems the desired control parameter is measured in the diluted product stream and output from this sensor regulates the flow rates of the two feed streams. Automatic carbonation can also be achieved using a similar control system. More detailed descriptions of some commercially available deaerating and high-gravity beer blending systems may be found in the following papers: Rubio *et al.* (1987), Andersson and Norman (1997), Koukol (1997), Anon. (1997).

## 2.6 Glossary of brewing terms

**Abv** Strength of beer as expressed as alcohol by volume. Units are % (v/v).
**Adjunct (*syn.* secondary brewing agent** Any carbohydrate source, other than malted barley, which contributes sugar to wort. In Germany, any ingredient used in the brewing process other than water, malted barley, yeast and hops.
**Ageing** Process of maturation in which green beer is stored, usually at low temperature, and in which final flavour adjustments are made and colloidal stability is achieved.
**Ale** Originally beer brewed from malted barley but not bittered with hops. Latterly, beer brewed from malt and hops by top fermentation at relatively high temperature. Now any beer brewed in this style but not necessarily by top fermentation.
**Aroma hops** Varieties with a high content of hop oils used for imparting flavour and aroma, as opposed to bitterness.
**Attenuation** Decrease in specific gravity of wort during fermentation. *Attenuation*

*rate* = rate of decline in gravity (fermentation rate); *attenuation gravity* = specific gravity at the end of primary fermentation; *attenuation limit* – see **Fermentable residue**.

**Auxiliary finings**  Silica-based preparations often added to primary fermenter to promote protein flocculation and thereby facilitate action of isinglass finings in later processing.

**Barm**  Yeast with entrained beer (barm ale) collected from fermenter.

**Barrel**  Measure of liquid capacity:

    USA barrel – 31.5 US gallons (119.344 litres)

    UK barrel – 36 Imperial gallons (163.65 litres)

**UK volume measures**

| | |
|---|---|
| Pin | 4.5 gallons |
| Firkin | 9 gallons |
| Kilderkin | 18 gallons |
| Barrel | 36 gallons |
| Tierce | 42 gallons |
| Hogshead | 54 gallons |
| Puncheon | 72 gallons |
| Butt (Pipe) | 108 gallons |
| Tun | 210 gallons |

**Beer**  Beverage, usually alcoholic, produced by fermentation of an aqueous extract of a malted cereal and usually flavoured with hops or other bitter and aromatic substances.

**Bitterness**  Bitter astringent flavour in beer derived from isomerised hop iso-alpha acids.

**Break**  Precipitate, predominantly protein and polyphenol formed during wort boiling ('hot break') and wort cooling ('cold break'). *Syn.* hot and cold 'trub'.

**Carrageenan**  Negatively charged polysaccharide from the marine alga *Chondrus crispus* (Irish moss). Used to promote formation of hot break during the copper boil. Has also been used as a support medium in some immobilised yeast reactors.

**Centrifuge**  Used primarily for removal of yeast and some suspended solids at the end of fermentation, often referred to as green beer centrifuge. Centrifuges may also be used for removal of hot break.

**Cereal cooker**  Mashing vessel used for solid adjuncts such as rice and maize grits, which require higher mashing temperatures than malted barley, for starch gelatinisation and amylolysis.

**Chill-proofing**  Treatments to remove polyphenol and protein components of beers, which may form permanent hazes when beer is cooled.

**Collection**  Process of filling fermenting vessels with wort.

**Coloured malts**  Speciality malts used as adjuncts with standard malts, which impart colour and flavour to beer. Usually produced by high-temperature kilning. The severity of the heat treatment determines the colour, as in chocolate, black, brown malt etc.

**Conditioning**  Synonymous with ageing. 'Warm conditioning' is performed in fermenter and synonymous with 'VDK stand' or 'diacetyl rest'. 'Cold conditioning' is the low-temperature storage for flavour maturation and attainment of colloidal stability, *syn.* 'lagering'. Also 'condition' (noun) referring to degree of carbonation in beer, particularly after secondary fermentation.

**Continuous fermentation**  Process in which a fermenting vessel is fed continuously with wort and a concomitant continuous stream of beer is produced. Continuous maturation/secondary fermentation is a variation.

**Cool ship**  Vessel, now very rare, used to collect hot wort from copper for holding prior to delivery to fermenter.

**Copper (*syn.* kettle)**  Process vessel in which sweet wort is boiled, usually with hops.

**Copper fining**  Process aids added to the copper to promote coagulation. Typically tablets of Irish moss (Carrageenan) or, occasionally silicates.

**Cropping (*syn.* yeasting, skimming)**  Removal of excess yeast from fermenter.

**Crystal malt**  Flavoured speciality malts produced by wetting a fully modified lightly kilned malt and then subjecting it to a further heat treatment which hydrolyses and liquefies the endosperm. After a further higher temperature kilning the grain interior becomes crystalline.

**Diacetyl stand (rest)**  See VDK stand.

**Dry hopping**  Addition of aroma hops to fermenter or cask for conferring intense hop flavours to beers.

**Enzymes**  Specifically in brewing, enzymes are preparations, often relatively impure, which contain particular activities which catalyse reactions whose effects are beneficial to the brewing process.

**Extract**  The concentration of dissolved solids in clarified wort. The percentage of water-soluble extract derived from a material using a standard test is termed the 'theoretical yield'. Extract of liquid adjuncts given in terms of their contribution to specific gravity and quoted as litre degrees per kg or barrel of adjunct. The total volume of wort containing all additions and ready for transfer to fermenter is termed the 'standard reference wort' for a particular product. The measured gravity of this wort is termed the 'original extract'. The proportion that is utilised during fermentation is termed the 'fermentable extract'.

**FAN**  Free amino nitrogen, a measure of nitrogen content of wort or beer measured using a test based on the ninhydrin reaction (see **TSN**).

**Fermentability**  Proportion of wort carbohydrate that may be converted to ethanol by yeast during fermentation. Usually given as ratio of fermentable to total solids content of wort.

**Fermentable residue**  Fraction of specific gravity at completion of primary fermentation consisting of fermentable carbohydrate that is unused because yeast metabolism has been arrested by the conditions in the fermenter. 'Non-fermentable residue' is the fraction of specific gravity at completion of primary fermentation representing solutes, which cannot be utilised by yeast.

**Fining**  Any agent (or the process) used to promote sedimentation of suspended particles. (see **Carrageenan, Copper fining, Irish moss, Isinglass, White finings, Auxiliary finings**).

**Green beer**  Immature beer before conditioning.

**Grist**  Charge of grain and other ingredients fed to the mill or the flour devolving from the mill.

**Gyle**  Any batch of beer or wort as it proceeds through the brewing process.

**Hanging fermentation (*syn*. sticking fermentation)**   Defective fermentation that slows or stops before the wort is fully attenuated.

**High-gravity brewing**   Production of concentrated wort that requires dilution, usually after fermentation, to produce finished beer.

**Hop back (*syn*. hop jack, montjeus)**   Vessels used for removing whole hop cones from wort after the copper boil.

**Hops**   Whole hops are dried cones of female plant *Humulus lupulus* added to beer to impart bitterness and flavour. Rarely used as dried cones, now usually in form of an extract of which many varieties are available.

**Hop isomerisation**   Conversion of hop alpha acids during the copper boil into the bitter tasting iso-alpha acids.

**Hopping rate**   Concentration of dried hops used.

**Isinglass (*syn*. white finings)**   Preparation of collagen with net positive charge prepared by treating fish swim bladders with acid. Used as a fining agent for beer, particularly for removal of yeast cells.

**Krausen**   'Rocky' or 'cauliflower' head formed at the surface of the wort during active primary fermentation. At the most active stage of primary fermentation is referred to as 'high krausen'. Also to 'krausen' is to add a proportion of actively fermenting wort to maturing beer in lager tanks to provide yeast and fermentable extract for secondary fermentation.

**Lager**   Style of beer produced by bottom fermentation usually at relatively low fermentation temperature and originally subject to lengthy lagering process.

**Lagering**   Period of cold-temperature storage for bottom-fermented beers during which flavour maturation and clarification occurs.

**Late hopping**   Hop varieties, or extracts, which are added to the copper relatively late in the boil so as to retain more of the hop oil fraction.

**Lauter tun**   Vessel with false bottom used for separating sweet wort from spent grains.

**Liquor**   Synonym for water.

**Mash (mashing, mashing-in)**   Mixture of grist and water. Process of mixing the grist with water and heating to prepare an aqueous extract.

**Mash copper**   Vessel used for boiling portion of mash in traditional decoction mashing regime.

**Mash mixer**   Stirred vessel used for preparing and heated mash prior to transfer to lauter tun.

**Mash press**   Plate and frame filter press used for separating sweet wort from spent grains.

**Mash tun**   Vessel used primarily in UK ale breweries for preparing sweet wort by a constant temperature infusion process and fitted with false bottom for separating sweet wort from spent grains.

**Maturation**   Process step in which green beer is treated to develop its finished flavour and aroma (see **Conditioning, Lagering** and **Ageing**).

**Mill**   Device used for grinding malted barley and other ingredients to form grist.

**Modification**   Extent to which the structural cell wall material of the endosperm of barley is broken down during malting. Measure of the quality of wort predictive of the amount of extract that will be liberated during mashing.

**Original gravity**   Gravity of wort measured at 20°C before fermentation has commenced. This parameter can be calculated at any stage during the fermentation by measurement of present gravity and alcohol concentration and back calculation using standard tables.

**Paraflow**  Wort paraflow or wort cooler, a plate heat exchanger used for cooling hot wort prior to delivery to the fermenter.

**Pitching**  Inoculating wort with yeast to achieve a desired suspended cell count (pitching rate). In USA, *syn.* 'brink yeast', 'brink rate'.

**Present gravity**  Gravity measurement made at 20°C. See also **Original gravity**.

**Primary fermentation**  Initial period of fermentation in which fermentable carbohydrate is utilised by yeast and converted to ethanol and carbon dioxide with the concomitant decrease in specific gravity.

**Priming**  Addition of ('priming') sugar to beer for flavour but usually to provide fermentable sugar for formation of carbon dioxide during secondary fermentation.

**Process aids**  Any substance used in brewing which is not a primary ingredient but facilitates the processes occurring during brewing and which may produce residues which persist in the finished beer. Examples include enzymes, filter aids, fining agents and salts.

**Racking**  Separation of green beer from yeast. Used to describe process of emptying fermenting vessels also to fill casks. Final specific gravity achieved at the end of primary fermentation is referred to as the 'racking gravity'.

**Reinheitsgebot**  German beer purity laws, which prohibit the use for brewing of any ingredient other than water, malted barley, hops and water.

**Ruh**  German brewing term. See **VDK stand**.

**Run-off**  Removal of sweet wort from mash or lauter tun.

**Secondary fermentation**  Any process stage after primary fermentation that requires the activity of yeast.

**Sparging**  Spraying bed of settled grains in lauter or mash tun with (sparge) liquor to wash out entrained wort. Occasionally used to describe the bubbling of gases (especially air or oxygen) through wort in fermenter.

**Trub**  See **Break**.

**TSN**  Total soluble nitrogen, a measure of the nitrogen content of wort. Now rarely used, the fraction of nitrogen in malt which remains dissolved after a standard period of boiling is referred to as the permanently soluble nitrogen ('PSN') (taken to be TSN × 0.94).

**Tun**  Originally a brewing vessel made by coopering, now any vessel used in brewing.

**Underback**  Vessel in which sweet wort is collected from mash tun and held before delivery to the copper.

**Underletting**  Method of raising temperature of contents of mash tun by pumping hot liquor in from the base.

**VDK stand (rest)**  Period of warm storage in fermenter after completion of primary fermentation during which yeast assimilates vicinal diketones (VDK) and reduces them to less flavour-active metabolites. In German, **Ruh** storage. One of primary functions of warm conditioning.

**Wort**  Aqueous extract of malted barley and other sources of extract ('sweet wort') boiled with hops to produce bittered wort, which forms the starting material for fermentation.

**Whirlpool**   Vessel into which hot wort is discharged from the copper and held (whirlpool stand) thereby allowing hot break to form a sediment via a hydrocyclone action.

**Yeast food**   Proprietary preparations containing salts, vitamins and amino acids used as supplements where nutrient deficiencies in wort are suspected

# 3 Biochemistry of fermentation

## 3.1 Overview

The biochemistry of brewery fermentation is complex and many of its aspects remain to be fully elucidated. The reactions – which underpin the growth of yeast during fermentation and the concomitant conversion of wort to beer – touch on almost every facet of cellular metabolism. Many of the pathways involved are those which may be found described in any standard biochemistry text and to duplicate this information here is both needless and beyond the scope of this book. Instead the intention is to concentrate on those aspects which are peculiar to brewery fermentation. In particular, the factors which control the flow of carbon and other nutrients between biomass formation, ethanologenesis and the formation of metabolites that contribute to beer flavour.

Three general areas of discussion may be considered which influence the biochemistry of fermentation. These are wort composition, the genotype of the yeast strain and phenotypic expression of the genotype as influenced by fermentation practice.

Wort composition is complex and it follows that, as with growth of microorganisms on any uncharacterised medium in a batch culture, the biochemical reactions which contribute to the assimilation and metabolism of its individual components will be equally convoluted. Wort provides a complete growth medium for yeast. Indeed, it should be remembered that brewery fermentation is nothing more than a manifestation of yeast growth and beer is merely the by-product of that activity. The yeast strain used will have been chosen because it has desirable fermentation properties and has the potential to produce beer with a suitable composition. The object of fermentation management and control is to regulate conditions such that the by-products of yeast growth and metabolism are produced in desired quantities and within an acceptable time.

The precise composition of wort is unknown, although provided similar materials and methods are used in its preparation the gross analysis should be relatively constant. All malt wort prepared with brewing liquor with an appropriate ionic composition provides a medium with the potential to produce new yeast biomass, ethanol and flavour components in balanced and desired quantities. To realise this potential it is necessary to control other parameters such as temperature and yeast inoculation (pitching) rate. Although wort composition is somewhat uncharacterised, it is possible to establish relationships between the assimilation by yeast of the major classes of wort nutrients and the formation of products of yeast metabolism. This equilibrium can be disturbed if wort composition is modified by, for example, changing the ratio of carbon to nitrogen in the wort by increasing the fermentable extract with sugar syrup adjuncts. Conversely, manipulation of wort composition can be a deliberate strategy for producing desired changes in fermentation.

The genotype of the particular strain of yeast used is critical to the outcome of fermentation. For example, the spectrum of flavour-active metabolites produced is as much determined by the yeast as by the conditions established during fermentation. Of fundamental significance is the response of brewing yeast strains to sugars and oxygen. All brewing yeast strains have limited respiratory capacity and are subject to carbon catabolite repression. In a brewery fermentation, irrespective of the presence of oxygen, metabolism is always fermentative and derepressed physiology never develops (see Section 4.3.1 for further discussion). Thus, the major products of sugar catabolism are inevitably ethanol and carbon dioxide. Respiration, in the true sense of complete oxidation of sugars to carbon dioxide and water, coupled to ATP generation via oxidative phosphorylation does not occur. The maximum ethanol concentration that may be generated during fermentation is also determined by the yeast genotype. In this regard strain-specific tolerances to ethanol and reduced water activity are important. Flavour considerations apart, this parameter may serve to limit the maximum concentration of sugar that can be used with no detrimental effect to the yeast.

Phenotypic expression of the genotype of the yeast is modulated by the conditions experienced during fermentation. In large vessels, in particular, the yeast is subjected to multiple stresses such as high hydrostatic pressure, elevated carbon dioxide concentration, low pH and reduced water activity. All of these have the potential to elicit specific biochemical responses by the yeast. In this regard channelling of a proportion of wort carbohydrate into accumulation of the disaccharide trehalose (see Section 3.4.2.2) may be implicated in the ability of yeast to withstand stress.

Aspects of yeast handling peculiar to the brewing industry are crucial to understanding the biochemistry of the process. In many fermentation industries it is normal practice to use an inoculum, which has been specifically cultivated for the purpose. In order to ensure rapid onset, starter cultures are often employed in the exponential phase of growth. This has the additional twin benefits of providing a means for introducing an inoculum which is of consistent physiological condition and which may be relatively small compared to the total new biomass generated during the fermentation.

In traditional brewing fermentations, such as Belgian lambic and many native (ethnic) beers, the inoculum may arise via contamination from the environment. However, most often and certainly in the modern process it is taken from yeast cropped from a previous fermentation. In consequence, the inoculum consists of stationary phase cells which are depleted in membrane lipids, sterols and unsaturated fatty acids. Replenishment of the pools of these essential lipids and restoration of membrane function is dependent on the supply of oxygen provided with wort at the start of fermentation. Commencement of fermentation and movement of yeast cells from stationary to growth phases is therefore associated with simultaneous assimilation of wort nutrients and a transition from aerobic to anaerobic conditions.

The practice of serial cropping and re-pitching requires that relatively high inoculation rates are used and this together with control of wort oxygen concentration limits subsequent yeast growth to modest levels. An inherent part of this practice is that the cells in the inoculum form a significant proportion of the total population in the fermenter, and, furthermore, there is an opportunity for these cells to persist through several generations of serially cropped and re-inoculated fermentation. Unlike bacteria, yeast cells have a finite life span and undergo an ageing process. Like any other mortal cell, both phenotypic and genotypic modifications are possible due to the effects of ageing and the onset of senescence (see Section 4.3.3.4).

The practice of serial fermentation necessitates having facilities for yeast handling during the interval between cropping and re-pitching. In consequence there is an opportunity for further modification to physiology depending on the time the yeast is held and the conditions under which it is stored. A feature of this aspect of fermentation management is that the yeast is subjected to periods of growth in fermenter interspersed with intervals of starvation during storage. It follows that whilst the yeast is in fermenter it must achieve a physiological condition in which it is capable of withstanding starvation during storage. In this respect regulation of carbon flow during fermentation between glycolysis and gluconeogenesis is of significance. Thus, carbohydrate reserves accumulated during fermentation provide a source of maintenance energy during the storage phase and possibly carbon for lipid synthesis during the aerobic phase of fermentation when utilisation of exogenous carbon is limited by lack of membrane function. The ability of yeast to channel carbon into gluconeogenic pathways during fermentation is favoured by the high ratio of carbohydrate to other wort components.

## 3.2 Mass balance

The gross changes which occur during the course of a typical high-gravity lager fermentation are illustrated in Fig. 3.1. In this case the initial temperature was 11°C. This was allowed to increase to 12°C and held at this value throughout primary fermentation. There was an initial lag phase, which lasted for 12–24 hours. During this period there was little or no observable change in specific gravity, yeast count and ethanol concentration. However, the oxygen concentration rapidly decreased, falling to undetectable levels within the first 24 hours. The concentration of free amino nitrogen began to fall as soon as the fermentation started and this was accompanied by a rapid decline in pH.

The fermentation rate, as judged by decline in wort specific gravity, gradually accelerated and reached a maximum after 24–36 hours. The decline in specific gravity was inversely related to increase in ethanol and yeast biomass. The minimum value of specific gravity was achieved after approximately 100 hours and this coincided with maximum observed values for ethanol and yeast biomass. Patterns of decline in free amino nitrogen and pH mirrored each other and minimum values of each were achieved after about 80 hours, some 20 hours before full wort attenuation was achieved. During the phase of active fermentation, total yeast biomass and the suspended yeast count were coincident. The maximum suspended yeast count was reached after about 100 hours and after this time this parameter declined with the deceleration in fermentation rate. This is a reflection of the lack of mechanical agitation in the vast majority of brewery fermenters.

During the course of the fermentation approximately 50° of gravity were fermented. This was accompanied by an increase in yeast dried biomass from roughly

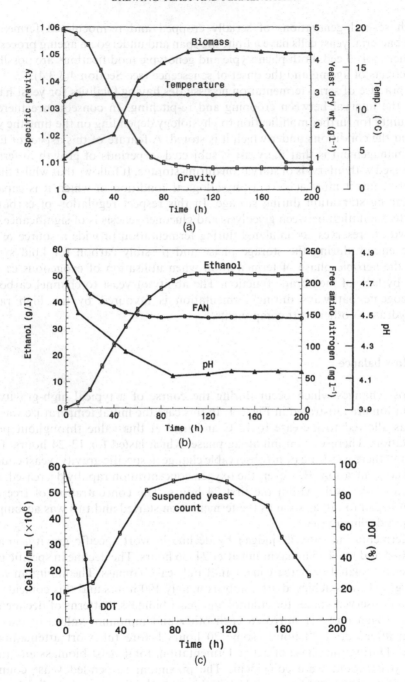

**Fig. 3.1** Some features of the progress of a typical high-gravity fermentation (Boulton, unpublished data).

$1 \, \text{g} \, l^{-1}$ to just over $4 \, \text{g} \, l^{-1}$. The terminal ethanol concentration was approximately $50 \, \text{g} \, l^{-1}$.

A crude mass balance for a fermentation of the type described above may be written as follows:

$$\begin{array}{cccccccc}
\text{Sugars} & + & \text{Free amino nitrogen} & + & \text{Yeast} & + & \text{Oxygen} & \rightarrow \\
(150 \text{ g } l^{-1}) & & (150 \text{ mg } l^{-1}) & & (1 \text{ g } l^{-1}) & & (25 \text{ mg } l^{-1}) &
\end{array}$$

$$\begin{array}{ccccc}
\text{Ethanol} & + & CO_2 & + & \text{Yeast} \\
(45 \text{ g } l^{-1}) & & (42 \text{ g } l^{-1}) & & (5 \text{ g } l^{-1})
\end{array}$$

Conversion of sugar to ethanol is approximately 88% of the theoretical, the shortfall being that proportion which is utilised by the yeast to generate additional biomass and to a lesser extent the formation of other metabolic by-products of yeast growth. Ethanol and carbon dioxide are, of course, produced in equimolar amounts; however, the yield of the latter is slightly reduced since a small proportion is utilised by the yeast cells for anabolic carboxylation reactions (Oura et al., 1980). The quantity of oxygen supplied at the beginning of fermentation is a key determinant of the regulation of the proportion of wort sugars used for new yeast biomass and that which is dissimilated to ethanol (Section 3.5).

## 3.3 Assimilation of wort nutrients

Brewing yeast strains are heterotrophic organisms capable of utilising a wide variety of nutrients to support growth and generate energy. A property of such organisms is that they must be capable of selective uptake. Thus, assimilation of individual nutrients from wort is made complex by the response of yeast to the mixture of components present. As with all cells, specific systems exist in brewing yeast strains to accommodate the uptake of individual or related classes of nutrients. Of particular relevance to brewery fermentation, assimilation of carbohydrates and nitrogenous compounds are highly regulated processes. When presented with a choice of nutrients, yeast cells tend to use first those that are most easily assimilated. Not only are some components utilised in preference to others but also the presence of some nutrients inhibits the utilisation of others. In consequence, uptake of carbohydrates and the various sources of nitrogen present in wort are ordered processes.

### 3.3.1 *Sugar uptake*

Brewing yeasts can utilise a wide variety of carbohydrates; however, there is some variability between individual strains (see Section 4.2.3). Ale strains of *S. cerevisiae* are able to ferment glucose, sucrose, fructose, maltose, galactose, raffinose, maltotriose and occasionally trehalose. Lager strains of *S. cerevisiae* are distinguished by being able to also ferment the disaccharide melibiose. *S. cerevisiae var. diastaticus* can utilise dextrins.

The patterns of uptake of sugars during an ale fermentation of starting specific gravity of approximately 1.040 are shown in Fig. 3.2. Sucrose is utilised first and the resultant hydrolysis causes a transient increase in the concentration of fructose. Fructose and glucose are taken up more or less simultaneously, in the case of the fermentation illustrated, disappearing from the wort after about 24 hours. Completion of assimilation of glucose is followed by uptake of maltose, the major wort sugar. Maltotriose is utilised last after all assimilation of maltose. Higher polysaccharides, the dextrins, are not utilised by brewing strains and these contribute to beer flavour by way of imparting fullness. Attempts have been made to utilise dextrins via two different strategies. First, through introduction of appropriate enzymes into brewing yeast strains by genetic manipulation (Tubb *et al.*, 1981; Goodey & Tubb, 1982; Vakeria & Hinchliffe, 1989; Lancashire *et al.*, 1989; Hansen *et al.*, 1990). Further discussion can be found in Section 4.2.4. Second, dextrins may be hydrolysed to assimilable sugars by addition to wort of commercial dextrinase enzymes.

The sequential uptake of wort sugars reflects the genotype of the yeast and ways in which this is expressed by repression and induction and by carbon catabolite inac-

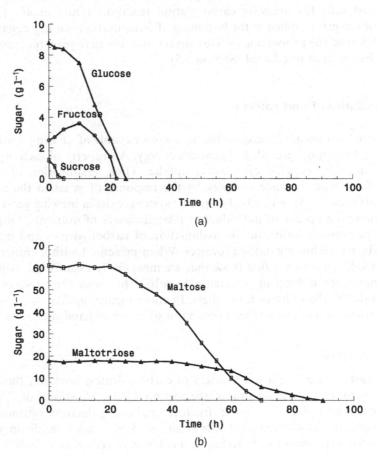

**Fig. 3.2** Utilisation of sugars during fermentation of an ale wort of original gravity 1.040 (Clutterbuck and Boulton, unpublished data).

tivation. The global effects on yeast metabolism of the presence of exogenous glucose and the mechanisms by which its effects are exerted are discussed in detail in Section 3.4. Some of these effects impinge on sugar uptake. Thus, there are specific and often multiple carriers for individual sugars. The activity of individual carriers is modulated by the spectrum and concentration of sugars present in wort. In particular, glucose appears to be the preferred substrate and when present in the medium its presence inactivates or represses carriers for the uptake of other sugars (Lagunas, 1993).

At least two glucose uptake systems have been recognised: low affinity and high affinity types, which apparently operate by facilitated diffusion (Bisson & Fraenkel, 1983a). The high affinity carrier requires the presence of a kinase although phosphorylation of glucose during uptake has been discounted in view of the evidence that the non-phosphorylable analogue, 6-deoxyglucose, had similar uptake kinetics to glucose (Bisson & Fraenkel, 1983b; Kruckenberg & Bisson, 1990). Both transporters are active with glucose and fructose. The low affinity system is constitutive, whereas the high affinity transporter is repressed in the presence of high glucose concentrations (Bisson & Fraenkel, 1984; Neigeborn et al., 1986). It follows that the role of the high affinity system is to provide an efficient scavenging mechanism in the event of competition for low concentrations of glucose. Repression of the high affinity system is associated with the general catabolite repression phenomenon and has been shown to occur only in fermentative yeast strains (Does & Bisson, 1989). It has been suggested that the low affinity system is merely passive diffusion. However, Gamo et al. (1995) refuted this on the basis that actual uptake rates using this transporter were 2–3 orders of magnitude higher than could be accounted for simply by passive diffusion.

The glucose carriers are influenced by components of the medium other than glucose itself. Thus, it is reported that exhaustion of nitrogen sources during batch growth brings about an irreversible inactivation of the glucose (and other) sugar transporters (Lagunas et al., 1982). This inactivation is apparently due to proteolysis of the carrier molecules (Busturia & Lagunas, 1986; Lucero et al., 1993). The consequences of these effects to brewery fermentation are unclear.

Maurico and Salmon (1992) concluded that differential patterns of loss of glucose transporters due to nitrogen starvation formed the basis of differences in ethanol productivity in two enological strains of S. cerevisiae. Thus, the strain that was capable of the greatest ethanol productivity had a putative second low affinity hexose transporter not subject to inactivation by nitrogen starvation. Presumably this phenomenon could be of significance to brewery fermentation depending on which component of the wort actually limits yeast growth. Should the sugar transporters be inactivated it follows that they would need to be switched on again during the initial lag phase in early fermentation. This would add further support to the contention that the carbon required for sterol synthesis during the aerobic phase of fermentation is supplied by dissimilation of endogenous glycogen reserves (Quain & Tubb, 1982). This could be due to the inability of yeast to assimilate sugars because of inactivation of the transporters, or equally to lack of membrane competence because of sterol depletion.

Sucrose is assimilated via the mediation of an invertase which is secreted into the cell periplasm. In S. cerevisiae the enzyme is encoded by the SUC2 gene and it hydrolyses both sucrose and raffinose (Carlson & Botstein, 1982). Once hydrolysed, the released fructose and glucose are taken up via glucose transporters. In the presence of high glucose concentrations the invertase is repressed via binding of a component Mig1p to the SUC2 gene promoter (Neigeborn & Carlson, 1994). The same group have suggested that low levels of glucose (0.1%, w/v) are actually required for maximum transcription of the SUC2 gene (Ozcan *et al.*, 1997).

Maltose utilisation is accomplished using the products of a multigene (MAL) family that occurs at several loci in the yeast genome and is not restricted to a single chromosome. Each locus consists of three genes: MALT which encodes for a maltose permease; MALS encoding for a maltase (α– glucosidase) and MALR which encodes for a post-transcriptional activator of the MALS and MALT genes (Needleman *et al.*, 1984; Michels & Needleman, 1984; Cohen *et al.*, 1985). Both the latter two genes are induced by maltose and repressed by glucose (Busturia & Lagunas, 1985; Cheng & Michels, 1991). The maltose uptake system is an active process requiring cellular energy. Uptake is via a proton symport system in which potassium $(K^+)$ is exported to maintain electrochemical neutrality (Serrano, 1977). As with glucose uptake it has been reported that there are also low and high affinity uptake systems for maltose (Busturia & Lagunas, 1985; Cheng & Michels, 1991). More recently Benito and Lagunas (1992) concluded that the low affinity component was due to non-specific binding of maltose to the yeast cell wall.

Jiang *et al.* (1997) investigated the inactivation of maltose uptake by glucose. They concluded that in maltose grown yeast cells, two specific signalling pathways exist for sensing the presence of extracellular glucose and that these were responsible for inhibition of maltose uptake. The first pathway was independent of glucose transport and caused proteolysis of maltase permease. The second pathway required glucose transport into the cell and resulted in proteolysis of the maltose permease and rapid inhibition of maltose transport. Wanke *et al.* (1997) reported that glucose inactivation of the maltose uptake system was modulated via the RAS/adenyl cyclase, protein kinase A signalling system.

The maltose uptake system shows no activity with maltotriose, instead there is a specific constitutive facilitated diffusion carrier (Michaljanicova *et al.*, 1982). Stewart *et al.* (1995) investigated the reasons for the observation that ale strains are frequently less effective than lager types at fully attenuating high gravity worts. They suggested that in some cases the strain-specific differences were due to the lack of a maltotriose permease. However, environmental effects were also important. Thus, elevated osmotic pressure inhibited uptake of maltose and maltotriose to a greater degree than glucose. High ethanol concentrations inhibited uptake of glucose, maltose and maltotriose. However, at modest concentration (5%, w/v) uptake of maltose and maltotriose were stimulated, an effect attributed to ethanol-induced changes in membrane configuration.

### 3.3.2   *Uptake of wort nitrogenous components*

Wort nitrogenous components are heterogeneous in nature. In a Canadian lager wort, Ingledew (1975) reported a rough distribution as: protein, 20%; polypeptides, 30–40%, amino acids, 30–40% and nucleotides, 10%. Of these the amino acid fraction is of most significance to fermentation performance and beer quality.

The uptake of wort amino acids uses a number of permeases, some specific for individual amino acids and a general amino acid permease (GAP) with a broad substrate specificity. Horak (1986) recognised 16 different amino acid transport systems in yeast. Of these 12 are constitutive and the remaining 4 are subject to regulation by the nitrogen sources present in the growth medium, a phenomenon termed nitrogen catabolite repression (Grenson, 1992). In other words, the presence of an exogenous supply of certain nitrogenous nutrients abolishes the utilisation of others by repressing the enzymes responsible for their assimilation. Uptake is an active process requiring energy (Hinnebusch, 1987). The patterns of uptake are complex, several regulatory mechanisms being evident. Thus, the spectrum of permeases present, their specificity, competition for binding to individual permeases and feedback inhibition of specific permeases by amino acids in the intracellular pool and other nitrogenous components are all influential.

The GAP permease is a high-affinity type, which is one of the group subject to nitrogen catabolite repression. Thus, maximum activity of this carrier is only expressed when nitrogen is limiting. Olivera *et al.* (1993) studied activity of amino acid permeases in chemostat cultures of *S. cerevisiae*, a technique which allows the effects of various nutrient limitations to be studied. These authors concluded that the specific permeases were likely to be involved in uptake of amino acids for anabolic pathways, notably protein synthesis, whereas the GAP permease and the others, which are subject to nitrogen catabolite repression, had catabolic roles. This provides an explanation as to why, for example, some amino acids are used in preference to ammonia when supplied as a mixture (see Table 3.1). Thus, although nitrogen is a preferred nitrogen source for catabolic reactions, certain amino acids may be used first for direct incorporation into proteins.

**Table 3.1** Classes of wort amino acids in order of assimilation during fermentation (Pierce, 1987).

| Class A | Class B | Class C | Class D |
|---------|---------|---------|---------|
| Arginine | Histidine | Alanine | Proline |
| Asparagine | Isoleucine | Ammonia | |
| Aspartate | Leucine | Glycine | |
| Glutamate | Methionine | Phenylalanine | |
| Glutamine | Valine | Tyrosine | |
| Lysine | | Tryptophan | |
| Serine | | | |
| Threonine | | | |

Investigations into nature of the specific permeases have relied on the use of mutant strains in which only the carrier of interest is functional. For example, Garcia and Kotyk (1988) studied L-lysine uptake in a double mutant of *S. cerevisiae*. They concluded that, in the strain used, a specific L-lysine permease was present, which was not active with any other naturally occurring amino acid. The maximum transporting activity of the carrier was more than an order of magnitude lower than the GAP enzyme with the same substrate. Tullin *et al.* (1991) considered that specific permeases are either high capacity, low affinity or low capacity, high-affinity. In the same communication these authors described a high affinity system in yeast active with the branched chain amino acids, L-lysine, L-leucine and L-valine. The correlation of substrate similarity and permease specificity has been noted for other specific amino acid carriers (Grenson, 1992).

Environmental conditions also exert modulating effects. Thus, Iglesias *et al.* (1990) reported that during the course of a wine fermentation the activity of the general amino acid permease decreased in parallel with increase in ethanol concentration. Slaughter *et al.* (1987) showed that increased pressure and carbon dioxide concentration resulted in altered patterns of amino acid uptake and that this could be related to changes in the formation of flavour volatiles.

In general, brewery worts made according to any given procedure have a relatively constant spectrum of amino acids (O'Connor-Cox & Ingledew, 1989). Pierce (1987) classified wort amino acids on the basis of patterns of assimilation during fermentation (Table 3.1). Those in the first group are assimilated immediately after the yeast contacts the wort. Those in class B are assimilated more slowly, whereas those in class C are not utilised until class A amino acids have disappeared from the wort. Proline is the sole member of class D and its dissimilation requires the presence of a mitochondrial oxidase not present under the repressed and anaerobic conditions of fermentation (Wang & Brandriss, 1987).

Although the amino acid spectrum of all malt wort is relatively constant, use of low nitrogen sugar adjuncts can perturb this balance. In this case an altered and unbalanced amino acid spectrum may arise. In order to provide a method of assessing the significance of changes in wort composition brought about by altered brewing practices, Pierce (1987) further categorised wort amino acids on the basis of their importance as regulators of the development of beer flavour metabolites (Table 3.2). In this case the significance of the concentration of individual amino acids was whether they could be synthesised *de novo* from wort carbohydrates, or if assimilation from the pre-formed pool in wort was critical to the formation of certain flavour-active metabolites. In this regard, the initial concentration in wort of class 1 amino acids was considered relatively unimportant since they could be either assimilated from wort or synthesised from sugar catabolism and transamination reactions. Deficiencies in class 2 and 3 amino acids were of greater significance. For example, relationships between valine availability and diacetyl production as well as utilisation

**Table 3.2** Wort amino acid classes based on effect on fermentation performance and beer analysis.

| Class 1 | Class 2 | Class 3 |
|---|---|---|
| Aspartate | Isoleucine | Lysine |
| Asparagine | Valine | Histidine |
| Glutamate | Phenylalanine | Arginine |
| Threonine | Glycine | Leucine |
| Serine | Tyrosine | |
| Methionine | | |
| Proline | | |

of various amino acids and higher alcohol production have been demonstrated (Inoue & Kashihara, 1995).

Individual amino acids are more readily assimilated than single sources of homo-peptides (Ingledew & Patterson, 1999; Patterson & Ingledew, 1999). The authors concluded that in simple mixtures the presence of amino acids, ammonia, allantoin and urea inhibited the utilisation of dipeptides. With more complex mixtures of three amino acids and three dipeptides, simultaneous assimilation was observed. In defined media, ammonium ions apparently inhibited peptide utilisation, whereas leucine enhanced peptide uptake.

Calderbank et al. (1985) demonstrated that peptides are utilised during the fermentation of malt worts. L-stereo isomers were taken up more readily than D-forms and peptides containing basic residues more easily utilised than those with acidic constituents (Becker et al., 1973; Becker & Naider, 1977). Marder et al. (1977) concluded that peptides containing no more than five amino acid residues were transported into the cell. Induction of peptide transporters was apparently regulated in complex but positive manner by the presence of trace concentrations of amino acids (Island et al., 1987, 1991).

### 3.3.3 Uptake of lipids

Lipids consist of a diverse group of molecules linked only by their properties of sparing solubility in water but being readily soluble in organic solvents such as chloroform. Of interest to brewing is the possibility of assimilation from wort of those lipids which would otherwise require oxygen for their synthesis. Thus, uptake of sterols and unsaturated fatty acids from wort has the potential to reduce the requirement for wort oxygenation (Section 3.5). For example, Lentini et al. (1994) analysed the lipid fraction of trub from wort and reported that 18:2 unsaturated fatty acids were the most abundant components. Furthermore, fermentations with trub-rich wort were associated with faster rates and increased yeast growth compared to bright worts.

Saccharomyces yeasts are able to take up fatty acids at low concentration using a facilitated diffusion system. At high concentrations entry is via simple diffusion (Finnerty, 1989). It is assumed that diffusion of such molecules into cells is facilitated by the lipophilic nature of the membrane (van der Rest et al., 1995). Free fatty acids are powerful detergents and immediately after entry into the cell they are esterified to coenzyme A, to reduce their potential for non-specific enzyme inactivation.

Salerno and Parks (1983) reported that sterols were taken up by yeast only when growing under aerobic conditions. The process was described as being passive by Lorenz et al. (1986) However, no uptake occurs in stationary phase cells, under anaerobic conditions. This phenomenon has been termed aerobic sterol exclusion (Lewis et al., 1988). It is related to the ability of the yeast to synthesise haem such that cells which are haem competent are not capable of sterol uptake (Shinabarger et al., 1989). Bourot and Karst (1995) isolated a gene (SUT1) from yeast and provided evidence that its product was involved in sterol uptake. These authors suggested that this formed part of a system used by yeast for sterol uptake under conditions of anaerobiosis, or in the absence of haem biosynthesis. This would suggest that in a brewery fermentation exogenous sterols could only be assimilated during the anaerobic phase before yeast growth ceases.

### 3.3.4    Metal ion uptake

Uptake of metal ions by yeast cells can be considered from two standpoints. First, a supply is required of several metal ions since they are essential nutrients. Absence of these from growth media has global adverse effects on cellular metabolism, mainly because of the number of enzymes which require metal ion co-factors for full activity. Of particular note in brewing are zinc, magnesium and calcium (Lie & Jacobsen, 1983; Lentini et al., 1990; Walker et al., 1996; Bromberg et al., 1997; Walker & Maynard, 1997). Second, in common with other microbial cells, yeasts are capable of concentrating many metal ions, which have no positive physiological function. Indeed uptake of this latter group may have toxic effects on the yeast cells. Where mixtures of metals are presented competition is possible between uptake of essential and nonessential metals. Yeast cells are particularly adept at removing metal ions from the external environment, irrespective of whether or not they are required as trace nutrients. This property has led to proposals that they could be of utility for the removal of heavy metal pollutants from contaminated waters or as a procedure for recovery of metal ions of economic value (Avery & Tobin, 1992).

Uptake of metal ions may be passive or active processes or a combination of both. Two mechanisms can be recognised, termed 'biosorption' and 'bioaccumulation'. The first of these is a passive process in which the metal ions become attached principally to the cell wall. This procedure is reported as being rapid, insensitive to temperature, not requiring metabolic energy and not inhibited by metabolic inhibitors. Biosorption does not require cell viability. Several mechanisms for binding have been suggested including ion exchange, adsorption, precipitation and complexation (for review see Blackwell et al., 1995). This non-specific binding has been suggested as having a physiological role in that it allows the cell wall to act as a filter, binding ions which could potentially be toxic to cells.

Bioaccumulation is a slower process, which has all of the hallmarks of a regulated process. Thus, it is temperature dependent and can be blocked by the use of metabolic inhibitors. Genes have been identified which appear to be associated with the uptake of specific metal ions. For example, FRE1 codes for a protein linked to reduction and uptake of ferric ions (Anderson et al., 1992). Dancis et al. (1994) described a high affinity copper uptake system under the control of another gene, CTR1. Yet another carrier is responsible for the uptake of calcium and magnesium (Borst-Pauwels, 1981). The kinetics of uptake may be complex. Thus, in the case of the latter transporter, uptake of strontium ions resulted in displacement of calcium, magnesium and protons (Avery & Tobin, 1992).

Suggested mechanisms for metal ion uptake include lipid peroxidation, permeation, involvement of specific carriers, endocytosis and ion channels. These systems make use of the pools of ions concentrated within the cell envelope. Once inside the cell, metal ions may be utilised by incorporation into proteins or wherever else they are required. Alternatively, ions may be sequestered within yeast vacuoles. This latter option appears to be a mechanism whereby cells can partition ions such that their cytotoxic effects are ameliorated although it may also serve as a temporary store of useful ions (Blackwell *et al.*, 1995).

Since metal ions can bind to yeast cells via electrostatic interactions it follows that this may also occur with other charged wort species and this may render essential metal ions unavailable for assimilation. In this respect, Lie *et al.* (1975) pointed out that with regard to yeast trace metal nutritional requirements it was necessary to consider both the absolute concentrations present in wort and also the presence of metal chelating agents. For example, Lentini *et al.* (1994) reported that worts with a high trub content were associated with reductions in available zinc due to non-specific binding. However, more recently, Kreder (1999) demonstrated that yeast can assimilate zinc bound to trub during fermentation.

## 3.4   Carbohydrate dissimilation

The principal pathways used by yeast for the dissimilation of glucose are shown in Fig. 3.3. All yeasts predominantly utilise the Embden-Myerhoff glycolytic pathway for generation of ATP via substrate level phosphorylation. A proportion of the carbon flow devolving from glucose, or other sugars, is dissimilated via the hexose monophosphate shunt. This pathway is of importance for generation of NADPH for use in anabolic metabolism such as lipid synthesis. Bruinenberg *et al.* (1983) calculated that at least 2% of glucose metabolism had to be via the hexose monophosphate pathway where ammonium was the nitrogen source in order to satisfy anabolic requirements. A much greater proportion was required with certain carbon sources, such as pentoses. Since in the case of a brewery fermentation anabolic requirements are modest it seems likely that this pathway is of relatively small significance.

The product of glycolysis, pyruvate, occupies a major branch-point in metabolism. With respect to sugar catabolism, carbon flow may be directed towards acetyl-CoA and subsequent oxidation via the TCA cycle and oxidative phosphorylation, or into ethanol formation via acetaldehyde. Lagunas (1986) estimated that in *Saccharomyces* yeasts most carbon was dissimilated via the glycolytic pathway in order to generate ATP and thence to ethanol via the fermentative route. Thus, this author presented data indicating that the proportions of glucose, maltose and galactose utilised for ATP production were 73%, 69% and 53%, respectively. Of these, only 3%, 4% and 14%, respectively, were metabolised via the oxidative pathway. Most significantly, in the case of glucose and maltose, the fraction of carbon flux metabolised via fermentative or oxidative routes was independent of the presence of oxygen. Using the assumption that the yeast respiratory chain had two phosphorylation sites, Lagunas (1986) calculated that under aerobic conditions the yeast obtained 27%, 40% and 84% of the yield of ATP from respiration using glucose, maltose and galactose, respectively, the remainder deriving from substrate level phosphorylation.

It may be concluded from the foregoing discussion that the nature and concentrations of sugars in wort exert far-reaching effects on yeast metabolism. The effects on carbohydrate uptake of the spectrum of sugars present in wort has been discussed previously (Section 3.3.1). *S. cerevisiae*, in common with several other yeast genera, is usually described as being 'facultatively fermentative'. That is, it is capable,

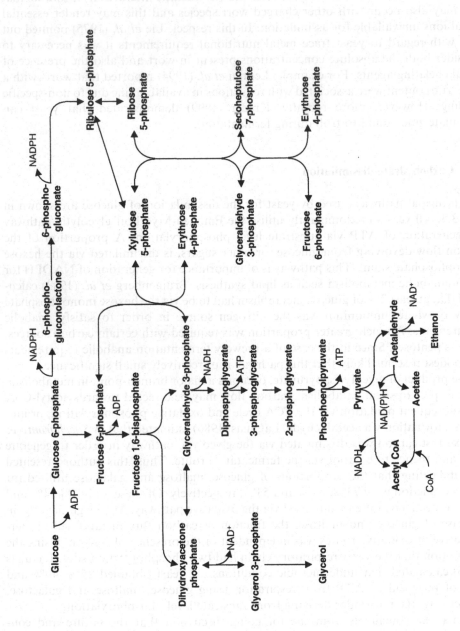

**Fig. 3.3** Pathways for dissimilation of glucose to ethanol in yeast via glycolysis and the hexose monophosphate shunt.

under the appropriate conditions, of fully oxidative metabolism in which sugars are respired to $CO_2$ and water and ATP is generated via oxidative phosphorylation. Alternatively, metabolism may be fermentative where the products of pyruvate metabolism are predominantly ethanol and $CO_2$ and energy is transduced by substrate level phosphorylation.

It may be assumed that the sole metabolic trigger, which switches metabolism between these two modes, is availability of oxygen. In fact, this is not necessarily the case and several other regulatory mechanisms occur in different yeast genera. These are summarised in Table 3.3. The Kluyver and the Custers effects are not of direct relevance to brewing in that they do not occur in *Saccharomyces* yeasts. For completeness brief mention is made here.

The physiological basis of the Kluyver effect has been elucidated by Barnett (Barnett, 1992b; Sims *et al.*, 1991; Sims & Barnett, 1991). The underlying observation is that certain yeasts cannot utilise certain disaccharides under anaerobic conditions even though assimilation under aerobic conditions is possible. In addition, anaerobic utilisation of one or more of the component hexoses may also occur. It is suggested that the effect is a result of reduced transport under anaerobiosis. Thus, because anaerobic conditions are associated with reduced availability of ATP due to blocked oxidative phosphorylation there may be insufficient energy to drive the proton pump system required for uptake of the sugar. In consequence, a reduction in glycolytic flux inactivates pyruvate decarboxylase.

**Table 3.3** Mechanisms for regulation of sugar metabolism in yeasts.

| Mechanism | Description |
|---|---|
| Pasteur effect | Inhibition of alcoholic fermentation in the presence of oxygen, or activation of glycolysis by anaerobiosis. |
| Crabtree effect – short term | Instantaneous aerobic ethanol formation following transition from carbon limitation to carbon excess. |
| Crabtree effect – long term | Aerobic ethanol formation at high growth rates under conditions of excess sugar or sugar limitation. |
| Custers effect | Stimulation by oxygen of glucose fermentation to ethanol. |
| Kluyver effect | Absence of alcoholic fermentation with certain sugars, particularly disaccharides even though glucose is fermented. |
| Carbon catabolite inactivation and repression | Suppression of respiratory metabolism by high sugar concentrations. |

The Custers effect is observed, among others, in members of the genus *Brettanomyces*, which although not a brewing yeast may be implicated in secondary fermentation in some traditional UK cask fermented ales (see Section 8.1.3.2). The phenomenon is exemplified by the observation that fermentation of glucose is more rapid under aerobic as opposed to anaerobic conditions. The explanation is reportedly related to altered redox requirements in these yeasts depending on the availability of oxygen. For example, yeasts belonging to this group may produce large amounts of acetic acid via an $NAD^+$-linked aldehyde dehydrogenase. The reduced co-factor may be re-oxidised by the respiratory chain in the presence of oxygen. Under anaerobic conditions this is less possible and the resultant high ratio of $NADH:NAD^+$ may result in inhibition of glycolysis at the triose phosphate level (Gancedo & Serrano, 1989).

The Pasteur effect is defined as an increase in the rate of glycolysis brought about by a shift from aerobic to anaerobic conditions or suppression of ethanol formation and reduced glycolytic rate in the presence of oxygen. The rationale for these related effects is that under aerobic conditions the possibility of highly efficient ATP generation via oxidative phosphorylation produces a reduced requirement for sugar utilisation. Conversely, under anaerobic conditions where respiratory activity is precluded glycolytic flux is increased to allow greater rates of substrate level phosphorylation.

Contrary to popular dogma, the Pasteur effect is of little importance in *Saccharomyces* and is only expressed under specific conditions (Lagunas, 1979, 1981, 1986; Lagunas *et al.*, 1982; Lagunas & Gancedo, 1983). Lagunas contended that the observations of Pasteur, which form the basis of the effect with his appellation, that biomass yields in aerobic yeast cultures were greater than under anaerobiosis was a misinterpretation of the data. Thus, these authors demonstrated under conditions of growth, rates of consumption of sugars and growth yields were similar under both aerobic and anaerobic conditions. The observations of Pasteur could be explained in terms of the possibility of greater aerobic biomass yields because of diauxic growth on glucose and ethanol.

In *S. cerevisiae*, the Pasteur effect can only be demonstrated either in a chemostat operating under carbon limitation or in resting cells exposed to sugar in the absence of a nitrogen source. Of fundamental significance is the observation that *Saccharomyces* yeasts have a limited respiratory capacity for oxidative growth, although this is disputed, as discussed later in this section. Respiration can, of course, be expressed only in the presence of oxygen but this expression is modulated by the concentrations of sugar and nitrogen in the medium. In resting cells deprived of nitrogen, sugar transporters are inactivated and because of this uptake of sugars occurs very slowly, fermentation rates decline but rates of respiration remain unaltered. Thus, under these conditions there is a proportional increase in respiration.

A similar effect occurs in carbon limited chemostat cultures, although in this case the reduction in sugar uptake is a result of lack of availability of these nutrients. Lagunas (1986) suggested that reduced sugar uptake results in a concomitant reduction in the steady state concentration of pyruvate. Since the $K_m$ for pyruvate dehydrogenase is much lower than that for pyruvate decarboxylase, a reduction in pyruvate concentration, under aerobic conditions, will progressively favour carbon flow into the respiratory pathway at the expense of fermentative metabolism.

The short-term Crabtree effect is also related to carbon flux through pyruvate. This phenomenon occurs in certain yeast strains when carbon limited aerobic chemostat cultures operating at low to medium dilution rates are pulsed with an excess of sugar and there is an instantaneous production of ethanol. It has been suggested that the explanation for this effect also resides in the limited respiratory capacity of these yeasts (Rieger *et al.*, 1983). Thus, under aerobic carbon limitation glucose is fully oxidised via pyruvate dehydrogenase and the respiratory pathway. A sudden increase in glucose concentration results in saturation of the respiratory pathway and as a result the overflow of carbon is channelled into pyruvate decarboxylase and ethanol formation.

Van Urk et al. (1988, 1990) compared the effects of transient glucose pulses in aerobic carbon limited chemostat cultures of a Crabtree positive (S. cerevisiae) and negative yeast (Candida utilis). These authors concluded that respiratory capacity of the two yeasts was not significantly different. After the transition from carbon limitation to carbon excess, the Crabtree positive yeast generated extracellular ethanol, pyruvate and TCA cycle intermediates but growth was unaffected. In the case of the Crabtree negative yeast no ethanol was produced but some pyruvate was generated. In addition, there was an increase in biomass, due to synthesis of reserve carbohydrates. It was demonstrated that Crabtree positive yeasts had more active pyruvate decarboxylases compared to the negative yeasts, whereas the latter group had increased activities of acetaldehyde dehydrogenase and acetyl coenzyme A synthetase. It was suggested, therefore, that in the Crabtree positive yeasts the ethanol produced in response to the glucose pulse was not due to saturation of the respiratory pathways but rather an increased carbon flux through the more active pyruvate decarboxylase. In the Crabtree negative yeasts ethanol does not arise because carbon metabolised through pyruvate decarboxylase can be channelled back into acetyl-CoA production via acetaldehyde dehydrogenase and acetyl-CoA synthetase. However, since respiratory capacities are similar, in both groups of yeasts some of the increased carbon flux in the Crabtree negative yeasts is directed towards glycogen synthesis.

The short term Crabtree effect is typified by production of ethanol under aerobic conditions when the glucose concentration is higher than a critical value. Postma et al. (1989) reported that in chemostat cultures of S. cerevisiae operated at low dilution rates under glucose limitation growth was entirely oxidative. With increase in dilution rate a critical value was reached at which alcoholic fermentation ensued and there was a concomitant decline in the biomass concentration. At higher dilution rates there was an increase in rates of oxygen consumption and carbon dioxide production and this was accompanied by production of exogenous pyruvate and acetate. With further progressive increase in dilution rate aerobic ethanol production ensued with simultaneous high rate of respiration. This phenomenon was termed the long-term Crabtree effect.

Postma et al. (1989) suggested that the organic acids produced under these conditions produced an uncoupling effect which caused the observed increase in rates of respiration. The effects due to the progressive increase in dilution rate could be explained in terms of overflow metabolism. It was proposed that at dilution rates where no ethanologenesis was observed, pyruvate decarboxylase was active; however, the resultant acetaldehyde re-entered the oxidative pathway via acetaldehyde dehydrogenase and acetyl-CoA synthetase. At higher dilution rates acetate appeared due to insufficient activity of the acetyl-CoA synthetase. At even higher dilution rates, ethanol was produced because the activity of the acetaldehyde dehydrogenase was insufficient to contain the carbon flow devolving from pyruvate decarboxylase.

In subsequent work, the effects on glucose metabolism were reported in a mutant strain of S. cerevisiae deficient in pyruvate dehydrogenase (Pronk et al., 1994). It was demonstrated that in aerobic carbon limited chemostat culture at medium dilution rates, metabolism was fully respiratory. In the absence of pyruvate dehydrogenase this was achieved by necessity using the pyruvate decarboxylase, acetaldehyde dehydrogenase, acetyl-CoA synthetase by-pass. However, biomass yields were reduced compared to growth of the wild type strain under similar conditions. This was attributed to the net lower ATP generation because of the increased proportion used by the acetyl-CoA synthetase in the mutant strain.

### 3.4.1  *Carbon catabolite repression*

The mechanisms described so far are examples of some of the ways in which carbon flow in yeast is directed at enzyme level. In this respect, the mechanisms are explainable in terms of the relative capacities of the various pathways involved and, therefore, they do not represent regulation in a true sense. The latter implies modulation of enzyme activity in response to the requirements of the yeast under a particular set of environmental conditions. Thus, examples of metabolic regulation would be induction or repression of enzymes synthesis, or covalent modification of preformed enzymes, resulting in altered activity.

There is disagreement as to whether or not glycolysis is regulated. Induction of various enzymes in the pathway has been claimed, although this is disputed by others (Wills, 1990). What is certain is that glucose, and to a lesser extent other sugars, exert far-reaching effects on yeast metabolism, and this is crucial to the outcome of brewing fermentation. When *S. cerevisiae* is grown in the presence of glucose the transcription of many genes is repressed and several proteins and transport systems are rapidly activated or inactivated at the level of post-transcription. These phenomena are termed carbon catabolite repression and catabolite inactivation, respectively (Gancedo, 1992; Trumbly, 1992; Wills, 1990, 1996). In respect to the global effects on yeast metabolism, Thevelein (1994) has likened the effects of glucose as being similar to those associated with hormones in mammalian systems. In other words glucose acts as a metabolic signal. Sugars other than glucose are also effective, particularly those which may be rapidly fermented such as fructose and maltose. Growth on sucrose does not usually elicit the response and *ipso facto* neither do non-fermentable sugars such as galactose.

The effects of catabolite repression and inactivation on sugar uptake have been described already (Section 3.3.1). The net result is that sugars are taken up in an ordered fashion. The sum of the other effects is to switch metabolism from a respiratory to a fermentative mode. In addition, gluconeogenic pathways may also be inhibited. These effects are independent of the availability of oxygen. Repressed cells exhibit no respiratory activity because of lack of a complete electron transfer chain due to cytochrome deficiency. The TCA cycle is disrupted and becomes a two-branched pathway because of the absence of 2-oxoglutarate decarboxylase. These biochemical changes are accompanied by morphological modifications, most notably a failure of repressed cells to develop functional mitochondria (Polakis & Bartley, 1965; Chapman & Bartley, 1968; Neal *et al.*, 1971; Wales *et al.*, 1980; Fiechter *et al.*, 1981). See Section 4.1.2.3 for a fuller discussion of 'mitochondria'. Carbon catabolite inactivation is most studied with respect to gluconeogenesis. This is seen when yeast grown on a gluconeogenic substrate such as ethanol is transferred to a medium containing glucose, or another repressing sugar. There is an immediate inactivation of enzymes associated with gluconeogenic metabolism such as malate dehydrogenase and isocitrate lyase which are associated with the glyoxylate cycle (reviewed in Wills, 1990).

It may be appreciated that the visible manifestations of glucose repression are the same as those seen in the Crabtree effect with the highly significant difference that repressed cells are not capable of respiratory growth because parts of the necessary metabolic machinery are absent. The significance of glucose as a metabolic signal has been reviewed by Thevelein (1994). Yeast cells possess regulatory mechanisms in which fundamental functions such as cell proliferation are initiated or terminated in response to external signals. In the case of relatively simple cells such as yeasts the initial signal is a nutrient. This is a complicating factor because in addition to provoking the signal response the nutrients are also utilised in normal metabolic pathways.

In the case of glucose, simple responses such as activation of glycolysis by induction of enzymes by glucose, or the products of other induced enzymes, is possible as has been discussed earlier. More complex responses may arise in which glucose constitutes an initial messenger which activates a signal transduction pathway, involving a cascade of sequentially modulated enzymes such as that seen in the RAS system. The RAS system was originally identified in mammalian tumour cells where they are involved in the regulation of cellular proliferation and homologous systems have been found in yeast cells (Tamanoi, 1988; Wiesmuller & Wittinghofer, 1994).

Two RAS (RAS1 and RAS2) genes are present. An external nutritional signal results in activation of the RAS genes by conversion of GDP to GTP. Thus, the RAS genes have an intrinsic GTPase activity and are active in a GTP-bound form but not in a GDP-bound form. The products of other genes control the GTPase activity and hence the activity of RAS. These other genes appear to be the target sites for the initial signal. The activated RAS genes in turn stimulate an adenylate cyclase leading to a transient increase in cAMP. In response a cAMP-dependent protein kinase is activated. This acts at several sites activating and inhibiting specific proteins and hence multiple metabolic pathways may be either induced or inhibited.

Several signal transduction pathways are believed to exist which regulate all aspects of metabolism. These pathways are arranged in a hierarchical fashion. For example, there is an oxygen induction pathway which controls expression of mitochondrial activity. One of the genes (CYC) involved in this pathway is subject to glucose repression (Forsberg & Guarente, 1989). This provides the explanation for glucose repression of respiration under aerobic conditions.

The reasons for glucose repression are obscure. It is apparent that yeast is capable of growth using purely fermentative pathways although this is an inefficient method of energy transduction. It has been suggested that when glucose is plentiful ethanol is produced since it may act as a natural biocide reducing competition from other less resistant organisms. In this respect the ethanol may be viewed as a temporary carbon store such that if glucose becomes exhausted and oxygen is available the ethanol may itself be utilised as a source of carbon and energy.

In a brewery fermentation, utilisation of ethanol is not an option since metabolism is fermentative at all times. When oxygen is present during the aerobic phase of fermentation the presence of high sugar concentrations ensures that yeast cells have a repressed physiology. When sugar concentrations fall to non-repressing levels anaerobiosis precludes respiration. Consequently, during the aerobic phase of fermentation, respiration – in its true sense – of oxidative phosphorylation using an electron transport chain with oxygen as the terminal electron acceptor does not occur. Undoubtedly, as some other authors have asserted, some of the features associated with derepression may be observed. The relevance of this debate to yeast metabolism during brewery fermentation remains to be resolved.

It is possible to obtain mutant strains, which do not exhibit catabolite derepression, and it has been suggested that these could be of value in brewing. Jones *et al.* (1986) described a procedure in which the technique of sphaeroplast fusion was used to produce hybrids of *S. diastaticus* and *S. cerevisiae* (see Section 4.3.4.3). Such strains had the potential for increased fermentation efficiency due to their possession of a glucoamylase which allowed utilisation of wort dextrins, a carbohydrate fraction not usually assimilated during fermentation. Using these hybrids a selective procedure was employed which allowed the isolation of derepressed mutants in which expression of the maltose uptake system and production of glucoamylase were constitutive. Compared with parental strains these mutants exhibited increased fermentation rates.

The effects of catabolite repression account for the patterns of ordered uptake of wort sugars as described in Section 3.3.1. The effects can be complex, particularly where worts are used that have high proportions of non-malt adjuncts. Armitt and Healy (1974) described a lager fermentation using a wort in which up to 30% of the sugar was obtained from cane sugar. In this case glucose and fructose did not disappear from the wort until approximately half-way through the fermentation, some 50 hours after pitching. These authors noted that disappearance of glucose and fructose was associated with an abrupt lowering of fermentation rate, as judged by decrease in gravity and $CO_2$ evolution. In addition, diacetyl levels showed an increased rate of formation at this time. This was ascribed to a 'half-way stress' and was assumed to be associated with the required adjustment needed to be made by yeast when changing from assimilation of glucose and fructose to maltose.

Phaweni *et al.* (1992, 1993) also reported that with some yeast strains high adjunct levels were associated with abnormal patterns of sugar uptake and altered levels of production of flavour compounds. In this case it was reported that with the particular yeast strain used, maltose and maltotriose uptake was not repressed by the presence of glucose. When adjuncts with high glucose concentrations were used fermentation rate and yeast growth were inhibited and there was a disproportionate increase in the formation of acetaldehyde and sulphur dioxide. These effects were amplified when the physiological status of the pitching yeast was compromised. It was concluded that this yeast strain was not subject to conventional glucose repression. However, there was an effect, defined as 'glucose block'. This phenomenon did not influence uptake of other carbohydrates; however, at high exogenous glucose concentrations utilisation of glucose was reduced and this was a consequence of increased intracellular glucose concentration. The implication was that elevated intracellular glucose levels may inhibit further uptake.

Under some circumstances sudden exposure to excess maltose can have adverse effects on yeast. Postma *et al.* (1990) reported that maltose-limited continuous cultures of *S. cerevisiae* subjected to a sudden maltose pulse resulted in an abrupt loss of viability. This phenomenon was termed substrate-accelerated death. No effect was noted with glucose pulses. The effect was ascribed to the fact that unlike glucose, maltose was taken up by an active proton symport system. Under conditions of limitation a sudden excess of maltose led to high rates of uptake and subsequent death due to osmotic burst.

It is apparent that uptake of sugars and their utilisation by strains of *S. cerevisiae* is complex and several mechanisms may be involved depending on the actual strain used. In particular, care must be taken when alterations to wort composition are made that influence the sugar spectrum. Thus, although the effects on the development of flavour compounds of altering the ratio of carbon to nitrogen is well described, other effects due to changing the spectrum of sugars are perhaps less recognised.

### 3.4.2 Storage carbohydrates

*S. cerevisiae* accumulates two classes of storage carbohydrates which reportedly have roles in brewery fermentation, namely glycogen and trehalose. Glycogen apparently serves as a true energy reserve, which may be mobilised during periods of starvation. Trehalose may also fulfil a similar role, although as will be discussed later, there is compelling evidence that it has another function and that is as a protectant used by yeast to help withstand imposed stresses.

#### 3.4.2.1 Glycogen.

Glycogen is a polymer of $\alpha$-D-glucose with a molecular weight of the order of $10^8$. It has a branched structure containing chains of 10–14 residues of $\alpha$-D-glucose joined by $1 \rightarrow 4$ linkages. Individual chains are connected by $(1 \rightarrow 6)$-$\alpha$-D-glucosidic linkages. In brewing yeasts considerable glycogen may accumulate. Thus, under appropriate conditions up to 40% (typically 20–30%) of the yeast dry weight may consist of glycogen. This implies that as much as 4% of the fermentable wort sugar may be used for synthesis of this reserve polymer (Quain & Tubb, 1982).

Glycogen is synthesised from glucose, via glucose 6-phosphate and glucose 1-phosphate (Fig. 3.4). The pathway uses uridine diphosphate (UDP) as a carrier of glucose units. Glycogen synthase (UDP-glucose: glycogen 4-$\alpha$-D-glucosyltransferase) catalyses chain elongation by successive transfer of glucosyl units from UDP-glucose to the growing $\alpha$-$(1 \rightarrow 4)$-linked polyglucose polymer. A second enzyme, branching enzyme [1,4-$\alpha$-D-glucan: 1,4-$\alpha$-D-glucan 6-$\alpha$-D-$(1,4$-$\alpha$-glucano)-transferase], forms the $\alpha$-$(1 \rightarrow 6)$-glucosidic bonds which form the branch points in the growing polymer.

Dissimilation of glycogen uses another enzyme system. Glycogen phosphorylase, in the presence of phosphoric acid, repeatedly removes successive glucose molecules from the non-reducing ends of glycogen chains, liberating molecules of glucose 1-phosphate. A debranching enzyme, which is a hydrolytic amylo-$\alpha(1 \rightarrow 6)$glucosidase, cleaves the $\alpha$-$(1 \rightarrow 6)$-glucosidic bonds such that in combination with the phosphorylase permits complete utilisation of glycogen.

Accumulation or degradation of glycogen is apparently controlled by yeast growth rate. Thus, accumulation is signalled by nutrient limitation in the presence of excess

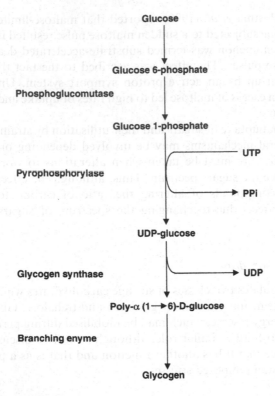

Fig. 3.4   Glycogen biosynthetic pathway.

sugar. For example, sulphur, nitrogen and phosphorus limitation (Lillie & Pringle, 1980). Glycogen accumulation may also occur under carbon limitation, for example, during diauxic growth on glucose. In this case when glucose falls to a concentration below that required to saturate the uptake system growth rate is restricted and glycogen accumulation is triggered. Accumulation continues until the exogenous sugar becomes exhausted. This glycogen store provides an energy source for induction of the respiratory and gluconeogenic systems, which are required for utilisation of ethanol (Gancedo & Serrano, 1989).

Regulation of glycogen synthase and phosphorylases activities is complex, as would be predicted from the number of external signals which can influence glycogen levels in yeast. Simultaneous accumulation and utilisation are possible, overall flux being controlled by co-ordinate regulation of glycogen synthase and glycogen phosphorylase. Wills (1990) has reviewed some of the mechanisms involved. Glycogen phosphorylase is activated by phosphorylation and deactivated by dephosphorylation. Glycogen synthase is activated by dephosphorylation. Two active forms of the enzyme occur, one of which is inhibited by glucose 6-phosphate. Apart from covalent modification to the enzymes, activity is also modulated by intracellular adenine nucleotide concentrations. Phosphorylation and dephosphorylation of the glycogen synthases and phosphorylase is regulated by signal transduction cascade pathways involving cyclic AMP and non-cyclic dependent kinases.

In the context of brewery fermentation it has been suggested that glycogen fulfils two vital roles. First, it provides the carbon and energy for synthesis of sterols and unsaturated fatty acids during the aerobic phase of fermentation and, second, energy for cellular maintenance functions during the stationary phase of fermentation and in the storage phase between cropping and re-pitching (Quain & Tubb, 1982). When yeast is pitched into aerobic wort there is an immediate mobilisation of glycogen reserves and this is accompanied by synthesis of sterol (Fig. 3.6). The period of rapid glycogen dissimilation is terminated by the disappearance of oxygen from the wort. The subsequent phase of rapid fermentation and yeast growth is associated with glycogen accumulation. Maximum glycogen concentrations are reached towards the end of primary fermentation after the point where yeast growth has ceased. In the final stationary phase when primary fermentation is complete, glycogen levels decline slowly.

It may be surmised that oxygen is the primary trigger for glycogen dissimilation immediately after pitching and that this is necessary to provide energy and carbon for sterol synthesis since exogenous sugars cannot be utilised due to lack of membrane function in the sterol-depleted pitching yeast. During the latter part of active primary fermentation, glycogen assimilation is favoured due to the presence in wort of the high ratio of sugars to other components. Exhaustion of nutrients other than sugars will gradually reduce yeast growth rate and progressively favour greater glycogen accumulation as primary fermentation progresses. When fermentable sugars become exhausted from the wort and anaerobiosis precludes utilisation of ethanol, glycogen dissimilation will be favoured in order to allow the yeast to withstand the starvation conditions.

The apparent importance of glycogen in fermentation has implications for storage of pitching yeast. Since glycogen is utilised for maintenance functions during storage it is important to remove yeast crops from fermenter as soon as is practicable in order to prevent excessive glycogen degradation in the period between the end of primary fermentation and cropping. Cooling of fermenter contents is useful in this respect. In storage vessels low temperatures also reduce yeast metabolic activity and conserve glycogen stores. Even so, the need to have sufficient glycogen to fuel sterol synthesis during subsequent fermentation serves to limit the time for which pitching yeast can be safely stored without compromising subsequent fermentation performance. In practice, this is usually no longer than 3 days at 2–4°C, under an inert gas atmosphere. For an exhaustive review of the impact of yeast storage conditions on glycogen turnover see Section 7.3.

The observation that rapid glycogen dissimilation may be triggered in anaerobic yeast suddenly exposed to oxygen implies that this may occur in stored pitching yeast exposed to air. In this case limited sterol synthesis may ensue and failure to correct for this by reducing wort oxygen concentrations will result in excessive yeast growth and loss of fermentation efficiency on re-pitching. For this reason yeast should be stored under an atmosphere of nitrogen or carbon dioxide. Alternatively, the ability of yeast to couple glycogen dissimilation to sterol synthesis in a controlled process which has the potential to remove the need for wort oxygenation has been proposed (Boulton et al., 1991; Masschelein et al., 1995). This is described in detail in Section 6.4.2.2.

**3.4.2.2** *Trehalose.* Trehalose is a disaccharide (α-D-glucopyranosyl-1,1-α-D-glucopyranoside) which contains two molecules of D- glucose. Like glycogen it is also synthesised in reactions which utilise uridine diphosphate as a carrier of glucose molecules (Fig. 3.5). The key enzyme is trehalose phosphate synthase (UDPG-glucose 6-phosphate transglucosylase) which catalyses the transfer of a glucose residue from uridine diphosphate glucose to glucose 6-phosphate. A phosphatase liberates the phosphate group forming trehalose. Synthesis of trehalose may derive from glucose or from glucosyl residues derived from degradation of glycogen. Furthermore, it is suggested that trehalose phosphate synthase has a greater affinity for UDP glucose than does glycogen synthase (Gancedo and Serrano, 1989). Mobilisation of trehalose occurs using another enzyme, trehalase, of which two occur in yeast, one which is associated with vacuoles and another which is cytosolic (Panek & Panek, 1990).

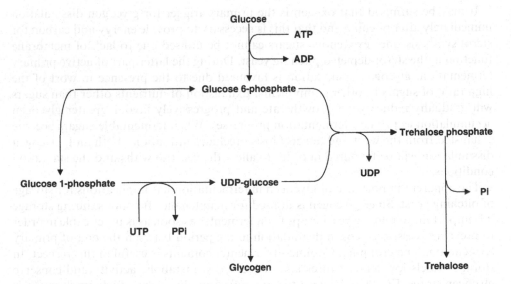

**Fig. 3.5**  Trehalose biosynthetic pathway.

It has been suggested that trehalose 6-phosphate may have a regulatory role in glycolysis in yeast by feedback control of hexokinases (Blázquez *et al.*, 1993). This perhaps suggests that in some circumstances it may act as a potential energy reserve. Both trehalose phosphate synthase and trehalase are subject to complex regulation in a similar manner to glycogen. Thus, cyclic AMP-dependent phosphorylation and dephosphorylation reaction are responsible for activation and deactivation of the enzymes leading to trehalose accumulation and degradation.

Like glycogen, trehalose also accumulates in yeast under conditions of nutrient limitation and therefore it was logical to conclude that it also served as a storage carbohydrate (Lillie & Pringle, 1980). However, the apparent redundancy of having two storage systems together with the element of interconvertability is perhaps indicative of differing roles for each polysaccharide. In this respect the same authors observed that although trehalose accumulation occurred under conditions of nutrient limitation this took place after glycogen formation and indeed there was evidence that glycogen was mobilised to provide glucose residues for trehalose formation. This was taken to indicate that trehalose did not behave like a typical storage polysaccharide. The low observed trehalose concentration in yeast during exponential growth is a consequence of glucose repression and inactivation. Thus, under repressed conditions the presence of glucose results in elevated levels of cyclic AMP due to activation of adenylate cyclase. This in turn activates protein kinase phosphorylase and thence activation of trehalase.

Simultaneously, a cyclic AMP-dependent phosphorylase converts the trehalose 6-phosphate synthase into an inactive phosphorylated form. In derepressed cells, such as is the case in the stationary phase of the growth cycle, these effects are reversed and trehalose accumulation is favoured. Glycogen synthase is regulated in a similar manner to trehalose synthase; however, the repressing effects of glucose are quantitatively greater in the case of trehalose synthetic pathway (Van der Plaat & van Solingen, 1974; Entian & Zimmermann, 1982; Uno et al., 1983; Londesborough & Varimo, 1984; Thevelein, 1984; Mittenbuhler & Holzer, 1988; Panek & Panek, 1990; Winkler et al., 1991). This – in combination with the already made observation that trehalose synthase has a greater affinity for UDP-glucose compared to glycogen synthase – provides an explanation for the observation that in the stationary phase of growth accumulation of glycogen occurs before that of trehalose.

Trehalose is known to confer resistance to heat and desiccation in a diverse range of organisms such as insects, plants, yeast and higher fungi. In addition, it is associated with spore formation (Elbien 1974; Crowe et al., 1984; Neves & Francois, 1992; de Virgilio et al., 1994). Colaco et al., (1992) described the ability of trehalose to stabilise protein structure such that in the presence of this disaccharide many enzymes exhibit startling resistance to heat and desiccation. Iwahashi et al. (1995) used whole cell NMR analysis of yeast and concluded that trehalose protected cells from temperature extremes by stabilising membrane structure. Further circumstantial evidence for a role for trehalose distinct from that of simple storage carbohydrate is provided by the fact that its synthesis requires metabolic energy, whereas no ATP is generated in its dissimilation. Trehalose has been shown to be a most effective agent for preventing damage to membranes by its ability to prevent phase transition events in lipid bilayers. The mechanism of action appears to be via binding of the hydroxyl groups of the sugar to the polar head groups of phospholipids in locations otherwise occupied by water. For maximum effectiveness, trehalose requires to be present at both the inner and outer surfaces of the membrane (Crowe et al., 1984).

It follows that in order for trehalose to exert its protective effects it must be transported from the site of synthesis in the cytosol to the membrane. Kotyk and Michaljanicova (1979) demonstrated the presence in S. cerevisiae of a transporter which rendered the cells capable of taking up exogenous trehalose. Eleutherio et al. (1993) postulated that the same carrier was responsible for transporting intracellular trehalose to the periplasm and inner membrane. The presence of this carrier was essential for the protective effects of trehalose to be expressed. Thus, mutant strains with no carrier could not withstand dehydration although trehalose accumulation was unimpaired. The same mutants could be afforded protection from dehydration by the addition of exogenous trehalose. As with trehalose accumulation, the activity of the carrier was repressed by glucose and was only active in stationary phase cells.

Yeast, in common with other eukaryotes, exhibits a heat shock response. This phenomenon is triggered when cells are exposed for a short period to a high but non-lethal temperature at which growth is not permitted. Such cells returned to a growth-permitting temperature exhibit an increased but transient tolerance to subsequent exposure to lethal temperatures. The effect is associated with the synthesis of a large number of so-called 'heat shock proteins' (Lindquist & Craig, 1988; Schlesinger, 1990). Acquisition of thermotolerance by expression of the heat shock response provides protection against other stresses. For example, increased tolerance to high osmotic pressures and ethanol (Piper, 1995). Indeed exposure to ethanol above a threshold concentration induces a similar response to heat shock (Piper et al., 1994).

There is much evidence to suggest that increased tolerance to stress is associated with accumulation of trehalose. For example, Sharma (1997) reported that in a strain of S. cerevisiae increased osmotic pressure brought about by elevated levels of sodium chloride was accompanied by increased accumulation of trehalose and increased tolerance to ethanol. A similar correlation between trehalose content and stress resistance in yeast was reported by Attfield et al. (1992). Elliott et al. (1996) isolated a mutant yeast strain which was unable to acquire heat shock resistance. The strain did not accumulate trehalose. The evidence suggests that induction of heat shock proteins and trehalose accumulation are related (Hottiger et al., 1989; Fujita et al., 1995).

The observation that trehalose accumulation occurs during the late stationary phase can be explained in that it represents a physiological response to cells under-going transition. Thus, in resting cells glycogen provides a readily utilisable source of carbon for maintenance energy. A proportion of this carbon, supplemented with exogenous sugar, is utilised for trehalose synthesis. The trehalose pool provides protection for cells during the resting phase. A variation of the transition from growth to resting state is differentiation into spore formation. In this case trehalose is the major carbon store and is essential for germination.

Rapid accumulation of trehalose in response to environmental changes may be taken as evidence that trehalose synthase acts as a stress protein. The mechanism of this adaptive response is believed to involve a general stress responsive promoter element (STRE) which has been identified in the upstream regions of several yeast genes. These include the TPS2 gene, which codes for a subunit of the trehalose phosphate synthase complex. The element is derepressed by nitrogen starvation in stationary phase cells and by applied stresses (Mansure et al., 1997).

The role of trehalose in brewing fermentation is unclear. It is accepted that high trehalose levels are important in commercial preparations of bakers' yeast, particu-larly active dried yeast (Trivedi & Jacobsen, 1986). In this case, trehalose provides protection during the drying phase. It also plays a role in providing cryoprotection when the yeast is used in frozen doughs. Bakers' yeast fermentations are conducted in such a way that trehalose is maintained at high levels, typically 15–20% of the dry weight, with 10% being considered critical (Gelinas et al., 1989). In brewing yeast cropped from fermenter trehalose concentrations tend to be modest, usually less than 5% of the dry weight. Higher levels may accumulate in yeast cropped from very-high-gravity fermentations.

Majara *et al.* (1996a and b) investigated the accumulation of trehalose in lager yeast fermentations using worts of varying gravities between 11 and 25°Plato. The higher-gravity worts were obtained by supplementation of low-gravity worts with glucose. It was demonstrated that trehalose accumulation occurred in late fermentation at a concentration proportional to the gravity of the wort. Thus, trehalose accounted for no more than 2–3% of the yeast dry weight in an 11°Plato wort, whereas this increased to 20–25% of the cell dry weight in yeast removed from a 25°Plato wort. The effect was independent of the ethanol concentration; however trehalose accumulation was favoured in the presence of sorbitol and could be correlated with the formation of glycerol. For these reasons the authors concluded that trehalose accumulated in response to osmotic stress. Other applied stresses such as freezing, exposure to ethanol, oxygenation, drying and carbonation all resulted in trehalose accumulation. It was concluded, therefore, that the detection of unexpectedly high concentrations of trehalose in pitching yeast should be taken as a sign that the yeast had been subject to stress.

It was contended that trehalose may be relatively unimportant as a stress protectant during early primary fermentation, as the presence of glucose would tend to mitigate against its synthesis and even when formed the repression of the carrier would not allow transport to the membrane for trehalose to exert a protective effect. In this regard, where very-high-gravity fermentation is practised it would be sensible to avoid high concentrations of glucose to promote early trehalose synthesis and transport. However, trehalose is important during storage of pitching yeast since these cells are derepressed and therefore the carrier is fully active. In this case, the conditions should be controlled to minimise trehalose degradation. In other words, short storage times, low temperatures and the absence of oxygen.

The optimum concentration of trehalose required by brewing yeast for it to exhibit resistance to applied stresses remains to be identified. Presumably, when sufficient is available on each side of the membrane for the stabilisation effect to occur no additional pool is required unless there is a continual dynamic turnover of synthesis and degradation. It would be interesting to know what this minimum required concentration is. In brewing fermentations, as discussed already, trehalose accumulation can be correlated with starting gravity, an effect ascribed to the increasing osmotic stress associated with elevated solute levels. It could also be argued that the increased accumulation of trehalose simply reflects the greater availability of carbon (or high carbon to nitrogen ratio) in very-high-gravity fermentations. After all, under the conditions of production brewing even a 10°Plato wort must present a considerable stress to yeast. However, under these conditions trehalose typically accounts for only 2–4% of the yeast dry weight even when the repressing effects of glucose are alleviated in late fermentation. During a study of sugar assimilation from a 300 hl 10°Plato ale fermentation, it was observed that a transient peak occurred when growth had ceased, which was tentatively identified as trehalose (C.A. Boulton, unpublished data). Presumably, loss of trehalose to the medium could be of more general occurrence, and overall biosynthetic rates may be greater than assumed, even in worts of moderate concentration.

An alternative to the usual practice of supplying wort with oxygen at the start of fermentation, is direct oxygenation of pitching yeast (see Section 6.4.2.2). It may be demonstrated that this process promotes sterol synthesis at the expense of glycogen dissimilation (Boulton *et al.*, 1991). Callaerts *et al.*, (1993) reported that the oxygenation process also resulted in trehalose accumulation in a pattern which showed a positive correlation with sterol formation. Presumably this was also a response to stress, in this case due to oxygen. The carbon for trehalose accumulation must have been provided by glycogen dissimilation.

### 3.4.3  *Fermentable growth medium induced pathway*

The phenomenon of glucose catabolite repression does not require growth for expression. Many of the metabolic events associated with yeast growth on wort are clearly growth related. For example, regulation of carbon flow between glycolysis and accumulation of polysaccharide reserves such as glycogen and trehalose (Section 3.4.2). Thevelein & Hohmann (1995) have developed a concept which they have termed the 'fermentable growth medium induced pathway' (FGMIP). This is defined as one or more signal transduction pathways, which are activated by a specific combination of nutrients, readily fermentable sugars and all other nutrients required for growth. The authors consider that this is distinct from the general glucose repression and RAS-adenylate cyclase pathways.

The characteristics and a summary for the evidence of the existence of this pathway is provided in Thevelein (1994). When yeast is starved for an essential nutrient such as nitrogen, sulphate or phosphorus in the presence of glucose the cells arrest in the G1 phase of the cell cycle and then enter the resting phase, $G_0$ (see Section 4.3.3.1). This is associated with accumulation of glycogen and trehalose, induction of heat shock proteins and repression of ribosomal protein genes. Subsequent addition of the missing nutrient induces growth and a rapid loss of stationary phase characteristics. Thus, there are rapid post-translationally induced changes in enzyme activity and gene expression which reverse the changes described above. Similar changes may be observed when exponentially growing fermentative cells are made to undergo a sudden increase in growth rate. This does not happen with cells growing on non-fermentable carbon sources.

Cells arrested in $G_0$ phase by starvation for a nutrient other than sugar are triggered to activate phosphofructokinase, glycogen synthase, glycogen phosphorylase and inactivate trehalose 6-phosphate synthase and trehalose 6-phosphate phosphatase by the addition of the missing nutrient. However, this occurs only in the presence of glucose. Inhibition of protein synthesis does not prevent the occurrence of these events. Addition of the missing nutrient does not produce a sudden spike in, cyclic AMP, which was taken to indicate that this metabolite was not acting as a secondary messenger. Instead it is argued that FGMIP in some way activates free catalytic subunits of cyclic AMP-dependent protein kinases.

Individual pathways are still to some extent overlapping. Thus, cells starved for nitrogen in the presence of glucose are glucose repressed. Addition of nitrogen induces FGMIP but the cells remain repressed. Presumably this is similar to the effects of oxygen in brewing fermentation being modified by the presence of repressing concentrations of sugar. The continued effects of repression were taken to indicate that expression of FGMIP did not require glucose-induced activation of the

Ras adenylate cyclase pathway since the latter is itself glucose repressible. Confirmation was provided by the observation that activation of FGMIP does not require glucose phosphorylation whereas this is mandatory in the RAS adenylate cyclase pathway. However, it was tentatively concluded that the initial glucose sensing mechanism for the RAS and FGMIP systems could be the same, although the different requirements for phosphorylation cast doubt on this premise. The search for the glucose sensor continues.

The reverse of a fermentation growth medium induced pathway is that which might come into play in stationary phase under starvation conditions. This has been termed a stress-induced (STRE) pathway and is described as a stationary phase co-regulated gene induction which is a response to starvation stress and is distinct from the heat shock response. Thevelein (1994) has also discussed this possibility within the context of FGMIP. The STRE pathway hypothesis states that under conditions of growth certain genes are not expressed. It has been reported that exhaustion of glucose, transfer to a non-fermentable carbon source or starvation of an essential nutrient results in the expression of the STRE genes. Thus, this is a general stress-sensing pathway and nutrient limitation would only be one of many triggers of the induction of it. In this sense a mechanism must exist for sensing sub-optimal concentrations of pertinent substrates.

It was argued that the existence of such a general stress-sensing pathway which involved both nutrient limitation and other stresses was unproved. Instead different stimuli may produce similar phenotypic effects by the use of the same target sites. For example, FGMIP and RAS adenylate cyclase pathway both interacting with cyclic AMP-dependent protein kinase. As the author points out, new signal transduction systems in yeast are being regularly discovered. Further elucidation of the structure and function of the yeast genome together with the advent of proteomics (see Section 4.3.2) will no doubt bring further clarification to this complex but fascinating aspect of yeast metabolism.

## 3.5 Requirement for oxygen

Failure to provide oxygen at the start of fermentation results in slow fermentation rate, incomplete attenuation and poor yeast growth. Oxygen is required in brewery fermentation to allow yeast to synthesise sterols and unsaturated fatty acids. These lipids are essential components of membranes (Parks, 1978; Brenner, 1984; Weete, 1989; Nes et al., 1993). Thus, S. cerevisiae is capable of growth under strictly anaerobic conditions only when there is an exogenous supply of sterols and unsaturated fatty acids (Andreason & Stier, 1953, 1954). Under aerobic conditions sterols and unsaturated fatty acids may be synthesised de novo from carbohydrates. The ability to grow under strictly anaerobic conditions is relatively rare among yeasts. Indeed, Visser et al. (1990) concluded that S. cerevisiae was a positive exception in this respect. These authors studied the oxygen requirements of type species from 75 genera of yeasts. In oxygen-limited shake flasks using a complex medium supplemented with ergosterol and Tween 80 (a source of unsaturated fatty acids), all stains tested were capable of fermenting glucose to ethanol. However, only 23% actually grew under anaerobic conditions. *S. cerevisiae* alone was capable of rapid growth at low oxygen tensions. It was suggested that these differences reflected the importance of some mitochondrial functions in growth during anaerobiosis and that *S. cerevisiae* was less reliant on these than other facultative anaerobes.

The quantity of oxygen required for fermentation is yeast strain-dependent. Kirsop (1974) classified ale strains into four groups based on the oxygen concentration, which produced satisfactory fermentation performance. The required initial oxygen concentrations were half-air saturation, air saturation, oxygen saturation and more than oxygen saturation (see Section 6.1.2). Jacobsen and Thorne (1980) observed similar strain-specific oxygen requirements for lager yeasts. Notwithstanding these strain-specific differences the proportion of oxygen used for lipid synthesis during fermentation is comparatively small. It has been estimated that in an air saturated wort, pitched with lipid depleted yeast, that approximately 10% of the dissolved oxygen is used for sterol synthesis and up to 15% for unsaturated fatty acid synthesis (Aires *et al.*, 1977; Kirsop, 1982). Although a proportion of the oxygen is consumed in oxidation of wort components, it is apparent that the bulk is utilised in reactions other than for synthesis of wort lipids.

Pitching yeast derived from a previous fermentation has an anaerobic (repressed) physiology. Sudden exposure to oxygenated wort provides an opportunity for synthesis of unsaturated lipids and possibly other advantageous aerobic reactions but it also presents a potentially lethal stress in the form of reactive oxygen radicals such as superoxide, hydroxyl and peroxide. In this sense sudden exposure to oxygen presents both threat and opportunity.

Provided the requirement for sterol and unsaturated fatty acid synthesis has been met, alcoholic fermentation and growth proceed under anaerobic conditions. As discussed in the previous section, under aerobic conditions and in the presence of repressing concentrations of sugars, metabolism is also fermentative. Other than lipid synthesis, therefore, there is no apparent role for oxygen in fermentation. However, it has long been recognised that respiratory deficient yeast strains – 'petites' – (see Section 4.3.2.7) produce poor fermentation performance (Ernandes *et al.*, 1993). In addition, fermentation efficiency, as judged by rates of ethanol formation, has been reported to increase under microaerophilic conditions (Grosz & Stephanopoulis, 1990).

As discussed already, *Saccharomyces* yeasts cannot grow under anaerobic conditions for an indefinite period but can tolerate anaerobiosis. This suggests that the natural environment of these organisms would be microaerophilic/fully aerobic with occasional periods of anaerobiosis. The additional implication is that yeast cells must be capable of responding rapidly to changes in oxygen tension both to ensure survival and to gain selective advantage over those organisms, which are either obligate aerobes or anaerobes.

Yeast cells have several mechanisms for nullifying the damaging effects of oxygen radicals (Krems *et al.*, 1995). The metalloenzymes superoxide dismutases convert the superoxide radical to hydrogen peroxide (Fridovich, 1986). The latter may then be dissimilated by catalase. Yeast cells have two superoxide dismutases (SOD), a Cu,Zn-SOD which is cytosolic and a Mn-SOD which is located in mitochondria. The cytosolic enzyme is constitutive whereas the mitochondrial activity is inducible. Two catalases are present, catalase T, which is cytosolic and catalase A, which is associated with peroxisomes. It has been suggested that catalase is not significant in protection against oxidative stress since the peroxisomal enzyme is associated with growth on fatty acids and is repressed by the presence of glucose (Lee & Hassan, 1987). See Section 4.1.2 for further information on the location of these enzymes.

Reduced glutathione reacts with superoxide and hydrogen peroxide and may represent another protective mechanism, as may sequestration of radicals by transition metals such as copper and iron. Finally, in humans it has been reported that squalene may serve as a scavenger of free radicals and prevent lipid peroxidation in the skin (Kohno et al., 1995). Since yeast cells accumulate squalene in the absence of oxygen a similar mechanism would be plausible.

The effects of free radicals are far-reaching and are associated with degenerative processes such as mutagenesis, transformation of cell lines to malignancy and ageing (see Sections 4.3.2.6, 4.3.2.7 and 4.3.3.4 for implications in yeast). In respect of these effects, no eukaryotic cells are immune, regardless of the presence of the multiplicity of protective mechanisms. Brewing yeast is no exception. It was demonstrated that during an abrupt transition from anaerobiosis to aerobiosis there was a rapid increase in the specific activity of the CuZn-superoxide dismutase. Increase in the specific activity of the mitochondrial Mn superoxide dismutase occurred only after a lag of several hours. During the transition and before induction of the Mn-superoxide dismutase there was an immediate (5–7%) drop in the viability of the culture. Similar losses of viability were demonstrated when anaerobically-grown yeast was exposed to 0.25 mM potassium superoxide. Aerobically grown cells were unaffected by similar exposure. It was concluded, therefore, that the CuZn-superoxide dismutase was responsible for protection against oxygen radicals in anaerobic cells suddenly exposed to oxygen. The Mn-superoxide dismutase was protective only in aerobic cells (Clarkson et al., 1991).

Zitomer and Lowry (1992) discussed regulation of gene expression by oxygen in S. cerevisiae. These authors described three classes of genes responsive to oxygen tension. First, respiratory growth under aerobic conditions which requires the expression of more than 200 genes. A list of these may be found in Tzagoloff and Dieckmann (1990). Second, there is a class of genes which are expressed only under anaerobic conditions, the functions of which are as yet unknown. Third, the so-called 'hypoxic class of genes' which respond to decreases in oxygen tension and are required for efficient utilisation of low concentrations of oxygen.

The signalling pathway by which molecular oxygen exerts its effects on metabolism remains to be fully elucidated. There is added complexity in that the effects due to oxygen may be modified by other external stimuli such as the overriding of induction of respiration by the presence of a repressing sugar (Section 3.4.1). Haem would appear to be a key intermediate in the oxygen sensing pathway. This molecule is a prosthetic group in cytochromes and also in oxygen-binding enzymes such as catalase. In addition, it serves as an effector metabolite in many of the pathways that involve the utilisation of molecular oxygen (Padmanaban et al., 1989). Oxygen is required for haem synthesis and anaerobically grown yeast contains all the necessary enzymes for haem synthesis. Therefore, the cellular concentration of haem is directly related to oxygen tension. Genes which are induced by haem include those which are involved in respiratory function and a second class which are protective against oxygen radicals. Hypoxic genes are repressed by haem and their products include enzymes involved in the synthesis of sterols, unsaturated fatty acids, haem itself and parts of the electron transport chain (Zitomer & Lowry, 1992). With some exceptions, repression by haem involves activities that are redundant under anaerobic conditions.

In the context of brewing yeast and fermentation, the evidence suggests that apart from lipid synthesis oxygen is required for certain activities which may be induced by haem and are not repressed by glucose and other sugars. These activities are implicated in the partial development of mitochondrial function, which occurs in the repressing but aerobic phase of fermentation. Anaerobic yeast cells contain promitochondria (see Section 4.1.2.3) which develop into fully functional organelles on exposure to oxygen and during derepression (Plattner et al., 1971). Since the enzymes for some essential anabolic reactions are located in the mitochondria it is assumed that in the absence of oxidative phosphorylation another mechanism must exist to generate energy for these reactions and to power transport of precursors and products between mitochondria and the cytosol.

The evidence suggests that during fermentative growth, mitochondrial ATP is derived from substrate level phosphorylation which occurs in the cytosol. Passage of adenine nucleotides between cytosol and mitochondrion is catalysed by an ADP/ATP translocase. Three genes have been identified which encode for this enzyme. One enzyme is constitutive, another is expressed under aerobic conditions and the third is induced by anaerobiosis (Kolarov et al., 1990). Zitomer & Lowry (1992) suggest that the anaerobic enzyme catalyses the reverse of the normal translocase reaction and actually imports ATP from the cytosol into mitochondria. Evidence for this has been provided by the observation that bongkrekic acid, a specific inhibitor of the ATP/ADP translocase, arrested growth under anaerobic conditions (Subik et al., 1972; Gbelská et al., 1983). O'Connor-Cox et al. (1993), investigating the roles of oxygen in brewery fermentation, also concluded that the ATP/ADP translocase was essential for mitochondrial energy generation. These authors noted that when the translocase was inhibited, also using bongkrekic acid, that yeast growth was reduced, assimilation of wort nitrogen was low and there was over-production of VDK, acetaldehyde, $SO_2$ and dimethylsulphide.

### 3.5.1  Synthesis of sterols and unsaturated fatty acids

In brewery fermentations, a proportion of the requirement for unsaturated fatty acids is obtained from wort. This may be vanishingly small particularly in the case of high-gravity bright worts made with a high level of non-malt adjunct. A small quantity of sterol may be present in wort, although under the conditions of brewery fermentation brewing yeast may not be able to assimilate it, as discussed later. In consequence, both sterol and unsaturated fatty acids must be synthesised during the initial phase of fermentation when oxygen is made available. In the case of sterols, the quantity of oxygen supplied regulates the quantity synthesised. Growth of yeast during the anaerobic phase of fermentation dilutes the pre-formed sterol pool between mother cells and their progeny. Cells divide until sterol depletion limits growth and therefore is responsible for the requirement for more oxygen when the yeast is re-pitched.

The aerobic synthesis of sterol during fermentation is accompanied by dissimilation of glycogen. A linear relationship may be demonstrated between the quantities of sterol formed and glycogen utilised. It is suggested that because of lack of membrane competence in pitching yeast, mobilisation of glycogen provides the metabolic fuel for sterol synthesis (Quain et al., 1981; Quain & Tubb, 1982). The profiles of changes in the concentrations of glycogen and total sterol synthesis during fermentation are shown in Fig. 3.6. The total quantities of sterol synthesised during fermentation are modest, typically increasing from 0.1% of the dry weight in yeast cropped at the end of fermentation up to approximately 1% at the end of the aerobic phase of fermentation. Greater concentrations of sterol are accumulated in derepressed cells, typically up to 5% of the yeast dry weight (Quain & Haslam, 1979).

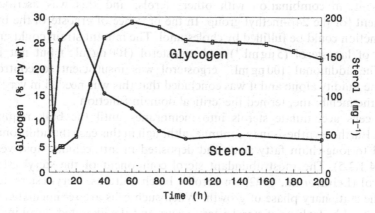

Fig. 3.6  Changes in levels of glycogen and sterol during the course of fermentation.

Several sterols may be found in yeasts but ergosterol is the most prominent. An excellent review of the role and diversity of sterols in yeast can be found in Parks (1978). By way of example, Behalova et al. (1994) examined the effect of growth rate on sterol synthesis in S. cerevisiae. They concluded that at low growth rates total sterol and ergosterol content of the yeast were c. 6.5% and 2–2.5%, cell dry weight, respectively.

Sterols have several putative roles in cells, both structural and functional. Sterols are important structural lipids, which play a key role in regulating membrane fluidity. Related to this, they exert other membrane effects such as regulating membrane permeability, influencing the activity of membrane-bound enzymes and affecting cell growth rate (Lees et al., 1995). The same authors describe a 'hormonal' role of specific yeast sterols, at low concentration (10 nM), in which they are vital to progression through the cell cycle. In addition to the role in membrane structural function, Parks et al. (1995) implicated sterols in amino acid and pyrimidine transport, in resistance to anti-fungal agents and some cations, as well as being required for the development of respiratory function.

With respect to membranes, sterols are considered to fulfil a vital role in maintaining membrane fluidity. Presence of sterols in membranes are reported to fulfil two roles, described as 'sparking' and 'bulk' functions (Rodriguez & Parks, 1983; Lorenz *et al.*, 1989). These authors suggested that, provided small 'sparking' quantities of ergosterol were present in the membrane, then other sterols could substitute for ergosterol and fulfil a bulk role. The critical feature of the sparking sterol was the presence of the 24 β-methyl group of ergosterol and it was essential that at least a small quantity of this lipid was present in membranes in order to maintain proper fluidity.

Rodriguez *et al.* (1985) recognised four groups of sterols – described as sparking, critical domain, domain and bulk, based on work with a sterol auxotroph of *S. cerevisiae* termed 'RD-5'. Ergosterol could fulfil all of these roles. The minimum concentration of free sterol below which cell proliferation could not occur was defined as the domain function. No growth was possible unless a small quantity of ergosterol was present, in combination with other sterols, and this was ascribed to the requirement for the 24β-methyl group. In the presence of ergosterol the bulk membrane function could be fulfilled by cholestanol. The mutant strain could grow in the presence of lanosterol ($5\,\mu g\,ml^{-1}$) and ergosterol ($100\,ng\,ml^{-1}$) but not lanosterol alone. The additional $100\,ng\,ml^{-1}$ ergosterol was insufficient to control overall membrane fluidity alone and it was concluded that this was needed in very restricted areas of the membrane, termed the 'critical domain function'.

Yeast cells accumulate sterols into membranes until the bulk requirement is satisfied. Further synthesis may continue, although in this case the additional sterol is esterified to long-chain fatty acids and deposited in intracellular lipid vesicles (see Section 4.1.2.5). The most abundant sterol component of the steryl ester pool is zymosterol (Leber *et al.*, 1992). In aerobic batch cultures, steryl esters accumulate during the stationary phase of growth. When such cells are re-inoculated into fresh medium rapid hydrolysis of steryl esters occurs and the liberated sterol incorporated into growing membranes (Quain & Haslam, 1979; Lorenz & Parks, 1991). In this sense, the steryl esters may be viewed as 'a store of sterols'.

Lewis *et al.* (1987) concluded that yeast cells must maintain a low level of free sterol, which is essential for growth. Provided there is sufficient sterol available, a further expandable free sterol pool may be maintained. However, as this pool of free sterol increased in size this was accompanied by a progressively increased rate of sterol esterification. The extent of esterification could be correlated with activity of acyl-CoA ergosterol acyltransferase.

The importance of steryl ester accumulation in brewing yeast strains during fermentation is not known. Thus, significant esterified sterol accumulation is associated with derepressed aerobic growth. This pool accounts for the five-fold difference in total sterol concentration of repressed and de-repressed cells (Quain & Haslam, 1979). In this regard it might be predicted that esterification would be of relatively small significance in brewing yeast. However, investigation of the biochemistry underlying the differences between high and low oxygen requiring brewing yeast strains has revealed that sterol esterification could be implicated. Thus, a comparison of sterol spectra revealed that the high-oxygen-requiring strain contained a relatively high concentration of zymosterol. This sterol is not readily incorporated into membranes and is usually esterified. Therefore, this implies that the high oxygen requiring strain would expend more oxygen to produce the same quantity of free membrane sterol as the low-oxygen-requiring strain (S.C.P. Durnin, personal communication).

In addition, when pitching yeast is deliberately exposed to oxygen under non-growing conditions some limited sterol synthesis may occur. This phenomenon has been used in a process designed to improve fermentation performance consistency (see Section 6.4.2.2). Here it was observed that during oxygenation of freshly cropped pitching yeast the total intracellular sterol content increased approximately five-fold from 0.2% to 1% of the total cell dry weight. In other words, a similar increase in total sterol concentration as is seen in a conventional oxygenated wort fermentation. However, analysis of the sterol spectrum revealed that ergosterol and zymosterol were synthesised in roughly equal proportions, and, furthermore, as a proportion of the whole, zymosterol increased at the expense of ergosterol (see Fig. 6.24). How the pool of steryl ester so formed would then influence subsequent fermentations remains unclear.

Sterols can be assimilated by yeast from the medium, however regulation is complex. Salerno and Parks (1983) reported that exogenous sterol could only be accumulated by anaerobic cells or mutant strains that are auxothrophic for sterol. Conversely, aerobic yeast capable of sterol synthesis did not utilise exogenous sterol. Parks' laboratory christened this phenomenon 'aerobic sterol exclusion' and has demonstrated that the process is associated with haem metabolism. Thus, several of the enzymes involved in the sterol biosynthetic pathway are haemoproteins. Not surprisingly therefore, haem deficient mutants are auxotrophic for sterol. Haem sufficiency, which requires aerobiosis, apparently derepresses sterologenesis and simultaneously inhibits sterol uptake (Lorenz & Parks, 1987; Shinabarger et al., 1989). The precise mechanisms for this phenomenon remain to be fully elucidated; however, it would suggest that during a brewery fermentation, sterol uptake could only occur after the initial aerobic phase was over.

Sterols are synthesised using carbon devolving from glycolysis via acetyl-CoA as part of the general pathway leading to the formation of branched isoprenoids. The first part of the synthesis is an anaerobic process, which involves the conversion of acetyl-CoA to squalene. An intermediate of this pathway is farnesyl pyrophosphate, which forms a branch-point between sterol synthesis and the formation of haem and ubiquinone, both essential components of the electron transport chain (Fig. 3.7).

In the terminal part of the sterol biosynthetic pathway devolving from squalene, molecular oxygen is used to form 2,3-epoxysqualene, followed by cyclisation to form the first sterol, lanosterol. Other sterols, including ergosterol are formed from lanosterol in a complex pathway in which the precise reactions vary from one yeast strain to another (Fig. 3.8). Apart from the initial epoxidation of squalene, molecular oxygen is involved in desaturation reactions involving cytochrome P450 mono-oxygenases.

Predictably, the regulation of sterol synthesis in yeast is complex! Of particular note is the influence of intracellular compartmentalisation (see Section 4.1.3). The sites of sterol synthesis and ultimate deposition may be distinct and therefore, regulation of synthesis and intracellular transport must be co-ordinated processes. Free sterols may be found in greatest abundance in the plasma membrane but smaller quantities are also found in the membranes surrounding other intracellular organelles such as the

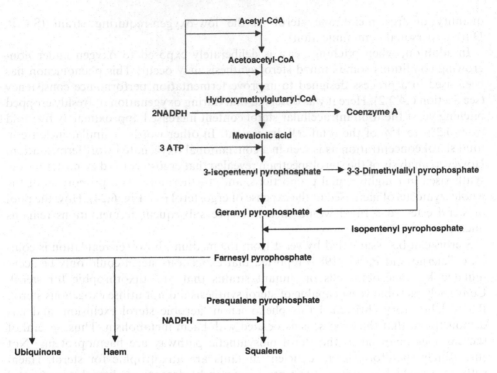

**Fig. 3.7** Pathway for the synthesis of squalene from acetyl-CoA.

mitochondria and vacuoles (see Section 4.1.1). Steryl esters, however, are associated with intracellular lipid particles. In mammalian systems sterol synthesis has been located in the endoplasmic reticulum (Reinhart, 1990). In yeast, however, sterol synthesis (at least the early stages) apparently occurs in mitochondria (Shimizu et al., 1973; Trocha & Sprinson, 1976). Clearly, this requires an intracellular transport system which must link the sites of synthesis of sterols with those of eventual deposition (possibly with the intermediary of esterification in lipid particles).

Alternatively, distinct biosynthetic pathways may provide sterols in individual cellular compartments. Casey et al. (1992) studied the control of the general iso-prenoid biosynthetic pathway in mutant strains of S. cerevisiae. They examined the effects of ergosterol and palmitoleic acid on the key enzyme, 3-hydroxy-3-methylglutaryl CoA reductase (HMG-CoA reductase) which catalyses the conversion of 3-hydroxymethylglutaryl CoA to mevalonate (Fig. 3.7). S. cerevisiae contains two isozymes of HMG-CoA reductase, coded for by the structural genes, HMG1 and HMG2. It was concluded that both isozymes directed equal amounts of carbon into sterol synthesis in haem-competent cells, despite a 57-fold difference in specific activity of the reductases. In mutant strains containing only HMG1 it was discovered that palmitoleic acid was a rate-limiting positive regulator, whereas ergosterol was an inhibitor of sterol production. In strains containing only the HMG2 gene, sterol production was inhibited by palmitoleic acid and to a lesser extent by ergosterol. The reductases were regulated differentially by haem but not by ergosterol and unsatu-

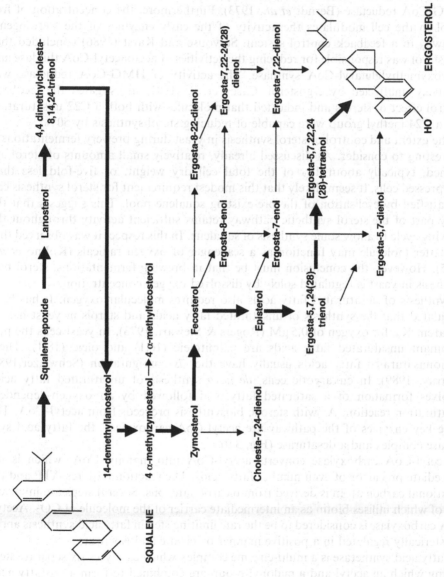

Fig. 3.8  Biosynthesis of ergosterol from squalene (adapted from Parks, 1978).

rated fatty acids. This led the authors to speculate that in this yeast there might be separate partitioned sterol biosynthetic pathways.

There is an abundance of evidence in the literature suggesting that sterol bio-synthesis is controlled at the level of hydroxymethylglutaryl-CoA reductase. The activity of this enzyme is low in anaerobically grown yeast. Supply of oxygen permits sterol synthesis and this is accompanied by a concomitant increase in the activity of HMG-CoA reductase (Berndt et al., 1973). Furthermore, the concentration of free sterol in the cell modulates the activity of the early enzymes of the sterologenic pathway in a feedback control system. Servouse and Karst (1986) concluded that ergosterol was responsible for reducing the activities of acetoacetyl-CoA thiolase and hydroxymethylglutaryl-CoA synthase. The activity of HMG-CoA reductase was relatively unaffected by ergosterol. Casey et al. (1991) investigated the feedback control effect of sterols and indicated that molecules with both a C22 unsaturation and a C24 methyl group were capable of reducing sterol synthesis by 50%.

The extent and control of sterol synthesis in yeast during brewery fermentations is interesting to consider. As discussed already, relatively small amounts of sterol are formed, typically about 1% of the total cell dry weight, or five-fold less than derepressed cells. It seems likely that this modest requirement for sterol synthesis can be satisfied by cyclisation of the pre-existing squalene pool. This suggests that the early part of the sterol synthetic pathway retains sufficient activity throughout the brewing cycle to allow some synthesis of squalene. In this respect it was observed that the latter molecule may function as a scavenger of oxygen radicals (Kohno et al., 1995). However, the conclusion must be that in brewing fermentations, sterol bio-synthesis in yeast is regulated solely by dissolved oxygen concentration.

Synthesis of unsaturated fatty acids also requires molecular oxygen. It has been calculated that the synthesis of unsaturated fatty acids and sterols in yeast has an apparent $K_m$ for oxygen of 0.5 μM (Rogers & Stewart, 1973). In yeast cells the pre-dominant unsaturated fatty acids are palmitoleic (16:1) and oleic (18:1). These monounsaturated fatty acids usually have the cis configuration (Schweizer,1989; Rattray, 1989). In eukaryotic cells, de novo synthesis of unsaturated fatty acids involves formation of a saturated fatty acid followed by an oxygen-dependent desaturation reaction. As with sterols, biosynthesis proceeds from acetyl-CoA. The three key enzymes of the pathway are acetyl-CoA carboxylase, the fatty acid syn-thetase complex and a desaturase (Fig. 3.9).

Acetyl-CoA carboxylase converts acetyl-CoA into malonyl-CoA, which is the immediate precursor of even number fatty acids. The reaction requires ATP and the additional carbon atom is derived from bicarbonate ions. Several steps are involved, one of which utilises biotin as an intermediate carrier of the molecule of $CO_2$. Acetyl-CoA carboxylase is considered to be the rate limiting step in fatty acid synthesis and is allosterically regulated in a positive manner by citrate and isocitrate.

Fatty acid synthetase is a multi-enzyme complex which catalyses six separate steps during which an acetyl and a malonyl group are combined to form a $C_4$-fatty acid. The additional carbon atom of the malonyl group is released in the form of $CO_2$. This latter carbon atom is the same as that which was fixed in the initial acetyl-CoA carboxylase reaction and, thus, the bicarbonate ion has a catalytic role. In E. coli, the fatty acid synthetase includes an acyl carrier protein (ACP) which contains the same

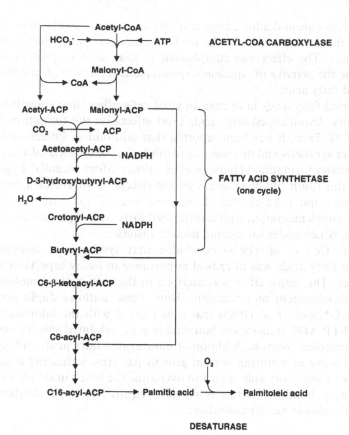

**Fig. 3.9** Biosynthesis of unsaturated fatty acids.

functional 4′-phosphopantotheine group as coenzyme A and fulfils a similar role. It is assumed that the yeast enzyme complex also used the same acyl carrier protein. Chain elongation proceeds via sequential condensation reactions between the product of the fatty acid synthetase and additional malonyl groups. Thus, palmitic acid results from seven cycles of fatty acid synthetase activity. Odd numbered fatty acids are synthesised in the manner described above, but in this case the initial activation reaction is between acyl carrier protein and propionyl-CoA. After the fatty acid has been released from the synthetase complex, desaturation may proceed, providing molecular oxygen is available.

The roles of sterols and unsaturated fatty acids in conferring specific structural and functional properties on membranes is described in Section 4.1.2.2. In some respects, the effects of these two classes of lipid are complementary and synergistic. Using model membrane systems, Hapala *et al.* (1990) concluded that membrane phospholipids were influential in regulating intramembrane sterol transfer. In particular the nature of the phospholipid was important. Butke *et al.* (1988) demonstrated that starving a sterol biosynthetic mutant of unsaturated fatty acids resulted in accumulation of squalene. Addition of exogenous unsaturated fatty acids restored the activity of squalene epoxide and allowed sterol synthesis to proceed. The most effective fatty acids were those which reduced the medium chain saturated fatty acid content of phospholipids. The effect was complicated in that *de novo* protein synthesis was essential for the activity of squalene expoxidase to be restored in cells deprived of unsaturated fatty acids.

Unsaturated fatty acids have roles in yeast cells other than contributing to membrane fluidity. Unsaturated fatty acids exert effects on the formation of esters (see Section 3.7.3). Thus, it has been reported that addition of linoleic acid resulted in reduced ester synthesis and this was due to inhibition of the alcohol acetyl transferase involved in ester formation (Thurston *et al.*, 1982). More recently, Fuji *et al.* (1997) duplicated this result but concluded that the effects of unsaturated fatty acids were more complex and were exerted at the gene level. These authors concluded that aeration or supplementation with unsaturated fatty acid repressed expression of the ATF1 gene, which codes for alcohol acetyltransferase.

O'Connor Cox *et al.* (1993) concluded that synthesis of adequate levels of unsaturated fatty acids was of critical importance in many aspects of fermentation performance. The major effect was ascribed to the role of these molecules in mitochondrial development and function. Thus, these authors duplicated the earlier findings of Gbelská *et al.* (1983) that cells treated with the inhibitor of the mitochondrial ATP/ADP translocase, bongkrekic acid, exhibited slow growth compared to unsupplemented controls. Addition of unsaturated fatty acids and ergosterol was partially effective at restoring normal growth patterns, indicating that these metabolites were in some way able in part to overcome the block in supply of energy from cytosol to mitochondria. Further research is required to clarify further the roles of unsaturated lipids in yeast metabolism.

## 3.6 Ethanol tolerance

The ability of yeast to withstand high concentrations of ethanol is obviously key to successful fermentation, particularly in the case of high gravity brewing using very concentrated worts (see Section 2.5). A number of factors may be considered as contributors to ethanol tolerance:

(1) Genetic components
(2) Toxicity of ethanol (or intermediates)
(3) Influence of physical environment
(4) Influence of physiological condition of yeast
(5) Influence of wort composition

It is generally accepted, although usually not proven, that there are genotypic differences in yeast strains, which predispose some but not all to withstand high ethanol concentrations. Two components may be recognised in this regard: the maximum concentration that may be generated during fermentation that does not compromise the viability of the crop and the maximum rate of ethanol production. These two elements are reflections of, first, the ability of the yeast to withstand the stress imposed by exogenous ethanol and, second, the susceptibility of the pathways leading to ethanol formation to feedback inhibition by the product. Brewing strains are considered to be moderately tolerant of ethanol, typically being used in fermentations in which up to 8% v/v ethanol is formed. Strains of *Saccharomyces* used for wine production are considered as being more tolerant of ethanol, performing fermentations which generate 10–15% v/v ethanol; in some extreme cases saké yeasts produce more than 20% v/v ethanol.

Such classifications assume that ethanol is actually inhibitory or toxic to yeast cells, or at least some aspects of their metabolism which relate to fermentation performance. Of course, if this is so then it is equally possible that intermediates of ethanol production or other products of fermentation are exerting deleterious effects in addition to, or instead of, ethanol. Conversely, ethanol tolerance may reflect differences in the ability of yeast strains to ferment very concentrated worts. In this regard the controlling factor may be ability to grow at reduced water activity and not in the presence of elevated ethanol concentrations, although these may be related. Assuming that ethanol exerts toxic effects on yeast cells it is probable that physical environmental conditions such as temperature, pressure, and pH will modulate the severity of the effects.

The purely genetic elements of ethanol tolerance, if they exist, will be modified by phenotypic factors. The condition of the pitching yeast is known to influence fermentation performance and therefore it is likely that ethanol tolerance will also be affected. In particular, the ability of yeast to withstand stresses is modulated by physiological condition. In this regard the conditions of storage of yeast in the interval between cropping and re-pitching will be influential. Further modifications to the yeast phenotype are possible due to the composition of the wort.

### 3.6.1 *Ethanol formation during fermentation*

During batch fermentation there are progressive changes in the rates of ethanol production. Thus, there is an initial lag phase, which corresponds with the passage of the yeast from lag to exponential growth. During the latter period the rate of ethanol formation reaches a maximum. Soon after yeast growth ceases the rate of ethanol production declines (see Fig. 3.1). It is assumed that the causes of the decline in rates of ethanol production are a consequence of a combination of nutrient depletion and toxic effects of ethanol.

Dombek and Ingram (1987) measured rates of ethanol production during batch fermentation of 20% glucose, which resulted in the formation of more than 10% v/v ethanol. The authors reported that specific activities of glycolytic and ethanologenic enzymes, measured *in vivo*, remained high throughout fermentation. However, rates of ethanol production, assessed off-line in terms of rates of $CO_2$ evolution, declined progressively as ethanol accumulated. Since removal of the ethanol did not restore rates of ethanologenesis it was concluded that the effects had other physiological causes, possibly related to irreversible ethanol-induced damage. In a later paper the same authors (Dombek & Ingram, 1988; Alterthum *et al.*, 1989) concluded that there was a decline in glycolytic activity and this was due to changes in the pool sizes of adenine nucleotides. In particular, hexokinase was strongly inhibited by increased AMP concentration.

Others have reported control of glycolysis at other parts of the pathway. For example, the possibility of regulation of phosphofructokinase by mechanisms such as adenylate energy charge as well as other metabolites such as citrate and isocitrate has been reported (Sols, 1981). Others have disputed this thesis. The necessity for phosphofructokinase for ethanol formation has been questioned (Breitenbach-Schmitt *et al.*, 1984, Heinisch & Zimmermann, 1985). The same group demonstrated that mutant strains with increased expression of glycolytic genes did not permit rates of ethanol production greater than the wild type (Schaaf *et al.*, 1989).

The control of carbon flux at the level of pyruvate has perhaps more plausibly been implicated as being influential in ethanol accumulation. Sharma and Tauro (1987) considered that strains of *S. cerevisiae* could be classified on the basis of their ability for rapid and slow ethanol production. These authors reported that the rapid ethanol producers had greater specific activities of pyruvate decarboxylase and lower aldehyde dehydrogenase compared to the slow producing class. The importance of control of carbon flow through pyruvate has been discussed with regard to the Crabtree effect (Section 3.4). However, it is interesting to note that during brewery fermentation pyruvate is excreted into the beer (Coote & Kirsop, 1974). The production of exogenous pyruvate is transient and the peak appears during the period of maximum ethanol production (Fig. 3.10). This implies that in brewing fermentations, at least for most of the time, dissimilation of sugars to the level of pyruvate is more rapid than carbon flow from pyruvate to ethanol and other products. Since ethanol accumulation occurs simultaneously with formation of extracellular pyruvate this also suggests that carbon flux via pyruvate decarboxylase is more rapid than that through other pyruvate-utilising pathways.

### 3.6.2   *Ethanol toxicity*

Several other authors have reported inhibitory effects due to ethanol. For example, Thatipamala *et al.* (1992) found that in batch culture an increase in ethanol con-

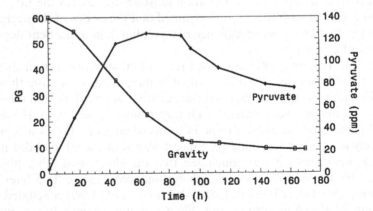

**Fig. 3.10**   Pattern of formation of extra-cellular pyruvate during the course of a high-gravity (1060) lager fermentation. The pitching rate was $12 \times 10^6$ cells ml$^{-1}$, the initial dissolved oxygen concentration was 18 ppm and the temperature was maintained at a constant 15°C throughout (Boulton and Box, unpublished data).

centration was responsible for an instantaneous decrease in biomass yield. Product inhibition was found to occur when the substrate concentration was higher than $150\,g\,l^{-1}$. In order that ethanol may exert its inhibitory effects, it has been suggested that intracellular ethanol accumulation takes place during fermentation. Nagodawithana and Steinkraus (1976) reported that ethanol generated by yeast was more cytotoxic than exogenous ethanol added at the same concentration. These authors suggested that this was due to an increased intracellular ethanol concentration which accumulated during rapid fermentation. Several other authors confirmed accumulation of ethanol (Novak *et al.*, 1981; Loureiro & Ferrera, 1983; Strehaiano & Goma, 1983; Legmann & Marglith, 1986).

Others have concluded that there is no accumulation of ethanol during fermentation (Dombek & Ingram, 1986; Jones, 1988). D'Amore *et al.* (1988) reported that up to 3 hours after the start of fermentation the intracellular concentration of ethanol was greater than that measured in the growth medium. However, after 12 hours both intracellular and extracellular concentrations were similar. It now appears that the plasma membrane is freely permeable to ethanol and intracellular accumulation only occurs when the rate of fermentation is very rapid and rates of production exceed those of diffusion out into the medium. This occurs rarely, usually only in early fermentation.

The explanation for the observation that ethanol generated intracellularly is more toxic than that added exogenously has been ascribed to the effects of other metabolites related to ethanol formation and which may exert toxic effects in addition to ethanol. Viegas *et al.* (1985) provided evidence that the powerful detergent properties of short chain fatty acids produced during fermentation exerted synergistic toxic effects with ethanol on yeast cells. Okolo *et al.* (1987) reported that higher alcohols increased the toxicity of ethanol. Others have described inhibitory effects due to acetate, which were not due simply to pH (Pampulha & Loureiro, 1989; Pampulha & Loureiro-Dias, 1990; Phowchinda *et al.*, 1995). Perhaps most convincingly it has been argued that acetaldehyde, the immediate precursor of ethanol, is an order of magnitude more toxic than ethanol (Jones, 1987, 1989).

Several mechanisms for ethanol toxicity have been described, both non-specific and those in which specific cellular sites of action have been identified. D'Amore *et al.* (1988) concluded that ethanol toxicity was a non-specific effect in which increased intracellular ethanol produced increased osmotic pressure. However, these authors concluded that nutrient limitation was responsible for the observed reduction in growth and fermentation rates. Jones and Greenfield (1987) also considered that non-specific osmotic effects were also contributory to ethanol toxicity. However, specific intracellular targets for ethanol have also been reported. In particular, ethanol is reported to exert toxic effects on cell membranes. Salgueiro *et al.* (1988) described the ethanol-induced leakage of amino acids and other UV-absorbing cellular components. It was concluded that ethanol tolerance and resistance to leakage could be correlated; furthermore, supplementation of growth media with the leaked material resulted in an improvement of alcoholic fermentation. Lloyd *et al.* (1993) described inhibition of transport systems in yeast by ethanol. Jimenez and Benitez (1987) reported that ethanol tolerant enological yeast strains were able to adapt membrane structure such that the toxic effects of ethanol could be withstood. Less tolerant strains were unable to undergo this adaptation and growth in the presence of ethanol resulted in increased but reversible sensitivity. The reversible effect required an energy source and was abolished by cycloheximide. This was taken to imply that protein components of the membrane might contribute to ethanol tolerance.

More recent work has shown that it is the lipid components of membranes that confer ethanol tolerance. Novotny *et al.* (1992) demonstrated that in chemostat cultures manipulation of carbon to nitrogen ratios which resulted in an increased membrane content of 5,7-unsaturated sterols was accompanied by increased resistance of the population to ethanol-induced death. Similarly, Alexandre *et al.* (1994) concluded that ethanol tolerance was associated with altered membrane fluidity. This was achieved by an increase in the proportion of ergosterol and unsaturated fatty acids, together with maintenance of phospholipid biosynthesis. Presumably, this would explain why elevated temperature increases the toxic effects of ethanol, since the former is associated with greater saturation of membrane lipids. A further response of yeast to ethanol may be accumulation of trehalose. The ability of this disaccharide to stabilise membranes has been described (Section 3.4.2.2).

Another target for ethanol toxicity is the mitochondrion (see Section 4.1.2.3). Aguilera & Benitez (1985) reported that yeast cells lacking mitochondria were more susceptible to ethanol inhibition. It is widely recognised that ethanol is a mutagen of the mitochondrial genome (see Section 4.3.2.7). Thus, Bandas and Zakharov (1980) noted that 24% v/v ethanol resulted in a five-fold increase in the generation of petite mutants in cultures of *S. cerevisiae* compared to the spontaneous mutation rate. This mutation is the most commonly occurring type in yeast during brewing fermentations, and, furthermore, such strains exhibit poor fermentation properties (Ernandes *et al.*, 1993).

Costa *et al.* (1997) provided evidence that ethanol tolerance in *S. cerevisiae* required the presence of an active Mn-superoxide dismutase. This is the mitochondrial form of this enzyme and this was taken to indicate that the formation of oxygen radicals were implicated in the mechanism of ethanol toxicity. It is notable that the stress response of yeast (see Section 3.4.2.2), which may be triggered by heat shock, is associated with induction of enzyme systems which are involved in protection against oxidative stress (Stephen *et al.*, 1995). Furthermore, there is overlap between cellular responses to heat and ethanol shock (Piper, 1995). Interestingly, the mitochondrial superoxide dismutase is not present in brewing yeasts under anaerobic conditions and requires several hours' exposure to oxygen before it is induced (Clarkson *et al.*, 1991). The implication of the findings regarding oxygen radicals is that ethanol should be less toxic to yeast under anaerobic conditions. Furthermore, this aspect of ethanol tolerance should not be of significance in brewing fermentations since aerobiosis occurs only when ethanol concentrations are low. Certainly, the oxygenation process described in Section 6.4.2.2 subjected yeast, suspended in beer, to prolonged aerobiosis. There was no observed decrease in viability during the course of this treatment (Fig. 6.23). Furthermore, under these conditions the Mn superoxide dismutase would not be expressed (see Clarkson *et al.*, 1991). Therefore if the toxic effects of ethanol are associated with the formation of oxygen radicals then the cytosolic and constitutive Cu-Zn superoxide dismutase, or other mechanisms, must have been active and effective.

Ethanol tolerance can be influenced by wort composition. The effect of wort oxygen concentration on the formation of sterols and unsaturated fatty acids and how these may then modulate membrane structure and ethanol tolerance are obvious from the previous discussion. Of course, the implication is that ethanol will be present in fermenter at high concentration when growth has ceased and unsaturated fatty acid and sterol concentrations will be minimal. Again the apparent toxicity of ethanol during the oxygenation process alluded to above, suggests that lack of sterol is not a critical factor with all yeast strains. Nevertheless, the synthesis of trehalose associated with the oxygenation process perhaps indicates both the level of stress associated with the process and the mechanism for dealing with it.

Several reports indicate that metal ions can moderate the inhibitory effects of ethanol. Dombek and Ingram (1986) observed that supplementation of a glucose medium with 0.5 mM magnesium resulted in prolonged exponential growth and a smaller reduction in fermentation rate compared to an unsupplemented control. Ciesarova et al. (1996) made similar observations but extended the investigations to include the effects of magnesium and calcium. They concluded that, under non-growing conditions in the presence of 10% v/v ethanol, both metal ions exerted protective effects. In fermentation, growth was stimulated by magnesium in the presence and absence of added ethanol, to a greater extent than calcium when added singly. Ethanol fermentation was more efficient when both metal ions were present.

It is suggested that some worts may contain less than optimal concentrations of these metals. Two groups of workers (Walker et al., 1996; Rees & Stewart, 1997) concluded that the ratio of magnesium to calcium was influential. With six yeast strains high magnesium to calcium ratios favoured rapid initial fermentation rates and improved ethanol yields. If the ratio was reversed ethanol production declined and uptake of maltotriose uptake was reduced. In the case of lager strains, maltose uptake was also affected. Conversely, Bromberg et al. (1997) noted stimulation of fermentation only by the addition of zinc; manganese, calcium and magnesium addition had no effect.

The mechanism by which metal ions may relieve ethanol toxicity is not known. However, magnesium in particular is a co-factor in numerous enzymes, including several involved in glycolysis. Furthermore, other divalent cations can compete for the magnesium binding sites. The metal ion contents of worts will depend upon the mineral composition of the brewing liquor, as well as those metals released from other raw materials. In addition, chelating agents in worts will serve to modulate the concentrations of metals available to yeasts. It may be readily appreciated, therefore, that metal ions may not always be limiting, depending on the particular wort under investigation.

## 3.7 Formation of flavour compounds

A multitude of compounds contribute to beer flavour. Many of these are derived directly from the raw materials used to produce the wort. In this respect the blend of malt and hops are influential. However, fermentation has the most significant impact on flavour development. Both ethanol and carbon dioxide contribute to beer flavour, imparting 'warming' and 'mouth tingle' characters, respectively. The essential character of any beer is determined by the plethora of other yeast metabolites, that arise during fermentation. Many of these are flavour-active at the concentrations found in beer. The balance of flavour metabolites formed is largely a consequence of the combination of yeast strain and wort composition which are used. In addition, a prime aim of fermentation management is to control conditions so as to ensure that these metabolites are produced in desired quantities.

Yeast metabolites, which contribute to beer flavour, are diverse chemically and include organic acids, medium chain-length aliphatic alcohols ('fusel alcohols'), aromatic alcohols, esters, carbonyls and various sulphur-containing compounds. Many hundreds of components may be detected in beers, although many remain to be positively identified. Some of these make a positive contribution; others impart undesirable notes. Meilgaard (1974) reported on the flavour thresholds and flavours of more than 200 beer components. Flavour thresholds varied to an enormous extent. Thus, the threshold detection concentration for ethanol was $13\,g\,l^{-1}$, whereas, for tertiary amyl mercaptan a value of $0.07\,ng\,l^{-1}$ was recorded! Flavour impact was shown to be influenced by the chain-length of the molecule. Thus, for aliphatic components, the greatest intensity was generally shown by compounds with 8–10 carbon atoms. Molecules of longer or shorter chain-lengths were less flavour-active.

Meilgaard (1975) ranked, in order of importance, the positive contributions to beer flavour as: ethanol, hop bitterness, carbon dioxide, 'banana esters' (isoamyl acetate), 'apple esters' (ethyl acetate) and 'fusel alcohols'. In terms of flavour defects the corresponding ranking was: 'sulphury' (dimethyl sulphide + hydrogen sulphide), 'toffee/butterscotch' (diacetyl) and 'stale' (2-trans-nonenal).

Predictably with such a wide range of components deriving from the results of yeast metabolism it is not possible to ascribe a single underlying rationale that explains their appearance in beer. Potential metabolic mechanisms are responses to reduced water activity, cellular redox balancing reactions, maintenance of intracellular pH, shock excretion, stress responses and counter ion nutrient uptake. Some of these are discussed in more detail in the following sections of this chapter. In addition, the significance of overflow metabolism should not be overlooked. With regard to carbon flow from assimilation of wort nutrients to formation of extracellular metabolites, it is noteworthy to consider the crucial importance of pyruvate and acetyl-CoA as major branch points (Fig. 3.11).

The significance of carbon flow from sugars and through pyruvate with respect to phenomena such as the Crabtree effect has been discussed previously (see Section 3.4.1). It is suggested that carbon flow from pyruvate to ethanol and acetyl-CoA is regulated only by the relative affinities of pyruvate dehydrogenase and pyruvate decarboxylase and that ethanol formation represents overflow metabolism. Furthermore, acetyl-CoA can arise from acetaldehyde, via the intermediary of acetate, thereby bypassing pyruvate dehydrogenase.

The evidence suggests that this bypass represents the major route for acetyl-CoA formation in brewery fermentations. Thus, the lipoamide dehydrogenase, which is an essential component of the multi-enzyme pyruvate dehydrogenase complex, is subject to glucose repression (Roy & Dawes, 1987). The absence of this enzyme, which is also the glucose repressible component of the 2-oxoglutarate dehydrogenase complex, is

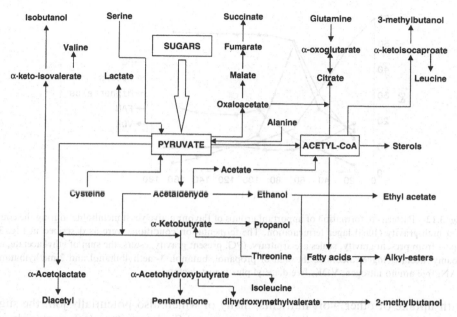

**Fig. 3.11** Central roles of pyruvate and acetyl-CoA in pathways leading from the utilisation of wort nutrients and formation of flavour-active metabolites.

responsible for the lack of a complete citric acid cycle in anaerobically grown *S. cerevisiae*. From another viewpoint, Flikweert *et al*. (1996) demonstrated that mutant strains of *S. cerevisiae* lacking pyruvate decarboxylase could not grow on glucose, even under non-fermentative conditions. This suggests that cytosolic formation of acetyl-CoA via the intermediary of acetaldehyde is essential to yeast. On the other hand, Wenzel *et al*. (1993) detected activity of pyruvate dehydrogenase during growth of yeast under anaerobic conditions and aerobic growth on ethanol, where it has no apparent catabolic role. These authors suggested that this may indicate an anabolic role in supplying acetyl-CoA for mitochondrial amino acid synthesis.

It seems reasonable to assume, therefore, that pyruvate dehydrogenase does not play a role in sugar catabolism in brewery fermentations. It follows that the very large quantities of sugars present in brewery worts must be channelled largely though pyruvate decarboxylase and thence, via either alcohol dehydrogenase or aldehyde dehydrogenase to generate ethanol or acetate, respectively. Presumably the inability of these enzymes to contain the carbon flux accounts for the appearance of pyruvate in beer (Fig. 3.10). Similar overflow mechanisms may account for the appearance of other metabolites in beer. Many important groups of flavour-active yeast metabolites arise in beer during early to mid-fermentation when the yeast is actively assimilating both sugars and other nutrients (Fig. 3.12). In the data given it may be seen that ester levels attained peak values coincident with the time at which uptake of wort free amino acids ceased. Both VDK and higher alcohols achieved maximum concentrations before this, when the wort was approximately two-thirds attenuated.

The period of maximum production of extracellular by-products of yeast metabolites occurs when there is a high rate of sugar uptake and dissimilation, coupled

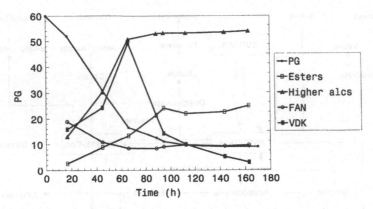

**Fig. 3.12** Patterns of formation of important groups of flavour active yeast metabolites during the course of a high-gravity (1060) lager fermentation. The fermentation conditions were as described in Fig. 3.9. Apart from present gravity, scales are arbitrary. (PG, present gravity; esters, the sum of ethyl acetate, and isoamyl acetate; higher alcohols; the sum of propanol, butanol, 3-methylbutanol and 2-methylbutanol; FAN, free amino nitrogen; VDK, free diacetyl plus α-acetolactate).

with uptake of other wort nutrients, many of which also potentially join the sugar catabolising pathways at the level of pyruvate. Since yeast growth extent during fermentation is relatively modest and there is no opportunity for complete oxidation of sugars, it is perhaps not surprising that many of the partial oxidation products are excreted from yeast cells, to appear as potential flavour-active beer components.

### 3.7.1    *Organic and fatty acids*

Meilgaard (1975) reported the presence of some 110 organic or short-chain fatty acids in beer. Many of these are derived from wort; however, the concentrations of some are modulated during the course of fermentation indicating their participation in yeast metabolism. Organic acids in general have sour flavours and they contribute to the lowering of pH that occurs during fermentation. In addition to sourness, individual organic acids are reported to have characteristic flavours. For example, succinate has a salty/bitter taste (Whiting, 1976). Short-chain fatty acids have a negative impact on beer flavour. For example, caproic and caprylic acids have unpleasant 'goat-like' aromas.

Organic acids include pyruvate (100–200 ppm), citrate (100–150 ppm), malate (30–50 ppm), acetate (10–50 ppm), succinate (50–150 ppm) lactate (50–300 ppm) and 2-oxoglutarate (0–60 ppm) (Coote & Kirsop, 1974; Whiting, 1976; Klopper *et al.*, 1986). The comparatively wide ranges of organic acids arising in various beers are in part due to variability in wort composition but are mainly a consequence of yeast strain-specific differences. The majority of organic acids are derived directly from pyruvate or from the branched tricarboxylic acid cycle which is characteristic of the repressed, anaerobic physiology of brewing yeast during fermentation (Wales *et al.*, 1980). Their excretion into beer can be explained by the lack of any mechanism for further oxidation, the need to maintain a neutral intracellular pH and the fact that they are not required for anabolic reactions. There is some evidence that levels of some organic acids fluctuate during the course of fermentation. For example, it has been reported that the formation of extracellular pyruvate, which occurs during active fermentation, may be re-assimilated and replaced by acetate (Coote & Kirsop, 1973, 1974 and see Fig. 3.10). Presumably this reflects changes in carbon flux through glycolysis during the course of fermentation. The contribution that they make to lowering the pH during fermentation could confer a selective advantage in mixed microbial populations.

Two oxo-acids, $\alpha$-acetolactate and $\alpha$-acetohydroxy acids, which are excreted by yeast during brewing fermentation, are of particular importance to beer flavour. These are the precursors of diacetyl and 2,3-pentanedione, respectively. The significance of these is discussed in Section 3.7.4.

On the whole fatty acids are undesirable components of beers both from the point of view of taste and their potential for adversely affecting foam performance. Chen (1980) reported that nearly 90% of the free fatty acids in worts were accounted for by palmitic (16:0), linoleic (18:2), stearic (18:0) and oleic (18:1). In beers 75–80% of fatty acids were caprylic (8:0), caproic (6:0) and capric (10:0). In three beers that results were given for there was a net increase in the total fatty acid contents of worts and beers of between 13% and 65% (5–7 mg l$^{-1}$ in wort increasing to 6–12 mg l$^{-1}$ in beers). The author suggested that these results indicated that long-chain fatty acids were assimilated by growing yeast and incorporated into structural lipids. The shorter-chain saturated fatty acids in beers were released as by-products of *de novo* lipid synthesis.

This result would infer that any change in fermentation conditions, which promoted the extent of yeast growth, would also favour increased levels of short-chain fatty acids found in beer. Thus, higher temperature, increased wort oxygenation and possibly elevated pitching rates would all be expected to be effective in this respect. However, caution must be exercised in the interpretation of such results. Short-chain fatty acids, particularly $C_8$–$C_{14}$, are toxic to yeast cells (Nordström, 1964). This is probably due to non-specific detergent effects disrupting cell membranes, and therefore it is unlikely that they would be excreted into beers. Possibly free fatty acids could be excreted if there was a restricted supply of coenzyme A. However, it is more plausible that short-chain fatty acids appear via an autolytic mechanism or through ethanol induced membrane leakage. Such a possibility has been proposed for the formation of 'yeasty' or stale off-flavours developing during beer maturation (Masschelein, 1986).

### 3.7.2 *Higher alcohols*

More than 40 higher alcohols have been identified in beers (Engan, 1981). Those that have organoleptic importance because they occur at concentrations above flavour thresholds are: n-propanol, iso-butanol, 2-methylbutanol and 3-methylbutanol. Their contribution to beer flavour is by a general intensification of alcoholic taste and aroma and by imparting a warming character. In addition, the aromatic alcohol 2-phenylethanol, which has a rose/floral aroma, is considered a desirable character (Meilgaard, 1974). Apart from being of significance to beer flavour and aroma in their own right the higher alcohols have the important secondary role of providing precursors for ester synthesis. Typical concentrations of higher alcohols in beers are of the order of 100–200 mg $l^{-1}$. Usually, 3-methylbutanol and 2-methylbutanol are the most abundant. In addition to these, glycerol may be found in beers at concentrations of the order of 1–2 g $l^{-1}$ (Quain & Duffield, 1985).

Higher alcohols may be synthesised by two metabolic routes. First, *de novo* synthesis from wort carbohydrates via pyruvate, the anabolic route, and, second, as by-products of amino acid assimilation, the catabolic or Ehrlich pathway (Ayrapaa, 1965, 1967a, b, 1968). In both cases, the immediate precursors are 2-oxo ($\alpha$-keto) acids. These are decarboxylated to form an aldehyde which is then reduced to the corresponding alcohol. In the anabolic route the 2-oxo acid derives from pyruvate or acetyl-CoA as part of amino acid biosynthetic pathways. In the catabolic route, the 2-oxo acid is formed by transamination of an amino acid (Fig. 3.13). Biosynthesis of higher alcohols important to beer flavour and their relation to amino acid metabolism is shown in Fig. 3.14.

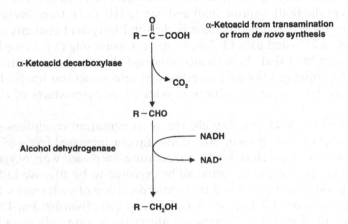

**Fig. 3.13** Generalised scheme for higher alcohol synthesis.

The relative contribution made by the anabolic and catabolic routes for higher alcohol synthesis is influenced by a number of factors. In the case of n-propanol, the anabolic route is the only one possible since there is no corresponding amino acid. Chen (1978) reported that with increased chain-length of the alcohol the catabolic route was progressively used. The wort amino acid concentration is important. Thus, where only low levels of assimilable amino nitrogen are available the anabolic route is predominant. Conversely, high concentrations of wort amino nitrogen favour the catabolic pathway due to feedback inhibition of the amino acid synthetic pathways by their respective products. In brewing fermentations Schulthess and Ettlinger (1978) concluded that yields of higher alcohols were approximately equal from both the anabolic and catabolic pathways. However, the catabolic pathway is of greater importance during the early phase of fermentation when wort amino acids are plentiful whereas the anabolic pathway is increasingly used as wort amino acids are assimilated (Inoue, 1975).

The metabolic explanation for the formation of higher alcohols is obscure.

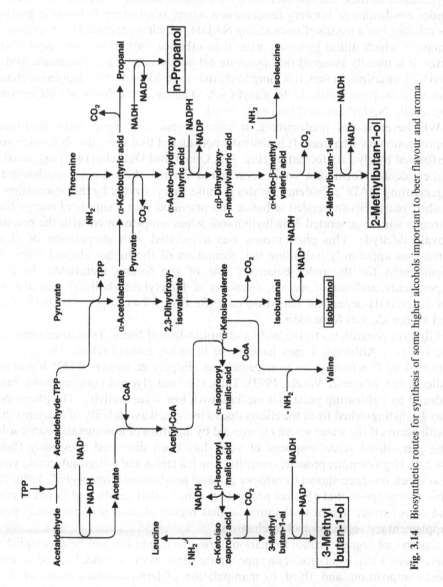

Fig. 3.14  Biosynthetic routes for synthesis of some higher alcohols important to beer flavour and aroma.

Although the proportion of wort carbon which appears in beers as higher alcohols is modest compared to ethanol it is still ostensibly profligate. Since yeasts, in common with all cells, are never wasteful of metabolic resources it suggests that there must be a reason for higher alcohol formation. Quain and Duffield (1985) argued that the appearance of these metabolites could be a manifestation of cellular redox control. Under conditions of brewery fermentation where respiratory function is precluded the cell requires a means of re-oxidising NADH, which is generated by glycolysis, and reactions which utilise pyruvate other than ethanol formation. With regard to the latter, it is usually assumed that pyruvate dehydrogenase is the significant route for $NAD^+$ formation. In fact, it is more likely that acetaldehyde dehydrogenase channels carbon flow from pyruvate to acetyl-CoA. The cytosolic form of this enzyme is apparently NADP linked (Dickinson, 1996).

Whichever routes predominate, it remains true that yeast must maintain an appropriate redox balance. It is generally considered that the redox-balancing role is performed by glycerol formation (Fig. 3.3). Quain and Duffield (1985) suggested that higher alcohol formation could represent a further 'fine tuning' mechanism for regenerating $NAD^+$. Evidence for this premise was provided by the observation that washed yeast cells suspended in buffer and provided with a supply of energy but no nitrogen source generated 3-methylbutanol when supplemented with the precursor, isovaleraldehyde. This phenomenon was associated with suppression of glycerol formation apparently indicating that formation of the higher alcohol relieved the requirement for the redox-balancing role of the former metabolite. In further experiments, addition of acetoin, a product of diacetyl metabolism which also arises via an NADH-dependent dehydrogenase (Section 3.7.4), suppressed both glycerol and higher alcohol formation.

Glycerol contributes to the body and mouth-feel of beers. Typical concentrations are $1–2\,g\,l^{-1}$. Although it may have a role in redox control (Oura, 1977), it is also reported to function as an osmoprotectant (Edgely & Brown, 1983; Blomberg & Adler, 1989; Mager & Varela, 1993). Thus, elevated glycerol concentrations may be induced by cultivating yeast in a medium with low water activity. The phenomenon may be distinguished from the effects due to increased availability of nutrients in that it still occurs if the water activity is reduced by addition of non-metabolisable solutes. The generalised stress response of yeast has been discussed previously (Section 3.4.2.2). High osmotic pressures constitute such a stress and indeed subjecting yeast to heat shock has been shown to enhance glycerol production (Omori et al., 1996, 1997). This also suggests that glycerol production is not solely implicated in redox control and adds further weight to the premise that higher alcohol formation may provide supplementary redox control mechanisms.

Control of higher alcohol formation during fermentation can be accomplished in three ways. First, by choice of an appropriate yeast strain, second, by modification to wort composition, and, third, by manipulation of fermentation conditions. Several authors have concluded that the yeast strain is the most significant factor. For example, Szlavko (1974), Engan (1978) and Romano et al. (1992) all observed variability in higher alcohol concentrations between individual strains cultivated under identical conditions greater than that seen with the same strain cultivated under a variety of different conditions. It has been reported that ale strains produce greater concentrations of higher alcohols compared to lager strains (Hudson & Stevens, 1960). Manipulation of the concentrations of individual higher alcohols is possible via genetic modification of yeasts. For example, Rous and Snow (1983) noted that mutants of Montrachet wine yeasts which were auxotrophic for specific branched chain amino acids produced reduced concentrations of the corresponding higher alcohols.

Wort composition influences higher alcohol formation, in particular the amino nitrogen content. This is of great significance in the case of high-gravity brewing where adjuncts may alter the ratio of carbon to nitrogen. As a general rule any change in fermentation conditions which increases the extent of yeast growth produces a concomitant increase in higher alcohol concentration. Thus, excessive wort amino nitrogen tends to elevate higher alcohol concentrations. However, very low amino nitrogen also produces excessive higher alcohols (Szlavko, 1974). In this instance it may be surmised that lack of nitrogen will promote enhanced synthesis of amino acids and that this will stimulate the anabolic higher alcohol synthetic route.

High wort oxygen concentration also favours increased yeast growth and this also enhances higher alcohol production. Increase in temperature has a similar effect (Barker et al., 1992). The effects of temperature are puzzling. This parameter influences yeast growth rate but not growth extent, and therefore it should not affect yields of metabolites which are growth-related. A similar effect is observed with respect to esters. In this case it has been suggested that alterations in membrane fluidity and diffusion rates from the cell might be implicated (Peddie, 1990). Control of excessive higher alcohol concentration in beer can be achieved by applying top pressure during fermentation. Thus, Rice et al. (1976) observed an inverse correlation between the concentrations of higher alcohols, other volatiles and yeast growth extent and applied pressure. It was demonstrated that this could be used to produce beers with similar volatile spectra fermented at 15°C with no applied pressure and at 22°C with 8 psig top pressure. The applied pressure had no effect on fermentation rate and therefore the advantage of using the higher temperature was retained.

### 3.7.3 Esters

Esters comprise possibly the most important set of flavour-active beer components which arise as a result of yeast metabolism. In the region of 100 distinct esters have been identified in beers (Drawert & Tressl, 1972; Meilgaard, 1975; Engan, 1981). Esters impart floral/fruity flavours and aromas to beers. Those whose concentrations in beer are considered crucial to product quality include ethyl acetate ('fruity/solvent'), isoamyl acetate ('banana/apple'), isobutyl acetate ('banana/fruity'), ethyl caproate ('apple/aniseed') and 2-phenylethylacetate ('roses/honey/apple'). Concentrations in beers are typically less than 1 ppm for minor components and 10–20 ppm for ethyl acetate (Table 3.4). Ethyl esters are the most abundant presumably because ethanol is the most readily available substrate.

Formation of esters by reactions between free fatty acids and alcohols is unlikely since the rates of such esterification would be too slow to account for the kinetics of their appearance during fermentation. Instead esters may potentially arise via two routes. First, from reactions between an alcohol, either ethanol or a higher alcohol

**Table 3.4**  Ester composition of a typical lager beer (OG 1030–1040).

| Ester | Concentration (ppm) |
|---|---|
| Ethyl acetate | 8–12 |
| Isobutylacetate | 0.03–0.05 |
| Ethyl butyrate | 0.04–0.06 |
| Isoamylacetate | 1–1.5 |
| Ethyl hexanoate | 0.12–0.18 |

and a fatty acyl-CoA ester (Nordström, 1962, 1963, 1964). Second, by esterases working in a reverse direction (Soumalainen, 1981). Schermers *et al.* (1976) contended that a positive correlation existed between reverse esterase activity and the concentrations of ethyl acetate and isoamyl acetate found in beers, thereby implicating these enzymes in ester synthesis. Nevertheless, it is generally accepted that the weight of evidence favours the first of these routes for ester synthesis in brewing yeasts. Biosynthesis involves two enzymes, an acyl-CoA synthetase and an alcohol acetyl transferase.

$$R_1COOH + ATP + CoASH \rightarrow R_1COSCoA + AMP + PPi$$

**Acyl Coenzyme A synthetase**

$$R_1COSCoA + R_2OH \rightarrow R_1COOR_2 + CoASH$$

**Alcohol acyl transferase**

It will be appreciated that the process requires energy and this suggests that esters serve a metabolic role. It is frequently stated in the literature that the acetyl Co-A precursor of esters arises from the action of pyruvate dehydrogenase. In fact, as discussed previously this enzyme is almost certainly inactive under the conditions of fermentation, and therefore acyl-CoA synthetase must be responsible. Evidence for the involvement of the alcohol acetyltransferase has been provided by genetic studies. The gene coding for the enzyme in *S. cerevisiae*, ATFI, has been cloned (Tamai, 1996) and it has been demonstrated that ester formation is dependent on its expression (Lyness *et al.*, 1997).

Much work has been performed in order to establish a unifying mechanism explaining the formation of esters during fermentation and to provide an insight into how their concentrations in beer may be regulated. In fact, the complete mechanism is still far from clear. It has long been brewing dogma that availability of acetyl-CoA exerts a controlling influence on ester synthesis. This premise precludes any role for reverse esterase activity. Changes in fermentation conditions, which influence the size of the pool of acetyl-CoA, should also affect the extent of ester synthesis. In particular, an inverse correlation between extent of yeast growth and ester synthesis has been demonstrated.

Thurston *et al.* (1981) have argued that ester synthesis may have metabolic significance in regulating the relative cellular concentration of acetyl-CoA and its precursor, CoASH. Certainly, at the mid-point of fermentation, the specific rate of ester synthesis increases more than two-fold at the same time that acetyl-CoA consumption through lipid synthesis ceases. Further evidence for the importance of maintaining an appropriate ratio of CoASH and acetyl-CoA was provided by the demonstration that, at the same time as the induction of ester synthesis, an acetyl-CoA hydrolase was induced (Thurston *et al.*, 1981). Although it is tempting to propose that ester synthesis enables 'fine tuning' of the cellular concentration of acetyl-CoA and CoASH, definitive evidence required the measurement of these metabolites during fermentation. Unpublished work (Pat Thurston and David Quain) reviewed by Quain (1988) confirmed that the intracellular concentration of acetyl-CoA increased in response to the shutting down of lipid synthesis. Indeed, the levels of this metabolite changed in tandem with the specific rate of ester synthesis. During the decline in both the rate of ester synthesis and in the level of acetyl-CoA, CoASH increased significantly.

The hypothesis that acetyl-CoA availability controls ester formation may be explored by examining the effects of fermentation conditions. With respect to wort composition, an increase in amino nitrogen concentration leads to increased ester synthesis. This may be interpreted as increased availability of acyl-CoA esters due to high rates of dissimilation of wort amino nitrogen. Conversely, where wort with low levels of assimilable nitrogen are used, such as might be the case with a high-gravity wort containing a large proportion of adjunct, yeast growth is restricted and in consequence the acetyl-CoA pool is also depleted.

Wort components, which promote yeast growth, tend to decrease ester levels. Thus, high trub levels are often accompanied by low ester levels. This can be explained in that trub may exert general stimulatory effects on yeast growth by favouring removal of carbon dioxide due to its ability to provide nucleation sites. In addition, trub may contain lipids and zinc, both of which promote yeast growth.

Increased provision of wort oxygenation encourages yeast growth and this is also accompanied by a reduction in ester levels. Again a plausible explanation is that because of the increase in yeast growth a large proportion of the acetyl-CoA pool is utilised for biosynthetic reactions and thereby the quantity available for ester formation is reduced. Changes in yeast pitching rate may also influence ester formation. In this instance the effects are less predictable. Maule (1967) reported that ester levels were reduced by large increases in pitching rate. A large increase in pitching rate would be expected to result in an overall reduction in yeast growth extent due to a restricted supply of oxygen available to each yeast cell and this should lead to elevated ester levels. However, the presence of high cell concentrations early in fermentation would be predicted to lead to rapid utilisation of wort amino nitrogen and oxygen for biosynthetic reactions. In which case the acetyl-CoA pool may be depleted and ester synthesis reduced.

Fermentation temperature and ester formation are directly related. The explanation for this effect is unclear. Peddie (1990) considered that higher temperatures could make cell membranes more fluid and this could possibly modulate the activity of the membrane-bound alcohol acetyltransferase or simply increase diffusion rates of esters from cells into the beer. However, elevated temperature also promotes increased formation of higher alcohols and this would provide greater concentrations of precursor for ester synthesis. In fact, the explanation for the effects of fermentation variables on ester formation is more likely to reside in control of the activity of alcohol acetyl transferase, as opposed to regulation via availability of substrate. In particular, the effects of lipids on ester formation appear to be crucial.

Alcohol acetyl transferase has proven a difficult enzyme to study since it is unstable in cell-free extracts. The instability is a consequence of its association with membranes. Thus Minetoki et al. (1993) reported that the enzyme is associated with the microsomal fraction of the yeast cell. It is well recognised that such proteins tend to be unstable when solubilised. Minetoki et al. (1993) purified the enzyme from a strain of S. cerevisiae used for saké production. In this beverage isoamyl acetate is the main flavour-active ester and it was reported that isoamyl alcohol was the best substrate, with acetyl-CoA for the alcohol acetyl transferase. The $K_m$ values for acetyl-CoA and isoamyl alcohol differed by nearly 200-fold, indicating, in the opinion of the authors, that supply of isoamyl alcohol and not acetyl-CoA would be rate-determining. It was suggested that varying substrate affinities could account for altered patterns of ester formation associated with different yeast strains. It was also demonstrated that the enzyme was strongly inhibited by unsaturated fatty acids.

Several reports (Thurston et al., 1982; Nakatani et al., 1991) have observed that ester synthesis occurs at low rates during the period of active yeast growth and then proceeds at high rates when yeast growth rate declines in mid to late fermentation. These authors claimed a correlation between the time of onset of rapid ester formation and cessation of lipid synthesis. It was speculated that the ratio of acetyl-CoA to free CoA could exert a regulatory influence on control of carbon flux between ester formation and lipogenic pathways. Crucially it was demonstrated that supplementation of worts with linoleic acid ($50 \, \mathrm{mg \, l^{-1}}$) resulted in an 80% decrease in ester formation. This was attributed to a possible inhibition of alcohol acetyl transferase.

Yoshioka and Hashimoto (1982a, b; 1984) partially purified alcohol acetyl transferase from brewers' yeast and observed inhibition by unsaturated fatty acids. Furthermore, activity could be correlated with the unsaturated fatty acid concentration of the cell membrane. It is suggested, therefore, that ester formation and lipid synthesis are related in an inverse manner. This allows a re-interpretation of the mechanism by which changes in fermentation conditions influence ester formation. Notably, the decrease in ester formation due to aeration can be explained in terms of inhibition due to increased synthesis of unsaturated fatty acids.

In fact the real effects of lipids on ester synthesis appear to be more complex than simple inhibition at the enzyme level. Thus the apparent regulatory effects of fatty acids could be interpreted as non-specific inhibition because of the ability of these compounds to inhibit enzyme activity due to non-specific detergent effects (Pande & Mead, 1968). Several authors have reported that oxygen and unsaturated fatty acids exert their effects by repressing the expression of the alcohol acetyl transferase gene. Thus, Malcorps et al. (1991) concluded that the step increase in ester synthesis, which occurred at the end of the yeast growth phase, was due to induction of alcohol acetyltransferase activity. The process required de novo protein synthesis and the induction was prevented by linoleic acid and oxygen. The same regulatory mechanism was considered to be involved for the synthesis of both ethyl and acetyl esters. Fuji et al. (1997) used Northern analysis to show that the alcohol acetyl transferase gene

(ATF1) was repressed by oxygen and unsaturated fatty acids. They concluded that the degree of repression exerted was probably the mechanism by which activity of alcohol acetyl transferase was controlled. Lyness *et al.* (1997) concluded that ATF1 gene expression and ethyl acetate concentration showed a close correlation; however no similar relationship was demonstrated with isoamyl acetate. These authors observed no effect on ATF1 expression and the addition of linoleic acid.

These apparently conflicting results may be explained in that there is good evidence that yeast cells possess more than one alcohol acetyl transferase and that these have differing substrate specificities. Dufour and Malcorps (1994) concluded that there may be a single enzyme which shows activity exclusively for ethyl acetate formation and another with broader specificity which can produce both acetyl and isoamyl acetate. Working with ATF1 mutants, Fuji *et al.* (1996) also concluded that the differing patterns of ethyl and isoamyl acetate formation could be explained by the presence of more than one enzyme.

It is apparent that regulation of ester synthesis is complex and intimately related to other pathways that utilise acetyl-CoA, in particular lipid formation. The evidence suggests that alcohol acetyl transferase is a key activity and several distinct enzymes may be present in the cell, each responsible for the formation of one or a few different esters. The activity of some, but not all, of these enzymes is regulated at the gene level, in a negative fashion, by oxygen and/or unsaturated fatty acids. Alcohol acetyl transferases may also be regulated by the availability of the alcohol substrate. In addition, it is not possible to exclude the role of esterases either acting in a reverse synthetic direction or selectively hydrolysing specific pre-formed esters. The characteristic patterns of esters formed by different yeast strains largely reflects the spectrum of alcohol acetyl transferases present, their substrate specificity and the precise manner of their regulation *in vivo*.

The metabolic role for ester formation is unknown. A plausible explanation is that it provides a route for nullifying the toxic effects of fatty acids. Thus, when lipid synthesis ceases, for example, due to oxygen depletion, the cell continues to produce medium-chain length fatty acids from acetyl-CoA. These are potentially toxic and the cell counters this threat by esterification and removal by diffusion into the medium.

### 3.7.4 *Carbonyls*

Some 200 carbonyl compounds have been detected in beers (Berry and Watson, 1987). Of importance to beer flavour and aroma and influenced by yeast metabolism are acetaldehyde and several other aldehydes and VDK.

Aldehydes have flavour threshold concentrations significantly lower than the corresponding alcohols. Almost without exception they have unpleasant flavours and aromas, variously described as 'grassy', 'fruity', 'green leaves' and 'cardboard', depending on the actual compound (Meilgaard, 1975). An aldehydic note is characteristic of the aroma of wort. This character is lost during a normal fermentation. In the case of low or zero alcohol beers that are made by limited fermentation, the 'worty' notes may be retained and this is considered an undesirable feature of such beers. Apart from aldehydes which are formed during wort mashing and boiling, others may arise as part of the catabolic and anabolic routes for higher alcohol formation, as described in Section 3.7.2. In beer, carbonyls form addition compounds with sulphur dioxide. In this form they may not be available for enzymatic reduction.

Reduction of aldehydes by yeast during fermentation is complex and apparently involves several enzyme systems with differing substrate specificities. Alcohol dehydrogenases are involved. Debourg *et al.* (1994) demonstrated that yeast with fermentative metabolism was more efficient at reducing aldehydes than respiratory sufficient cells. This was explained in that in fermentative cells the ADHI gene is fully expressed and the alcohol dehydrogenase coded for by this gene is the one responsible for aldehyde reduction. By implication the products of ADHII and ADHIII, the respiratory alcohol dehydrogenases, are not involved in aldehyde reduction.

In other reports the same authors have confirmed that alcohol dehydrogenase is active in reducing wort aldehydes, in particular pentanal and pentenal. In addition, several other enzyme activities have been detected. There were two NADPH reductases specific for 3-methylbutanal and one for pentanal. Aldo-ketoreductases with broad specificity were also detected (Colin *et al.*, 1991; Debourg *et al.*, 1993; Laurent *et al.*, 1995).

Acetaldehyde must be considered separately to other longer chain aldehydes because of its importance as an intermediate in the formation of ethanol or acetate. In some circumstances acetaldehyde may persist in beers at concentrations above its flavour threshold value of 10–20 ppm. In this case it produces an unpleasant 'grassy' flavour and aroma (Meilgaard, 1975).

Acetaldehyde formation in beer occurs in early to mid-fermentation during the period of active yeast growth. Later in the stationary phase acetaldehyde levels usually decline (Pessa, 1971; Geiger & Piendl, 1976). Accumulation of acetaldehyde is dependent on the kinetic properties of the enzymes responsible for its formation and dissimilation, pyruvate decarboxylase and acetaldehyde, and alcohol dehydrogenase, respectively. Alcohol dehydrogenase (ADH1), the fermentative enzyme, exhibits highest specific activity during the period of active primary fermentation. In late fermentation specific activity usually declines (Fig. 3.15).

Two acetaldehyde dehydrogenases are found in *S. cerevisiae*. One is mitochondrial,

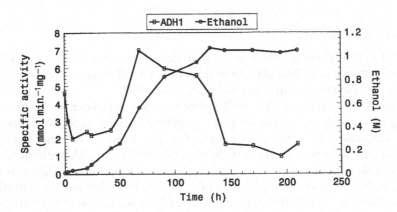

**Fig. 3.15** Specific activity of alcohol dehydrogenase (ADH1) and ethanol concentration during a high-gravity lager fermentation (W. Tessier, unpublished data).

activated by $K^+$ and $NAD^+$ or $NADP^+$-linked, and is reportedly implicated solely with oxidative growth (Jacobsen & Bernofsky, 1974). The other is cytosolic, activated by magnesium and $NADP^+$-linked (Dickinson, 1996). As discussed previously, acetaldehyde dehydrogenase probably forms the major route for generation of cytosolic acetyl-CoA in brewing yeast under the conditions of fermentation, suggesting that the cytosolic acetaldehyde dehydrogenase would be involved in this role. However, since this enzyme is $NADP^+$-linked it would not participate in the cycle of $NAD^+/NADH$ redox-balancing reactions in which pyruvate dehydrogenase is implicated.

There is evidence that the mitochondrial aldehyde dehydrogenase is active under brewery fermentation conditions (Fig. 3.16). The data presented shows the activity of the cytosolic and mitochondrial acetaldehyde dehydrogenases during the course of a high-gravity lager fermentation. Surprisingly, the cytosolic enzyme is apparently rapidly induced during the aerobic phase of fermentation, reaches a peak of activity in early fermentation and then declines to undetectable levels after some 130 hours. In contrast, the mitochondrial activity could be detected throughout the fermentation and would appear to be implicated with the fall in beer acetaldehyde concentration which is characteristic of the later stages of fermentation. This again serves to emphasise the common theme that certain mitochondrial functions are essential under conditions of repression and anaerobiosis.

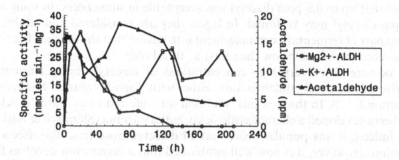

**Fig. 3.16** Specific activities of the cytosolic (Mg2+-ALDH) and mitochondrial (K+-ALDH) acetaldehyde dehydrogenases, together with the concentration of acetaldehyde in beer during a high-gravity lager fermentation (W. Tessier, unpublished data).

High levels of acetaldehyde in beer at the end of fermentation are associated with non-standard performance. This could take the form of premature separation of yeast from wort due to induced changes in yeast flocculence, such that metabolism of the pre-formed acetaldehyde pool is restricted. It may be indicative of a loss of fermentation control. Use of too high a temperature, over-oxygenation of worts and high pitching rates are all reported to result in high beer acetaldehyde levels (Geiger & Piendl, 1976). More commonly it is indicative of poor yeast quality. In the latter case it is not easy to ascribe simple cause and effect. Thus, a failure to provide adequate control of fermentation conditions may have stressed the yeast such that acetaldehyde metabolism is impaired and in consequence it may accumulate to abnormally high concentrations in beer.

Conversely, it may be the formation of acetaldehyde that provides the initial stress to the yeast. Jones (1989) has argued that acetaldehyde is highly toxic to cells. Thus its ability to form Schiff bases with amino residues can lead to inactivation of enzymes and disruption of synthesis of proteins and nucleic acids. These effects may be manifested as disruption of cell growth, cell death or mutagenesis. This has led Jones (1989) to suggest that acetaldehyde may account for some of the toxic effects ascribed to ethanol. Furthermore, toxic effects of acetaldehyde may explain why ethanol generated intracellularly is apparently more harmful than ethanol added extra-cellularly at the same concentration (Section 3.6.2). Stanley and Pamment (1993) concluded that the toxic effects of acetaldehyde might be magnified since cells accumulated this metabolite. Thus ethanol was shown to exit from yeast more rapidly than acetaldehyde.

3.7.4.1 *Vicinal diketones.* Vicinal diketones (VDK) have progressively higher flavour threshold concentrations with increase in molecular weight (Meilgaard, 1975). Most are considered to have unpleasant flavours. With respect to beer the two most important members of this group are diacetyl (2,3-butanedione) and 2,3-pentanedione. Both have the flavour and aroma of butterscotch but the threshold concentration of diacetyl is almost ten times lower than 2,3-pentanedione, 0.15 ppm and 0.9 ppm respectively (Mielgaard, 1975). In some beers, for example UK top-fermented ales, VDK contributes to the overall palate and aroma. Wainwright (1973) reported that up to 0.6 ppm diacetyl was acceptable in some beers. In some wines up to 8 ppm diacetyl may be found. In lagers they are considered undesirable and an essential part of fermentation management is to ensure that the finished beer contains VDK at concentrations below their flavour thresholds.

The occurrence of very high concentrations of diacetyl during fermentation is indicative of microbial contamination, either with *Lactobacillus* or *Pediococcus* spp (see Section 8.1.2). In the past this condition was referred to as 'Sarcina sickness' in which beers developed a characteristic sickly buttery aroma (Shimwell & Kirkpatrick, 1939). Indeed, it was popular belief that all diacetyl arose from the effects of con-tamination. However, it is now well established that a proportion devolves from the activity of brewing yeast and is part of the normal fermentation process. Peak con-centrations of diacetyl and diacetyl precursor (α-acetolactate) produced by yeast during the course of fermentation are within the range 1–5 ppm, depending on the type of beer being produced.

Diacetyl and 2,3-pentanedione arise as an indirect result of yeast metabolism. It is generally accepted that the pathway for the formation of VDK and their subsequent dissimilation is that shown in Fig. 3.17. The precursors are α-acetohydroxy acids, which are intermediates in the pathways in the biosynthesis of valine and isoleucine. During the early to middle parts of primary fermentation some of the intracellular pools of these α-acetohydroxy acids are excreted into the fermenting wort where they undergo spontaneous oxidative decarboxylation to form diacetyl and 2,3-pentanedione. From middle to late fermentation extracellular VDK is metabolised by yeast cell reductases to form acetoin and 2,3-butanediol from diacetyl and 2,3-pentanediol from 2,3-pentandione. These metabolites persist in beers but since they are much less flavour-active than VDK, their presence can be tolerated.

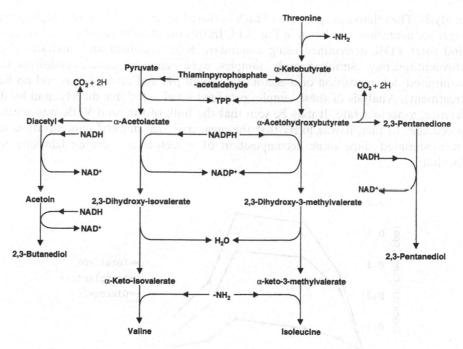

**Fig. 3.17** Pathways leading to the formation and dissimilation of the vicinal diketones, diacetyl and 2,3-pentanedione.

An alternative synthetic route has been suggested (Chuang & Collins, 1968, 1972). As a result of studies with radiolabelled substrates these authors concluded that in *S. cerevisiae*, acetoin and diacetyl arose from condensation reactions between hydroxyethylamine pyrophosphate and acetaldehyde and hydroxyethylamine pyrophosphate and acetyl-CoA, respectively. Although the evidence supported the existence of this pathway the majority of work performed by others suggests that it is likely to be of minor importance and that most diacetyl arises via the route depicted in Fig. 3.17. Thus, all the enzymes in the pathway have been identified and characterised. Mutant strains lacking acetohydroxy acid synthase, and which therefore cannot generate acetolactate, do not form diacetyl. Furthermore, supplementation of non-growth media with valine or α-ketoisovaletate does not increase diacetyl formation, indicating that the reverse pathway is apparently not operative (Wainwright, 1973).

The biochemistry of VDK and the influence of fermentation conditions can be considered in three phases: steps leading to the formation of α-acetohydroxy acids, factors affecting the spontaneous decarboxylation of α-acetohydroxy acids and the subsequent reduction of VDK. As discussed later in this section, the evidence suggests that under most circumstances the rate-determining step for VDK formation is the non-enzymic decarboxylation of α-acetohydroxy acids. Furthermore, the subsequent yeast catalysed reduction of VDK is usually very rapid. Therefore, analysis of fermenting worts is usually referred to as 'total VDK' and represents the sum of α-acetohydroxy acids and diacetyl. Practically, this requires subjecting samples to a heat treatment, which ensures that all α-acetohydroxy acids are converted to VDK prior to analysis. The relative proportions of each of these formed in a laboratory high-gravity lager fermentation are shown in Fig. 3.18. In this experiment samples were removed and total VDK determined using a standard heat treatment and analysis by gas chromatography. Simultaneously samples were removed under conditions that minimised decomposition of α-acetolactate (high pH and anaerobiosis and no heat treatment). Analysis of these samples provided a value for free diacetyl and by difference, α-acetolactate. It may be seen that the bulk of the total VDK was actually precursor. In fact, it was likely that the measured free diacetyl concentration was over-estimated since some decomposition of α-acetolactate during handling was inevitable.

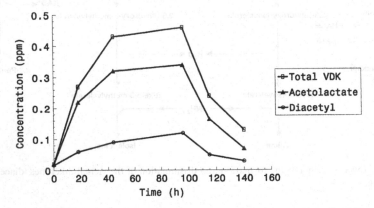

**Fig. 3.18** Proportions of diacetyl and α-acetolactate formed during the course of a 2-litre stirred high-gravity (1060) wort fermentation (Box & Boulton, unpublished data).

The formation of α-acetohydroxy acids is intimately related to amino acid metabolism. In this respect the total wort amino acid content and amino acid spectrum are influential, as are parameters which regulate yeast growth and by inference effect amino acid utilisation. Nakatani *et al.* (1984a) presented data from a large number of fermentations performed under a variety of conditions. They concluded that at high levels of wort free amino nitrogen (FAN) the VDK profile took the form of an extended peak, which appeared early in fermentation. At low wort FAN levels, the profile was transformed into two peaks. The first was small and coincided with the appearance of the broad single peak seen in the high FAN condition, the second peak was sharper and larger. It was demonstrated that a correlation existed between the maximum achieved total diacetyl concentration (T-VDK$_{max}$) and the minimum value of wort free amino nitrogen (FAN$_{min}$). The hyperbolic relationship was described by the following equation:

$$T\text{-VDK}_{max} = \frac{0.161}{FAN_{min} - 3.87} + 0.415$$

From this, it was concluded that in order to minimise the magnitude of the VDK peak it was necessary to provide a wort with sufficient FAN to ensure a given minimum concentration at the end of fermentation. Since FAN utilisation and yeast growth are positively correlated it was also possible to demonstrate relationships between initial fermentation conditions and the size of the VDK peak. Thus, high wort oxygen concentrations, trub levels, pitching rates and fermentation temperature, all of which favour rapid and extensive yeast growth, also promoted high VDK levels, presumably due to increased FAN utilisation.

In subsequent work (Nakatani et al., 1984b) the same authors confirmed the results of others (Wainwright, 1973), that the valine and isoleucine contents of wort exert specific effects on the formation of α-acetohydroxy acids. Valine and isoleucine inhibit the pathways leading to their biosynthesis by inhibition of the α-acetohydroxy synthetase enzyme and this removes the precursors of diacetyl and 2,3-pentanedione formation. These amino acids are assimilated in the early to middle phase of fermentation and this suppressive effect accounts for the delayed onset of VDK formation. With high wort FAN levels the suppressing effect persists longer because of the increased availability of valine and isoleucine. With low wort FAN levels the requirement for amino acid biosynthesis in middle to late fermentation explains the late large peak of VDK formation.

Amino acids other than valine and isoleucine also modulate the patterns of VDK formation. Barton and Slaughter (1992) concluded that supplementation with valine, leucine, alanine, threonine, glutamate and, surprisingly, ammonium chloride, all decreased VDK concentration in fermentation. These effects were show to be due to inhibition of α-acetohydroxy synthetase. Presumably these effects relate to the complexities of control of amino acid uptake and metabolism in yeast cells. It may be concluded that both total and spectrum of wort FAN affects VDK formation. Since there is little practical control available over FAN spectrum this observation is likely to be of academic but not practical interest.

The non-enzymic oxidative decarboxylation of α- acetohydroxy acids occurs in the wort and (with some caveats, as discussed subsequently) does not require the activity of yeast. Decomposition of α-acetohydroxy acids is rapid under aerobic conditions but the process proceeds under anaerobiosis where copper, ferric and aluminium ions may function as alternative electron acceptors (Inoue et al., 1968a, b). The reactions are also favoured by acidic pH values. Fermentation conditions that produce rapid yeast growth are also associated with a rapid fall in pH and this promotes VDK formation (Inoue et al., 1991). Inoue (1992) reported that non-oxidative decarboxylation of α-acetolactate to acetoin may also occur under appropriate conditions. This may be encouraged by heating at 60 to 70°C under strictly anaerobic conditions but it is suggested that it may also occur at fermentation temperatures providing a sufficiently low redox state pertains. These authors contended that maintenance of low redox during warm conditioning to enhance the proportion of acetolactate non-oxidatively decarboxylated to acetoin could be another essential role of yeast cells in VDK metabolism.

Andries et al. (1997) described a process in which the spontaneous non-oxidative decarboxylation of α-acetolactate to acetoin could be encouraged by passage of green beer through a column containing zeolite resin by a mechanism which was not described. Several different grades of zeolite were tested with very variable results. Conversion rates were temperature dependent; up to 90% at a temperature of 20°C but only 2% at 5°C. The process was reportedly not affected by the presence of oxygen or the redox state of the beer although no results were provided to support this. No other effects on beer analysis were observed and this appears to be a promising approach for use at production scale.

The reduction of VDK occurs in the late stages of fermentation or during conditioning (see Chapter 6) and requires the presence of viable yeast. The enzymology of the reactions is not well characterised. Hardwick *et al.* (1976) demonstrated that yeast alcohol dehydrogenase (ADH1) reduced methyl glyoxal, diacetyl and 2,3-pentanedione but at much lower rates than acetaldehyde. No activity with acetoin was observed and this suggested that other specific enzymes might be involved. Several authors have reported the presence of specific diacetyl reductases in yeast (Louis-Eugene *et al.*, 1988; Legeay *et al.*, 1989; Heidlas & Tressl, 1990; Schwarz & Hang, 1994). These enzymes were either NADH or NADPH dependent. In a later study, Murphy *et al.* (1996) investigated diacetyl reductases in a number of strains of brewing yeasts. A number of different enzymes were detected on the basis of differing thermal lability curves in extracts and mobilities on polyacrylamide gel electrophoresis. Two groups of yeasts could be differentiated on the basis of patterns of dehydrogenases present. Yeasts in one group were primarily bottom fermenting ale types and the other were lager varieties. The lager strains all contained a heat stable acetoin reductase as well as alcohol dehydrogenases which showed no activity with acetoin. The ale types lacked the same acetoin reductase activity but did possess another dehydrogenase which was active with acetoin and diacetyl. These strains also had alcohol dehydrogenases activity different to the lager strains.

Yeast cells are usually capable of rapid diacetyl reduction. The data presented in Fig. 3.19a–e shows the effect of addition of exogenous diacetyl to stirred laboratory high-gravity lager fermentations. In this trial, five identical fermentations were performed. Samples were removed at appropriate intervals and analysed for total VDK using a gas chromatographic procedure. To all but the control fermentation (Fig. 3.19a) diacetyl (1 ppm) was added at the times indicated by the sudden appearance of a peak in the VDK profiles. As may be seen in all cases, the yeast was capable of rapidly reducing the added diacetyl. However, the profiles suggested that there was a progressive decrease in the ability to reduce added diacetyl as a function of time (Fig. 3.19b–d). Interestingly the yeast was capable of reducing exogenous diacetyl during the early part of fermentation where the metabolism was still directed towards production of endogenous VDK (Fig. 3.19e). These data confirm that diacetyl reduction is not the rate-limiting step in the VDK cycle although the apparent decline in activity with time would suggest that the yeast's physiological condition may be of some relevance to diacetyl reduction. In this respect the role of membrane competence could be of importance. It must be assumed that prior to reduction the diacetyl is taken up by the yeast. Since membrane condition will deteriorate during the course of fermentation due to sterol depletion this could in part explain the decline in reduction rates seen in late fermentation. However, it is also entirely possible that there may be a diminution in the activity of the diacetyl reductases during fermentation in a manner analogous to the decline in activity of alcohol dehydrogenase (Fig. 3.15).

An essential aspect of the warm conditioning phase of lager fermentations, if practised, is the necessity to achieve a minimum total VDK concentration (see Section

6.5.3). It is noteworthy that the time taken to achieve this minimum VDK concentration does not necessarily correlate with the magnitude of the peak value. Frequently it may be observed that there is a decline in the rate of decrease in VDK concentration in late fermentation. The inflexion point between rapid and slow decrease in VDK concentration usually occurs near to the threshold value at which cooling is applied to terminate the fermentation. Variability in this slow phase of VDK uptake can be a major cause of inconsistency in vessel turn-round time.

This is illustrated in Fig. 3.20, which shows the terminal phase of decline in total VDK for a number of pilot scale lager fermentations. Fermentation conditions were identical apart from a different batch of wort being used in each case. In the case of these fermentations the specified maximum VDK concentration at which cooling could be applied was 0.1 ppm. As may be seen the most rapid fermentation achieved the VDK specification in approximately 150 hours, whereas the slowest took more than 220 hours. It must be emphasised that the observed variability was not due to differences in suspended yeast count since similar biphasic profiles were also seen with small scale stirred fermentations. Furthermore, the wort gravity attenuation profiles were essentially identical in all cases (data not shown). The reasons for this variability are unclear but appeared to be related to differences in wort composition.

From the previous discussion it is apparent that within the confines of normal fermentation practice little may be done to accelerate VDK removal or to improve the consistency of VDK metabolism other than to apply rigorous control to all aspects of wort production and fermentation management. Other more positive approaches to accelerating VDK removal or reducing its formation have been proposed. Rapid VDK removal using continuous maturation processes is described in Section 5.7.2.

The possibility of spontaneous non-oxidative decarboxylation of α-acetolactate directly to acetoin occurring during fermentation has been discussed previously. In some bacteria this reaction is catalysed by an enzyme, α-acetolactate decarboxylase. Preparations of the bacterial enzyme are available commercially for use as process aids in fermentation (Aschengreen & Jepsen, 1992). Addition of this enzyme to wort at the start of fermentation was shown to reduce diacetyl formation to the extent that it abolished the requirement for a conventional VDK rest. No other effects on fermentation performance or beer quality were reported. A modification to this method was suggested by Dulieu *et al.* (1997) in which the α-acetolactate decarboxylase preparation was encapsulated in a polymer made from a combination of sodium cellulose sulphate and poly (diallyl dimethyl ammonium) chloride. The claimed advantage was that it would be possible to retain the enzyme within a bioreactor column, which unlike the free enzyme would lend itself to multiple use.

Several reports describe attempts to eliminate or reduce VDK stand-times by genetic modification of brewing yeast strains (see Section 4.3.4). The gene coding for the enzyme α-acetolactate decarboxylase has been cloned from certain bacterial species – *Klebsiella terrigena, Enterobacter aerogenes* and *Acetobacter aceti* – and transformed into brewing yeast (Sone *et al.*, 1987, 1988a, b; Suihko *et al.*, 1989; Shimizu *et al.*, 1989; Blomqvist *et al.*, 1991; Tada *et al.*, 1995). In the latest of these studies it is reported that the gene has been integrated into the yeast chromosome to produce a stable transformant in which α-acetolactate decarboxylase expression is regulated by the promoters for the yeast phosphoglycerokinase (PGK1) or alcohol

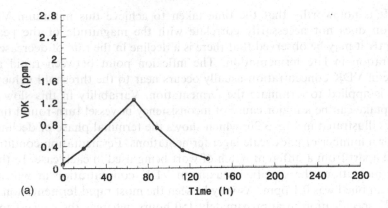

(a)

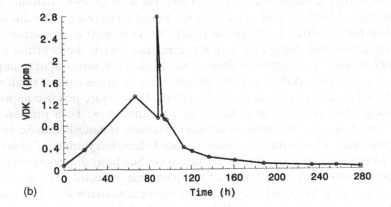

(b)

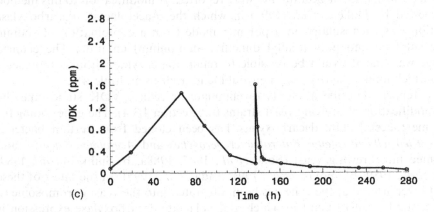

(c)

**Fig. 3.19** *Contd on next page.*

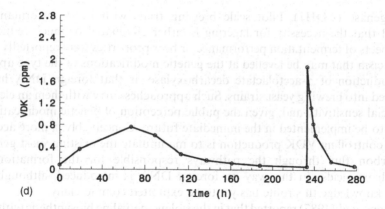

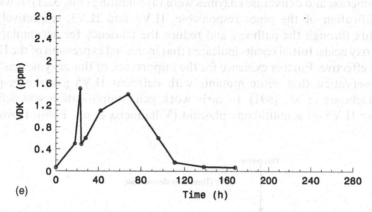

**Fig. 3.19** Effect of addition of exogenous diacetyl to VDK profiles of 2-litre stirred laboratory high-gravity wort fermentations. Five identical fermentations were performed and 1 ppm diacetyl was added to fermentations (b) to (e) at the times indicated by the appearance of the transient peak (Box & Boulton, unpublished data).

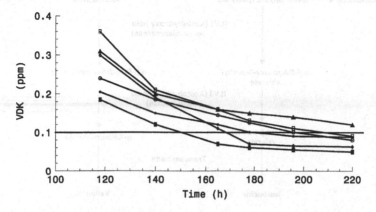

**Fig. 3.20** End profiles of VDK concentration for a number of pilot scale (8 hl) lager fermentations – see text for explanation (Besford & Boulton, unpublished data).

dehydrogenase (ADH1). Pilot scale brewing trials with the transformants have indicated that the necessity for lagering is either eliminated or much reduced. No other aspects of fermentation performance or beer properties were reportedly altered.

A criticism that may be levelled at the genetic modifications of the type applied to the introduction of α-acetolactate decarboxylase, is that foreign DNA has been introduced into brewing yeast strains. Such approaches carry with them an element of commercial sensitivity and, given the public perception of genetic modification, it is unlikely to be implemented in the immediate future. Conceivably, a more acceptable route to controlling VDK production is to manipulate the existing yeast genome to alter carbon flux through the pathways responsible for the formation of α-acetohydroxy acids. In this way no foreign DNA is introduced, although to the authors' knowledge this route has yet to be exploited commercially.

Goossens *et al.* (1987) reported that in the isoleucine/valine biosynthetic pathway the reductoisomerase and dehydrase enzymes were rate-limiting (Fig. 3.21). It was argued that amplification of the genes responsible, ILV5 and ILV3, respectively, would increase flux through the pathway and reduce the tendency for accumulation of α-acetohydroxy acids. Initial results indicated that increased expression of the ILV5 gene were most effective. Further evidence for the importance of this enzyme was provided by the observation that petite mutants with deficient ILV5 genes over-produced diacetyl (Debourg *et al.*, 1991). In early work gene amplification was achieved by introducing ILV5 on a multi-copy plasmid (Villanueba *et al.*, 1990). However, this

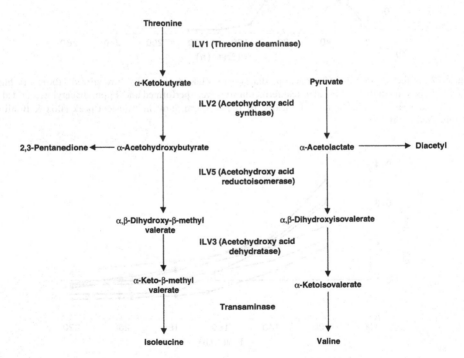

**Fig. 3.21**  Activities corresponding to the ILV genes and their relationship to amino acid and VDK formation.

proved unstable and in order to overcome this problem transformants were produced in which the gene was integrated into the chromosome and controlled by the constitutive GDP2 (glyceraldehyde 3-phosphate) promoter (Goossens *et al.*, 1991, 1993). Such transformants were stable and in brewing trials reduced diacetyl formation by 50% compared to wild-type strains. No adverse effects on beer quality were reported.

No metabolic functions for the VDK pathway have been proposed. However, the convoluted pathway of excretion of α-acetohydroxy acids and subsequent assimilation of VDK suggests that some metabolic role must be being fulfilled. As with the formation of higher alcohols the only obvious function would be yet another route for maintenance of cellular redox using the NAD(P)H-linked dehydrogenases involved in the terminal steps of VDK reduction.

### 3.7.5 *Sulphur compounds*

A number of sulphur containing compounds, both inorganic and organic, contribute both directly and indirectly to beer flavour. Many of these derive directly from wort and persist unchanged in beers but some are influenced by yeast metabolism. Those of particular significance, which are influenced by yeast metabolism during fermentation, are hydrogen sulphide ($H_2S$) and sulphur dioxide ($SO_2$). At low concentrations, these may make a positive contribution to beer but at higher levels they may impart undesirable tastes and aromas. For example, a low but detectable concentration of $H_2S$ is acceptable and indeed characteristic of some top-fermented ales; however, at high concentrations it would be considered a most undesirable flavour defect.

Indirect flavour effects may arise from the ability of sulphur-containing compounds such as $SO_2$ to form reversible adducts with carbonyl compounds. In this way, for example, high sulphite concentrations, which may arise during fermentation, can stabilise beer flavour by binding compounds associated with beer flavour staling such as acetaldehyde and trans-2-nonenal. In addition, sulphite acts as a natural anti-oxidant. In general there is a positive correlation between the quantity of sulphite formed during fermentation and the concentration of the wort. In low alcohol beers, which may be produced by fermentation of dilute worts or by partial fermentation of normal worts, there may be an insufficiency of sulphite formed and this may affect adversely the staling potential of such beers. Although the shortfall may be corrected for by addition of sulphite to packaged beers, this represents an additional expense and requires specific labelling in some countries. This has provided the impetus for several groups to perform research work into the biochemical and genetic basis of sulphite formation so that methods of maximising the formation of this metabolite can be identified.

Dufour (1991) discussed the significance of carbonyl-sulphite adduct formation during fermentation. This author pointed out that the degree of binding of each type of carbonyl would be a function of the magnitude of the adduct's equilibrium constants. In this regard, the highest affinities would be shown for acetaldehyde, whereas the vicinal diketones would have medium affinities. It was concluded that during fermentation the rate of sulphite formation would regulate the proportion of carbonyls bound as adducts and that fraction reduced by yeast. Carbonyl adducts persisting in beer are flavour negative; however, under some circumstances they may be released to exert their staling effects. The total concentration of sulphite present will correlate with flavour stability. Thus, if sulphite concentrations are low carbonyls can be released due to competitive irreversible reactions of sulphite with other beer components such as polyphenols and quinones.

As described in Clarke *et al*. (1991), during the early part of fermentation $H_2S$ accumulates and then declines during the later stages. Under some circumstances, the phase of decline may not occur and indeed a further increase is possible, resulting in undesirable levels of this metabolite persisting in the beer after fermentation is completed. The quantity of $H_2S$ formed during fermentation is much influenced by the yeast strain (Romano & Suzzi, 1992). However, excessive concentrations may arise through deficiencies in wort composition, inappropriate control of fermentation or use of yeast with stressed physiology.

The influence of wort composition on the formation of $H_2S$ and other sulphur volatiles is discussed later in this section. Considerable $H_2S$ may be lost during active fermentation because of the effects of gas stripping. Should the vigour of fermentation be lessened for any reason, there will be a concomitant reduction in gas purging effects. Such a situation can arise, for example, due to a failure to provide adequate wort oxygenation. Sticking fermentations may also be a consequence of the use of pitching yeast that is in poor condition. In this event $H_2S$ may arise as a result of yeast death and autolysis with subsequent degradation of sulphur-containing amino acids.

Traditionally, excessive $H_2S$ was controlled by treatment with copper sulphate solution, thereby forming a precipitate of insoluble copper sulphide. This approach must be treated with caution to avoid poisoning yeast. In addition, use of such potentially dangerous additives is now rightly frowned upon. Interestingly, replacement of copper fermenting vessels with stainless steel types has been associated with an increase in beer $H_2S$ (R. Wharton, personal communication). King *et al*., (1990) reported that supplementation of worts with pantothenate (0.01 ppm) suppressed $H_2S$ formation. Conversely, addition of certain amino acids, particularly serine, to cask beers during secondary fermentation was effective at reducing sulphury odours. Supplementation of worts with pantothenate also decreases concentrations of $SO_2$ and acetaldehyde during fermentation (Lodolo *et al*., 1995). These authors suggested that this would improve flavour stability since acetaldehyde binds to $SO_2$ in preference to more flavour staling longer-chain aldehydes. It was speculated that for some wort and yeast combinations there is a deficiency of pantothenate which may lead to a shortfall of coenzyme A and this results in overproduction of acetaldehyde.

$SO_2$ formation during fermentation is influenced principally by wort composition Dufour (1991). Dufour *et al*. (1989) demonstrated that a positive correlation existed between sulphite concentration and wort OG. Conversely, increase in wort oxygenation, or elevated wort lipid (in the form of trub) was associated with reduced $SO_2$. The yeast's physiological condition was also shown to be influential. In this case it was demonstrated that starvation of yeast prior to pitching was associated with increased $SO_2$ accumulation. Thus, a positive correlation between starvation time and $SO_2$ concentration formed during fermentation was reported.

These relationships can be explained in terms of yeast growth extent during fermentation and how this influences assimilation of wort nutrients, in particular amino acids and sources of inorganic sulphur such as sulphate. Biosynthesis of $SO_2$ from sulphate during fermentation requires metabolic energy, and therefore a source of fermentable sugar is needed. It follows that more concentrated worts offer the possibility of increased accumulation. As a general rule, amino acids inhibit the biosynthetic route from sulphate, and therefore the ratio of fermentable carbohydrate to amino nitrogen will exert a modulating effect. However, products of carbohydrate catabolism and specific amino acids exert additional specific effects, as discussed subsequently, adding further levels of complexity. Factors which influence the extent of yeast growth and thus utilisation of amino nitrogen will also influence the formation of $SO_2$, hence the effects of increasing wort oxygenation and lipids. Conversely, yeast in poor physiological condition grows to a limited extent during fermentation and amino acid assimilation will also be affected.

Beer sulphur-containing components, which are influenced by yeast metabolism, can arise from two routes. First, from the dissimilation of complex organic molecules such as sulphur-containing amino acids and vitamins and, second, from assimilatory reactions involving inorganic sulphur-containing nutrients. The pathway of assimilation of sulphate, its reduction and incorporation into sulphur-containing amino acids is shown in Fig. 3.22. Sulphate enters the cell via a specific permease and is reduced to sulphite and then sulphide in a sequence of energy-requiring reactions. Sulphide is incorporated into the sulphur-containing amino acids cysteine, methionine and S-adenosylmethionine.

Regulation of sulphur metabolism is complex involving feedback inhibition of enzyme activity and repression of gene expression. In particular, S-adenosylmethionine represses transcription of all the enzymes involved in sulphate uptake and reduction to sulphide, as well as those which catalyse S-adenosylmethionine synthesis. Thus, growth in the presence of high concentrations of methionine inhibits sulphite production. The presence of threonine increases sulphite formation, reportedly by feedback inhibition of aspartokinase such that there is a depletion of the pool of O-acetylhomoserine, and hence methionine levels fall and relieve the repression effects exerted by this metabolite. Isoleucine brings about feedback inhibition of threonine utilisation and hence has the same outcome as the presence of threonine alone (Gyllang et al., 1989).

Korch et al. (1991) demonstrated a correlation between sulphite production and concentration of glucose in wort. Since there was a concomitant increase in ethanol and acetaldehyde these authors proposed that an intermediate of glucose catabolism could be implicated. They suggested that pyruvate and acetaldehyde formed adducts with sulphite and this deprived the methionine synthetic pathway of sulphite. As a result the sulphite synthetic pathway becomes increasingly derepressed.

Dufour (1991) proposed four stages of sulphite production. In the first during very early fermentation the presence of high levels of methionine and threonine repress the sulphite synthetic pathway. In the second phase which equates to the period of active yeast growth the decline of wort methionine and threonine derepresses the sulphite synthetic pathway but the pool so formed is fully utilised for the synthesis of sulphur-containing amino acid required for biomass formation. Hence, no extracellular sulphite accumulates. In the third phase (mid to late fermentation), yeast growth ceases, the sulphite reductase declines to a low level but the continued availability of fermentable sugar allows sulphite biosynthesis to continue unabated and extracellular

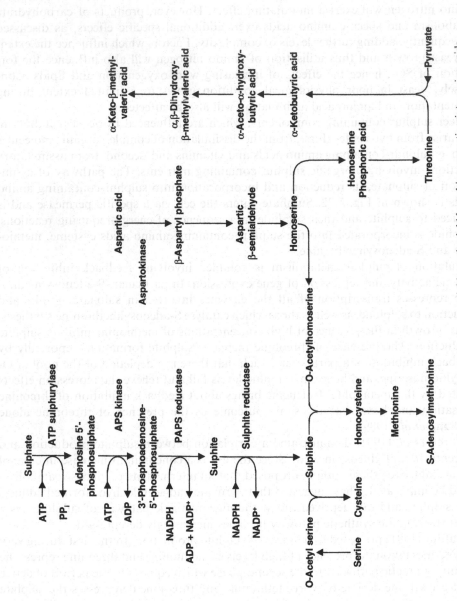

Fig. 3.22  Pathways of assimilation of sulphate, reduction to sulphite and sulphide and incorporation into amino acids.

accumulation occurs. In the fourth phase, sulphite synthesis and accumulation ceases due to depletion of the carbohydrate source.

Several reports (Dufour et al., 1989; Korch et al., 1991; Hansen & Kielland-Brandt, 1995) describe genetic approaches to produce yeast transformants capable of increased sulphite formation. Although details differ, the aim is to produce strains in which the genome has been modified so that the suppressing effects of amino acids on the sulphite producing pathway are relieved. An essential aspect of these methods is to ensure that over-production of sulphite is not accompanied by an equal surfeit of $H_2S$. Fuller details of these experimental approaches are provided in Section 4.3.4.

Another important sulphur-containing beer flavour component is dimethyl sulphide (DMS). At high concentrations this compound has an unpleasant flavour and aroma described as 'cooked sweet-corn' or 'cooked vegetable'. In top-fermented ales its concentration is not usually above its flavour threshold of approximately 30 ppb (Harrison & Collins, 1968) and it is not important in these beers. Higher concentrations are found in lager beers. Here within the range 30–100 ppb, it is considered an essential flavour component contributing to the distinctive flavour and aroma of lagers. At concentrations above 100 ppb it is considered objectionable (Anness & Bamforth, 1982).

DMS arises in beer via two routes. First, from S-methylmethionine (SMM) which decomposes to DMS on heating and, second, via the reduction of dimethyl sulphoxide (DMSO). The latter reaction is catalysed by yeast during fermentation. The common precursor of all DMS arising in beer is SMM and this derives from green malt. Both DMS and DMSO are formed during the various steps involved in the conversion of green malt to finished malt. The proportions of each are dependent on the conditions used for each step in particular, the temperature of kilning. Thus, significant quantities of DMSO are only formed where a kilning temperature greater than 60°C is used (Anness et al., 1979; Parsons et al., 1977).

DMS is volatile and consequently most of that present in malt is lost during mashing and wort boiling. Conversely, DMSO is heat stable and persists unchanged through the mashing and wort boiling stages. SMM is converted to DMS during the wort boil. Much of this is also lost; however, thermal decomposition of SMM continues during the copper cast and throughout cooling, such that wort in fermenter contains a mixture of SMM, DMS and DMSO. The proportions of each vary depending on the raw materials, conditions of wort production and the nature of the brewery plant employed (Anness & Bamforth, 1982; Dickenson, 1983).

No further conversion of SMM to DMS occurs during fermentation since the temperature is too low for thermal decomposition. SMM is assimilated by yeast where it is converted to methionine. DMSO in wort is reduced to DMS by yeast during fermentation (Anness et al., 1979; Gibson et al., 1985). In a second paper (Gibson et al., 1985), the latter authors demonstrated that the specific activity of dimethyl sulphoxide reductase increased when yeast growth had stopped. The increase in specific activity was associated with nitrogen limitation. In trials using defined media it was demonstrated that DMSO reduction was most rapid in cultures where methionine was the limiting nitrogen source. Furthermore, in experiments using 14[C]methionine, radiolabelled methionine sulphoxide was identified in cells.

From these results it was concluded that DMSO reductase was subject to nitrogen catabolite repression and that it was involved in the uptake of methionine.

Anness and Bamforth (1982) suggested that the reduction of DMSO by yeast is performed by methionine sulphoxide reductase. This complex enzyme utilises thioredoxin and thioreoxin reductase to transfer electrons from NADPH. These authors conclude that reduction of DMSO may be a fortuitous side reaction.

Anness and Bamforth (1982) summarised the effects of fermentation variables other than wort amino nitrogen on DMSO reduction by yeast. Strains designated as *S. cerevisiae* were generally more effective than strains of *S. uvarum* (presumably 'lager' strains). Low temperatures favour formation of DMS by yeast and this may partly explain why levels in lagers are generally higher than ales. Wort concentration and DMS formation by yeast are positively correlated. The relationship is not linear and at very high gravities there is a disproportionate increase in the yield of DMS. Formation of DMS during fermentation is favoured by high pH. High capacity deep fermenting vessels are associated with high DMS levels.

The multiplicity of factors both in wort production and during the fermentation stage allows for great variability in the relative importance of SMM and DMSO as the immediate sources of beer DMS. Thus, in many fermentations the yeast DMSO reductase route appears to be negligible, for example, Dickenson (1983), whereas others have reached a totally opposite conclusion (Leemans *et al.*, 1993).

### 3.7.6 Shock excretion

When yeast is suspended in water it responds by exuding protons. The effect is magnified by the presence of a fermentable sugar (Sigler & Hofer, 1991). The ability to acidify the external medium is utilised in a rapid method for assessing brewing yeast quality, the acidification power test (see Section 7.4.2.2).

It has been observed that under such conditions brewing yeast also excretes a variety of metabolites. This phenomenon has been christened 'shock excretion' (Lewis & Phaff, 1964, 1965). These authors showed that the bulk of the released metabolites were amino acids, derived from the intracellular pool, and a variety of other compounds which absorbed at 260 nm. The process occurred over a period of 1–2 hours and required the presence of exogenous sugar. Glucose was the most effective sugar although maltose could substitute providing the yeast had been grown in the presence of this sugar. Within 4 hours most of the excreted amino acids were re-absorbed; however, other metabolites, including nucleotides, remained in the medium. Differences in the patterns of excretion and re-absorption were observed with individual yeast strains.

It must be assumed that this phenomenon may also occur when yeast is pitched into wort. Presumably, it may reflect a temporary depolarisation of the membrane allowing release of some cell constituents followed by a period of re-equilibration in which some of the released material is again assimilated. It is possible, however, that some of the material that is not re-absorbed, in particular, the nucleotide fraction, might make a subtle contribution to beer flavour.

# 4 Brewing yeast

## 4.1 Morphology, cytology and cellular function

It has been said that yeast is known more for what it can do than what it is (Robinow & Johnson, 1991). Until comparatively recently this was certainly the case and, thus, early brewers had by and large to be content with gauging yeast condition on the basis of observed relationships between behaviour and macro-morphological appearance. For example, assessing fermentation progress based on the appearance of the yeast head in an open square fermenter. In this sense yeast was (and perhaps still is by some practitioners of a less sensitive nature) treated as a bulk ingredient rather than a population of individuals.

Nevertheless, brewers were also relatively early converts to recognising the value of direct observation of the appearance of individual yeast cells as evidenced by the fact that as early as the late nineteenth century a microscope was considered an essential piece of laboratory apparatus. For example, Lindner (1895) discussed the application of microscopy to the study of pure yeast culture, yeast morphology and the theory of infection. In a later publication (Hatch, 1936) stressed the value of microscopic examination of pitching yeast under bright field. He provided the description 'healthy, vigorous yeast will show cells uniform in size and shape with walls clearly defined but not thickened. Few or no granular cells should be present and the number of bacteria should not exceed 2 or 3% of the total number of yeast cells'.

Direct examination of yeast cells using a light microscope continues to provide a method for detecting morphological abnormalities in the population of yeast used in the brewery. For example, an abrupt shift in microscopic appearance which might suggest that a process failure or change in practice was having a deleterious effect on the yeast. Similarly it provides a means of detecting gross contamination with bacteria and possibly other yeast strains, providing the latter have sufficiently different morphology to render them distinguishable from the native population.

A detailed elucidation of cellular ultrastucture requires microscopic tools with a precision and discrimination greater than that afforded by simple optical methods. Several such methods have been developed and yeast has been used frequently in these studies as a convenient model eukaryotic cell. Similarly, the brewing industry has been at the forefront in adopting new approaches for examining yeast cell morphology and cytology. Thus, Mitchison *et al.* (1956) using interference microscopy measured changes in cell dry mass, volume and number in growing and dividing *Schizosaccharomyces pombé*. Royan and Subramaniam (1956) used phase contrast and dark field microscopy to investigate the appearance of the nucleus and vacuole in yeast cells growing in wort. In the same period reports of the use of the electron microscope begin to appear, for example Bradley (1956) described the use of a carbon replica shadowing method for visualising the surface of yeast cells.

Further refinements in light microscopy continue to provide ever more powerful tools for studying cell ultrastructure. For example, the use of fluorescence microscopy (Robinow, 1975; Pringle *et al*., 1989) and confocal microscopy (Bacalfao & Stelzer, 1989). The latter method allows the production of three-dimensional images of yeast cells using a laser scanning device in which the resultant fluorescence light intensity is detected via a photomultiplier. Coincident with advances in light microscopy have been the development of ever more elegant incarnations of the electron microscope. Some of these as applied to yeast cells are described by Kopp (1975) and Wright and Rine (1989).

The development of the methods outlined in the previous discussion has allowed a detailed picture of yeast ultrastructure to be drawn up. Although this has been a rewarding exercise in itself, it is, of course, all the more valuable to be able to link structure with function. In this regard the development of a battery of additional techniques for the isolation and fractionation of individual yeast cell organelles has been of great importance. Thus, disruption of yeast cells under carefully controlled conditions and the subsequent collection of individual organelles followed by appropriate analysis has allowed the intracellular location of a multitude of biochemical pathways to be identified. Such studies have demonstrated that compartmentalisation plays a key role in the regulation of metabolism of individual cells. Furthermore, changes in physiological condition of yeast cells may be accompanied by visible changes in cytology. In this respect the effects on yeast of catabolite repression, exposure to oxygen, starvation and cellular ageing are all pertinent to brewing.

Apart from the localisation of groups of pathways into individual cellular compartments there is strong evidence that metabolism is ordered both biochemically and spatially in a co-ordinated fashion within cellular compartments. Several studies suggest that glycolytic enzymes are physically associated with the cytoskeleton in an ordered fashion such that products and substrates of individual enzymes are channelled in a controlled manner through the pathway (Shearwin & Masters, 1990; Masters, 1992). Recently, Götz *et al*. (1999) provided evidence that changes at the enzyme level associated with shifts from glycolytic to gluconeogenic modes of metabolism, in yeast, could be mediated by the cytoskeleton.

Several review articles describe procedures for sub-cellular fractionation and collection of yeast organelles, for example, Walworth *et al*. (1989) and Lloyd and Cartledge (1991). More specific methods are described for isolating protoplasts (Kuo & Yamamoto, 1975); nuclei (Duffus, 1976); vacuoles (Wiemken, 1976); plasma membranes (Henschke & Rose, 1991) and mitochondria (Linnane & Lukins, 1976; Deters *et al*., 1978; Guérin, 1991).

As alluded to already, a considerable body of work is now available in the literature providing great detail of the relationships between yeast cell structure and function. Much of this work has been performed on brewing or closely related yeast strains. However, there is much less literature describing how the conditions to which yeast is exposed during brewing result in modifications to cellular morphology and ultrastucture. In particular the effects of serial fermentation with associated transient aerobiosis and growth followed by intermittent periods of storage under starvation conditions. Some of the effects of brewing on yeast morphology are summarised in subsequent sections of this chapter. In some cases a lack of published data has necessitated a degree of speculation.

### 4.1.1 Cell morphology

Lodder (1970) describes the cells of S. cerevisiae as being spheroidal, subglobose, ovoid, ellipsoidal or cylindrical to elongate, occurring singly, in pairs, occasionally in short chains or clusters. Cells may be grouped into three classes on the basis of size. A large type, 4.5–10.5 × 7.0–21.0 µm (microns); a small-cell type falling within the range 2.5–7.0 × 4.5–11 µm and an intermediate group with cells measuring 3.5–8.0 × 5.0–11.0 µm. Some yeasts may form filaments which may be up to 30 (m in length. Brewing yeast cells fit into any of these categories; however, they tend to be quite large cells, a consequence of polyploidy. Vágvölgyi et al. (1988) demonstrated the effect of ploidy on yeast cell size, thus, the mean diameters of hapoid, diploid and triploid S. cerevisiae cells being 4.2, 5.2 and 5.9 µm, respectively (see also Section 4.3.2.5). The relationship between volume and surface area for an ellipsoidal yeast cell (e.g. S. cerevisiae) has been given as 45 and 71 µm$^2$ for haploid and diploid cells with volumes of 29 and 55 µm$^3$, respectively (Hennaut et al., 1970). See Section 4.3.2.5 for a fuller discussion on ploidy and cell size.

The mean cell size of a particular yeast strain is not a constant but varies according to the stage in the growth cycle, the growth conditions and the age of the individual cell. Changes in cell size associated with age (see Section 4.3.3.4 for a full discussion on ageing.) were described by Woldringh et al. (1995) using a centrifugal elutriation system which allowed the collection of new-born daughter cells. These authors reported that the volume of these increased from a mean value of 17 µm$^3$ to 34 µm$^3$ after five generations and 81 µm$^3$ at 15 generations (see also Section 4.3.3.4). Hartwell and Unger (1977) concluded that the increase in cell volume occurred during the phase in the cell cycle when budding has finished. In fact, the mean cell volume decreases by approximately a third during the budding phase. Apart from volume changes, predictably, there are also cell cycle-associated oscillations in cell mass. Baldwin and Kubitschek (1984) demonstrated that the cellular density reached a peak during the mid-growth cycle and this was attributed to a loss of water with a con-comitant increase in dry mass. Such observations serve to illustrate the fact that caution should be exercised in interpreting experimental data, expressed as percentage dry mass, regarding fluctuation in the concentrations of individual cellular compo-nents during growth.

The growth conditions also influence cell size. Robinow and Johnson (1991) reviewed the effects of incubation temperature on cell size and reported a variable response depending on the yeast. Thus, most yeast types, probably including Sac-charomyces, showed a temperature dependent increase in cell size. However, Schizosaccharomyces pombé cells increased in size at temperatures higher and lower than the optimum. Brewing yeast cells growing fermentatively on maltose are sig-nificantly bigger than the same cells growing oxidatively on ethanol (C. A. Boulton, unpublished data).

Short-term perturbations in cell size may also occur, presumably as a result of osmotic effects, when yeast is transferred between different media. For example, the effects of pitching yeast suspended in barm ale into wort. Thus, Quain (1988) observed that in the first 4 hours after pitching yeast, at laboratory scale, into wort there was a transient increase in cell size. The mean cell volume increased from approximately 170 $\mu m^3$ to 200 $\mu m^3$. The volume changes occurred before the onset of budding and cell proliferation and were independent of cell dry weight. These results suggest that the degree of turgor of the plasma membrane exerts an influence on cell size and that the cell wall is sufficiently flexible to accommodate such short-term fluctuations.

Changes in cell size over longer periods have also been recorded. Cahill *et al.* (1999b) reported the application of image analytical techniques as a method for improving the control of pitching rate. The apparatus was used to monitor cell size of stored pitching yeast. It was observed that for both ale and lager yeast the mean cell volume reduced by 19% (302 to 244 $\mu m^3$) and 7% (208 to 194 $\mu m^3$), respectively during storage over a period of 14 days at 4°C. The changes in mean cell volume were correlated with glycogen content, the latter being utilised for maintenance energy during storage.

A diagrammatic representation of a section through an idealised yeast cell indicating the major organelles is shown in Fig. 4.1. Not all of these features are visible within the cell at all times; indeed some may not be present in yeast at all, as discussed later. A brief discussion of the structure and function of some of these organelles follows. The description is not comprehensive; however, an attempt has been made to draw out those features which might be expected to be characteristic of brewing yeast.

4.1.1.1    *Cell composition.*    Most published data for yeast composition relates to bakers' yeast which predictably will have significant differences in some cellular

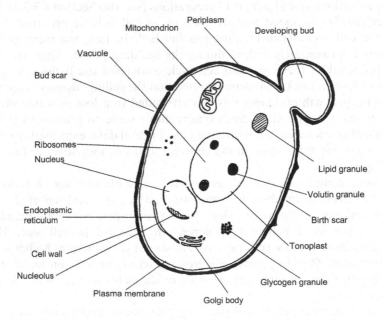

**Fig. 4.1**   A diagrammatic cross section of a yeast cell.

components compared to brewing yeast. In particular, levels of trehalose are engineered to be high in bakers' yeast to ensure good resistance to drying (see 3.4.2.2). In addition, sterol levels are high in bakers' yeast, a consequence of catabolite derepression (see Section 3.5.1).

The molecular composition of dried wine yeast is given in Table 4.1. From these data Rosen (1989) gave a rough molecular formula for this yeast as:

$$C_{4.02}H_{6.5}O_{2.11}N_{0.43}P_{0.03}$$

Table 4.1 Molecular composition of dried wine yeast (Rosen, 1989).

| Component | Percentage of dry weight |
| --- | --- |
| Carbon | 48.2 |
| Hydrogen | 6.5 |
| Oxygen | 33.8 |
| Nitrogen | 6.0 |
| Phosphorus | 1.0 |
| Magnesium | 0.1 |
| Calcium | 0.04 |
| Potassium | 2.1 |
| Sulphur | 0.01 |
| Iron | 0.005 |

Reed and Nagodarithana (1991) gave the gross macromolecular composition of bakers' yeast as shown in Table 4.2. The same authors also provided a more detailed mineral composition of bakers' yeast and this is reproduced in Table 4.3, together with earlier data for brewers' yeast as given by Eddy (1958). Quite large differences may be observed in the concentrations of individual minerals between different yeast samples. Apart from the differences in growth conditions, alluded to already, it would be suspected that the materials of construction of the plant used for handling the yeast, particularly the fermenters in the case of the brewing types were influential.

The carbohydrate fraction of yeast includes structural components, especially those associated with the cell wall (see Section 4.4.2) and those with a storage and/or stress resistance function such as glycogen and trehalose (see Sections 3.4.2.1 and 3.4.2.2). Yeast lipids are mainly structural components of the plasma membrane and other

Table 4.2 Macromolecular composition of bakers' yeast (Reed & Nagodarithana, 1991).

| Component | Percentage |
| --- | --- |
| Moisture content | 2–5 |
| Protein | 42–46 |
| RNA and DNA | 6–8 |
| Minerals | 7–8 |
| Total lipids | 4–5 |
| Carbohydrates | 30–37 |

**Table 4.3**   Mineral composition of yeast (Eddy, 1958; Reed & Nagodarithana, 1991).

| | μg/g yeast dry weight | | | |
|---|---|---|---|---|
| Component | Bakers' | Brewers' 1 | Brewers' 2 | Brewers' 3 |
| Aluminium | – | 3.0 | 2.0 | 1.0 |
| Calcium | 0.75 | | | |
| Chromium | 2.2 | | | |
| Copper | 8.0 | 37.0 | 104 | 34 |
| Iron | 20.0 | 17.0 | 17.0 | 25.0 |
| Lead | | 2 | 14 | 100 |
| Lithium | 0.17 | | | |
| Magnesium | 1.65 | | | |
| Manganese | 8.0 | 4.0 | 5.0 | 11.0 |
| Molybdenum | 0.04 | 0.1 | 0.04 | 2.7 |
| Nickel | 3.0 | 3.0 | 4.0 | 3.0 |
| Phosphorus | 13 | | | |
| Potassium | 21 | | | |
| Selenium | 5.0 | | | |
| Silicon | 30.0 | | | |
| Sodium | 0.12 | | | |
| Sulphur | 3.9 | | | |
| Tin | 3.0 | 3.0 | > 100 | 3.0 |
| Vanadium | 0.04 | | | |
| Zinc | 170.0 | | | |

intracellular membrane-bound organelles, as described in a later section of this chapter.

Spent yeast is used as a nutritional supplement because it is an excellent source of vitamins. Yeast does not contain vitamin C or the lipophilic types A and D. Nevertheless, other vitamins, particularly those of the B complex, are present in relative abundance (Table 4.4).

The gross macromolecular constitution of brewing yeast is much influenced by the genotype of the organism and the conditions to which it is exposed during the brewing process. The method of expressing the relative concentrations of individual cellular components requires careful interpretation. It is common practice to use yeast dry (or wet) weight as the denominator in expressions of macromolecular concentration.

**Table 4.4**   Vitamin content of brewers' yeast (from Eddy, 1958).

| Vitamin | μg/g dry yeast |
|---|---|
| P-aminobenzoic acid | 9–102 |
| Biotin | 0.5–2.0 |
| Choline | 4850 |
| Folic acid | 3.0 |
| Inositol | 2700–5000 |
| Niacin | 310–1000 |
| Pantothenic acid | 100 |
| Pyridoxine | 25–100 |
| Riboflavin | 36–42 |
| Thiamine | 50–360 |

Clearly, the results of metabolism may be reflected by simultaneous changes in the concentrations of several classes of macromolecules as well as the water content of the cell. Lagunas and Moreno (1985) pointed out that intracellular volume, cell dry weight, protein content and cell number change independently of each other. For these reasons direct comparison of published data can be difficult.

Ogur *et al.* (1952) described the RNA and DNA contents of a laboratory *S. cerevisiae* cells as influenced by ploidy. In this case, the results were expressed as a function of cell number (Table 4.5). Similar proportionalities between RNA and DNA can be observed in brewing yeast (D.E. Quain and P.A. Thurston, unpublished data). In this case the contents of RNA, DNA, glycogen and protein content of brewing yeast were monitored, together with dry weight and cell number, during fermentation an all-malt wort. Cell number increased during fermentation to a maximum value of approximately $80 \times 10^6$ per ml. This was accompanied by an increase in dry biomass. The latter parameter continued to increase after the cessation of cell proliferation from 4.7 to 6.0 mg ml$^{-1}$. RNA and DNA increased in parallel with cell number. The concentration of RNA, expressed as a function of cell dry weight, was approximately 5%. This amounted to 50 times the concentration of DNA. The protein concentration increased in a linear fashion throughout fermentation, eventually accounting for approximately 30% of the cell dry weight. The increase in biomass observed in the stationary phase, when cell proliferation had ceased, was a result of the accumulation of glycogen. In the subsequent maintenance phase prior to removal from fermenter, the dried biomass decreased and this was in large part due to glycogen dissimilation (see Section 3.4.2.1).

**Table 4.5** Nucleic acid content of *Saccharomyces cerevisiae* in cells of different ploidy (from Ogur *et al.*, 1952).

| Ploidy | RNA ($\mu$g/cell $\times 10^9$) | DNA ($\mu$g/cell $\times 10^9$) |
|---|---|---|
| Haploid | 55.6 $\pm$ 16.6 | 2.26 $\pm$ 0.23 |
| Diploid | 96.5 $\pm$ 22.4 | 4.57 $\pm$ 0.60 |
| Triploid | 164.0 $\pm$ 33.6 | 6.18 $\pm$ 0.54 |
| Tetraploid | 172 $\pm$ 6 | 9.42 $\pm$ 1.77 |

### 4.1.2 *Cytology*

4.1.2.1 *Plasma membrane.* As with all living cells the plasma membrane of yeast forms the barrier between the cytoplasm and the external environment. Several distinct roles for the plasma membrane can be recognised. These are to present a barrier to free diffusion of solutes, to catalyse specific exchange reactions, to store energy in the form of transmembrane ion and solute gradients, to regulate the rate of energy dissipation, to provide sites for binding specific molecules involved in metabolic signalling pathways and to provide an organised support matrix for the site of enzyme pathways involved in the biosynthesis of other cell components (Hazel & Williams, 1990).

The yeast plasma membrane is not smooth but the outer surface has a series of deep invaginations (Moor & Mühlethaler, 1963). These features are more prominent in stationary phase cells as compared to those in the exponential phase (Walther *et al.*, 1984). The role of the invaginations is not known. Takeo *et al.* (1988) observed that the furrows could form connections between specialised parts of the plasma membrane and the cell wall. Thus, the membrane is trapped within the wall, thereby possibly stabilising these regions of the plasma membrane. These authors concluded that the invaginations were retained in areas of membrane where no growth was occurring, indicating that in these regions the membrane was relatively static.

The membrane consists principally of lipid and protein in approximately equal proportions. Rank and Robertson (1983) isolated about 150 polypeptides from a yeast membrane preparation. Bearing in mind the large array of tasks performed by the membrane, it must be assumed that majority of membrane proteins are functional as opposed to structural. Thus, membrane proteins are responsible for regulating solute transport, they include the enzymes which catalyse cell wall synthesis, some receptors and, of course, the ATP-ase responsible for maintaining the plasma membrane proton-motive force (van der Rest *et al.*, 1995). Perhaps surprisingly, the membrane lipid biosynthetic enzymes are located elsewhere and indeed the initial steps of sterol synthesis up to mevalonate reportedly occur in the mitochondria in yeast (Coolbear & Threfall, 1989; Henschke & Rose, 1991). Adenylate cyclase, the enzyme responsible for initiating cyclic AMP-dependent cascade systems, is reportedly a membrane protein, attached to phospholipid components (Field *et al.*, 1988). However, Mitts *et al.* (1990) suggested that it might only be membrane-associated in a loose peripheral fashion.

The lipid components of membranes are amphipathic and confer hydrophobic properties. The major membrane lipids are phospholipids and sterols. The principal phospholipids are phosphotidylinositol, phosphotidylserine, phosphotidylcholine and phosphotidylethanolamine. Ergosterol is the principal sterol, although trace quantities of others are also found (see Section 3.5.1).

The current view of membrane structure is that of Singer and Nicholson (1972), termed the 'fluid mosaic model'. This holds that the amphipathic lipids are arranged spatially in a bi-layer such that the hydrophobic moieties of molecules are situated towards the interior of the membrane whereas the hydrophilic polar groups are arranged outermost in contact with the aqueous environment. Transmembrane and peripheral proteins are associated with the lipid bi-layer. Both lipids and proteins undergo lateral diffusion within the plane of the membrane. In the case of sterols, the combination of the polar hydroxy groups and hydrophobic sterol skeleton and alkyl side-chain allows them to be orientated in a perpendicular manner between the hydrophobic chains of other lipids.

Studies using artificial pure phospholipid monolayers have concluded that sterols modulate membrane fluidity. Thus, such artifical membranes, when cooled to a temperature less than a critical value (the transition temperature) undergo a phase change between gel-liquid and crystalline states (Brenner, 1984). The value of the transition temperature is dependent upon the composition of the lipid component of membranes. It is assumed that in the case of real biological membranes correct function requires appropriate fluidity. The process of regulating membrane fluidity in response to changes in environment, such as fluctuating temperature, is accommodated by changes in membrane lipid composition (Brenner, 1984).

With respect to sterols, the group led by Parks has recognised four functions that are essential for membrane function. These were termed sparking, critical, domain, domain and bulk. All of these functions can be satisfied by ergosterol. The sparking function is the most critical and is highly structure specific. It was demonstrated using a sterol auxotroph; growth on cholestanol was precluded unless trace amounts of ergosterol ($1$–$10 \, \text{ng ml}^{-1}$) were present. In the presence of traces of ergosterol, cholestanol could fulfil the bulk sterol function. Growth was observed in the presence of lanosterol ($5 \, \mu\text{g ml}^{-1}$) and ergosterol ($100 \, \text{ng ml}^{-1}$) but not lanosterol alone. Park's group concluded that the additional ergosterol was required for maintaining fluidity in very restricted areas of the membrane and that this was the critical domain function. The domain function was considered to be the minimum concentration of free sterol required for cell proliferation (Parks, 1978; Bottema et al., 1985; Lorenz et al., 1989; Rodriguez & Parks, 1983).

The plasma membrane of brewing yeast would be predicted to be modified in several ways during the brewery cycle. The effects of carbon and nitrogen catabolite repression and inactivation on the spectrum of plasma membrane transporters have been discussed in Chapter 3. Clearly, during most of primary fermentation, the membranes will be highly active sites of growth in terms of both individual cell size and proliferation. Superimposed on this will be the continual dynamic changes in the patterns of active and inactive transporters. Furthermore, the pre-formed pools of sterols and unsaturated fatty acid components of membrane lipids will be progressively diluted between mother cells and their progeny. The effects on plasma membrane integrity of ethanol, low water activity, high hydrostatic pressures and other stresses, as experienced by yeast during fermentation, together with the possible ameliorating effects of trehalose, have been discussed elsewhere.

4.1.2.2   *Periplasm.*   The periplasm is the space between the outer surface of the plasma membrane and the inner surface of the cell wall. The periplasm is not a continuous space because of the interrupting presence of invaginations in the plasma membrane and irregularities in the inner surface of the cell wall, as described in the previous section. It is perhaps not an organelle as such; however, it is worthy of comment since it is the location where a number of specific yeast enzymes may be found.

The contents of the periplasm of a living yeast cell will clearly be something of a metabolic cross-roads. There will be a dynamic mixture of soluble components passing into the cell from the external medium mixing with those moving out of the cell as a result of metabolism. However, there are a few enzymes that are presumably synthesised intracellularly then transported into the periplasmic space where they are retained because of the limited porosity of the cell wall (Scherrer et al., 1974).

Arnold (1991) defines periplasmic enzymes as being those which may be assayed in intact cells without disruption of the plasma membrane. Enzymes associated with the periplasmic space are most notably invertase and acid phosphatase. The latter hydrolyses phosphate esters in the medium; however, it may have an additional role in regulating other periplasmic enzymes via dephosphorylation. Other enzymes include a variety of binding proteins and melibiase (Wainer et al., 1988; Arnold, 1991). Panek (1991) considered that the fraction of yeast glycogen termed 'water-insoluble' is located in the periplasmic space where it is attached to the inner surface of the cell wall by alkali-resistant bonds.

The role of the periplasm as a distinct cellular compartment, rather than as a mere boundary area between the wall and the plasma membrane, is open to discussion. It is possible that the retention in the periplasm of a spectrum of enzymes which are required to utilise particular substrates, for example sucrose, could afford a competitive advantage over those organisms which fully secrete similar enzymes into the growth medium. Arnold (1991) commented that the protein concentration in the periplasm is very high, indeed possibly higher than needed for purely catalytic purposes, and suggested that the resultant gel-like nature of the periplasm provides a protective layer for the plasma membrane.

4.1.2.3 *Mitochondria.* The literature describing the structure and functions of mitochondria is vast and here it is possible only to provide a brief outline. For more details the reader should consult the authoritative texts of Lloyd (1974), Stevens (1981) and Guérin (1991). Of particular note here are the effects on the ultrastructure, modifications to the complement of enzyme activities and the metabolic roles of mitochondria in yeast as employed in brewing fermentations.

In yeast cells growing oxidatively under derepressed conditions, the total volume due to mitochondria (termed the 'chondriome') occupies about 12% of the total cell volume. Mitochondria have two distinct membranes, the inner of which has well-developed cristae which project into the interior of the organelle (Guérin, 1991). Four distinct compartments or locations can be recognised: the outer membrane, the intermembrane space, the inner membrane and the matrix. Specific enzyme activities are associated with each of these locations, as discussed later.

The membranes of mitochondria, like the yeast plasma membrane, contain both phospholipids and sterols. Sperka-Gottlieb *et al.* (1988) studied the composition of the inner and outer mitochondrial membranes and concluded that there were differences. The outer membrane contained predominantly phosphatidylcholine, phosphotidylethanolamine and phosphotidylcholine. Only a modest amount (5% total phospholipid) of cardiolipin was present in the outer membrane (compared with 13% of the phospholipid content of the inner membrane). The ratio of ergosterol to phospholipid in the outer membrane was shown to be lower than that of the inner membrane. The phospholipid to protein ratio of the outer membrane was significantly higher than that of the inner membrane.

Mitochondria contain self-replicating DNA and protein synthetic systems. The genome is comparatively small and it codes for just 5% of all mitochondrial proteins (Guérin, 1991). The remaining proteins are coded for by nuclear genes. The obvious implication is that many functional and structural components have to be transported into mitochondria. The interrelationships between the mitochondrial and nuclear genome and the regulation of the relevant genes which lead to the synthesis and assembly of sub-components during mitochondrial biogenesis is reviewed in Grivell (1989). Transport of some solutes between individual compartments of the mitochondrion and the cytosol involves channels or pores which contain a protein termed porin (Dihanich *et al.*, 1989; Palmieri *et al.*, 1990; Mannella, 1992; Szabó *et al.*, 1995).

Mitochondria are perhaps best known as the site of the respiratory chain leading to

ATP formation using the electron transport chain. The process is catalysed by five complexes: NADH CoenzymeQ reductase; succinate CoenzymeQ reductase, CoenzymeQH$_2$-cytochrome C reductase, cytochrome C oxidase and ATP synthase (De Vries & Marres, 1987). Two further electron carriers, ubiquinone and cytochrome C, are also involved. The electron transport chain generates ATP as described by the chemi-osmotic theory of Mitchell (1979). In essence this states that the components of the electron transport chain are located in the inner membrane of the mitochondrion such that protons are translocated from the inner to outer side as electrons pass down the chain. The ATPase uses energy stored in a transmembrane proton electrochemical potential to generate the high-energy bond associated with ATP.

Several other enzyme systems are also located within mitochondria as discussed by Guérin (1991). The enzymes of the sterol biosynthetic pathway responsible for mevalonate biosynthesis from acetyl-CoA are located here, as are steryl ester hydrolases involved in sterol degradation. The outer membrane contains some of the enzymes associated with phosphlipid biosynthesis. The intermembrane space contains adenylate kinase. Several enzymes of the Krebs cycle are found in the matrix (Robinson & Srere, 1985). In addition, the matrix contains enzymes which are involved with the biosynthesis of some amino acids and nucleotides and also the Mn-superoxide dismutase. The inner membrane contains the electron transport chain and also a number of transport carriers.

It may be appreciated from the previous discussion that transport between mitochondria and the cytosol is an essential prerequisite to normal metabolic activity. Most notable is the adenine nucleotide carrier, which under conditions of oxidative growth exchanges ADP from the cytosol for ATP generated from oxidative phosphorylation and the catabolite repressible proton symport system for importing phosphate into mitochondria (Guérin, 1991).

In other respects, however, the transport system is perhaps complicated by virtue of the fact that many metabolites are not accessible to the mitochondrial membranes and, therefore, indirect methods involving other intermediates are required. For example, since the mitochondrial and cytosolic pools of nicotinamide nucleotides are distinct, methods are required for producing reducing equivalents in both intracellular compartments. Systems such as the malate/aspartate shuttle fulfil this role in many cells. In *S. cerevisiae* this system apparently does not occur and instead there are two NADH:ubiquinone-6 oxidoreductases (De Vries *et al.*, 1992). One is bound to the inner membrane of the mitochondrion and the other faces the mitochondrial matrix. The first enzyme is involved in the oxidation of NADH produced in the cytosol, the other oxidises NADH arising from the tricarboxylic acid cycle and mitochondrial alcohol dehydrogenase. Similarly, since acetyl-CoA cannot pass through the mitochondrial membrane its passage is accomplished, at least in the case of *S. cerevisiae*, as citrate or acetate, as discussed following.

With regard to the responses of brewing yeast to the conditions experienced during fermentation the effects of anaerobiosis and repression on mitochondrial structure and function are of obvious importance. It is now accepted that yeast grown under anaerobic and/or repressed conditions contains 'undeveloped' mitochondrial bodies termed promitochondria (Criddle & Schatz, 1969). In yeast, growing under repressed conditions Stevens (1981) describes mitochondria as assuming simple spherical or long, sinuous branching shapes. Individual mitochondria possess only a few cristae. The same author (Stevens, 1977) estimated that the chondriome of glucose-repressed yeast occupied just 3–4% of the total cell volume. Damsky (1976) studying mitochondrial biogenesis in *S. carlsbergensis* observed that during aerobic adaptation of anaerobically-grown cells there were increases in density and development of cristae.

In brewing yeast during the fermentation cycle, the conditions for fully functional mitochondria to arise do not exist. Thus, when oxygen is present during the aerobic phase of fermentation, the high concentrations of sugar in the wort maintain repressing conditions. When the bulk of the sugar has been utilised by yeast in late fermentation, the repressing conditions are lifted but the prevailing conditions of anaerobiosis again prevent full mitochondrial biogenesis. The question, therefore, is what metabolic roles are played by promitochondria in brewing yeast?

In a gross sense, it is well recognised that yeast cells with impaired mitochondrial function, such as petite mutants (see Section 4.3.2.7), do not perform well in wort fermentations. In particular, $\rho^-$ mutants, which have lost the mitochondrial genome, appear to be especially disadvantaged. For example, Ernandes *et al.* (1992) compared the fermentation performance of spontaneously- arising $\rho^-$ mutants of both ale and lager strains and demonstrated that, although attenuation rates of high-gravity rates were similar, the deficient strains produced beers with altered volatile spectra and increased diacetyl. The same group (Good *et al.*, 1993) also concluded that the respiratory deficient strains were less tolerant to stress.

These results are explainable in that it is apparent that many of the metabolic activities associated with fully developed mitochondria, other than the electron transport chain, are fully functional in promitochondria. Furthermore, these other functions are required for normal brewing performance. Bandlow *et al.* (1988) demonstrated the presence of both cytosolic and mitochondrial adenylate kinases in yeast. The latter was situated in the intermembrane space and was required for growth under non-fermentative conditions. There is now compelling evidence that the mitochondrial ATP-ADP translocase can be used to transport ATP into either cell compartment, depending on the mode of growth. Thus, during anaerobiosis, or under repressing conditions, ATP is generated in the cytosol via substrate level phosphorylation and a proportion is transported into the mitochondria where it is used for the multitude of essential biosynthetic reactions associated with this organelle and as described already in this section.

An admirable review of the evidence describing the role of mitochondria in brewing yeast during beer fermentations has been supplied by O'Connor-Cox *et al.* (1996). Of particular note would appear to be the importance of the regulation of carbon flow devolving from glycolysis. Specifically, the formation and utilisation of acetyl-CoA in respect to cellular compartmentalisation.

Acetyl-CoA is an essential precursor of numerous pathways with enormous impact on both yeast growth during fermentation and beer flavour. For example, the formation of sterols, fatty acids, some amino acids, esters, higher alcohols and diacetyl. Some of these processes are cytosolic/microsomal and some are mitochondrial (see Section 4.1.3). Clearly, the mitochondrial pathways must be active in promitochondria for growth during fermentation and development of flavour compounds.

Under aerobic conditions acetyl-CoA derived from sugar metabolism would be generated in mitochondria, via the activity of pyruvate dehydrogenase. However, as described in Section 3.4, there is evidence that pyruvate dehydrogenase is not active in brewery yeast at any stage during fermentation. Therefore, supplies of cytosolic acetyl-CoA, from *de novo* synthesis probably arise from pyruvate via pyruvate decarboxylase, acetaldehyde dehydrogenase and acetyl-CoA synthetase. Both cytosolic and mitochondrial aldehyde dehydrogenases and acetyl-CoA synthetases have been detected (Jacobsen & Bernofsky, 1974; Satyanarayana *et al.*, 1980; Meaden *et al.*, 1997). It would thus appear that during fermentation, mitochondrial acetyl-CoA could be derived from the import of cytosolic acetate and, thence, conversion to acetyl-CoA using mitochondrial acetyl-CoA synthetase. Conversely, the acetate units may be transported into mitochondria utilising the acetyl-carnitine shuttle which has also been reported in *Saccharomyces* yeasts (Atomi *et al.*, 1993). However, this enzyme was associated with high activity of yeast grown on oleic acid. Most of the enzyme was peroxisomal; however a mitochondrial fraction was also detected.

The essentiality of a degree of mitochondrial function for normal fermentation performance is apparently an ineluctable truth. No doubt further elucidation of the interactions between promitochondria and other cellular compartments will be forthcoming.

### 4.1.2.4 *Vacuoles.*

Vacuoles are the most obvious features of yeast cells viewed under the light microscope. However, Schwenke (1991) makes the point that the single large visible vacuole is part of a complex dynamic membrane system which includes smaller vacuoles, the endoplasmic reticulum, secretory vesicles, endocytic vesicles and the Golgi complex.

Vacuoles have a plastic structure and their number and size vary during the growth cycle. With respect to brewing yeast the presence of extensive vacuoles has been taken as a sign of stress. Thus, in an early report, Aswathanarayana (1958) observed the presence of extensive vacuolation in yeast cells subject to starvation. The vacuoles rapidly disappeared when the same cells were transferred to fresh medium. In other reports the degree of vacuolation has been shown to fluctuate throughout the cell cycle. For example, Wiemken *et al.* (1970) observed fragmentation of large vacuoles into greater numbers of smaller structures coincident with the commencement of budding.

The vacuolar membrane, or tonoplast, is similar to that of the plasma membrane, although Schwenke (1991) points out that there are some differences since the latter is considerably more elastic than the former. Compared to the plasma membrane the tonoplast contains more phospholipids and unsaturated fatty acids but less sterol (Kramer *et al.*, 1978). Presumably, these differences confer the elasticity which allows the vacuole to enlarge or decrease in volume, as required.

Vacuoles have two principal functions. First, they serve as dynamic stores of nutrients. In so doing they regulate the concentration of nutrients in other cellular compartments. Second, they provide a site for the breakdown of some macromolecules, in particular proteins, to release intermediates for other metabolic pathways. It may be readily appreciated that the fluctuations in these requirements account for the dynamism of vacuolar dimensions.

Amino acids are major soluble constituents of vacuoles, in particular basic types such as arginine. These amino acids function as a nitrogen reserve and they arise as a result of proteolysis. Achstetter and Wolf (1985) identified at least seven distinct proteinases in vacuoles. These included proteinases, carboxypeptidases, aminopeptidases and dipeptidyl aminopeptidases.

In a later report the same group (Rendueles & Wolf, 1988) concluded that the major role of these proteinases was to provide amino acid moieties during sporulation under conditions of nitrogen deficiency. This would suggest little relevance to brewing yeast; however many worts, particularly of very high gravity, may contain limited nitrogen and under such conditions protein turnover may be of importance.

Control of the activity of proteinases and sorting of proteins is considered to require regulated acidifcation of the vacuole. It is proposed that this is accomplished using a proton-translocating membrane bound ATP-ase (Nelson, 1987; Anraku et al., 1989; Klionsky et al., 1992). Other vacuolar membrane transporters have been identified which are responsible for the active transport of amino acids, $Ca^{2+}$ and other cations including iron (Wada et al., 1987; Bode et al., 1995).

Vacuoles also serve as a principal store of inorganic phosphate distinct from the pools, which have been identified, associated with the plasma membrane and nucleus (Schuddenat et al., 1989). Vacuolar phosphate deposits occur as a linear polymer of polyphosphate linked by high-energy phospho-anhydride bonds. As well as serving as a reserve material, polyphosphate has been implicated in the sequestration of basic amino acids such as arginine (Durr et al., 1979). Schwenke (1991) has reviewed the evidence that the 'volutin' granules visible in bright-field light micrographs of yeast vacuoles consist of an association between polyphosphate and S-adenosyl-L-methionine which are capable of interacting with basic amino acids and cations in a complex and as yet uncharacterised manner.

Vacuoles also contain a trehalase (Mittenbühler & Holzer, 1988). This enzyme is distinct from the cytosolic enzyme and unlike the former is not subject to complex regulation by phosphorylation/dephosphorylation reactions (Londesborough & Varimo, 1984). The role of acid trehalase is not known, although the relative lack of regulation would suggest that it is not involved with the response to stress ascribed to the neutral trehalase.

4.1.2.5  *Other cytoplasmic inclusions.*　Robinow and Johnson (1991) comment that apart from the specific organelles described already and ribosomes, the cytoplasm of yeast is not greatly populated by other defined structures. There is relatively little endoplasmic reticulum and certainly in the case of *S. cerevisiae*, microbodies are few. From the perspective of brewing yeast two features are of note. These are glycogen bodies and lipid granules.

Brewing yeast during fermentation can accumulate high concentrations of glycogen (see Section 3.4.2.1). Apart from the 'structural' glycogen that is apparently associated with the periplasm (see Section 4.1.2.2), the pool of glycogen that is used as a storage carbohydrate is located in the cytoplasm. It may be readily be visualised under bright field microscopy using iodine staining (Quain & Tubb, 1983). The glycogen is borne in clumps of spherical bodies each with a diameter of approximately 40 μm (Matile et al., 1969).

Lipid particles occur in the cytoplasm of yeast. The particles consist of triacylglycerols and steryl esters in approximately equal quantities with small amounts of phospholipid (Schaffner & Matile, 1981; Zinser et al., 1993). In a more recent paper, Leber et al. (1994) purified lipid particles from yeast and concluded that they consisted of a hydrophobic core which contained triacylglycerols and steryl esters. The core was surrounded by surface membrane which consisted of phospholipids and protein. These authors demonstrated that the lipid particles contained $\Delta^{24}$-methyltransferase activity and considered this to be evidence that the particles were involved in sterol biosynthesis. Behalova and Vorisek (1988) observed that when *Saccharomyces cerevisiae* was grown under conditions of aerobiosis, low growth rate and nitrogen limitation, sterols accumulated. This accumulation was accompanied by the formation of small vacuoles which contained lipid. It was concluded that these had a storage function.

It must be assumed that these sterol esters in the lipid particles function as a sterol store which may be utilised in times of need. Conversely, the lipid particles may simply represent temporary sterol-handling organelles which are implicated in the biosynthetic pathway as carriers transporting the lipids from the sites of synthesis to the sites of deposition into membranes. Leber et al. (1995) studied the effects of various metabolic inhibitors on steryl ester intracellular transport and hydrolysis. They reported that lipid particles did not contain steryl ester hydrolase activity and that this enzyme was found in highest activity in the plasma membrane. They concluded that steryl ester transport did not involve the micotubule-dependent vesicle system but that active membrane synthesis provided the driving force for release of free sterols.

The Golgi complex consists of a series of stacked membranes and vesicles whose presence in eukaryotic cells seems to be ubiquitous. Although it has been oberved in numerous yeast genera, its presence in *S. cerevisiae* has been doubted (Robinow & Johnson, 1991). However, in another chapter of the same publication, Schwenke (1991) comments that in this yeast both endoplasmic reticulum and Golgi bodies are highly dynamic bodies which do not accumulate, and therefore their presence may be difficult to demonstrate. In this respect it is perhaps similar to promitochondria in repressed yeast which are also difficult to visualise. Schwenke (1991) states categorically that the Golgi complex is a normal cellular constituent of *S. cerevisiae*.

The organelle is part of a yeast secretory pathway involving an ordered sequence passing from endoplasmic reticulum to Golgi complex to vesicle and thence to the vacuole or cell surface. In this pathway, the Golgi complex appears to function as a regulator of destination of intracellular protein trafficking. It is in this organelle that proteins are sorted and directed to the vacuole or to the plasma membrane.

Peroxisomes are a class of membrane bound bodies which have been observed in the cytosol of *S. cerevisiae*. They have been associated with activities involving the glyoxylate cycle, ß-oxidation of fatty acids and catalase A (Kunau & Hartig, 1992). This spectrum of activities indicates the role of this organelle, that is, as an essential pre-requisite of growth on lipid carbon sources. Thus, Kunau and Hartig (1992) demonstrated that peroxisome proliferation was encouraged by growth on oleic acid but repressed during growth on glucose. In consequence, these organelles would seem to have small relevance to brewing yeast.

### 4.1.3  Intracellular location of enzymes

Intracellular compartmentalisation provides the cell with both opportunities and problems. Discrete locations provide microenvironments where conditions may be manipulated to suit the particular aspects of metabolism on-going at that particular location, insulated from adjoining cellular compartments. For example the acidic environment of vacuoles compared with the cyctosol (Section 4.1.2.5). Conversely, the presence of intracellular barriers requires a complex system of transport to ensure effective communication between cellular compartments and possibly duplicated enzyme pathways where particular metabolites are restricted to cellular locations. For example, the discrete aldehyde dehydrogenases associated with the cytosol and mitochondria (see Section 4.2.1.3).

The function of particular organelles is obviously strongly indicated by the complement of enzymes present. By way of a summary to the preceding discussion, the locations of some enzymes/pathways of importance to brewing yeast are given in Table 4.6. Clearly, the list is not exhaustive.

## 4.2  Taxonomy and differentiation

### 4.2.1  Taxonomy and the Saccharomyces

Although taxonomy has been defined as 'the art of classification' (Thorne, 1975), it is apparent that biological systematics are something of a minefield. For the non-specialist this is in part due to the intimidating complexity of the various organisa-

**Table 4.6**  Intracellular locations of enzymes.

| Enzyme | Location | Comments | Reference |
| --- | --- | --- | --- |
| Invertase | Periplasm | | Carlson and Botstein (1982) |
| Melibiase | Periplasm | | Arnold (1991) |
| Steryl-ester hydrolase | Plasma membrane | | Leber *et al.* (1995) |
| Hexose transporters | Plasma membrane | | Ozcan and Johnston (1999) |
| Invertase | Endoplasmic reticulum | Localised by e.m. – evidence of yeast secretory pathway | Brada and Schekman (1988) |
| Alcohol acetyl transferase | Microsomes | | Minetoki *et al.* (1993) |
| Catalase A Glyoxylate cycle enzymes Fatty acid β-oxidation | Peroxisomes | Induced by growth on lipid substrates | Kunau and Hartig (1992) |
| Sterol Δ$^{24}$-methyltransferase | Cytosolic lipid particles | | Leber *et al.* (1994) |
| Adenylate cyclase | Plasma membrane and cytosol | Probably peripheral membrane protein | Mitts *et al.* (1990) |

*Contd*

**Table 4.6** *Contd.*

| Enzyme | Location | Comments | Reference |
|---|---|---|---|
| Glycolytic pathway Hexokinase Phosphoglucose isomerase Phosphofructokinase Fructose 1,6-bisphosphate aldolase Triose phosphate isomerase Glyceraldehyde 3-phosphate dehydrogenase Phosphoglycerate kinase Phosphoglyceromutase Enolase Pyruvate kinase Pyruvate decarboxylase | Cytosol | Possible association with cytoskeleton | Shearwin and Masters (1990), Masters (1992) |
| Glycogen synthase Glycogen branching enzyme | Cytosol | | Panek (1991) |
| NAD$^+$-glycerol 3-phosphate dehydrogenase | Cytosol Mitochondria | Two isozymes in yeast, cytosolic enzyme involved in osmotic response in yeast | Nevoigt and Stahl (1997) |
| Neutral trehalase | Cytosol | Regulatory enzyme with putative involvement in yeast stress responses | App and Holzer (1989) |
| Alcohol dehydrogenase | Cytosol | Not active in aerobically-adapted yeast, presumably involved in ethanol production during fermentation | Wales *et al.* (1980) |
| Acetyl CoA carboxylase Fatty acid synthestase | Cytosol | Initial enzymes of fatty acid synthesis | Ratledge and Evans (1989) |
| Cu,Zn-superoxide dismutase | Cytosol | | Berminham *et al* (1988) |
| Mg$^{2+}$ acetaldehyde dehydrogenase | Cytosol | | Meaden *et al.* (1997) |
| Acetyl-CoA synthetase | Cytosol and mitochondria | | Satyanarayana *et al.* (1974) |
| Acetohydroxy synthase | Mitochondria | ILV2 gene product involved in isoleucine valine biosynthetic pathway Implicated in diacetyl metabolism | Falco and Dumas (1985) |
| Acetohydroxy acid reductoisomerase | Mitochondria | ILV5 gene, see above | |
| Alcohol dehydrogenase | Mitochondria | ADH3 gene product, probably active in depressed cells only | |

*Contd*

**Table 4.6** *Contd.*

| Enzyme | Location | Comments | Reference |
|---|---|---|---|
| NAD(P)-aldehyde dehydrogenase | Mitochondria | | Jacobsen and Bernofsky (1974) |
| Mn-superoxide dismutase | Mitochondria | | Van Loon *et al.* (1986) |
| Cytochrome c oxidase | Mitochondria – inner membrane | | Wright *et al.* (1995) |
| Adenylate kinase | Mitochondria and cytosol | | Bandlow *et al.* (1988) |
| $F_1$-ATPase | Mitochondria | Promitochondrial activity: | Criddle and Schatz (1969) |
| Succinate dehydrogenase | | 4-fold reduction | |
| NADH-ferricyanide reductase | | 2-fold reduction | |
| Succinate-ferricyanide reductase | | 3-fold reduction | |
| Succinate-cytochrome c reductase | | absent | |
| Succinate oxidase | | absent | |
| NADH oxidase | | 375-fold reduction | |
| Cytochrome c oxidase | | 550-fold reduction | |
| Fumarase | Mitochondrial matrix and cytosol | | Boonyarat and Doonan (1988) |
| TCA cycle enzymes | Mitochondrial matrix | 2-oxoglutadehydrogenase absent under anaerobiosis/repressed conditions (Wales *et al.*, 1980) | Robinson and Srere (1985) |
| Hydroxymethyl-CoA synthetaseglutaryl Hudroxymethyl-CoA reductaseglutaryl | Mitochondria | Initial steps of squalene synthesis in sterolgenesis | Coolbear and Threfall (1989) |
| NADH kinase | Mitochondria (inner membrane) | Involved in formation of NADPH from NADH | Iwahashi and Nakamura (1989) |
| Proline oxidase | Mitochondria | Requires aerobiosis | Brandriss (1983) |
| NADH-ubiquinone-6 oxidoreductase | Mitochondria | Functions in place of malate/aspartate shuttle in *S. cerevisiae* | De Vries *et al.* (1992) |
| Pyruvate dehydrogenase | Mitochondria | | Van Urk *et al.* (1989) |
| Carboxypeptidase Aminopeptidase Dipeptidyl aminopeptidase | Vacuole | | Rendueles and Wolf (1988) |
| α-Mannosidase | Vacuole | Marker of vacuolar membrane | Yoshihisa *et al.* (1988) |
| Phosphohydrolase | Vacuole | | Lichko and Okorokov (1990) |
| ATP-ase | Vacuolar membrane | | Okorokov and Lichko (1983) |
| Acid trehalase | Vacuole | | Mittenbühler and Holzer (1988) |

tional levels together with the myriad of seemingly different criteria for inclusion. In the case of brewing yeast it is a member of the fungi kingdom which has been estimated to consist of 100 000 species (Anonymous, 1996b). The 'first cut' of classification ('division') graphically demonstrates the diversity of the fungal kingdom (Table 4.7) ranging from brewing yeast, edible mushrooms and *Penicillium* through to 'thrush' (*Candida albicans*) and a host of moulds, mildews and rusts. The complexity of classification through descending hierarchy of taxa can been seen (Fig. 4.2) through tracing the lineage of brewers' yeast – *S. cerevisiae* – down through the kingdom to the yeast species.

For most, the practical fruits of classification are at the levels of *genus*, *species* and *strain*. Definitions are difficult, debatable and, occasionally controversial. For simplicity, 'popular' definitions are presented here which, although not precise, provide a 'feel' of the scope of each taxa. A *genus* is a 'group of species closely related in

**Table 4.7** Fungal taxonomy.

| Fungal division | Comments | Examples |
| --- | --- | --- |
| Oomycota | Frequently resemble algae, often parasitic to plants | Water mould, white rust, downy mildew |
| Zygomycota | Filamentous, often insect parasites | Black bread mould |
| Ascomycota | Sac fungi, yeasts with endogenous spores, usually hyphal | *Saccharomyces*, powdery mildew, red bread mould (*Neurospora*) |
| Basidiomycota | Yeast, exogenous spores, some plant parasites, some saprophytic | Corn smut, edible mushroom, sulphur fungus, black stem rust, puffball, *Cryptococcus* |
| Deuteromycota | Fungi imperfecti, mixed group, do not produce spores | Thrush (*Candida albicans*), *Aspergillus*, *Bretanomyces*, *Verticilium*, *Fusarium*, *Penicillium* |

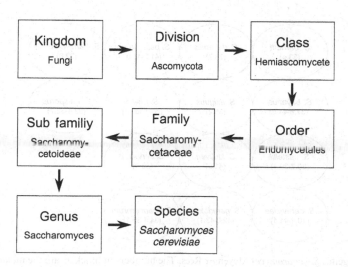

**Fig. 4.2** The taxonomic hierarchy.

structure and evolutionary origin' whereas *species* can be defined as 'individuals, which interbreed but are unable to breed with other such groups' (Anonymous, 1996b). The final level of classification is at the level of *strains*. Here the phenotypic and genotypic differences between strains within a species are – taxonomically – comparatively minor.

As noted above, brewing yeast is of the genus '*Saccharomyces*' and the species '*cerevisiae*' resulting in the specific name *S. cerevisiae* (the species is always coupled with the genus and has a lower case initial letter). The names of strains of *S. cerevisiae* often relate to the brands they produce, brewery they originate from or, if from a culture collection, are simply described numerically.

The changing complexities of yeast taxonomy have been reviewed at length in recent years (Barnett, 1992; Campbell, 1996). Perhaps not surprisingly, the classification of yeasts has provoked much lively debate with phases of taxonomic expansion and reduction. This is ably demonstrated by considering the number of yeast genera and species reported over the years in *The Yeasts, A Taxonomic Study*. Between the second (Lodder, 1970) and third (Kreger-van Rij, 1984) editions of this major text, the number of genera increased from 39 to 60 and the number of species from 349 to 500. In the fourth edition (Kurtzman & Fell, 1998) there are now 100 genera representing over 700 species. This, as noted by Walker (1998) 'represents only a fraction of the yeast biodiversity on this planet'.

The genus *Saccharomyces* Meyan *ex* Rees has been subject to both contraction and slight expansion. The 41 species identified in 1970 (Lodder, 1970) were reduced to ten in 1990 (Barnett *et al.*, 1990; Barnett, 1992a) and increased to 14 in 1998 (Kurtzman & Fell, 1998) (Fig. 4.3). The ten 'core' species described by Barnett (1992a) have been subdivided into three groups based on ribosomal RNA sequence (Vaughan-Martini & Martini, 1993) and the size and stability (see Section 4.3.2.7) of the mitochondrial genome (Piskur *et al.*, 1998). The *Saccharomyces sensu stricto* group consists of *S. cerevisiae* and three other closely related 'sibling' species *S. bayanus*, *S. paradoxus* and

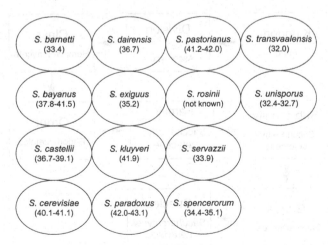

**Fig. 4.3** The genus *Saccharomyces* Meyen *ex* Rees. The numbers in brackets are the molar percentage of guanine + cytosine in nuclear DNA (see Section 4.2.1.1).

*S. pastorianus.* Although closely related, the four species split into two 'clusters' (Table 4.8) (Montrocher *et al.*, 1998) consisting of (i) *S. bayanus* and *S. pastorianus* and (ii) *S. cerevisiae* and *S. paradoxus*. The two clusters differ fundamentally in their response to temperature. Compared to the *'cerevisiae'* cluster, the *'bayanus'* cluster grows at a lower optimum and maximum temperature (see Section 4.2.3), is able utilise melibiose (Naumov, 1996) and transports fructose via an active proton symport (Rodrigues de Sousa *et al.*, 1995). Despite these differences, it is worth noting that although separated into clusters, the yeasts within the *Saccharomyces sensu stricto* group consist of 'very closely related yeasts when the whole genus *Saccharomyces* is considered' (Montrocher *et al.*, 1998). In contrast, the *Saccharomyces sensu lato* group is more diverse consisting of *S. dairensis*, *S. castelli*, *S. exiguus*, *S. servazzii* and *S. unisporus*. This group is genetically heterogeneous both in terms of chromosome number (7–16) and genome size. Accordingly and somewhat inevitably, these yeasts can be further subdivided into four groups (Petersen *et al.*, 1999). The third group consists of one species, *S. kluyveri*, which increasingly is being consolidated within the *Saccharomyces sensu lato* group (Marinoni *et al.*, 1999; Petersen *et al.*, 1999). Whatever, *S. kluyveri* is the most divergent member of the genus with only 5 or 7 chromosomes (Vaughan-Martini & Martini, 1998; Petersen *et al.*, 1999) and, within the *Saccharomyces*, a unique fatty acid profile (Augustyn *et al.*, 1991).

**Table 4.8** Split of the *Saccharomyces sensu stricto* – commonality and differences between the species.

| | *S. cerevisiae* | *S. paradoxus* | *S. pastorianus* | *S. bayanus* |
|---|---|---|---|---|
| Habitat | Domesticated – brewing (ale), baking, enology | Not domesticated | Domesticated – brewing (lager) | Domesticated – enology |
| Genome size | 1 | – | 1.5–1.6 | 1.2 |
| Maximum growth temperature (°C) | ≥ 37 | ≥ 37 | ≤ 34 | ≤ 34 |
| Fructose transport | Facilitated | Facilitated | Active | Active |
| Melibiose utilisation | Yes | Yes | No | No |
| rDNA spacer sequences | 'Cerevisiae' cluster | 'Cerevisiae' cluster | 'Bayanus' cluster | 'Bayanus' cluster |

The genus *Saccharomyces* is without doubt the most commercially exploited yeast Of the 14 species (Vaughan-Martini & Martini, 1998), only *S. cerevisiae* contributes to the three major yeast-based industrial processes, (i) industrial alcohol and beverages (both wine and beer), (ii) baking and (iii) the production of biomass, autolysates and flavours. Although of obvious industrial importance, the bottom-fermenting hybrid, *S. pastorianus* (syn. *S. carlsbergensis*) is exclusive to lager fermentations. *S. bayanus* remains a 'niche' industrial yeast being the junior partner (to *S. cerevisiae*) in wine making. *S. bayanus* is found in low-temperature wine fermentations (Naumov, 1996) and predominates, because of its high ethanol tolerance, toward the end of natural wine fermentations (Subden, 1990; Guidici *et al.*, 1998; Naumov *et al.*, 1998). The remaining species of the *Saccharomyces sensu stricto* group, *S. paradoxus*, is not 'domesticated' and is found in natural habitats such as broad-leaved trees, soils and insects (Vaughan-Martini & Martini, 1993; Naumov *et al.*, 1998). It is noteworthy that *S. cerevisiae* (and by implication *S. bayanus* and *S. pastorianus*) are considered (Martini, 1993) not to be 'natural' as they are rarely found in vineyard soil or the surface of ripe grapes. The generally accepted view (Martini, 1993) is that wine and brewing strains of *S. cerevisiae* have evolved in industrial environments via selective pressure through being better equipped to ferment grape musts and wort. Although beyond the scope of this book, fascinating insights into the complex and varied ecology of 'natural' yeast genera can be found in expert reviews (Phaff & Starmer, 1987; Spencer & Spencer, 1997).

Practically, the *Saccharomyces* species are differentiated from each other via a collection (Table 4.9) of physiological tests (Barnett, 1992; Vaughan-Martini & Martini, 1993) although the most recent incarnation (Vaughan-Martini & Martini, 1998) includes criteria based on chromosome size. Unfortunately, this 'churn' of species seems inevitable and is likely to accelerate as molecular techniques reveal new interrelationships between yeasts. The conclusion from the excellent review of Barnett (1992) offers a slightly despairing view of taxonomy:

'there is no correct classification of *Saccharomyces* species, except in terms of rigorous and arbitrary taxonomic rules. These rules have always varied with the techniques used, nutritional, serological, genetic or molecular biological, and the current prejudices of taxonomists. The changes would be of little interest and no consequence to experimental biologists were they kept as esotericisms and did not lead to confusions of nomenclature'.

**Table 4.9** A taxonomic key for the genus *Saccharomyces* (after Vaughan-Martini & Martini, 1993).

| | | Test | Go to Test |
|---|---|---|---|
| 1 | a | Growth in the presence of 1000 ppm cycloheximide | 2 |
| | b | No growth in the presence of 1000 ppm cycloheximide | 3 |
| 2 | a | Ethylamine. HCL, cadaverine and lysine assimilated – *S. unisporus* | |
| | b | Ethylamine. HCL, cadaverine and lysine not assimilated – *S. servazzii* | |
| 3 | a | Growth in the presence of 1000 ppm cycloheximide – *S. exiguus* | |
| | b | No growth in the presence of 1000 ppm cycloheximide | 4 |
| 4 | a | Ethylamine. HCL, cadaverine and lysine assimilated – *S. kluyveri* | |
| | b | Ethylamine. HCL, cadaverine and lysine not assimilated | 5 |
| 5 | a | D-ribose assimilated – *S. dairensis* – *S. castellii* | |
| | b | D-ribose not assimilated | 6 |
| 6 | a | Growth in vitamin-free medium – *S. bayanus* | |
| | b | Growth in vitamin-free medium | 7 |
| 7 | a | Active transport mechanism for fructose present, maximum growth temperature 34°C or below – *S. pastorianus* | |
| | b | Active transport mechanism for fructose not present | 8 |
| 8 | a | D-mannitol assimilated, maximum growth temperature at least 37°C – *S. paradoxus* | |
| | b | D-mannitol not assimilated, maximum growth temperature variable – *S. cerevisiae* | |

Similar concern has been expressed by Naumov (1996), who in passing notes that applied microbiologists are 'confused by the repeated revisions of the genus *Saccharomyces*'.

4.2.1.1 *Identification of yeasts.* The fourth edition of *The Yeasts, A Taxonomic Study* details current approaches to the identification of yeasts. As with previous editions of this text, 'traditional' identification to family and genus revolves around reproduction and morphology. As noted above for the *Saccharomyces*, physiological tests are used to describe and identify yeast species.

Taxonomically (Yarrow, 1998), the first clue to identity is through the mode of sexual reproduction. The ascomycetes form *asci* and basidiomycetes form *basidia*. Where sporulation is not seen – the so-called 'imperfect yeasts' – staining with Diazonium Blue B is used to differentiate between the ascomycetes and basidiomycetes. In 'perfect yeasts', major clues as to the genus of an unidentified yeast can be gleaned from visual analysis of the asci or basidia. Despite this, the mode of asexual or vegetative reproduction provides a more universal insight into possible identity. The majority of yeasts reproduce via budding or, to a lesser extent, via fission, and a few form conidia on short stalks. Similarly, much can be gleaned from the morphology of vegetative cells. For example, the size and shape of cells, their organisation (chains, pseudohyphae) together with colony colour and texture on solid media.

Yeast species are identified via an array of tests including fermentation and assimilation of various carbon sources, growth on a selection of nitrogen sources and growth at different temperatures and osmotic pressures. The number of tests required to achieve identification hampers the use of such 'keys'. For example, the key of Payne *et al.* (1998) involves 73 discrete tests, which would be expected to limit routine 'casual' use. However, the need for identification of yeasts, particularly clinically important yeasts, has led to the commercial development of paired-down, miniaturised systems. Typically, the number of tests is reduced to 20–30 in microtitre plates or strips of microwells or cupules. Turnaround time ranges from 2–3 days where growth or metabolism is assessed manually or automatically via a pH indicator or turbidometrically, through to as little as four hours using enzyme hydrolysis of chromogenic substrates. Identification is achieved through characteristic fingerprint or 'microcode' which is interrogated either manually, or preferably electronically, against a database. This results in a putative identity and measure of confidence for the accuracy of the identification.

Of the numerous commercial approaches available for the identification of yeasts (for a review see Deak & Beuchat, 1996), the bioMerieux API (Analytical Profile Index) series is the most widely used. In particular the API 20C consisting of 19 assimilation tests has become the benchmark method by which new, more rapid (4 h) approaches for clinically significant yeasts are assessed (Espinel-Ingroff *et al.*, 1998; Heelan *et al.*, 1998; Kellog *et al.*, 1998). Application of this and other approaches to food-borne yeasts has been less successful (Deak & Beuchat, 1996) and is presumably a consequence of the clinical focus of the species databases. Despite this, these test strips usually have the capability to successfully identify isolates of *S. cerevisiae*, which is often included as control for non-clinical yeasts. More exhaustive test strips such as the API 50CH (Ison, 1987) or the ATB 32C (Michelle Schofield and David

Quain, unpublished results) may be more appropriate for the identification of yeasts commonly found in alcoholic and soft drinks.

A more detailed phenotypic approach has been described (Duarte *et al.*, 1999) that exploits enzyme polymorphism between species. Electrophoretic fingerprinting of glucose-6-phosphate dehydrogenase and three other enzymes enabled the clear and reproducible differentiation of the four species that comprise *Saccharomyces sensu stricto*. Similarly, this approach is reported (Duarte *et al.*, 1999) to generate distinct profiles for yeasts as diverse as *Pichia membranifaciens* and *Zygosaccharomyces bailii*.

Genetic approaches to yeast taxonomy provide 'the bigger picture' as classification reflects the genome rather than the activity of, potentially, a single gene. Kurtzman & Phaff (1987) have reviewed developments in molecular taxonomy, which invariably revolve around measures of DNA 'relatedness'. For example, the molar percentage of guanine + cytosine (% G + C) in nuclear or mitochondrial DNA has been used with some success, although, of course, a similar G + C does not always mean similar DNA sequences. Indeed, as shown in Fig. 4.3 the range of '% G + C' for the 14 species that currently make up the *Saccharomyces* genus (Vaughan-Martini & Martini, 1998) is relatively limited with significant overlap between them. Other approaches, which include testing the homology of DNA-DNA (Kurtzman, 1998) or ribosomal RNA to DNA (Kurtzman & Blanz, 1998), are also finding application in determining phylogenetic relationships. Recent developments (see Section 4.2.6) in DNA 'fingerprinting' are increasingly providing more targeted molecular approaches to the identification and differentiation of yeast taxa.

### 4.2.2 Taxonomy of ale and lager yeasts

For most in the industry, *ale* and *lager* is a sufficient description of production yeasts. If pushed many would recognise *S. cerevisiae* as an ale yeast and *S. carlsbergensis* as lager yeast. Unfortunately, this happy state of affairs is taxonomically incorrect. Long ago in 1970 *S. carlsbergensis* was consolidated within *S. uvarum*, which itself was repositioned as *S. cerevisiae* in 1990 (Barnett *et al.*, 1990). To make matters worse, *S. cerevisiae var. carlsbergensis* is now classified as *S. pastorianus* (Barnett, 1992a; Pederson, 1995; Vaughan-Martini & Martini, 1998). However, there is now good evidence that in terms of DNA relatedness, bottom-fermenting lager yeasts are more correctly classified as *S. pastorianus* than *S. cerevisiae* (Rodrigues de Sousa *et al.*, 1995). In the light of this taxonomic confusion, it has been argued that despite taxonomic consolidation, there is a case for 'special purpose taxonomy' (van der Walt, 1987). This would allow the retention of industrially applied (and understood) names such as *S. cerevisiae* and *S. carlsbergensis* within the brewing industry.

It can be argued that the debate about the taxonomy of lager yeasts is merely one of semantics. Yet it does obscure the reality that although related, ale and lager strains are fundamentally distinct when viewed from a genetic or physiological standpoint. A clue to explaining this (see Fig. 4.4), was provided by Vaughan-Martini and Kurtz-man (1985) who demonstrated that *S. pastorianus* (syn. *S. carlsbergensis*) exhibits a relatively high DNA homology with both *S. cerevisiae* (53%) and *S. bayanus* (72%). However, tellingly, *S. cerevisiae* and *S. bayanus* have little commonality in their DNA. These observations have provided compelling evidence that lager strains are a

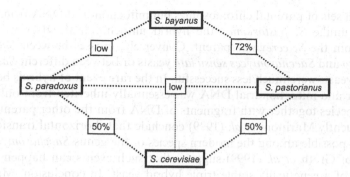

**Fig. 4.4** The *Saccharomyces sensu stricto* group.

hybrid of *S. cerevisiae* and either *S. bayanus* (Vaughan-Martini & Kurtzman, 1985; Turakainen *et al.*, 1994) or the closely related *S. monacensis* (Pedersen, 1994; Hansen & Kielland-Brandt, 1994). Although originally isolated in 1882 by Hansen as 'bottom fermenting yeast no. II' (Kielland-Brandt *et al.*, 1995), *S. monacensis* is a synonym of *S. pastorianus* (Vaughan-Martini & Martini, 1998) but, confusingly, is considered as *S. bayanus*.

Not surprisingly, given its hybrid nature, the genome of *S. pastorianus* is 50 (Vaughan-Martini & Kurtzman, 1985) to 60% (Rodrigues de Sousa *et al.*, 1995) bigger than that of *S. cerevisiae*. There is also growing evidence that the lager yeast genome has inherited more of its genetic material from *S. bayanus* than *S. cerevisiae* (Montrocher *et al.*, 1998). Further, there are reports that the mitochondrial DNA (Piskur *et al.*, 1998) and ribosomal DNA (Montrocher *et al.*, 1998) in *S. pastorianus* is derived from *S. bayanus* rather than from the other parent, *S. cerevisiae*. Extending the theme, analysis of chromosomes in bottom-fermenting isolates of *S. pastorianus* has shown the coexistence of chromosomes from both parents (Tamai *et al.*, 1998). This is in keeping with an earlier report (Hansen & Kielland-Brandt, 1994) that *S. carlsbergensis* (syn. *S. pastorianus*) contained two genes (MET2) encoding homeoserine acetyltransferase. One gene was from *S. cerevisiae* and the other MET2 gene showed high homology for sequences in type strains of *S. monacensis* and *S. bayanus*. Similar observations (Masneuf *et al.*, 1998) have been made with commercial isolates of a cider and wine yeast which have both been shown to have *S. cerevisiae*-like and *S. bayanus*-like chromosomes and both forms of the MET2 gene. Groth *et al.* (1999) have reported further genetic studies on the French cider yeast (CID1) of Masneuf *et al.* (1998). Intriguingly, CID1 is a hybrid of three yeasts, *S. cerevisiae*, *S. bayanus* and a *Saccharomyces* species (IFO 1802) contributing the mitochondrial DNA which has only been found in Japan.

A possible mechanism for the evolution of hybrid yeast species has been proposed by Marinoni *et al.* (1999). In model laboratory experiments, crosses were made between a number of *Saccharomyces* species, some which were closely related (*Saccharomyces sensu stricto*) and others (*Saccharomyces sensu lato*) with less in common. Perhaps not surprisingly, crosses between *S. cerevisiae* and *S. bayanus* resulted in viable and stable genetic hybrid yeasts. Like *S. pastorianus*, the hybrid variant contained both sets of parental chromosomes but mitochondrial DNA from one parent. However, unlike *S. pastorianus*, the hybrid mitochondrial DNA was preferably sourced from the *S. cerevisiae* parent. Conversely, crosses between *Saccharomyces sensu stricto* and *Saccharomyces sensu lato* yeasts or between different *Saccharomyces sensu lato* yeasts were much less successful. In the rare event of hybrids being formed, the nuclear and mitochondrial DNA was essentially inherited from only one of the parental species together with fragments of DNA from the other parent.

Consequently Marinoni *et al.* (1999) conclude that 'horizontal transfer of genetic material is possible among the modern species of the genus *Saccharomyces*'. Indeed, the work of Groth *et al.* (1999) suggests that such events can happen twice in the formation of a genetically stable triple hybrid yeast. In conclusion, Masneuf *et al.* (1998) round things off nicely by commenting that:

> 'it appears that among the *Saccharomyces* yeasts used in the fabrication of wine, cider and beer, stable interspecies hybrids are quite common. Whether such hybrids originate from events having taken place in the production environments or in nature is not known'.

Although perhaps logical, the outcomes of the above discussion are too unwieldy and taxonomically unstable for everyday use in the brewing industry. The take-home message is simple. Irrespective of nomenclature and classification, top-fermenting ale yeasts – *S. cerevisiae* – and bottom-fermenting lager yeasts – *S. pastorianus/carlsbergensis* – are taxonomically and genetically different (see Section 4.2.3). This accepted, most will feel comfortable with 'ale or lager strains' of *S. cerevisiae* – an approach used here. Alternatively, some will sympathise with Pedersen's (1995) proposal 'that the species name *S. carlsbergensis* is reintroduced as a specific name for lager brewer's yeast.' Whatever, for simplicity and clarity, throughout this book brewing yeasts are described as *S. cerevisiae*. Although taxonomically more precise, lager strains are only described as *S. pastorianus* where explicitly necessary.

### 4.2.3   *Diversity and differences between ale and lager yeasts*

*S. cerevisiae* is well placed to produce alcoholic beverages as the *Saccharomyces* are – almost uniquely – capable of the rapid anaerobic (i.e. in the absence of oxygen) fermentation of simple sugars to produce ethanol (Visser *et al.*, 1990; Barnett, 1992a). Why this should be is not clear but it has been proposed that anaerobic growth was facilitated by the duplication of the *Saccharomyces* genome during evolution (Wolfe & Shields, 1997). Indeed, some of the duplicated genes (including sugar transporters) are regulated differently under aerobic and anaerobic conditions (see Section 4.3.2.1). Table 4.10 provides an overview of the various sugars that *S. cerevisiae* can grow on anaerobically and aerobically. From the viewpoint of the development of alcoholic beverages, it can be seen with hindsight why *S. cerevisiae* has evolved to be so successful a micro-organism. Clearly, its physiological credentials of rapid anaerobic growth with concomitant production of ethanol offer a selective advantage for both yeast and man.

Although taxonomically closely related – or even indistinguishable – strains of *S. cerevisiae* have evolved through selection and have become adapted to specific and

**Table 4.10**  Growth of *S. cerevisiae* on different carbon sources.

| Carbon source | Anaerobic growth | Aerobic growth |
|---|---|---|
| D-glucose | ✓ (Yes) | ✓ |
| D-galactose | Variable | Variable |
| Methyl-α-D-glucopyranoside | Variable | Variable |
| Maltose | Variable | Variable |
| Sucrose | Variable | Variable |
| Trehalose | Variable | Variable |
| Melibiose | Variable | Variable |
| Lactose | × (No) | × |
| Cellobiose | × | × |
| Melezitose | Variable | Variable |
| Raffinose | Variable | Variable |
| Inulin | × | × |
| Starch | Variable | Variable |
| L-sorbose | × | × |
| DL-lactate | × | Variable |
| Glycerol | × | Variable |
| D-glucosamine | × | × |
| D-glucitol | × | × |
| D-mannitol | × | Variable |
| Succinic acid | × | Variable |
| D-xylose | × | × |
| Xylitol | × | × |
| Ethanol | × | Variable |
| Methanol | × | × |

mutually exclusive applications. For example, strains of *S. cerevisiae* used in baking or winemaking may produce splendid bread or wine but quite unacceptable beers. Similarly, there are many environmental 'wild' yeasts, which, although taxonomically *S. cerevisiae*, result in the 'spoilage' of beer through the formation of unpleasant flavours or aromas. An excellent example of this is those strains of *S. cerevisiae* able to ferment non-fermentable oligosaccharides or starch (formally described by the synonym of *S. diastaticus*). These strains cause spoilage through fermenting further than brewing strains as well producing an unpleasant 'phenolic' type aroma. To complicate matters further, *S. cerevisiae* is not exclusively an industrial micro-organism but is perhaps the most popular laboratory organism used world wide as a 'model eukaryote'. These strains differ significantly from brewing strains of *S. cerevisiae* in that they are genetically simpler and are well defined. Such strains lack the robustness and genetic complexity of their taxonomic relatives and are incapable of producing anything remotely like beer!

Whilst the taxonomic debate lumbers on, there is compelling evidence that ale and lager strains are different. As noted above, the genomes of *S. cerevisiae* and *S. pastorianus* have some commonality (e.g. the '*cerevisiae*' genome) and fundamental differences (e.g. the *S. bayanus* element in the *S. pastorianus* genome). A fundamental point of difference well known to taxonomists and yeast physiologists is the capability of lager strains to utilise the disaccharide melibiose (α-D-galactose-(1→6)-α-D-glucose). Genetically, this capability is complex (Turakainen *et al.*, 1994) involving up to ten *MEL* genes, which are exclusive to strains fermenting melibiose (Naumov *et al.*, 1995). Lager yeasts do not grow on melibiose as such but on the products of its hydrolysis – galactose and glucose – via α-D-galactosidase (for a review see Barnett, 1981).

This has provided a simple route to the differentiation of ale and lager strains of *S. cerevisiae* (see Sections 4.2.5.2 and 4.2.5.6). The activity of the α-D-galactosidase is highly regulated being induced by galactose but repressed by glucose (Gadgil *et al.*, 1996). In passing, 'melibiose utilisation' has been one of the targets of the genetic modification of bakers' yeast. Beet molasses, the feedstock for bakers' yeast, contains up to 3% raffinose which is normally unavailable to the 'highly specialised strains' used in the baking industry (Vincent *et al.*, 1999).

Lager yeasts also exhibit a greater affinity for galactose than ale strains. In aerobic culture, lager strains metabolise galactose and maltose simultaneously whereas ale strains ferment the maltose preferentially (Crumplen *et al.*, 1993). Lager strains also appear to form more sulphite than ale yeasts in wort fermentations (Crumplen *et al.*, 1993). Other differences have been reported in the utilisation of the wort sugar, maltotriose which lager strains seem able to use more rapidly than ale strains (Stewart *et al.*, 1995). As noted earlier (Section 4.2.1), bottom-fermenting yeasts (*S. pastorianus*) differ from ale strains (*S. cerevisiae*) in that lager yeasts transport fructose via a proton symport mechanism (Rodrigues de Sousa *et al.*, 1995).

However, process development and change have undermined some of the more empirical distinguishing features between ale and lager yeasts. For example, as rule of thumb, ale yeasts were 'top-fermenting yeasts' and lager yeasts were 'bottom-fermenting yeasts'. Today, such a distinction has become blurred. The use of large cylindroconical fermenting vessels has resulted in ale yeasts – classically top fermenting – becoming through selection essentially bottom fermenting. Similarly, it is a fundamental 'rule' that lager yeasts perform ideally at low temperatures (8 to 15°C) whereas ale yeasts operate best at higher temperatures (approx. 20°C). Physiologically this is borne out by laboratory studies (Walsh & Martin, 1977) that show ale strains to have a higher maximum growth temperature (37.5 to 39.8°C) than lager strains (31.6 to 34.0°C). Similarly, there are differences in optimal growth temperature ($T_{opt}$) with lager yeasts (*S. pastorianus*) having a $T_{opt}$ below 30°C and ale strains (*S. cerevisiae*) a $T_{opt}$ above 30°C (Guidici *et al.*, 1998). These differences are used taxonomically to type yeast species within the *Saccharomyces*, where *S. pastorianus* never grows above 34°C whereas strains of *S. cerevisiae* happily grow at 37°C (Vaughan-Martini & Martini, 1998). Further, a detailed comparison (Walsh & Martin, 1977) of growth rates between 6 and 12°C, showed a lager strain to grow significantly more quickly than an ale strain. Consequently, it would appear that lager strains are better 'equipped' than ale strains to grow at lower temperatures. It is tempting to conclude that this reflects the contribution to the lager yeast genome of *S. bayanus* which is noted to be 'cryophillic' (Naumov, 1996). However, the view that lagers can only be fermented at low temperatures and are unacceptable when produced at ale temperatures (Lewis, 1974) has been challenged. Perfectly acceptable, and, in terms of product matching, indistinguishable lagers can be produced at ale fermentation temperatures (unpublished observations, David Quain & Chris Boulton).

As noted above, lager yeasts are by no means as diverse as ale strains. In a fascinating account Casey (1990) notes that top-cropping yeasts had been in use – mostly unknowingly – for at least 3000 years (Samuel, 1997). Conversely, bottom-cropping lager yeasts were used exclusively by Bavarian brewers until the 1840s when they were smuggled to Czechoslovakia and Denmark. With increasing trade and travel, lager yeasts were soon being used worldwide. Consequently, compared to ale yeasts, the diversification of lager strains is relatively new. Indeed, Casey (1996) – using chromosomal fingerprinting (see Section 4.2.6.3) – has shown that a selection of lager strains from throughout the brewing world have essentially one of two basic finger-prints ('Tuborg' and 'Carlsberg') with small but reproducible differences between strains. Conversely, similar analysis of numerous ale yeasts failed to show any form of common fingerprint. Casey (1996) concludes that ale yeasts 'lack a common origin' with 'breweries in different regions' selecting 'unique yeast strains for each location'.

Casey's observations are notable as they strike at the heart of strain diversity. The lager strains originate from one or two common sources but although closely related are genuinely different. This clearly points at genetic change through environmentally driven adaptation or through ongoing chromosomal rearrangement. Whatever the driver, it is increasingly recognised that genetic change or 'instability' is far more common than previously thought in brewing yeast (see Section 4.3.2.6). Further complexity is introduced when considering the contribution of the environment to strain diversity. Not only would the indigenous population of *S. cerevisiae* be expected to be influenced by geography but – overlaid on top – are genetic changes that may (or may not) offer some selective advantage. Musing on the genetic diversity of brewing yeast strains is inevitably speculative. Perhaps, in the future, Casey's initial work will be developed and the genealogy of ale yeasts will begin to be unravelled.

There is no universal view as to the commercial need for a diversity of brewing yeast strains. Some breweries derive their portfolio of products from one or two yeasts. Other breweries maintain that a diversity of different brewing yeasts are necessary to produce a diversity of products. If anything, the 'reductionists' have gained the upper hand. Brewery closures and consolidation of brewery production within large efficient breweries have tended to result in the culling of yeast strains servicing relatively minor brands. Although doubtless necessary, this is unfortunate as the rich diversity of the currently unfashionable or process-difficult yeasts are being lost forever.

4.2.3.1 *Culture collections of brewing yeasts.* One solution to this loss of diversity is to purchase suitable yeast strains from a culture collection. Here, yeasts have been deposited – often anonymously – which are being, or more often, have been used in commercial fermentations. Most collections also maintain 'type' strains, genetic strains, genetically modified strains and general strains 'of interest'. Invariably, a small biography is attached which details what is known about the yeast strain (source, use, flocculence, genetic markers etc.). It is usual practice to store strains in, or above, liquid nitrogen or, occasionally, freeze-dried to assure long-term survival and minimise genetic change.

Arguably, the best-known brewing yeast culture collection is the UK's National Collection of Yeast Cultures (NCYC) which currently holds almost 700 strains. Deposited strains date back to 1910 but were presumably in use before that (Anonymous, 1995). Although selection of strains can be made conventionally via a hard-copy catalogue, the NCYC database can be searched via the Internet according to the desired strain criteria (attenuation, head formation, flocculence, rate of fermentation etc.). Other important collections that include brewing yeasts (often local) are the American Type Culture Collection (ATCC), the Centraalbureau voor Schimmel-cultures (CBS) in Holland, VTT Culture Collection in Finland and BCCM in Belgium. Details of how to contact these collections are given in Table 4.11.

**Table 4.11**  Commercial collections of yeasts.

| Collection | Address |
|---|---|
| American Type Culture Collection (ATCC) | 12301 Parklawn Drive, Rockville, MD 20852, USA<br>*Tel:* (+1) 301 881 2600<br>*Fax:* (+1) 301 816 4365<br>www.atcc.org/general.html |
| Centraalbureau voor Schimmelcultures (CBS) | Yeast Division, Julianalaan, 2628 BD Delft, Netherlands<br>*Tel:* (+31) 15 278 7466<br>*Fax:* (+31) 15 278 2355<br>www.cbs.knaw.nl/www/cbshome.html |
| National Collection of Yeast Cultures (NCYC) | Institute of Food Research, Research Park, Colney, Norwich, NR4 7UA, UK<br>*Tel:* (+44) 1603 255 000<br>*Fax:* (+44) 1603 458 414<br>www.ifrn.bbsrc.ac.uk/NCYC/Brew.html |
| VTT | VTT Biotechnology and Food Research, PO Box 1500, FIN-02044 VTT, Finland<br>*Tel:* (+358) 9 456 5133<br>*Fax:* (+358) 9 455 2103<br>www.vtt.fi/bel/mib/prosmik/ekokelm.htm |
| Belgian Co-ordinated Collections of Micro-organisms (Agro)industrial fungi & yeasts collection | BCCM™/MUCL, Mycothèque de l'Université Catholique de Louvain, Place Croix du Sud 3, B-1348, Louvain-la-Neeuve, Belgium<br>*Tel:* (+32) 1047 3742<br>*Fax:* (+32) 1045 1501<br>http.www.belspo.be/bccm/mucl.htm |

### 4.2.4  Differentiation – an introduction

Most – if not all breweries – need to identify with some certainty *their* production yeasts. The need is usually proportional to the complexity of brewery operations. For example, breweries who use many brewing yeasts (or mixed strains), or who franchise brew, or regularly propagate new lines of yeast will periodically require the assurance of strain identity. Unfortunately providing a brewing strain with a unique 'fingerprint' via some laboratory test has for many years frustrated micro-biologists the world over. Strains that are so easily characterised in fermenter or in final product are difficult – seemingly impossible – to identify unambiguously via laboratory methods. Consequently, a multitude of approaches has, over the years, been developed to differentiate brewing production yeast strains. Many of the older methods, whilst not providing definitive identity, have found routine application in differentiation between strains. For many this was and is sufficient! However, where a more precise identification is required, the new approaches described in recent years at last offer the potential of unambiguously identifying and differentiating brewing yeasts.

For convenience, the methods for the differentiation of brewing yeast strains can be divided into 'traditional' and 'modern' methods. *Traditional* methods are generally based on what might be described as conventional brewing microbiology. Conversely, the *modern* methods owe little to brewing microbiology but to developments elsewhere. For example, the dramatic developments in the molecular biology of yeast have allowed the genetic differences between yeasts to be probed and 'fingerprinted'. A second fledgling route to differentiate brewing yeasts involves the application of sophisticated instrumentation/data handling which can be used to probe differences in cell composition between strains.

Unfortunately, despite much fevered activity, the take-up of the modern approaches to differentiation has been relatively poor. Such approaches still remain in the domain of 'research' with only a few reported applications in routine brewing QA (Wright *et al.*, 1994; Quain, 1995; Casey, 1996). There are at least three reasons why traditional methods of differentiation have not been superseded by modern methods: (i) the capital and revenue costs of implementing modern methods are high, (ii) the required 'skill sets' are not normally found in brewery QA laboratories and (iii) the overall cost/risk/benefit analysis is not attractive.

Developments in the differentiation of brewing yeast strains have been subject to review in recent years (Quain, 1986; Casey *et al.*, 1990; Meaden, 1990; Gutteridge & Priest, 1996; Russell & Dowhanick, 1996).

### 4.2.5 *'Traditional' methods*

The 'traditional' methods are tried and trusted approaches that offer some degree of strain identification and differentiation. They are not definitive or unequivocal. However these methods are generally simple, affordable and are 'fit for the purpose'. Such methods focus on the morphological, physiological and biochemical differences between strains of *S. cerevisiae*. Indeed some of these approaches exploit taxonomic differences whereas others relate more directly to brewing performance.

#### 4.2.5.1 *Plate tests.*   Isolation of micro-organisms by growth on solid agar plates remains of pivotal importance in microbiology (see Section 8.3.3.2). Despite this, our understanding of how yeast cells grow on agar plates has been scanty. This is changing with a series of publications that have begun to shed light on the growth of yeast colonies on solid media. For example a gene (*IRR1*) has been identified that is required for the formation of yeast colonies on solid media but has no role on growth in liquid culture (Kurlanzka *et al.* (1999). Elegant work by Meunier and Choder (1999) has quantified colony size such that 'solitary' colonies (an average of five colonies per 9 cm plate) contain greater than $10^9$ cells. Plates with 250 (more typical) and 5000 (heavily loaded) colonies per plate contain $5 \times 10^7$ and $10^6$ cells respectively. Surprisingly, the initial logarithmic growth rate resembles that in liquid culture and in 'solitary' colonies growth is biphasic with growth being focused on the periphery of the colony. Although this resembles the classic diauxie of growth on glucose in liquid culture, the second growth phase on plates does not reflect a transition from

fermentative to oxidative growth. Consequently, this biphasic growth pattern is observed with cells growing oxidatively on ethanol or with respiratory deficient petite mutants.

Colony morphology and shape is highly organised. Although typically circular, and, in older colonies, pyramidal, there are reports of quite atypical colony shape. For example, pseudohyphal outgrowth of colonies (also implicated in the penetration of human tissues by pathogenic isolates of *S. cerevisiae* (see Section 4.3.2.2) has been reported under specific conditions of nitrogen starvation (Wright *et al.*, 1993). Perhaps more intriguing is the report (Engelberg *et al.*, 1998) of stalk-like structures from the few yeast cells that survive UV irradiation of a lawn of cells on plates. These stalks can be as long as 3 cm and as wide as 4 mm. On resuspension in liquid and replating, the cells from the stalks grew as conventional colonies. This suggests that the cells in these fascinating structures are not mutants caused by the UV irradiation.

Unlike shaken liquid media, cells on plates are subject to a complexity of nutrient and metabolite gradients. It is interesting to speculate how nutrients are transported to cells at the top of 'multicellular' colonies, which are far removed from the agar surface. One mechanism – passive capillary movement – has been proposed (Meunier & Choder, 1999) although 'more active feeding mechanisms cannot be ruled out'. Indeed, the issue of cell and colony 'communication' has been enlivened by the report that ammonia acts as a signal between individual colonies of *Saccharomyces* and other yeast genera (Palkova *et al.*, 1997). Colonies pulse ammonia toward neighbouring colonies, which, in turn, reciprocate pulsing ammonia in the return direction. This results in growth inhibition of facing parts of both colonies, which thereby orientate their growth towards areas that minimise the competition for nutrients.

The giant (or 'solitary') colonies have a 'highly organised morphology specific for a particular yeast strain' (Palkova *et al.*, 1997). Indeed in brewing microbiology, the giant colony method is the most traditional of the 'traditional' methods of strain characterisation! Its beginnings can be traced back to 1893 (Hall, 1954) when Lindner reported that up to ten single yeast colonies on thick wort-gelatin plates gave rise to characteristic 'giant colonies' (Fig. 4.5) when incubated for *c*. three weeks at 18°C. Subsequent embellishments (Hall, 1954; Richards, 1967) added consistency and robustness to the method. Indeed, colonies could be classified into six groups (Richards, 1967) by the detailed analysis of colony appearance. With hindsight, the giant colony method seems a little archaic, the method had many devotees and was used successfully in breweries worldwide for many years. However, despite the method's virtues, the lengthy turnaround time for results is now clearly unacceptable in today's 'real time' world.

Chromogenic media is increasingly finding application in the differentiation and identification of micro-organisms, most notably for the 'rapid' differentiation of coliforms in water. In brewing, the use of colony colour to exploit differences in media acidification for the differentiation of yeasts has been somewhat limited. The most common application is in the use of WLN (see Section 8.3.3.2) agar – a general purpose 'green' media used in the detection of aerobic yeasts and bacteria. Colonies on WLN typically range from light lime green through to dark green. In experienced hands this can offer some differentiation between brewing strains. For example, Lawrence (1983) noted that increasing the concentration in WLN of Bromocresol green exaggerates any differences in colony colour such that closely related brewing yeasts can be differentiated on colony colour and size.

Lager and ale strains of *S. cerevisiae* can be most effectively differentiated by colony colour after growth in media containing a chromogenic substrate ('X-α-gal', 5-bromo-4-chloro-3-indoyl-α-D-galactoside) (Tubb & Liljeström, 1986). Lager yeast expresses a α-galactosidase, which cleaves X-α-gal to form an insoluble blue-green dye. Consequently lager yeast colonies are blue whereas ale colonies remain cream. Unfortunately, for this is a useful and effective method, subsequent health and safety concerns about the solvent used to prepare X-α-gal (N,N'-dimethy-formamide) have relegated the method from routine usage. However, the demise of approach is premature and the identification of an alternative, food-grade solvent, propanediol, has given the method a new lease of life. This modified method can also be used for the 'real time' detection (through the blue coloration) of very low levels of lager yeast contamination in ale yeast slurries (but unfortunately not *vice versa*) (Wendy Box, unpublished observations).

The antibiotic cycloheximide (or 'actidione') inhibits cytoplasmic protein synthesis in yeasts and other eukaryotes. Resistance in yeasts is associated with a single amino acid (proline to glutamine) modification of the ribosomal protein, L41 (Kawai *et al.*, 1992). In brewing microbiology (see Section 8.3.3.2), the inclusion of cyclohcximide makes growth media selective for bacteria and 'wild' yeasts, as brewing strains are sensitive to low levels of the antibiotic (1 mg/l). There have been reports (Harris & Watson, 1968; Gilliland, 1971a; Lawrence, 1983) that resistance to cycloheximide can be used to differentiate between brewing strains. However, as noted by Gilliland (1971a), differentiation of strains at such low concentrations of cycloheximide (0.06–0.4 mg/l) requires care (cycloheximide is heat sensitive). Gilliland (1971a) also expressed further reservations about this approach as strains can vary in their resistance to cycloheximide.

A more comprehensive approach has been described (Simpson *et al.*, 1992) which quantifies the inhibition of growth by six dyes and antibiotics (including cycloheximide). Discs impregnated with growth inhibitors are positioned around a ring, which is then placed on microbiological media seeded with a lawn of yeast. The degree of growth inhibition is quantified by measuring the zone of clearing around each inhibitor. Although many brewing strains exhibit the same inhibition 'fingerprint', some can be clearly differentiated. Like many methods, this approach is perhaps more useful for characterising 'wild' *Saccharomyces* and non-*Saccharomyces* yeasts.

4.2.5.2 *Flocculation tests.* Flocculation tests remain one of the most popular approaches to brewery yeast differentiation (see Section 4.4.7 for a detailed review of 'flocculation'). This is hardly surprising as – even in closed vessels – *flocculence* remains one of the handful of characters used by brewers to describe the process performance of a yeast. However, despite very real differences in flocculence in fermenter, laboratory methods to differentiate strains on the basis of differing flocculence remain based on empirical approaches devised many years ago (Burns, 1941; Gilliland, 1951; Hough, 1957).

The various flocculation tests are described in Table 4.12. Of them, perhaps only the 'Gilliland' method remains in widespread use. This approach was developed some

(a)

makes them able selective for bacteria and wild yeasts, as the enumeration is
sensitive to low levels of the total count ($1 \text{ mg/l}$). There have been reports (Harris &
Watson, 1968; Gil *et al.*, 1978; Lawrence, 1983) that resistance to ethanol, and so
are used to differentiate between brewing strains. However, as noted by Graham
(1970), differentiation of strains of such low concentrations as etched nitro Ole
0-4 (m/l) requires test flocculmeasile is very sensitive. Gilliland (1971) also
expressed further reservations about this approach as strains can vary in their char-

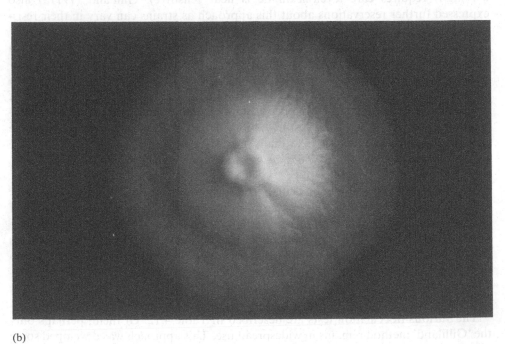

(b)

(c)

**Fig. 4.5** Examples of giant colonies: (a) ale, (b) lager and (c) wild yeast (all × 10 magnification).

years earlier (Donnelly & Hurley, 1996) by J.W. Tullo to monitor the proportion of the two strains used in Guinness fermentations. Indeed, in 1932 these methods revealed that the yeast consisted of four distinct varieties of yeast ranging from non-flocculent or 'dispersed' (Class I) to the heavily flocculent or 'filamentous' chain forming strains (Class IV). The typical fermentation performance of the four classes is shown in Fig. 4.6.

Although these methods (Gilliland, 1951) are inevitably retrospective, they have found routine application for monitoring pitching yeast purity (Gilliland, 1971a). Indeed the Gilliland method is still is use today at Guinness (Donnelly & Hurley, 1996) and in Bass Brewers. In the later case, the Gilliland method is used to track the proportion of two strains used in ale fermentations where one strain is Class I and the other is Class II. Remarkably, the Gilliland classification has even penetrated the very different world of advertising. A poster for 'Draught Bass' in the late 1980s noted that '... our two yeasts – named Gilliland I and II, after the chemist who classified them – have been together now for over 200 years'!

4.2.5.3 *Fermentation performance.* Although fraught with difficulty, using 'fermentation performance' to characterise brewing yeasts is potentially a most rewarding approach. Not only is a yeast strain presumably identified but a whole host of useful data can be gleaned in terms of process performance and product quality. It is perhaps no surprise that the methods described to assess 'fermentation performance' are fundamentally the same! This reflects the difficulties encountered in attempting to mimic production fermentations in the laboratory. Although stirred

**Table 4.12** Flocculation tests used to differentiate brewing yeasts.

| Author | Method |
|---|---|
| Burns (1941) | <ul><li>Laboratory wort fermentations.</li><li>Flocculation one of a number of fermentation tests.</li><li>Flocculation assessed in 'own beer' or distilled water.</li><li>Flocculation in sodium acetate-acetic acid buffer (pH 4.6) is of 'cardinal significance'. Clumped yeast is quantified.</li></ul> |
| Gilliland (1951) | <ul><li>50 colonies are inoculated into $50 \times 5\,ml$ wort cultures.</li><li>After three days the cultures are examined and the appearance of the yeast sediment noted after (i) gently swirling the culture and, where the yeast remains compact, (ii) decanting all but 0.5 ml of the supernatant and then shaking the sediment.</li><li>Four classes can be identified.</li><li>Class I – compact sediment which on resuspending in 0.5 ml is completely homogeneous.</li><li>Class II – compact sediment which on resuspending is distinctly granular.</li><li>Class III – sediment peels away in large flakes/dense round clumps which will not disperse.</li><li>Class IV – sediment consists of loose flakes/clumps of yeast.</li></ul> |
| Hough (1957) | <ul><li>Yeasts are characterised according to whether they flocculate (or not) under the below defined conditions.</li><li>Washed yeast (0.5 g) is suspended in 10 ml calcium chloride (0.1%, w/v).</li><li>One aliquot (0.5 ml) is adjusted to pH 3.5 and another to pH 5. Flocculence is assessed.</li><li>Where flocculation is observed at pH 3.5, maltose is added (final concentration 10%, w/v). In some strains maltose disperses flocculant cells.</li><li>Where flocculation is not found at pH 3.5, ethanol is added (final concentration 2%, v/v). Some strains then flocculate.</li><li>Where flocculation is not observed at pH 5, a second strain of *S. cerevisiae* is added and co-flocculation assessed. Those coflocculating with NCYC 1108 are described as 'type I', those with NCYC 1109 as 'type II' and those failing to flocculate as 'type III'.</li></ul> |

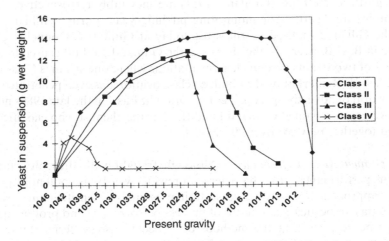

**Fig. 4.6** The fermentation performance of yeasts from the four Gilliland flocculation classes (redrawn from Gilliland, 1951).

fermentation systems have their place, the two-litre 'tall tube' fermenter – based on the dimensions of a cylindroconical fermenter – has found particular application in studies of fermentation performance (Walkey & Kirsop, 1969; Gilliland, 1971a; Thorne, 1975; Bryant & Cowan, 1979). Although a fruitful approach, tall tube fermentations are regarded by many as 'user hostile'. The Carlsberg Research Centre has overcome such concerns via the development of automated systems such as the 'multiferm' (Sigsgaard, 1996) and Octo-Duo-Ferm (Skands, 1997) which, respectively, consist of 60 and 16 two-litre tall tube fermenters. Other developments have focused on data handling (Bryant & Cowan, 1979) and product analysis (Gilliland, 1971; Thorne, 1975).

There is consensus (Burns, 1941; Walkey & Kirsop, 1969; Gilliland, 1971; Thorne, 1975; Bryant & Cowan, 1979) that the fermentation performance of brewing yeast can be described broadly by five criteria. These include (i) the formation of a yeast head, (ii) deposit formation (flocculation), (iii) rate of fermentation, (iv) extent of fermentation and (v) clarification after fining. Although clearly adding value, analysis of beer flavour through tasting panels or 'analytically' (Gilliland, 1971; Thorne, 1975) offers little diagnostic benefit.

Unfortunately, 'fermentation performance' lacks sufficient sensitivity to differentiate between closely related brewing yeasts. Indeed, in a major survey, Walkey and Kirsop (1969) used the above 'five criteria' to characterise 153 brewing strains in tall tube fermentations from the National Collection of Yeast Cultures. This approach was extended by Bryant and Cowan (1979) to include a further 82 strains from the NCYC but with sophisticated data treatment of all 235 strains by principal coordinates analysis.

Although essentially similar conclusions can be drawn from both studies, Bryant and Cowan (1979) identified five distinct groups dependent on head or deposit formation (see Table 4.13). Group 1 consisted of low head (score of 1) and high deposit (score of 5) formers. Conversely group 5 consisted of head forming, low deposit strains. Groups 2–4 catered for strains falling between the extremes. Further subdivision was effected by 'A' (high clarification) or 'B' (higher extent of attenuation) subgroups.

**Table 4.13** Classification of 235 brewing strains by fermentation performance (from Bryant & Cowan, 1979).

| Group | Number of strains | Head formation | Deposit formation | Degree of attenuation | Rate of attenuation | Clarification |
|-------|-------------------|----------------|-------------------|-----------------------|---------------------|---------------|
| 1A | 42 | 1.0 | 5.0 | 4.0 | 4.3 | 4.0 |
| 1B | 2 | 1.0 | 5.0 | 5.0 | 5.0 | 1.0 |
| 2A | 37 | 2.1 | 4.0 | 3.8 | 4.1 | 3.8 |
| 2B | 4 | 1.5 | 3.5 | 5.0 | 4.8 | 1.0 |
| 3A | 15 | 2.9 | 2.9 | 4.5 | 4.0 | 3.5 |
| 3B | 15 | 1.7 | 1.7 | 4.9 | 4.3 | 1.1 |
| 4A | 7 | 4.7 | 2.7 | 3.3 | 4.3 | 5.0 |
| 4B | 24 | 3.4 | 1.4 | 4.9 | 4.0 | 1.3 |
| 5A | 29 | 5.0 | 1.0 | 1.7 | 4.3 | 4.6 |
| 5B | 60 | 5.0 | 1.0 | 4.4 | 4.1 | 2.6 |

Although useful in broadly classifying brewing yeast strains, the real value of this approach is in selection of strains that are 'fit for the purpose'. Strains in group 1 (19% of 235 strains tested) are clearly best suited to fermentation in conical vessels (non-head forming and flocculent) whereas strains in group 5 (38%) are most suitable in traditional open vessels where cropping is through skimming the yeast head.

4.2.5.4 *Assimilation/fermentation*. Taxonomists have long used fermentation and assimilation tests as a route to typing yeasts (Barnett *et al.*, 1990). Typically, *fermentation* is judged by the formation of carbon dioxide (usually in a Durham tube) from a carbon source under anaerobic or 'semi-anaerobic' conditions. Conversely, *assimilation* of a carbon source is an aerobic process that is usually judged by measurement of yeast growth. Although normally performed in liquid culture (5–10 ml), assimilation can also be assessed on solid media. For example, *auxanograms* where pour plates of yeast nitrogen base (without a carbon source!) are seeded with crystals (approx. 5 mg) of each test compound on the agar surface. Where assimilation occurs growth is observed. Another approach – *replica plating* – involves the transfer of colonies from a master plate to a variety of agar plates containing different carbon sources. Transfer is achieved by placing a sterile velvet pad on the master culture, which is then used to 'inoculate' the test plates.

Although seemingly simple and effective, both fermentation and assimilation tests have been found to be technically wanting! The methods' very simplicity can be their undoing inasmuch that results from such tests are frequently unreliable and often ambiguous. For example, Barnett *et al.* (1990) note that the interpretation of fermentation tests in Durham tubes can be complicated by any aerobic metabolism of the test compound, background growth on contaminating substrates, flocculation and protracted incubation time. Similarly, Visser *et al.* (1990), in a wide ranging study of 75 yeast genera, noted the need for genuine anaerobic conditions and that visible gas production was unreliable as a criterion for fermentation – the measurement of ethanol being a more reliable approach.

Liquid culture assimilation tests require even more care! As noted by Barnett *et al.* (1990), the limited oxygen transfer compromises the use of static, tilted test tubes to test the aerobic assimilation of carbon sources. Although rocking/agitation will undoubtedly improve matters, the preferred route is to move from test tubes into triangular/conical flasks that are agitated in orbital incubators or shaking baths. Indeed, *aerobic* assimilation tests can only reliably be performed under *truly* aerobic incubation conditions.

These and other complications, were graphically demonstrated by Quain and Boulton (1987a) who studied the growth and metabolism of the sugar alcohol – mannitol – by over 40 strains of *S. cerevisiae*. The assimilation of mannitol was shown to be obligately aerobic, with growth simply stopping under anaerobic conditions and restarting when switched back to aerobiosis. Of the 20 or so strains capable of growth on mannitol (5%, w/v), some (but not all) were *unable* to grow at low substrate concentrations (1–2%, w/v). This is important as, typically, in taxonomic assimilation tests the concentration of the carbon source is less than 1% (Lodder, 1970; Barnett *et al.*, 1990). Further, the composition of the media can have a dramatic impact on assimilation. Growth on mannitol could not be demonstrated in fully defined yeast nitrogen base but required the presence of yeast extract and 'salts'. As the taxonomic approach is to use yeast nitrogen base it is perhaps no surprise that mannitol utilising yeasts have been reported as being unable to use the sugar alcohol when subject to 'blind' typing by the National Collection of Yeast Cultures (David Quain, unpublished).

Despite these reservations, the fermentation and/or assimilation of various carbon sources by strains of *S. cerevisiae* has been described in the taxonomic texts as being 'variable' (Lodder, 1970; Barnett *et al.*, 1990, Table 4.10). Although fermentation based tests have little application in the differentiation of brewing yeast strains, assimilation of various carbon sources may provide simple differentiation between strains. For example, Kirsop (1974a) was able to segregate 130 strains of *S. cerevisiae* into 25 groups dependent on their assimilation of melizitose, α-methyl D-glucoside, lactic acid, glycerol, ethanol and succinic acid.

One fermentation test that does find application is as noted in Section 4.2.5.2, the differentiation of lager and ale strains of *S. cerevisiae* by exploiting the presence of periplasmic α-galactosidase in lager yeasts. These strains are able to hydrolyse melibiose (galactosyl-α-1,6-sucrose) to the fermentable products, sucrose and galactose. Without the α-galactosidase, ale strains exhibit no growth on melibiose.

4.2.5.5 *Immunology.* Over the years, there has been sporadic interest in using immunological methods to identify and differentiate *Saccharomyces* strains. More recently, such methods have perhaps found greater application in the identification and detection of beer spoilage bacteria (Eger *et al.*, 1995; Russell & Dowhanick, 1996; Whiting *et al.*, 1999).

Fundamental to immunology is, of course, the antigen–antibody reaction (see Muller, 1991 for a review). An *antibody* (or immunoglobulin) is a protein synthesised by an animal in response to an *antigen* or 'foreign' macromolecule. Either antibodies can be highly specific for an antigen (monoclonal) or, alternatively, cheaper 'soups' of polyclonal antibodies can be raised against pure immunogens. In laboratory studies, antibodies are raised from rabbits by immunisation with immunogens (typically in this example, whole yeast cells or isolated cell walls will contain a number of antigens). After about five weeks, the antisera is recovered and treated with ammonium sulphate to isolate the immunoglobulin antibodies. In *S. cerevisiae* the immunodominant yeast antigens appear to be primarily the cell wall mannan side chains and, arguably, a group of glucan-free proteins (for a review see Fleet, 1991). Clearly, the complexity of the antigenic determinants potentially limits the application of immunology to differentiate brewing yeast. It would appear that there are common antigens as *S. cerevisiae* can be differentiated from other *Saccharomyces* species by two universal antigens (Umesh-Kumar & Nagarajan, 1991), whereas Cowan and Bryant (1981) reported 19 different antigens in a study of 43 brewing strains.

Methods differ in how antigen–antibody reactions are visualised and quantified. A range of approaches have been used in brewing microbiology including immunofluorescence (Cowland, 1968; Campbell, 1971; Gilliland, 1971a; Thompson & Cameron, 1971), immunofluorescence coupled with flow cytometry (Eger *et al.*, 1995), immunoelectrophoresis (Cowan & Bryant, 1981). The advent of ELISA (enzyme-linked immunosorbent assay) (Russell & Dowhanick, 1996) has led to the rejuvenation of immunoassay-based methods. This is because ELISA-based protocols are comparatively simple, hugely flexible and, importantly, allow quantification.

It is telling that recent developments such as ELISA have found little application in the differentiation or identification of brewing yeasts. This perhaps reflects the success of other approaches together with fundamental problems inevitable to any phenotypic method of brewing yeast differentiation – the contribution of growth physiology and the close relatedness of brewing yeasts. For example, Cowland (1968) using immunofluorescence, clearly demonstrated that the antigenic determinants of brewing yeast are determined by cell physiology and age. Thompson and Cameron (1971) drew similar conclusions and also reported that the application of immunofluorescence was limited to the differentiation of top and sedimentary yeast strains. Attempts to differentiate within these classes were unsuccessful. Subsequent work using the more sensitive immunoelectrophoresis (Cowan & Bryant, 1981) extended earlier work (Bryant & Cowan, 1979) which had used fermentation characteristics to classify brewing yeasts. Building on this initial work, Cowan and Bryant (1981) raised antisera against representative strains of the five brewing groups (Bryant & Cowan, 1979) which revealed a total of 19 different antigens distributed amongst the 43 strains examined in this study. As with previous work (Thompson & Cameron, 1971) the 'best fit' was between specific antigens and the fermentation properties of top and sedimentary strains (Cowan & Bryant, 1981).

### 4.2.5.6   Other approaches.

- *Temperature* – Lager and ale strains of *S. cerevisiae* can be simply differentiated by incubation at $37 \pm 0.5°C$ (Lawrence, 1983). Ale strains (and wild yeasts!) have a higher maximum growth temperature (37.5 to 39.8°C) than lager strains (31.6 to 34.0°C) (Walsh & Martin, 1977). The application of this seemingly simple method has been hampered by what is frequently poor temperature control of laboratory incubators. Ideally, to be meaningful, such studies should be performed in gradient heat blocks (Walsh & Martin, 1977) or gradient water baths (McCusker *et al.*, 1994).
- *'Volatiles'* – yeast metabolism during fermentation is responsible for the vast majority of esters and higher alcohols found in beer. Measurement of volatiles – either directly or organoleptically – has been described as part of broad approaches directed toward typing or differentiating brewing strains (Gilliland, 1971a; Thorne, 1975). Typically, analysis of volatiles would play a role in the sort of approaches previously described under 'Fermentation performance' (Section 4.2.5.3). Although something of an aside, the ratio of the higher alcohols, 2- and 3-methyl-1-butanol can be used to differentiate between lager and ale strains of brewing yeast (lager $= 0.23 \pm 0.026$ [$n = 11$ strains], ale $= 0.14 \pm 0.012$ [$n = 4$ strains], David Quain, unpublished).
- *Organic acids* – brewing yeast produces a variety of organic acids as by-products of metabolism. Coote and Kirsop (1974) tracked the formation of organic acids during fermentation and noted significant differences between brewing yeast strains. This potential approach to strain differentiation was extended in a series of publications (Bell *et al.*, 1991a–c) which concluded that

the organic acid profile could be used to discriminate after growth on the assimilatory carbon sources described by Kirsop (1974a, see Section 4.2.5.4) but not after growth on glucose.

- *Oxygen requirement* – the amount of oxygen necessary to support a successful standard fermentation varies between brewing strains (Kirsop, 1974b; Jacobsen & Thorne, 1980). Although of significance in terms of process performance and of interest in terms of yeast physiology, differing oxygen requirements are too cumbersome for the differentiation of brewing yeast strains.

### 4.2.6 'Modern' methods

In recent years great strides have been made in the differentiation of brewing strains. Unlike the 'traditional' methods, the 'modern' approaches have succeeded in the unequivocal identification and differentiation of individual brewing strains. This success has primarily been a consequence of the explosion of activity in the molecular biology of the *Saccharomyces*. The development of techniques to probe the yeast genome has spawned spin-off techniques that have been enthusiastically applied by geneticists to the differentiation of brewing yeasts. Although a simple premise that different strains are genetically different, the challenge has been to employ methods that reveal these differences. The three approaches to DNA fingerprinting described below have all been shown to achieve the differentiation of closely related brewing strains.

#### 4.2.6.1 *Genetic fingerprinting – RFLP*. Arguably one of the most successful approaches to yeast differentiation has been the RFLP (restriction fragment length polymorphism) DNA fingerprinting approach (for applied reviews see Meaden, 1990, 1996; Walmsley, 1994). The application of RFLP DNA fingerprinting to the differentiation of commercial brewing strains has been described (Schofield *et al.*, 1995; Wightman *et al.*, 1996), as has the its role in the routine QA of yeast supply to breweries (Quain, 1995). Not surprisingly, this approach is one of a number of fingerprinting methods that has found favour in characterising and tracking clinical isolates (see Section 4.3.2.2) of *S. cerevisiae* (Clemons *et al.*, 1997; McCullough *et al.*, 1998).

The RFLP approach to DNA fingerprinting is neither simple nor quick! Indeed, the introduction of RFLP into a brewing laboratory can be costly (hardware) and, from a zero base, can involve a steep learning curve. Although RFLP uses what are now standard molecular protocols, it is prudent to seek hands-on training with subsequent access to guidance and support. This can be important, as experience has shown that the interpretation of routine genetic methods by microbiologists can differ significantly from that of molecular biologists. This can lead to a variety of unfathomable problems and complications that can often be labyrinthine to unravel.

RFLP DNA fingerprinting can be viewed as a series of modular steps (Fig. 4.7) which take in all about five days from start to finish. In outline (see Meaden, 1996 for more detail) DNA is extracted and digested with a restriction enzyme. The enzymes – of which over 100 are now commercially available – are bacterial endonucleases, which cut DNA into fragments. Different restriction enzymes cut DNA at different

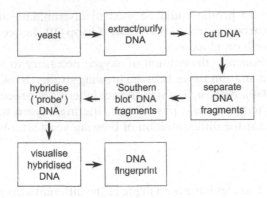

**Fig. 4.7**  The steps in RFLP DNA fingerprinting.

recognition sequences producing a different set of fragments. This soup of restriction fragments is separated by size (range 200–20 kilobases) using gel electrophoresis and is then subject to 'Southern blotting'. This step – named after its inventor, Ed Southern – transfers the separated DNA fragments from the gel to a nylon membrane.

This matrix enables the DNA fragments to be 'probed' with another piece of yeast DNA. This is necessary as restriction patterns alone support little or no differentiation between strains (Casey *et al.*, 1990). The choice of DNA probe (and restriction enzyme) is of critical importance to the differentiation of closely related brewing yeasts. Two multi-locus probes – poly GT (Walmsley *et al.*, 1989; Walmsley, 1994) and Ty 1 (Schofield *et al.*, 1995, Wightman *et al.*, 1996) – have been used which bind, respectively, to the many poly GT tracts or complimentary 'Ty elements' to be found in the yeast DNA fragments. This 'hybridisation' is visualised by radiolabelling or, preferably, by labelling the probe with a plant steroid, digoxigenin that is itself detected by a subsequent colour reaction.

The elegance of the colorimetric reactions that finally result in a 'fingerprint' is worthy of explanation (see Fig. 4.8). The digoxygenin (DIG) labelled probe is bound to an antibody ('ant-DIG') which in turn is conjugated with the enzyme alkaline phosphatase. This enzyme forms an insoluble blue dye when the chromogenic 'X-phosphate' is added.

The complexity of RFLP DNA fingerprints is shown in Fig. 4.9. Here three restriction enzymes – *Hind*III, *Eco*RI and *Pst*I – have been used to digest three different ale strains. It is apparent that each enzyme creates a different fingerprint for each yeast strain which, in turn, can differentiate between strains. This approach – *Hind*III and Ty1-15 probe – is routinely used to differentiate 24 commercial production yeasts (Quain, 1995) and periodically to successfully identify strains 'blind' (Philip Meaden, unpublished observations).

**4.2.6.2**  *Genetic fingerprinting – PCR.*  The polymerase chain reaction is a powerful molecular tool that exploits the way that DNA is naturally replicated. It is increasingly widely used in food microbiology (Candrian, 1995), particularly in the detection

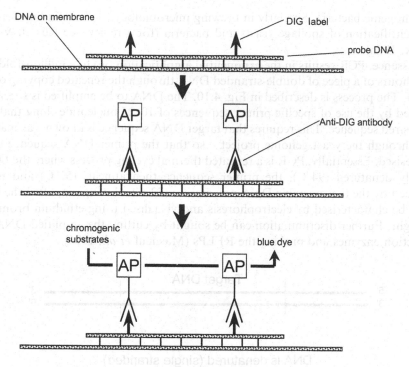

Fig. 4.8 The colourimetric 'cold' method for visualising an RFLP fingerprint.

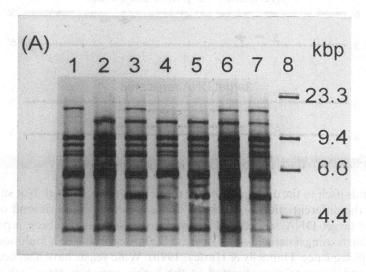

Fig. 4.9 A typical RFLP DNA fingerprint (kindly provided by Philip Meaden, ICBD, Heriot-Watt University).

of pathogenic bacteria. Similarly in brewing microbiology, PCR is finding favour in the identification of spoilage yeasts and bacteria (for a review see Russell & Dowhanick, 1996).

In essence, PCR results in the exponential amplification (up to a million-fold) over a few hours of a piece of double-stranded DNA through the repeated copying of each strand. The process is described in Fig. 4.10. The DNA to be amplified is specifically targeted by the use of specific primer sequences of 10–25 nucleotides long that flank the desired sequence. This requires that target DNA sequence is known – as many are now through the yeast genome project – so that the primer DNA sequence can be synthesised. Essentially, PCR is a repeated thermal cycling process where the DNA is initially denatured (94°C), the primer sequence then anneals (55°C) and is then extended by the thermostable DNA polymerase (72°C). Having amplified the DNA, it can be characterised by electrophoresis and visualised using ethidium bromide or UV light. Further discrimination can be sought by cutting the amplified DNA with restriction enzymes and probing the RFLPs (Masneuf *et al.*, 1996).

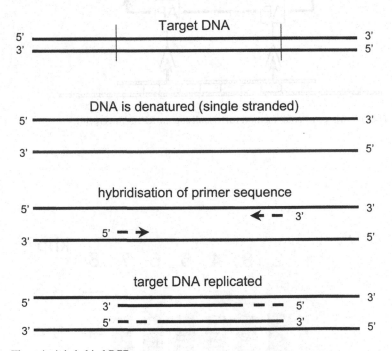

**Fig. 4.10**   The principle behind PCR.

PCR lends itself to the differentiation at the genus or species level. Not surprisingly, its use in the differentiation of strains is limited, as this would depend on changes within the target DNA sequence. However, some success has been reported with brewing yeasts using a defined set of probes (Walmsley, 1994) or a multi-locus primer sequence (δ sequence, Donnelly & Hurley, 1996). Wine yeasts have also been broadly differentiated using primers directed at the highly variable intron splice sites (de Barros Lopes *et al.*, 1996). Subsequent work with this approach (de Barros Lopes *et al.*, 1998) has shown the differentiation of the closely related cluster of species, *Saccharomyces sensu stricto* (see Section 4.2.1). However, in another study the use of hypervariable repetitive sequences was unable to successfully differentiate wine, brewing or baking strains of *S. cerevisiae* (Lieckfeldt *et al.*, 1993). Similarly, mixed success has been reported (Yamagishi *et al.*, 1999) for differentiation of brewing and non-brewing yeasts using PCR targeted to a fragment of the putative flocculation gene, *FLO1*. Much improved discrimination was reported by digestion of the PCR products with specific restriction enzymes and RFLP analysis (see Section 4.2.6.1). This hybrid approach of PCR and RFLP enabled the differentiation of brewing strains from wild *Saccharomyces* and non-*Saccharomyces* yeasts.

An alternative to the targeted probes is the use of randomly designed non-specific DNA probes ('random amplified polymorphic DNA' or RAPD PCR). This approach has the benefit of not requiring any prior knowledge of DNA sequence and can – if the primer sequence is 'right' – result in several PCR products as the random primer bind at several sites. These products are separated and visualised to give a fingerprint. This approach has been successful in the differentiation/identification of brewery bacteria (Tompkins *et al.*, 1996) or ale from lager yeasts (Laidlaw *et al.*, 1996).

Compared to the complexities and hardware of RFLP DNA fingerprinting, PCR techniques are simple, straightforward and rapid. However PCR has suffered somewhat from a 'bad press' inasmuch that 'contamination' can be an issue as can the presence of interfering/inhibitory beer components such as polyphenols (Dowhanick, 1995). In conclusion, the development of PCR in the brewing industry is presently more likely to be driven by the detection and identification of spoilage organisms rather than the differentiation of brewing yeast strains.

### 4.2.6.3 *Genetic fingerprinting – karyotyping.*

'Karyotyping' is the determination of chromosomal size and number (Meaden 1990). This approach has arguably been applied more successfully to the differentiation of brewing yeast strains (Casey *et al.*, 1988, 1990; Casey, 1996; Donhauser, 1997; Oakley-Gutowski *et al.*, 1992; Pedersen, 1994; Sheehan *et al.*, 1991) than any other method.

*S. cerevisiae* is particularly amenable to karyotyping as it has 16 chromosomes (more than other yeast!) which range dramatically in size from 230 kb (chromosome I) to approximately 1.5 Mb (chromosome IV). The size distribution of the 16 chromosomes is shown in Fig. 4.19 (Goffeau *et al.*, 1996). Such complexity is both a strength (karyotypes are likely to vary widely) and a weakness (electrophoretic separation of both big and small chromosomes at the same time is difficult).

The problem (or 'opportunity') of resolving yeast karyotypes was resolved with the development in 1984 of pulsed field gel electrophoresis (PFGE) (Carle & Olson, 1984; Schwartz & Cantor, 1984). This approach supports the separation of chromosomes ranging from 50 kb to 10 Mb in agarose gels when electric fields are applied in different directions. Different pulse times are used to separate only the small chromosomes ( < 500 kb, Chromosomes I, III, VI and IX) or all of the chromosomes (Sato *et al.*, 1994; Casey, 1996). Once separated the chromosomes are stained with ethidium bromide and then visualised under UV light. As with RFLP, karyotyping yields a barcode-like fingerprint (Fig. 4.11) which can be used to differentiate between strains.

Inevitably, the PFGE approach has been developed primarily in the area of elec-

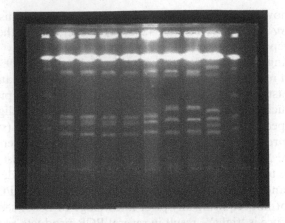

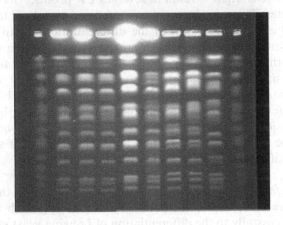

**Fig. 4.11** A typical karyotype DNA fingerprint (kindly provided by Miles Schofield, BRi).

trode design and consequent electrical field. Two approaches have found particular favour in the brewery applications of karyotyping – CHEF (contour clamped homogeneous electric field) and TAFE (transverse alternating field electrophoresis) (for details see Russell & Dowhanick, 1996).

Although the original PAGE approach has been used (Donhauser, 1997), the majority of reports on brewing yeast karyotytping have favoured the CHEF (Sheehan et al., 1991; Oakley-Gutowski et al., 1992, Pedersen, 1994; Wright et al., 1994) or TAFE (Casey, 1996; Casey et al., 1988, 1990) approach.

Unlike the RFLP approach, karyotyping offers more than strain fingerprinting and differentiation. In a notable paper, Casey (1996) was able to show that changes in chromosomal size and dosage can be associated with changes in cell colony morphology ('smooth' v. 'rough'), culture purity and long-term culture stability. Karyotyping was also able to track the genetic evolution of American lager strains which were shown to have originated 150 years ago from either the Tuborg or Carlsberg breweries in Denmark.

As is described in Section 4.3.2.5, Casey's (1996) observations fit the growing theme that brewing yeasts can be genetically unstable. Instability can be beneficial or detrimental. These so-called 'chromosome-length polymorphisms' (Zolan, 1995) are well-recognised in fungi and can be explained by movements of 'bits of DNA' either through deletions, insertions or translations within and between chromosomes.

An interesting extension of karyotyping has been the use of CHEF to study the ecology of wild *S. cerevisiae* flora in wine musts in Spain (Briones *et al.*, 1996). In a survey of 14 vats from three cellars, 392 discrete colony types were recovered. Remarkably, after CHEF analysis, 174 different karyotypes were identified which when subject to cluster analysis reduced to four major groups. Intriguingly, 'the majority of *S. cerevisiae* strains in cellars A and C possess the karyotype characterizing Groups 1 and 2; while in cellar B the strains correspond to three Groups: 1, 3 and 4' (Briones *et al.*, 1996).

### 4.2.6.4 *Genetic fingerprinting – AFLP.*

Amplified fragment length polymorphism (AFLP) is described (Savelkoul *et al.*, 1999) as the 'newest and most promising method' for the identification and typing of organisms at the DNA level. First described by Vos *et al.* (1995), AFLP can simply be viewed as a hybrid method comprising RFLP (see Section 4.2.6.1) and PCR (see Section 4.2.6.2). The extracted DNA is digested with two restriction enzymes, one with average cutting frequency (e.g. *Eco*RI) and one with a higher cutting frequency (e.g. *Mse*I). The fragments are selectively amplified by PCR which, after gel electrophoresis, results in 'highly informative' (Savelkoul *et al.*, 1999) fingerprints with 40–200 bands. As with all molecular fingerprinting approaches, the complex AFLP patterns require measurement and software driven data capture of the digitised images. For an insight into the many commercial packages used in fingerprinting, see Savelkoul *et al.* (1999).

AFLP has been used with a diversity of yeast species including those of the *Saccharomyces sensu stricto* cluster (see Section 4.2.1) (de Barros Lopes *et al.*, 1999). As the yeast strains used in this work mirror those used previously by this group with PCR (de Barros Lopes *et al.*, 1998), it is possible to compare the performance of the two approaches. The previous work (de Barros Lopes *et al.*, 1998), had been unable to differentiate two environmental isolates of *S. bayanaus* which could be differentiated using AFLP. This and other observations suggest that although more labour intensive, AFLP offers greater sensitivity, is more reproducible and is more 'discriminatory' than PCR.

However, as with many molecular approaches, AFLP has its limitations. Whilst particularly effective in demonstrating genetic relatedness and accordingly differentiation of domesticated industrial strains of *S. cerevisiae*, AFLP cluster analysis failed to identify the known genomic relationships in the closely related species found in the *Saccharomyces sensu stricto* group.

### 4.2.6.5 *Pyrolysis mass spectroscopy.*

The demands of clinical microbiology have spawned a variety of approaches that examine aspects of cell composition. Initially, pyrolysis gas chromatography (PyGC) and, now the more sophisticated, pyrolysis mass spectroscopy (PyMS) are increasingly being used to identify and differentiate micro-organisms to the level of genus, species and subspecies (for reviews see

Goodacre, 1994; Goodacre & Kell, 1996; Gutteridge & Priest, 1996). Perhaps the secret of PyMS's success is its simplicity and speed. Typically, sample time is less than two minutes with preparation limited to the careful transfer of a colony from a plate or drying down a few microlitres of a liquid culture. Indeed, Goodacre (1994) noted that a British daily newspaper had reported that PyMS was 'so simple that a chimpanzee could be trained to do it!'

In essence, pyrolysis involves the thermal degradation in an inert atmosphere of – in this case – microbial biomass. Typically the 'Curie point' pyrolysis temperature is 530°C which has been shown to give an ideal balance between the fragmentation of microbial polysaccharides and proteins (Goodacre, 1994). The ensuing volatile, low-molecular-weight fragments are typically separated using a mass spectrometer (PyMS) (see Fig. 4.12). This powerful approach has superseded separation using a gas chromatograph.

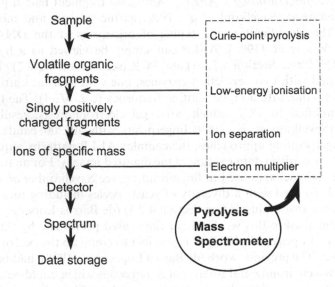

**Fig. 4.12** The main steps in PyMS (redrawn from Goodacre, 1994).

PyMS mass spectra provide relatively crude differentiation of micro-organisms if limited to fragment mass/charge ratio ($m/z$) and intensity. Typically, a range of fragments with a $m/z$ between 50 and 150 provide a quantitative spectral fingerprint (see Fig. 4.13). However, interpretation is not easy and the spectra are frequently qualitatively similar for different genera or species. The small but significant differences in mass intensities require sorting and 'reduction' using multivariate statistical techniques such as principal components analysis (PCA) together with the closely related, canonical variates analysis (CVA). These treatments enable samples to be grouped and differentiated between. Further analysis enables the construction of a 'percentage similarity matrix' and hierarchical classification to form dendrograms (Fig. 4.14) that enable the relationships between different strains to be visualised.

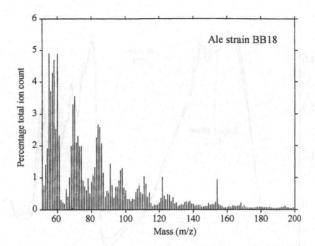

**Fig. 4.13** A typical normalised PyMS spectra (redrawn from Timmins *et al.*, 1998).

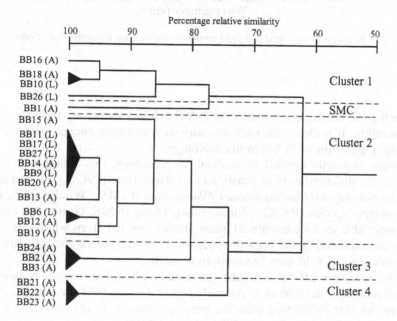

**Fig. 4.14** Dendrogram based on PyMS data showing percentage similarity between 22 brewing strains of *S. cerevisiae* (redrawn from Timmins *et al.*, 1998) Ale (A) and lager (L) strains are identified by their 'BB' coding.

A relatively new and exciting development in PyMS is the application of 'artificial neural networks' (AAN) to uncover complex relationships in multivariate data. Goodacre and Kell (1996) have described the use and benefits of 'supervised learning' through AANs over the 'unsupervised learning' of PCA and CVA. The chief advantages for PyMS of such 'trained' AANs are in routine discrimination of closely related biological samples, quantification and screening of microbial cultures and

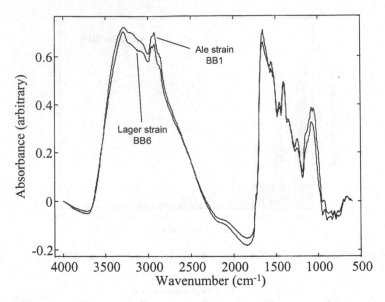

**Fig. 4.15** FT-IR spectra of two unrelated yeast strains (redrawn from Timmins *et al.*, 1998).

overcoming the fundamental issue that PyMS exhibits poor long-term (> 30 days) reproducibility. It is clear that such developments can only encourage the development and application of PyMS in microbiology.

Although primarily applied to medical microbiology, this technology has been applied to the differentiation of yeasts such as *Ascomcetes* (PyGC – Viljoen & Kock, 1991), *Saccharomyces/Rhodosporidium/Filobasidium* (PyMS – Windig *et al.*, 1981/2), *Saccharomyces* species (PyGC – Jones, 1984). Using PyMS, Gutteridge and Priest (1996) were able to differentiate 51 yeast strains isolated from what was clearly a heavily contaminated 'home-brew product', into brewing yeasts, contaminant ('wild') *Saccharomyces* and wild non-*Saccharomyces* strains.

Despite its rapidity, PyMS has yet to be 'taken on board' as a route to yeast strain differentiation and identification. An early report (Quain, 1988) briefly described the opportunities that PyMS may offer the brewing industry. In the first detailed study, Timmins *et al.* (1998), demonstrated the power of PyMS in broadly discriminating 22 production brewing yeasts into four main clusters and a 'single member cluster' (Fig. 4.14). As these strains have also been characterized by RFLP DNA fingerprinting (Schofield *et al.*, 1995; Wightman *et al.*, 1996), this work has allowed the direct comparison of phenotypic (including FT-IR below) and genotypic approaches to the differentiation of strains. Intriguingly, Timmins *et al.*, (1998) conclude 'that these phenetic approaches mirror the known genotype (and brewing phenotype) of these organisms'. However, despite such promising observations, the cost and limited availability of PyMS is likely to hamper development and application of this technology within brewing.

4.2.6.6   *Other approaches.*

- *Fourier transform infrared spectroscopy* – like PyMS, FT-IR is a physico-chemical method that provides a 'snap-shot' of the whole cell phenotype (DNA/RNA, proteins, carbohydrates, etc.). The wave number $\nu$, the reciprocal of the wavelength, is used as the physical unit for FT-IR spectroscopy. Typically the 'middle' infrared ($\nu = 4000$–$200$ cm$^{-1}$) is used to differentiate and identify yeasts. Like PyMS, FT-IR data should be used with multivariate statistical techniques, as direct analysis of spectra reveals few differences (see Fig. 4.15 for spectra of two distinct strains of *S. cerevisiae*). The various peaks in the spectra represent the vibrations of bonds within functional groups, for example the peak at 1000–100 cm$^{-1}$ is mainly due to carbohydrate C–O vibrations.

   FT-IR is finding growing application in the differentiation of yeasts. An early study used FT-IR to reveal the interrelationships between eight yeast species and hybrid crosses between seven of them and *S. diastaticus* (Hopkinson *et al.*, 1988). FT-IR was also used by Timmins *et al.* (1998) in their study of the differentiation of 22 production brewing yeasts. In addition to successfully discriminating between strains, FT-IR shows the same clusters of related strains as Py-MS, see Fig. 4.16.

   A substantial paper from Kummerle *et al.* (1988) generated a FT-IR spectral library from 332 different food related yeasts representing 74 species of 18 genera. When challenged with 772 unknown yeasts from the food industry, FT-IR identified 699 yeasts (97.5%) correctly as validated by conventional physiological and morphological tests. These results suggest that FT-IR may well find wider application in the future, particularly as identification is achieved within 24 hours of colony isolation.

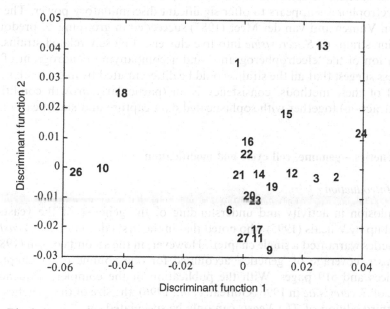

**Fig. 4.16** Discriminant analysis bioplot based on FT-IR data for 22 strains of *S. cerevisiae*. These same strains were subject to PyMS analysis in Fig. 4.14 (redrawn from Timmins *et al.*, 1998).

- *Fatty acid profiles* – FAME (fatty acid methyl ester) profiling has proved particularly successful in the typing of clinical bacterial genera and species (Gutteridge & Priest 1996). However, in the case of yeasts, it is generally held that the profiles of saturated and unsaturated fatty acids in *S. cerevisiae* are essentially the same, although the profile is sufficiently distinctive and consistent to differentiate *S. cerevisiae* from other species of yeast encountered in the wine industry (Tredoux *et al.*, 1987). Augustyn and Kock (1989) have proposed the exciting possibility that fatty acid profiles may be used to differentiate strains of *S. cerevisiae*. Central to this proposal are prescriptive protocols for the aerobic growth of each strain, defined conditions of extraction/methylation and analysis using sensitive capillary GLC columns together with mass spectroscopy. Although up to 14 different fatty acids could be identified, statistical analysis of ten fatty acids (as a relative percentage) enabled the differentiation of 13 strains of *S. cerevisiae*. However, 'as the number of strains to be differentiated increases, it is quite possible that the data generated by a study of 10 fatty acids used here will prove to be inadequate' (Augustyn & Kock, 1989).

- *Protein fingerprinting* – It has been estimated that there are more than 2000 different proteins in the microbial cell (Gutteridge & Priest 1996). Indeed, in yeast, the genome project has suggested almost 6000 'potential' proteins (Goffeau *et al.*, 1996). With so much diversity it is perhaps not surprising that the extraction and electrophoretic separation of cellular proteins has provided a successful route for the identification and characterisation of, particularly, medically important bacteria. Within brewing microbiology there has only been limited application of protein fingerprinting, notably the identification of lactic acid bacteria (Gutteridge & Priest 1996). However, with *S. cerevisiae*, protein electrophoresis appears to offer significant discriminatory power. The work of van Vuuren and van der Meer (1987) succeeded in grouping 29 predominately wine strains of *S. cerevisiae* into five clusters of closely related strains. Examination of the 'electropherograms' and accompanying dendrogram of relatedness suggest that all the strains could be differentiated from each other. As with all of these methods, consistency is all (particularly growth conditions and extraction) together with sophisticated data capture and statistical treatment.

## 4.3  Genetics – genome, cell cycle and modification

### 4.3.1  *Introduction*

The explosion in activity and understanding of the genetics of the yeasts is ably summed up by Wheals (1995) who noted that in the first edition of *The Yeasts* (1969–71) 'genetics warranted a single chapter'. However, in the second edition (1987–95) of *The Yeasts*, coverage of 'genetics' accounted for one volume of 11 chapters, four appendices and 619 pages. With the publication of the complete sequence of the genome of *S. cerevisiae* in 1996 (Goffeau *et al.*, 1996), the size of the 'genetics' element of any third edition of *The Yeasts* can only be speculated on.

Yeast genetics is considered to have commenced in the 1940s with the pioneering work of Winge in the Carlsberg Laboratories in Denmark and Lindegrin in Illinois, USA. Since then *S. cerevisiae* has become very much the organism of choice in the study of eukaryote genetics and cell biology. However, for the worldwide brewing industry, yeast genetics came of age in the 1980s with the development of gene cloning and the recognition that genetic manipulation offered the possibility to tailor yeasts to meet desired needs. In the succeeding years, a number of 'foreign' genes have been successfully introduced into brewing yeast so much so that, technically, genetic manipulation could almost be described as being 'routine'! However, as yet, labelling demands, regulatory hurdles and, more importantly, consumer negativity or concern have mitigated against the commercial use of a genetically modified yeast (see Section 4.34).

Although genetic manipulation remains 'too hot to handle', the endeavour of many committed geneticists has resulted in numerous valuable spin-offs and insights. In particular, genetic methods are now used to routinely fingerprint and differentiate brewing yeast strains (see Sections 4.2.6.1–3) and have provided hitherto unexpected insights into genetic instability and change (see Section 4.3.2.5). In the more rarefied atmosphere of applied brewing research, genetic methods increasingly feature in the routine technical armoury that is used to unravel biological function and interaction.

Although the detailed consideration of yeast genetics is beyond the scope of this book, there is room, and a need, to review the special and applied case of *brewing* yeast genetics. This is especially necessary as – for reasons detailed below – brewing yeast strains differ significantly from laboratory isolates of *S. cerevisiae*. Consequently, the important points of brewing yeast genetics are usually lost in a welter of detail and understanding that surrounds the laboratory strains. For specialist, authoritative reviews of brewing yeast genetics see Kielland-Brandt *et al.* (1995) and Hammond (1993, 1996). For detailed reviews of the wider vista of yeast genetics the interested reader is referred to volume 6 of *The Yeasts* edited by Rose *et al.* (1995).

4.3.1.1 *Genetic nomenclature and definitions.* Genetics is not alone in being impenetrable for all but the professional. The intention of this section is to provide the reader with an understanding of the 'hot' topics in brewing genetics. Inevitably, there is much jargon, which, not surprisingly can confuse and frustrate. In an attempt to minimise this pain, Table 4.14 provides a glossary of some of the terms used in this chapter.

### 4.3.2 *The genome*

4.3.2.1 *Yeast genome project.* Work on the sequencing of the genome of *S. cerevisiae* began in 1989 and finished with the publication of the sequence on 24 April 1996. The work was lauded as a milestone in biology as (i) it was the first eukaryotic genome to be sequenced and (ii) it represented the collective efforts of some 633 scientists from 96 laboratories worldwide. The project was very much of its time, using 'modern infomatics technology' (Goffeau *et al.*, 1996) and the Internet to facilitate communications and project management. Indeed, the sequence of the genome was first 'published' on the World-Wide Web and is now easily sourced and interrogated on the Internet (for a review see Brown, 1998).

**Table 4.14**  Glossary of terms and abbreviations used in genetics.

| Abbreviation | Definition |
| --- | --- |
| ($\rho^-$) or [rho$^-$] | Petites – gross deletion of mitochondrial DNA – no growth on non-fermentable substrates. |
| [rho°] | No mitochondrial genome. |
| allele | One of the alternative forms of a gene, found on the same place on the chromosome. |
| amino acids | The building blocks of proteins. They all have the same carbon backbone structure but differ from one another according to the individual side chain. Only 20 of the naturally occurring amino acids are commonly found in proteins. |
| aneuploid | A polyploid whose chromosomes are not a perfect multiple of ploidy. |
| ascospore (spore) | A specialised haploid cell produced during meiosis. |
| ascus | A sac-like structure containing four spores in *S. cerevisiae*. Plural: asci. |
| auxotrophic | A yeast strain that requires nutritional supplements to grow. |
| base pair (bp) | Two bases, linked by noncovalent forces, that pair in double-stranded DNA or RNA molecules. |
| bases | The variable part of DNA. The nitrogenous bases of DNA are divided into two groups: purines [adenine (A) and guanine (G)] and pyrimidines [thymine (T) and cytosine (C)]. In RNA, thymine is replaced by uracil (U). |
| cdc | Cell division cycle mutants. |
| cDNA (complementary DNA) | A DNA molecule usually obtained by reverse transcription of a mRNA molecule. |
| centromere | The attachment site of chromosomes to the mitotic or meiotic spindle. |
| CHEF | Contour clamped homogeneous electric field. |
| CHRs | Cluster homology regions. |
| chromosomes | Discrete physical units carrying genetic information. Each chromosome contains a long duplex DNA molecule and associated proteins. Usually, chromosomes are visible only during cell division. The number of chromosomes varies widely among different species. |
| CLP | Chromosome-length polymorphism. |
| codon | A sequence of three nucleotides (in a DNA or mRNA), that encodes a specific amino acid to be incorporated into a protein. |
| diploid | A term used to describe an organism having two sets of identical chromosomes (2*n*). *S. cerevisiae* is chimeric, in that haploid and diploid forms exist. |
| DNA (deoxyribonucleic acid) | The primary genetic material of all cellular organisms. It is a polymeric macromolecule composed of a repeating backbone of phosphate and sugar subunits to which different bases are attached. DNA is arranged in two opposing strands (the Watson-Crick double helix) in which the complementary bases form hydrogen-bonded basepairs across the two strands. The sugar backbone of DNA is composed of deoxyribose subunits. |
| DNA-replication | Process of synthesis of a new DNA strand by a mechanism in which a pre-existing strand (parental strand) is used as a template. Each new strand is complementary to the parent strand. |

*Contd*

**Table 4.14**  *Contd.*

| Abbreviation | Definition |
|---|---|
| eukaryote | An organism whose genetic information is, in contrast to prokaryotes (such as bacteria), contained in a separate cellular compartment: the nucleus. Besides algae, fungi and protozoa, all multicellular cell-differentiating organisms including plants and animals are eukaryotes. In addition to their nuclear genome, all eukaryotic cells contain small additional, extranuclear genomes, which are contained in mitochondria (in all eukaryotes) and in plastids (only in eukaryotes that can perform photosynthesis). |
| EUROFAN | European Functional Analysis Network. |
| exons | Segments of a eukaryotic gene that encodes mRNA. In DNA, exons are adjacent to non-coding DNA segments, called introns. |
| Flo⁻ | A strain which is non-flocculent. |
| *FLO*⁺ | All wild-type genes controlling flocculation. |
| Flo⁺ | A strain which is flocculent. |
| *FLO1* | A locus or dominant allele. |
| *flo1* | A locus or recessive allele. |
| flo1-Δ1 | Partial or complete deletion of the *FLO1* gene. |
| *FLO1-1* | A specific dominant allele or mutation. |
| *flo1-1* | A specific recessive allele or mutation. |
| Flo1p | The protein encoded by *FLO1*. |
| gene | A unit of heredity. A section of DNA coding for a single polypeptide chain; a particular species of tRNA, rRNA; or a sequence that is recognised by and interacts with regulator proteins. |
| gene designation | Three letter italicised name that typically gives an indication of the function of the gene or the major phenotype associated with a defect in that gene. |
| genome location | YAR050w where Y = yeast, A = chromosome number (A = chromosome I, B = chromosome II, etc), R = right arm (of the chromosome, L = left arm), 050 = designated ORF number and w = which DNA strand the ORF is on (w = Watson, c = Crick). |
| haploid | The state of having only one set of chromosomes. See diploid. |
| introns | Noncoding regions of eukaryotic genes, which are transcribed into mRNA but are then excised by a process called RNA-splicing. See exon. |
| kb | Kilo bases – 1000 bases. |
| locus | The position of a gene, mutation on a chromosome. |
| long-terminal repeats (LTRs) | Identical DNA sequences, several hundred nucleotides long, found at both ends of transposons. They are thought to have a role in integrating the transposons into the host DNA. Solo LTRs and remnant LTRs indicate former integration sites of transposable elements. |
| mating type | The sex of a yeast cell. In *S. cerevisiae* three types of cells can be distinguished: a, α and a/α. The haploid a and α cells can mate with each other. During mating, cellular and nuclear fusion of the two cells of opposite mating type occurs, forming the third cell type, the diploid a/α cell. The a/α cell cannot mate but, unlike a and α cells, can be induced by external signals to enter meiosis and undergo sporulation. |

*Contd*

**Table 4.14** *Contd.*

| Abbreviation | Definition |
| --- | --- |
| meiosis | A special process of nuclear division during which spores are produced. Meiosis involves a diminution (by half) in the amount of genetic material; it consists of two successive nuclear divisions with only one round of DNA replication producing four haploid daughter cells (the spores) from an initial diploid cell. |
| MIPS | Munich Informatioan Centre for Protein Sequences. |
| mitochondria | Semi-autonomous and self-reproducing organelles located in the cytoplasm of eukaryotic cells. These organelles are responsible for the energy conversion of most of the cellular energy metabolites into adenosine triphosphate (ATP) by oxidative phosphorylation. |
| mitosis | The process of nuclear division in eukaryotic cells. |
| monosomic | A diploid strain lacking one of a pair of chromosomes is monosomic in that chromosome. |
| mRNA (messenger RNA) | A single-stranded RNA that acts as the template for the amino acid sequence of proteins. |
| nfs | Nonfermentable substrate. |
| open reading frames (ORFs) | DNA stretches that potentially encode proteins. They always have a start codon (ATG) at one end and a translation-terminating stop codon at the other end. |
| orphan genes | Genes without any recognised function. |
| PCR | Polymerase chain reaction. |
| PFGE | Pulsed field gel electrophoresis. |
| phenotype | The observable characteristics of a yeast strain. |
| polymorphism | Typically variability in chromosome size that enables differentiation of closely related strains of yeast. |
| polyploid | Generic term that typically means that the ploidy of the yeast is greater than diploid. |
| proteins | Polymers composed of a linear string of amino acids. |
| proteome | The complete set of proteins that yeast can form. |
| prototrophic | A yeast strain that has no obvious requirements for nutritional supplements. |
| RAPD | Random amplified polymorphic DNA. |
| RFLP | Restriction fragment length polymorphism. |
| RNA (ribonucleic acid) | Polymer composed of a repeating backbone of phosphate and sugar subunits to which different bases are attached: adenine (A), cytosine (C), guanine (G), and uracil (U). The sugar backbone of RNA is composed of ribose subunits. RNAs can be distinguished by their different properties: mRNA, tRNA, and rRNA. |
| rRNA (ribosomal RNA) | A structural and functional component of ribosomes. Ribosomes are the cellular machinery responsible for the translation of the genetic information into proteins. They are composed of rRNA and ribosomal proteins. |

*Contd*

**Table 4.14** *Contd.*

| Abbreviation | Definition |
|---|---|
| *Saccharomyces cerevisiae* | Taxonomic classification: Eukaryota; Plantae; Thallobionta; Eumycota; Hemiascomycetes; Endomycetales; Saccharomycetaceae. *Saccharomyces* is a genus of ascomycetes. They are normally diploid unicellular fungi that reproduce asexually by budding. Asci, containing four haploid ascospores, develop directly from the diploid vegetative cells by meiosis. After germination of the ascospores the haploid cells can reproduce vegetatively, or haploid cells of different mating type can fuse to form a diploid zygote. Most laboratory strains used are, in contrast to wild-type yeasts, stable haploids. |
| SGD | *Saccharomyces* Genome Database. |
| snRNA | A set of RNAs that are typically smaller than 300 nucleotides and function in the nucleus in the form of small nuclear ribonucleoprotein particles (snRNPs). The function of snRNPs is to mediate and regulate post-translational RNA processing events. |
| spindle | A network of fibrous microtubules and associated molecules formed during mitosis between the opposite poles (centromeres) of eukaryotic cells. It mediates the movement of the duplicated chromosomes to opposite poles. |
| sporulation | The process of spore development. Sporulation can be induced by external signals, such as absence of nitrogen. |
| TAFE | Transverse alternating field electrophoresis. |
| telomere | The terminal part of a eukaryotic chromosome, consisting of a few hundred base pairs with a defined structure. Telomeres are important for maintaining chromosomal structure and stability, as they permit replication of the ends of the linear DNA molecule. |
| tetraploid | Four copies of each chromosome ($4n$). |
| transcription | The process by which DNA is used as a template for the synthesis of an RNA molecule. |
| translation | The process of protein synthesis from a mRNA template, occurring at the ribosome. |
| transposable element (transposon) | A mobile DNA sequence that can move from one site in a chromosome to another, or between different chromosomes. The transposable elements in yeast are the Ty elements. |
| triploid | Three copies of each chromosome ($3n$). |
| trisomic | A diploid strain carrying three copies of chromosome III is trisomic for chromosome III. |
| tRNA (transfer RNA) | A set of RNAs that act during protein synthesis as adaptor molecules, matching individual amino acids to their corresponding codon on a mRNA. For each amino acid, there is at least one tRNA. |
| Ty elements | Tys are members of a widely distributed family of eukaryotic elements called LTR-containing retrotransposons. They have the same sequence organisation as retroviruses. The complete retrotransposons are 5 to 6 kilobases long. They are bracketed by long-terminal repeats (LTR), which are 300 to 400 basepairs long. |
| zygote | The diploid cell resulting from the union of two haploid cells of complementary mating types. |

Although undeniably a major success, the DNA sequence of *S. cerevisiae* is relatively small with a genome of 13.5 Mb. Indeed the second eukaryote to have its genome sequenced – the nematode worm (*Caenorhabditis elegans*) – has a genome of 97 Mb. This sequence published in December 1998 (*C. elegans* Sequencing Consortium, 1998) has enabled the comparison of two highly diverged eukaryotes, one a unicellular micro-organism and the other a multicellular animal (see Section 4.3.2.2). Of course, this is the ultimate prize although the genome of man (*Homo sapiens*) is – not surprisingly – much larger, being in the order of 3500 Mb (Oliver, 1996)! Inevitably there is already astonishing progress on the Human Genome Project such that the sequence of one of the smallest of the 23 pairs of chromosomes, chromosome 22, was published in December 1999 (Dunham *et al.*, 1999).

There are two major web sites: (i) the *Saccharomyces* Genome Database (SGD) at http://genome-www.stanford.edu/Saccharomyces/ and (ii) the European Union funded database at MIPS (Munich Information Centre for Protein Sequences) at http://www.mips.biochem.mpg.de/mips/yeast/MIPS. As detailed elsewhere in this section, these web sites are a splendid and continually updated resource. The review by Cherry *et al.* (1998) breaks the SGD down into easily digested chunks: (i) (predominately) DNA sequence information, (ii) structural information of proteins (iii) underpinning literature and abstracts and (iv) maintenance of gene nomenclature. Such is the success of the SGD web site that it is 'hit' about 45 000 times a week (Cherry *et al.*, 1998)!

The yeast genome project required the sequencing of some 12 million nucleotide bases (for an overview see Mewes *et al.*, 1997). This encodes about 6217 potential proteins ('open reading frames' – ORFs) which account for almost 70% of the total sequence. Just how revolutionary this project is can be gleaned from our current knowledge of sequence and function. Only about half the proteins are 'known', being well-characterised biochemically or genetically. Of the rest, understanding ranges from 'some indication of their function *in vivo*' to 'similarities to other uncharacterised proteins or (show) no similarities at all' (Mewes *et al.*, 1997). The 30% or so of genes without any recognised function – so called 'orphan genes' – represent virgin territory to geneticists and protein chemists. To understand the function of these genes new ways and approaches have had to be devised that frequently challenge traditional thinking.

In principle, knowledge of the sequence of the genome should enable the theoretical synthesis of a complete set of proteins (the so-called 'proteome', Goffeau *et al.*, 1996) that yeast can form. This fascinating opportunity arrives through computer-driven comparisons of the anticipated amino acid sequences with proteins of known function. However, only 50% or so of the predicted proteins that meet the required criteria for 'similarity score' can be functionally identified. Consequently, the distribution of proteins by functional role (Fig. 4.17 ) is not yet 'carved in stone'. It might be expected that, with time and new effort, the proportion of ORFs with assigned function would increase. Where known, the functions of individual ORFs are given in the 'overview' of Mewes *et al.* (1997). In some instances, this analysis has revealed gene products in yeast whose existence was previously in doubt.

For example, the classical 'function first' approach (Oliver, 1997) proceeds from understanding biological function to identification of a DNA sequence. Conse-

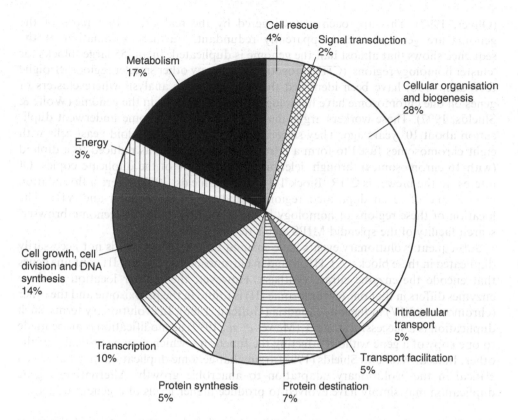

**Fig. 4.17** Yeast genome sorted by eleven functional categories (redrawn from Mewes *et al.*, 1997).

quently, the 'orphan' genes (without any recognised biological function) would be missed using this approach. This has led to the concept of performing 'genetic analysis in the reverse direction' (Oliver, 1997). Here, each unknown protein is selectively deleted and its biological function determined by detailed screening of the mutant's phenotype. Hampsey (1997) has described – in a useful 'hands-on' review – the screening over 70 distinct phenotypes in yeast. A more sophisticated route (Oliver, 1997; Oliver *et al.*, 1998) couples two-dimensional gel electrophoresis with the analytical power of mass spectrometry (see Section 4.2.6.4) – to achieve a 'metabolic snapshot' of each deletant. This concept has been extended to enable direct analysis of more than 100 proteins (e.g. the yeast ribosome complex) via a combination of liquid chromatography with mass spectrometry (Link *et al.*, 1999).

A more direct approach to understanding the function of yeast genes is to systematically create deletion mutants of all 6000 ORFs found in the yeast genome. As with the original genome sequencing project, this phase of work is being addressed by a consortium of laboratories in the USA, Canada, Japan and in Europe (European Functional Analysis Network or EUROFAN).

This project is also enabling gene functions to be probed in other ways. Work is ongoing to construct a 'minimalist' genome where every gene is an essential one

(Oliver, 1997). This approach was triggered by the realisation that parts of the genome are gene poor and apparently redundant. Further examination of the sequence shows that almost half the genome is duplicated. In all, 55 large 'blocks' or 'cluster homology regions' (CHRs) together with many other smaller regions (Seoighe & Wolfe, 1998) have been identified through computer analysis where clusters of genes on one chromosome have homologues somewhere else in the genome (Wolfe & Shields, 1997). These workers argue that the entire yeast genome underwent duplication about $10^8$ years ago. They suggest that two ancestral diploid yeast cells with eight chromosomes fused to form a tetraploid which then was reduced to a diploid (with 16 chromosomes) through deletion of about 85% of the duplicate copies. Of interest to the brewer is CHR 'Block 1' (Wolfe & Shields, 1997) where a flocculation gene is one of seven duplicated regions found on chromosome I and VIII. The location of these regions of homology is easily visualised via the 'genome browser' search facility of the splendid MIPS web site.

Subsequent evolutionary events have ensured that gene function is not necessarily duplicated in these blocks. A CHR found in chromosomes XIV and III contains genes that encode the enzyme citrate synthase. However, the cellular location of these enzymes differs in that one (chromosome III) is found in the peroxisome and the other (chromosome XIV) in the mitochondria (Goffeau, 1996). In evolutionary terms, such duplication is a successful strategy (Mewes *et al.*, 1997) as modifications can be made to one copy of a gene without affecting the function – which may be critical – of the other. Indeed, Wolf and Shields (1997) argue that genome duplication may have been critical in the evolutionary adaptation to anaerobic growth. Alternatively, gene duplication may simply have evolved to produce higher levels of a gene product.

4.3.2.2  *Yeast genome project, human disease and pathogenicity.*  One of the more effecting themes from the genome project is the view that an understanding of the yeast proteome is 'a prerequisite for understanding the more complex human proteome' (Goffeau, 1996). One of the 'big wins' is that 'nearly half of the proteins known to be defective in human heritable diseases show some amino acid sequence similarity to yeast proteins' (Goffeau, 1996). Although perhaps oversold – 'about 30% of human disease associated genes significantly match yeast genes' (Foury, 1997) – the ease with which yeast genes can be manipulated and modified can be anticipated to provide considerable insight into understanding the molecular basis of human disease. By way of example, some of the yeast genes with a high sequence similarity to human disease genes are detailed in Table 4.15. Such 'comparative genomics' are clearly of major academic and commercial significance, which inevitably will extend beyond *S. cerevisiae* to include other sequenced model organisms. For example Chervitz *et al.* (1998) have compared the yeast genome with that of the multicellular nematode worm, *C. elegans*. From this it is clear that most of the core biological functions (intermediary metabolism, DNA and RNA metabolism etc.) are carried out by orthologous proteins in both *S. cerevisiae* and *C. elegans*. These 'orthologous proteins' are proteins of different species that can be traced back to a common hypothetical ancestor. However, those genes (e.g. signalling) involved in multicellularity have no yeast orthologs. Taken together these are powerful insights into eukaryote biology and provide strong support for the basic assumption that model

**Table 4.15**  Yeast homologues (with high sequence similarity) of human disease-associated genes (from Bassett *et al.*, 1996; Foury, 1997).

| Yeast gene | Function in yeast | Human gene | Human disease |
|---|---|---|---|
| MSH2 | DNA repair protein | MSH2 | Colon cancer |
| MLH1 | DNA repair protein | MLH1 | Colon cancer |
| YCF1 | Metal resistance protein | CFTR | Cystic fibrosis – defect in chloride ion transport, leads to gastro-intestinal insufficiency |
| CCC2 | Probable copper transporter | WAD | Wilson disease – copper accumulation in the liver and brain. Leads to liver failure, dystonia, dementia, haemolytic anaemia, reduced renal function, jaundice |
| SGS1 | helicase | BLM1 | Bloom syndrome – genomic instability in somatic cells, leads to telangiectatic facial erythema, photosensitivity, dwarfism and other abnormalities |
|  |  |  | Werner's syndrome – DNA helicase Q-related protein; premature ageing and strong predisposition to cancer |
| PAL1 | Peroxisomal membrane protein | ALD | X-linked adrenoleukodystrophy – accumulation of very-long-chain fatty acids, leads to demyelination of the central nervous system and death, or in a milder form to adrenal insufficiency |
| TEL1 | P13 kinase | ATM | Ataxia telangiectasia – cerebellar degeneration and mental retardation, immunodeficiency, chromosomal instability, cancer predisposition, radiation sensitivity |
| GEF1 | Voltage-gated chloride channel | CLCN5 | Fanconi syndrome – nephrolithiasis: renal deficiency |
| SOD1 | Superoxide dismutase | SOD1 | Amyotrophic lateral sclerosis – decreased removal of superoxide radicals, leads to a progressive loss of motor function |
| YPK1 | Serine/threonine protein kinase | DM | Myotonic dystrophy – myotonia, muscular dystrophy, cataracts |
| YIL002C | Putative IPP-5-phosphate | OCRL | Lowe syndrome – deficient inositol phosphate metabolism, leads to mental retardation, renal abnormalities, cataracts and glaucoma |
| IRA2 | Inhibitory regulator protein | NF1 | Neurofibromatosis (type 1) – fibromatous skin tumors, skin marks |
| CCH1 | Component of a yeast $Ca^{2+}$ channel that may mediate $Ca^{2+}$ uptake in response to mating pheromone, salt stress, and $Mn^{2+}$ depletion | CACNL-A4 | Migraine – calcium channel; familial hemiplegic migraine and episodic ataxia |
| BTN1 |  | CLN3 | Batten's disease – CLN3 gene product may function as a chaperone involved in the folding/unfolding or assembly/disassembly of other proteins, specifically subunit c of the ATP synthase complex |

organisms will provide 'reliable functional annotation of the human DNA sequence' (Chervitz *et al.*, 1998). As befits a relatively young and visible field, the Internet plays a central role in cross-referencing and homology searching of genome sequences via a publicly accessible database (see Bassett *et al.*, 1997; Foury, 1997; Chervitz *et al.*, 1998).

The homology of yeast sequences with those implicated in human heritable diseases provides, if it was needed, the primary justification for the yeast genome project. Remarkably, this single celled 'primitive' eukaryote provides an experimental vehicle for probing the genetics and biochemistry of human diseases with the ultimate goal of control or avoidance. Comparative genomics provides the experimental scenarios that are already enabling yeast to be used as a screen for antitumour and antiviral drugs. The construction of the 'minimalist yeast' described above is anticipated to be an important next step in such a programme (Oliver, 1996).

The power of comparative genomics has been brought to life from work on Werner's syndrome and Batten's disease (Table 4.15) in yeast. The former results in premature ageing (see Section 4.3.3.4), whereas Batten's disease is an inherited neurodegenerative disease in children with an incidence of 1 in 12 500 live births in the USA (Pearce & Sherman, 1998). Batten's disease is characterised by a decline in mental abilities, untreatable seizures, blindness, loss of motor skills and premature death. Comparative genomics have shown that, in yeast, the product of the non-essential *BTN1* gene is 39% identical and 59% similar to the protein from the human gene (*CLN3*) for Batten's disease. Compelling work from Pearce and co-workers has shown that yeast is an excellent experimental model to determine the function of the *CLN3* gene. The human protein can be expressed in yeast and complements functions lost by deletion of the *BTN1* gene (Pearce & Sherman, 1998). Subsequent work (Pearce *et al.*, 1999) has suggested that Batten's disease is caused by a defect in vacuolar pH control. Tellingly, the authors conclude that their work 'draws parallels between fundamental biological processes in yeast and previously observed characteristics of neurodegeneration in humans' (Pearce *et al.*, 1999).

Coincidentally, yeast may also throw some light on another, more high-profile neurodegenerative disease, Creutzfeldt-Jakob disease, that has been associated with BSE (bovine spongiform encephalopathy) or 'mad cow disease' in cattle. Intriguingly these disorders are believed to be brought about by infectious proteins or 'prions' transmitted via a 'protein only' mechanism which is not coded for genetically. It is thought that the protein – which is encoded in the usual manner but is free of DNA or RNA – flips its shape to form an abnormal insoluble protein (prion) which, consequently, loses its normal function. The prion is infectious and multiplies by converting other copies of the normal protein into prions by inducing the benign molecules to change shape (see Fig. 4.18). Accordingly, Wickner *et al.* (1999) have described prions as 'genes composed of altered proteins rather than nucleic acids'. Given this challenging notion it is not surprising that the maverick protein concept remains controversial. However the originator of the prion concept, Stanley Pruisner, – was awarded the Nobel Prize in 1997 (see review by Pruisner, 1995).

Inevitably, given its increasing use in probing human disease, work with *S. cerevisiae* (Wickner, 1997) is making a substantial contribution to unravelling the mammalian prion hypothesis. However, this is an increasingly complex and per-

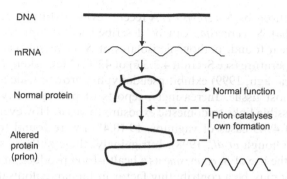

DNA

mRNA

Normal protein → Normal function

Prion catalyses
own formation

Altered
protein
(prion)

**Fig. 4.18** The 'prion hypothesis' (redrawn from Wickner, 1997).

plexing field and the interested reader looking for a fuller explanation is directed to the excellent review of Wickner *et al.* (1999). From a brief perspective, work in the late 1960s and early '70s identified two infectious proteins (prions), which were able to convert the normal form of the protein into an abnormal form. [PSI] is the prion form of the SUP35 protein (involved in protein translation) and [URE3] is the prion form of the URE2 protein that switches off the utilisation of a poor nitrogen source in the presence of an alternative 'good' source. In her memorably titled review – 'Mad cows meet mad yeast: the prion hypothesis' – Lindquist (1996) argued that the work with yeast would provide 'new insights and approaches that will synergise with mammalian studies'. As if a premonition, Liu and Lindquist (1999) have reported that the conversion of the native protein (SUP35p) to [PSI] is triggered by increasing the number of characteristic short, oligopeptide repeats in the protein. The prion then propagates forming characteristic amyloid filaments that can be visualised using congo red staining and microscopy (Wickner *et al.*, 1999). Conversely, deletion of the repeated sequences eliminates conversion to the prion format. As mammalian prion diseases are associated with an expansion in oligopeptide repeats, this work in yeast has identified a general mechanism in which mutation can lead to a protein misfolding disease. Further work (Ma & Lindquist, 1999) has focused on protein trafficking and the degradation of misfolded proteins. Rather than being degraded *in situ* within the secretary pathway, misfolded proteins are delivered by retrograde transport to the cytoplasm where they are deglycosylated and degraded. Occasionally, for reasons that have yet to be established, the abhorrent misfolded protein is not degraded and aggregates to become a prion-like protein. In conclusion, whatever the eventual outcome of the 'prion story', it seems likely that the yeast model will provide important insights into understanding and, ultimately, controlling and eradicating an important high profile human disease.

Unlike *Candida* and *Cryptococcus* yeast species (for a review see Hurley *et al.*, 1987), *S. cerevisiae* has little clinical presence. Indeed, in a survey of pathogenic yeasts found in women, 80 distinct strains were found, consisting of six *Candida* species and a *Saccharomyces*. Of the 80 strains found, 59 isolates were *C. albicans* with *S. cerevisiae* accounting for only two isolates. This is in keeping with the view that *S. cerevisiae* is not normally recognised as being pathogenic to man although there are occasional clinical reports (notably in immunocompromised AIDS and transplant patients) of infections by *S. cerevisiae*. A recent review (Murphy & Kavanagh, 1999) has suggested that *S. cerevisiae* can be described as an 'opportunistic pathogen'. Significantly, when found, pathogenic isolates of *S. cerevisiae* can grow at the atypically high temperature (see Section 4.2.5.6) of 42°C (McCusker *et al.*, 1994) and can (Murphy & Kavanagh, 1999) exhibit pseudohyphal growth which may facilitate the penetration of host tissue. Increasingly, reports of pathogenicity have focused on vaginitis and possible links to domestic exposure to yeast. However cases are rare, so much so that of 4943 cases of vaginitis, 19 (0.4%) were found to be caused by *S. cerevisiae* (McCullough *et al.*, 1998). Intriguingly, these authors speculate that 'the proliferation of the use of *S. cerevisiae* as a health-food product, in home baking, and in home brewing may be a contributing factor in human colonisation and infection with this organism'. Similarly, Posteraro *et al.* (1999) reported a strong correlation between instances of yeast associated vaginitis and the 'frequent domestic use of yeast'. However, molecular fingerprinting of implicated commercial bakers' yeasts failed to show any epidemiological link with the clinical isolates.

Baking and brewing strains of *S. cerevisiae* have also been implicated in Crohn's disease (Walker, 1998), the chronic inflammation of the gut which is also associated with an increased risk of intestinal cancers. Sufferers exhibit an immune hypersensitivity to dietary yeast, with a specific response to specific cell wall mannan antigens (Young *et al.*, 1998) (see Section 4.4.2). Persuasive evidence for the role of yeast in this disease comes from analysis of occupational mortality that shows bakers to have the highest odds ratio for Crohn's disease (Alic, 1999).

4.3.2.3 *Yeast genome project and brewing yeast.* It is telling – and a little disappointing – that such a major project as sequencing the yeast genome has provoked so little comment in the brewing press. Meaden (1996), in a brief review, succinctly captured the key facts. He noted that 'the overall genetic picture will be similar for brewers' yeast, whether an ale or lager strain'. However, inevitably, there will be differences between the sequenced yeast and brewing strains. The genome project involved a number of closely related haploid strains derived from the progenitor strain S288C (Philip Meaden, personal communication; Cherry *et al.*, 1998). The most notable difference derives from the genealogy of S288C as reported by Mortimer and Johnston (1986). Although a complex and fascinating story, EM93 – the main progenitor strain of S288C which is estimated to contribute 88% of the genome of this strain – was originally isolated from a rotting fig in California in the late 1930s!

Fundamentally, the most obvious difference between the yeast whose genetic sequence we now have and brewing strains is one of ploidy. As is discussed in Section 4.3.2.4, brewing strains are polyploid and as such have three or four copies of each chromosome. The yeast used in the genome project is haploid and, by definition, has only one copy of each chromosome. Fundamentally, significant differences have been shown between specific chromosomes of laboratory haploid strains (such as those in the genome project) and the chromosomes in commercial strains such as distillers' 'M' strain (Wicksteed *et al.*, 1994) and Carlsberg lager yeast (Casey, 1990). Consequently, it is no great surprise that the fine detail of the genetic 'snapshot' of a haploid strain of *S. cerevisiae* is anticipated to be different to brewing strains. As reported by Meaden (1996), the sequenced yeast has one copy of the genes responsible for maltose fermentation whereas brewing strains typically contain ten or more sets of *MAL* genes. Similarly, some genes (such as the second alcohol acetyltransferase, Lg-*ATF1*) that are found in brewing yeast cannot be found in the genome sequence. Conversely, approaching from the other viewpoint, Goffeau *et al.* (1996) have noted that seemingly redundant genes may be 'more apparent than real'. Indeed, many apparently redundant genes may be required to deal with natural non-laboratory 'real world' environments such as rotting figs, grapes or more pertinently, the brewery (Goffeau *et al.*, 1996; Oliver, 1996).

Clarification of just how significant these 'differences' are awaits the sequencing of a brewing strain of *S. cerevisiae*. As it seems likely that this will be a lager strain, there will be the dual challenge of sequencing the hybrid genome of *S. monacenis* and the more familiar *S. cerevisiae* (see Section 4.2.2). If and indeed when this might happen is not clear. Initial signs that a consortium of laboratories might tackle this challenge have come to nothing. A more likely scenario is that a single laboratory might undertake the task (Morten Kielland-Brandt, personal communication).

4.3.2.4 *Chromosome number.* Between 1960 and 1997, 12 genetic maps of *S. cerevisiae* have been published (Cherry *et al.*, 1997). Although as recently as 1981 there were thought to be 17 chromosomes (Mortimer & Schild, 1981), by 'Edition 11A' (Mortimer *et al.*, 1995) and the twelfth 'and probably the last' map (Cherry *et al.*, 1997) the yeast genome consisted of 16 chromosomes. These range dramatically in size from the 230 kb of Chromosome 1 to 1532 kb of Chromosome IV (see Fig. 4.19 ). These provide a benchmark for chromosomal size, as chromosomal polymorphism is increasingly recognised (Section 4.3.2.5) and probed using karyotyping (Section 4.2.6.3).

It will come as no surprise that the genome sequencing project has played an

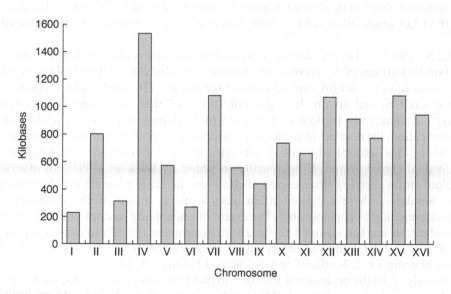

**Fig. 4.19** The size of the 16 yeast chromosomes of the sequenced *S. cerevisiae.*

important role in defining the genetic map of *S. cerevisiae*. This physical approach has complemented and extended the traditional techniques of genetic mapping. A detailed analysis of chromosomal composition can be found in Goffeau *et al.* (1996) and as updated on the MIPS web site. Options here (Heumann *et al.*, 1996) range from a 'top-down' overview of a chromosome to the detailed chromosome sequence and, as noted previously, use of the MIPS genome browser to visualise common sequences on chromosomes. Table 4.16 presents details of chromosome size, number of ORFs, hypothetical proteins and Ty sequences. Data on Ty elements is included as these sequences can be probed for in DNA fingerprinting (Section 4.2.6.1). Ty elements can move from one site in a chromosome to another or move between chromosomes. These transposable elements can trigger genetic rearrangements (Hammond, 1996).

There is a wealth of information to be gleaned from the overview of Mewes *et al.* (1997). As noted previously, the location of the 5800 proteins is reported systematically and diagrammatically, chromosome by chromosome. Groups of genes that encode proteins with a common function (e.g. amino acid metabolism) are further sub-divided (regulation, transport, degradation and so on) (see Fig. 4.17 ). Although undeniably a remarkable and awe inspiring document, the 'gazetteer' of Mewes *et al.* (1997) is fascinating but not particularly user-friendly. For the genetic 'tourist', the gazetteer is a milestone document of a milestone project and, as such, will meet their need. However, the professional geneticist will access the interactive web sites of MIPS or the SGD where chromosomes and genes can be browsed and searched (these and other sites are listed in Brown, 1998). As noted above, these extraordinary web sites allow the yeast genome to be searched interactively. For example, searching the entire genome via MIPS for references to the flocculation gene – *FLO* – delivers a 28 page report detailing 31 ORFs. The level of detail together with the capability to drill down to further strata of complexity is awesome. An edited list of some of the information that can be gleaned from such a search is found in Table 4.17. In addition to the FLO genes, other ORFs whose disruption affects flocculation are reported.

4.3.2.5 *Ploidy*. The number of 'sets' of chromosomes is described as the 'ploidy'. Laboratory strains of *S. cerevisiae* are usually *haploid* or *diploid*. Haploid strains of *S. cerevisiae* have one set (*n*), diploid strains have two sets (2*n*), and triploid strains have three sets (3*n*) and so on. It is generally accepted that 'domesticated' yeasts are *polyploid*, frequently triploid or tetraploid (4*n*) (Hammond, 1996). To add a little complication, the number of copies is not necessarily a perfect multiple of the haploid number. This 'aneuploidy' allows for extra or indeed reduced copy numbers of individual chromosomes. Consequently, as noted by Wickner (1991) 'an otherwise diploid strain carrying three copies of chromosome III is trisomic for chromosome III'. Similarly, where a diploid strain lacks a one of a pair of chromosomes it is monosomic in that chromosome. The extent of aneuploidy in domesticated yeasts is revealing. In marked contrast to the above assumptions of tetraploidy, a report by Codon *et al.* (1998) showed the ploidy of 16 industrial strains to vary widely (wine 1.9*n*, brewing 1.6–2.2*n*, distilling 1.3–1.6*n* and baking 1.3–3*n*).

Naively, it might be assumed that in polyploid or aneuploid strains, each copy of each chromosome is identical. The reality is somewhat different. The advent of

Table 4.16 Distribution of genes and Ty elements in the sequenced yeast genome (from the MIPS web site: http://www.mips.biochem.mpg.de/mips/yeast/MIPS).

| Elements | Chromosomal number | | | | | | | | | | | | | | | | Total |
|---|---|---|---|---|---|---|---|---|---|---|---|---|---|---|---|---|---|
| | I | II | III | IV | V | VI | VII | VIII | IX | X | XI | XII | XIII | XIV | XV | XVI | |
| Sequenced length | 230 | 813 | 315 | 1532 | 577 | 270 | 1091 | 563 | 440 | 765 | 666 | 1078 | 924 | 784 | 1091 | 948 | 12067 |
| ORFs (n) | 106 | 424 | 171 | 818 | 292 | 136 | 573 | 291 | 231 | 388 | 335 | 550 | 488 | 422 | 571 | 500 | 6296 |
| Hypothetical proteins | 103 | 393 | 160 | 752 | 279 | 130 | 516 | 280 | 220 | 356 | 315 | 509 | 458 | 399 | 534 | 463 | 5867 |
| Ty elements | 1 | 3 | 2 | 9 | 2 | 1 | 6 | 2 | 1 | 3 | 0 | 6 | 4 | 3 | 4 | 5 | 52 |

**Table 4.17**   Edited digest of a MIPS database search for *FLO* genes.

| Gene | Chromosome | Location | MIPS information |
|---|---|---|---|
| *FLO1* | I | YAR050w | • Synonyms – *FLO2, FLO4, FLO5, FLO8*<br>• Cell wall protein involved in flocculation<br>• Dominant, one copy causes flocculation<br>• Null mutants fail to flocculate<br>• May function to activate a lectin rather than being a lectin itself<br>• Location – cell wall |
| *FLO5* | VIII | YHR211w | • Strong similarity to Flo1p<br>• Putative transcriptional activator of Flo1p<br>• Probable pseudogene, similar to *FLO1*<br>• Location – nucleus |
| *FLO8* | V | YER108c | • Probable transcriptional activator of Flo1p<br>• Nonsense mutation in the haploid S288C prevents flocculation<br>• Location – nucleus |
| *FLO9* | I | YAL063c | • Strong similarity to Flo1p<br>• May be involved in the flocculation process<br>• Location – cell wall |
| *FLO10* | XI | YKR102w | • Similarity to Flo1p<br>• Location – extracellular, cell wall |
| *FLO11* | IX | YIR019c | • Synonyms – *MUC1, STA4, DEX2, MAL5, STA1*<br>• Extracellular α-1→4-glucan glucosidase<br>• Flocculin<br>• *Flo11* deletion mutants do not flocculate<br>• Location – extracellular |
| | I | YAR062w | • Putative pseudogene<br>• Strong similarity to Flo1p |
| *GTS1* | VII | YGL181w | • Transcriptional factor<br>• *GTS1*-induced flocculation is hardly sensitive to mannose in contrast to *FLO1*-determined flocculation<br>• Overexpression results in constitutive flocculation<br>• Location – nucleus |
| *MIG1* | VII | YGL035c | • Transcriptional repressor<br>• Overexpression causes flocculation<br>• Location – nucleus |

chromosomal fingerprinting or karyotyping (see Section 4.2.6.3) has shown differences in chromosome size (polymorphism) to be common. Although usually a function of chromosomal translocation, chromosomal differences may be due to more fundamental reasons. As noted in Section 4.2.2, lager strains of *S. cerevisiae* are a species hybrid of *S. monacensis* and top fermenting *S. cerevisiae* (Pedersen, 1995). Consequently, lager strains contain two distinct but functionally similar populations of chromosomes that resemble either *S. monacensis* or *S. cerevisiae*.

The ploidy in brewing strains is difficult to determine, particularly where it is based on measurement of cellular DNA and comparison with the DNA content of a haploid reference strain (Hammond, 1996). A comparatively crude approach to measuring DNA was used by Aigle *et al.* (1983) who showed two industrial yeasts to be tetraploid. A more recent application (Guijo *et al.*, 1997) with enological isolates of *S. cerevisiae* (see below) showed strains to be diploid, triploid and tetraploid. The work of Codon *et al.* (1998) reported above moved ploidy analysis forward by using flow cytometry to determine DNA content. However, for most, even a crude measurement of ploidy is of little relevance! It is sufficient to accept that production-brewing yeasts are, inevitably, polyploid or aneuploid. However, polyploidy is of concern for geneticists as such strains are difficult or impossible to work with using what are now traditional genetic approaches. Such strains do not lend themselves to the classical method of 'tetrad analysis' (for a review see Wickner, 1991) as they sporulate poorly and where sporulation occurs the spores are frequently inviable. This was important as tetrad analysis was very much at the core of haploid yeast genetics and, as such, was instrumental in much of the genetic mapping programme (Cox, 1995). Consequently, yeast breeding programmes and analysis of brewing strains was hampered until the arrival of new genetic approaches in the 1970s and '80s (see Section 4.3.4).

The sophisticated molecular approach reported by Hadfield et al. (1995), quantified the chromosome copy number by the use of 'ploidy probes'. Without going into detail, this approach hinges on the insertion of a DNA fragment into a specific chromosome. The fragment contains a known yeast gene that is found as a single copy in the haploid genome. The DNA probe recognises its homologous counterpart on the chromosome and replaces it. Inclusion of a 'selectable marker' (frequently resistance to an antibiotic) in the DNA probes enables the copy number of the probed chromosome to be determined.

The 'ploidy probes' approach has been used with two related top (Y1) and bottom-fermenting (Y9) ale strains (Hadfield et al., 1995). Analysis of Y1 suggested this strain to be triploid with three copies of chromosomes IV, V and XV. However, in Y9 there was evidence of aneuploidy, having three copies of chromosomes IV and XV but two copies of chromosome V. It is not immediately obvious how this change can bring about changed flocculence. Of the probed chromosomes, only chromosome V has been associated with flocculation through *FLO8* (see Table 4.17). However, although necessarily tentative, it is tempting to conclude that the change in flocculation is related to the change in chromosomal copy number. Indeed Hadfield et al. (1995) noted that 'this may be a mechanism by which yeast can develop new characteristics in an alternative manner to gene mutation'.

An alternative, Guijo et al. (1997) reported a more traditional genetic approach to determining chromosome copy number in *S. cerevisiae*. Here, six enological strains of *S. cerevisiae* used in the fermentation and 'flor' film ageing of sherry-type wines were found to contain two, three or four copies of chromosomes. Indeed all six strains were aneuploid, with four being predominately diploid, one triploid (but disomic and tetrasomic for certain chromosomes) and one tending toward the tetraploid but trisomic for certain chromosomes.

There is general acceptance that polyploidy is advantageous for industrial domesticated yeasts. However, specific evidence of the benefits (or disadvantages) of polyploidy in yeast is a little thin on the ground! Inevitably, the advent of the yeast genome sequence and associated analytical tools has enabled the implications of ploidy to be systematically unravelled. In a keynote paper, Galitski et al. (1999) constructed a series of isogenic or genetically identical strains that differed only in ploidy ranging from haploid ($n$) to tetraploid ($4n$). The implications of increased ploidy were far-reaching and not always predictable. Although cell size increased with increasing ploidy (see Table 4.18), $3n$ and $4n$ strains grew more slowly in aerobic

**Table 4.18**  The impact of yeast ploidy on cell morphology (from Galitski *et al.*, 1999).

| Ploidy | Cell volume ($\mu m^3$) | Cell length/width ratios |
|---|---|---|
| Haploid | $72 \pm 1$ | $1.20 \pm 0.01$ |
| Diploid | $111 \pm 2$ | $1.24 \pm 0.01$ |
| Triploid | $152 \pm 3$ | $1.29 \pm 0.01$ |
| Tetraploid | $289 \pm 6$ | $1.39 \pm 0.02$ |

culture with a greater lag phase than the corresponding haploid and diploid strains. Although undeniably interesting, the real core of the work reported by Galitski *et al.* (1999) was the effect of ploidy on the expression of all yeast genes during aerobic exponential growth. Using DNA chip technology, they searched for genes whose expression, relative to total gene expression, increased or decreased as the ploidy changed from haploid to tetraploid. Using a stringent cut-off of a ten-fold difference in gene expression between *n* and 4*n*, ten genes were ploidy induced and seven genes were ploidy repressed. Of these 17 genes, 10 encode known functions. One (*CLN1*, which encodes a cell cycle protein) was ploidy repressed, which may explain the greater cell size associated with higher ploidy. Similarly and intriguingly *FLO11*, 'whose molecular function is unknown', and which determines invasive growth into agar plates was ploidy repressed. Two of the genes that were ploidy induced were of interest: *CTS1*, which is involved in the separation of mother and daughter cells, and *CTR3*, which is involved in copper transport. However, despite these fascinating insights the vast majority of yeast genes were unaffected by an increase in ploidy.

Accordingly, from the work of Galitski *et al.* (1999), the 'jury is still out' as to the identifiable benefits (or not) of polyploidy. A more empirical 'real world' but admittedly soft argument for polyploidy is the realisation that – as noted above – domesticated strains *are* at least diploid. Perhaps the real selective benefit arises from aneuploidy and chromosomal rearrangements and translocations. This suggests that there is selective advantage in maintaining a number of copies of most (if not all) chromosomes that benefit the cell in terms of gene dosage. Simplistically, increasing the copy number of important genes is of benefit and is selectively advantageous. Certainly there is evidence that increasing the dosage of genes responsible for utilisation of maltose increases the rate of maltose fermentation (Hammond, 1996). A further argument is that polyploid strains are innately more stable as genetic change will require that each copy of a gene is effected by a 'change event' (Casey, 1990; Reed & Nagodarithana, 1991; Hammond, 1996). Evidently in evolutionary terms, such genetic 'buffering' (Soltis & Soltis, 1995) is a successful strategy as the genome project suggests that chromosomal tracts have been duplicated in the haploid genome (Section 4.3.2.1). However, we may have been lulled into a false sense of security, as genetic instability in production brewing strains is increasingly being reported (see Section 4.3.2.6). Perhaps it is more appropriate to view the greater stability of polyploid strains as being relative to the more susceptible haploid isolates.

4.3.2.6  *Chromosomal instability.*  In recent years chromosomal instability in brewing yeast has had a 'good press'. The growing use of DNA fingerprinting (Sections 4.2.6.1–4.2.6.3) techniques has provided evidence that the brewing yeast genome can and does change. Karyotyping has shown changes in chromosome size (Sato *et al.*, 1994; Casey, 1996) whereas RFLP fingerprinting has succeeded in associating changes in process performance of a variant strain with genetic changes (Wightman *et al.*, 1996). For many this is, and remains, somewhat surprising. However, the view that by nature of their polyploidy, brewing yeasts are somehow immune to genetic change is now increasingly accepted as outdated and incorrect.

Perhaps it is more surprising that there is any need to debate the concept of genetic instability or change in brewing yeast. How else can the diversity of brewing yeasts (Section 4.2.3) be explained other than through genetic change and evolution? An excellent case history is the evolution of lager strains of *S. cerevisiae*. As noted by Casey (1996),

> 'over the last one hundred and fifty years, each strain has shown tremendous ability to adapt to environmental and nutritional selection pressures unique to its propagation, maintenance and utilisation – as reflected by chromosome copy number and size differences in virtually every chromosome region.'

Without wishing to prolong the argument, it is both ironic and contradictory to accept the principle of selecting stable strains with different process characteristics and not to recognise that this must be a consequence of a genetic change. Perhaps it is a matter of semantics. Genetic evolution through rearrangement is expected, genetic instability is not!

The development of karyotyping (Section 4.2.6.3) provided the tools by which chromosomes are separated and characterised by size. With it came the 'striking discovery that most (fungal) species exhibit chromosome-length polymorphism (CLP)' (Zolan, 1995). The CLPs, are detected by their change in size, and are caused by chromosomal rearrangements, which can occur via a variety of routes. Zolan (1995) has described five scenarios where DNA can be exchanged, added or lost which leads to chromosomes increasing or decreasing in size. An even more radical rearrangement can result in the complete loss of a chromosome. The evidence that such events happen is indisputable. At a fundamental level, the genome project has shown that almost half the haploid genome is duplicated with a host of 'cluster homology regions' (see Sections 4.3.2.1). Further, the many observations of chromosomal length polymorphism and the occurrence of aneuploidy (Section 4.3.2.4), unequivocally argue that the yeast genome is labile and is prone to rearrangement.

The potential frequency of adaptive mutations in yeast was first reported by Paquin and Adams (1983). Using a glucose limited chemostat, the frequency of adaptation over time for haploid and diploid strains was scored against parameters such as resistance to cycloheximide, canavanine and 5-fluorouracil. The authors concluded that 'genetic changes in both haploid and diploid populations occur surprisingly frequently'.

Subsequent work by Adams and co-workers (Adams *et al.*, 1992) extended this approach to demonstrate the sheer scale of chromosomal rearrangement that can occur during continuous cultivation of yeast. After about 50 generations in a phosphate-limited chemostat, both haploid and diploid strains of *S. cerevisiae* exhibited occasional physiological changes. These were interpreted as being 'population replacements' where 'the predominate clone in the population is replaced by one with a greater adaptation to the environment' (Adams *et al.*, 1992). Analysis of these populations by electrophoretic karyotyping together with chromosome-specific DNA probes, showed these changes to be genetic–either translocations, deletions or additions. The scale of these genetic rearrangements was such that nine of the sixteen chromosomes showed changes. Perhaps the most dramatic change was the appearance, in one population, of a new large (1100 kb) chromosome made up of sequences homologous to chromosomes II and XVI. Another population gained an extra copy of chromosome V containing a 50 kb deletion. Numerically, the most changes were seen in chromosome XII which, subsequently, has been shown to be subject to size changes in the ribosomal DNA cluster region (Chindamporn *et al.*, 1993). Although chromosomal deletions were limited to no more than 50 kb, the more common additions ranged from 10 to a substantial 390 kb. The magnitude of these additions is put into perspective by the realisation that chromosome I is only 230 kb!

The approach of Adams (Adams *et al.*, 1992; Paquin & Adams, 1983) has been extended to good effect by Ferea *et al.* (1999). Here, a related diploid strain of *S. cerevisiae* was cultured continuously in three glucose-limited aerobic chemostats for between 250 and 500 generations. All three populations changed similarly by becoming more efficient in the use of the limiting carbon source, glucose. Compared to the parental strain, the evolved strains produced at steady state *c.* four-fold more cells, *c.* three-fold more biomass and at least an order of magnitude less ethanol. Analysis of gene expression showed that at least 3% of the yeast genome – 184 genes – exhibited a two-fold difference in expression between parental and evolved strains. Of these genes, 88 could be identified via the *Saccharomyces* Genome database. Two groups could be identified as exhibiting decreased expression (glycolysis) or increased oxidative phosphorylation, TCA cycle and mitochondrial structural proteins. This clearly suggests that the phenotype is not carved in stone but, in this case, adapts presumably via 'a small set of mutations' so that the metabolism of glucose is energetically more effective. This is perhaps not the complete picture, as Ferea *et al.* (1999) note that a similar number of uncharacterised genes also changed patterns of expression in evolved strains.

Since the work of Adams *et al.* (1992) there have been numerous observations of chromosomal changes in both laboratory and brewing strains of *S. cerevisiae*. Table 4.19 summarises the various reported changes across the yeast genome. Currently only three chromosomes (VII, IX and XV) in *S. cerevisiae* remain unscathed as to reports of chromosome polymorphism. What triggers such changes remains a matter of conjecture and debate. Globally polymorphisms are reported to arise through major events such meiosis and sporulation (Yoda *et al.*, 1993) and vegetative growth (mitosis) (Longo & Vezinhet, 1993). Whether such changes can be bracketed under the umbrella of being 'selectively advantageous' (Adams *et al.*, 1992) remains to be seen. Whatever, it is noteworthy that in these 'academic' studies, chromosomal changes are anything but rapid, requiring 275 (Longo & Vezinhet, 1993) to upwards of 1000 generations.

Prior to the introduction of genetic fingerprinting methods, evidence for what was assumed to be genetic change was inferred from observations of altered behaviour. As remains the case today, such observations frequently revolve around changes in

**Table 4.19** Reports in the literature of genetic change in *S. cerevisiae* (sorted by chromosome).

| Chromosome | Size (kb) | Ploidy | Change | Reference |
|---|---|---|---|---|
| I | 230 | 2n | 140 kb addition | Adams *et al.* (1992) |
| | | 2n | size polymorphism | Longo and Vezinhet (1993) |
| | | polyploid | size polymorphism | Sato *et al.* (1994) |
| | | polyploid | size polymorphism | Casey (1996) |
| II | 813 | 1n | 10 kb addition | Adams *et al.* (1992) |
| | | 1n | 20 kb addition | Adams *et al.* (1992) |
| | | 1n | 250 kb addition | Adams *et al.* (1992) |
| | | 2n | 10 kb addition | Adams *et al.* (1992) |
| III | 315 | 2n | 40 kb deletion | Adams *et al.* (1992) |
| | | 2n | size polymorphism (350, 450, 530, 630 kb) | Yoda *et al.* (1993) |
| | | polyploid | size polymorphism | Sato *et al.* (1994) |
| IV | 1532 | polyploid | size polymorphism | |
| V | 577 | 1n | Extra copy with 50 kb deletion | Adams *et al.* (1992) |
| | | 2n | 30 kb deletion | Adams *et al.* (1992) |
| | | polyploid | size polymorphism | Sato *et al.* (1994) |
| | | polyploid | size polymorphism | Pedersen (1993) |
| VI | 270 | 2n | size polymorphism | Longo and Vezinhet (1993) |
| | | polyploid | size polymorphism | Sato *et al.* (1994) |
| | | polyploid | size polymorphism | Casey (1996) |
| VII | 1091 | | | |
| VIII | 563 | 2n | size polymorphism | Longo and Vezinhet (1993) |
| IX | 440 | | | |
| X | 765 | polyploid | size polymorphism | Casey (1996) |
| XI | 666 | 1n | 50 kb deletion | Adams *et al.* (1992) |
| | | 1n | 10 kb addition (× 2) | Adams *et al.* (1992) |
| | | polyploid | size polymorphism | Sato *et al.* (1994) |
| | | polyploid | size polymorphism | Casey (1996) |
| XII | 1078 | 1 and 2n | size polymorphisms (× 9) | Adams *et al.* (1992) |
| | | 2n | size polymorphism | Chindamporn *et al.* (1993) |
| | | polyploid | size polymorphism | |
| XIII | 924 | 1n | 60 kb addition | Adams *et al.* (1992) |
| XIV | 784 | 2n | 390 kb addition | Adams *et al.* (1992) |
| XV | 1091 | | | |
| XVI | 948 | 1n | 50 kb addition | Adams *et al.* (1992) |
| | | 1n | 20 kb addition | Adams *et al.* (1992) |

flocculation. Whether this reflects some predilection of brewing yeast or, perhaps more likely, the visual nature of such changes is not clear.

Over the years, workers at Guinness have reported changes in flocculation and utilisation of maltotriose, characteristics 'which tend to change spontaneously' (Donnelly & Hurley, 1996). Early observations originate from two doyens of brewing yeast physiology, V.E. Chester (1963) and R.B Gilliland (1971a) working, respectively, in London and in Dublin. Both demonstrated that flocculation was prone to instability. Notably, cultures derived from a single cell became less flocculent on subculturing (Chester, 1963). Gilliland (1971a) reported practical observations of an ale yeast switching from flocculent (Class II) to non-flocculent (Class I) (see Section 4.2.5.4). Opposite but less common was the process observation that a lager strain became more flocculent. Gilliland (1971a) also noted 'the second serious mutation was the loss of ability to ferment maltotriose and this has been found on a number of occasions, both in top fermentation breweries and in lager breweries'.

This theme was extended by Hammond and Wenn (1985), who reported sluggish fermentation problems that 'disappeared' after seven to ten fermentations together with flocculation changing from flocculent Class II to non-flocculent Class I (Gilliland, 1971a). Intriguingly, the 'new' problem yeast grew well in laboratory cultures on an oxidative carbon source (glycerol) but exhibited much weaker growth on the wort trisaccharide, maltotriose. Conversely the established 'normal' yeast showed strong growth on maltotriose and weak growth on glycerol. The case for these observations being linked to genetic instability coupled with selective advantage was reinforced by laboratory observations where maltotriose utilisation could occasionally be induced in the problem yeast by repeated subculturing on maltotriose. In conclusion, Hammond and Wenn (1985) argued that these events are caused by some subtle mitochondrial malfunction, which evades detailed genetic analysis of mitochondrial DNA.

A similar but seemingly less frequent observation on the switching of flocculation in production yeast strains has been made in Bass Brewers. Here, a major lager yeast strain becomes dramatically more flocculent which results in cropping difficulties. Typically, the pumped yeast solids increase from 40–45% to 60–75%. As is shown in Fig. 4.20, the timing of the switch is not consistent but, once made, is stable. Two

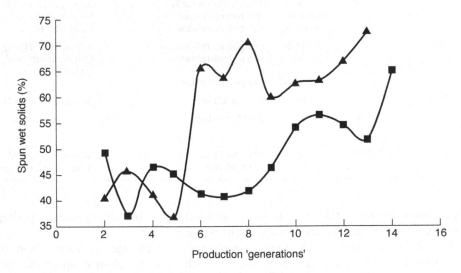

**Fig. 4.20**  Pumped solids for two production yeasts exhibiting a change in flocculence.

pieces of evidence suggest that the atypically flocculent population is genuinely different to the 'parental' population and that such observations are not some artefact of the process. Firstly, the populations differ in their response to calcium, which is required for flocculation (see Section 4.4.6.1). Washed cells of the atypical yeast are more susceptible to calcium-promoted flocculation than the normal yeast (see Fig. 4.21). Furthermore, there are small but definite genetic differences between the two yeasts. RFLP fingerprinting (Section 4.2.6.1) of the two strains (Wightman *et al.*, 1996) using two different restriction enzymes has shown there to be three bands which differ in intensity between the two yeasts (Fig. 4.22). Although by no means definitive, these observations strongly suggest a genetic explanation for changes in flocculation in a production environment. What is not clear is how such events seemingly result in a step change in phenotype rather than a progressive but slow shift in behaviour.

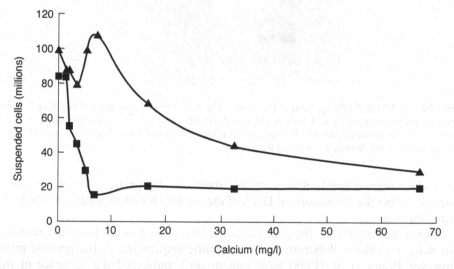

**Fig. 4.21** Role of calcium in the flocculation of the yeasts with normal (▲) and atypical (■) flocculence.

Casey's (1996) observations on long-term culture stability are particularly interesting. Retrospective chromosomal analysis of Stroh Brewery yeast samples deposited on a yearly basis between 1958 and 1985 revealed seven different karyotypes. However, the changes were restricted to four chromosomes, I (230 kb), VI (270), X (745) and XI (667). Casey argues that positive selective pressures may drive these changes as these chromosomes carry genes that are important in yeast performance (I/flocculence, VI/glycolyis, XI/maltose utilisation and X/diacetyl production).

4.3.2.7 *Mitochondrial instability.* There is general acceptance about one genetic change in brewing yeast, the 'petite' ($\rho^-$) mutation first identified in *S. cerevisiae* in 1949 by Ephrussi (Cox, 1995) by their atypically small colony size on agar plates. These mutants are respiratory deficient and, consequently, are unable to grow on oxidative substrates such as ethanol and glycerol. Correspondingly, unaffected cells

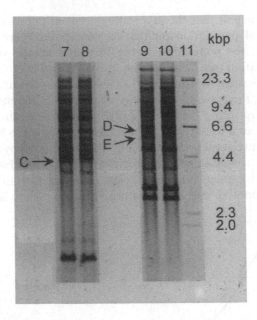

**Fig. 4.22**   RFLP DNA fingerprinting of the yeasts in Fig. 4.20. Normal yeast (lanes 7 and 9) and yeast with atypical flocculence (lanes 8 and 10) digested with different restriction enzymes (*Pst*I lanes 7, 8 and *Sal*I lanes 9, 10). Differences in the RFLP fingerprint identified by arrows. From Wightman *et al.* (1996), with permission from the Society for Applied Bacteriology.

($\rho^+$) which form normal colonies are described as being 'grande'. This genetic change does not effect the chromosomal DNA of the nucleus but the specialised DNA of the mitochondria.

Mitochondrial DNA is the poor relation of the yeast genome inasmuch that it was not subject to the collaborative and systematic sequencing of the genome project. However Foury *et al.* (1998) have subsequently published the sequence of mitochondrial DNA from the same haploid strain of *S. cerevisiae* as used in the genome project. At only 86 kb, mDNA is significantly smaller than the smallest chromosome (chromosome I at 230 kb). Although representing only 0.5% of the yeast genome (as measured by length or 'hypothetical' proteins), the mitochondrial genome is enriched (*c.* ten-fold) by genes that code for transfer RNA. Indeed, the mDNA codes for only about 25 identified proteins and seven hypothetical potential proteins (ORFs – see Section 4.3.2.1) as the vast majority (95%) of mitochondrial proteins are encoded by the nuclear genome (Dujon, 1981) and imported into the mitochondria.

The petite mutation is typically caused by loss ($\rho^0$), gross alterations or, in the case of lager yeast petites, extensive deletion of mitochondrial (mt) DNA (Good *et al.*, 1993). In the later situation, the remaining mtDNA is amplified so that the petite contains the same amount of mtDNA as the wild type yeast (Good *et al.*, 1993; Piskur *et al.*, 1998). The mutation is frequently described as being 'spontaneous' and 'natural' occurring in *S. cerevisiae* at a frequency of *c.* 1% (Silhankova *et al.*, 1970a). However, in the laboratory, mutagens such as ethidium bromide (Meyer & Whittaker, 1977) can be used to generate petites (Piskur *et al.*, 1998).

Classically, petites are identified by the 'tetrazolium overlay' plate test of Ogur *et al.* (1957) as white colonies whereas grandes are red. Alternatively, and preferably, petites can be estimated by more functional approaches. For example, replica plating from a fermentative (glucose) to an oxidative (glycerol) carbon source or directly by classification as large ($\rho^+$) or small ($\rho^-$) colonies on media containing a limiting amount of fermentable carbohydrate (Piskur *et al.*, 1998).

As ever, it is unwise to make sweeping generalisations about the genetics and physiology of relatively closely related species of yeast. For example, the three groupings that make up the genus *Saccharomyces* (see Section 4.2.1) differ in the size and stability of their mitochondrial genomes (Piskur *et al.*, 1998). Both the *Saccharomyces sensu stricto* and *Saccharomyces sensu lato* form petites spontaneously but differ in susceptibility to ethidium bromide. The *Saccharomyces sensu stricto* group that contains *S. cerevisiae* has the biggest mtDNA (64–85 kb), and petite formation is induced by treatment with ethidium bromide. The mitochondrial genome is smaller (23–48 kb) in the *Saccharomyces sensu lato* group and ethidium bromide does not induce significant formation of petites. The third group, consisting of *S. kluyveri*, is quite distinct in not giving rise to petites, neither naturally nor after treatment with the mutagen, ethidium bromide.

To add further complexity, within *S. cerevisiae*, the frequency of the petite mutation can vary. For example, Donnelly and Hurley (1996) have reported that a population with 2–4% petites is 'acceptable for brewing purposes'. Conversely Morrison and Suggett (1983) reported that in one production lager strain, petites accounted for almost 50% of the population. In this case, petites were seemingly triggered by protracted yeast storage. Others have reported the induction of petites by the use (and presumably carryover in fermenter) of formaldehyde in the brewhouse to improve colloidal stability (Cowan *et al.*, 1975). Finally, there may be more subtle reasons for any variation in susceptibility to petite formation. As discussed at length earlier in this chapter (Section 4.2.2), ale yeasts are exclusively *S. cerevisiae* whereas lager yeasts are a genetic hybrid between *S. monacensis* and *S. cerevisiae*. Interestingly, as shown by Piskur *et al.* (1998), the mtDNA of lager strains is exclusively derived from the *S. bayanus* parent which, although similar, differs from the mtDNA of *S. cerevisiae*. Consequently, differences in the nature and frequency of the petite mutation might be anticipated in lager and ale isolates of *S. cerevisiae*.

Arguably, the petite mutation should be of little consequence in brewery fermentations. Afterall, respiration – as noted elsewhere (see Section 3.4.2) – plays no direct role in brewery fermentation. Despite this, there is substantial evidence that petite mutants are undesirable. It is increasingly clear that the large primitive mitochondria ('promitochondria') found in anaerobic cells (Visser *et al*, 1995) are vital to cell metabolism during anaerobiosis (for a review see O'Connor-Cox *et al.*, 1996). For example numerous enzymes are located in mitochondria, notably some of those involved in the citric acid cycle, sterol biosynthesis and amino acid synthesis (Visser *et al.*, 1994). Further compelling evidence for the contribution of promitochondria to anaerobic metabolism of yeast comes from studies with bongkrekic acid, which inhibits ATP transport into mitochondria. Under anaerobic conditions, this is necessary as ATP cannot be supplied via respiration but is required to fuel transport of metabolites in and out of the mitochondria. On addition of bongkrekic acid, growth rate is reduced in anaerobic cultures, both continuous (Visser *et al.*, 1994) and batch cultures (O'Connor-Cox *et al.*, 1993). Other work has shown that a functional mitochondrial genome is important in tolerance to ethanol, with petites being more sensitive to the growth inhibitory effects of ethanol (Brown *et al.*, 1984; Hutter & Oliver, 1998). Furthermore, exposure to high concentrations of ethanol (e.g. 18%,v/ v) can induce the formation of petites. Chi and Arneborg (1999) have extended this theme to show a relationship between petite formation and ethanol tolerance and membrane lipid composition. More ethanol-tolerant strains are less prone to the formation of petites. Compared to ethanol-sensitive strains, ethanol-tolerant yeasts have comparatively higher levels of phosphatidlycholine, ergosterol and long-chain unsaturated fatty acids. These differences, Chi and Arneborg (1999) argue, would reduce membrane fluidity, which in turn would protect against the fluidising effects of ethanol on membranes.

The applied significance of mitochondria is borne out by more applied work in fermentations with petites where beer quality and process performance are markedly affected (Silhankova *et al.*, 1970b; Morrison & Suggett, 1983; Ernandes *et al.*, 1993). Elevated levels of petites result in sluggish fermentation rates, reduced yeast growth and changes in flocculation. Corresponding beers have a skewed flavour with elevated levels of diacetyl and higher alcohols together with reduced concentrations of ethyl acetate.

### 4.3.3 *Cell cycle*

Like the rest of this chapter, the focus here is on polyploid brewing yeast strains and vegetative asexual reproduction. Sexual reproduction (*mating* or *conjugation*) and *sporulation*, which apply respectively to haploid and diploid cells, are not considered here. The interested reader is directed to various reviews for a description of these processes (Byers, 1981; Pringle & Hartwell, 1981; Wheals, 1987; Cox, 1995; Sprague, 1995).

Whether laboratory, baking or brewing strains, *S. cerevisiae* undergoes asexual reproduction via an asymmetric form of cell division (see Section 4.3.3.1) called 'budding'. One cell gives rise to two 'daughter' cells that are genetically identical to the original 'mother' cell. This simple relationship is central to the success of the yeast cell, be it industrially in a brewery fermenter, in a laboratory culture or environmentally on the surface of a grape. Of course, cell division is an ongoing process in that the progeny of cell division – the 'virgin' daughter cells – themselves divide becoming mother cells and so on (Fig. 4.23 ). However, cell division is not a linear, never-ending process. Cell division slows or stops ('arrests') when growth nutrients become limiting (see Section 4.3.3.3 Stationary phase), when cells age and become senescent (see Section 4.3.3.4 Ageing) or when cells die (see Section 4.3.3.5 Death).

#### 4.3.3.1 *Cell division.*
Other than a rather academic review by Duffus (1971), the yeast cell cycle has attracted little interest in the brewing literature. However, recent work on cell ageing (Section 4.3.3.4) in brewing yeast may well renew interest in this fundamentally important area.

The biology of the cell cycle of the budding yeast *S. cerevisiae* and, particularly, the

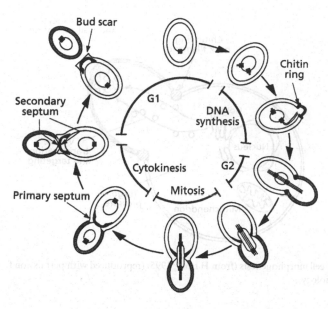

**Fig. 4.23** The cell cycle in *S. cerevisiae* (from Harold, 1995) (reproduced with permission from the Society for General Microbiology).

fission yeast *Schizosaccharomyces pombe*, continue to provide rich pickings for cytologists, morphologists and geneticists. Consequently, an astonishing amount of information has been – and continues to be – published about the yeast cell cycle. As ever, the driver for this research is that *S. cerevisiae* is a model system for understanding eukaryote cell division. As noted by Jazwinski (1990), 'it has become abundantly clear that virtually any phenomenon involving individual eukaryotic cells can be profitably studied in yeast'. In this case, the key elements of the cell cycle appear to be highly conserved throughout evolution, so what happens in yeast is likely to find application in higher eukaryotes from plants to man. Here, coverage is limited to an overview of the fundamental steps of this process. A comprehensive review can be found by Wheals (1987) in *The Yeasts*.

The cell cycle is divided into a series of events or 'landmarks' that describe this remarkable process that creates a new genetically identical cell. Although inevitably an overview of this sort cannot do justice to the wonders of the cell cycle, the diagram of Harold (1995) (Fig. 4.23) succinctly captures the cycle landmarks together with cell morphogenesis. Normally, unless synchronised in some way, an actively growing population will contain cells at all stages of this cycle. The unbudded gap period or 'GI' phase represents the time prior to starting a new round of cell division. This commences at a point toward the end of GI termed 'START', a decision point that represents a collection of cellular events necessary for cell division. Although complex (see review by Sherlock and Rosamond, 1993), one of the questions posed before committing to the cell cycle is whether the environment contains adequate levels of nutrients. If there is, the next step is that of DNA synthesis ('S' phase), during which time the daughter cell 'bud' starts to develop. The post-synthetic G2 phase coincides

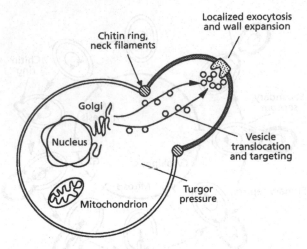

**Fig. 4.24**  Yeast cell morphogenesis (from Harold, 1995) (reproduced with permission from the Society for General Microbiology).

with significant bud growth, which then culminates in mitosis ('M' phase), and nuclear division. The final steps of the cell cycle are (i) cytokinesis where daughter and mother cells are separated by septa and (ii) cell separation. The sites of attachment on mother and daughter cells are seen (under the microscope) as, respectively, bud and birth scars.

The physiology of bud growth continues to excite much interest. Not surprisingly, this complex sequence of events (see Fig. 4.24) requires numerous checkpoint controls. Even the site of a new bud is controlled, being either axial (haploid) or bipolar (diploid) (Chant, 1994). As noted by Harold (1995) 'expansion of the wall is patterned in space and time: to begin with, growth takes place chiefly at the tip of the forming bud, then it becomes uniformly distributed, and eventually expansion halts'. These events are driven through secretory vesicles that deliver the appropriate enzymes and precursors for wall growth. Intriguingly, there is a view that 'cables' of actin microfilaments track through the cytoplasm into the bud, both to facilitate and to target the vesicles (Harold, 1995). The chitin (Section 4.4.3.3) ring elaborated early bud development remains after separation as a bud scar. These are usefully visualised using fluorochromes such as primulin or Calcofluor (Streiblova, 1988).

Not surprisingly given its complexity, cell division is 'resource hungry'! According to Wheals (1987) there are about 70 cdc (cell-division cycle) mutations. The yeast genome project extends this commitment and allocates about 14% of the 3167 functional ORFs (see Fig. 4.17 ) to 'cell growth, cell division and DNA synthesis' (Mewes *et al.*, 1997). Much of what is known about the cell cycle, although undeniably fascinating and worthy of discussion, is beyond the scope of this book. However, the role of yeast cell size in the cell cycle warrants comment.

At the end of the cell cycle, on separation, daughter cells are usually smaller than their mother cells (Fig. 4.25). Work in the late 1970s (Johnston *et al.*, 1977) showed that the daughter cell had to increase its cell volume before it was capable of budding.

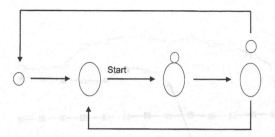

**Fig. 4.25**   The role of yeast cell size in the cell division cycle.

Accordingly, the G1 period of the daughter cell was longer than the mother cell whose size was more appropriate for entering the cell cycle. This asymmetric division has been crystallised in terms of the need to achieve a 'critical cell size' before a new round of division can start. Typically, there is no definitive 'critical cell size'! The work of Tyson and colleagues (Tyson *et al.*, 1979) showed the mean cell size in aerobic cultures to increase 2.5 fold with increasing growth rate (doubling time from 450 to 75 minutes). Correspondingly, the 'critical size' varied with growth rate, being (at slow growth rates) almost half the volume of cells dividing quickly.

As something of an aside, mutants have been isolated which have been used to probe the relationship between cell size and division. These mutants – *wee* (*S. pombe*) (Nurse, 1975) and *whi* (*S. cerevisiae*) (Sudbery *et al.*, 1980) – start budding at only half the cell size of wild type cells. Consequently, because of their small size, the mutants were named *wee* and *whi*. However, this is only part of the story. The *wee* designation also reflects on the Scottish heritage of this mutant (the work was done in Edinburgh) whereas *whi* stems from the whisky that was consumed in celebration of obtaining the mutant (Andrew Goodey, personal communication)!

**4.3.3.2**   *Cell division and brewery fermentation.*   Cell cycle events in brewery fermentations have received little attention. This is a little surprising given the widespread and high-profile interest in the cell cycle over the last 25 years. Perhaps, in comparison to the genetic improvement of brewing yeast strains, it is seen to offer little obvious added value. Another likely explanation is the comparative technical complexity of brewery fermentations. This is a two-edged sword! On the one hand, there is the need to replicate as precisely as possible a brewery fermentation in a laboratory culture without, on the other hand, so many compromises that the conclusions cannot be extrapolated to the real world. This, together with the ever-changing cultural environment (notably oxygen) in brewery fermentation, is perhaps sufficient to dissuade substantial work in this area.

This is a pity as even laboratory 'brewery' fermentations can provide fascinating insights into cell cycle events. By definition having been recovered from a previous fermentation, pitching yeast is in stationary phase (see Section 4.3.3.3). On pitching into a single batch of air-saturated wort, the critical cell cycle events on which successful fermentation depends can be seen to unravel step by step (Quain, 1988). The shift from the effectively 'starved' conditions of yeast storage to fresh wort results in early events being seemingly synchronised (Fig. 4.26). The initially unbudded

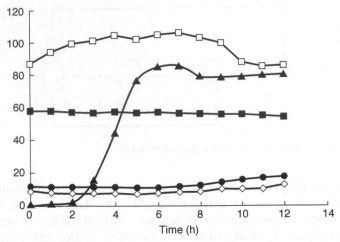

**Fig. 4.26** Events during the first 12 hours of a stirred laboratory fermentation (redrawn from Quain, 1998). Units are biomass ($\Diamond$, mg.ml$^{-1}$ – for real figure $\div$ by 10); cell number ($\bullet$, $10^6$.ml$^{-1}$); budded cells ($\blacktriangle$, %), cell volume ($\square$, $\mu m^3$ – for real value $\times 2$) and present gravity ($\blacksquare$).

stationary phase ('G0') cells presumably 'sense' the presence of nutrients and shift into the G1 phase and START. During these first two hours cell volume increases perceptibly but PG and cell number remain static whereas total biomass declines by almost 20%. Arguably, glycogen breakdown during this time fuels the G0 to G1 transition and, specifically, new lipid synthesis. As pitching yeast is 'lipid depleted', the formation of sterols and unsaturated fatty acids is a critical requirement for cell division to occur in brewery fermentations. Entry into S phase is apparent with the onset of budding at 2 hours, which peaks at almost 90% of the population within 6 hours of pitching. By now the population is asynchronous as, against a budding index of 80%, new daughter cells begin to accumulate and the cell number increases from 7 hours onwards. At this point changes are detectable in biomass and sugar uptake (PG). As would be predicted, these early events in fermentation are accompanied by an increase in cell size, which peaks 20% higher at 6 hours. Presumably, this reflects the need for individual cells to achieve a critical cell size before entering the cell cycle.

This frenetic early period of activity prepares the cell for its future role in fermentation. During active fermentation (Fig. 4.27) there are about 3.5 rounds of cell division as cell numbers increase from 10 to 100 $\times 10^6$ cells ml$^{-1}$. Other indicators of cell cycle activity (budding index and cell volume) decline.

Of course, such measures are very much an average of what is a changing and heterogeneous cell population with, typically, 50% virgin daughter cells, 25% single budded mothers and 25% multi-budded mother cells (Deans *et al.*, 1997). Interestingly (Fig. 4.28), cell division does not continue throughout fermentation but 'arrests' at about the 'mid-point'. This is reciprocated by the budding index declining to basal level. It is generally accepted that division is not arrested by nutrient deficiency in the fermenting wort but by dilution by cell division of essential lipids. As noted in more detail elsewhere the sterols and, to a lesser extent the unsaturated fatty acids, determine yeast growth and the rate (and extent) of fermentation. Consequently, brewery

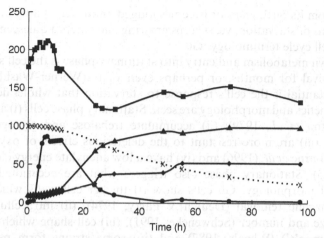

**Fig. 4.27** Events during a stirred laboratory fermentation (see Fig. 4.26) (unpublished observations of Fintan Walton and David Quain). Units are biomass ($\blacklozenge$, mg.ml$^{-1}$ – for real figure $\div$ by 10); cell number ($\blacktriangle$, $10^6$.ml$^{-1}$); budded cells ($\bullet$, %), cell volume ($\blacksquare$, $\mu$m$^3$) and present gravity ($\times$, as a % of the original gravity, 1060).

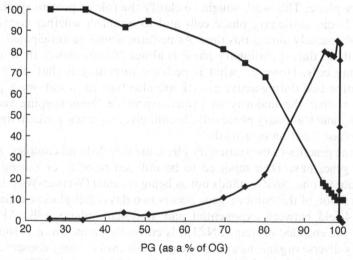

**Fig. 4.28** Relationship between cell division and fermentation (same experiment as Fig. 4.26) (unpublished observations of Fintan Walton and David Quain). Units are budded cells ($\blacklozenge$, % of total cells) and cell number ($\blacksquare$, $10^6$.ml$^{-1}$).

fermentations are in a sense nutrient limited! The point of difference being that the nutrient is supplied and consumed some days before 'deficiency' takes effect.

### 4.3.3.3 *Stationary phase.*

In the 'real world', nutrient starvation is often the 'norm' for micro-organisms. As noted by Werner-Washburne *et al.* (1996), 'yeast cells can be translocated from relatively nutrient-rich environments, e.g. ripe grapes and plant leaves (especially wilted plant leaves), to nutrient-poor environments as rapidly as a cell is blown from its birth place or becomes lodged on the leg of a fly'. Accordingly, under conditions of starvation, yeast stops growing and enters a quiescent 'stationary phase' or, in cell cycle terminology, G0.

Shutting down metabolism and entry into stationary phase is the cell's strategy for long-term survival for months, or perhaps even years (Werner-Washburne *et al.*, 1996). So substantial is the cell's response to starvation that wholesale changes in physiology, genetics and morphology are seen. Stationary phase cells (i) are more heat resistant (Walton *et al.*, 1979), (ii) accumulate trehalose and glycogen (Lillie & Pringle, 1980), (iii) are more resistant to the deleterious effects of oxygen radicals (Werner-Washburne *et al.* (1996) and (iv) have a low adenylate energy charge (Ball & Atkinson, 1975). Stationary phase also triggers what are occasionally dramatic changes in cell morphology. G0 cells show (i) thicker cell walls which are more resistant to enzymic removal (Deutch & Parry, 1974), (ii) intracellular vacuoles increased in size and number (Schwencke, 1991), (iii) cell shape which can become distorted ('schmoo's') (Wheals, 1987) and (iv) some strains form pseudohyphae (Kuriyama & Slaughter, 1995).

Most of the contemporary studies on the stationary phase in yeast focus on cell biochemistry and genetics. A notable publication by Fuge *et al.* (1994) overcame some of the criticism of some of the work in the area by monitoring cells that were genuinely in stationary phase. This work sought to clarify the role of protein synthesis over a protracted 23-day stationary phase cells and to establish whether certain proteins were formed uniquely during this time. As perhaps would be anticipated, the rate of protein synthesis during stationary phase is about 300-fold lower than in exponentially growing cells. However, what is perhaps surprising is that the portfolio of proteins synthesised during active growth are also formed in stationary phase cells. This suggests that the majority of proteins provide 'housekeeping functions' in exponentially and stationary phase cells. Seemingly, only a few proteins are unique to stationary phase function or growth.

As ever, the genetics of the stationary phase are detailed and complex. Although a number of genes have been reported to be induced prior to or during entry into stationary phase, one, *SNZ1*, stands out as being unusual (Werner-Washburne *et al.*, 1996). Expression of this 'snooze' gene occurs two days after glucose exhaustion and increases 14-fold between exponential and stationary phase cells. Although the function of the encoded protein (SNZ1p) is currently unknown, it is found in phylogenetically diverse organisms and is one the most evolutionary conserved proteins, which suggests that it has a bigger role in biology (Braun *et al.*, 1996).

Although subject to a formidable amount of work, our understanding of stationary phase in yeast is based almost exclusively on aerobic cultures of haploid strains growing on glucose in defined laboratory media at 30°C. Little work has been published on stationary phase in brewery fermentations. Consequently, as conditions in brewery fermentations are almost diametrically opposed to the laboratory model, it is something of an act of faith that stationary phase is similar in brewing yeast. A further rider is that the trigger for entry into stationary phase is not solely exhaustion of a (usually fermentable) carbon source but includes nitrogen and sulphur limitation. The onset of stationary phase in (anaerobic) brewery fermentations has not been explicitly studied. Whether entry into stationary phase at the end of fermentation is triggered by exhaustion of fermentable carbohydrate or some other nutrient is not entirely clear. Indeed, the presence of residual fermentables (notably maltotriose) in green beer possibly suggests a more complex interpretation than simply the apparent lack of external nutrients. In this case, it may be more appropriate to consider that sugars become 'unavailable' because of changes (brought about by the effects of lipid depletion on membrane function) in the affinity of transport proteins or through cell flocculation. Certainly, brewery fermentations can 'arrest' or 'stick', in the presence of large amounts of fermentable carbohydrate, when lipid synthesis is limited by the addition of insufficient oxygen at the beginning of fermentation. Of course, with so little hard evidence, the simple view that the unavailability of fermentable carbohydrate triggers entry into stationary phase does not preclude the possibility that the exhaustion of some other nutrient may initiate this transition.

Much of what is known about stationary phase in brewery fermentations relates to the consequences of yeast storage and its subsequent 'fitness for purpose'. Physiologically, although apparently quiescent, brewing yeast in stationary phase continues to 'tick over' – the rate of which is dependent on temperature. In gross terms, stationary phase cells are under threat and metabolism is focused on survival in the hope that the cells' environment will eventually improve. In common with laboratory observations under aerobic conditions (Lillie & Pringle, 1980), stored (stationary phase) brewing yeast dissimulates accumulated storage carbohydrates such as glycogen and, to a lesser extent, trehalose (Quain et al., 1981). Quantitatively, the contribution of glycogen to survival in stationary phase is substantial. As if in recognition of this, glycogen accumulation during active brewery fermentation is substantial, accounting for up to 40% or more of the cell biomass. The extent of glycogen turnover during stationary phase in fermenter or in storage vessel can be extensive depending on duration and temperature. Presumably, glycogen breakdown provides energy for ongoing cell maintenance during stationary phase. Although the significance is not clear, the rate of glycogen breakdown during stationary phase in brewing strains is found to vary by up to three-fold (Quain, 1988).

The advent of the concept of 'vitality' and 'vitality testing' (see Section 7.4.2) has resulted, indirectly, in an improved understanding of the physiology of cells in stationary phase. Numerous measures have been advocated to assess the vitality of stored yeast. Most focus on metabolic response post-feeding with glucose, such as uptake of oxygen, evolution of carbon dioxide, intra- or extracellular changes in pH. Whatever their utility in establishing a population's 'fitness' for pitching-on, these tests provide a useful benchmark for the varying physiological status of batches of yeast in store. However, interpretation of 'cause and effect' is difficult, as these measures are unable to differentiate between the impact of yeast storage and physiological changes brought about by stationary phase.

In passing, although stationary phase cells are physiologically better equipped to withstand the rigours of yeast storage and handling, such cells will be more resistant to killing through pasteurisation.

4.3.3.4 *Ageing.* Understanding the phenomena of cellular ageing has a universal and popular appeal. Together with the nematode *C. elegans* (whose genome is being sequenced, Section 4.3.2.1) and the fruit fly, Drosophila, *S. cerevisiae* is at the forefront of high-profile studies that seek to model and unravel the mechanisms of cell ageing. The profile is such that national newspapers and press agencies report developments in this field. For example, work at the Massachusetts Institute of Technology (MIT) created a furore with 'Yeast gene yields clues to the ageing process' (Reuters, 28/8/97) and 'Yeast cells may hold key to human ageing' (*The Times*, 26/12/97). Simplistically the media see the prize of such work as being understanding and even control of human ageing. Inevitably, this ever-changing area is continually subject to review. Recent general reviews include those by Sinclair *et al.* (1998) and Jazwinski (1999) together with more applied reviews by Smart (1999) and Powell *et al.* (2000).

The yeast cell has a finite life span, measured by the number of divisions and not by chronological age. On stopping division, the cell becomes senescent and eventually dies. In addition to the usual arguments about being a model eukaryote, *S. cerevisiae* lends itself to experimental studies on ageing. The reasons for this are numerous but includes: (i) the cell and organism are one and the same, (ii) the cell is thought not to be affected by extracellular factors (hormones, other cells), (iii) cells can be isolated singly or in bulk and (iv) bud scars act as a 'biomarker' for cell age (Jazwinski, 1990). Although currently fashionable, the study of the life span of yeast is by no means new, dating back to work by Mortimer and Johnston (1959). This important paper concluded that the life span of a diploid strain of *S. cerevisiae* was an average of 24 generations, with a maximum of about 40. Some decades later and using the same methodology as Mortimer and Johnston (1986), Egilmez and Jazwinski (1989) reported remarkably similar data for a haploid strain, which had a mean life span of 24 ± 9 (standard deviation) generations. Other strains have a shorter life span. Work with a polyploid lager strain (Barker & Smart, 1996) reported a mean life span of 17 generations and a maximum of 29 generations. Similar data was reported for a haploid strain (mean = 17, maximum = 31) (Austriaco, 1996). In passing, the approach first described by Mortimer and Johnston (1959) requires long hours of tiring microscopic examination. Given this, it is not surprising that yeast cells under study require to be refrigerated overnight so as to slow division and to 'provide relief for the investigator' (D'mello *et al.*, 1994)! Such treatment has no bearing on determinations of life span (Egilmez & Jazwinski, 1989).

The physiology of ageing in yeast is fascinating. Morphologically, the cell takes on an aged appearance, becoming granular, crenellated and wrinkled (Mortimer & Johnston, 1959; Jazwinski, 1990; Barker & Smart, 1996) and accumulates lipid granules (Austriaco, 1996). Further, as first noted by Mortimer and Johnston (1959), yeast cell size and, to an extent, generation time, increase with successive generations. The increase in cell size is linear with age (Jazwinski, 1990). With lager yeast NCYC 1166 there is a six-fold difference in cell volume (Fig. 4.29 ) between young new mothers (163 $\mu m^3$) and senescent cells (950 $\mu m^3$) at the end of their life span (Barker & Smart, 1996). The relationship between generation time and cell age is less pleasing to the eye! Egilmez and Jazwinski (1989) noted that a 'pattern of moderate increase in generation time between the ages of 10 and 20 generations was always followed by a sharper increase after generation 20'. Barker and Smart (1996) (Fig. 4.30) reached broadly similar conclusions although the generation time was stable for almost 70% of the cell's life span before increasing dramatically during the last two or three divisions.

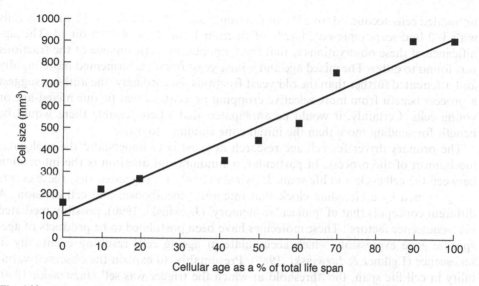

**Fig. 4.29** Relationship between cell volume and age (redrawn from Barker and Smart, 1996).

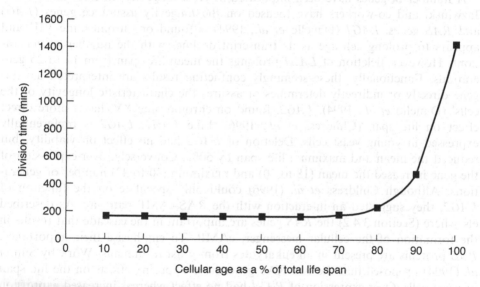

**Fig. 4.30** Relationship between cell division time and cellular age (redrawn from Barker and Smart, 1996).

The implications of yeast cell ageing in brewery fermentations were studied by Deans *et al.* (1997). In this elegant work, the yeast cone from a 2000 hl fermenter was fractionated into 5 hl fractions. Each fraction was characterised by age (bud scar analysis and Calcofluor staining, see Section 4.3.3.1) and fermentation performance in tall tube fermenters. As would be anticipated, the age distribution is skewed across the cone, ranging from predominately older cells in early (bottom) fractions to predominately young cells in last cuts at the top of the cone. For example, virgin unbudded cells accounted for 35% of fraction 1 and 72% of fraction 12 whereas cells with 1–2 bud scars represented 42% of fraction 1 and 21% of fraction 14. The significance of these observations is that the fermentation performance of the fractions was found to differ. The mixed age and young yeast fractions fermented more rapidly and attenuated further than the old yeast fractions. Accordingly, the authors suggest a process benefit from more selective cropping procedures that favour mixed-age or young cells. Certainly, it would be anticipated that where feasible there would be benefit for sending more than the initial cone runnings to waste.

The primary driver for cell age research in yeast is to understand the molecular mechanism of the process. In particular, a fundamental question is the interaction between the cell cycle and life span. Jazwinski (1990) has suggested that these events are governed by a circadian clock that integrates metabolism and cell division. A different concept is that of 'molecular memory' (Jazwinski, 1990), possibly mediated via 'senescence factors'. These molecules have been postulated to be products of age-specific gene expression, which accumulate in ageing cells resulting eventually in senescence (Egilmez & Jazwinski, 1989). Presumably, to explain the observed variability in cell life span, the 'threshold at which the trigger was set' (Jazwinski, 1990) would vary between yeast cells and strains.

A number of genes have been implicated in yeast longevity. In Louisiana, USA, Jazwinski and co-workers have focused on the *l*ongevity *a*ssurance *g*enes (*LAG*) and *RAS* genes. *LAG1* (D'mello *et al.*, 1994) is found on chromosome VIII and appears to prolong cell age as its transcript declines with the number of generations. However, deletion of *LAG1* prolongs the mean life span from 17 to 25 generations. Functionally, these seemingly conflicting results are interpreted as 'this gene directly or indirectly determines or assures the characteristic longevity of the cells' (D'mello *et al.*, 1994). *LAG2*, found on chromosome XV, has a more direct effect on life span (Childress *et al.*, 1996). Like *LAG1*, *LAG2* is preferentially expressed in young yeast cells. Deletion of *LAG2* had no effect on viability, but reduced the mean and maximum life span by 50%. Conversely, over-expression of the gene increased the mean (18 to 20) and maximum (30 to 42) number of generations. Although Childress *et al.* (1996) could only speculate on the function of *LAG2*, they suggested an interaction with the RAS-cAMP pathway. As described elsewhere (Section 3.4.2) the *RAS* genes are important in the cascade that results in the formation of the cellular messenger, cAMP. To emphasise their importance, *RAS* proteins are present in all eukaryotes from yeast to humans. Work by Sun *et al.* (1994) proposed that the two RAS genes have opposing effects on the life span of yeast cells. Over-expression of *RAS1* had no effect whereas increased expression of *RAS2* extended the life span and postponed the senescence-related increase in generation time. Disruption of *RAS2* decreased the life span whereas deletion of *RAS1* prolonged it! Although an attractive concept, Sun *et al.* (1994) found that elevated levels of cAMP did not extend but curtailed life span.

Guarente and co-workers at MIT isolated mutants in the wonderfully named *UTH* ('youth') genes with increased stress resistance and longevity (Austriaco, 1996). More recently the group have used yeast as a vehicle to probe a human disease, Werner's syndrome, that causes premature ageing. This is caused by a recessive mutation to the *WRN* gene, which has a homologue in yeast, *SGS1* (Sinclair *et al.*, 1997). The *SGS1* gene is important in yeast cell ageing as its deletion reduced the average life span to about 40% of the wild type strain.

Both the *WRN* and *SGS1* genes code for helicase proteins that are involved in the transcription of the highly reiterated ribosomal DNA in the cell nucleolus. Any mutation that impacts on ribosome formation or structure would potentially effect mRNA translation and protein synthesis. Intriguingly the work of Sinclair *et al.* (1997) has shown that deletion of the SGS1 gene results in the nucleolus becoming enlarged and fragmented. These events, it is argued, prevent cell division in yeast and, consequently, senescence and death. A subsequent paper (Sinclair & Guarente, 1997) has developed the theme that the nucleolus is the cell's 'Achilles' heel' in ageing. In some respects, this work harks back to the 'senescence factor' theory of Egilmez and Jazwinski (1989). Fragments of DNA (extrachromosomal ribosomal DNA circles or 'ERCs') accumulate as cells become older. These ERCs, which are believed to be by-products of the cell's ongoing DNA repair processes, may act as a 'clock' for cell ageing.

4.3.3.5 *Death and autolysis.* Death and autolysis are the final events in the life cycle of yeast. Although cell death (or loss of viability) and, to extent, autolysis can be measured, the triggers for these events are at best empirically understood. In the case of cell viability, programmed cell death through ageing and senescence only account for a fraction of the dead cell population found in fermenter. Clearly, as is well known, the environment of the yeast cell can have a direct impact on cell death. Most notable of the numerous environmental factors is the concentration of the narcotic, ethanol (see Section 3.6.2). The complexity surrounding the assessment of cell death and the mystery of 'vitality' are discussed at length elsewhere (see Sections 7.4.1 and 7.4.2).

It will come as no surprise that yeast is proving a useful vehicle to probe the phenomena of cell suicide in animal systems. Apoptosis (for a review see Matsuyama *et al.*, 1999) or 'programmed cell death' describes the process where in adult humans, some 50–70 billion cells are eradicated on a daily basis! Although the apoptotic pathway found in animal species has not been identified in yeast, *S. cerevisiae* is being used to gain important insight into these events in higher eukaryotes. Conversely, this vibrant area may throw light on cell death in yeast. For example, a recent report (Madeo *et al.*, 1999) links the accumulation of reactive oxygen species (peroxide, superoxide radicals) with induction of apoptosis in yeast.

Although less glamorous, yeast autolysis has received its fair share of attention. This reflects the impact of cell autolysis on beer flavour and appearance together with its importance as a process in the production of yeast extracts and spreads for human consumption. Autolysis was first observed accidentally by Salkovski in 1989 (Joslyn, 1955). Although considered a process of 'self digestion', the focus of studies on autolysis has been the end result rather than the cellular mechanisms. Suffice to say autolysis is caused by intracellular events mediated by hydrolytic enzymes such as proteinases and glucanases (see Thorn, 1971).

Within brewing, autolysis is triggered by a variety of environmental factors such as temperature, pH, ethanol concentration and osmotic pressure (Lee & Lewis, 1968; Thorn, 1971; Chen *et al.*, 1980; Yamamura *et al.*, 1991). As ever, autolysis is prevented by good practice in fermentation and yeast handling. The impact of yeast autolysis on beer flavour is dependent on the extent of autolytic damage but, at it worst will give rise to a 'yeast bitten' palate. Autolysis products include low-molecular-weight substances (nucleotides, fatty acids) (Lee & Lewis, 1968; Chen *et al.*, 1980) and high-molecular-weight cell wall material (glucans and mannans) (Lewis & Poerwantaro, 1991). Depending on the concentration, the low-molecular-weight materials affect beer flavour and appearance whereas the cell wall materials can cause haze.

### 4.3.4  Genetic modification

4.3.4.1 *Drivers for strain development.*  Fundamentally, two drivers have fuelled research into the genetic modification of brewing strains. First, to modify strains to do what they do better, and, second, to modify strains to do what they currently cannot do. Depending on your point of view, the distinction between these two drivers can be narrow. Suffice to say the goal of genetic modification is one of strain improvement enabling yeast to be better equipped or more capable to perform its task of (usually) beer production.

Ironically, brewing yeast strains are poorly equipped to take advantage of what brewing technology in the twenty-first century can offer. Indeed, the physiology, biochemistry and genetics of brewing strains are such that they are now 'rate-limiting' in the introduction of innovative fermentation processes. In other words, we have gone as far as we can with fermentation technology using the current portfolio of brewing strains. To achieve a significant incremental or step change two approaches can be envisaged. First, finding existing strains through natural selection which are better equipped to achieve a desired task, or, second, the introduction of desired genes through genetic manipulation. Although the former approach has undoubtedly been successful in the selection of micro-organisms for the commercial production of enzymes, chemicals, vitamins and pharmaceuticals (Steele & Stowers, 1991), it is the latter approach that this section is focused on. Arguably, this route is quicker and more targeted than the serendipity of natural isolation and selection.

From the mid-1980s, there have been quite stunning developments in the manipulation or modification of yeasts, usually *S. cerevisiae*. Over this period yeast genetics has galloped ahead through substantial exponential developments in funding, insight and protocols, culminating in the sequencing of the yeast genome (Section 4.3.2.1). Nowadays genetic modification of (primarily) laboratory strains of *S. cerevisiae* is a routine event such that it is frequently performed by university undergraduates. The focus has moved on to understanding gene function, regulation and integration within cell metabolism.

Inevitably against this background, there have been significant developments in the genetic manipulation of brewing strains. Indeed, a number of generic targets can be identified that have been subject to the attention of brewing yeast geneticists. Articles by Tubb (1981, 1984) at what is now Brewing Research International (BRi), whetted the industry's appetite by painting an appealing picture of what might be possible. Three major targets were identified: (i) enabling yeast to use different carbon sources,

(ii) increasing the efficiency and productivity of fermentation and (iii) improving the control of fermentation and beer quality. With hindsight, the strategy of yeast genetics in brewing has been built around the premise of 'things we currently cannot do' (see Table 4.20). Although relatively easy to identify the 'targets' for genetic change, the 'solutions' are often compromised or incomplete. As with the vastly more significant prizes of understanding cell ageing and human diseases, genetic insight requires to be underpinned by a solid understanding of yeast physiology. This situation was recognised as early as 1991, when Steele and Stowers (1991) noted that 'industry leaders see trained microbial physiologists as being the limiting factor in the development of biotechnology in the coming decade'. Closer to home, Hammond (1995) observed that for brewing yeasts the 'range of improvements possible is still restricted by our lack of knowledge of yeast physiology'.

**Table 4.20** General objectives for brewing strain development.

| Driver | Objective | Target | Comments |
|---|---|---|---|
| Vessel utilisation | Ferment very-high-gravity worts at the normal rate without compromise to beer quality, yeast viability or serial repitching | Ethanol tolerance Osmotic pressure General yeast physiology | Collection gravity > 1100 Final abv > 10% v/v |
| Efficiency | Ferment non-fermentable wort dextrins | Introduce amylolytic enzymes | Hydolyse linear/branched oligosaccharides (> G3) to fermentable sugars |
| Vessel utilisation | Ferment at elevated temperatures without compromise to beer quality, yeast viability or serial repitching | Thermotolerance General yeast physiology | Lager fermentations at 20 to 30°C |
| Yeast handling | Modify non-flocculent 'powdery' strain to flocculent strain | Flocculence gene(s) | Move from yeast cropping ex-centrifuge to conventional cone cropping |
| Vessel utilisation | Avoidance of diacetyl 'rest' or 'stand' in fermenter or downstream in maturation/conditioning tank | Introduce foreign enzyme that circumvents the formation of diacetyl or increases flux through pathway | Depending on the fermentation process, diacetyl stands range from a few days to weeks |
| Beer quality | Modify beer flavour | Manipulate specific target genes | Capability to control the concentration of flavour substances, e.g. esters, $H_2S$, $SO_2$ |
| Cost | Avoid the need for addition of exogenous enzymes (processing aids) | E.g. β-glucanases, proteases and amylases | |

4.3.4.2  *Approaches to strain development.*   It is beyond the scope of this section to consider in any detail the various techniques in use, pre-recombinant DNA technology, by geneticists to modify brewing yeast. Without exception, these approaches are characterised by limitations. The most common and the most problematic was the inability of these methods to selectively effect a single specific change. A good example of the frustrations that this could bring was the introduction of the capability to ferment dextrins into brewing strains. This particular theme has been a popular target of yeast genetics as a 'demonstrator project' for the technology. Early attempts to introduce the glucamylase gene (*DEX1* or *STA2*) from the dextrin utilising wild *Saccharomyces* (see Section 8.1.3.1), *S. cerevisiae var. diastaticus* were technically successful in that dextrins were fermented. Unfortunately, *S. cerevisiae var. diastaticus* has the gene for phenolic off flavour (POF) which, as luck would have it, was adjacent to the gene for glucoamylase. As the early genetic modification methods lacked precision about the bits of DNA being introduced into the recipient strain, the modified yeast produced beers that were phenolic and consequently unacceptable. Although such frustrations could be overcome, the approaches of the early 1980s were rapidly superseded with the advent of recombinant DNA technology. For a user-friendly introduction to the world of the 'early approaches' such as 'mutation and selection', 'hybridisation', 'rare mating', 'cytoduction', and 'spheroplast fusion' the interested reader should consult *Brewing Microbiology* Hammond (1996).

Recombinant DNA technology brought the vision (Tubb, 1984) of the genetic manipulation of brewing yeast to life. As noted by Meaden (1986) this technique had the major advantage of introducing 'specific genes only', thereby avoiding 'contamination by unwanted or undesirable material from the donor organism'. The details of recombinant DNA technology and application with brewing strains are well beyond the scope of the chapter. General user-friendly overviews are to be found in the 'popular' brewing press (see Lancashire, 1986; 2000; Meaden, 1986; Vakeria, 1991). The interested reader wishing to get into the 'nitty gritty' of genetic modification should consult Walker (1998) or one of John Hammond's excellent reviews on the subject (Hammond, 1996).

The success of recombinant DNA technology with yeast owes much to the relative simplicity and, most importantly, precision with which a desired gene can be inserted into the genome of the target strain. The process is described simplistically below, and by no means does justice to what is an elegant and increasingly sophisticated process. As noted above, the reader is urged to read one of the recommended reviews.

The early part of the process is described diagrammatically in Fig. 4.31 (from Meaden, 1986). The donor DNA is cut into fragments with a restriction enzyme (see Section 4.2.6.1) which is also used to introduce what is usually a single break in a plasmid. These plasmids are small circular DNA molecules, which are often based on the so called '2 μm plasmid' from yeast. The DNA fragments from the donor are mixed with the open plasmid and, the DNA is 'stitched-up' or ligated to form a bank of recombinant plasmids. Transformation of the target strain is achieved by introduction of the plasmid into the yeast cell. The plasmid is then introduced into the cell by removal of the cell wall ('spheroplasting'), treatment with lithium salts or application of an electric current. If transformation is successful, a 'transformant' carrying the desired gene is recovered through a selection process. The stability and the level of

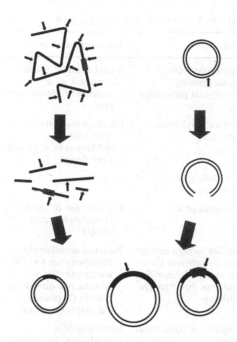

**Fig. 4.31**  Construction of recombinant DNA plasmid (redrawn from Meaden, 1986).

expression of the 'foreign' DNA depends on whether or not the DNA becomes integrated into the chromosomes or remains within the plasmid. Although substantially more stable when part of the genome, the copy number (and consequently gene expression) can be low. Conversely if maintained on the plasmid, the expression can be much higher but with poor long-term stability within the cell population.

The advent of recombinant DNA technology matched the then ambitions of brewing geneticists with capability to achieve hitherto impossible goals. Although the vagaries of yeast physiology continued to obstruct some of the more far-fetched concepts, the decade from 1985 proved particularly fruitful in delivering some of the early promise of genetic modification. Table 4.21 links some of the achievements of genetic modification with the 'drivers' and 'targets' established in Table 4.20. The list is by no means exhaustive and, consequently, fails to capture the many excellent reports that detailed similar achievements or were important steps on the way to technical success. Although some modified strains have been subsequently rebadged as 'demonstration GM' projects (Lancashire, 2000), there is no doubting the potential process benefits of, for example, a brewing yeast that circumvents the formation of diacetyl.

With the winds of change and the realisation that commercial exploitation of transformed brewing strains was increasing unlikely, development of new strains has effectively stopped. Indeed, the 1999 EBC contained just one paper on yeast transformation, in this case the construction of a lager yeast which is unable to form sulphite (Francke Johannesen *et al.*, 1999). Today 'genetics' in brewing is about

**Table 4.21**  Examples of genetic transformation of brewing yeast.

| Driver | Target | Transformation of yeast | Reference |
|---|---|---|---|
| Vessel utilisation | Ethanol tolerance<br>Osmotic pressure<br>General yeast physiology | Introduction of maltose permease gene increased fermentation rate | Kodama *et al.* (1995) |
| Efficiency | Introduce amylolytic enzymes | Glucoamylase from *S. cerevisiae var. diastaticus* (also cloned from fungi) | Perry and Meaden (1988) |
| Vessel utilisation | Thermotolerance<br>General yeast physiology | | |
| Yeast handling | Flocculence gene(s) | Introduction of yeast flocculation genes (*FLO1*) | Watari *et al.* (1994) |
| Vessel utilisation | Introduce foreign enzyme that circumvents the formation of diacetyl or increases flux through pathway | Bacterial acetolactate decarboxylase (ALDC) which converts α-acetolactate directly to acetoin (bypassing diacetyl formation) | Tada *et al.* (1995)<br>Yamano (1995) |
| Beer quality | Manipulate specific target genes | Increasing $SO_2$ production by brewing yeast by elimination of the *MET10* gene<br>Capability to manipulate ester synthesis by increasing alcohol acetyltransferase | Hansen and Kielland-Brandt (1996)<br>Fuji *et al.* (1994) |
| Cost | E.g. β-glucanases, proteases and amylases | β-glucanase ex *Bacillus subtilis*<br>Secretion of protease | Lancashire *et al.* (1989)<br>Young and Hosford (1987) |

differentiation of yeast strains (Section 4.2.6) and identification of bacteria (Section 8.3.4.2). Elsewhere, the technology continues to be exploited, if only in the laboratory. Various publications have described the transformation of non-brewing strains of *S. cerevisiae* to express a fungal phytase (Han *et al.*, 1999), antifreeze peptides (Driedonks *et al.*, 1995), and to overproduce glycerol (Remize *et al.*, 1999).

4.3.4.3  *Legislation, public perception and commercial implementation.*  In 1986, the *Journal of the Institute of Brewing* published a series of Centenary Reviews. One by Graham Stewart and Inge Russell titled 'One hundred years of yeast research and development in the brewing industry' captured the excitement of what genetic engineering might offer the brewing industry. In the epilogue to this admirable review, Stewart and Russell (1986) noted that the prediction of future developments was 'a difficult, if not foolish, pastime'. However, as with many applied yeast scientists at the time, the future prognosis for genetic manipulation in the brewing industry was decidedly rosy. Indeed Stewart and Russell (1986) commented that 'the use of manipulated yeast strains in brewing will become commonplace within the next decade with yeast strains specifically bred for such characteristics as extracellular amylases, β-glucanases, protease, β-glucosidase production, pentose and lactose utilisation, carbon catabolite derepression (higher productivity) and production of intracellular heterologous proteins (value added spent yeast).'

Given this upbeat introduction it is disappointing that what, at the time, was perfectly reasonable prediction should be so far from the truth. Other than the innate conservatism of the brewing industry, it is not immediately obvious why a genetically modified strain was not used commercially. Certainly, the current media interest and consumer reticence about genetically modified food was not nearly as public or articulate in the early to mid-1990s. Whatever, it is clear that today the odds are stacked against the commercial exploitation of a genetically modified yeast in the brewing industry. Against a background of consumer concern coupled with in a highly competitive market, such an action would be tantamount to commercial suicide!

Within the UK, one genetically modified brewing yeast has been cleared for food use by the appropriate authorities. Hammond (1995) and Walker (1998) have described this daunting and seemingly complex process of approval. The approved yeast was developed by BRi as a 'demonstrator' for the GM technology. The interesting story of the construction and evaluation of this amylolytic brewing yeast can be found in Hammond (1998). Although not a commercial product, BRi periodically produce 'Nutfield Lyte' which continues to be positively received by tasters.

In conclusion, despite the availability of the technology, the numerous hurdles (consumer, media, regulatory) make it almost inconceivable that genetically modified yeast will be used in the brewing industry. The focus is subtly shifting to consideration of exploiting the natural genetic variants that are likely to exist in the yeast population. Although an attractive solution to genetic improvement, the success of such initiatives will depend on the availability of suitably sensitive screening techniques for variants present at very low levels.

## 4.4 Cell wall and flocculation

### 4.4.1 The cell wall – an introduction

The brewing yeast cell wall is a hugely important and frequently underestimated organelle. Primarily it is made up of an array of carbohydrates (80–90% of the wall) with proteins embedded within it. As memorably described by Stratford (1994) it is not an 'inorganic egg shell' but a living organelle whose properties and functions change during the cell's lifetime. As if to emphasise its importance the wall accounts for 15–25% of cell dry weight. The yeast cell wall has been subject to a number of general reviews over the years (MacWilliam, 1970; Ballou, 1982; Fleet, 1991; Kreutzfeldt & Witt, 1991; Stratford, 1994; Cid et al., 1995; Lipke & Ovalle, 1998; Smits et al., 1999).

Structurally the yeast cell wall has been likened to reinforced concrete (Stratford, 1994). Where 'steel reinforcing rods are represented by enmeshed alkali-insoluble β-(1→3)-glucan fibrils, comprising some 35% of the wall. The reinforcing is surrounded by concrete, pebbles in a sand/cement matrix: secreted mannoproteins represent pebbles, some 25–50% of the wall, encased and bonded to the reinforcing fibrils by a matrix of amorphous β-glucan and chitin'.

An idealised representation of the cell wall and its various strata is presented in Fig. 4.32 (Kreutzfeldt & Witt, 1991). In reality, the micro-architecture of the cell wall is ill defined. As noted by Cabib *et al.* (1982) the layers are amorphous as the polymers interweave and overlap.

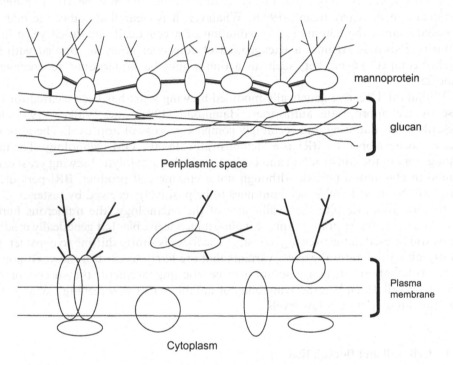

**Fig. 4.32** A diagrammatic cross section the cell wall of yeast (redrawn from Kreutzfeldt and Witt, 1991).

Increasingly, much is known about the cell wall and its many roles, functions and responsibilities. In particular, the cell wall provides osmotic protection, a point well made when on its removal the cell lyses! The cell wall is analogous to a 'thick overcoat' providing general protection from the environment as well as limiting (through its permeability) what can enter or leave the cell. Further, it is instrumental in determining cell shape and morphology. The cell wall also provides a matrix for a variety of enzymes involved in wall maintenance and development together with hydrolytic proteins. Somewhat inevitably it is also involved in attachment to surfaces and, importantly in brewing, molecules from the outside world are attached to the wall. Finally, the role of the cell wall in flocculation has attracted perhaps more than its fair share of attention from brewing scientists (see Section 4.4.4). This, of course, is to be expected and encouraged. Despite this, it is fair to say that the global impact of brewery fermentations and serial repitching on the cell wall and its physiological roles are poorly understood. This special case would benefit from some structured investigation.

### 4.4.2 *Composition*

The yeast cell wall is a carbohydrate chemist's dream! As can be seen from the review by Fleet (1991), the fractionation and structural analysis of cell wall carbohydrates is usually a harsh affair involving acid and alkali extraction. Consequently, although gross interactions are understood, 'judgements about the molecular organisation of the wall become difficult if not impossible' (Fleet, 1991).

#### 4.4.2.1 *Glucans.*

The 'glucans' are the major polymers in the cell wall, accounting for between 30–60% of the wall. Although they cover the entire cell in the form of a microfibrillar net they are restricted to the inner layer of the wall (Fig. 4.32). Fleet (1991) has described three classes of glucans: (i) alkali insoluble-acetic acid insoluble β-(1→3) (35% of the cell wall), (ii) alkali soluble β-(1→3) (20%) and (iii) highly branched β-(1→6) (5%). The β-(1→3) glucans are branched via β-(1→6) linkages whereas the β-(1→6) glucans are branched through β-(1→3) linkages. With so much glucan in the wall it is no surprise to find that these polymers play an important role in the cell architecture. Fleet (1991) proposed that the alkali insoluble-acetic acid insoluble β-(1→3) glucan maintains wall rigidity and shape, whereas the alkali soluble β-(1→3) glucan 'may confer flexibility to the wall'. More recently, Stratford (1994) has reported that the two types of glucan are one and the same, differing only in the extent of cross linkage with chitin (see Section 4.4.2.3). The role and location of the β-(1→6) glucan is less clear. Cabib *et al.* (1982) have speculated that it forms a layer between mannoprotein and β-(1→3) glucan. More recently (Kollar *et al.*, 1997) have developed this theme and have shown the β-(1→6) glucan to play a central role in the organisation of the cell wall. Essentially the β-(1→6) glucan interconnects with the cell wall polysaccharides, β-(1→3) glucan, mannoprotein (Section 4.4.2.2) and chitin (Section 4.4.2.3).

How these polymers might confer rigidity and strength is perhaps explained by considering β-(1→3) glucan as 'rope' rather than 'steel reinforcing rods'. Ballou (1982) has proposed that two chains of β-(1→3) glucan wind around each other to form a rope-like double helix. These, presumably, form the microfibrillar net seen on removal of the mannoprotein outer layer of the cell wall. If Ballou (1982) is right, elaboration of new wall through cell growth could simply be achieved through the chains sliding along each other.

#### 4.4.2.2 *Mannoprotein.*

The outer layer of the cell wall is composed of mannoprotein. This glycoprotein is a major player, accounting for 25–50% of the cell wall. There is evidence that the 'structural' mannoproteins are anchored to the cell wall through linkage with β-(1→6) glucan (Fleet & Manners, 1976; Cid *et al.*, 1995). In addition, extracellular enzymes such as invertase (that hydrolyse sucrose to fructose and glucose) are mannoproteins.

Although, their function is not clear, cell wall mannoproteins are essential for cell survival (Fleet, 1991) but play no role in cell shape or rigidity. Not surprisingly, given their location, mannoproteins appear to have an 'interactive' role acting as antigenic determinants (Section 4.3.3.2), as receptors for 'killer toxins' (see Section 8.1.3.1) and sexual agglutination. More importantly in the brewing context, mannoproteins are the receptors in the flocculation process (Section 4.4.6.2).

Mannoproteins have also been implicated in cell wall porosity. As noted by Nobel and Barnett (1991), the size and shape of molecules passing through or out of the cell wall has been subject to much debate. Early work in the 1950s and '60s, which suggested that only relatively small molecules could permeate the cell wall, has been overturned by the realisation that proteins are easily secreted out of yeast cells. Indeed, as described elsewhere (Section 4.3.4), yeast is an excellent host for the genetic manipulation and subsequent secretion of 'foreign' heterologous proteins. The current picture of cell wall porosity remains somewhat speculative. Nobel and Barnett (1991) suggest that the confusion over what can and cannot diffuse may be explained by a mix of predominately small pores in the cell wall with a few 'large' holes that allow the diffusion of big heterologous and homologous proteins. The mannoproteins in the cell wall are thought to obstruct diffusion through ionic interactions and the web of mannan side chains.

The structure of mannoprotein is complex. The review of Fleet (1991) notes that mannoprotein consists of about 90% mannose and 10% protein together with small amounts of phosphate located on the outside of the molecule (Cawley *et al.*, 1972). In passing, it is noteworthy that this phosphate is the major source of negative charge on the cell surface (see Sections 4.4.6.5 and 4.4.6.6). The polysaccharide part of the complex consists of $\alpha$-(1$\rightarrow$6)-linked mannose with side chains of $\alpha$-(1$\rightarrow$2) and $\alpha$-(1$\rightarrow$3) linked residues (see Fig. 4.33). Although at first glance beyond the scope of this book, it is appropriate to dwell a little on the structure of these side chains because of their involvement as flocculation receptors. The side chains on the much-repeated 'outer chain' are all linked $\alpha$-(1$\rightarrow$2) to the backbone. Characterisation after cleavage and fractionation shows there to be four types of side chain: (i) mannobiose ($M^{2-1}M$), (ii) mannotriose ($M^{2-1}M^{2-1}M$), (iii) mannotriose ($M^{2-1}M^{3-1}M$) and (iv) mannotetraose ($M^{2-1}M^{2-1}M^{3-1}M$). The mix of the side chains varies within the repeating unit of the outer chain. It is likely that immunochemical differentiation of strains and

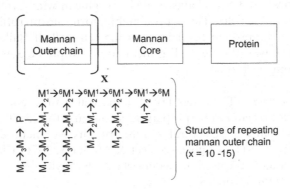

**Fig. 4.33** The molecular structure of a cell wall mannoprotein.

species is due to the polymorphism of yeast mannoproteins. Numerous mannoprotein mutants (*mnn*) have been isolated and characterised. Amongst other studies, these have allowed the unravelling of minimum receptor structure that will support flocculation (Section 4.4.6.2).

4.4.2.3 *Chitin.* As well as the yeast cell wall, chitin is found in the exoskeleton of crustaceans and insects. However, it is quite distinct from the other polysaccharides found in the cell wall. Structurally it is comparatively simple, being a linear polymer of $\beta$-(1→4) linked N-acetylglucosamine. Although chitin is believed to play no role in the structure of the cell wall, more recent evidence with chitin mutants has suggested it may be important in wall structure (Stratford, 1994). Chitin is found, almost exclusively (*c.* 90% of total), in scars left on the mother cell surface after the cell has undergone reproduction through budding (see Section 4.3.3.1). These 'bud scars', which resemble volcanic craters, are left every time a cell buds. The chitin-rich scar is all that is left of the primary septum between mother and daughter cell (Cabib *et al.*, 1982; Fleet, 1991).

Robinow and Johnson (1991) noted somewhat tartly that 'considering how truly unimportant yeast scars are, it is remarkable that they have a literature all their own'. With this admonishment in mind, a few comments about bud scars are appropriate! First, the view that the cell's surface limits the number of bud scars is not the case (Robinow & Johnson, 1991). Second, the number of bud scars provides a useful measure of cell age. Although a seemingly trite comment, *S. cerevisiae* is a model organism to study cellular senescence (see Section 4.3.3.4) which has triggered some valuable insights into yeast cell age during brewery handling (Barker & Smart, 1996; Deans *et al.*, 1997).

The use of microscopy (conventional light or, occasionally electron microscopy) (Fig. 4.34) to visualise bud scars is limited by the '2D nature of the image'. Work at

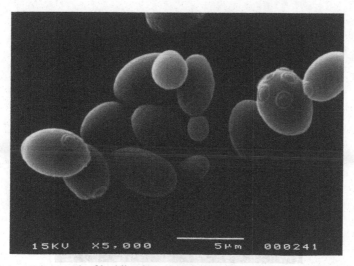

**Fig. 4.34** Electron micrograph of budding brewing yeast (kindly provided by Katharine Smart, Oxford Brookes University).

Oxford Brookes University in Oxford, UK has explored the use of confocal microscopy together with wheat germ agglutinin to generate 3D images of yeast cells. Some these stunning images (Christopher Powell, in preparation) are presented in Fig. 4.35).

4.4.2.4 *Proteins.* Protein accounts for 5–10% of the cell wall. The cell wall 'proteins' are represented by (i) structural proteins (ii) enzymes and (iii) surface receptor proteins (see Fleet, 1991). The structural proteins include the heterogeneous man-

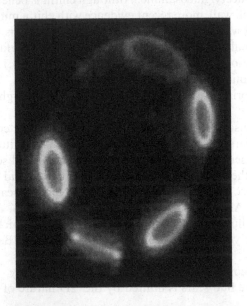

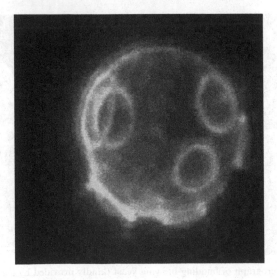

**Fig. 4.35** *Contd on next page*

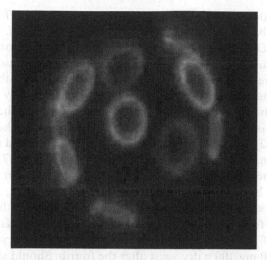

**Fig. 4.35** Confocal micrographs of budding yeast cells yeast – bud scars stained with wheat germ agglutinin (kindly provided by Christopher Powell, Oxford Brookes University).

noprotein population as well as proteins that are believed to interact with cell wall glucans (Fleet & Manners, 1976). The enzymes divide into two groups; those which are involved in the metabolism of nutritional substrates and those involved in cell wall morphogenesis. The former class that includes the mannoproteins invertase and acid phosphatase are located in the 'never-never land' (Robinow & Johnson, 1991) of the periplasm – an ill defined region between the plasma membrane and the cell wall (see Fig. 4.1). The wall degrading enzymes, of which the $\beta$-$(1 \rightarrow 3)$-glucanase is by far the best characterised, appears to have a roving brief around the cell wall. Finally, the lectin-like proteins represent the surface receptor proteins. These are discussed at length in Section 4.4.6.1.

### 4.4.3 Cell wall and fermentation

The cell wall polysaccharides – glucan and mannan – each account for $c$. 5% of the yeast biomass during fermentation (Quain *et al.*, 1981). Although somewhat low key compared to the arguably more exciting storage polysaccharide glycogen, the cell wall and its components play an important role in brewery fermentations.

Although its role in flocculation has received the plaudits (see below), the general contribution of the yeast cell wall to brewery fermentation has yet to be fully appreciated! One facet, which draws little attention, is the role of the yeast cell wall as a protective layer to the ever-changing environment in which the cell finds itself. In particular, brewery worts contain a melange of suspended particles that associate with the cell wall. Depending on production methods, wort can contain as much as 4% (v/v) particulates or 'trub' (Lentini *et al.*, 1994). As a rule of thumb, the more turbid the wort, the more trub it contains. Trub is a miscellany of substances that include insoluble proteins, complex carbohydrates, lipids, polyphenols and hop substances. There is no doubt that trub associates intimately with the cell wall. Trub lipids such as linoleic acid are adsorbed by yeast and can have a beneficial effect on fermentation, yeast growth and beer flavour (Schisler *et al.*, 1982; Siebert *et al.*, 1986; Lentini *et al.*, 1994). However, O'Connor-Cox *et al.* (1996) have argued that the brewer should strive for bright and not cloudy worts. In their view, fermentation performance and product quality are compromised in cloudy worts. Intriguingly they propose that the 'yeast cells become coated with the "cloudy" material thereby preventing optimal activity' (O'Connor-Cox *et al.*, 1996). Extending this concept, the association of trub with the cell wall is recognised as potentially interfering with the fining out of yeast (see Section 4.4.3.2). There is no doubt that hop substances bind to the yeast cell wall (Hough & Hudson, 1961; Dixon & Leach, 1968; Laws *et al.*, 1972) – so much so that significant 'bitterness' is lost because of fermentation, which can simply be recovered from the cell wall by washing with water (Kieninger & Durner, 1982).

The impact of serial repitching on the binding of trub and hop iso-α-acids is of interest. A model for what may happen has been reported by Kieninger and Durner (1982). Binding of iso-α-acids increased from 16.9 mg/100 g dry yeast after the first fermentation to 70 mg/100 g dry yeast after the fourth. Should similar accumulation occur with wort proteins, polyphenols and other substances, it is tempting to assume that such events may well influence flocculation, substrate transport and other wall functions. There is evidence that the charge of the cell wall becomes progressively more negative with serial repitching (Lawrence *et al.*, 1989). Further, Bowen and Ventham (1994) have speculated that there is electrostatic attraction between ale yeasts and trub, which may enhance flocculation (see Section 4.4.6.6).

### 4.4.3.1 *Acid washing.*

Of the many abuses yeast is exposed to during its life in the brewery, acid washing (see Section 7.3.3) to kill contaminating bacteria (but not contaminating yeasts!) is arguably the most extreme. 'Washing' yeast at pH 2.1 for two hours or so in the cold would be expected to damage the cell wall and result in generalised stress to the cell. Work by Simpson and Hammond (1989) has shown that although brewing yeasts are generally resistant to acid washing, the cell wall exhibits changes. In particular, surface blistering ('blebbing') was observed particularly at elevated temperature (25°C) (Fig. 4.36). Further, acid washing removes associated trub-like material (Fig. 4.37) and makes the yeast sticky (Fig. 4.38).

### 4.4.3.2 *Fining.*

The need to separate liquid and solids is a common theme in the brewing process. In an ideal world, the removal of yeast from beer is primarily achieved through flocculation (see Section 4.4.4). Green beer centrifuges are used on run down from fermenter to 'trim' yeast counts further or, where required, to remove powdery, poorly flocculent strains. However, inevitably, green beer will still contain yeast usually in the range of $1-3 \times 10^6$/ml. This loading is reduced significantly by the addition of isinglass finings either in conditioning tank (chilled and filtered products) or final package (cask beer). These materials interact with yeast cells to form increasingly large flocs that sediment or 'fine' out of solution (for a review see Leather, 1994).

It is intriguing to speculate about the origins of 'white finings' or 'isinglass' as they are sourced from the swim bladders of tropical fish! Their success in removing yeast cells is down to the presence of collagen. This large linear protein carries a positive

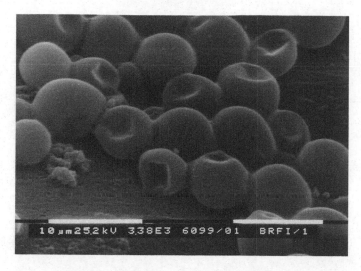

**Fig. 4.36** 'Blebbing' of yeast cells after acid washing (from Hammond and Simpson (1989), with permission from the Institute of Brewing).

charge which, it is believed, interacts with the negatively charged yeast cell. The contribution of the yeast cell wall to fining is widely accepted, but it must be assumed that the phosphate moiety within the mannoprotein matrix (see Section 4.4.2.2) is the key determinant of successful fining. Mindful of comments above about interactions with trub (4.4.3), recent work by Leather *et al.* (1997) has shown that increasing loadings of positively charged 'non-microbiological particles' (NMP) has progressively deleterious effects on isinglass fining performance (see Section 4.4.3.3). Presumably, NPM associates with cell wall phosphate, which is then unable to interact with isinglass.

4.4.3.3 *Commercial applications.* The yeast cell wall has been identified as a source of interesting and commercially useful polymers. In particular, glucan has 'functional characteristics' being able to hold water and to provide thickening properties (Johnson, 1977). More recently, mannoproteins from *S. cerevisiae* have been identified as being 'bioemulsifiers' (Cameron *et al.*, 1988) and foam stabilisers (Kunst *et al.*, 1996). A protein-rich fraction from yeast cell walls has been advocated for use in clarification processes such as fining of beer (Jackson, 1996).

4.4.4 *Flocculation – an introduction*

Yeast flocculation is a physical process and is of fundamental importance in brewery fermentations. Contrary to popular opinion, flocculation is not about sedimentation but the aggregation of single yeast cells into multicellular 'flocs'. Perhaps 'aggregation' (Miki *et al.*, 1982) or, less elegantly, 'clumping' are more useful descriptions than flocculation. Whatever, a good and popular definition of this process is 'the phenomenon wherein yeast cells adhere in clumps and sediment rapidly from the medium in which they are suspended or rise to the medium's

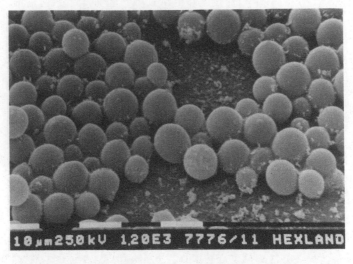

(a)

(b)

**Fig. 4.37** Yeast cells (a) before and (b) after acid washing (from Hammond and Simpson (1989), with permission from the Institute of Brewing).

surface' (Stewart & Russell, 1981). In passing, flocculation is unrelated to another form of yeast cell aggregation – the small cell flocs formed by certain *chain forming* brewing strains.

Flocculation is critical to yeast recovery be it through the skimming of open fermenters ('flotation') or, more typically, the cropping of closed cylindroconical vessels ('sedimentation'). Because of this, it is tempting to conclude that such yeasts have been selected as strains which flocculate at the appropriate time so as to facilitate yeast recovery/pitching-on as well as controlling downstream yeast cell counts to racking or conditioning tank. Whether, in addition to making the brewer's life a little easier, flocculation offers some benefit to the yeast cell remains to be seen! Johnson (Stewart

**Fig. 4.38** Sticky yeast post acid washing (from Hammond and Simpson (1989), with permission from the Institute of Brewing).

& Russell, 1981) and Iserentant (1996) have, quite reasonably, suggested that the formation of cell flocs is one of the many responses of yeast to stress. 'Multicellular' flocs, it is argued, offer greater protection to the external inclement environment.

Although not usually routinely monitored or measured, yeast flocculation is a high profile fermentation parameter second only to fermentation rate (PG drop). This is borne out by the readiness with which any in-process changes in yeast flocculence are seen through yeast solids or head formation. Such changes are thought to reflect genetic changes (see Section 4.3.2.6).

Ironically, despite its importance, yeast flocculation in the brewing industry remains a 'given' with little real practical understanding of the factors that determine yeast flocculation. This is despite the efforts of legions of yeast physiologists, biochemists and geneticists who have exhaustively studied flocculation. Indeed, it has been estimated that in the last 20 years, 10–15 papers per year have been published on 'flocculation' (Speers & Ritcey, 1995). This 'hit rate' may well increase as, in recent years, flocculation has acquired a new interest and input from biotechnologists studying the removal of yeast cells from non-brewing fermentation broths.

### 4.4.5 *Overview*

Yeast flocculation has inspired numerous and lengthy reviews. In particular the reader is referred to the reviews of Stratford (1992a, 1996) and Speers *et al.* (1992) which ably detail the complexity and occasional ambiguity that has characterised work on flocculation. Although the majority of studies focus on brewing strains of *S. cerevisiae*, flocculation is found in other yeast genera (Stratford, 1996) such as *Candida* and *Kluyeromyces*.

Athough the debate continues, there appears to be general agreement that the physiology and biochemistry (but perhaps not genetics) of yeast flocculation are now broadly understood. From the viewpoint of brewery fermentations, the aggregation of cells into clumps ('flocs') is an unusual, possibly unique process insomuch that individual yeast cells interact directly with each other. In summary, yeast flocculation occurs toward the end of fermentation (typically early stationary phase) and involves the interaction of cell wall proteins on one cell to carbohydrate 'branches' on the cell wall of another. Calcium is necessary for the 'activation' of the cell wall protein. As would be anticipated the process is reversible! On repitching flocculent yeast is 'deflocculated' by the presence of simple sugars or, in the laboratory, by the removal of calcium.

### 4.4.6  *Mechanism*

Given the history of 'flocculation' it seems inevitable that the present 'best bet' for the mechanism of this process will be subject to further development and embellishment. Certainly, as noted by Stratford (1996) 'several pieces of evidence have recently come to light that cast doubt on the (below) simple picture of yeast flocculation'. However, despite these slight reservations, the currently accepted theory of flocculation fits the majority, if not all, the facts and observations made on this process in the last 20 years or so.

The current concept of binding of surface proteins to carbohydrate receptors on neighbouring cell walls (Stratford, 1996) dates back to Eddy and Rudin (1958). However, what has now become, the 'lectin theory of flocculation' stems from a report by Miki *et al.* (1982). This seminal paper describes flocculation in terms of the recognition and interaction of cell wall factors. It is argued, through convincing experimental evidence, that the cell to cell interactions of flocculation are driven through 'lectin-like' proteins bonding with outer mannan chains.

4.4.6.1  *Lectin-like proteins.*   The involvement of lectin-like proteins or 'adhesins' in flocculation was first proposed by Miki *et al.* (1982). The interaction of the lectin-like proteins on flocculent cell surfaces (Fig. 4.39) with mannose receptors (Fig. 4.40) (see Section 4.4.2.1) is now accepted as being at the heart of the flocculation (Fig. 4.41) theory. The involvement of either party can be simply demonstrated by the inhibition

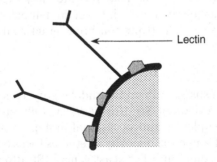

- Require Ca for activity
- Determine flocculation

**Fig. 4.39**  Diagrammatic representation of cell wall lectins.

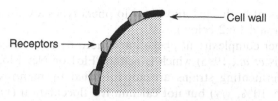

Receptors are present throughout fermentation.
Receptors require at least 2–3 mannose residues.

**Fig. 4.40**   Diagrammatic representation of cell wall receptors.

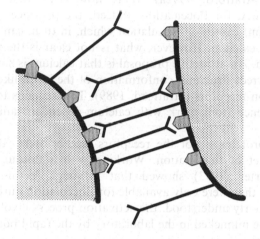

**Fig. 4.41**   Diagrammatic representation of yeast flocculation.

of flocculation by the presence of excess but related molecules. For example, a plant lectin (Concanavilin A) is able to competitively inhibit flocculation (Miki *et al.*, 1982). Similarly, specific simple sugars (mannose, maltose etc.) are able to reversibly inhibit flocculation (Miki *et al.*, 1982; Stratford & Assinder, 1991).

The inhibition of flocculation by sugars is worthy of fuller consideration. Although early work (Eddy, 1955) showed sucrose, mannose, maltose and glucose to be effective 'deflocculating agents', more recent work (Stratford & Assinder, 1991) with 42 strains of *S. cerevisiae*, has identified two distinct phenotypes. One group, Flo1, were inhibited only by mannose which – as this sugar is not present in wort – implies such strains are typically heavily flocculent throughout fermentation. Conversely, mannose, maltose, sucrose and glucose inhibited the 'NewFlo' phenotype. Such strains are 'typical brewing strains'. To overcome any metabolic interference or confusion, sugars were added to heat killed but flocculent cells. A ready explanation of these observations is that there are two types of lectin-like proteins in yeast. In addition to differences in sugar inhibition (Stratford and Assinder, 1991), the two lectins were shown to differ in response to pH, protease digestion and inhibition by cations. These differences, it is argued, reflect the NewFlo proteins' greater suscept-ability or availability to these treatments. However, despite such differences, the preferred receptors of both Flo1 and NewFlo phenotypes are the outer branches of mannan (see Section 4.4.6.2 below).

Typically, further complexity has been introduced by the identification of a third phenotype (Dengis *et al.*, 1995) which is neither Flo1 or NewFlo. 'Mannose insensitive' (MI) top-fermenting strains are not inhibited by mannose or sucrose and require ethanol (5–10%, v/v) but not calcium for flocculation (Dengis & Rouxhet, 1997). These MI strains may not obey the current rules of flocculation, being 'governed by non-specific interactions solely, or by non-lectin specific interactions (e.g. protein-protein)' (Dengis & Rouxhet, 1997).

The involvement of inorganic salts in yeast flocculation has been recognised for many years (see Stratford, 1992a). It is now clear that calcium ions are unequivocally required for flocculation. Indeed, the presence of chelating agents (that remove calcium) causes deflocculation which, in turn, can be reversed by the addition of further calcium. However, what is not clear is the precise role of calcium in flocculation. An attractive proposal is that calcium is somehow involved in maintaining the correct structural conformation of the lectin-like protein for bonding with flocculation receptors (Stratford, 1989). This remains to be seen but there is supporting evidence from work with calcium-dependent animal lectins (Stratford, 1992b).

The lectin-like proteins – not the receptors (see below) – determine both the capability and onset of flocculation. Working with a flocculent brewing strain, Stratford and Carter (1993) showed that although lectins were synthesized throughout growth they were only available for flocculation during early stationary phase. Although poorly understood, the activation process involves no new protein synthesis and can be mimicked in the laboratory by the rapid boiling and cooling of 'non-flocculent' cells.

Isolation and characterisation of the lectin-like protein would clearly add credibility to its proposed role in flocculation. However, as if in support of Stratford's (1996) clouding of the 'simple picture', the reports detailing various putative lectin-like proteins have not fully supported the expectations of this protein. For example, two reports have described surface proteins from yeast with lectin-like properties (Straver *et al.*, 1994; Shankar & Umesh-Kumar, 1994).

One report (Straver *et al.*, 1994a) describes the recovery of an 'agglutinin' from cell walls of both flocculent *and* non-flocculent cells. Intriguingly, the protein mimics flocculation in being sensitive to mannose, pH and is calcium dependent. Further addition of the partially purified protein to the same strain of yeast stimulated flocculation in flocculating (stationary phase) and in non-flocculating (exponential) cells. However, in the latter case, stimulation although measurable was 'weak' and required significantly more agglutinin to trigger a response. Although clearly part of the flocculation process, this lectin-like protein differs from that postulated in the model in that it is easily released from yeast cells by agitation during a laboratory flocculation protocol. This would suggest that there are other anchored lectin-like proteins, that release through agitation is artefactual or, fundamentally, that the model is flawed. Although a conclusion has not been reached, the authors present an alternative model where the released agglutinin cross links a non-lectin glycoprotein ('flocculin') found only in flocculent cells (Straver *et al.*, 1994b). As the authors conclude, there is a need for 'more insight into the interplay between flocculin, fimbriae-like structures and agglutinin' (Straver et al., 1994a).

The putative lectin-like protein isolated by Shankar and Umesh-Kumar (1994) is perhaps a better contender for the hypothetical protein. Unlike the Straver et al. (1994b) agglutinin, this protein is *not* found in non-flocculating yeast. Further, in a particularly elegant experiment, the protein was shown to bind to inert surfaces coated with yeast mannan but only in the presence of calcium. Similarly the protein was shown to bind to a non-flocculent 'mutant'. As noted by the authors, 'this is probably the first report on the isolation of active lectins associated with flocculation' (Shankar & Umesh-Kumar, 1994).

4.4.6.2 *Receptors.* The identity of the flocculation receptors has proved less controversial. The inhibition of flocculation by mannose is consistent with the lectin-like protein interacting with this sugar. As mannan – a polysaccharide consisting of mannose – is a major component of the yeast cell wall it is hardly surprising that mannan became the major candidate for the flocculation 'receptor'. Further evidence came from Miki et al., (1982) who showed inhibition of the process by a plant lectin (Concanavilin A) which binds selectively to α-mannan.

Stratford (1992c) using mannoprotein mutants elegantly demonstrated the involvement of mannoproteins in flocculation. Although structurally complex (see Fig. 4.33), only the 'outer chain' of the mannoprotein molecule is involved in flocculation. Using a selection of *mnn* mutants, Stratford (1992c) was able to probe the requirements for receptor structure by coflocculation of the non-flocculent mutants with flocculent strains and by aggregation using concanavalin A (a plant lectin that binds to the flocculation receptors). Two mutants – *mnn2* and *mnn5* – were unable to coflocculate. As *mnn2* mutants lack the side chains on the outer chain and *mnn5* contain truncated side chains with only one mannose unit, it is clear that the receptors are the di/trisaccharide side chains of the outer mannan chain.

Given the apparent structural importance of mannoproteins in the yeast cell wall, it is perhaps not surprisingly that functional receptors are present throughout growth and, consequently, do not determine flocculence. Stratford (1993) demonstrated this for 11 strains of *S. cerevisiae* using the simple but powerful coflocculation and concanavalin A methods used in the above work on receptor structure.

4.4.6.3 *Interaction between receptors and lectin-like proteins.* The timing of the flocculation event is of clear importance. Early flocculation will result in 'stuck', poorly attenuated fermentations whereas late flocculation will lead to over-attenuated beers and difficulties in yeast cropping. Typically however, flocculation occurs in laboratory aerobic fermentations when cell division has stopped (Smit et al., 1992) and the culture has entered stationary phase (Amri et al., 1982). It seems likely that this is governed by the inhibitory presence of sugars as apparently non-flocculent exponentially growing yeast can be flocculated by washing and resuspension in calcium containing buffer (Stratford, 1992a). This is in keeping with the above observations that receptors (Stratford, 1993) and the lectin-like proteins (Stratford & Carter, 1993) are present throughout fermentation.

The vast majority of flocculation studies are performed aerobically in shake flask cultures. Although technically more convenient, the physiology of aerobically grown yeast is often different and distinct from yeast grown anaerobically. In this case it is heartening to report that in anaerobic fermentations of wort, flocculation of (presumably) NewFlo brewing strains begins when cell division stops (Kempers *et al.*, 1991; Straver *et al.*, 1993) at the so-called 'mid-point' of fermentation (Thurston *et al.*, 1981). Although probably true of many production brewing strains the work of Gilliland (1951) clearly demonstrated at least four different classes of flocculence. It is tempting to conclude that the differentiation of brewing strains into four groups (see Section 4.2.5.2) reflects a non-flocculent strain (class 1), a top-fermenting NewFlo strain (class 2), a bottom-fermenting NewFlo strain (class 3) and an early, heavily flocculent Flo1 strain.

Although seemingly contradictory, flocculation requires agitation! The early 'fortuitous' observations of Stratford and Keenan (1987) have been further developed into a persuasive argument (for a review see Stratford, 1992a). Initial laboratory studies (Stratford & Keenan 1987) showed that the harder flocculent yeasts were shaken the better they flocculated. Indeed, without agitation, a flocculating culture was unable to flocculate. Further, the rate of flocculation increased in parallel with increasing mechanical agitation.

On the face of it, such observations are in conflict with the real world of brewery fermentations where *mechanical* mixing is either unavailable or ineffectual. However, to put 'agitation' into perspective, Stratford and Keenan's (1987) experiments involved relatively gentle mixing (70–120 rpm) to trigger flocculation. Consequently, as brewery fermentations are indirectly 'mixed' though the release of carbon dioxide, it is easier to integrate the involvement of agitation in flocculation. Indeed, at the 'mid-point' of fermentation, flocculation starts against a background of highly active 'mixed' fermentation.

The bridge-like bond between the lectin-like protein and receptors confers the strength and stability of the flocs. There is evidence that hydrogen bonding is important in floc stability. Agents which disrupt hydrogen bonding – such as urea or elevated temperature (50°C) – disrupt flocs (Speers *et al.*, 1992) or the interaction in vitro between isolated (putative) lectin and mannose (Shankar & Umesh-Kumar, 1994).

It would be anticipated that the number of interactions between flocculating cells would be substantial. As perhaps a reflection of the technical challenge, relatively few attempts have been made to quantify these interactions. A preliminary report noted that up to five contacts per individual cell have been estimated in a single plane (Miki *et al.*, 1982). What this means in terms of cell–cell interactions is not clear but may be gauged from observations quantifying the number of lectin sites per cell. Elegant work with fluorescent mannose or galactose probes (Masy *et al.*, 1992) suggests the number of lectins per cell ranges from $4 \times 10^6$ (Flo1 strain) to $2 \times 10^7$ (NewFlo strain). Other work (Speers, unpublished) using fluorescent avidin (that contains dimannose) suggests comparable lectin numbers ($4 \times 10^6$ per cell). It is not clear how many lectin molecules on a cell actively bind to receptors on other cells. The qualitative and quantitative contribution of mannan receptors (as 'mannoproteins' – see Section 4.4.3.2) might suggest that the stability of flocculation is due to the number of cell to cell interactions.

4.4.6.4 *Genetics.* The genetics of flocculation (for a review see Jin & Speers, 1998) in many ways exemplifies our understanding of the flocculation process in general. Like the measurement of flocculation, the genetics of the process is subject to great but confusing activity. The MIPS search reported in Table 4.17 is revealing in both the number of *FLO* genes and the debate as to their role. Although by no means definitive, there is growing evidence that the product of *FLO1* is a cell wall protein involved in flocculation. The work of Teunissen and Steensma (1995) is persuasive in that this gene confers flocculence when introduced into a non-flocculent strain. Further, flocculation was strongly correlated with the amount of *FLO1* protein that was expressed. More recently, Javadekar *et al.* (2000) have provided further fuel for the role of this gene in flocculation. These workers isolated a cell surface lectin that had at least 70% homology with the predicted N-terminal sequence of the putative *FLO1* as well as *FLO5* gene products. This protein, which was isolated from a highly flocculent strain of *S. cerevisiae*, was shown to play a role in flocculation and to bind to the branched trimannoside cell wall receptor. These collective observations suggest that the riddle of flocculation genes is getting ever closer to resolution!

4.4.6.5 *Premature flocculation.* Although rarely monitored, gross changes in yeast flocculence are observed through increased cell counts at beer run-down (less flocculent) or, conversely, elevated racking gravity (more flocculent). Early or 'premature' flocculation has received greater attention, perhaps because it impinges more on product quality or, simply, is more common!

Although genetic instability may play a role in premature flocculation (see Section 4.4.4), there is growing evidence that wort polysaccharides adhere to yeast cell walls and trigger early flocculation. To explain this, Stratford (1992a) has proposed that large 'multivalent' wort polysaccharides overcome sugar inhibition by binding to cell wall lectins and, consequently, flocculation occurs early.

Work by Herrera and Axcell (1991a and b) showed premature flocculation to be triggered by a lectin-like, gum type polysaccharide present in the malt husk. Addition of the semi-purified polysaccharide to normal fermentations resulted in early flocculation (Herrera and Axcell, 1991a). In an accompanying publication (Herrera & Axcell, 1991b), use of enzyme-linked immunosorbent assays (ELISA) showed the polysaccharide to be present in normal and *problem* worts. However, more (65%) was found in the wort causing premature flocculation. More recent work (Nakamura *et al.*, 1997) has shown the 'premature yeast flocculation' factor to be present in barley husks. A screening method is described where the propensity of barley to trigger premature yeast flocculation can be determined in four days.

These observations are in keeping with the growing realisation that the subtleties of wort composition have a much greater impact on fermentation performance than previously believed. It is tempting to speculate that such 'premature yeast flocculation' factors already contribute a further layer of complexity to the flocculation story in commercial fermentations. Further work in this area should be encouraged!

4.4.6.6 *Hydrophobicity.* One of the few certainties in the great flocculation debate is that the cell wall plays an essential role! However, in addition to providing 'lectin-like proteins' and 'receptors', the hydrophobicity of the cell surface has been implicated in flocculation. The key determinant of cell hydrophobicity appears to be the phosphate content of the outer layer of the cell wall (Mestdagh et al., 1990), hydrophobicity being associated with low phosphate and vice versa. Another view is that hydrophobicity is related to the number of bud scars and, by implication, cell age (Akiyama-Jibiki et al., 1997).

Straver and Kijne (1996) have proposed that 'a high level of cell surface hydrophobicity may facilitate cell-to-cell contacts in an aqueous medium, leading to a more specific interaction between the cells, i.e. calcium-dependent lectin-sugar binding'. Certainly, in some strains of S. cerevisiae, there is a step increase in hydrophobicity at the end of logarithmic growth just prior to the onset of flocculation in aerobic (Smit et al., 1992) or anaerobic cultures (Straver et al., 1993). However, no changes in hydrophobicity were seen during the anaerobic growth of a non-flocculent mutant (Straver et al., 1993). In a later paper, the same group, working with five strains of varying flocculence, demonstrated a positive correlation between hydrophobicity and flocculation (Straver & Kijne, 1996). Similar conclusions were drawn by van der Aar (1996) working with two strains and Azeredo et al. (1997) working with four strains. Particularly convincing evidence of a relationship between cell surface hydrophobicity and flocculation was reported by Akiyama-Jibiki et al. (1997) working with 75 lager strains.

However, others (Garsoux et al., 1993; Dengis et al., 1995; Dengis & Rouxhet, 1997) have been unable to demonstrate any significant differences in hydrophobicity between flocculent and non-flocculent cells. Nor have they been able to associate increased hydrophobicity with the onset of flocculation (Wilcocks & Smart, 1995).

It is not clear why there should be such confusion as to whether (or not) cell surface hydrophobicity is involved in flocculation. Perhaps a simple explanation is the diversity of methods used to quantify hydrophobicity. This ranges from 'water contact angle determinations' (Smit et al., 1992; Garsoux et al., 1993; Dengis et al., 1995; Azeredo et al., 1997) through cell adhesion to polystyrene petri dishes (Amory et al., 1988; Smit et al., 1992) or magnetic beads (Straver & Kijne, 1996) to hydrophobic interaction chromatography (Akiyama-Jibiki et al., 1997). Perhaps more telling is the 'old chestnut' of yeast strain differences. This, of course, can be exacerbated where the conclusions are drawn from work with only one or two strains – a point that applies to key papers from both camps (e.g. Smit et al., 1992; Dengis et al., 1995)!

Cell surface hydrophobicity however may explain why some yeast strains are 'top-fermenting'. These strains are more hydrophobic than bottom-fermenting strains (Hinchliffe et al., 1985; Mestdagh et al., 1990; Dengis et al., 1995; Dengis & Rouxhet, 1997) and, as a consequence, are more able to adhere to carbon dioxide bubbles and to form yeast heads at the top of the fermenter. Inevitably these observations are linked to flocculation as the strains used by Dengis and co-workers are those unusual yeasts described as being 'mannose insensitive' (see Section 4.4.6.1).

4.4.6.7   *Zeta potential.*   Measurement of zeta potential of brewing yeast and its involvement (or not) in flocculation has, in recent years, attracted much attention. Zeta potential is a measure of surface charge (for a full explanation see Lawrence et al., 1989) which is primarily determined by phosphate found in cell wall mannoproteins (Section 4.4.3.2) (Speers *et al.*, 1992) and the pH of the surrounding medium (Lawrence *et al.*, 1989). Indeed, a curvilinear relationship has been found between increasing phosphate concentration and progressively more negative zeta potential (Amory *et al.*, 1988; Mestdagh *et al.*, 1990).

During fermentation, there some confusion as to whether the zeta potential declines (i.e. becomes more positive) (Lawrence *et al.*, 1989; Bowen & Ventham, 1994) or increases during active fermentation only to decline again at the end (Brown, 1997b). There is general agreement however that 'such a lessening in zeta potential would reduce electrostatic repulsion between individual cells and so favour flocculation, even if it is not the cause of flocculation' (Bowen & Ventham, 1994). The same workers also note that at the end of fermentation, trub has a positive zeta potential that they suggest would be attracted to the negatively charged yeast cells. Bowen and Ventham (1994) also speculate that trub could enhance flocculation and cite as evidence work on wort proteins triggering premature flocculation (see Section 4.4.6.5).

### 4.4.7 *Measurement*

Unfortunately, there is no agreement on a 'standard' flocculation test. The vast majority of workers in this field appear driven to develop their own method, which frequently involves a minor modification of an existing method. This frustrating state of affairs was succinctly captured by Speers and Ritcey (1995) who noted that between 1990 and 1995, '17 different flocculation methods (or variations thereof) have been proposed in 30 articles written by 21 authors'. Indeed, the methods issue is now so big that it has spawned a number of reviews that seek to make sense of the diversity of approaches (Stratford, 1992; Speers *et al.*, 1992; Speers & Ritcey, 1995; Soares & Mota, 1997). Arguably, real progress in understanding yeast flocculation awaits the advent of a standard and universally accepted method of measuring yeast flocculation.

Cynically, there are two drivers for the development of a 'new' flocculation test. First, the current approaches simply do not work or are strain specific and, second, the methods are so simple that 'improvements' by fine-tuning are almost irresistible! It may well be that the existing raft of methods, which fundamentally are very similar, will never (irrespective of fine tuning) provide a definitive, standard method.

The vast majority of methods are derived from the test of Burns (1941) (see Section 4.2.5.4) who simply washed yeast and resuspended (at 5% w/v) in an acetate buffer at pH 4.6. After 10 minutes standing, flocculation (or as Burns noted 'clumping power') was determined by measurement of the sediment. As with all subsequent flocculation methods, this test tracks the transition from a population of exclusively single cells to a mixed population of single cells and flocculated clumps (see Fig. 4.42). However, as noted by Stratford (1992), quantification of flocculation requires two of the three possible measurements from the relationship:

Free cell fraction + Flocculated fraction = Total cell count

Many of the methods used today operate with a standard total cell count and quantify the 'free cell fraction' after a period of flocculation. Typically, for convenience, cell counts are derived from measurements of absorbance in a spectro-

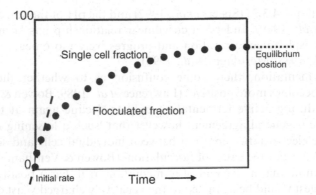

**Fig. 4.42** Phase diagram during flocculation.

photometer. Flocculation is then expressed as a percentage of settled or flocculated cells. Alternatively, some publications describe methods where flocculation is assessed *in situ* in a spectrophotometer. In such circumstances, flocculation is quantified by measurement of the rate of *change* in absorbance.

4.4.7.1  *Current methods.*   Although there are a host of variants, there are essentially three methods used to measure flocculence (see Table 4.22). The 'Helm' sedimentation test (Helm *et al.*, 1953) – which is an extension of the Burns method – is perhaps the most popular method, having been subject to numerous modifications and 'improvements' (see Speers *et al.*, 1992; D'Hautcourt & Smart, 1999). As perhaps a measure of the method's universal acceptability, the Helm test is 'recommended' in its original subjective form by the Institute of Brewing (1997) and, as a more objective absorbance method, by the American Society of Brewing Chemists (Bendiak, 1994, 1996). The 'Stratford' approach although by no means as popular has been used in support of an impressive body of work on the physiology of yeast flocculation (Stratford, 1989; Stratford & Assinder, 1991; Stratford, 1993; Stratford & Carter, 1993).

Both of the above methods operate at the end of an *in vitro* flocculation process, when equilibrium has been achieved between the single-cell fractions and flocs (see Fig. 4.42). The third approach to the measurement of flocculence (Miki *et al.*, 1982) determines the (initial) rate of flocculence directly by change in absorbance in a spectrophotometer. This approach has also been subject to numerous (minor) modifications and has found wider usage in the more fundamental, academic studies on yeast flocculation.

Ironically, given the seemingly never-ending debate about the measurement of flocculation, the above three approaches have surprisingly much in common! As can be seen in Table 4.22, the three approaches agree on the need to wash and resuspend (at a pH of 4.6) cells, to add calcium and to mix to promote the flocculation event. Essentially the differences in protocol are restricted to volume, incubation time and, as noted above, philosophy of measurement at the beginning or at the end of the flocculation process.

**Table 4.22**  Current approaches to the quantification of flocculation.

| | Helm (as Soares & Mota, 1997) | Stratford (as Soares & Mota, 1997) | Miki et al. (as Smit et al., 1992) |
|---|---|---|---|
| Recover cells | Centrifugation | Centrifugation | Centrifugation |
| Wash | EDTA (250 mM) | EDTA (250 mM) | Water |
| Resuspend in buffer | Sodium chloride (250 mM), pH 4.5, 1 × $10^8$ cells/ml | (a) Deionised water, 4 × $10^9$ cells/ml (b) 1 ml of 'a' + 39 ml citrate buffer (50 mM), pH 4.5 | Sodium acetate (50 mM), pH 4.5, $OD_{620}$ = 2.5, 30 minutes 'acclimatisation' |
| Calcium concentration | 4 mM | 4 mM | 9 mM |
| Final volume | 25 ml in measuring cylinder | 40 ml in 100 ml conical flask | 0.65 ml in 1 ml cuvette |
| Mix/agitate | 18 inversions | Orbital incubation at 120 rpm | Whirlimix (20 seconds), invert cuvette (5 ×) |
| Incubation time | ≤ 7 minutes | ≤ 4 hours | 5 minutes |
| Aliqouts removed | 0.2–1 ml from '20 ml' | Stop shaking, after 30 seconds remove 0.2–1 ml from just below the meniscus | No |
| Cell count | Disperse in sodium chloride (250 mM) and determine $OD_{620}$ | Disperse in EDTA (250 mM) and determine $OD_{620}$ | Monitor decrease in $OD_{620}$ |

An interesting paper that seeks to define more precisely the conditions necessary for flocculation (and by implication for a 'test') is that of van Hamersveld et al. (1996). Here, flocculation was investigated in situ, in fermenter at the end of fermentation. Using some innovative techniques, factors such as medium composition, calcium concentration, pH, temperature and shear rate (both laminar and turbulent) were evaluated in terms of floc size, settling rate, bond strength and fraction of single cells. This work 'adds value' in demonstrating the implications of changing environmental variables, and, importantly, notes that temperature, which is frequently ignored, is ideally 15°C (van Hamersveld et al., 1996).

An important contribution to the methods debate is the detailed work of Soares and Mota (1997) who compared their absorbance-based version of the Helm test with a composite Stratford method. Essentially, the two methods yielded results that where indistinguishable in terms of flocculation or, indeed, precision although the Helm method was significantly quicker to perform than the Stratford approach. Soares and Mota (1997) plump for their variant of the Helm test as a 'rigorous and an objective method for quantification of yeast flocculation'. So perhaps despite earlier pessimism, the possibility of an agreed standard test is now not so far away.

4.4.7.2  *New approaches.*  After more than 50 years of wrangling and development of the Burns test, it is timely to consider the opportunities for the development of alternative approaches to assess flocculence. The advent of 'bead'

technology perhaps provides new routes to the isolation and separation of flocculent cells.

Although it is not clear whether or not cell surface hydrophobicity is a universal
factor in flocculation (see Section 4.4.2.6), bead technology has been applied successfully to its measurement (Wilcocks & Smart, 1995; Rhymes & Smart, 1996) and
isolation (Straver & Kijne, 1996). Essentially hydrophobic cells adhere to hydrophobic beads, either 1–2 μm polystyrene-coated latex (Straver & Kijne, 1996) or
0.8 μm latex (Wilcocks & Smart, 1995). The larger beads of Straver and Kijne (1996)
are paramagnetic, enabling their removal with a magnet.

A similar approach that potentially offers better targeting of cells with the capability to flocculate is the use of 'Dynabeads' (Anonymous, 1996a). These super-
paramagnetic polymer particles (2.8 μm) can be coated with ligands that enable
selective isolation of cells with a magnet. This is because the beads are coated with
streptavidin, a protein with a strong affinity for the B vitamin, biotin. Ligands
such as antibodies, proteins, lectins, sugars or nucleic can be attached after being
tagged with biotin ('biotinylated'). Once customised, the beads can then be used to
probe suspensions for target cells, which are then selectively recovered via a magnet (Fig. 4.43).

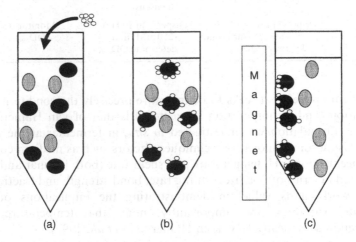

**Fig. 4.43** Diagrammatic representation of lectin magnetic bead technology.

Preliminary studies suggest that customised Dynabeads may be useful to probe or
quantify flocculence. Unpublished work (Wendy Box, Andrew Prest and David
Quain) has shown Dynabeads with an attached snowdrop lectin to bind to brewing
yeast cells with almost 100% efficiency (Fig. 4.44). This lectin binds to mannose and,
presumably, interacts with the cell wall mannose receptors. In agreement with
Stratford (1993), receptors appear to be constitutive, as binding was effectively 100%
throughout fermentation. Unfortunately attempts to probe flocculence lectins with
biotinylated dimannose on beads have been unsuccessful. It is assumed that although
only two mannose residues are necessary for lectin binding (Section 4.4.6.2), bio-

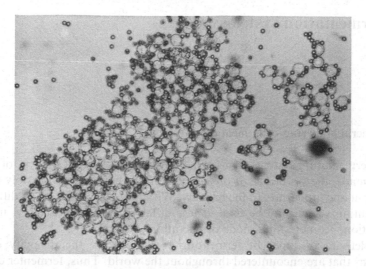

**Fig. 4.44** Attachment of magnetic beads to flocculent yeast.

tinylation results in some sort of steric hindrance. It is anticipated that trimannose may be a better substrate for biotinylation and subsequent bead-cell binding. In the event of Dynabeads being able to probe for flocculence lectins, it is anticipated that a more functional flocculence test will be devised.

# 5 Fermentation systems

## 5.1 General properties of fermentation vessels

Several fermentation systems are used in brewing. These reflect the type of beer which is being made, the traditions of the country of production and possibly the volume throughput and modernity of the brewery. Within these systems many different types of fermenting vessel may be used. These differ in terms of capacity, materials of construction, geometry and mode of operation.

The plethora of types of fermenting vessel mirrors the diversity of brewing operations that are encountered throughout the world. Thus, fermenter design must cater for the requirements of the small unsophisticated cottage industry, brewing beer for domestic consumption, through to the very large ultramodern brewing factory producing brands for the international market place. In between these two extremes is the traditional brewery using vessels with a design pedigree that is centuries old. Other breweries use vessels made to a similar traditional design but constructed from new materials, which are better suited to the modern industry. In addition, the efficiency or ease of use of traditional vessels may be improved by the introduction of new methods for monitoring and controlling the processes occurring within them.

A recent phenomenon is the micro brewery, producing small volumes of often specialist beer types for one or a few retail outlets. The fermenting vessels used are normally of traditional design, but the relatively small volumes involved allow the use of materials such as plastics that have hitherto not been utilised for this purpose in the brewing industry. A further rather niche area requiring fermentation vessels of a special design is that of research and development. Academic and commercial institutions involved in brewing research use laboratory and pilot scale fermentation facilities, which are designed to act as model versions of their production-scale counterparts. Several different small-scale fermentation systems are in common use.

Vessel design, the method of operation, the variety of yeast used and the type of beer produced are all intimately interrelated. In the case of many traditional fermentation vessels, these relationships have been derived empirically using the technology, scientific knowledge and engineering expertise available at some time in the past. Brewing is an innately conservative industry and naturally there is little will to change any part of the process that could result in a reduction in product quality. This applies particularly to fermentation since this stage of brewing is critical to the development of beer flavour. Consequently, there can be resistance to replacing vessels of traditional design with modern substitutes, designed purely on functional grounds to take full advantage of current knowledge. Thus, traditional vessels continue to be used and replaced, some shortcomings being accepted as the price to be paid to maintain a perceived product quality.

On the other hand, the conservatism of parts of the brewing industry has always coexisted with a more radical element, which has sought improvement by innovation. For example, the use of aluminium cylindroconical fermenting vessels together with a method for producing clear worts was first applied to lager brewing at the end of the nineteenth century (Nathan, 1930a, b). This author demonstrated that his fermentation system allowed the production of lager in approximately 12 days compared to the then conventional open fermenter and cellaring system, which took several months to complete. However, whether or not the beers produced by the rapid process were a precise organoleptic match for their slower brewed counterparts was the subject of some controversy. This debate continues; thus, German wheat beers are traditionally produced in large open top-cropping fermenters at relatively high temperature (25 to 30°C). In many breweries this practice continues, whereas in others closed, bottom-cropping cylindroconical fermenters are employed. It has been claimed that the change in vessel type has resulted in wheat beer which has a blander character compared to the traditionally produced product (Hook, 1994).

In many countries in recent years, there has been a trend towards fewer and larger breweries. To meet commercial needs these breweries have tended to use ever-larger individual fermentation batch sizes. These very large fermentations represent a considerable business risk should the resultant beer be in any way atypical. In almost all cases, these very large vessels are of a closed cylindroconical or a related type because of their superior hygiene, ease of monitoring, ability to control and excellent utilisation of available ground area. The trend towards the increasing use of cylindroconical vessels seems set to continue (Maule, 1986).

It may be readily appreciated from the preceding discussion that there is no single ideal fermenting vessel, merely one that is fit for the purpose to which it is put. However, there are some common process and performance criteria. These may be considered in the following categories:

(1) Materials of construction
(2) Hygiene
(3) Capacity
(4) Vessel geometry
(5) Facilities for monitoring and control.

### 5.1.1 Materials used for vessel construction

The materials from which fermentation vessels are fabricated must have sufficient mechanical strength and rigidity to withstand the forces exerted upon them by the volume of fermenting wort. Conversely, they may require to be sufficiently light or malleable, or have other properties which make them easily formed into the required shape and facilitate installation into the brewery. The internal surfaces of the vessel must be capable of being cleaned with ease and be durable enough to withstand sterilisation regimes and the rigours of long-term use. In addition, the internal surfaces should be smooth so as not to present potential sites for microbial colonisation. The material in contact with the fermenting liquid must be inert; first, to ensure that no beer or wort components are adsorbed onto its surface, and, second, such that no potentially toxic and/or flavour tainting materials are leached out into the product.

Fermentation is an exothermic process and it is necessary to apply cooling to maintain a desired temperature. In addition to the need for providing a method of attemperation, the materials used for fermenter construction must have appropriate properties of thermal conductivity. Thus, where attemperation is achieved via external cooling jackets the inner surfaces of the fermenter should be made from materials with high thermal conductivity to ensure rapid and efficient heat transfer from wort to coolant. The exterior walls of the vessel, which enclose the cooling jackets, should then be provided with low-thermal-conductivity lagging material to prevent heat pick-up from the environment. Conversely, where attemperation is achieved via devices submerged in the fermenting liquid, the walls of the vessel should be largely constructed from materials with low thermal conductivity to minimise heat pick-up from the fermenting room.

Many materials have been used for the construction of fermenting vessels and these are summarised in Table 5.1. Frequently the chosen material had much to do with what could be obtained in a particular locality, the availability of a large work force skilled in the use of such materials and, by inference, cost. The tendency to use locally

**Table 5.1**  Materials used for the construction of fermenting vessels.

| Material | Comments |
| --- | --- |
| **Natural materials** | |
| Hollowed gourds, animal skins | Used for domestic brewing, particularly 'native' beers. |
| Stone, slate | Durable, non-porous, inert, low thermal conductivity, very heavy, only suitable for comparatively small square or rectangular vessels |
| Wood | Durable, relatively light, easily worked and shaped, must be well-seasoned, some timbers require inert lining, low thermal conductivity, now expensive, needs manual cleaning. |
| **Fabricated materials** | |
| Plastics | Light, can be formed into any shape, inexpensive, easily deformed, may not be resistant to heat and chemicals, low thermal conductivity, must be inert, not easily cleaned. |
| Ceramics | Inert and non-porous (particularly if glazed), easily worked, inexpensive, low thermal conductivity, fragile and only suitable for small scale domestic brewing. |
| Glass | Light, can be formed into any shape, inert, non-porous, relatively inexpensive, easily cleaned, medium thermal conductivity, very fragile and only suitable for laboratory or domestic vessels. Used in form of vitreous lining with some production scale vessels. |
| Concrete | Durable, strong, relatively inexpensive, can be formed into any shape, relatively heavy but can be used for large vessels, low thermal conductivity, requires inert lining material. |
| Aluminium | Light, durable, relatively inexpensive, easily worked, can be used as light lining material or as main fabric of vessels, inert, easily cleaned, some alloys susceptible to attack by alkalis, high thermal conductivity, subject to electrochemical corrosion. |
| Iron, steel | Strong, relatively inexpensive, easily worked, durable, high thermal conductivity, subject to rusting so must have inert lining fitted. |
| Copper | Strong, easily worked, durable, very expensive, easily cleaned, very high thermal conductivity. |
| Stainless steel | Strong, very durable, expensive, difficult to machine, excellent cleaning properties, austenitic types very corrosion resistant, can be formed into any shape, high thermal conductivity. |

available materials inevitably resulted in vessels which were peculiar to a particular geographical region. In more recent times, and in particular with vessels of non-traditional design, it has been more usual to choose construction materials purely on the basis of cost and functionality.

Historically, fermenting vessels were most commonly constructed from wood, frequently using a circular design and employing the same skills required to cooper casks. In many ways wood is an ideal material for the task. It is, or perhaps was, universally available at reasonable cost, it is durable and strong yet easy to work with. It is not toxic to yeast, it does not impart flavour to the beer and it permits low rates of heat transfer. It does suffer the disadvantage that it is always porous to a certain extent thereby providing potential sites for microbial infection.

The problem of porosity can be minimised by choosing an appropriate wood. Ideally the timber should be hard, close grained, as far as possible free from knots and well seasoned. Favoured woods were oak, kauri pine, red deal and cypress (Lloyd Hind, 1940). The first two of these can be used in vessels where the wood is in direct contact with the fermenting wort. Some of the disadvantages of using wood for fermenter construction can be eliminated by providing an impermeable and inert coating layer between the timber carcase and the product. Materials which have been used include waxes, enamel, pitch and varnish. Lloyd Hind (1940) recommended that lining materials such as metal sheeting should be avoided since this arrangement may promote rotting of the underlying timber. However, there are reports of wooden vessels lined with copper which had shown no deterioration after 70 years of service (Lowe, 1991).

Slate or stone are natural materials which have been used for fermenter construction. They share many of the advantages of wood, for example, inertness, durability and low thermal conductivity and in addition, they are totally non-porous. On the other hand, they are heavy, not easily worked and suitable only for use in vessels with square or rectangular cross-section. They have tended to be used where local supplies were available and in vessels of particular traditional design as typified by the Yorkshire stone square. Although slate is eminently suitable for these fermenters and many remain in use, where they have been replaced the new vessels have been built to a similar design but using more modern materials such as stainless steel (Griffin, 1996). Wood and slate are suitable only for vessels of limited capacity, typically less than 80 hectolitres. Wooden vessels with metal linings may be used for larger vessels with volumes of 150–300 hectolitres. For even larger vessels, it is necessary to use fabricated construction materials such as concrete or various metals.

Reinforced concrete is durable, relatively inexpensive and can be formed into any desired configuration. It has been used for the construction of fermenting vessels, usually of cylindrical design and of any desired capacity. It has high mechanical strength but is relatively heavy, and therefore has tended to be used in new installations where the vessels form part of the fabric of the fermentation room. As in the case of some woods, concrete vessels need to have an inert lining material in contact with the fermenting wort. Lloyd Hind (1940) describes two patented linings, Australite and Ebon, specifically developed for bonding to the inner surface of concrete fermenting vessels. In addition to these materials, aluminium liners have been used.

Metals are the most advantageous of all materials. They are strong, durable, can be fabricated into any shape, inert, non-toxic and easily cleanable. Metals have high coefficients of thermal conductivity and consequently where internal or jacket attemperation is used the external walls of vessels must be lagged. Several metals are used for fermenter construction, notably aluminium, copper, iron, steel and stainless steel. The metal may be used as comparatively thin sheets forming a liner, with other materials used to give structural strength. Alternatively, the entire vessel may be constructed from thicker gauge metal. As in the case of the other materials already described, inner coatings may be used to form an inert barrier between the metal and the fermenting liquid. The coating may be added purposely or it may take the form of a naturally occurring layer of metal oxide.

Aluminium is light, relatively cheap and easily welded and formed, and in the past has been much favoured for vessel construction, for example, the original cylindroconical vessels introduced by Nathan (1930b). It may be used as a lining material with wooden or concrete vessels, as already described, or as the main fabric of the vessel. Aluminium alloys have greater strength than the pure metal but are more susceptible to corrosion. To overcome this it is common to use alloys that have a thin coating of pure aluminium. The major disadvantage of aluminium is that it is attacked by concentrated alkalis and consequently caution is required in the choice of cleaning materials. More significantly, it can be subject to electrochemical corrosion. This can be a particular problem where aluminium vessels are fitted with internal copper attemperators (Schoffel, 1970; Junger & Kruse, 1972). Thus, the charged species in the fermenting wort allow an electrochemical potential to be set up between the two metal components. Severe corrosion to aluminium may occur, especially if any microscopic fissures are present. The effect must be prevented by not using different metals in the same vessel or by insulating the aluminium with an inert layer of lacquer.

Copper has been used widely for small open square ale fermenters in many traditional UK breweries. Many of these vessels have been in constant use for a hundred or more years. Copper has excellent properties with respect to inertness, strength, durability, ease of working and cleaning. It is interesting to consider that relatively low concentrations of copper ions are toxic to many brewing yeast strains (see Section 8.3.3.2) and yet copper fermenting vessels have been used to no apparent detriment. It must be assumed that the layer of copper oxide, which is always present, exerts a protective effect. Use of copper in new vessels is now precluded on economic grounds. Where it has been replaced with the now more usual stainless steel, deleterious changes in beer flavour have been observed, for example, increase in the concentrations of sulphidic components (R. Wharton, personal communication). The explanation is that the presence of copper ions removes some sulphur-containing beer constituents in the form of insoluble copper sulphide.

Iron and steel have both been used for fermenter construction. They possess most of the advantages of metals in general and are relatively inexpensive. Both suffer the obvious drawback of rusting and they must be lined. As in the case of concrete, several proprietary inert resins have been used. In addition, glass lined vessels are common. The vitreous lining is fused to the metal vessel body by means of a heat treatment. The lining is fragile and great care must be taken to ensure that it is not damaged. Epoxy lined steel vessels have a very smooth finish which is claimed to be superior to polished stainless steel for cleaning and efficient removal of yeast crops from conical fermenters (Hollis *et al.*, 1989).

The most favoured material for fermenter construction is undoubtedly stainless steel and the majority of new vessels are now made from this. It has three times the strength of copper and relatively thin gauge sheets (6–10 mm thickness) can be used to fabricate very large vessels. It is totally inert and not subject to corrosion under the conditions encountered in brewery fermentations, provided the correct grade of stainless steel is used and installations are made to appropriate engineering standards. With some caveats, as discussed below, it can be welded with ease and formed into any desired shape. It can be polished to a high degree thereby providing an excellent cleanable and well-draining surface.

Stainless steel is a low-carbon alloy steel containing at least 11.5% chromium. There are three types of stainless steel, termed 'ferritinic', 'martensitic and 'austenitic'. A summary of the alloy composition of these is given in Table 5.2. The brewing industry uses austenitic stainless steels since these are the most resistant to corrosion (Gregory, 1967). The increased resistance is achieved by the addition of nickel and in some grades molybdenum, apart from chromium, which is present in all types of stainless steel. The resistance to corrosion requires oxidising conditions since it is dependent upon the formation of a surface oxide layer. Under reducing conditions and in the presence of chloride ions the oxide layer may be disrupted and not allowed to reform thereby allowing corrosion to occur. To avoid these conditions it is important to ensure that during critical operations such as cleaning oxygen is present.

**Table 5.2** Composition of some stainless steel alloys (Perry *et al.*, 1963). Types 304 and 316 are commonly used for brewing fermenter construction.

| Alloy | Composition |
| --- | --- |
| Ferritinic | 13–20% chromium; less than 0.1% carbon |
| Martensitic | 10–12% chromium; 0.2–0.4% carbon; up to 2% nickel |
| Austenitic | 18–20% chromium; more than 7% nickel |
| Type 304 | Cr, 19%; Ni, 10%; C, 0.08% max. |
| Type 316 | Cr, 16.5–18.5%; Ni, 10% min.; C. 0.08% max.; Mo, 2.25–3.0%; Si, 1.0% max. |

Welding can increase the susceptibility of stainless steel to corrosion since heat-induced chromium carbide precipitation causes localised depletion of the protective chromium. In austenitic grades of stainless steel, this problem is ameliorated by stabilising the chromium by the addition of other components such as titanium, tantalum or columbium. Alternatively, or in addition, the carbon content may be reduced to a minimum concentration. No high-carbon steel components must be used for fittings directly welded to fermenting vessels because of the potential for chromium depletion and thereby promotion of corrosion due to weld penetration.

Although austenitic stainless steels are durable, they can suffer from a defect termed stress corrosion cracking. This phenomenon occurs when steel with an applied tensile stress is exposed to a combination of soluble chlorides and oxygen. Under these conditions, the protective surface oxide layer on the stainless steel may be attacked and over a relatively long period pits may develop. This introduces points of weakness such that the tensile stresses cause sudden cracking to occur. Fortunately, this type of corrosion is not appreciable at temperatures below 50°C, and therefore should not be a problem in the case of fermenting vessels. Although fermenters may be exposed to hot chlorine-containing chemicals in the presence of oxygen during the clean cycle, these conditions are transitory and there is time for the protective oxide layer to be reformed. Nevertheless many stainless steel vessels have not been in service for many years and regular inspection to screen for early signs of developing corrosion would seem prudent.

Although stainless steel has been almost universally adopted for the construction of new vessels, the high-quality grades required are expensive. In small installations cheaper alternatives have been occasionally used, for example plastics such as polypropylene (Anonymous, 1991). Polypropylene lacks the strength to be used as the sole material for the construction of large vessels. It has found application as a new lining material where traditional open square fermenters have been refurbished. It is also used to replace wooden headboards which are fitted to square fermenters to contain the rising yeast head and for temporary covers used during cleaning-in-place operations (Haworth, 1983). It is essential that where materials such as plastics are used they are of a type that is totally inert and must not have components that can leach out into the product. This has been reported to be a problem elsewhere. For example, domestic brewing of African beers traditionally used fermenting vessels made from ceramics or gourds. In recent years recycled metal containers have sometimes been substituted and this has resulted in contamination of the beers with metal ions such as zinc, copper and iron (Reilly, 1972).

### 5.1.2  Vessel hygiene

The hygiene requirements of fermenting vessels are:

(1)  All vessels must be capable of being cleaned in between individual fermentations to remove soiling and avoid the possibility of taints being introduced into the product.

(2)  It may be necessary to disinfect the vessel, prior to filling with wort, to minimise the risk of subsequent microbial contamination and to ensure that all yeast from the previous fermentation is removed.

(3)  In most cases but not all, after the fermentation has commenced the vessel must present a microbiological barrier to the external environment. This is to prevent microbial contaminants gaining entry to the vessel and to confine the yeast within the vessel to minimise the risk of cross-contamination where several yeast strains are used in a single brewery.

In practice, the rigour with which these requirements are pursued depends upon the sophistication of the fermenter and the type of beer that is being produced. Thus, although all vessels must be capable of being cleaned, it is probable that no fermenting vessel is sterilised in the literal meaning of the word. Instead the somewhat paradoxical concepts of 'near sterility' or 'commercial sterility' are used, as with pasteurisation, to denote a condition in which cleaning and sterilisation procedures reduce microbiological loading to an acceptable level. Treatments which together give physical cleanliness and reduction in microbial counts are sometimes described as 'sanitisation' and the chemical cleaning agents, disinfectants and biocides as 'sanitisers' (Brennan *et al.*, 1976). See Section 8.2.1.1 for a full discussion of cleaning-in-place (CiP) operations in the brewing industry.

In terms of fermenter hygiene, CiP achieves a hygienic status that is 'fit for purpose'. Additional protection against adventitious microbial contamination is achieved through environmental and product parameters such as closed vessels, anaerobiosis, low pH and the antimicrobial activity of hop components (see Section 8.1.1). Where open fermenters are used it is essential to add the pitched wort as soon as possible after cleaning and disinfection. The blanket of carbon dioxide produced by the yeast presents a natural barrier to contamination. Nevertheless, to ensure good hygiene the design of the fermentation room must prevent materials, microbial or otherwise, inadvertently dropping into open vessels. On the other hand, worts used to produce Belgian lambic beers are allowed to become contaminated with airborne yeast and other bacteria to start the fermentation (see Section 8.1.4.2). In the case of these products, vessel sterilisation is not necessary and great care is taken to ensure that the microbiological ecology of the fermenting rooms are undisturbed (De Keersmaecker, 1996). At the other end of the spectrum, because of the financial investment associated with very large batch sizes, modern fermenters require a rigorous regime for cleaning and disinfection.

The importance of the materials used for the internal surfaces with regard to ease of cleaning and draining has been discussed in Section 5.1.1. The geometry of the vessel should also facilitate good hygienic practice. In this respect, some traditional fermenters are less than ideal. Those that have square or rectangular cross-section and possibly have internal attemperators are not free draining and have angles that can retain soil. Such vessels may have to be cleaned manually.

Modern vessels tend to be of enclosed design, which assists good hygienic operation. Larger fermenters normally have curved internal geometry that aids both cleaning and draining, and are provided with automatic CiP systems. The latter are permanent plumbed-in systems which automatically clean, rinse and disinfect vessels (see Section 8.2.1.1). They are very effective but require careful setting up. The spray balls must be positioned so that all interior surfaces are cleaned and there must be no shadow areas. All fittings such as sample cocks, valve seals, probes, agitators if fitted, manway door seals etc. must be cleansed properly during the CiP process. The chemicals used must be compatible with all the materials they meet. They must be of an appropriate concentration and the rate at which they are delivered must be sufficient to clean the vessel but not so rapid that draining is impeded.

## 5.1.3   *Fermenter capacity*

It is important to size fermenting vessels in relation to the volume of wort that can be produced. In practice, fermenters tend to have a greater capacity than the output of the brewhouse, such that two or three batches of wort are required to fill a vessel. A more disproportionate mismatch between the capacities of fermenter and brewhouse is undesirable since the time taken to produce sufficient wort to fill the vessel would be unacceptable.

The capacity and number of fermenting vessels must be appropriate for the production requirements of the brewery. Where many beer qualities are produced within a single brewery, it is convenient to have a large number of comparatively small fermenters to foster flexibility. Thus, a multiplicity of vessels allows simultaneous brewing of several beer qualities and in addition, it provides a convenient means of fine tuning production in response to seasonal demand in sales. Conversely, if one or a small number of beer qualities are produced, a few large vessels may be more convenient, or alternatively a continuous fermentation process could be considered. However, some degree of compromise is inevitable since few but large vessels and continuous systems are inherently inflexible and cannot respond easily to fluctuating demands for beer volume.

If new vessels are to be installed in a brewery to increase fermentation capacity, the most cost-effective option is fewest and largest, within the constraints already discussed. The economic factors to be considered are capital costs of installation and fabrication versus revenue costs associated with operation. A comparison of a small multi-vessel and single large vessel installation is given in Table 5.3.

**Table 5.3** General comparison of installation and running costs of four medium sized (4 × 400 hl) and one large (1600 hl) cylindroconical vessel.

| | 4 × 400 hl vessels | 1 × 1600 hl vessel |
|---|---|---|
| Volume output | 1600 hl (4 × 400 hl batches). | 1600 hl (1 × 1600 hl batch). |
| Flexibility | Good – system has ability to ferment four simultaneous product lines. | Poor – only one product line can be fermented at a time. |
| Vessel space utilisation | Poor | Good |
| Capital costs | Cheaper individual vessels but four required.<br>More complex pipe-work, valving etc.<br><br>Costs of monitoring/control equipment for four vessels | More expensive vessel but only one required.<br>Simpler pipework but may need heavier duty pumps.<br>May need higher capacity brewhouse. |
| Revenue costs | Large number of unit operations – filling, cropping, racking, CiP.<br><br>Coolant jacket surface area of four small vessels approximately 1.6 × jacket surface area of one large vessel. | Fewer unit operations for same volume throughput but longer time required for each process step.<br>Greater individual coolant load for larger diameter vessel. |

In modern breweries, the trend towards larger batch sizes has provided the impetus for the almost universal adoption of cylindroconical fermenters. Typically, vessels have a capacity in the range 1500–4000 hl although vessels of 12 000 hl are in use. Probably this is close to, or greater than, the practical maximum for all but the largest breweries. Very large vessels are frequently used for both fermentation and conditioning ('dual-purpose vessels' – see Section 5.4.3). To accommodate these dual functions, modifications are required, particularly to the provision for cooling. The influence of capacity on fermenter performance is intimately related to vessel aspect ratio and geometry. These topics are discussed together in Section 5.1.4.

Very-large-capacity fermenters are employed where high-volume throughputs are required. An alternative approach, particularly where a single beer quality is produced, is to consider the use of continuous fermentation (Masschelein, 1994). Continuous beer fermentation systems are described in Sections 5.6 and 5.7.

### 5.1.4   Fermenter geometry

The geometry of the fermenting vessel must facilitate the operations that have to be performed within it, be suitable for the fermentation characteristics of the chosen yeast strain and be appropriate for the quality of beer produced.

The method of addition of wort should be arranged such that the risk of contamination is minimised and foam generation is not favoured. Conversely, filling of vessels should produce sufficient turbulence to ensure that the yeast is evenly distributed throughout the wort. Brewery fermentation vessels are not usually provided with mechanical stirrers and carbon dioxide bubble formation provides a natural method of agitation. Vessel geometry should be such that gas evolution maintains homogeneous conditions during active fermentation so as to ensure consistent yeast count throughout the body of the wort and to assist with attemperation. However, uncontrolled foam generation should be avoided, which would result in loss of product. Similarly, gas evolution rates should be controlled to minimise stripping of volatile flavour components and evaporation of ethanol. On economic, environmental and safety grounds, it is desirable to provide a means for collecting the evolved carbon dioxide.

At the end of fermentation and possibly during its course, it is necessary to separate yeast from the green beer. The geometry of the vessel should facilitate this operation and be appropriate to the type of yeast that is being used. Vessel emptying should be rapid but not cause turbulent flow and the vessel should drain completely to minimise losses. Before the vessel can be re-used it must be cleaned and disinfected as discussed in Section 5.1.2. These processes should be facilitated by the design of the vessel.

Vessels may be open or closed. Originally, the majority of shallow types were of open construction. However, modern variants tend to be of the closed construction. Deep vessels are invariably closed. The requirement for improved hygiene provided the impetus for the introduction of closed fermenters. This has on occasion produced unexpected results. For example, it has been reported that closed fermenters have increased susceptibility to infection by lactic acid bacteria, possibly a result of the very low oxygen tensions associated with these vessels (Ulenberg et al., 1972; Hoggan, 1978). Cleaning and sterilisation of closed vessels is facilitated by their design; however, anti-vacuum devices must be fitted to eliminate the possibility of collapse. Shuttlewood (1984) identifies a number of circumstances under which a partial vacuum may arise. These are:

(1)   emptying an unvented vessel under gravity;
(2)   cooling down after hot cleaning or sterilisation;

(3)  chemical reactions occurring in the head space (particularly conversion of gaseous carbon dioxide to sodium carbonate when using sodium hydroxide as a cleaning agent); and

(4)  siphoning of vessel contents from a top-mounted down pipe.

Other advantages accrue with closed vessels; for example, collection of carbon dioxide is possible. Vessels of closed design may be pressurised. This can be used to regulate yeast growth and thereby provide a practical measure for controlling the development of yeast-derived organoleptic volatile components of beer (see Section 6.4.1.4).

There are two general geometries used for fermentation vessels. Low-aspect-ratio types, as exemplified by open and closed squares or horizontal cylindrical types, and the high-aspect-ratio variety, as in the cylindroconical or related vessel. Historically, shallow vessels pre-date high-aspect-ratio designs. In the authoritative text of de Clerck, first published in 1948 and then in translation in the UK in 1954, it was recommended that fermenters should be at least 1 metre in depth but not greater than 2 metres. This author advised that settling of yeast and solids would be unacceptably slow and cleaning too difficult in vessels of greater depth.

The move towards deeper and enclosed vessels was pioneered by Nathan (1930a, b) by his introduction at the end of the nineteenth century of the aluminium cylindroconical design. Advances in hygienic design and ease of construction fostered by the use of construction materials such as mild steel and latterly stainless steels promoted the use of larger vessels. Nevertheless progress was slow and the widespread use of large fermenters did not begin until the late 1950s. Initially there was much resistance to changing vessel geometry because of the fear of adversely affecting beer quality. Consequently, capacities were increased simply by building larger vessels to a traditional design but using modern construction materials. For example, the St James Gate Brewery of Guinness in Dublin, Ireland, installed, in 1957, enlarged stainless steel rectangular vessels with capacities of up to 5200 hl and depths approaching 7 m.

During the following three decades, a body of experience was gained which demonstrated that deep vessels of novel design were not a bar to successful brewing. The benefits of economies of scale overrode the misgivings of the conservative and many breweries installed tall high-volume vessels. For example, the Whitbread Brewery in the United Kingdom installed cylindroconical fermenters with capacities of 1200 hl and wort depths of 6 m. Using these vessels, ales were produced that were considered analytical and organoleptic matches for similar beer fermented in traditional shallow open vessels (Shardlow & Thompson, 1971; Shardlow, 1972). Mackie (1985) reported similar findings when cylindroconical vessels where used for the production of British cask-conditioned ales. Ulenberg et al. (1972) described the construction and application for lager production at the Heineken Brewery at Zouterwoude in the Netherlands of a giant cylindroconical vessel of 4800 hl capacity and overall height of 25 m. Many other tall vessel designs were introduced during the 1960s and '70s. These include dished-ended cylinders, cylinders with cones of varying included angles, cylinders with sloping bases and even spheres with conical bases. In many cases, these new vessels were designed for the dual purposes of fermentation and conditioning. The construction and operation of these is described later in this chapter.

Although tall deep vessels are now the norm it is still of benefit to compare their performance with shallower types. Low-aspect-ratio vessels are generally associated with the use of top-cropping yeast strains and high-aspect-ratio vessels for use with bottom-cropping yeast strains. The physiological basis of top- and bottom-fermenting yeast strains is discussed elsewhere (Sections 4.2.2 and 4.4.6.6). The aspect ratio of the fermenter, coupled with the method of yeast cropping, has several ramifications. In top-fermenting systems, there is an opportunity to purify the yeast used in subsequent fermentations. The first yeast head, which contains much entrained wort solids, is discarded. The second yeast head, which forms as the result of vigorous growth during active fermentation, is cropped and retained for subsequent re-pitching. Other non-yeast solids and mainly dead yeast cells sediment. Thus, top-cropped yeast tends to be very clean and of high viability. Many traditional breweries have used the same culture of top-cropped yeast repeatedly pitched and cropped for many tens of years with no apparent decline in performance. However, there are several disadvantages. The method of cropping may be labour intensive or require elaborate and difficult-to-clean plant. Hop utilisation is poor in top fermenters since the bittering components, iso-homulones, selectively bind to yeast cell walls (Section 4.4.2) and hydrophobic species present in foams and these are selectively removed during cropping. Overall efficiencies in terms of product yields are low because of beer losses associated with the cropping regime. In addition, compared to bottom-cropping systems, there is more opportunity for the yeast to be exposed to air. Apart from the increased risk of microbiological contamination, this also can result in reduced fermentation efficiency. Thus, excessive yeast growth in subsequent fermentations may be promoted due to sterol synthesis in response to oxygen exposure during cropping from the previous fermentation.

Removal of yeast crops from bottom-fermenting systems such as cylindroconicals is efficient, can be automated and beer losses are minimal. Since the vessel is closed there is little possibility of exposure to atmospheric oxygen, and therefore maintaining high growth efficiencies is facilitated. On the other hand, unless very bright worts are used, the crop will contain relatively high levels of entrained non-yeast solids and there may be further enrichment with each subsequent repitching. In top-cropping systems, the yeast is removed relatively early in the fermentation when it is in good condition. With bottom-cropping vessels the yeast is not usually removed until the fermentation is completed. Consequently, the yeast may remain in the cone of a deep vessel for quite long periods of time, subjected to the physiological stresses of high hydrostatic pressure, temperature hotspots and high concentrations of carbon dioxide and ethanol. It is possible to crop at intervals during the fermentation (see Section 6.7.2); however, this introduces the possibility of selecting for yeast variants with increased flocculence characteristics. In addition, if multiple cropping is practised, the advantages of ease of yeast handling associated with closed vessels are obviously diminished. The combination of solids enrichment and deterioration due to additive stresses or selection of non-standard variants mitigates against continued cropping and repitching. For this reason where deep enclosed vessels are employed it is usual to limit the number of serial fermentations and periodically introduce newly propagated yeast (see Section 7.2).

The flocculence characteristics of the yeast are important with regard to the aspect ratio of the vessel. Very flocculent yeast, if no remedial action is taken, can settle out before the fermentation is complete with the result that the desired degree of wort attemperation is not achieved. Some traditional shallow vessels were designed specifically to be used with highly flocculent yeast strains, for example, Yorkshire stone squares and the Burton Union system. With these systems the design and mode of operation ensures that yeast remains in contact with the fermenting wort (see Section 5.3). Where flocculent yeast strains are used in other vessel types it may be necessary to provide some means of agitation other than carbon dioxide evolution, to prevent premature sedimentation. It is easier to keep yeast in suspension in shallow vessels of comparatively small volume, compared to their larger, high-aspect-ratio counterparts. In the latter case, it may be necessary to provide a mechanical agitator or to pump the fermenting wort from the base of the vessel back up to the top using a recirculation loop. Obviously, this represents additional capital and revenue costs. Alternatively, when fermentations are scaled up and there is a change from shallow to deep vessels it may be necessary to select a less flocculent yeast strain.

The mixing characteristics of shallow and deep vessels are significantly different. Contrary to expectation, agitation produced by carbon dioxide evolution is much more vigorous in tall thin vessels compared to shallow types. Bishop (1938) reported a detailed study of his observations on carbon dioxide evolution in open square fermenters. He noted that gas bubbles form only at the bottom of fermenters. Sedimented solid particulates but not yeast cells acted as nucleating sites. As the result of a *tour de force* of meticulous observation, Bishop concluded that gas bubbles arose only from particles which had rough or creviced surfaces. Presumably, this explained why the relatively smooth yeast cells did not act as nucleation sites. Once formed, the stream of bubbles rose through the fermenting wort carrying yeast cells within the bubble film. At the surface of the wort the small bubbles collapsed and coalesced into a loose mesh of larger bubbles with entrapped yeast cells held within.

In deep vessels using bottom fermenting yeast, gas bubbles have also been reported to form on sedimented particulates but in this case no yeast cells were observed within the bubble film. Instead yeast cells were simply re-suspended in the wake of rising bubbles (Delente *et al.*, 1968; Ladenburg, 1968). These authors confirmed the earlier observations of Bishop (1938) that no gas bubbles were formed directly from yeast cells. They further noted that the gas bubbles increased in diameter as they rose through the wort and that all bubbles were roughly the same size at any given height within the fermenter. The following interpretation was advanced. Gas bubbles break away from the nucleation site and rise due to buoyancy and in so doing re-suspend yeast cells. Carbon dioxide is evolved at the same rate throughout the entire body of fermenting wort. However, within the body of the vessel the evolving carbon dioxide diffuses directly into the ascending bubbles thereby causing them to grow, an effect further magnified by the decreasing hydrostatic pressure. In growing the bubbles become more buoyant, and their rate of ascent and power to drag both increase. The power of agitation due to the gas bubbles varies in proportion to the rate of fermentation and the logarithm of the height of the fermenter (Fig. 5.1). As may be seen,

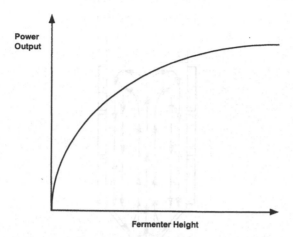

**Fig. 5.1** Relationship between power output due to carbon dioxide evolution and fermenter height (adapted from Delente *et al.*, 1968).

since the power output progressively flattens out, the advantages of good mixing in large-aspect-ratio vessels are eventually lost. The superior mixing characteristics of tall cylindrical vessels are further enhanced by the provision of wall cooling jackets and a conical base. With this configuration, the yeast cells are dragged up through the central core of the vessel with the ascending gas bubbles. At the top, the liquid flow is directed towards the walls. Here the cooling effect increases the density of the fermenting wort and the yeast cells are carried back down towards the base of the vessel there to be returned into the path of the central rising gas stream. Eventually when the rate of gas evolution slows due to depletion of fermentable sugars, the yeast sediments in the base of the cone, thereby facilitating cropping. The patterns of agitation are illustrated in Fig. 5.2. Further discussion of patterns of agitation as a function of vessel geometry may be found in Section 5.4.2.

The use of high-aspect-ratio vessels has been linked with perturbations in beer flavour. Vrieling (1978) reported that the concentrations of the important flavour compounds, ethyl acetate and iso-amyl acetate, formed during fermentation were inversely related to vessel height. Other authors have observed that changing fermenting vessels from low- to high-aspect-ratio types resulted in faster fermentation rates, greater attenuation of worts, increased concentrations of higher alcohols, increased bitterness, increased yeast growth, reduced beer nitrogen, increased dissolved carbon dioxide and lower beer pH (Shardlow, 1972; Ulenberg *et al.*, 1972; Hoggan, 1978; Masschelein, 1986a, 1989)

The alteration in dissolved carbon dioxide is obviously a consequence of the increased hydrostatic head. The improved hop utilisation is a result of the reduced loss rate of bittering compounds associated with top cropping of yeast. Masschelein (1994) identified the high agitation rate as being the principal cause of the other changes. High rates of carbon dioxide evolution promote loss of 'volatiles' due to gas stripping. More significantly, good mixing increases the suspended yeast count and this results in high fermentation rates and enhanced yeast growth. Increases in biomass yields are associated with greater utilisation of wort nitrogen, elevated levels of

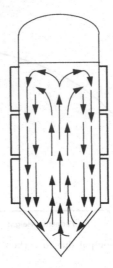

**Fig. 5.2**  Pattern of mixing in a cylindroconical fermenter with wall cooling jackets.

higher alcohols and reduced pH. In addition, a greater proportion of the intracellular pool of cytosolic acetyl-CoA is used to fuel growth at the expense of ester formation. Apart from increased agitation, tall vessels can influence yeast growth because the hydrostatic head increases oxygen solubility. The dissolved oxygen concentration provided in wort at the start of fermentation is the primary mechanism by which the extent of subsequent yeast growth is controlled.

Other factors may be implicated in vessel-size-associated differences in fermentation performance. Large vessels can take a long time to fill and this may require several batches of wort. It is common practice to pitch all the yeast with the first portion of wort. If the filling stage is very prolonged there is an opportunity for the yeast to start budding before wort addition is completed. Should oxygen be supplied to all the wort, a higher biomass yield per unit volume will be achieved in a large vessel as compared to that obtained with a smaller vessel capable of being filled with a single batch of wort. Other flavour anomalies may arise where wort collection is prolonged. Masschelein (1981) observed that vicinal diketone concentrations were three-times higher where a 1200 hl fermenter was filled over a 24 hour period, compared to a similar vessel filled in 4 hours.

In large vessels, during active fermentation the circulating yeast is subject to a continually changing hydrostatic pressure. When the turbulent 'active phase' of fermentation is complete, the yeast sediments into the base. This yeast will be exposed to a combination of high hydrostatic pressure, elevated carbon dioxide and possibly increased temperature 'hot spots'. The latter is due to the difficulty of applying cooling to packed yeast which is still generating heat via exothermic metabolism. In addition, the beer surrounding the packed yeast may have a much higher ethanol concentration than that in the upper parts of the fermenter. This is caused by a localised build up of ethanol produced via fermentation of residual sugar and by dissimilation of yeast glycogen reserves (Quain & Tubb 1982). Masschelein and van der Meersche (1976) reported that there may be a degree of stratification in large vessels. In consequence the beer in the bottom may have a different composition to that in the top, due to the presence of yeast excretion products, including amino acids, peptides, nucleotides and organic phosphates. Prolonged residence times lead to the formation of off- flavours and these have been associated with the release of various short-chain fatty acids from the yeast cells (van de Meersche *et al.*, 1979).

Apart from matching vessel size to the needs of the brewery, the optimum capacity and aspect ratio of fermenting vessels is largely governed by the costs of construction and operation. Within certain limits, larger vessels are the most economical since a doubling in volume attracts a cost increase of just 50–60%. Lindsay and Larson (1975) identified three primary contributors to vessel costs: size, geometry and operating pressure. They calculated that for all vessels the logarithm of relative cost was directly proportional to the logarithm of the volume. Vessels of all configurations followed this rule producing a family of parallel lines each with a slope of approximately 0.65. For any vessel type of the same geometry there is a correlation between volume and surface area (surface area $\propto$ volume $\cdot$ 0.67). Therefore, the relationship can be further simplified to cost being proportional to surface area raised to the power of 0.97. In other words, there is an almost direct correlation between these two parameters. It follows that in terms of materials costs the most effective geometry for fermenting vessels are those which enclose the maximum volume within the minimum surface area.

The optimum geometry is therefore a sphere and the least effective is one of square or rectangular cross-section. Cylindrical vessels with low aspect ratios are more economical than tall thin varieties. Spherical vessels with conical bases were introduced in the 1970s (Martin *et al.*, 1975). The fermenters were constructed from stainless steel, had a diameter of 10 metres and a capacity of 5000 hl. The geometry favours the easy application of cooling because of the high surface area relative to the volume, and pressurisation is favoured. Costs were reported as being 12% less than all other vessel geometries. However, Lindsay and Larson (1975) considered that spherical vessels were difficult to construct and they were associated with high material wastage.

In practice spherical vessels have found little favour and there has been a trend towards the almost exclusive use of tall cylindroconical vessels, irrespective of the type of beer being produced or the strain of yeast employed. The fears that beer quality would suffer as a result of the change to tall vessels appear to be unfounded. It is interesting to note that when cylindroconical vessels were introduced to the Whitbread brewery (Shardlow, 1972) it was felt necessary to select a new bottom-cropping yeast strain for production of ales. In later trials a top-cropping variety, hitherto used in shallow fermenters, was tested in a deep vessel. It was noted that in this case the fermentation was normal, as was the beer, but the yeast formed a satisfactory bottom crop.

The effects on fermentation performance and beer quality of higher agitation rates associated with tall vessels were, in fact, recognised by Nathan (1930a, b) and he recommended that this should be counteracted by reducing the fermentation temperature. There are other strategies, apart from temperature, which may be considered when attempting to match performance in large-volume deep vessels with that of shallow types. Thus, wort gravity, dissolved oxygen tension and yeast pitching rate can all be modulated independently to obtain a desired outcome. It is usual to have to reduce hop addition rates to correct for the lower loss rate and clearly this is advantageous. Conversely, because of the hydrostatic head, deep vessels inevitably produce beers with high dissolved carbon dioxide concentrations. This may be acceptable with lagers but many beer qualities will require adjustment down-stream of the fermenter.

Although vessels with high aspect ratios are used with success to produce many beer qualities, a ratio of height to diameter of 3:1 is most commonly encountered. In terms of volume, 1640 hl is the most usual, giving a vessel with dimensions of approximately 4 metres in diameter and 12 metres in height (Shuttlewood, 1984). This represents the optimum compromise between economy of fermenter scale and batch production requirements. Larger vessels tend to be dual-purpose types used for both fermentation and conditioning. Regardless of the pros and cons of traditional versus deep fermenters, the industry has made a decision, and in the vast majority of commercial breweries beer is fermented in large-volume vessels. Thus, it is reported (Derdelinckx & Neven, 1996) that only 5% of total beer production is now produced by top fermentation. This is the almost exclusive preserve of Europe. The distribution is illustrated in Fig. 5.3.

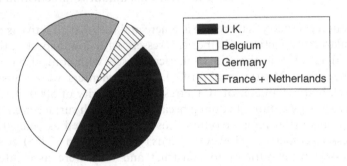

**Fig. 5.3** Percentage share of European production of beer by top fermentation (Derdelinckx & Neven, 1996).

### 5.1.5  Monitoring and control

The design of fermentation vessels must facilitate control of the processes occurring within them. Provision must be made for monitoring the progress of the fermentation, in order that the completion of key stages may be recognised and, if necessary, to identify deviation from normal behaviour and thereby take remedial action. As batch sizes have grown ever bigger, the need to improve the consistency of fermentation performance has provided the impetus for devising improved means for monitoring and control. Whereas the 1960s and '70s were characterised by the widespread adoption of deep closed-fermenting vessels, the intervening period has focused on the development of new and often automatic systems for controlling and monitoring fermentation.

There are three elements to controlling and monitoring brewery fermentations:

(1)  establishment of a desired set of initial conditions;
(2)  monitoring the progress of the fermentation and applying control to maintain an appropriate rate;
(3)  identification of the completion of key stages during fermentation and a point at which the contents of the vessel are deemed to be in a state suitable for transfer to the next stage of processing.

With respect to the initial conditions, the pertinent parameters are wort volume, present gravity, dissolved oxygen concentration, yeast pitching rate and temperature. All of these, with the possible exception of wort volume and present gravity, are established upstream of the fermenter and therefore do not impact directly on fermenter design. However, it may be necessary to make some adjustments to these parameters in the fermenter and appropriate provision must be made to facilitate this. For example, in the event of an unacceptably slow fermentation a possible remedial action would be to add additional oxygen. Vessels should be provided with a means of achieving this aim, either via direct injection of gas into the base of vessels or via an external circulation loop.

In many cases, there may be a requirement to make additions during the course of the fermentation, for example, process aids such as enzymes, fining agents or additional yeast and wort. Vessels must be designed to cater for these needs. There should be a means of measuring the volume of wort within the vessel, either via metered addition or the use of sight-glasses or dipsticks. In some countries, this may be a legal requirement for the purpose of excise duty payments. It may be a practical requirement if the wort collection gravity requires adjustment by dilution with brewing liquor.

During fermentation, it is necessary to monitor progress and usually this is achieved by periodic measurement of the specific gravity of the wort. The achievement of a desired minimum specific gravity marks the end of primary fermentation and signals the commencement of the next stage in the process. In the vast majority of cases, the specific gravity is measured off-line on samples removed from the vessel. These samples may also be used to measure the suspended yeast count and the concentration of flavour metabolites such as vicinal diketones which are also relevant to the progress of some fermentations (Section 3.7.4.1).

In open fermenters, samples of fermenting wort may be obtained with a dropping can or similar implement. Top-opening manway doors in deep vessels allow access to perform the same operation; however, it is more satisfactory to provide a hygienic, cleanable sample point. Where sample cocks are used they must be located in a position which will provide a sample which is representative of the whole of the contents of the vessel. As an alternative to manual sampling and off-line measurement, several commercial systems have been developed which can be fitted into vessels and which provide an automatic measure of specific gravity or other parameters. These are described in Section 6.3.2.3. Apart from providing a continuous measure of the fermentation rate, such systems have the advantage that they provide an output which can be used in automatic feedback control loops.

During fermentation, the primary and usually sole method of controlling rate is via the application of cooling to counterbalance the heat produced by the exothermic metabolism of the growing yeast. Fermenting vessels must be designed to facilitate the measurement and control of temperature. In unsophisticated traditional vessels, temperature measurement, like gravity, may be performed off-line. Conversely, modern vessels are provided with wall-mounted thermometers. These are usually of a type which provide an output that can be used to provide automatic attemperation. In deep vessels several temperature probes are required and these must be positioned carefully to ensure that readings are representative of the whole vessel.

It is not usual to provide 'external' means of increasing temperature in fermenting vessels, although it is not unknown. Thus, the cylindroconicals described by Shardlow (1972) had facilities for the application of both cooling and heating. However, the latter was found by experience to be superfluous and not used in subsequent installations. Normally the yeast and wort are added at a temperature slightly below that at which the primary fermentation is conducted. When yeast activity commences, the temperature is allowed to rise until a desired set-point is reached and control is then applied.

The degree of cooling which is capable of being generated must be appropriate for the fermentation system and type of beer being produced. Cooling loads may be comparatively slight, for example, in the case of top-cropping ale fermentations carried out in small open squares. These are performed at relatively high temperatures (18 to 22°C) and at the end are subjected to modest chilling (8 to 10°C). Fermentations such as these typically use internal attemperators through which cold water is circulated. Conversely, lager fermentations performed in large deep vessels at lower temperatures (6 to 15°C) and culminating in rapid cooling to between 2 and 4°C to sediment yeast have much greater cooling duties. In these cases, external cooling jackets are the norm and refrigerants such as glycol are required. Dual-purpose vessels require the greatest cooling capacities since in addition to attemperation during primary fermentation they must also cater for cold conditioning at sub-zero temperatures. In the case of these vessels there are four distinct cooling duties: attemperation during primary fermentation, crash cooling at the end of primary fermentation to promote yeast sedimentation, cooling to the conditioning temperature and attemperation at the conditioning temperature.

Occasionally, cooling is achieved by circulating the contents of the vessel through an external heat exchanger. This has some drawbacks since it introduces additional complex and difficult-to-clean components. However, it can be a useful in improving mixing, where a particularly flocculent yeast strain is employed. Nonetheless, wall cooling is an expensive undertaking where large numbers of vessels are employed. An alternative approach is to control the temperature of the fermenting room and use unlagged vessels. Historically, this was standard practice, hence the siting of vessels in cellars or caves, particularly those used for the low-temperature secondary fermentation stage of lager production. Latterly this approach is used as a means of augmenting the cooling applied directly to the vessels.

An interesting case history was reported by de Witt and Hewlett (1974) from a brewery in New Zealand. The authors described a situation in which the brewery required additional capacity for both fermentation and cold conditioning. For the sake of economy of scale, it was decided that large vessels were required and to maximise flexibility they should be dual purpose, suitable for both fermentation and cold conditioning. Three options were considered: outdoor weatherproof insulated tanks fitted with cooling jackets; jacketed, insulated tanks contained in a light construction building; unjacketed and uninsulated tanks contained within an insulated and refrigerated building. It was concluded that the third option was the most economic and this plan was implemented with success. However, this approach has not seen widespread adoption.

In addition to controlling fermentation rate, attemperation is also required to minimise the stresses imposed on the yeast crop. In top-cropping systems the yeast is removed comparatively early in the fermentation and is transferred to the relative security of chilled storage. In the case of deep fermenters, the total residence time of the yeast may be prolonged, and, furthermore, at the end the bulk of the yeast will be sedimented in the base of the vessel. Here local heat generation can occur owing to the combination of exothermy of the packed yeast and lack of convection currents. To avoid deterioration of the crop and improve attemperation, large cylindroconicals are frequently fitted with cone cooling jackets.

The rate of fermentation also can be controlled by the application of pressure. This is generated by restricting the dissipation of evolved $CO_2$. Elevated pressure also has been used to modulate levels of beer esters (see Section 6.4.1.4). Obviously this approach is only feasible in closed vessels and, apart from the need to measure and control the pressure, they must be designed to meet appropriate engineering standards. The mandatory requirements relating to pressure vessels cover materials, their nature and thickness, vessel design and method of construction. The regulations apply to the vessels themselves and to associated fittings such as cooling jackets. Vessels classified as pressure types are liable to inspection either by officers of the Government or representatives of the insurer. The most commonly used codes are those applicable in the United States and Canada (American Society of Mechanical Engineers, Boiler and Pressure Vessel Code, ASME Section VIII, Division 1) and their equivalents in the United Kingdom (BS 5500) and Germany (*A.D. Merkblatter*). The British Standards Institution publication *Boilers and Pressure Vessels* (1975) details the legislation relevant to 76 political jurisdictions.

## 5.2  Fermentation rooms

The outcome of fermentation is crucial to the success of the entire brewing process. Since wort is an ideal growth medium for a multitude of micro-organisms apart from brewing yeast, it is essential that every precaution is taken to prevent infection and to ensure that controlled conditions are maintained within fermentation vessels. In this respect, the fermenting vessels and the room in which they are located should be viewed as an integrated whole. Thus, the vessel represents the primary barrier between the fermenting wort and the immediate environment whereas the fermentation room (hall or cellar) is the secondary barrier between the vessels and the external environment. It follows that the design of fermentation room and fermenting vessels should be complementary.

The principal functions of the room may be summarised as follows:

(1)  to enclose the fermenters and ancillary plant within a contained and hygienic environment which minimises the risk of microbiological (or other) contamination of the contents of the fermenter;
(2)  to assist in attemperation of fermenting vessels;
(3)  to provide a safe working environment with respect to concentrations of atmospheric $CO_2$.

In many respects, the degree of sophistication required for the fermenting room is inversely related to the level of containment and control achievable in the fermenting vessel. Vessels of very simple design with rudimentary control facilities rely on the room to compensate for their deficiencies. Conversely, large closed vessels are self-contained and in many cases are constructed entirely outdoors with no dedicated fermentation room. Commonly, tall cylindrical fermenters are arranged in 'tank farms' with their lower parts contained within a fermentation room and the upper two-thirds projecting through the roof and exposed to the elements.

There are advantages and disadvantages to each type of installation. Open vessels, by their very nature, must be enclosed within a room. However, the cost of the room can be defrayed to some extent in that it is not necessary to weatherproof the individual vessels. Further cost savings may be made if the room is the sole source of attemperation since it is less expensive to refrigerate the room as opposed to supplying coolant to several vessels. On the other hand room refrigeration as a means of attemperating vessels is inefficient, very low temperatures cannot be achieved and it lacks flexibility.

Vessels with separate facilities for attemperation offer maximum flexibility. Individual vessels may be used simultaneously to produce several beer qualities, each requiring a different attemperation regime. Similarly, fermentation and cold conditioning may be performed concurrently. However, each vessel must be supplied with services for monitoring and control. Outdoor installations save costs because no building is required but this is partially offset by the need to insulate and weatherproof each vessel. However, there is no restriction on the size of vessels and groups of tall cylindrical tanks make very effective use of available floor space.

Tank farms that are totally outdoors can only be operated safely in areas with a mild and dry climate. In less temperate geographical locations it is necessary to enclose the base of vessels within a building. The building provides a physical support for the vessels and a contained, hygienic and weatherproofed area, which protects the mains, pumps, and other services.

### 5.2.1  Hygienic design of fermenting rooms

It is essential that fermenting rooms are constructed and operated with an appropriate regard to standards of hygiene. Frequently this important aspect of fermentation system design is neglected. Where new installations are made, it is common for budgets to not extend to the finish of the new fermentation room. Furthermore, rooms may be constructed to good standards but subsequent poor maintenance can entirely negate the initial economic investment. In fact, it might be argued that the only occasion where hygienic design of the room is not relevant is where the fermentation is spontaneous, as in the case of Belgian lambic beers (see Section 8.1.4.2). The rigour of the hygienic precautions extended to the room depends upon the microbiological susceptibility of the fermentation system and the scale of the potential economic loss due to contamination.

Historically in United Kingdom breweries, using open vessels to produce ales, it was considered uneconomic to brew during the summer months because of losses due to contamination (Anderson, 1989). Improved hygiene, in all respects, coupled with the introduction of effective attemperation has removed this bar to brewing! Nevertheless, open vessels can be prone to infection and they afford the possibility of cross-contamination where several yeast strains are used in the same fermentation room. Closed vessels are by design more microbiologically robust, and consequently the design of the fermenting room is less critical. The trend towards the increasing use of closed vessels perhaps explains why the finer nuances of room design are sometimes neglected. Fermentation systems that use immobilised yeast are now becoming more common and continuous fermentation, although rare, is a commercial reality. In these cases, the consequences of contamination are very severe because of the prolonged unproductive down-time. Although such systems are designed to present a microbiological barrier to the external environment, it is still important to minimise risks by locating them in hygienic surroundings.

The internal surfaces of the fermenting room should be covered with a material which is cleanable, crevice-free and impervious. Preferably the surface should be seamless and particular care should be taken to ensure that corners are sealed. The most suitable materials are synthetic resins, which provide a continuous membrane. Glazed tiles are satisfactory but it is essential to keep the grouting in good repair. Table 5.4 summarises the hygienic properties of commonly used materials, together with details of resistance to chemicals and wear.

**Table 5.4** Hygienic and durability properties of materials used for internal surfaces of fermenting rooms (Institute of Brewing, 1988).

| Material | Hygiene | Resistance to wear | Resistance to strong acids | Resistance to strong caustic |
|---|---|---|---|---|
| Concrete | Poor | Good | Not recommended | Poor |
| Granolithic concrete | Fair | Excellent | Not recommended | Poor |
| Cement | Poor | Good | Not recommended | Not recommended |
| Resin modified cement | Poor | Good | Not recommended | Poor |
| Rubber | Good | Good | Good | Excellent |
| Paviors | Good | Good | Excellent | Excellent |
| Ceramic tiles | Good | Good | Excellent | Excellent |
| Steel tiles | Poor | Good | Not recommended | Poor |
| Asphalt | Good | Good | Excellent | Excellent |
| Timber | Poor | Fair | Not recommended | Not recommended |
| Vinyl sheet | Good | Good | Good | Good |
| Impervious synthetic resins: | | | | |
| Epoxy | Excellent | Good | Good | Good |
| Polyester | Excellent | Good | Excellent | Excellent |
| Polyurethane | Excellent | Excellent | Good | Excellent |
| Methylmethacrylate | Excellent | Good | Good | Excellent |
| Vinylester | Excellent | Good | Excellent | Excellent |

The floors of fermenting rooms must be free draining and drainage channels should not contain stagnant water, which can be a source of odours that could potentially taint beers. Equipment within the room should be located, wherever possible, away from the walls to provide adequate access for cleaning and disinfection. Windows should be protected with grilles to prevent the entry of insects. Measures should be taken to prevent entry of rodents. Ceiling design is of particular importance where open vessels are used because of the potential for foreign materials falling into the beer. To minimise the risk, the material used for the ceiling must be smooth and non-peeling. If possible there should be no fittings, mains or trunking located directly above vessels. If this is unavoidable, a sealed false ceiling is advisable. Light fittings should be shrouded. Doors leading to the room must be kept closed to minimise the risk of atmospheric contamination and access should be restricted to essential personnel. If the consequences of contamination are considered to be very severe (e.g. immobilised yeast reactor or a continuous fermentation system), a slight positive pressure should be maintained to further reduce the likelihood of airborne contamination (see Section 8.1.4.2). Ideally but rarely, the room should be maintained at a constant temperature and humidity to prevent the formation of condensation. This is of particular importance where the room is refrigerated.

The most common hygiene problem encountered in fermenting rooms (and other areas within the brewery) is the development of mould growth. These micro-organisms are particularly adept at colonising surfaces, particularly where there is persistent moisture. Their habit of proliferating via airborne spores promotes rapid spread of any infection. They will grow readily on most surfaces, including stainless steel. In fact, the black mould, *Geotrichum candidum*, is often referred to as 'machinery mould' because of its propensity for producing slimy growth on metal surfaces (Phipps, 1990).

The consequences of mould growth in fermenting rooms can be severe and unexpected. Unchecked mould infection is associated with corrosion of metals and other materials. The by-products of their metabolism are associated with 'musty' off-flavours in beer. In one instance, a problem of over-attenuation in an unpasteurised beer was thought to be the result of contamination with amylases, introduced as the result of growth of the mould *Rhizopus oryzae* on an unclean flexible hose (D.E. Quain, unpublished data).

Prevention of microbial contamination of the room should be achieved via combination of good initial design followed by careful management. A number of paints, epoxy resins and mastics are available that contain biocides and these should be used where appropriate. In addition to moulds, these are effective against algae, bacteria and lichens. The surfaces of the room and the plant should be washed regularly with a disinfectant. Phipps (1990) recommends ammonium diproprionate as being particularly effective against moulds since it has good residual activity. Incipient mould infections may be recognised as an increase in the numbers of airborne spores. It is good practice to check for this using a suitable microbiological air-sampling device (see Section 8.1.4.2).

## 5.2.2 *Temperature control of fermenting rooms*

Where fermenters are enclosed totally within a room, the latter must assist with vessel attemperation. Traditionally, fermenters and conditioning vessels were located within cellars or caves to take advantage of the low and relatively constant ambient temperature. Vessels were constructed from materials such as wood, which have low thermal conductivities, and assist in minimising heat pick-up from the room. In the United Kingdom, ales tend to be fermented at relatively high temperatures (18 to 22°C) and presumably this reflects the limits of temperature control that were available before the advent of effective refrigeration. Traditional lager beers were fermented at temperatures lower than ales (6 to 12°C). The long secondary fermentation associated with these beers was performed at even lower temperatures (1 to 10°C). Historically, production of lagers evolved in the countries of central Europe and Scandinavia where the relatively colder winter climate facilitated attemperation. For example, the Pilsner Urquell brewery in the Czech Republic still uses wooden fermenting vessels of 25 hl capacity. The vessels have no internal attemperation but they are located in a cellar. The pitching temperature is 4.8°C and during the 12 day fermentation the temperature rises to no more than 8.5°C (Hlavacek, 1977).

Fermentation and conditioning could be practised throughout the entire year in the United Kingdom only when a means of artificial cooling was devised. Initially, this was achieved by placing blocks of ice in fermentation rooms (de Clerck, 1954). Of course, this was expensive and inefficient and when the requisite technology became available, rooms were chilled using cold air delivered via a refrigeration unit. Less ideally, fermenting vessels, or more often conditioning vessels, may still be attemperated using room refrigeration. This requires that the room be insulated but not the vessels. The latter should be constructed from materials with high coefficients of thermal conductivity. In addition, there must be adequate ventilation to avoid a build-up of dangerous concentrations of carbon dioxide and care should be taken to avoid condensation.

In modern installations, the use of refrigerated rooms has been almost entirely superseded by the use of insulated vessels with individual facilities for attemperation. This has been accompanied by the use of rooms of light construction with little or no specific facilities for temperature control.

## 5.2.3 *Control of carbon dioxide concentration*

Gaseous carbon dioxide produced during fermentation is a serious health hazard. The risk to brewery personnel is particularly grave where open fermenters are used and there is no facility to collect the gas as it is evolved. It is essential to provide sufficient ventilation within the fermentation room so that the gas may be dissipated safely. The carbon dioxide concentration within the fermenting room should be monitored at all times using an automatic monitor fitted with an audible alarm.

## 5.3   Traditional fermentation systems

### 5.3.1   *Open fermenters*

Figures 5.4 (a), (b) and (c) show stages in the evolution of traditional fermentation vessels, which are principally associated with top-fermenting yeasts. Figure 5.5 shows a photograph of square fermenting vessels. The most primitive vessel (Fig. 5.4(a)) is a shallow open coopered wooden cask which receives no attemperation other than that provided by the room in which it is housed. The vessel is mounted on legs to allow air circulation and promote cooling. Cooling is further assisted by the shallow depth and relatively large surface area. The low aspect ratio and large surface favours rapid and efficient formation of a yeast head. Vessels of this design are small by modern brewing

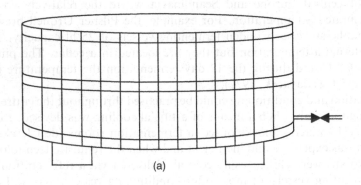

(a)

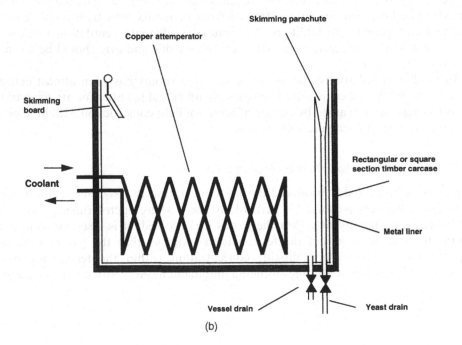

(b)

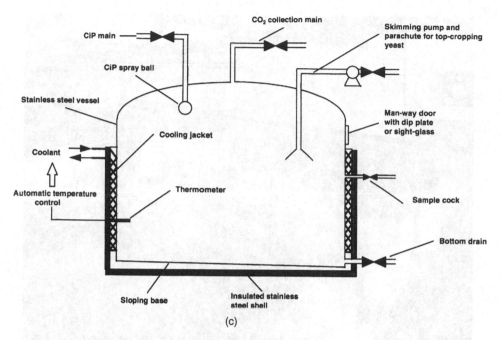

**Fig. 5.4** (a) Traditional wooden open fermenter with bottom drain. (b) Lined wooden open square fermenter. (c) Stainless steel closed square fermenter.

standards, typically less than 30 hl. Although used mainly for top fermentation, similar vessels are used with bottom-cropping yeasts, for example, traditional Pilsner lager production in the Czech Republic (Hlavacek, 1977). Belgian lambic beers (Section 8.1.4.2) are fermented in closed casks (with an aperture for the egress of carbon dioxide) made from oak or chestnut and originally used to hold wine (de Keersmaecker, 1996).

A slightly more sophisticated vessel, an open square fermenter, is shown in Fig. 5.4(b). These may also be constructed from wood, but in this case there is an inner metal liner, typically aluminium or copper. This is more easily cleaned than a wooden surface. Control of fermentation rate is facilitated by an internal attemperator. To ensure good heat transfer this is usually constructed from copper. Open square fermenters have modest capacities, most often within the range 30–50 hl. Like the primitive cask fermenter, the low aspect ratio of the open square favours the formation of a yeast head. Removal of this is via a wide mouthed drain ('parachute'), attached to the side of the vessel. The height of the parachute is adjustable to allow for different wort depths and thickness of yeast head. The skimming board consists of a length of wood or polypropylene, which allows the yeast head to be pushed towards the parachute inlet. Other methods for collecting the yeast head are used, for example, suction pumps. With this arrangement it is possible to pass the yeast through a filter press and return the beer filtrate to the vessel, thereby reducing losses associated with top cropping.

In operation, both the open square and the cask type vessels require a high degree of manual input. In particular, such vessels are normally cleaned by hand. It is

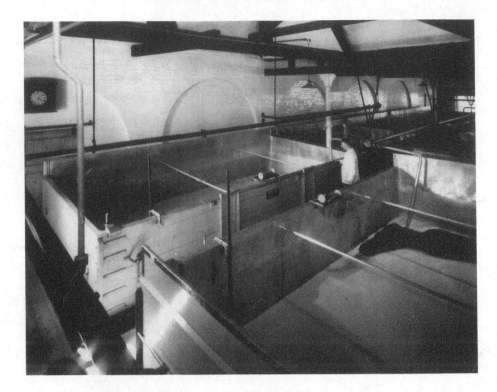

**Fig. 5.5**  Image of open 'square' FVs (courtesy of the Bass Museum, Burton-upon-Trent, England).

possible to automate the cleaning of open squares by providing a temporary cover, usually made of polypropylene. Apart from containing the process, the cover incorporates spray balls that distribute the cleaning agent over the surface of the vessel. The residues are collected via the vessel bottom drain. Even with this automatic system, it is still advisable to remove the attemperator for separate cleaning.

### 5.3.2  Closed square fermenters

The modern incarnation of the square fermenter is shown in Fig. 5.4 (c). These vessels are now usually constructed entirely from stainless steel and typically have capacities of up to 500 hl. Larger vessels have been used, for example, the Guinness brewery in Dublin, in the 1960s installed stainless steel rectangular fermenters with capacities of 5200 and 11500 hl (Lindsay & Larson, 1975). Closed square or rectangular fermenters are provided with a dedicated CiP system. Attemperation is achieved by wall mounted cooling jackets. Apart from being more efficient, these lead to much improved hygiene since the interior of the vessel is relatively uncluttered and therefore much more easily cleaned. Temperature is measured with a thermometer mounted in the wall of the vessel. If desired, output from this may be used to regulate the flow of coolant and thereby provide automatic attemperation. Since the vessels are of a closed design, it is possible to collect carbon dioxide. Modern closed square fermenters may be used for both top and bottom fermentation. If the former is used there are facilities for automatic skimming of the yeast head via a suction pump. The yeast head may be pushed towards the pump inlet using a gas jet. The base of the vessel slopes towards the outlet, which assists with removal of the yeast when used for bottom fermentation.

### 5.3.3 Ale dropping system

Traditional open square or cask type fermenters are not fitted with mechanical stirrers and compared to large cylindrical vessels natural agitation is limited. This can be a problem where particularly flocculent strains of yeast are used since sedimentation of yeast cells may cause premature slowing or even cessation of fermentation. Furthermore, several traditional ale fermentations use mixed yeast strains. If one strain is particularly flocculent it may sediment differentially and eventually disappears from the subsequent top crop. In the United Kingdom, several fermentation systems have been developed specifically to overcome these problems. In the dropping system, the fermentation is started in one open square fermenter and then after 24–36 hours, the partially fermented wort is transferred to another similar vessel. This resuspends the yeast, ensures thorough mixing and high fermentation rates. In addition, the transfer has a cleansing action. Thus, if carefully managed it is possible to transfer the partially fermented wort and yeast but leave much of the non-yeast solid material behind. However, there are some disadvantages. The process uses two vessels for each fermentation, and therefore is expensive in terms of labour and doubles the number of cleans required. In addition, the approach is inherently inefficient since the wetting losses associated with emptying vessels are also doubled.

### 5.3.4 Yorkshire square fermenters

The Yorkshire stone or slate square is a traditional open vessel designed specifically for use with flocculent ale strains (Fig. 5.6). It is a one-vessel system, and therefore overcomes the doubled cleaning disadvantage of the dropping system. As the name suggests, such vessels were originally constructed from slabs of stone or slate and were small, in the region of 30–50 hl. Modern versions are constructed from stainless steel or aluminium and have capacities of 300–400 hl.

Several variations have evolved; however, the characteristic of the vessel is that it is divided into two compartments by a partition or 'deck', positioned towards the top of the vertical walls. The deck slopes slightly, either from the centre or from one side to the other and is perforated by a central circular manhole. The latter has a raised rim about 15 cm high. In addition, the upper and lower compartments are connected by a number of tubes, known as 'organ pipes', which extend downwards almost to the bottom of the vessel. At the periphery of the deck, there is a drain for removal of yeast. A further aperture in the deck allows access for dipping. Vessels were originally cooled by circulating cold water through an external jacket; later these were superseded by submerged attemperators with the inlet and outlet fed through the manhole.

In operation, wort is added to a depth of a few centimetres above or below the

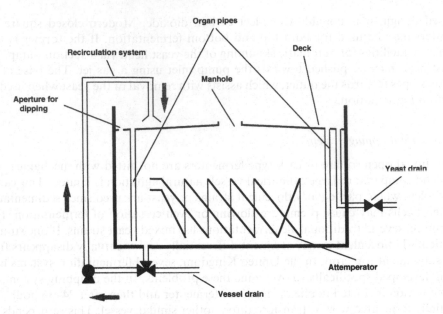

**Fig. 5.6**  Yorkshire slate square fermenter.

height of the deck. The optimum depth is judged by experience and depends on the properties of the yeast that is used. As the fermentation proceeds, yeast rises through the manhole where it is retained on the deck. Beer separates from the yeast head and drains back into the lower compartment via the organ pipes. If very flocculent yeast strains are used the contents of the vessel may be continuously roused using a pumped recirculation system. At the appropriate time, the rousing is stopped and the yeast settles onto the deck from where it may be collected by pushing it into the yeast drain. After the yeast has been skimmed the beer is removed via the bottom drain.

Yorkshire squares, in common with other traditional fermenters, are inefficient since they require considerable manual input. Beer losses are high due to the method of yeast cropping. However, regardless of these shortcomings they continue to be used since it is claimed that the beer is superior to the same product produced in more modern closed vessels. In one Yorkshire brewery, new squares have recently been fitted (Griffin, 1996). These were chosen in preference to the ubiquitous cylindro-conical vessels. The fermentation uses a mixture of two yeast strains and it is claimed that both are necessary to achieve beer with the desired qualities. The strains have differing flocculation characters and it was found that when used in cylindroconical fermenters this resulted in loss of one yeast type.

The new Yorkshire squares have a capacity of 880 hl and have been redesigned to improve efficiency and vessel hygiene. They are enclosed so that collection of carbon dioxide is possible. Automatic CiP systems are fitted and attemperation is automatic via glycol-containing wall cooling jackets. The yeast skimming procedure is automated with each vessel having two suction points set into the deck. These are connected to vacuum yeast collection tanks. After the bulk of the yeast has been removed, the remainder is pushed automatically towards the suction points by rows of water jets located above the deck. Flow through each row of jet heads is controlled sequentially as the yeast advances towards the exit drain. A similar arrangement in the lower compartment controls the removal of tank bottoms.

### 5.3.5 *The Burton Union system*

The Burton Union system of fermentation derives from central England where it was commonly used for the production of ales that required a non-flocculent, or 'powdery' yeast strain. It includes elements of both the dropping system and the Yorkshire square. Thus, the fermentation is started in a conventional open or closed square fermenter. After 24–36 hours, when the fermentation is proceeding vigorously, the contents of the square are transferred into the Burton union fermenter or 'set'.

The essential features of a Burton union set are shown in Fig. 5.7(a), (b) and (c). The actively fermenting wort is transferred to a number of pairs of attemperated wooden casks, termed unions. The casks are suspended below a central 'top trough' to which they are linked via stainless steel tubes known as swan necks. As the fermentation proceeds, yeast and entrained liquid moves upwards from the casks into the top trough. This is cooled using an internal attemperator and this causes the yeast to sediment. The top trough is slightly inclined and this allows beer to run into the feeder trough from whence it returns to the casks via the side rods. When the fermentation is complete most of the yeast is in the top trough and from here it is collected via a drain in the feeder trough. Beneath each row of casks is the bottom trough. This is used to collect the beer from the casks via bottom drain cocks and direct it towards the racking tank. The height of the drain cocks is adjustable so that there is some control over the quantity of yeast transferred with the beer.

The capacity of each union cask is approximately 150 imperial gallons (*c.* 7 hl). The number of casks varies from 24, up to between 50 and 60. Thus, a large set with say 56 casks would have a total capacity in the region of 326 hl. When this system was in common usage in the United Kingdom, a brewery such as Bass, at Burton-upon-Trent, had a hundred sets of this capacity in regular use (see Fig. 5.8). Now, only one brewery, Marstons, also of Burton-upon-Trent, continues to use the union system.

It has many disadvantages, in fact so many it is perhaps difficult to see how it ever arose, let alone survived! For instance, it requires both a conventional fermenter as the well as the union set and this is expensive in terms of plant utilisation. The union set is complex and difficult to clean. Much of the cleaning and routine operation needs a high degree of manual input. The large number of casks and the open trough arrangement presents a significant microbiological risk. The casks require the skills of a cooper to fabricate and maintain. The efficiency of the system is very poor since beer losses are high due to the multiplicity of vessels.

On the other hand, the beer produced from Burton union sets is of the highest quality, as is the yeast collected from the top trough. In the past, efforts were made to simplify the design of the system and improve the hygiene but retain the essential features. For example, replacement of the oak casks with a single long shallow stainless steel tank, fitted with CiP facilities. Such plant was installed at the Bass Brewery in Burton-upon-Trent but it was not considered a success. There were undesirable changes in both beer flavour and the quality of the yeast crop (Steve

Price, personal communication). The now sole user of the Burton union system, Marstons, has recently invested in new plant (Harvey, 1992). The design was totally traditional and the investment was justified since it was claimed that when using conventional fermenters it was not possible to match the existing beer flavour or collect sufficient good quality yeast for re-pitching purposes.

## 5.4  Large-capacity fermenters

In modern breweries, there has been a trend towards batch sizes larger than those achievable with traditional vessel designs. To meet this demand several fermenter

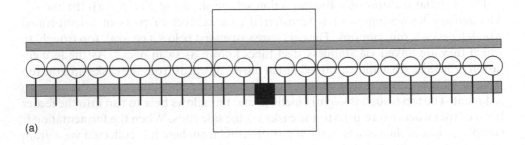

(a)

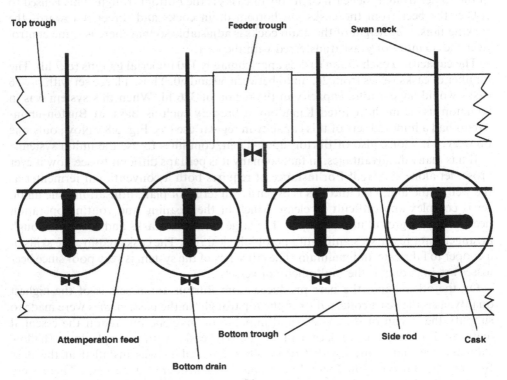

(b)

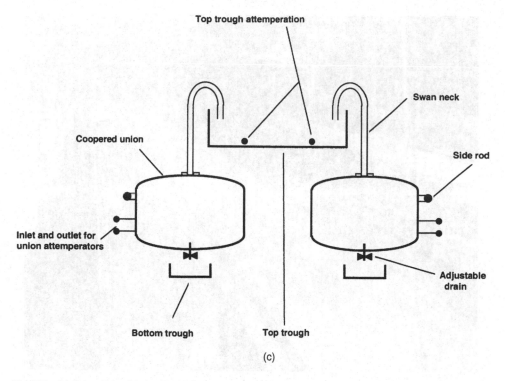

Fig. 5.7 (a) Schematic, (b) detail and (c) side elevation of a 44 cask double Burton Union set.

types have been developed. These are all of the closed design and now are invariably fabricated from stainless steel. Earlier forms were constructed from mild steel, aluminium or lined concrete. Apart from the need to be able to control and monitor the fermentation, a common design criterion is the need to fit numbers of tanks into a minimum of floor space. With one exception, the spheroconical fermenter, this has been accomplished by adopting a cylindrical geometry. In all cases, the yeast crop is harvested from the bottom of the vessel.

### 5.4.1 Cylindrical fermenters

The simplest vessel design takes the form of a cylinder with two dished ends. The vessel may be oriented in a horizontal (Figs 5.9(a) and (b)) or vertical plane (Fig. 5.10). Typical capacities are 500–2000 hl. Both vessel types are of a closed design and so they have good hygienic properties and collection of carbon dioxide is possible. Attemperation is via wall cooling jackets and temperature control is usually automatic. In order to insulate the vessels from the external environment they must be lagged, and, if located outdoors, weatherproofed. There is a single main, which is used for both addition of wort at the beginning of fermentation and removal of yeast and green beer at the end.

Yeast cropping is a more efficient operation from the vertical vessel since the yeast collects as a more compact mass in the smaller area of the dished base. In the case of

**Fig. 5.8** Images from the Burton Union room (courtesy of the Bass Museum, Burton-upon-Trent, England).

the horizontal vessel the yeast sediments along the entire long axis of the base. To facilitate its removal horizontal vessels are inclined slightly towards the exit main. Even so, very flocculent strains may be very difficult to crop and the use of such yeast with these vessels is not recommended.

In view of the different aspect ratios of horizontal and vertical vessels it would be predicted that mixing due to the natural agitation of fermentation would be much less in the former (see Section 5.1.4). In view of this, it might also be predicted that fermentation performance would be markedly different in each vessel when using similar worts and yeast strains. In particular, faster fermentation rates due to better mixing would be expected in the vertical vessel. Experience does not necessarily bear this out. The data in Table 5.5 shows residence times for similar high-gravity lager fermentations performed in either 800 hl horizontal cylindrical vessels or 1600 hl vertical cylindrical vessels. As may be seen, there were no significant differences in fermentation performance between the two types of vessel. However, in this case the yeast strain was a non-flocculent type, which possibly provides an explanation for the result. With a more flocculent variety, the horizontal vessel would be expected to give a slower fermentation. To overcome this shortcoming and improve circulation, horizontal cylindrical fermenters are often fitted with slow speed mechanical stirrers.

Both types of vessel can be arranged in groups, or tank farms, to maximise the utilisation of available floor space. From a civil engineering standpoint, this is more

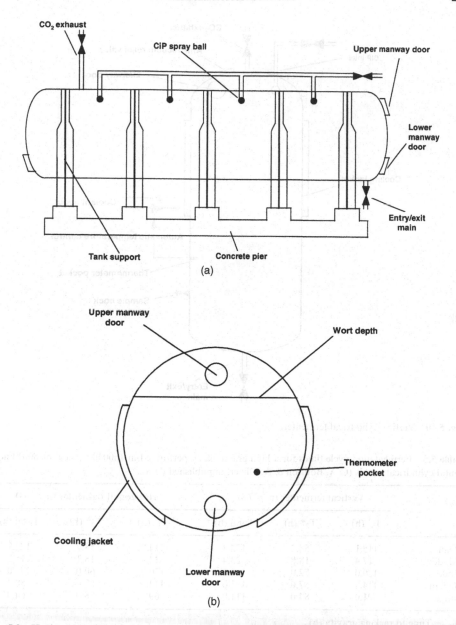

**Fig. 5.9**  Horizontal cylindrical fermenter. (a) Side elevation. (b) End-on.

difficult with horizontal vessels. If they are located within a building, the tanks will probably be arranged on a number of floors. If they are outdoors, the vessels may be stacked in vertical arrays but substantial and expensive supporting structures are needed, and, furthermore, a complex arrangement of service mains is required. On the other hand, vertical cylindrical tanks can be conveniently located in close standing groups with a much simpler supporting structure. If they are outdoors there is no

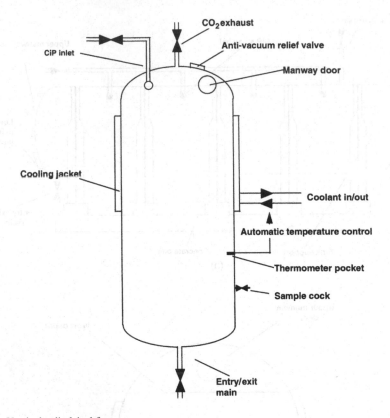

**Fig. 5.10**   Vertical cylindrical fermenter.

**Table 5.5**   Fermentation cycle times for a high gravity lager performed in 1600 hl vertical or 800 hl horizontal cylindrical vessels (C.A. Boulton & A. Oliver, unpublished data).

|  | Vertical fermenter (n = 68) | | | Horizontal fermenter (n = 64) | | |
|---|---|---|---|---|---|---|
|  | $T_1^\dagger$ (h) | $T_2^*$ (h) | Total (h) | $T_1^\dagger$ (h) | $T_2^*$ (h) | Total (h) |
| Mean | 115.1 | 58.2 | 173.3 | 114.8 | 58.7 | 173.7 |
| Std. dev. | 17.4 | 18.8 | 23.4 | 14.7 | 18.1 | 15.9 |
| Mode | 120.0 | 72.0 | 162.0 | 120.0 | 60.0 | 177.0 |
| Median | 114.0 | 57.0 | 171.0 | 114.0 | 60.0 | 88.0 |
| Range | 91.0 | 84.0 | 111.0 | 69.0 | 78.0 | 64.0 |

$^\dagger T_1$ = Time to racking gravity (h).
$^* T_2$ = Warm conditioning time (h).

height restriction. The majority of services are supplied to the base of vertical vessels in a simpler arrangement, which uses fewer valves. With very tall vertical vessels, the large hydrostatic head requires heavy-duty pumps to fill from the base. In this regard, horizontal vessels present less of a problem; however, even this advantage is lost where the vessels are stacked.

On balance, vertical cylindrical vessels are more advantageous than their horizontal counterparts and this is reflected in their relative use. As early as 1984, Shuttlewood reported that, of all new cylindrical vessel installations, less than 1 in 20 were of the horizontal type. The move towards vertical geometry seems set to continue.

### 5.4.2  Cylindroconical fermenters

This style of vessel was originally designed by Nathan (1930a, b) at the end of the nineteenth and introduced by him, initially to Switzerland, as a means of improving the efficiency of the traditional lager fermentation. After this time they slowly came into common usage in mainland Europe and throughout much of the rest of the world. The United Kingdom resisted wholesale adoption of cylindroconicals because of the belief that they were unsuitable for the production of top- fermented ales. This prejudice has now been largely dispelled (see Section 5.1.4) and cylindroconicals are now the most widely used high-capacity fermenters in the United Kingdom and elsewhere.

Installation of cylindroconical vessels (CCVs) was particularly prevalent during the 1960s and '70s. During this period, in many countries, there was a trend towards fewer and larger brewing companies, producing brands for both national and international markets. This produced a requirement for larger batch sizes. Simultaneously, the fabrication and material costs of large stainless steel vessels fell and this coincidence of demand and economic feasibility paved the way for the introduction of cylindroconicals and other related high-capacity vessel types. Since that time little has changed in terms of vessel design. Recent advances mainly relate to improved methods for monitoring and control and these are discussed in Chapter 6.

The original Nathan process improved the efficiency of lager production by combining primary fermentation and cold conditioning in a single tank. In modern parlance, this was a 'unitank' operation. Modern cylindroconical vessels are essentially the same as the original Nathan design, and, similarly, they may be used for both primary fermentation and cold conditioning. However, in the majority of breweries these operations are separate and involve an intermediate tank-to-tank transfer. Versions with a slightly simpler design may be used solely for primary fermentation, others just for cold conditioning. An idealised representation is given in Fig. 5.11 together with photographs of various CCVs in Fig. 5.12. The vessel as depicted would be suitable for carrying out both primary fermentation and cold conditioning and it shows some fittings, which may be considered as optional. The daunting challenge of delivery and installation is captured in Fig. 5.13.

The original Nathan vessels were constructed from aluminium. This was a disadvantage in that the propensity of this metal to corrode was not fully understood and this did little for their initial reputation. Subsequently, mild steel lined with epoxy resin was commonly used but this has now been superseded largely by stainless steel. The essential feature of the vessel is the replacement of the lower dished end by a cone. The interior surface of this is highly polished to reduce friction and thereby facilitate the collection and removal of the yeast crop. The vessels may be located within a building or outdoors. If the latter option is chosen, external weatherproofing must be provided. Frequently, the fermenters are grouped such that the cones and associated

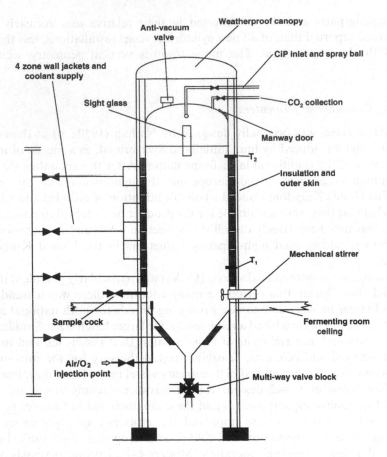

**Fig. 5.11**   A cylindroconical fermenter.

services are enclosed within a hygienically designed room. The ceiling of the room provides mechanical support for the vessels and access to the area at the base of the cylindrical portion, a convenient location for a sample cock. In one reported instance an installation was made in which the vessels were uninsulated and erected within a refrigerated room. The latter provided attemperation during primary fermentation. Circulation of the green beer through external heat exchangers provided a means of cooling to cold conditioning temperatures (de Witt & Hewlett, 1974).

In the majority of installations, attemperation is via wall cooling jackets. These are surrounded by an insulated outer skin to minimise heat pick-up from the environment. Thermometers ($T_1$, $T_2$, Fig. 5.11) are provided within the vessel and output from these is used to control temperature automatically via the supply of coolant to the wall jackets. The number and location of the thermometers are dependent on the duties to which the vessel is put. The natural agitation of fermentation may be augmented by the provision of a small mechanical agitator. Occasionally, a loop system may be used, where the vessel contents are circulated out from the base and back into an entry point at the top. The latter system may also incorporate an in-line heat

(a)

(b)

(c)

(d)

**Fig. 5.12** Cylindoconical fermenters (a) courtesy of Briggs PLC, (b) kindly supplied by Maxine Bellfield of Vaux Breweries, (c) kindly supplied by Harry White, Bass Brewers and (d) courtesy of Ian Dobbs, Bass Brewers.

(a)

(b)

**Fig. 5.13** Delivery (a), lifting (b) and installation (c) of new CCVs to Bass Brewers Burton Brewery (kindly supplied by Colin Turton, Bass Brewers).

exchanger to supplement or even replace the wall cooling jackets. In addition, post-collection aeration/oxygenation systems can be installed which will also facilitate mixing. Vessels are gauged to allow an accurate measure of wort volume and where appropriate, provide a means of calculating the rate of addition of dilution liquor. In some installations the traditional manual sight glass or dip plate arrangement is replaced by an in-line flow meter located up-stream of the fermenters. Other refinements in monitoring and control of fermentations have been introduced in the last five years and these are discussed in Chapter 6. Vessels are suitable for collection of evolved carbon dioxide and may be pressurised if designed to the appropriate specification.

Filling and emptying of vessels is via a single entry point at the base of the cone. This simplifies the design of the vessels but requires a complex valve block so that the process flow can be directed to the appropriate destination. The pipework is arranged so that a number of vessels may be serviced by a minimum number of common mains (Figs 5.14 and 5.15). More complex valving and duplication of service mains, although providing greater operational flexibility, will, of course, escalate capital and revenue costs.

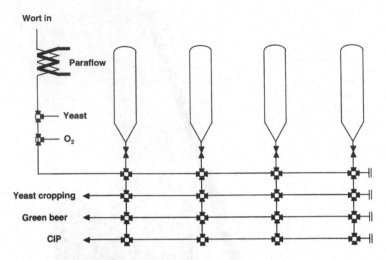

**Fig. 5.14** Arrangement of process pipework for cylindroconical tank farms.

**Fig. 5.15** CCV cones and associated mains (kindly supplied by Maxine Bellfield of Vaux Breweries).

In operation the vessels and associated mains are cleaned and sterilised automatically by a dedicated CiP system. Wort is delivered to the vessels already oxygenated and inoculated with yeast. Both of the latter operations are performed in-line and may be controlled automatically (Chapter 6). Frequently, the wort may be diluted in-line with brewing liquor using an automatic blending system, to achieve a desired gravity. Alternatively, dilution liquor may be added to the partially filled vessel. If the latter option is chosen, it is important to ensure that the vessel contents are mixed to homogeneity. As shown in Fig. 5.11, this may be achieved using a mechanical stirrer or by gas rousing. Of these two options, mechanical stirring is preferable. With gas rousing there is the inevitable risk of perturbing the initial wort dissolved oxygen concentration, and thereby influencing, in an uncontrolled manner, the outcome of the fermentation (Chapter 6). Thus, the initial oxygen concentration may be supplemented if compressed air is used, or reduced by gas stripping oxygen if nitrogen or carbon dioxide is used.

At the end of fermentation the yeast sediments in the cone from where it is cropped in the form of a concentrated slurry and usually is transferred to a storage vessel. Here it may be retained and used to seed a subsequent fermentation; alternatively, it may be disposed of after first recovering the entrained beer. If the fermenter is used solely for primary fermentation, the green beer is transferred to the next stage of processing. Most of the suspended yeast is removed by centrifugation and cold conditioning of the partially clarified beer is performed in a second chilled (frequently cylin-droconical) storage vessel. Alternatively, with a single tank operation both primary fermentation and cold conditioning are performed in a single vessel. In this mode of operation, the green beer is chilled to between 2 and 4°C at the end of primary fer-mentation (and possibly a period of warm conditioning) to sediment the yeast. After the yeast crop has been removed, the vessel contents are chilled to a 'conditioning' temperature (typically just below 0°C), and held for an appropriate period to allow flavour maturation and achievement of colloidal stability.

The duties for which cylindroconical vessels are used influences their design. Those used solely for primary fermentation tend to be of a high aspect ratio and have a steep sided cone. Thus, Shuttlewood (1984) reported that the most commonly supplied cylindroconical fermentation vessel is one of approximately 1500 hl with 12–13 m body height, a diameter of 4.0–4.2 m and cone with an included angle of 70°. This design favours good mixing and the relatively steep-angled cone facilitates efficient removal of the yeast crop. Similarly, these vessels have good draining properties. Lager fermentations generally make more efficient use of the vessel capacity com-pared to ales because the latter are usually performed at higher temperatures and this promotes greater fobbing. Consequently, it is necessary to leave a greater headspace (freeboard) to accommodate the foam. This problem can be ameliorated by sparing application of silicone antifoam or other surface active lipids, such as sorbitan monolaurate and glyceryl monoleate (Button & Wren, 1978). Preferably, the addition should be at vessel collection although antifoam can be sprayed directly onto the developing foam. Some brewers view the use of chemical antifoams with suspicion because of the fear of adversely affecting head retention of the finished beer. Mechanical foam breaking devices have not found much favour because they intro-duce components that are difficult to clean. A new and patented approach, at present only tested at pilot scale, suppresses foaming by the application of ultrasound radiation (Freeman et al., 1997). Whether or not this method sees widespread adoption will require developments in the technology and reductions in cost.

Vessels used for both primary fermentation and cold conditioning tend to have higher capacities, smaller aspect ratios and frequently a much shallower cone. These differences relate to costs versus capacity. Where a single tank process is operated, and therefore fewer vessels in total are needed, it is sensible to use the largest possible batch size, which is consistent with production requirements and the capabilities of the brewhouse. With a single large tank the material costs decrease in relation to the aspect ratio, and furthermore shallow cones are less expensive than the steep variety. On the other hand, a lower aspect ratio reduces the surface area available for wall cooling and shallow cones are less efficient for yeast crop removal and increase wetting losses. The cooling jackets assist in giving structural strength to vessels. Those used for brewing vessels are most commonly of dimpled or half-pipe design, depending on the type of coolant used (Shuttlewood, 1984). The former are used with isothermal refrigerants such as ammonia, whereas the latter are suitable for non-isothermal coolants such as propylene glycol (Wilkinson, 1991). The relative merits of each type of coolant have been discussed by Gerlach (1995). This author considered that ammonia was the most advantageous for several reasons: no glycol transformation stage, higher temperatures at the compressor, smaller pumps, an overall energy saving of 30–40%, smaller refrigeration ducts, more precise temperature control and better flexibility.

Single- and dual-purpose vessels must cater for different cooling requirements and this is reflected in the number and location of wall jackets.

The cooling duties are:

(1)  attemperation during the period of maximum exothermy (primary fermentation);
(2)  rapid cooling ('crash cool 1') to 2–4°C at the end of primary fermentation and attemperation at this temperature, to sediment yeast;
(3)  cooling ('crash cool 2') to the conditioning temperature; and
(4)  attemperation at the conditioning temperature.

Several factors determine how much cooling must be provided to fulfil the individual duties outlined above. A proper consideration of these is essential in order to ensure that cooling is efficient and cost effective.

All four duties are influenced by the geometry of the vessel, in particular, the aspect ratio, the capacity, the number, disposition and surface area of cooling jackets, the temperature and flow rate of the coolant. In addition, the thermal conductivity of the material used for vessel construction, the efficiency of the insulation and the temperature of the environment. During primary fermentation, the magnitude of exothermy is governed by the factors which control the rate and extent of yeast growth, principally wort composition, initial dissolved oxygen tension, yeast pitching rate and the required holding temperature. For crash cooling (both 1 and 2) the total temperature drop and the minimum acceptable time to achieve it are influential. With a two-tank arrangement, reduction of green beer to the cold conditioning temperature (crash cool 2) may be accomplished by in-line chilling during transfer. In this case, the second vessel requires only to be capable of maintaining the conditioning temperature. Conversely, with a single vessel operation, or where the beer is transferred without in-line chilling, the cooling capacity must be sufficient to reduce the beer to the conditioning temperature and then maintain it.

Transfer of heat from the beer to the coolant involves a combination of convection and conduction. This can be quantified using the heat transfer formula:

$$Q = UA \, \Delta T$$

where Q is the rate of heat transfer (Joules $s^{-1}$); U is the overall heat transfer coefficient ($J\,m^2\,s^{-1}\,K^{-1}$); A is the area of jacket ($m^2$); $\Delta T$ is the mean difference in temperature between the two liquids (K).

In addition to the transfer of heat from the beer to the coolant, there is also the possibility of heat transfer from the environment to the coolant and the magnitude of this effect will also be controlled by the heat transfer formula. For efficient cooling, vessels should be constructed to minimise the value of U with respect to heat exchange between the exterior and the coolant and to maximise the value of U with respect to heat exchange between the coolant and the beer.

Several processes combine to determine these values of U and these are illustrated in Fig. 5.16. Each phase may be viewed as a barrier to heat transfer. Thus, solid materials such as the tank wall, the outer jacket wall, the insulation layer and the outer skin have defined coefficients of thermal conduction. The values of these may be increased by surface soiling (fouling layers). Transfer of heat through the fluid layers is governed by convection. Fluids flowing over plane solid surfaces have a surface film, or boundary layer, where flow may be laminar.

Temperature

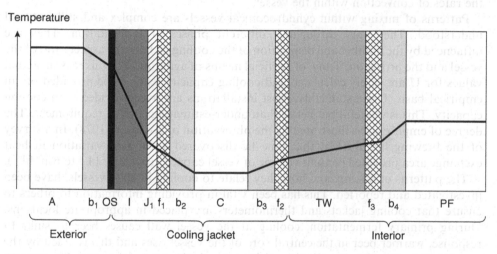

**Fig. 5.16** Schematic representation of the barriers to heat transfer between the exterior and interior of a jacketed and insulated fermenter:A, external atmosphere; $b_1$, air boundary layer; OS, vessel outer skin; I, insulation; $J_1$, outer jacket wall; $f_1$, outer jacket fouling layer; C, coolant; $f_3$, inner coolant fouling layer; TW, tank wall; $f_3$ beer fouling layer; $b_4$, beer boundary layer; PF, process fluid (beer/wort); $b_2$ outer coolant boundary layer; $b_3$, inner coolant boundary layer.

Surrounding this is a phase where flow is turbulent. The size of the boundary layer is influenced by the magnitude of the convection currents in the bulk of the fluid. Where two bodies of fluid of different temperature are separated by a solid layer, the biggest temperature differential occurs within the boundary layers since this region offers the most resistance to heat transfer.

In the case of a fermenter, the insulation should ensure that loss of cooling efficiency due to heat transfer from the environment to the coolant is minimal. This may not always be so. Shuttlewood (1984) estimated that up to 50% of the cooling capacity of fermenters could be dissipated in this way.

Most cylindroconical vessels are made from either stainless or mild steel and both of these have high coefficients of thermal conductivity. However, the metal thickness is governed by mechanical strength and therefore is not a variable with respect to cooling. In a properly designed and cleaned vessel, fouling layers both within the vessel and cooling jackets should be insignificant. The coolant temperature should be sufficiently low so that it is possible to achieve and control the lowest required beer temperature but with no ice formation. Thus, use of very low temperature coolant is an attractive means of increasing the overall heat transfer coefficient but in a well-insulated vessel the inevitable result will be that the beer will freeze. Formation of ice on the inner tank wall dramatically reduces the efficiency of heat transfer. If the coolant temperature is too high, the desired crash cool temperature will either never be achieved or not achieved within an acceptable time span. Coolant is pumped through the jackets and so the fluid flow rate is relatively high, therefore convection is forced and rates of heat transfer are high. The most significant factors which influence heat transfer (or increase the value of U) are the total coolant jacket surface area and the rates of convection within the vessel.

Patterns of mixing within cylindroconical vessels are complex and still not well understood. They vary throughout different phases of fermentation. They are influenced by the number and disposition of the cooling jackets, the aspect ratio of the vessel and the provision, if any, of artificial means of agitation. For this reason, actual values for U are rarely calculated and cooling capacity tends to be provided on an empirical basis. Understandably, most installations are over-provided with cooling capacity. This is wasteful but better than under-estimating cooling requirement. The degree of empiricism is illustrated by the observation of Knudsen (1978). In a survey of the brewing literature at that time he discovered a 100-fold variation in heat exchange area provided per unit volume of vessel capacity ($0.006\,m^2\,hl^{-1}$ to $6\,m^2\,hl^{-1}$).

The patterns of mixing, and how they relate to cooling in large vessels, have been investigated and reported. This has been vital in providing information to others to ensure that cooling jackets and thermometers are placed in appropriate locations. During primary fermentation, cooling at the vessel wall causes beer to sink. In response, warmer beer in the central core of the vessel rises and this is aided by the ascending stream of carbon dioxide arising from the bottom of the cone. This pattern of mixing is illustrated in Fig. 5.2. and is based on the observations of Ladenburg (1968). In a later report, Knudsen (1978) claimed that these observations were invalid since they were obtained from a model which used relatively thin wafers to observe patterns of agitation. These failed to take into account boundary layer effects which can obscure what is happening in the deeper recesses of vessels. In addition, this author cautioned that observations made with small-scale model systems do not always reflect faithfully what happens in larger tanks. Knudsen's approach, also using a scaled-down model, was to study hydraulic movements in saline solutions (to simulate wort) containing suspended organic solvent droplets, of the same density as the saline and stained with the lipophilic dye Sudan Black. Agitation was provided by sparging carbon dioxide from the base of the cone of the vessel.

The observed agitation patterns were as shown in Fig. 5.17. Two zones of mixing

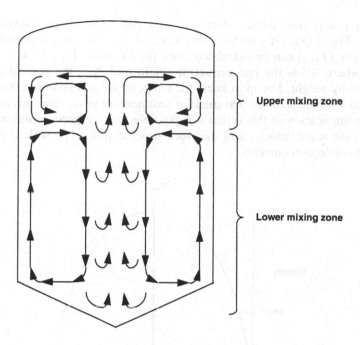

**Fig. 5.17** Convection currents in a cylindroconical fermenter with $CO_2$ sparging from the base of the cone (after Knudsen, 1978).

were distinguished and flow rates were always more rapid in the upper zone. Upward movement of the lower zone occurred at a velocity proportional to the size of the upper zone. Increasing the sparge rate increased the size of the lower zone but produced a smaller more vigorously mixed upper zone. Decreasing the vessel diameter had the same effect as increasing the sparge rate. The implications were considered that narrow vessels of high aspect ratio would be more prone to foaming, due to the high rates of turbulence in the upper zone. Cooling jackets for attemperation during primary fermentation should be located at the upper zone to maximise free convectional heat transfer. More recently, Schuch and Denk (1996) described another model cylindroconical system in which flow patterns were visualised using a laser beam arrangement. The results were similar to those of Knudsen (1978) but there was evidence of much greater complexity of fluid flow.

During the active phase of primary fermentation, there is usually sufficient agitation to ensure homogeneity of vessel contents, and therefore convectional heat exchange is efficient, attemperation is accurate and temperature readings measured at any location are valid. During crash cooling the situation is less satisfactory. At the end of primary fermentation, mixing due to carbon dioxide generation is minimal. In the absence of any other source of agitation, heat transfer is dependent on mixing due to convection currents generated by the formation of density gradients in beers of differing temperature. Weissler (1965) showed that for any beer there is a specific temperature where the density is maximal. As beer is cooled the density increases and hence it tends to sink. At temperatures less than that of maximum density (the inversion point) beer density decreases and the direction of convection flow is reversed (Fig. 5.18). This author demonstrated that the temperature of maximum beer density ($T_{MD}$) can be calculated from the formula, $T_{MD} = 4 - (0.65\ RE - 0.24A)$, where RE is the real extract in °Plato and A is the percent ethanol concentration by weight. For most beers the temperature of maximum density is in the range 2 to 4°C, the same as that used for sedimenting yeast after primary fermentation. The implication of this is that it is possible to have beers of different temperatures but the same density, and therefore stratification is possible if cooling relies solely on convection currents.

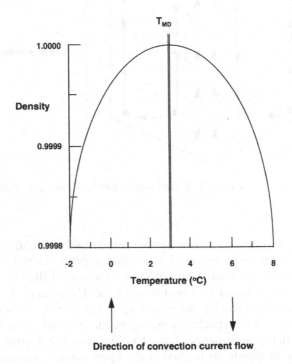

**Fig. 5.18**  Relationship between temperature and beer density (adapted from Wilkinson, 1991).

Temperature distribution and fluid flow patterns in large cylindroconical vessels during crash cooling have been studied by Brandon (Larson & Brandon, 1988; Reuther *et al.*, 1995). The studies used a combination of experimental observation in production-scale and model vessels. The ineluctable conclusion was that the mixing patterns are far more complex than is usually appreciated. It was observed that when a beer was being held at a temperature below the inversion point, the predicted upward fluid flow could be detected in beer adjacent to cooling jackets. At the same time, there was also a simultaneous and unexpected downward movement. Similarly, temperature measurements made at a point near the wall indicated rapid fluctuations over a period of a few minutes, indicating the presence of vortex motion. These authors hope for an outcome in which optimum cooling and vessel design may be achieved using predictive models. This seems still to be a distant prospect! Schuch

(1996) concluded that when using a single point thermometer, it was prudent to cover as much of the cone as possible with cooling surfaces and at least 50% of the surface area of the cylinder. Similarly, Manger and Annemuller (1996) reported that it was not possible to make general recommendations regarding the suitability of any one cooling system because of the multiplicity of factors involved. Appropriate cooling systems were best fitted in response to site-specific needs.

Notwithstanding the theoretical and perhaps still unresolved aspects of vessel cooling, some practical points may be made with regard to the optimum placing of cooling jackets for different duties. The much-quoted work of Maule (Maule, 1976; Crabb & Maule, 1978) is invaluable in this respect. In this study, a 1600 hl cylindroconical fermenter was fitted with a manifold device which allowed temperature measurement and withdrawal of samples at various locations within the vessel. From a study of a number of fermentations, both ale and lager, it was concluded that during primary fermentation the vessel contents were essentially homogeneous throughout. Satisfactory attemperation could be achieved for both lager and ale fermentations, held at temperatures of 8°C using a single wall jacket. This was best located high up the vertical sides of the vessel whereas the thermometer should be placed at the base of the cylinder. This arrangement promotes good convective mixing. Thus, the thermometer is located in the warmest part of the vessel. Output from this activates coolant flow in the jacket above and in response the cooled beer flows downwards due to the increased density. With very tall vessels, especially with high temperature fermentations performed at 21°C, a second jacket was considered advisable.

Crash cool 1 (2°C to 4°C) in a tall vessel required the use of two or more side wall jackets, one located near the top of the cylinder and the other abutting the junction with the cone. In this case the beer was cooled to near the inversion point and to overcome the concomitant lack of mixing, provision of cooling at both high and low level was necessary. In addition, if the beer was held in the same vessel for a long period, a cone-cooling jacket was shown to be necessary to avoid localised heating in this area.

A cone-cooling jacket was shown to be necessary for storage at conditioning temperatures to avoid localised heating of the yeast and beer in the cone. In this case, in the absence of cone cooling, the main body of the beer may be cooled to a temperature lower than the inversion point and thus, it tends to rise. Yeast sediments in the cone and generates metabolic heat, which cannot be dissipated because of the lack of downward convection. The positioning of thermometers was also important. If there was no cone cooling and temperature was controlled by a single thermometer located just above the cone, the rising warm stratum activated the wall cooling jackets leading to overcooling of the main body of the beer. In this case, it was better to have two wall-cooling jackets with the controlling thermometer located mid-way between them. Thermometers should not be located at the top of the vessel because of the possibility that they may be inadvertently exposed to the headspace gas. In this case, over-cooling and ice formation was possible. Thermometer probes must be long enough to project into the main body of liquid, and therefore avoid local cold currents close to the cooling jackets.

Where vessels were used solely for cold conditioning and were filled with beer that had been pre-chilled, satisfactory attemperation could be achieved with cooling jackets located at the base of the vessel. If the beer was delivered warm to the conditioning vessel a second jacket positioned mid-way up the vessel was recommended. In both instances, the controlling thermometer should be placed mid-way up the vertical wall. Obviously, if vessels are used for both primary fermentation and cold conditioning it is necessary to have a combination of wall and cone jackets as well as multiple thermometers in appropriate locations. Such vessels require complex control systems, which activate combinations of cooling jackets, and controlling thermometers, depending on which cooling duty is called upon. The number and positioning of jackets required for cylindroconical vessels with different cooling duties are summarised in Fig. 5.19.

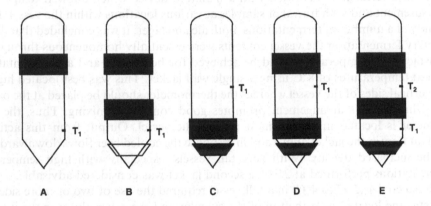

**Fig. 5.19** Positioning of cooling jackets and thermometers ($T_1$, $T_2$) for cylindroconical vessels: A, used for primary fermentation only (low-temperature lager-type fermentation); B, used for primary fermentation only (high-temperature ale-type fermentation); C, used for cold conditioning only (centrifuged beer, precooled on receipt); D, used for cold conditioning only (centrifuged beer, not pre-cooled on receipt); E, dual-purpose vessel suitable for all duties including single tank fermentation and conditioning.

The sole use of wall cooling jackets, particularly for cold conditioning, has obvious disadvantages because of the lack of mixing. This fundamental drawback can be ameliorated by providing forced agitation in the vessel. Small mechanical agitators, as illustrated in Fig. 5.11, may be fitted for this purpose; however, care must be taken to ensure that the surfaces are cleaned properly by the CiP system. A less invasive approach is to use gas agitation. Ladenburg (1968) reported that injection of liquid carbon dioxide into the base of a cylindroconical fermenter produced sufficient agitation such that the same rate of heat transfer was obtained as with a mechanical turbine impeller. Furthermore, cooling was assisted by evaporation of the liquid carbon dioxide. Similarly, Masschelein and colleagues (Lemer *et al.*, 1991) showed much reduced cooling times to conditioning temperatures in fermenters injected with gaseous carbon dioxide. In this case slow addition rates were used in a gas lift approach, which, it was claimed, produced circulation currents similar to those produced by thermal convection. Despite the attraction of this approach to improving temperature control, it is yet to see widespread adoption. It does have the disadvantage that it may lead to over-carbonation, thereby requiring further adjustment of carbon dioxide levels during subsequent processing.

### 5.4.3 *Dual-purpose vessels*

As discussed in the previous section, cylindroconical vessels with steep cones may be used for both primary fermentation and cold conditioning. Although this may be a single tank operation (Harris, 1980), this use is comparatively rare. There are, however, a small number of alternative large-capacity vessels that were designed specifically for combined fermentation and conditioning. These were developed during the 1960s and '70s to cope with the demand, at the time, for cost-effective increased capacity and accelerated throughput.

The single tank concept is contentious and many brewers consider that fermenting and conditioning should be distinct operations performed in dedicated vessels. Dual-purpose vessels are maybe over-engineered since they must cater for all duties. Thus, conditioning vessels are essentially very simple since they need do little more than hold the green beer at a low temperature. The presence of facilities for yeast cropping, carbon dioxide collection, monitoring and control associated with primary fermentation are unnecessary and potential hygiene hazards. Conversely, the cooling capacity of dual-purpose vessels must be sufficient to achieve and hold at sub-zero conditioning temperatures. This degree of cooling is not needed during primary fermentation and indeed the temperature differential between coolant and wort is so great that there is a risk of thermal shock to the yeast. Furthermore, the transfer between separate vessels allows greater flexibility. Thus, there is more opportunity for blending different batches of green beer, in-line additions may be made and in-line cooling is possible.

There are advantages to a single-tank operation, the most important of which is the reduction in total process time, compared to the two-tank approach. There is an element of flexibility in that it is not necessary to decide on what is the ratio of fermenting to conditioning vessels most appropriate to any particular brewery. Instead, calculations may be based simply on total capacity requirements and individual batch size. The single tank approach avoids the intermediate tank-to-tank transfer and therefore avoids the possibility of oxygen pick-up. Losses associated with vessel emptying are halved as are the number of cleans. Maule (1976) reported that cooling to conditioning temperature in a single tank was quicker and resulted in less temperature stratification than chilling of similar beer after transfer. Fricker (1978) considered this was due to beer stratification. In the single tank, convection currents are generated by evolution of carbon dioxide released from deeper saturated strata moving upwards to regions of lower pressure when cooled to below the inversion point. Presumably, during the tank-to-tank transfer the stratification would be destroyed, and, on subsequent cooling, convection current flow would be less dramatic.

In-line chilling during tank-to-tank transfer removes much of the cooling duty of the conditioning tank, as discussed in the previous section. However, it does make large demands on the brewery refrigeration plant. Luckiewicz (1978) described a brewery operation in which beer cooled in fermenter to 5°C was transferred to a conditioning tank via a heat exchanger. Transfer time was 4 hours, during which the beer temperature was reduced to 0°C. It was found that during this operation the refrigeration load for the brewery increased four-fold over all other duties. This necessitated a high capital investment to cope with a relatively transient peak demand. When the transfer time was extended to 8 hours, the differential between the average and peak refrigeration loading was reduced to a much more acceptable level.

Ultimately the choice between the single- or two-tank approach depends on the throughput and batch size requirements of the individual brewery. Undoubtedly the single-tank approach is more suited to large batch sizes and just a few beer qualities. In smaller breweries, or where there is a need to produce short runs of several different beer qualities, the dual-tank approach probably offers more flexibility.

5.4.3.1  *Asahi vessels.*  Asahi vessels, developed by the Japanese brewing company of the same name during the mid-1960s, were the first designed purposely for uni-tank operation (Takayanagi & Harada, 1967). Apart from Japan, Asahi vessels were installed in many United States breweries, also during the 1960s and '70s, where they were seen as a cost-effective method for increasing fermentation and conditioning capacity (Lindsay & Larson, 1975). These authors reported on the installation at the Genesse Brewing Company, New York State, of two Asahi-type vessels, each with a capacity of 10 600 hl. These were two of the largest known fermentation vessels ever to be constructed.

Vessels are made from stainless steel and are cylindrical. The total capacity is 5000 hl and the aspect ratio is close to 1, the internal diameter being 7.5 m and the height 11.8 m. The characteristic feature of the vessel is the internal base, which slopes towards the exit main to facilitate removal of yeast and other solids (Fig. 5.20). Attemperation is provided by jackets in the wall and base. The jackets are surrounded by insulation and an outer weatherproof skin. For single-tank operation additional cooling is obtained by circulating the vessel contents through an external heat exchanger. The circulation loop also includes a centrifuge by which means the yeast loading may be reduced. Circulation of the vessel contents by-passing the centrifuge and chiller is also possible. The point of beer re-entry consists of a stainless steel pipe, which is pivoted. This arrangement allows improved control of the dissolved carbon dioxide concentration.

Operation of the single tank operation was described by Amaha *et al.* (1977). Prior to filling the vessel, wort cold break was removed using a flotation treatment. All the yeast, which was a weakly flocculent and highly attenuating type, was pitched with the first wort addition. Vessel filling required five brewlengths and this took 20 hours. Fermentation and conditioning was of the traditional lager type, being lengthy and using comparatively low temperatures. The initial temperature was 6°C and during primary fermentation this rose to 9°C then decreased to 5°C. At the end of primary fermentation, which took 8 days, the green beer was pumped via the in-line chiller and centrifuge to both accelerate cooling and reduce the yeast count. This process was continued for 7 hours, with a pumping rate of 350 hl h$^{-1}$, and during this time the temperature fell between 2 and 3°C and the suspended yeast count from *c.* 55 to *c.* 25 million cells ml$^{-1}$. The green beer was then matured using a traditional lagering process for a further period of some 30 days. During this time the temperature was further reduced to $-1$°C. Mixing during the lagering period was improved by pumping the beer through the external loop by-passing the chiller and centrifuge. At the end of lagering approximately 80% of the yeast, which was suspended at the end

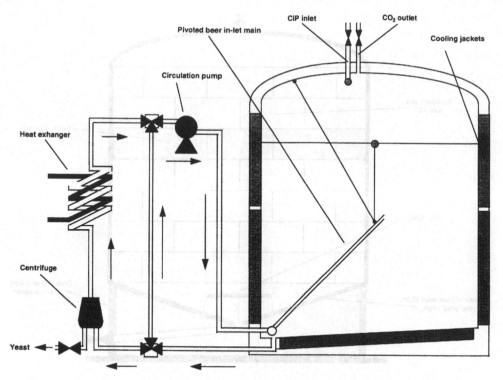

**Fig. 5.20**   Schematic representation of an Asahi vessel (adapted from Amaha *et al.*, 1977).

of primary fermentation, was shown to have formed a compact sediment that was easily removed from the vessel.

Amaha *et al.* (1977) reported that based on five years' experience, beer made using a single-tank method in Asahi vessels was indistinguishable from that made by the two-vessel approach. The single-tank method was shown to be the most cost effective. Thus, initial capital and subsequent running costs were reported to be 88% and 65% respectively of those incurred by the conventional two-tank method.

5.4.3.2   *The Rainier uni-tank.*   Like the Asahi vessel the uni-tank also arose during the 1960s, in this instance at the Rainier Brewing Company of Seattle, Washington State, USA. Similarly, they were also designed specifically for single-tank fermentation and conditioning, hence the name which is a contraction of 'universal tank' (Fig. 5.21).

The reasoning which led to the design of this vessel is described by Knudsen and Vacano (1972). The optimum tank size, taking into account production requirements and the capacity of the brewhouse, was considered to be 5500 hl. The intention was to enclose this volume within a tank of a geometry that used the minimum quantity of construction materials. A spherical configuration was discounted on the basis of prohibitive construction cost, and therefore a cylinder with an aspect ratio close to 1 was chosen. To maintain the low aspect ratio and again keep costs to a minimum, the vessel was fitted with a shallow conical base with an included angle of 12.5°.

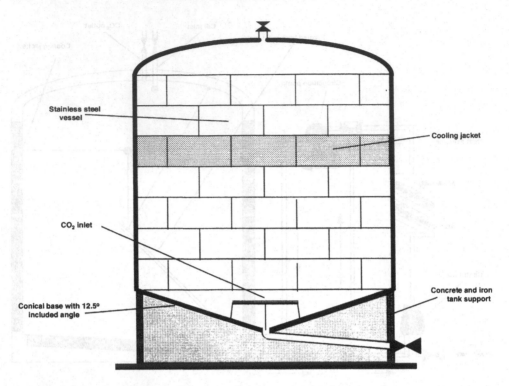

**Fig. 5.21**   Rainier uni-tank (redrawn from Knudsen & Vacano, 1972).

The vessel was constructed from readily available stainless steel plates, each 1.2 m wide. Using seven courses of these gave a tank with a straight side measurement of 8.5 m and the same internal diameter (8.5 m × 8.5 m). Cooling was provided by liquid ammonia fed to a dimpled jacket mounted towards the top of the vertical wall. Insulation was provided by a 15 cm layer of polyurethane foam, held in place by an outer weatherproof aluminium skin. The coolant capacity was capable of reducing the temperature by *c*. 5.6°C in 24 hours, starting from 13.3°C. This was shown to be adequate for attemperation during the most active period of the fermentation. Thus, although the aspect ratio of the vessel was chosen to minimise the use of construction materials, it was still apparently capable of providing an adequate cooling area, in this case *c*. 0.005 m$^2$ hl$^{-1}$. Cooling to the cold conditioning temperature was assisted by injection of carbon dioxide. This was achieved by directing the flow of gas upwards through a ring located in the centre of the tank at the base of the vertical wall. In addition to promoting convective cooling, the gas flow assisted in sedimenting the yeast in the cone and stripping unwanted volatile beer components. The exhaust gas was either liquefied and collected or purified and re-introduced into the tank.

When used as a uni-tank the vessel was filled with 10 brewlengths of wort, representing 87% of the total volume. On some occasions, vessels were filled to over 90% of the total capacity. This high-volume utilisation rate was possible because it was claimed that the good mixing characteristics of the uni-tank, presumably a function of the low aspect ratio, resulted in only small generation of foam. Primary fermentation lasted for 3–4 days and was performed at 13°C. After a further 2–3 days' warm conditioning to reduce vicinal diketone levels, the beer was cooled to −1.7°C over a period of 6 days. The yeast was removed after primary fermentation and was suitable for retention for future pitching. The relatively shallow cone had no adverse effect on yeast cropping, at least with this strain. After beer removal, vessels were cleaned with cold gluconated caustic and hypochlorite via an automatic CiP system.

### 5.4.3.3 *Spheroconical fermenters.*

Spherical vessels offer many advantages. The sphere is the optimal geometry for enclosing the maximum volume within a minimal surface area, and therefore it is the most cost-effective shape with respect to usage of construction materials. In addition, it is an ideal shape for withstanding pressure. Furthermore, it would be suspected that a spherical fermenting vessel with wall cooling would naturally generate strong convection currents, and thereby facilitate good mixing and attemperation. Lastly, it presents a relatively easily cleaned interior and, compared to other vessels, a minimum surface area. Despite the misgivings of some, for example Knudsen and Vacano (1972), that the construction of spherical vessels would be precluded by the cost and difficulty of construction, others have successfully followed this route to fruition. Thus, the Aguila Brewery in Madrid has designed and installed such vessels to be used in a uni-tank process (Martin *et al.*, 1975; Posada, 1978).

The principal features of the vessel are shown in Fig. 5.22. Since it would be very difficult to crop yeast from an entire sphere, a conical base is fitted. The vessels are constructed from stainless steel, surrounded by 220 mm thick foam insulation and an exterior coating of epoxy resin. Cooling is provided by wall jackets through which a 25% aqueous solution of propyleneglycol at −4°C is circulated. The jackets are arranged in four rings around the spherical part of the vessel. In addition, there is a cone cooling jacket. The total cooling surface is 150 m$^2$. The capacity of the vessels is 5000 hl. The diameter of the sphere is 10 m and the height, including the cone, is 11.95 m.

Martin *et al.* (1975) described a uni-tank process using spheroconical vessels. The wort had a gravity of 11.4°Plato, the initial dissolved oxygen concentration was 3–5 ppm, the starting temperature 12°C and the yeast pitching rate 30 × 10$^6$ cells ml$^{-1}$. During primary fermentation, which lasted for 4 days, the temperature rose to 14°C. After this time, the green beer was cooled to 8°C, over a period of 20 hours. The yeast was a highly flocculent type, which facilitated sedimentation in the cone such that, unlike the Rainier uni-tank, it was not necessary to use a centrifuge to reduce the cell count at the end of primary fermentation. After removal of the yeast crop, the beer was further cooled to 0°C and held for 21 days for maturation. During cold conditioning, the carbonation was adjusted to 5.5 g l$^{-1}$. This process was assisted by the ability to pressurise the vessel. It was noted that carbonation levels at the end of cold conditioning were, respectively, 5.3 and 5.7 g l$^{-1}$ at the top and bottom of the vessel. This relative homogeneity was taken as evidence of the excellent mixing characteristics of spheroconical vessels.

As with the Rainier uni-tank, beers produced in spheroconicals using a single-tank process were indistinguishable from similar beers produced in conventional fermenters and cellar tanks. However, it was observed that it was necessary to reduce hop

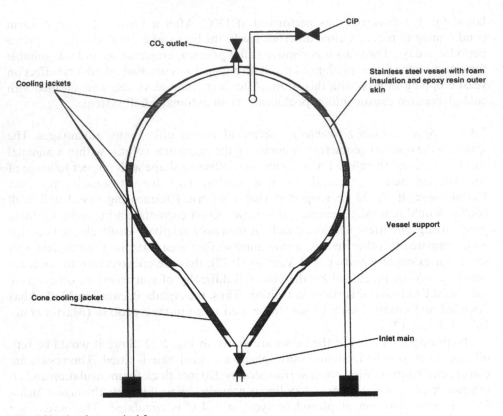

**Fig. 5.22**  A spheroconical fermenter.

rates by 12% in the uni-tank process. This improved utilisation was thought to be due to a reduction in the loss of bittering components on the yeast as a result of reduced foaming in the spheroconical vessel.

## 5.5  Accelerated batch fermentation

Several approaches have been suggested for increasing vessel productivity by manipulation of the physical and biochemical parameters that regulate the progress of fermentation. These may be loosely grouped, as others have done, under the heading of accelerated batch fermentation. The simplest methods require no special modification to vessels but merely changes to the way in which they are managed. In this respect, they are not distinct fermentation systems; however, mention is made here for the sake of completeness. A detailed discussion of the effects on performance of fundamental control parameters is given in Chapter 6. These simple methods are attractive economically since they offer the promise of increase in productivity from existing plant.

Batch fermentations, depending on the type, are made up of some combination of the stages given below:

(1)  vessel filling;
(2)  primary fermentation (lag phase, exponential phase and phase of decline);
(3)  warm conditioning;
(4)  warm yeast cropping;
(5)  chilling;
(6)  cold bottom yeast cropping;
(7)  vessel emptying;
(8)  CiP;
(9)  conditioning tank filling;
(10)  cold conditioning;
(11)  vessel emptying;
(12)  CiP.

Clearly, some of these operations may be simultaneous, or overlapping, for example, primary fermentation and yeast skimming from a top-cropping ale fermentation. However, since they all contribute to the total vessel residence time it is useful to consider, albeit briefly, how the time taken to accomplish each phase may be shortened.

Vessel filling, chilling and emptying are largely pre-determined by the design of the fermenter and associated plant and little improvement is possible to existing installations. Manual yeast cropping procedures may be replaced by automated methods that may be more rapid and efficient, as described in Chapter 6, although again the actual time savings are likely to be modest. Where a period of cold conditioning is required, the time-savings which accrue from the use of a uni-tank approach have been discussed previously (Section 5.4.3). By default, therefore, the most profitable approach to reducing vessel residence time is to shorten the time taken for primary fermentation (and warm conditioning, if required).

The parameters that influence the duration of primary fermentation are wort composition, temperature, initial wort dissolved oxygen concentration, yeast pitching rate, strain-specific yeast properties, pitching yeast physiological condition and physical factors pertaining to the vessel and its management. Yeast strain-specific properties include flocculence character, cell surface hydrophobicity and oxygen requirement, amongst others. Physical factors would include parameters such as pressure, dissolved carbon dioxide concentration and the extent of mixing during fermentation. All of these interact in a complex manner and together determine the outcome of fermentation.

It may be readily appreciated that most of these factors could not be classed as process variables. Manipulation of wort composition to effect improvements in vessel productivity, as in high-gravity brewing, is discussed in Section 2.5. Deliberate alteration of yeast strain-specific characteristics by, for example, genetic manipulation, to produce variants with more desirable brewing properties could be a route to accelerating batch fermentation (see Section 4.3.4). This leaves dissolved oxygen concentration, yeast pitching rate, pitching yeast physiology, temperature and other physical parameters as being amenable to manipulation.

Increasing pitching rate, temperature and dissolved oxygen concentration all result in a more rapid primary fermentation, mainly by increasing the rate of wort attenuation in the exponential phase, although also by shortening the duration of the initial lag phase. The reduction in primary fermentation time can be very dramatic providing the parameters are increased to a sufficient degree. However, the apparent benefits are somewhat illusory since significant rate improvements can usually be obtained only at the expense of unacceptable changes in beer flavour and other quality parameters. Thus, changes in pitching rate and dissolved oxygen concentration, in particular, influence the extent of yeast growth during fermentation. Concomitantly, this will also change the spectrum of flavour-active metabolites whose concentration in beer is related to yeast growth.

With some fermentations, it is possible to gain significant advantage by modest increases in temperature alone. This is because temperature exerts most effect on yeast growth rate and not growth extent. However, this strategy must also be treated with caution since some lager beers, traditionally fermented at low temperature, can lose much of the subtle flavour character if too high temperatures are used. It has been suggested that the use of increased pressure can ameliorate the negative flavour effects due to elevated temperature whilst retaining the gains in time saving (Kumada et al., 1975).

Primary fermentation can be shortened by the use of improving vessel agitation, either mechanically or by gassing with carbon dioxide. As with the other approaches some caution is required to avoid changes in beer flavour. Masschelein (Haboucha et al., 1967; Masschelein, 1986a) reported that forced agitation resulted in increased yeast growth and concomitant reduction in esters and elevated levels of higher alcohols. The same author (Masschelein et al., 1993) suggested that the problem could be overcome by using a fed-batch approach. In this pilot scale study, stirred batch growth was allowed to proceed until the wort was 40% attenuated, at which point the wort feed was started. Wort was added to the culture at an exponential rate to keep the residual fermentable extract at a constant 40% attenuation. This approach allows yeast growth rate to be regulated by the medium feed rate, as in a continuous system (see Section 5.6). In this case, yeast growth extent was reduced to that measured in a comparable unstirred batch fermentation. By inference the production of flavour-active metabolites dependent on yeast growth could also be manipulated by this method whilst retaining advantages in productivity. Application at commercial scale remains to be demonstrated.

## 5.6  Continuous fermentation

The use of very large combined fermentation and conditioning vessels in a uni-tank process is in essence a strategy which seeks to increase productivity by a combination of shortening process times and reducing capital costs. Thus, the uni-tank process is more rapid than the traditional separate fermentation and conditioning tank approach and with very large vessels there is an opportunity to minimise the ratio of surface area to volume. An alternative approach is continuous fermentation in which a single vessel (or group of linked vessels) produces a continuous beer stream. In this case, high productivity is possible because the downtime and costs associated with vessel filling, emptying and cleaning of the batch process are much reduced. Continuous fermentation should be very efficient since beer losses are low, conditions are arranged such that rates of fermentation are rapid but yeast growth is low and volume output is high relative to the capital costs of the plant. When fully operational with a stable flow rate, the beer should be of consistent quality. The degree of consistency should be better than that achievable with batch fermentations since the variability due to inconsistency in the physiological condition of individual batches of pitching yeast is eliminated.

Continuous beer fermentation is not a new concept. Kleber (1987) describes a system, that of Van Rijn, which was patented in 1906. Similarly, Ricketts (1971) referred to continuous beer fermentation systems which date from the end of the nineteenth century. However, the period of most intense interest in this approach was during the 1960s, coincident with the introduction of large uni-tanks. At this time several systems were developed, some to the point of commercial exploitation. Subsequent experience showed that few of the anticipated benefits of continuous fermentation were actually realised and most breweries, with some notable exceptions, have continued to use the traditional batch approach albeit using large-capacity vessels.

The reasons for the abandonment of continuous fermentation were given by Portno (1978) and with some caveats, they remain valid today. The major strength of the batch system, using several vessels, is that it is flexible. It is able to cope with both seasonal, or shorter-term fluctuations in demand and can easily be adapted to vary the spectrum of production of several beer qualities. Conversely, the benefits of continuous fermentation are only realised fully when the systems are operated for long periods with minimum downtime for changes in beer quality. In this respect, they are suited only to breweries producing a single or very limited portfolio of beer qualities. The inflexibility extends to rates of beer production since this parameter can only be varied within small limits with most continuous systems.

Most breweries have been designed to work within the confines of the batch system. Brewhouses are sized to provide sufficient wort to fill the largest fermenters within a convenient period of time. With a continuous system the rate at which wort is produced must be sufficient to supply the needs of the fermenter at all times. Inevitably, this requires some type of pre-fermenter wort collection vessel. Downstream of the fermenter the brewery must be capable of handling a continuous supply of green beer. The consequences of failure in a continuous system pose a serious threat to production. Thus, emptying, cleaning, start-up times and establishment of stable running conditions are lengthy procedures, possibly taking as long as two weeks. The prolonged nature of continuous fermentation has inherent risks. Extended running times increase the opportunities for microbial contamination, and, perhaps more seriously, yeast 'variants' (Section 4.3.2.6) may be selected for with the concomitant risk of undesirable changes in beer quality. Continuous systems are more sophisticated than many brewery batch fermenters, and therefore skilled personnel must be on-site, night and day, to provide technical support. The advantages and disadvantages of continuous fermentation compared to the batch process are summarised in Table 5.6.

Notwithstanding the disadvantages outlined previously, there has in recent years been a resurgence of interest in continuous fermentation. Where fermentation capacity is simply to be extended within an existing brewery, already using batch

**Table 5.6** Advantages and disadvantages of continuous beer fermentation compared to a batch process.

| Advantages | Disadvantages |
|---|---|
| • Rapid rates of conversion of wort to beer. | • Lack of flexibility with respect to number of beer qualities. |
| • High efficiency fermentations<br>　– Low yeast growth<br>　– High ethanol yield. | • Limited ability to vary output rate. |
| • Efficient vessel utilisation. | • Complex costly vessel(s) and ancillary plant. |
| • Only one or few fermenting vessels needed. | • Requirement for continuous skilled technical support. |
| • Consistent beer. | • Very long start-up times. |
| • Reduced beer losses. | • Serious consequences of break-down or microbial contamination. |
| • Fewer cleans. | • Possibility of selection of yeast mutants. |
| • Reduced usage of detergents/sterilants. | • Wort production and beer processing capability must match needs of fermenter. |
| • Reduced need for pitching yeast storage. | • May not be suitable for all beer qualities. |

vessels, there is still little chance that a continuous fermenter would be chosen. However, the approach may now be given serious consideration if a new brewery is to be constructed. The renaissance of interest is a result of advances in hygienic design and facilities for control. These allow a continuous system to be used with a fair degree of confidence that downtime due to plant failure or contamination will be virtually non-existent. More importantly, between the 1960s and the present day, there have been technical advances that have broadened the spectrum of fermentation-associated processes to which continuous systems may be applied. These advances allow improvements to process efficiency that are not possible to achieve with a batch system.

The crucial step forward in continuous technology has been the development of commercial immobilised yeast reactors. This approach has generated sufficient interest to form the subject of an entire European Brewing Convention Symposium – 'Immobilised Yeast Applications in the Brewing Industry' held in Finland in 1985. This technology is discussed in detail in Section 5.7; however, the principal advantages of immobilised reactors are that the very high yeast concentrations which are achievable allow very rapid process throughputs. This is of particular benefit when applied to rapid beer maturation. Thus, a single immobilised yeast reactor can eliminate the often very time-consuming warm conditioning step associated with lager fermentation. In many breweries, this can result in reduction in fermentation process times of a few to several days, with consequent improvements in the overall productivity of large capacity vessels. The application of immobilised yeast to beer maturation need not be restricted to new breweries; it may be retro-fitted within any site as an added process at the end of conventional batch primary fermentation.

Immobilised yeast reactors have also found use in new fermentation processes, for example, in the production of low- or zero-alcohol beers. Thus, immobilised reactors are particularly suited to the approach where primary fermentation is restricted to a brief contact between wort and yeast, which serves primarily to remove wort flavour character with the formation of little or no ethanol.

### 5.6.1   Theoretical aspects

Microbial cultures may be characterised as being of either 'closed' or 'open' type. In the former, as typified by a batch culture, some component that has a positive or negative influence on growth, is contained within the system. An example of such a critical component would be an essential nutrient, or a growth-inhibiting metabolite. In consequence, the conditions within a closed system are in a continual state of transition. The growth rate will always tend towards zero as the essential nutrient becomes exhausted or the concentration of a metabolite increases to a toxic level. In contrast, continuous fermentations are open systems in which by definition components may freely enter or leave. In consequence, it is possible, by the continuous replenishment of nutrients, or removal of potentially toxic metabolites, to produce conditions where the microbial growth rate and by inference the rate of production of metabolites, is constant.

Pirt (1975) describes two basic types of continuous culture, the plug flow system and the chemostat. In a plug flow system (Fig. 5.23(a)) the reactor takes the form of a coil or similar elongated form. Inoculum and growth medium are mixed at the point of entry and fed simultaneously and continuously into the reactor. Within the reactor, the conditions are arranged such that there is a minimum of backward and forward mixing, and thus batch growth proceeds within each discrete 'plug' as it travels through the reactor. The reactor may therefore be viewed as a continuum of batch cultures in which spatial location is related to culture age. The factors that regulate growth rate in a conventional batch culture such as temperature, inoculation rate and substrate concentration, are also influential in a plug flow continuous culture. In addition, the composition of the culture issuing from the reactor will also be a function of the flow rate. By careful regulation of all of these parameters, it is possible to establish a steady state where the product is of a constant and desired composition.

A further refinement that may be introduced to a plug flow reactor is biomass recycling. In this case, a means is provided for separating and concentrating some of the microbial cells from the product stream issuing from the reactor. This biomass is returned to the entry point of the reactor where it is used as inoculum. Used in this way the reactor requires only to be supplied with fresh medium.

A true plug flow continuous reactor is a theoretical entity only as there will always be some degree of backward and forward mixing. In order to turn the concept into practical reality some compromise is inevitable. This can be achieved by using multiple tanks, which provide physical separation of individual stages of the batch culture. The extent to which this arrangement approaches the ideal is a function of the number and size of each individual tank.

The second open continuous culture system is the chemostat (Fig. 5.23(b)). This consists of a stirred reactor to which medium is introduced by an entry pipe. The rate of medium addition is controlled by a variable speed pump. Culture is removed from the reactor via a second pipe, which is arranged in the form of a siphon or weir such

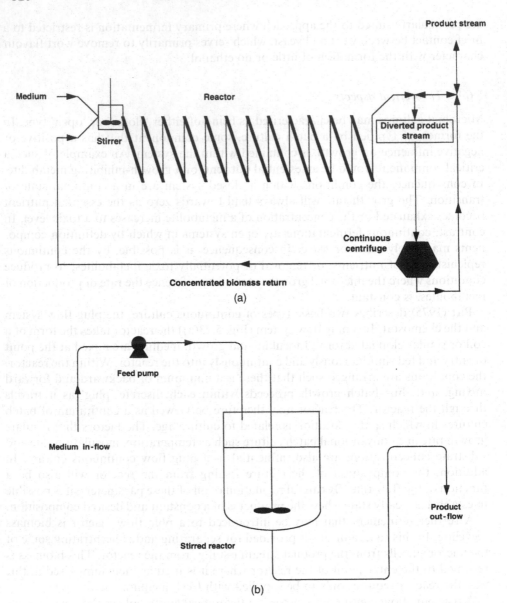

**Fig. 5.23** Continuous fermentation systems. (a) Plug flow reactor (shown with biomass feed-back). (b) Continuous fermentation systems – Chemostat.

that the culture volume within the reactor remains constant. Unlike the plug flow reactor, the chemostat vessel is inoculated once only and when batch growth has commenced the in-flow of fresh medium is initiated. It is assumed that mixing is perfect so that in-flowing medium is instantaneously and homogeneously distributed throughout the reactor.

The conditions within a chemostat are governed by two opposing effects: growth on the nutrients in the in-flowing fresh medium and dilution due to loss of culture through the exit tube and its replacement by fresh medium. Mathematically this may be explained as follows:

Growth in any culture is characterised by the general growth equation:

$$N_t = e^{\mu t} N$$

where $N_t$ is the population at time t; N is the population at t = 0; $\mu$ is the specific growth rate.

The specific growth rate ($\mu$) is defined as the rate of growth per unit amount of biomass and has the units of reciprocal time. The specific growth rate of a micro-organism in a culture is controlled by environmental factors such as nutrient supply and temperature. However, under ideal conditions, where growth is not limited by any external factor, each organism has a maximum specific growth rate ($\mu_{max}$) that is determined by the genotype.

Returning to the chemostat; from the general growth equation, the instantaneous growth rate of the population is given by:

$$\frac{dN}{dt} = \mu N$$

The rate of loss of cells being removed from the growth vessel is:

$$\frac{dN}{dt} = -DN$$

D is the dilution rate and is equal to f/V, where f is the rate of addition of medium and V is the volume of culture in the vessel. It also has the unit of reciprocal time.

The net change in population is therefore:

$$\frac{dN}{dt} = \mu N - DN \text{ or } N(\mu - D)$$

If the dilution rate (D) is greater than $\mu_{max}$ the above expression will have a negative value and the population in the chemostat will tend, with time, towards zero, termed 'wash-out'. If the dilution rate is less than $\mu_{max}$ the value of the expression is positive and the population size will increase. The increase in population density increases the rate of depletion of nutrients in the in-flowing medium and eventually the concentration of an essential nutrient will fall to a growth limiting value. In consequence, the specific growth rate falls until it is equal to the dilution rate. At this point dN/dt = 0 and a steady state is achieved, the population density (and composition of the spent medium) being constant. Different steady states can be achieved by varying the dilution rate between zero and some fraction of $\mu_{max}$. At each steady state the concentration of the growth limiting nutrient within the reactor will be zero. It follows, therefore, that apart from varying the dilution rate to regulate growth rate within the chemostat, it is also possible to exert control by varying the composition of the growth medium. The productivity of a micro-organism growing in a chemostat is always greater than that achievable using the same combination of organism and medium in a batch culture of equal volume. This is because, in the former, growth rate remains constant throughout, as there is no lag or stationary phase.

Some embellishments to the basic design and operation of chemostats are possible. One approach is the reverse of the conventional chemostat. Thus, a steady state may be established by automatic self-regulation of the dilution rate in response to an input signal from a sensor which measures biomass concentration or a growth-related parameter. Examples include the turbidostat where biomass concentration is determined using a light scattering spectrophotometric approach. Growth related parameters could be rate of carbon dioxide evolution, rate of oxygen consumption, exothermy or the concentration of any other growth related metabolite for which a suitable sensor exists. The advantage of the turbidostat, or related approach, is that it is possible to establish steady states where all essential nutrients are present in excess. This may provide a means of exerting greater control over the composition of the product stream than is possible with the simple chemostat.

It is possible to use a biomass concentration loop similar to that shown for the plug flow reactor shown in Fig. 5.23(a). In this way, the rate of formation of product and biomass can be increased to higher levels than are achievable with a simple chemostat. Another alternative is to use multiple chemostat vessels linked in series. In this arrangement, the outlet stream issuing from the first vessel may be augmented with additional nutrients as it is fed into the second vessel. Multi- stage chemostats allow the establishment of distinct steady states in each growth vessel. Like the single stage chemostat, the mutli-stage variety may also be fitted with a biomass concentration and feed-back loop system.

The types of continuous systems that may be used for brewing, have been classified by Portno (1970). Based on an earlier classification of Herbert (1961) these may also, somewhat confusingly, be differentiated into open or closed types. In this case, open types are those in which no attempt is made to retain yeast cells and the process flow issuing from the fermenter is of the same composition as that inside. Conversely, closed types have some method which separates the issuing beer from the yeast, the latter being restrained within the fermenter. Portno (1970) extended this classification to include another category, partially closed, defined as where the concentration of cells escaping is less than the concentration retained in the fermenter but greater than zero. Partially closed systems can be quantified by the retention (or closure) index (R %), defined as the restriction over escape of cells from systems in steady state:

$$R = \frac{(1 - y)}{x} 100 = \frac{(1 - \mu)}{D} 100$$

where y = concentration of cells in the effluent; x = concentration of cells in the fermenter; $\mu$ = specific growth rate; D = dilution rate.

A further division of systems is possible based on the extent of mixing within the fermenter. Where mixing is assumed to be perfect, as in the simple chemostat, the system is described as homogeneous. Conversely, where a gradient of conditions exists between the point of entry and exit, the system is defined as heterogeneous. As with the partially-closed category Portno (1970) also recognised partially-heterogeneous systems in which linked series of tanks could include elements of both homogeneity within an individual vessel yet maintaining a fermentation gradient throughout the entire system. Portno (1968a) also distinguished two classes of heterogeneous closed systems, namely, single and two-phase. The former describes a plug flow arrangement in which cells and substrates/products move simultaneously through the reactor. In the latter, yeast and substrate/product traverse the fermentation system at different rates. Closed and open, homogeneous and heterogeneous systems are, of course, not mutually exclusive.

The requirements of an ideal brewing continuous fermentation system were discussed by Portno (1978). They remain a useful set of criteria and are reiterated here. The first three points relate to performance of the system with respect to productivity and flexibility. Thus, a continuous system should be capable of maintaining a high yeast concentration within the reactor in order to ensure a constant fermentation rate close to the maximum rate of a batch fermentation. Preferably, the system should function with any yeast strain. Output from the fermenter should be variable within reasonable limits in order to retain at least some of the flexibility of the multi-tank batch approach.

The remaining criteria relate to the need for the sum of the conditions within the continuous system to be sufficiently similar to those that occur within a batch culture so that acceptable beer is produced. Many important beer flavour compounds are produced in proportions that are relative to the extent of yeast growth. It follows that should conditions within a continuous fermenter deviate greatly from those in a batch culture the resultant beer is likely to be equally non-standard. With a homogeneous system, particularly where the retention index is high, it is difficult to control yeast growth such that its extent is similar to that of a batch system. Therefore, the most important of this second group of criteria is the need for a fermentation gradient to be established within the continuous fermenter. In other words, heterogeneous continuous systems offer the greatest likelihood of success.

In a batch fermentation, yeast assimilates wort sugars in an ordered fashion with sucrose and glucose being used first followed by maltose, the major wort sugar (see Section 3.3.1). Utilisation of maltose is inhibited in the presence of glucose.

In a homogeneous continuous system the continual in-feed of fresh glucose-containing wort disrupts maltose utilisation, and, therefore, a requirement of a batch fermentation – that all fermentable sugars are assimilated – is not fulfilled. In partially closed systems, these effects are compounded. Such systems are characterised by low growth rates, indeed the latter is directly proportional to the retention index. Induction of α-glucosidase is a growth-related process, and therefore, in partially closed continuous systems, the presence of glucose inhibits maltose utilisation and the low growth rate restricts the synthesis of the enzyme responsible for maltose hydrolysis. Under these circumstances, the presence of glucose in the in-feed can cause a progressive increase in the concentration of maltose in the fermenter.

In some closed or partially closed continuous fermenters, flocculent yeast is used and it is this characteristic which retains the yeast. Maltose is a powerful deflocculating agent, and therefore where its assimilation is impaired its concentration may increase to a point where the resultant de-flocculated cells may be washed out of the fermenter. In both partially closed and open homogeneous continuous brewing fermenters the complex interactive effects of the varying spectrum of sugars may produce oscillations in the present gravity and yeast concentration (Portno, 1968a, b).

The lack of a fermentation gradient can have other undesirable effects. In a homogeneous continuous system, operating at a low dilution rate, the yeast has, in effect, to grow in the presence of beer. The resultant high concentrations of products and concomitant absence of wort substrates can affect yeast physiology, beer composition and produce instability in the fermenter. Thus, some yeast strains are pleomorphic under such conditions and may develop a pseudomycelial form (see Section 4.2.5.1). Such cells may not be able to flocculate and as in the case of the presence of maltose, wash-out may occur. Greater than usual degree of esterification of higher alcohols is possible where the concentrations of the latter are high, resulting in beer with abnormally high ester levels. In extreme cases, in the absence of wort sugars and under aerobic conditions, it is possible for the yeast physiology to become derepressed, again with consequent undesirable effects on beer composition. It is vital, therefore, to ensure that oxygen is provided at a point where sugar levels are still high.

### 5.6.2   Continuous fermentation systems

Several continuous brewing fermentation systems have been described in the brewing literature although the majority were restricted to the scale of the laboratory or occasionally pilot plant. A description of some of these, which may be termed experimental continuous systems, is given here since they appeared mainly during the late 1950s and '60s, the period of burgeoning interest in continuous fermentation. They are worthy of inclusion since they provide a record of the development stages of continuous brewing and they represent working models of some of the different classes of system as outlined in the previous section. The results of trials using these various models provided data on which the design of some of the later commercial systems were based.

Comparatively few systems were developed to production scale. For reasons of historical completeness, details of these are provided although most are now defunct. Only breweries in New Zealand still use a continuous process, albeit over a period of some 30 years and very successfully.

### 5.6.2.1   Experimental continuous fermentation systems.

Hough and Rudin (1958) described a simple system, shown in Fig. 5.24, which consisted of a series of linked stirred round-bottomed flasks. The first flask was connected to a wort reservoir via a reciprocating pump and the outflow from the final flask was directed towards a beer receiver. The contents of each flask were attempered by immersion in constant temperature water baths. This efficacy of this system was tested using either one flask or linked series of two or three flasks. Thus, it could be used either as a simple open homogeneous system, or as a partially-heterogeneous open system.

Results were presented of the effects of varying process parameters such as temperature, flow rate and wort gravity using a top-fermenting ale yeast strain. It was claimed that after a single inoculation with yeast, steady state conditions were achieved in each flask and these could be maintained over a period of several weeks.

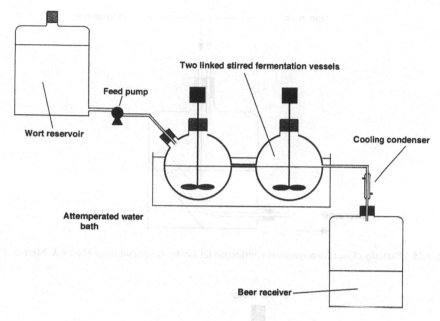

**Fig. 5.24**  Simple laboratory continuous fermenter (re-drawn from Hough & Rudin, 1958).

The resultant beer was similar in analysis and flavour to that derived from batch fermentations using similar wort and yeast. Two or three flask systems were found to be more efficient than the single flask arrangement. In a subsequent paper the effect of varying yeast strain was tested (Rudin & Hough, 1959). This study concluded that both top and bottom fermenting yeast strains could be used in the laboratory system to produce palatable beer. In some cases, where mixtures of strains were used, one strain became dominant over a period of time.

A simple partially-closed system was described by Harris and Merritt (1962). In this single vessel system, escape of yeast was restricted by placing the outlet tube within a secondary containing tube (Fig. 5.25). The contents of the flask were stirred magnetically, such that the bulk of the culture was homogeneous. However, in the region adjacent to the exit tube the lack of turbulence allowed yeast in this region to settle, and thereby be returned to the stirred culture. Provision was made for direct aeration of the culture.

A more sophisticated heterogeneous closed system was devised by Hough and Ricketts (1960). This apparatus (Fig. 5.26) may be viewed as the progenitor of the commercial tower continuous fermenter (Section 5.6.2.2). The fermenter consisted of two linked tubular sections, one vertical and the other inclined at an angle of approximately 25°. Wort is introduced at the top of the vertical tube and is mixed via a top-mounted motor and drive shaft arrangement. Fermenting wort and yeast move from the base of the vertical section into the adjacent inclined unstirred tube. The apparatus was designed specifically for use with a flocculent yeast strain. In the inclined tube section, the lack of mixing allows such yeast to settle out so that beer issuing from the exit line at the top of the inclined tube was shown essentially free of yeast.

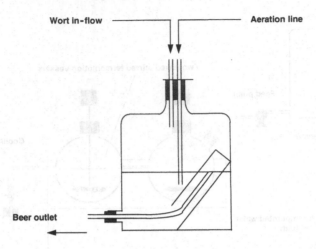

**Fig. 5.25**   Partially-closed homogeneous continuous fermenter (re-drawn from Harris & Merritt, 1962).

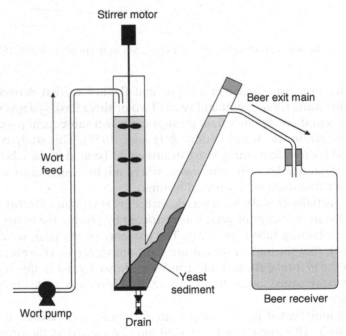

**Fig. 5.26**   Heterogeneous closed continuous fermentation system (re-drawn from Hough & Ricketts, 1960).

The system was tested with a number of yeast strains of varying flocculence. The time taken for steady state conditions to become established was directly proportional to the degree of flocculence of the yeast strain used. Similarly, increased flocculence was associated with the establishment of progressively higher yeast concentrations within the fermenter, less yeast in the effluent and higher rates of fermenter productivity. Productivity using a flocculent strain was significantly greater than that which could be obtained using a simple open homogeneous continuous system. Although the beers arising from the system were claimed to be palatable and a good match for a batch-made product, no detailed analytical data was provided. It was reported that both the concentrations of ethanol and nitrogen were elevated in the continuously made beer, perhaps suggesting that an exact match for the batch product had not been achieved.

Design of a pilot scale continuous fermenter was also provided. In essence this was a scaled-up version of that shown in Fig. 5.25. More detailed description of a pilot scale continuous facility was provided in a later paper (Hough *et al.*, 1962). It consisted of two stirred, attemperated cylindroconical vessels each fed with wort supplied from either of two stainless steel reservoir tanks at a rate controlled by two variable speed feed pumps (Fig. 5.27). Each wort feed pipe was fitted with a trap arrangement, which prevented back-growth from fermenter to wort reservoir. Each fermenter had a total working capacity of approximately 0.7 hl, but smaller volumes could be used by lowering the position of the exit main. If required the outflow from one fermenter could be coupled to the inflow of the second. Each fermenter was stirred by two electrically driven paddles. A third blade swept the liquid surface acting as a foam breaker.

If required, retention of yeast in the fermenters could be encouraged by the use of an adjustable sleeve, which fitted over the exit main and allowed the yeast to settle. The outflow from each fermenter was fed via a yeast separation vessel. This comprised a chilled stainless steel cylindroconical vessel fitted with an internal vertical partition. Beer and yeast was fed into the top of one compartment. Yeast settles in the

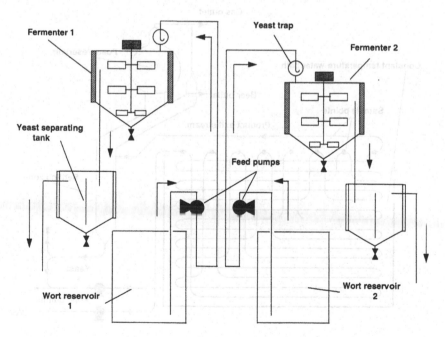

**Fig. 5.27** Pilot scale continuous fermentation system (re-drawn from Hough *et al.*, 1962).

conical base of the vessel and the beer passes under the base of the partition and out via an exit pipe located in the upper wall of the adjacent compartment.

Several laboratory continuous fermentation systems have been described in a series of papers by Portno (1967, 1968a, b). These were used to investigate fully the kinetics of continuous wort fermentation and form the conclusion that heterogeneous systems, which generate a fermentation gradient, offered the greatest likelihood of success, as described in Section 5.6.1.

The gradient-tube continuous fermenter was a plug-flow reactor and is illustrated in Fig. 5.28. It consisted of a tube, 70 metres in length, immersed in a constant temperature water bath. The tube had a diameter of $c.$ 0.5 cm, which was claimed to be sufficiently narrow to virtually eliminate back-mixing. Ideal plug flow behaviour was further assisted by the formation of carbon dioxide bubbles in the tube, which tended to partition the flow into a series of segments. Pasteurised wort, mixed with a desired proportion of yeast slurry, was fed into the tube and fermentation proceeded as the mixture passed through. Fermentation performance was gauged by analysis of process fluid removed from a number of sample points arranged at intervals throughout the length of the tube. Beer exiting from the tube passed into a device that separated yeast and beer into two streams. The yeast could be recovered from the base of the separating device and mixed with fresh wort and re-circulated back through the tube.

Performance of the gradient tube fermenter was governed, predictably, by temperature, flow rate and the proportion of yeast introduced with the in-flowing wort. Results, which were obtained from analysis of samples taken at points throughout the

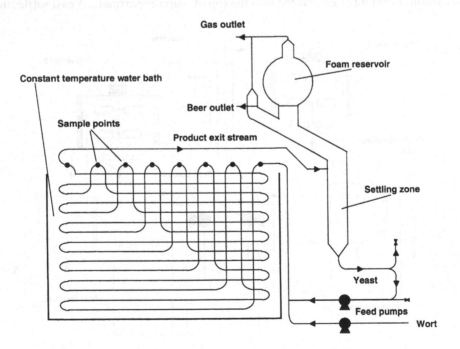

**Fig. 5.28**   A gradient-tube continuous fermenter (re-drawn from Portno, 1967).

length of the tube, confirmed that a gradient had been set up which mirrored the changes associated with a batch fermentation.

The centrifugal continuous fermenter was a partially closed type, which did not rely solely on flocculence for retention of yeast cells. In this case, yeast sedimentation could be controlled using an internal centrifugal rotor (Fig. 5.29). The fermenter consisted of a cylinder fitted with a variable speed stirrer and a bottom-located wort in-feed tube. The drive shaft of the stirrer was hollow and this provided a route for exit of beer. A centrifugal rotor was attached to the drive shaft, which contained a number of chambers, each contiguous with the hollow drive shaft.

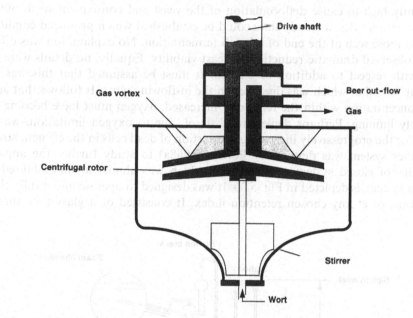

**Fig. 5.29**   Centrifugal continuous fermenter (re-drawn from Portno, 1967).

Beer exiting upwards through the drive shaft is subjected to a downward centrifugal force of a magnitude which is a function of the stirring speed. This force separates yeast from the beer and returns it to the fermenting vessel. It was observed that for any given yeast strain there was an optimal rotation speed at which yeast retention was maximal. At sub-optimal rotation speeds it was concluded that the centrifugal force was insufficient to retain yeast. At supra-optimal rotation speeds the yeast concentration in the outflow increased, due, it was considered, to yeast deflocculation caused by turbulence in the fermenter.

At very high yeast retention rates, it was not possible to achieve a satisfactory steady state. Thus, a trial was described in which the rotor speed allowed 96.5% of the yeast to be retained. After 2 days' batch growth on an all-malt wort of specific gravity, 10°Plato, continuous flow was initiated. The yeast concentration during this period increased from 10 to $20\,\mathrm{g\,l^{-1}}$ dry weight. During the first 4 days of continuous operation, the yeast concentration remained constant, after this time it progressively decreased, falling to approximately $8\,\mathrm{g}\,l^{-1}$ at day 11. The decrease in yeast concentration was accompanied by a progressive reduction in yeast viability, from close to 100% at the end of batch growth to less than 40% at day 11. As the yeast concentration decreased, there was a concomitant increase in levels of wort α-amino nitrogen. The specific gravity was constant and low throughout the first 4 days of continuous operation; after this time it increased to a value close to that of the in-flowing wort.

This instability was ascribed to the decrease in yeast growth rate and fermentative ability brought about by the high retention rate. As the growth rate decreased, the extent of assimilation of wort sugars also fell until the concentration of maltose was sufficiently high to cause deflocculation of the yeast and consequent wash out. At lower retention rates, a steady state could be established which produced conditions closer to those seen at the end of a batch fermentation. No explanation was offered for the observed dramatic reduction in yeast viability. Equally, no details were provided with respect to addition of oxygen. It must be assumed that this was at a constant rate, that which was dissolved in the in-flowing wort. It follows that as the yeast concentration within the fermenter increased, oxygen must have become progressively limiting. Perhaps depletion of sterol, due to oxygen limitation, was the reason for the progressively increasing proportion of dead cells in the effluent stream.

Another system was devised by Portno (1968a) to study further the apparent instability of closed systems. This device, which was almost an immobilised fermenting system, is depicted in Fig. 5.30. It was designed to operate under fully closed conditions, or at any chosen retention index. It consisted of a glass tube through

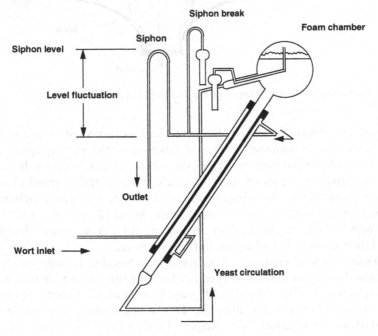

**Fig. 5.30**  Closed tubular homogeneous fermenter (re-drawn from Portno, 1968a).

which wort was circulated. Enclosed within the tube was a central resin-impregnated fibreglass tube, the end of which was attached to a glass bulb. The resin tube contained the yeast; however, pores, roughly 2 μm in diameter, allowed transfer of nutrients and metabolites between the inner and outer tubes. Yeast was circulated from the base of the inner tube and back into the upper bulb. The latter also served as a foam reservoir. In operation it was noted that a yeast deposit formed on the inner surface of the fibreglass tube. To prevent this, an automatic siphon back-flushing arrangement ensured that yeast was removed by the induced liquid washing action.

Using this apparatus, it was confirmed that it was not possible to achieve a stable steady state in a fully closed continuous system. Thus, with time, the yeast viability, as judged by methylene blue staining, gradually decreased. This was mirrored by a gradual increase in the specific gravity of the wort. Within these trends, both of these parameters also showed regular oscillations. Sugar analyses revealed that the oscillations in specific gravity were due principally to fluctuations in the concentration of maltose.

Since it appeared that open or closed homogeneous continuous systems were inherently unstable, Portno (1968b) devised a laboratory heterogeneous model fermenter to confirm that such an approach would be capable of producing satisfactory beer under stable steady state conditions. The apparatus, shown in Fig. 5.31, consisted of two linked stirred and attemperated flasks, the first of which was fed with a constant supply of pasteurised wort. The overflow from the second flask led to a chilled vessel in which the yeast was allowed to form a sediment. The overflow from the sedimentation vessel was attached to a further reservoir in which surplus yeast collected and from the top of which, beer was taken for analysis. The sedimentation vessel was also attached to another vessel in which yeast could be aerated and then recycled into the first fermenting vessel. The sedimentation, aeration and surplus

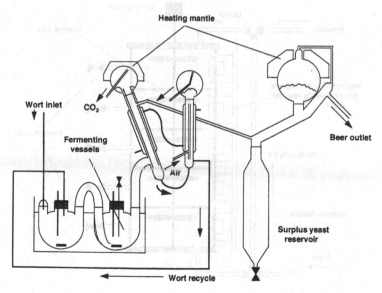

**Fig. 5.31**  Two-stage hterogeneous continuous fermenter (re-drawn from Portno, 1968b).

yeast vessels were each fitted with top-mounted spherical flasks to contain foam. Foaming was suppressed by applying heat to these flasks. Control of the dilution rate and the rate of yeast recycling provided a means of regulating the conditions within each fermentation flask.

It was demonstrated that steady states could be achieved, in each flask, over a wide range of dilution rates. The oscillatory behaviour seen in homogeneous systems was entirely lacking and high yeast viability was retained over several days. Furthermore, the resultant beers were similar in analysis to those obtained from batch fermentations using similar wort and yeast.

The benefits of a multi-vessel heterogeneous continuous fermentation system in terms of stability could be clearly demonstrated. However, such systems are complex and require considerable process pipework and valving. With an eye to future commercial exploitation, Portno (1969) also devised a single tank heterogeneous system, which retained the advantages of the multiple tank, without some of the complexities. This was termed a 'progressive continuous system' and is illustrated in Fig. 5.32. It consisted of a glass tube fitted at each end with stainless steel base plates to give a capacity of 1400 ml. Wort was fed in through the bottom and the beer outlet through a port in the top. The fermenting vessel was separated into five connected chambers by metal discs attached to a central rotating drive shaft. The gap between the discs and the vessel wall was 25 mm. The contents of each chamber were stirred by angled blades attached to the drive shaft. Silicone rubber membranes were attached to each rotating metal disc. These served as one-way valves so that flow was restricted to an upward movement from wort inlet to beer outlet. Good mixing within each chamber was further encouraged by wall mounted baffles, the angle of which could be adjusted by external magnets.

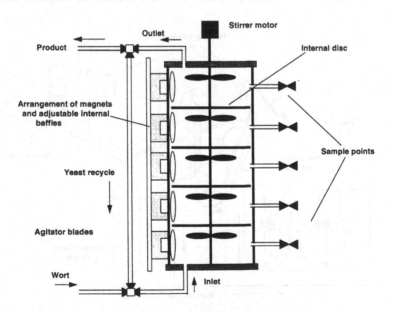

**Fig. 5.32**   Progressive continuous fermenter (re-drawn from Portno, 1969).

The apparatus was used in conjunction with the ancillary vessels used in the two-stage heterogeneous system shown in Fig. 5.31 and in a similar fashion yeast could be recycled. Using this apparatus it was demonstrated that steady states could be achieved at dilution rates varying between 10 and 100% of the fermenter volume per hour. Analysis of samples taken from each chamber revealed that, as predicted, the composition of the wort and the physiological condition of the yeast changed in the same ordered sequence as would be seen throughout the time course of a conventional batch fermentation. Beers obtained at all dilution rates were comparable with the batch product; however, it was observed that where a high yeast recycle rate was used, levels of esters were enhanced. This was attributed to the relatively high levels of fully attenuated beer introduced with the yeast into the first chamber. This mirrored the abnormally high ester levels that have been observed with homogeneous continuous systems.

The Wellhoener system (Wellhoener, 1954) was a pilot scale continuous fermentation and maturation system consisting of a series of six vessels. The first three vessels were maintained at 10 to 12°C and were used for fermentation. The process commenced by filling the first fermentation vessel with filtered wort and pitching with yeast slurry. When fermentation had started, continuous operation was established by initiating flow of aerated and filtered wort to the first vessel. A pressure differential, using carbon dioxide, was maintained between the three fermentation vessels. This was used to move wort between vessels and reportedly to control the rate of fermentation. Green beer exiting from the third vessel was filtered in-line and directed towards the first maturation vessel, all vessels being held in a cooled room at a temperature of 0°C. Maturation proper and adjustment of carbonation occurred in the second maturation vessel, the third vessel was used as a buffer tank.

The capacity of the fermentation vessels were 40, 30 and 20 hl respectively, and the conditioning vessels, 15 hl each. It was claimed that beer could be produced at a rate of 5 hl per day and the overall residence time was 27 days.

This lengthy residence time seems to have been the major drawback of the system, and, as with many others, it never progressed beyond the pilot scale. However, Wellhoener did introduce a successful discontinuous accelerated fermentation and maturation system as describd by Kleber (1987). The system was suitable for both top- and bottom-fermented beers and used pressure as a method of controlling yeast growth. The method claimed that increased carbon dioxide pressure inhibits yeast replication but has a smaller effect on fermentative metabolism and therefore, by inference, application of pressure can be used to increase fermentation efficiency. In the method described, bright aerated wort was introduced into a first vessel and pitched with yeast. After 6 hours, pre-fermentation pressure was allowed to build up to 0.3 bar by restricting the release of evolved carbon dioxide. At 50% wort attenuation, the pressure was allowed to increase to 0.7 bar, at 70% attenuation to 1.2 bar and 78% to 1.5 bar. After 2.75 days, the wort achieved 79% attenuation, at which point it was transferred to a maturation vessel. During fermentation, the pressure was partially released periodically to encourage mixing of the vessel contents and purge undesirable flavour volatiles and therefore further accelerate the process. During transfer to the maturation vessel the beer was cooled in-line. This efficacy of this system was confirmed by Kumada et al. (1975).

A complete experimental pilot scale continuous brewery was described by Hudson and Button (1968). This allowed continuous or batch production of hopped wort, which could be directed towards conventional batch or continuous fermenter. The latter was of a partially closed design as shown in Fig. 5.33. It consisted of an attemperated vessel of 10 litre capacity, the contents of which were mixed by an electrically driven stirrer. Aerated wort was fed into this vessel and this caused the fermenting liquid to rise through a vertical column to an upper vessel, the expansion chamber. Here the green beer exited through a shrouded outlet and the retained yeast was allowed to fall back into the main fermentation vessel. The upper surface of the expansion chamber was heated electrically to encourage foam collapse.

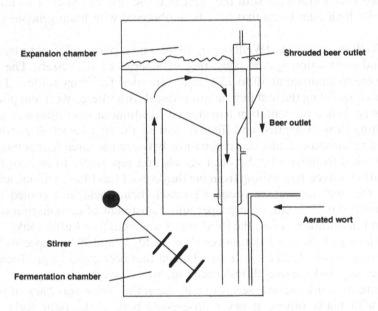

**Fig. 5.33**  Partially closed pilot scale continuous fermenter (re-drawn from Hudson & Button, 1968).

### 5.6.2.2  *Commercial continuous fermentation systems.*

Some of the more applied and largely pilot scale continuous systems were introduced in around 1900, and thus pre-date the laboratory models already described. In some respects the traditional brewing process has always incorporated some continuous elements, for example, the semi-conservative nature of yeast collection and re-pitching. Furthermore, some long established procedures which were introduced with a view to accelerating batch fermentation, such as mixing partially fermented and fresh wort, may be viewed as semi-continuous processes. It is perhaps unsurprising, therefore, that the concept of a fully continuous fermentation process should have come to the mind of these late Victorian brewers even though at that time the practice was not supported by a proper theoretical understanding.

Early semi- and fully continuous systems were reviewed by Green (1962) and Kleber (1987). For completeness, a brief mention is made here. Although all of these systems were intended for commercial use, many did not progress beyond the pilot scale.

A two-tank semi-continuous process was patented by Schneible in 1902. This involved pitching wort in the first vessel and then encouraging vigorous yeast growth by continuous aeration and rousing. The fermenting wort was then transferred into the second vessel. When the wort was attenuated the green beer was removed leaving the bulk of the yeast behind. Fresh wort was then added to this yeast and the process repeated.

A more complex process was patented by Schalk in 1906, which involved six connected vessels. The first vessel was filled with aerated wort and pitched at double the normal rate. After 24–48 hours, when the fermentation was actively proceeding, half the contents of the first vessel were transferred to the second vessel and then both were topped up with fresh wort. As soon as the contents of each vessel were again actively fermenting, half the volume of second tank was transferred to the third vessel and again both of these were topped up with fresh wort. This procedure was continued until all six vessels were filled with fermenting wort. By the time the final vessel was filled, attenuation of the wort in the first was complete. After the green beer was removed from the first vessel for further processing, the procedure was continued by transferring half the contents of the sixth vessel into the first. The sequence could be repeated *ad infinitum*, provided infection was kept at bay. The advantage of the Schalk process was, therefore, that the requirement for pitching yeast and its handling were much simplified. Of course, the consequences of microbial contamination would be severe in that there would be the potential for losing six batches of beer, as opposed to one in the conventional process. According to Kleber (1987) the process was not exploited commercially.

A contemporary of the Schalk process was that of van Rijn (1906). This was also a multi-tank approach but was truly continuous in that constant wort feed was required. The tanks were arranged in the form of a cascade, thereby allowing gravity feeding from overflow pipes fitted at the top of each vessel. The process was controlled by the rate at which wort and yeast were fed into the base of the uppermost tank. It was claimed that control of the rate of addition of wort allowed a constant yeast population to be maintained in each vessel. Therefore, this is an early example of an open single-phase heterogeneous continuous system.

A similar approach using a cascade of four linked fermenting vessels was patented by Williams and Ramsden (1963). This consisted of four linked jacketed tanks, each with a bottom located in-feed and top located overflow so arranged as to allow gravity flow (Fig. 5.34). Wort and yeast were added to the first vessel and a batch fermentation was allowed to proceed. At a suitable point, continuous fermentation was then initiated by addition of wort to the base of the first vessel. No further yeast was added to the system. The contents of the first and second vessels could be gas roused and provision was made for collection of carbon dioxide from the second. The arrangement of vessels as shown in Fig. 5.34 was specifically designed for use with top-fermenting yeast. In this case, the third tank, in which primary fermentation was essentially completed, was fitted with a top-mounted baffle, which contained the yeast head and prevented it from being carried over into the last vessel. The crop was pumped from the third vessel to a plate and frame filter from which yeast was recovered and the barm ale returned to the in-feed of the fourth vessel. The system could also be used with bottom-fermenting yeast; in this case the third tank was of

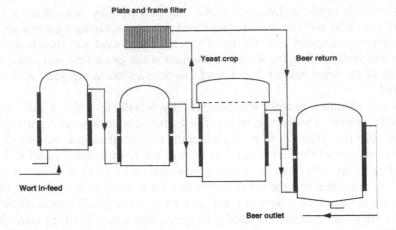

Fig. 5.34 Cascade continuous fermentation system.

cylindroconical design, and the top-cropping deck was omitted. The fourth vessel was essentially a conditioning tank in which the green beer was cooled and provision was made for adjustment of levels of carbonation.

In 1956, Mr Morton Coutts, Technical Director of the Dominion Breweries Company of New Zealand, patented a method for wort stabilisation and continuous brewing process, the contents of which are described in Coutts (1966). This was the forerunner of the only successful continuous brewing systems now remaining in use. The Dominion Group of New Zealand operates continuous brewing systems at several of its breweries, with two new ones commissioned in 1993 (Stratton *et al.*, 1994). The Coutts process is another multi-vessel approach. A diagrammatic representation of a modern incarnation is shown in Fig. 5.35(a).

High-gravity wort is cooled and stored in a receiving vessel where trub is separated out by gravity sedimentation. During transfer from the receiving vessel to the continuous fermenter, the wort is diluted with sterile water to a desired specific gravity. In the first tank of the continuous fermenter, the hold-up vessel, the wort is diluted 1:1 with process fluid recycled from the second vessel. In addition, a stream of yeast is also added which is recycled from the end of the process. The yeast strain is described as being highly flocculent and having a relatively high oxygen requirement. The hold-up tank volume is 6% of the total and its purpose is to improve the microbiological robustness of the system. Thus, blending back of partially fermented wort reduces the pH to less than 4.2, adds ethanol at a concentration of approximately 2% v/v and introduces yeast in the exponential phase of growth. All of these measures mitigate against adventitious contamination. The contents of the hold-up vessel are aerated vigorously. Provision of appropriate quantities of oxygen at this stage has been shown to have a critical influence on the extent of subsequent yeast growth and achieving a desirable balance of beer esters.

The second and third tanks are fermentation vessels occupying respectively 63% and 31% of the total fermenter volume. Both are stirred and passage from one to the other is via gravity feed. Beer then passes into a conical separator where the flocculent

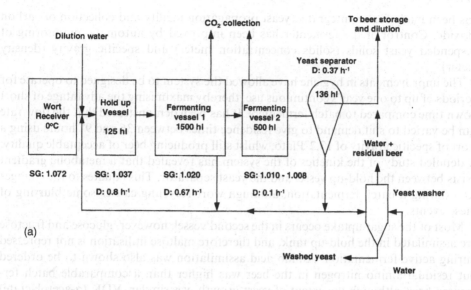

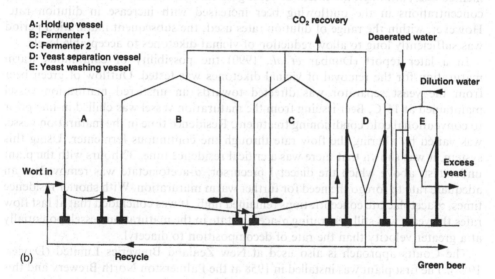

**Fig. 5.35** Coutts system of continuous fermentation. (a) Diagrammatic. (b) Schematic of Auckland Brewery continuous plant (from Dunbar *et al.*, 1988).

yeast is allowed to settle. The excess yeast stream is directed towards a washer where the beer is removed via a countercurrent of deaerated water. The resultant beer and water mixture is used for original-gravity adjustment of the green beer as it is transferred to conventional maturation vessels. Use of the recovered green beer in this way, it is claimed, reduces losses to a minimum.

Several refinements have been incorporated in the most modern installations (Stratton *et al.*, 1994). Notably, advances made with regard to hygienic design of conventional fermenters have been applied to the continuous fermenter. Provision has been made for an integrated yeast propagation facility and collection of carbon dioxide. Control of the fermenter has been improved by automatic monitoring of suspended yeast solids (solids concentration meter) and specific gravity (density meter).

The improvements in hygiene have allowed the system to be designed to operate for periods of up to one year's continuous use, thereby maximising the advantage of short down time compared to batch fermenters. It has been demonstrated that the flow rate can be varied to suit demand to give residence times between 36 and 97 hours using a wort of specific gravity of 13.2°Plato, whilst still producing beer of acceptable quality. A detailed study of the kinetics of the system has revealed that a metabolic gradient exists between the hold-up vessel and the yeast separator. This equates to the changes seen during a batch fermentation, although wort recycling causes some blurring of these events.

Most of the sugar uptake occurs in the second vessel; however, glucose and fructose are assimilated in the hold-up tank, and therefore maltose utilisation is not repressed during active fermentation. Amino acid assimilation was also shown to be ordered but residual amino nitrogen in the beer was higher than a comparable batch fermented beer, although the extent of yeast growth was similar. VDK ($\alpha$-acetolactate) concentrations in the outflowing beer increased with increase in dilution rate. However, within the range of dilution rates used, the subsequent maturation period was sufficiently long to allow reduction of vicinal diketones to acceptable levels.

In a later report (Dunbar et al., 1990) the possibility of continuous warm maturation for the removal of vicinal diketones was tested. Outflow of green beer from the yeast separator was directed towards an unstirred maturation vessel maintained at 15°C. Beer issuing from the maturation vessel was chilled in-line prior to conventional cold conditioning treatment. Residence time in the maturation vessel was varied by altering the flow rate through the continuous fermenter. Using this system, it was shown that there was a critical residence time, 32 hours with the plant under test, above which the diacetyl precursor, $\alpha$-acetolactate was removed at an adequate rate to forego the need for further warm maturation. With shorter residence times, $\alpha$-acetolactate concentrations remained high. It was concluded that at fast flow rates the yeast was still generating $\alpha$-acetolactate in the maturation vessel, potentially at a greater velocity than the rate of decomposition to diacetyl.

The Coutts approach is also used at New Zealand Breweries Limited (Davies, 1988). The first plant was installed in 1958 at the Palmerston North Brewery and this became the first in the world to be totally dependent on continuous fermentation. Similar installations were made at four other breweries, the most modern at the Christchurch brewery in 1976. This plant is similar to that shown in Fig. 5.35(b) and uses two rectangular stainless steel fermenting vessels with capacities of 1500 and 1100 hl. The throughput is 72 hl per hour.

In 1982, there was a requirement at the New Zealand Breweries' Christchurch plant to increase capacity up to 2 million hl per annum. Coincident with this, the method of payment of excise liability was changed from a levy on wort extract to an end-product duty. After a consideration of relative costs, it was concluded that the cheapest option was to generate the increased capacity by installing conventional cyclindroconical fermenters. Complete costing of the project indicated that obtaining similar volumes through the continuous approach would be 20–42% more expensive. This, coupled with a requirement to produce a much wider product portfolio would seem to have heralded the end of further investment in continuous fermentation by this company.

Another system contemporary with the Coutts process was the Canadian Labatt ABM fermentation process described by Geiger (1961). This patented approach (Geiger & Compton, 1961), used to produce lager beer, was a heterogeneous open system with yeast recycling (Fig. 5.36). The key characteristic of the system was the physical separation of yeast propagation and fermentation. Wort was collected twice each week and held in hygienic storage tanks for up to four days at 8°C. From the storage tanks, the wort was pumped into the first vessel in which growth of yeast was encouraged by the maintenance of highly aerobic conditions. Yeast plus wort was then pumped into the anaerobic fermentation vessel. The second tank was described as being a fluidised bed reactor in which very high yeast concentrations were maintained, thereby allowing rapid rates of fermentation to proceed. For some trials, a second fermentation vessel was used. The out-flowing beer passed through a yeast separator and thence was cooled in-line prior to transfer to a conditioning tank. Yeast from the separating vessels could be either recycled into the fermentation vessel(s) or directed to waste.

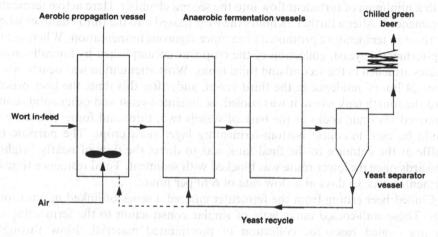

**Fig. 5.36** Labatt ABM continuous fermentation system.

In the pilot scale plant system described, the two main vessels each had capacities of 120 hl and the yeast separation vessel was of approximately 20 hl. When operated at 15°C, the overall residence time was 30–40 hours and beer was produced at the rate of 6 hl h$^{-1}$. It was claimed that yeast growth was some 25% less than that seen in a comparable batch fermentation. Beers were considered a close organoleptic match for those produced by conventional means. Analytically, matching was also close although esters and phenylethanol were significantly elevated in the continuous product. However, in some trials, a single homogeneous tank system was used and in this case the beers were considered unacceptable on the basis of both taste and analysis. The system was apparently quite resistant to contamination and on one occasion was run for four months without breakdown. It was used for commercial brewing albeit only briefly.

In the same period as the Labatt process, and also from Canada, O'Malley (1961) presented plans at both pilot and production scale for a system of continuous wort production, fermentation and maturation. Unfortunately this very innovative system was another casualty of the 1970s *volte-face* over continuous fermentation and it was never implemented at commercial scale.

Hopped wort was produced in a moving belt continuous combined mash lauter tun and a cylindrical multi-stage horizontal copper. The fermenter consisted of four linked rectangular tanks, the first with a sloping bottom and the other three with bottoms of triangular cross-section (Figs 5.37(a) and (b)). The combined tanks were covered by a dome, which contained manways, sight glasses and carbon dioxide outlets above the second and third compartments. Temperature control was via wall jackets and internal baffles in the fourth tank.

The fermenter was an example of an open heterogeneous single-phase type. Thus, as in a plug flow reactor, the stages of batch fermentation were intended to take place in the individual vessel compartments. To better mirror the batch process no provision was made for mechanical agitation. Pitched wort was fed into the first vessel where it remained for 8 hours, equating to the batch lag phase, and was then pumped with a minimum of turbulent flow into the second chamber. Here active fermentation commenced. After a further 18 hours the wort passed into the third chamber where an increase in temperature promoted even more vigorous fermentation. When used with top-fermenting yeast, collection of the crop was accomplished by laterally mounted chutes attached to the second and third tanks. Wort attenuation was nearly complete after 24 hours' residence in the third vessel, and, after this time, the beer proceeded into the fourth tank where it was cooled. Sedimented yeast and other solids could be removed via drain cocks at the base of vessels two, three and four. The same route could be used to collect bottom-fermenting lager yeast crops. The purpose of the baffle at the entrance to the final tank was to direct the flow of nearly bright beer upwards once the lower route was blocked with sediment. Total residence time in the fermenter was 3.5 days at a flow rate of 6 hl per hour.

Chilled beer exiting from the fermenter entered a series of linked maturation vessels. These wall-cooled tanks were of similar construction to the fermenting tanks, having angled bases for collection of precipitated material. Flow through the maturation vessels was slow with very little turbulence and a decreasing temperature gradient was maintained between the entrance and exit. Total residence time in the maturation vessels was three weeks at a flow rate of approximately 6 hl per hour, equivalent to a fluid flow rate of 17 cm per hour.

Williamson and Brady (1965) described another continuous system developed by the Carling Brewing Company (USA) and Canadian Breweries Ltd. A pilot scale facility was constructed at the Cleveland plant and a production scale system at Fort Worth. This was also a progressive multi-vessel system similar to the Coutts process.

The production scale facility (Fig. 5.38) consisted of an initial vertical rectangular pitching tank in which aerated wort was introduced after mixing with recycled yeast. This vessel was fitted with internal baffles designed to provide a long path length to ensure a sufficiently long residence time to complete the lag phase of fermentation.

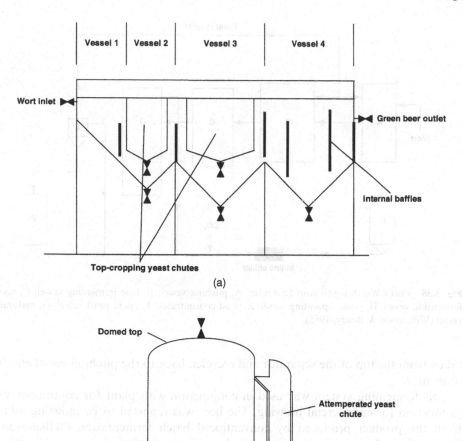

Vessel 1 | Vessel 2 | Vessel 3 | Vessel 4

Wort inlet

Green beer outlet

internal baffles

Top-cropping yeast chutes

(a)

Domed top

Attemperated yeast chute

Fermenting vessel

(b)

**Fig. 5.37** Continuous fermenter. (a) Side view. (b) Front view. (After O'Malley, 1969.)

There were two fermentation vessels, the first roughly double the capacity of the second. Both vessels were agitated and attemperated at the maximum temperature used for the batch process. High fermentation rates were achieved by maintaining high yeast concentrations. Beer exiting from the second fermenter entered a vertical conical-bottomed yeast separating vessel in which the specially chosen flocculent strain was allowed to settle. The relatively yeast-free beer was taken from the top of the separating vessel and chilled and carbonated in-line before transfer to a maturation tank. Yeast and beer taken from the bottom of the separation tank was pumped to a yeast concentrator. In this small conical tank yeast was separated from the beer by a pressure treatment and sent to waste. A mixture of beer and yeast was

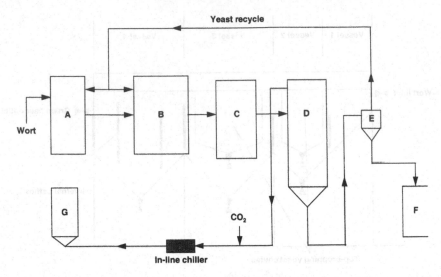

**Fig. 5.38** Forth Worth continuous fermenter: A, pitching vessel; B, first fermenting vessel; C, second fermenting vessel; D, yeast separating vessel; E, yeast concentrator, F, spent yeast vessel; G, maturation vessel (Williamson & Brady, 1965).

taken from the top of the separator and recycled back to the pitching vessel and first fermenter.

This fermenting system was used in conjunction with plant for continuous wort production for commercial brewing. The beer was reported to be indistinguishable from the product produced by conventional batch fermentation (Williamson & Brady, 1965). Apparently it was susceptible to bacterial infection; however, in spite of this, relatively long periods of uninterrupted use were achieved. Despite this success the plant was decommissioned and the breweries returned to batch fermentation.

Bishop (1970) was responsible for introducing another multi-vessel continuous system into four UK breweries of Watney during the late 1960s. Wort was produced by a conventional batch process and stored in chilled tanks for up to 14 days. For a commercial process it was recommended that three wort tanks were available since with a conventional brewhouse this provided a continuous supply of wort whilst allowing necessary down-time for tank cleaning and re-filling.

The continuous fermenter comprised an in-line wort steriliser, a unique oxygenation column, two fermentation vessels and a yeast separation tank (Fig. 5.39). The process was initiated by filling the first fermenter with wort and inoculating with a pure yeast culture. Wort was pumped into the system at an accelerating rate to match increased yeast growth. After some 2–3 days a steady state was established and a wort flow rate was set which was considered appropriate for the temperature regime used and provided beer of suitable quality. Supply of oxygen at an optimum concentration was considered to be critical to the success of the process. If too low, then fermentation rate would be diminished because of inadequate yeast growth. If too high, it was observed that yeast cells became elongated, although any effect of this pleomorphism on beer quality was not reported. Accurate dosing of oxygen into the in-

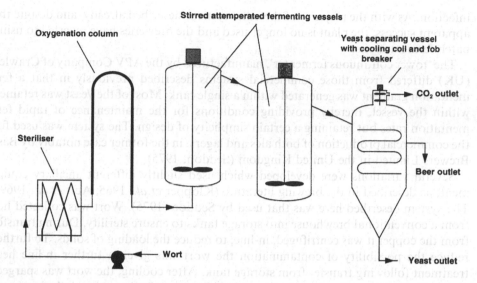

**Fig. 5.39** Bishop continuous fermentation system.

flowing wort was achieved using the column oxygenator. This consisted of an inverted U-tube. Wort and oxygen were introduced into the first ascending arm. When the wort passed into the second downward arm any undissolved oxygen bubbles would tend to rise and meet more incoming wort, therefore providing adequate time for complete solution.

Within the first fermenting vessel, at a desired steady state, it was observed that the wort was approximately 50% attenuated and the yeast achieved a concentration of 5 lb per barrel (*c.* 2–2.5 g l$^{-1}$ dry weight). In the second vessel the final gravity was achieved with a yeast concentration of 6 lb per barrel (2.5–3.0 g l$^{-1}$ dry weight). The yeast settling tank was cylindrical with a conical bottom and contained a centrally mounted helical spiral, cooled by circulating brine. Chilling the beer stopped fermentation and caused flocculated yeast to settle into the cone from whence it was recovered at intervals, pressed and retained. Recovered beer was returned to the outflowing product stream. The yeast was used as a source of pure culture but unlike the other continuous systems already described there was no recycling. Green beer flowing from the top of the separator contained less than 0.5 pounds per barrel yeast (0.1–0.15 g l$^{-1}$ dry wt). Carbon dioxide gas was recovered from the top of the separating vessel after passing through a mechanical fob breaker.

This system was shown suitable for all beer qualities, ales and lagers; indeed a strong beer (20–22.5°Plato original gravity) was produced by continuous fermentation in two days compared to nine months by a traditional procedure! All beers produced with the system were considered the equal of their batch-produced counterparts. When installed at all four breweries the maximum output was 1.6 million hl per annum. Each fermenter was rated at approximately 6500 hl per week. On one occasion, a fermenter was operated continuously for 13 months, which would seem to belie the oft-made criticism of continuous systems that they are prone to infection. As with the majority of the other systems described already, and despite the apparent success, the plant is no longer used and the breweries have reverted to using batch fermentations.

The 'tower continuous fermenter', manufactured by the APV Company of Crawley (UK) differed from those commercial systems described previously in that a fermentation gradient was generated within a single tank. Most of the yeast was retained within the vessel, thereby providing conditions for the maintenance of rapid fermentation rates but retaining a certain simplicity of design. The system was used for the commercial production of both ales and lagers, in the former case notably by Bass Brewers Limited in the United Kingdom (Seddon, 1975).

Several variations were developed which used slightly different ancillary equipment, as described in the brewing literature (Klopper et al., 1965; Ault et al., 1969). The system described here was that used by Seddon (1975). Wort was collected hot from a conventional brewhouse into storage tanks to ensure sterility. During transfer from the copper it was centrifuged, in-line, to reduce the loading of solids. To further reduce the possibility of contamination the wort was given a further in-line heat treatment following transfer from storage tank. After cooling, the wort was sparged, also in-line, with sterile air and carbon dioxide at a point immediately before entry into the base of the fermenter. The purpose of the carbon dioxide was to improve agitation at the bottom of the tower and prevent yeast compaction.

The tower consisted of a wall attemperated stainless steel vertical cylinder with a larger diameter head structure (Figs 5.40 and 5.41). A specially selected highly flocculent, low ester and diacetyl producing and slow growing strain was used in the fermenter. Such yeast tended to settle and form a plug in the base. Yeast concentrations, at the base of vessel, were in excess of 350 g wet weight per litre. Upward flowing wort passed through the yeast plug and was therefore fermented at a very rapid rate. Concentration and retention of the bulk of the yeast at the base of the tower in transient contact with wort allowed the formation of a gradient throughout the vertical axis of the vessel. The tendency for back mixing was reduced naturally by the density gradient due to progressive fermentation and it was further discouraged by the provision of a number of horizontally mounted perforated metal plates. This fermenter, therefore, may be classified as a heterogeneous, closed, two-phase type.

Yeast carried to the top of the fermenter completed the attenuation of the accompanying wort. In the top region of the tower an arrangement of baffles separated evolved carbon dioxide and allowed some of the residual yeast to be diverted into a reservoir. From here, the yeast could be recycled back into the base with the inflowing wort. Different types of separator could be used depending on whether the yeast was an ale or lager type (Klopper et al., 1965). Claimed advantages for the tower fermenter were very rapid rates of attenuation due to the high yeast concentration. Seddon (1975) described use of a battery of four towers each of nearly 1 m diameter and c. 8.5 m tall. Together these were capable of producing ale at the rate of 5000 hl per week. This represented rates of wort attenuation of the order of 3–4 hours. In addition to rapidity, the low rates of yeast growth led to improvements in the efficiency of ethanol yield. Use of several fermenters provided some degree of flexibility in terms of being able to vary production rates. It was also shown that if necessary it was possible to chill the contents of the tower and stop the flow of wort for periods of

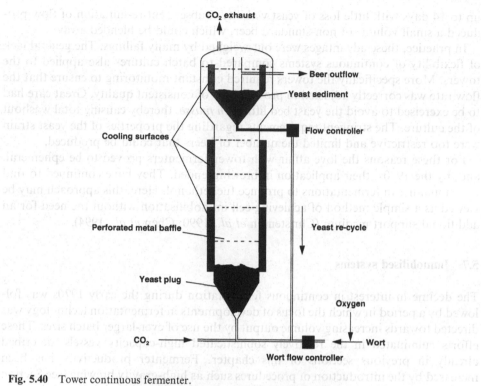

**Fig. 5.40**   Tower continuous fermenter.

**Fig. 5.41**   Tower fermenter (kindly supplied by Harry White, Bass Brewers).

up to 14 days with little loss of yeast viability. Subsequent re-initiation of flow produced a small volume of non-standard beer, which could be blended away.

In practice, these advantages were out-weighed by many failings. The general lack of flexibility of continuous systems compared to batch cultures also applied to the towers. More specifically the towers required constant monitoring to ensure that the flow rate was correctly adjusted to provide beer of consistent quality. Great care had to be exercised to avoid the yeast bed lifting *en masse*, thereby causing total washout of the culture. The stringent requirements regarding the properties of the yeast strain were too restrictive and limited the number of beers that could be produced.

For these reasons the love affair with tower fermenters proved to be ephemeral, and, by the 1970s, their application in brewing ended. They have continued to find occasional use in fermentations to produce fuel ethanol. Here, this approach may be viewed as a simple method of achieving cell immobilisation without the need for an additional support medium (Christensen *et al.*, 1990; Chen *et al.*, 1994).

## 5.7  Immobilised systems

The decline in interest in continuous fermentation during the early 1970s was followed by a period in which the focus of developments in fermentation technology was directed towards increasing volume output by the use of ever-larger batch sizes. These efforts culminated in the relatively sophisticated high-capacity vessels, described already in previous sections of this chapter. Fermenter productivity has been increased by the introduction of procedures such as high-gravity brewing (see Section 2.5). Some time savings have been made by the use of combined fermentation and maturation vessels (Section 5.4.3). However, these have been relatively modest and not of the order that was expected of continuous fermentation.

During the late 1970s through to the present, there has been a renaissance of interest in continuous brewing processes. This is because of the promise that many of the shortcomings of the early systems could be overcome by the use of relatively small vessels containing immobilised yeast cells. Such reactors have already found application at commercial scale for continuous flavour maturation and for the production of low-alcohol and alcohol-free beers by limited fermentation. As yet they have not been used for primary fermentation at production scale although several pilot scale systems have already been developed and commercial application in the near future would seem likely. Coupling of primary fermentation and warm maturation into a single continuous process catalysed entirely by immobilised yeast is of course feasible, providing the other upstream and downstream processes are capable of supporting the fermenters. In this case, it has been suggested that eventually it may be possible to produce beer within one day (Atkinson & Taidi, 1995)!

The major perceived advantages of the use of immobilised yeast reactors are:

(1)   very rapid process times because of the high yeast concentration present in the reactor coupled with the ability to use rapid flow rates that would cause washout in open continuous systems;

(2)   high efficiency of conversion of sugar to ethanol formation by restricting the extent of yeast growth (when used for primary fermentation);

(3) relatively small reactors giving large-volume continuous throughputs with a minimum of down-time;

(4) reduced risk of microbial failure because potential contaminants are unable to compete successfully with the existing yeast;

(5) rates of metabolism of immobilised yeast cells may be more rapid than free cells;

(6) no restriction on the choice of yeast strain;

(7) reduced requirement for yeast propagation;

(8) simplified yeast handling;

(9) simplified separation of product from yeast;

(10) improved product consistency;

(11) possibility of modular design allows improved flexibility compared to conventional continuous systems.

There are some disadvantages, particularly with respect to application to primary fermentation. It is advisable to use bright wort to avoid the possibility of clogging of support beds. In most breweries, this will require some form of pre-filtration. Brewhouses must be capable of a constant supply of wort, otherwise, as with conventional continuous fermentation, much of the volume productivity gain is lost. There is a requirement to store wort under sterile conditions. The large volumes of carbon dioxide evolved during primary fermentation may cause disruption of some mechanically fragile support media and excessive gas breakout can reduce the efficiency of mass transfer of solutes in packed beds. More significantly, as stated already, yeast growth is not a prerequisite of immobilised systems. This is an advantage with respect to efficiency of conversion of sugar to ethanol. However, the concomitant reduced uptake of wort amino acids results in beers which contain low levels of esters, and are therefore poorly matched to those produced by conventional batch fermentation. As will be described subsequently, one approach to circumventing this problem is the use of multiple tanks and a combination of free and immobilised yeast. In this instance, the supposed advantage of simplicity is perhaps brought into question.

Application of immobilised yeast reactors for primary fermentation within existing breweries at the expense of conventional fermenters and associated plant would be difficult to justify in terms of cost. However, it could be given serious consideration where the need arose for additional fermentation capacity, although obviously great care would need to be taken with respect to product matching. A method has been proposed whereby existing conventional fermenters could be modified for use as immobilised reactors by fitting retaining screens and using yeast entrapped in calcium alginate beads (Hsu & Bernstein, 1985). However, using existing brewery plant for duties for which it was not designed must at best be considered an unsatisfactory compromise.

Continuous warm maturation is an ideal application for immobilised yeast since it does not require yeast growth, involves little carbon dioxide evolution and can be fitted as an adjunct to conventional batch fermenters. At present, therefore, this application is a much more attractive proposition for fitting into existing breweries. For the same reasons production of low- or zero-alcohol beers by limited fermentation is suited to an immobilised process. In the long term, use of immobilised reactors for combined continuous primary fermentation and maturation seems destined only to be considered seriously where entirely new breweries are being planned.

### 5.7.1 Theoretical aspects

Radovich (1985) defined immobilisation as 'physical confinement or localisation of microorganisms in a way that permits economic re-use'. McMurrough (1995) gave a more specific definition, 'cells physically confined or localised in a certain defined region of space with retention of their catalytic activity, if possible or even necessary their viability, which can be used repeatedly and continuously'. Thus, using the classification described in Section 5.6.2, immobilised yeast reactors would be viewed as totally closed systems where the process fluid, whether it be beer or wort, makes transient contact with actively metabolising but retained yeast cells. Unlike batch systems or conventional continuous fermenters, it is not essential that any cellular growth occurs within an immobilised reactor, although growth is not precluded. In this sense, the yeast may be viewed simply as a 'biocatalyst'.

In theory, compared to a conventional continuous fermenter much greater volumetric productivity may be achieved using an immobilised yeast bioreactor. With a completely open continuous fermenter, as in a chemostat, the maximum achievable flow rate (critical dilution rate) is governed by the maximum yeast growth rate which may be obtained under the particular selected operating conditions. Use of a dilution rate greater than the critical value results in washout of the yeast culture since the growth rate becomes less than the rate at which fresh medium is added (Section 5.6.1). In the case of an immobilised system, the yeast is retained and cannot wash out and therefore, dilution rates greater than the critical value may be used. In practice, the volumetric productivity is limited by the biomass loading in the reactor and the mass transfer of nutrients and products from the liquid phase to the yeast. These parameters are dependent on the method of immobilisation and the design and operation of the reactor, as will be discussed subsequently.

*5.7.1.1 Methods of immobilisation.* Five methods for immobilising cells are recognised (Godia *et al.*, 1987; Scott, 1987; McMurrough, 1995). These are flocculation, entrapment within a polymeric matrix, adhesion to a solid surface, colonisation of porous materials and retention behind membranes.

The simplest method of immobilisation is to take advantage of the natural flocculent character of some yeast strains. This enables retention by sedimentation against an upward flow of process fluid. This is the operating principle of the continuous tower fermenter, described already (Section 5.6.2.2). It has the advantage that there is no requirement for an immobilising support medium. Disadvantages are that the choice of yeast strain is restricted, flow rates must be modest to avoid washout and evolved carbon dioxide may disrupt flocs.

Entrapment of cells in a porous matrix has been probably the most widely reported method of immobilisation. The matrix material must be non-toxic and retain entrapped cells but be sufficiently porous to allow passage of nutrients and metabolites. The materials which have been used include calcium alginate, κ-carrageenan, polyacrylamide, agarose, pectin, gelatin, chitin and locust bean gum (Godia *et al.*, 1987). Many of these, for example, calcium alginate, κ-carrageenan and chitin, are approved for food use. The carriers may conveniently and inexpensively be formed into spherical beads, which can contain very high loadings of biomass. The low cost removes the need for the beads to be regenerable.

Mass transfer of substrates and metabolites between the beads and the process fluid is of critical importance to the efficiency of any immobilised cell reactor (Radovich, 1985). It is of particular importance with entrapped yeast supports since reactants and products must traverse through the matrix. Thus, mass transfer is dependent on diffusion gradients and these are influenced primarily by the bead size, the density of the matrix, the biomass loading, the bead concentration and the velocity of fluid flow over the bead surface. In this respect beads with a diameter of 0.2–3 mm offer a suitable compromise between level of biomass loading and diffusional path length. Unfortunately, most types of entrapment bead have low mechanical strength. They may be disrupted by evolution of gaseous carbon dioxide, and cell outgrowth may be considerable. Furthermore, they are compressible and, consequently, reduced flow rates due to compaction can be a problem with some reactor types. On the other hand they can be of roughly the same density as the suspending fluid which can make for easier mixing. The mechanical strength of the beads can be increased by using a greater degree of cross-linking in the gel or by the use of more robust surface layers. Of course, this is at the expense of diffusivity.

Cell immobilisation by attachment to surfaces is proving to be one of the most effective approaches for brewing applications at commercial scale since the available supports offer both mechanical strength and reasonable levels of biomass loading. In addition, diffusion between cells and process fluid is obviously less restricted than with entrapment systems. A wide range of supports have been used. These include wood chips, ceramics, glass, cotton fibres, diatomaceous earth, stainless steel, various resins, polyvinyl chloride and cellulose (Anselme & Tedder, 1987; Godia *et al.*, 1987; Ryder *et al.*, 1995).

The physiological basis of adhesion to inert surfaces is unclear but would appear to involve a combination of electrostatic and hydrophobic effects (Mozes *et al.*, 1987). Yeast immobilised in this way may be by combinations of attachment to inert surfaces and cell-to-cell flocculation. These interactions are dependent on the nature of the surface, the state and type of cell and the conditions in the surrounding environment. Thus, the material and cell surfaces both have inherent charges and varying extents of hydrophobicity. These will also be influenced by environmental parameters such as pH, osmolarity, ionic composition and the presence of other solutes, which may either promote or block adhesion. Surfaces may be subject to chemical modification to promote adhesion, for example, addition of positively charged ligands. With regard to yeast, parameters such as starvation, mean cell age, cell size and cell concentration have effects on cellular morphology. These might be expected to influence adhesion and flocculation (Wood *et al.*, 1992; Smart *et al.*, 1995; Barker & Smart, 1996; Rhymes & Smart, 1996).

Attachment of cells to surfaces via covalent bonding is possible but has not been used widely since many of the procedures are cytotoxic. There have been some reports, for example, of use of glutaraldehyde cross-linked gels that contain covalently bound yeast cells coated onto glass beads (Godia *et al.*, 1987). These have been tested for the production of fuel alcohol but not for the production of alcoholic beverages.

For brewing applications, DEAE-cellulose has proven to be the favoured support. It is available in commercial quantities supplied under the trade name Spezyme® (Cultur, Finland) and described as granulated derivatised cellulose (GDC). The material has a positive charge and binds negatively charged yeast cells in a mono-layer. The material is robust and has a rough surface, which provides some protection against accidental dislodgement of cells; however, excessive shear forces can result in some shedding (Norton & D'Amore, 1994). The material is FDA approved, 0.4–0.8 mm granular size and regenerable by treatment with a 2% w/v sodium hydroxide solution at 80°C (Pajunen, 1995). In the same report, the yeast loading was given as being up to $500 \times 10^6$ cells per gram wet weight of carrier.

The fourth method of immobilisation, colonisation of porous materials, is another form of entrapment. The materials are usually provided with comparatively small surface pores and larger internal cavities. Single cells may bind to the surface of the carrier but the bulk enters the central matrix via the pores and grows in the internal cavities forming retained aggregates. Immobilisation is via a combination of surface adsorption and cell-to-cell aggregation. Most supports of this type are robust, regenerable, non-compressible, and, since most of the immobilised cells are protected within the internal matrix, losses due to shearing are minimal. For these reasons they may be used in reactors where there is a relatively high degree of agitation, to promote good rates of mass transfer. With some supports care must be taken to avoid excessive abrasion damage.

Biomass loadings are high and an open pore structure allows egress of carbon dioxide with minimal floc disruption. As with gel entrapment methods, solute mass transfer is restricted compared to surface adhesion. Norton and D'Amore (1994) concluded that with porous beads only a relatively shallow band of biomass, extending to a depth of 140 μm, took an active role in immobilised reactors. Similarly, Inoue (1988) argued, in this case with alginate beads although the point is the same, that cells in the core region exhibited limited metabolic activity because of oxygen starvation and high local ethanol concentrations. For this reason, it is important to use beads of an appropriate diameter to maximise the available surface area.

Porous supports which have been recommended for brewing applications include artificial sponges (Scott & O'Reilly, 1995), glass beads (Breitenbucher & Mistler, 1995; Hyttinen *et al.*, 1995; Yamauchi & Kashihara, 1995) and ceramic rods with silicon carbide matrices (Krikilion *et al.*, 1995). Of these, the latter two have been applied at commercial scale, and therefore received the most attention. Porous glass beads are sold under the trade name SIRAN®, produced by the Schott Engineering Company (Mainz, Germany). They are prepared from a mixture of glass powder and salt such that during the process of bead formation the salt is dissolved and removed leaving pores in the glass of 60–300 μm diameter. Beads of 2–3 mm diameter are recommended for fixed bed reactors and 1–2 mm diameter for fluidised bed types. Each bead contains approximately 55–60% pore volume. Yeast loading levels are reported to be in the region of $15 \times 10^6$ cells/g bead (Breitenbucher & Mistler, 1995). The beads are regenerable by treatment with hydrogen peroxide and steam sterilisation. Cleaning is more difficult than surface carriers because of the need to remove all organic matter from the matrix.

Reactors employing yeast entrapped within silicon carbide have been described in several recent publications (for example, van de Winkel *et al.* (1991, 1993, 1995), Krikilion *et al.* (1995)). The support consists of a ceramic cylinder, which contains a number of channels through which the process fluid traverses. The channels are embedded in a matrix of silicon carbide which has surface pores of roughly 8–30 μm diameter and internal pores of 100–150 μm diameter, depending on the grade used. Void volumes of up to 60% are achievable. This arrangement allows entrance of single cells via the small surface pores and provides space for extensive internal colonisation and yeast retention. The ceramic elements can be arranged in groups through which process fluid is circulated. Unlike fixed or fluidised reactors, there is little restriction to flow with these channelled elements, and therefore mass transfer efficiencies are high and there is no possibility of bed compression or abrasion damage. The elements are robust and the arrangement of channels allows easy egress of carbon dioxide. The elements may be regenerated and cleaned by successive forward and backward flushing with hot sodium hydroxide solution, detergent and hot water. This is followed by sterilisation with either steam or cold temperature treatment with a mixture of peracetic acid and hydrogen peroxide.

The fifth method of immobilisation is the use of a semi-permeable membrane that retains cells but allows exchange of soluble metabolites and gases. They may be used in many configurations. Cells may be attached to the membrane or freely suspended behind it. The former approach is commonly used in the special application of biosensors, the latter in membrane bioreactors (Scott, 1987). Many configurations of membrane reactor have been suggested, for example, membrane re-cycle, hollow fibre, dialysis membrane (Cheryan & Mehaia, 1984; Park & Kim, 1985; Godia *et al.*, 1987). Few have progressed beyond the laboratory scale. They have inherent disadvantages of restricted mass transfer, a problem that may be exacerbated by a tendency to clog. Application at commercial scale for brewing seems most improbable.

A comparison of the methods of yeast immobilisation, together with examples pertinent to brewing applications is given in Table 5.7. No attempt has been made to quantify the differences in biomass loading achievable by each type of support since no single unit is used in the literature and comparisons are difficult. In any case, direct comparisons are of little value since they do not entirely reflect the volumetric productivity of which each type of support is capable. Thus, this parameter is influenced by not only the type of carrier but also the design of the bioreactor and the conditions under which it is operated. From a practical standpoint, the gel encapsulation supports seem unlikely to be applied at commercial scale for brewing applications. Of the other supports described in Table 5.7, both the DEAE-cellulose and glass bead supports have advantages over gel entrapment methods and both are clearly suitable for application at commercial scale. The silicon carbide approach looks particularly promising, having a good combination of high biomass loading, reasonable rates of mass transfer and unrestricted flow.

### 5.7.1.2 *Effects of immobilisation on yeast physiology.*

There is now a considerable body of evidence suggesting that the physiology of immobilised yeast differs from that

**Table 5.7** Methods of yeast immobilisation used in brewing applications, advantages and disadvantages.

| Method | Advantages | Disadvantages | Carrier | Application | Reference |
|--------|-----------|---------------|---------|-------------|-----------|
| Flocculation | No requirement for support medium | Restricted use of yeast strain<br>Floc disruption by carbon dioxide evolution<br>Wash-out at high flow rates | None | Primary fermentation | Seddon (1975) |
| Entrapment | Inexpensive supports<br>High biomass loading<br>Usable with any yeast strain | Restricted mass transfer<br>Restricted yeast growth<br>Potential cell outgrowth<br>Fragile<br>Disruption by $CO_2$<br>Compression of reactor beds<br>Non-regenerable | Chitosan beads | Primary fermentation (lab. scale) | Shindo et al. (1994) |
| | | | κ-carrageenan | Diacetyl removal using immobilised *Bacillus polymixa* (laboratory scale) | Willetts (1988) |
| | | | κ-carrageenan | Primary fermentation (pilot scale) | Mensour et al. (1995) |
| | | | Calcium alginate | Primary fermentation (laboratory scale) | White & Portno (1978) |
| | | | Calcium alginate | Continuous alcohol-free beer process (pilot scale) | Dziondziak & Seiffert (1995) |
| | | | Calcium alginate | Primary fermentation (laboratory scale) | Shindo et al. (1994) |
| Attachment | High mass transfer rates<br>Robust<br>Non-compressible<br>Regenerable<br>Not affected by $CO_2$ | Expensive<br>Cells may shear off at high flow rates<br>Reduced biomass loading cf. other methods | Diatomaceous earth | Primary fermentation (production scale) | Narziss & Hellich (1972) |
| | | | DEAE-cellulose | Continuous maturation (production scale) | Pajunen (1995) |
| | | | DEAE-cellulose | Continuous wort acidification using immobilised lactic acid bacteria | Pittner et al. (1993) |
| | | | DEAE-cellulose | Continuous maturation and alcohol-free beer production | Mieth (1995) |
| | | | DEAE-cellulose | Continuous alcohol-free beer production | van Dieren (1995) |
| | | | DEAE-cellulose | Continuous maturation | Nothaft (1995) |

of free cells. With respect to brewing processes using such yeast, some of these alterations have beneficial effects and some are disadvantageous. The precise causes of the changes are obscure and many factors appear to be influential. A proper elucidation of cause and effect is difficult because of the heterogeneous nature of immobilised systems. In addition, it is difficult to unravel those effects due solely to immobilisation of cells and those which relate to restricted mass transfer. It is an important area of study since a proper understanding of the physiology of immobilised yeast is clearly a prerequisite to optimising the design and operation of bioreactors.

Factors influencing yeast physiology fall into two loose groupings, which roughly coincide with the positive and negative responses to immobilisation. Thus, as Ryder et al. (1995) have commented, yeast cells in nature are usually found attached either to each other and/or to a surface (see Section 8.1.4.1 Biofilms). Therefore, immobilisation may be viewed as a protected and indeed preferred habitat. In the discussion to the same paper, Iserentant makes the point that cell-to-cell contact has profound but as yet incompletely understood effects on yeast physiology and that immobilised cells may react as a colony and not as individuals. Again, the implication is that this is a natural state for yeast cells to be in, and, as a corollary, freely suspended cells are in an abnormal and disadvantageous situation.

The negative responses relate to the stresses that immobilisation imposes on yeast cells. The most important of these are undoubtedly the effects of restricted mass transfer of supply of nutrients, particularly oxygen, and the removal of potentially toxic products such as ethanol and carbon dioxide. The severity of these effects will depend on the biomass loading, the type of support and the degree of agitation present in the reactor. Thus, it would be predicted that surface attached cells in a rapidly stirred reactor would be the least affected. In contrast, cells entrapped in the core of large beads in a packed bed type reactor are subject to the greatest degree of stress and will exhibit the most negative response. In between these two conditions, a whole continuum of varying restrictions to mass transfer is possible. For example, Pilkington et al. (1999) using lager yeast entrapped in carrageenan gel beads noted that after six months' continuous fermentation the viability of the entrapped cells decreased to approximately 50%. Cells in the core of the beads were noted as having fewer bud scars and had an altered morphology compared with those at the outer surface.

Superimposed on these effects are responses, which are due to the use to which the reactor is put. Where it is being used for primary fermentation, and, consequently, where a complete aerobic growth medium is supplied, the response of the immobilised yeast would be expected to be different to that exhibited by yeast in a reactor used for continuous maturation, where the feed stock is beer with little or no nutritive content. Similarly, as with conventional continuous fermenters, stirred homogeneous reactors used for primary fermentation are likely to elicit different responses in yeast to those which use unstirred packed beds.

Bearing in mind the complexities which arise from the preceding discussion some general comments may be made. There is a consensus that overall metabolic rates are enhanced in immobilised yeast cells adding support to the contention that this is an advantageous state. For example, increased rate of glucose uptake, increased rate of ethanol formation, increased yields of ethanol, lower intracellular pH and alterations in the activities of glycolytic enzymes have all been reported (Doran & Bailey, 1986; Tyagi & Ghose, 1982; Aires Barros et al., 1987; Galazzo & Bailey, 1989, 1990). In contrast, specific growth rates are reduced, as are biomass yields, specific rates of oxygen uptake and carbon dioxide evolution (Tyagi & Ghose, 1982; Doran & Bailey, 1986). These alterations can be dramatic; thus, in the report of Doran and Bailey (1986) specific rates of ethanol production in immobilised cells were 40–50% greater than in free cells. The differential in rate of glucose consumption was two-fold in favour of the immobilised yeast, yet biomass yield only 30% of the suspended cell value.

In addition to increased ethanol production, both glycogen and trehalose pools are elevated compared to free cells (Doran & Bailey, 1986). These reserve polysaccharides are implicated in resistance to starvation and other stresses and perhaps provide a part explanation for the observation that immobilised cells are more resistant than free cells. For example, Jirko (1989) concluded that immobilised yeast was better able to withstand starvation than free cells and this was due to the former being able to maintain a high adenylate energy charge for longer periods – presumably at the expense of dissimilation of the larger pool of glycogen.

Increased resistance to starvation in immobilised cells is accompanied by improved ethanol tolerance (Holcberg & Margalith, 1981; Dror et al., 1988). The precise mechanism remains obscure. Dror et al. (1988) proposed that cross-linked gels could form a thin layer on the cell surface and this provided stabilisation in the presence of ethanol. Hilge-Rotmann and Rehm (1991) concluded that under anaerobic conditions and in the presence of ethanol, immobilised cells contained greater concentrations of long-chain saturated fatty acids, compared to free cells. A positive correlation between fermentation rate and the degree of fatty acid saturation in immobilised cells was demonstrated.

This is contrary to the more usual assertion that ethanol tolerance and the proportion of unsaturated fatty acids in membranes are directly related (Beavan et al., 1982). However, Alexandre et al. (1994) concluded that membrane fluidity is the key to ethanol tolerance and that different cells have varying ways of manipulating this parameter by altering membrane lipid composition. It is conceivable, therefore, that the micro-environment to which immobilised yeast is exposed may produce a response to ethanol different from that of free cells. This aspect of immobilised yeast physiology is intriguing in that such cells may have restricted access to dissolved oxygen and of course, this is required for the de novo synthesis of both unsaturated fatty acids and sterols. It might be expected, therefore, that immobilised yeast would have reduced tolerance to ethanol but this is apparently not the case.

A further characteristic of the immediate environment of immobilised cells may be reduced water activity, which represents yet another stress that the cells must cope with. Typically, yeast cells react to low water activity by excreting osmoprotective metabolites such as glycerol. Galazzo and Bailey (1989) observed that the rate of production of this metabolite by alginate-entrapped yeast was approximately double that of suspended cells, which perhaps supports this view. Of course, it is also true that glycerol production has a redox-balancing role and this could be implicated in these changes. It is known that when yeast is subject to sub-lethal heat stress a response is elicited, termed the heat shock response in which increased thermotolerance is acquired (Lindquist & Craig, 1988). The response is complex and involves the simultaneous induction of a large number of so-called heat shock proteins (Schlesinger, 1990). Stresses other than heat can also trigger the response and it has been demonstrated that increased tolerance to these other stresses is also acquired. For example, heat shock leads to enhanced ethanol tolerance (Watson & Caviccioli, 1983) and increase in levels of trehalose (van Laere, 1989). It is tempting therefore, to speculate that some of the physiological changes seen in immobilised yeast may be stimulated by applied stresses and manifest in a similar way to the heat shock response.

Several authors have reported altered morphology of immobilised cells, which suggest that cell proliferation and bud separation may be affected. Jirko *et al.* (1980) noted that yeast cells covalently linked to a modified hydroxyalkyl methacrylate gel proliferated in nutrient media without separation of daughter cells and new cells were of an elongated shape. It was assumed that the altered morphology was due to attachment disrupting normal growth. Koshcheyenko *et al.* (1983), also using gel entrapment, reported that cells in the surface layers were apparently of normal morphology but those deep in the matrix showed many abnormalities. With pro-longed incubation these cells lysed, presumably due to nutrient starvation. Crypto-genic growth of the outer viable cells on the lysis products was also evident. Doran and Bailey (1986) showed that immobilised yeast contained increased concentrations of structural polysaccharide and had higher ploidy than free cells, indicating that the normal cell cycle had been disrupted.

It would appear that growth, as measured by cell proliferation and/or increase in mean cell size, is restricted in some immobilised systems at least for some of the time. This is of particular relevance to application to primary fermentation since growth and the concomitant assimilation of wort nutrients other than sugars, is necessary for the production of an appropriate spectrum of flavour-active metabolites. The perti-nent question is whether growth and nutrient assimilation are restricted by immo-bilisation *per se*, or can the limitation be overcome by increasing rates of mass transfer?

The evidence suggests that restricted mass transfer is the more significant factor, particularly the supply of oxygen. Thus, most growth occurs during the initial start-up period at a rate governed by the type of reactor, physical parameters such as tem-perature and the nature of the process fluid. In this initial period the carrier is not fully loaded and as growth proceeds the new daughter cells occupy the spare capacity. This may be via attachment to unoccupied surface or by colonisation of a matrix, depending on the type of carrier. When surface attachment carriers are completely occupied new daughter cells are released into the medium, suggesting that growth is not limited by restricted mass transfer, providing sufficient agitation. For example, van Dieren (1995) described use of DEAE-cellulose immobilised yeast for the production of alcohol-free beer. The reactor was operated at a temperature of 0 to 1°C under anaerobic conditions using filtered acidified wort. The initial cell count was $50 \times 10^6$ cells per gram carrier, and during the initial start-up period this doubled. However, the biomass con-centration slowly increased throughout the entire operation to the extent that periodically excess yeast had to be removed to prevent blockages occurring. With this system, it would appear that a stable steady state was never achieved.

Yamauchi and Kashihara (1995) working with a primary fermentation system investigated the effect of varying the dissolved oxygen concentration of wort introduced into a packed bed reactor containing yeast immobilised in calcium alginate. They reported that increasing the oxygen concentration from less than 0.3 ppm to 6.0 ppm had no effect on the bound yeast concentration ($c.$ $1 \times 10^7$ cells $ml^{-1}$) but the suspended yeast count increased from $1 \times 10^6$ to $1 \times 10^7$ cells $ml^{-1}$. It was assumed that the increase in free yeast count was due to greater liberation of new daughter cells from the carrier. This was consistent with the observation that during prolonged use the viability of the bound yeast gradually declined. Again, this indicates that steady-state conditions are not achieved with these types of continuous reactors.

It is, of course noteworthy that yeast cells have a finite life span and can produce only a certain number of daughters. It would be predicted, therefore, that the mean age of the immobilised cells would gradually increase and eventually a state of general senescence would pertain. On the other hand, non-viable cells could either detach from surface supports or lyse within porous types leaving space for further colonisation. At present, these possibilities are poorly researched.

Where reactors have been designed to have high mass transfer characteristics, growth and assimilation of wort components such as amino nitrogen are comparable with that seen in conventional batch fermentations, as is flavour development in the resultant beers. See, for example Mensour *et al.* (1995) and Krikilion *et al.* (1995). However, this is perhaps an over-simplistic view of what are very complex interactions between yeast and the external environment. The influences of mass transfer, residence times and type of reactor are all intertwined. It has been discussed already (Section 5.6.1) that in a plug flow type continuous reactor there is an opportunity for ordered assimilation and metabolism of wort nutrients as in a batch culture. The opportunity for this to occur exists in a packed bed reactor with no back mixing. The extent and rate of uptake of metabolites will depend on flow rate, temperature, provision of oxygen and the efficiency of mass transfer in the reactor.

In a homogeneous reactor mass transfer may be improved by efficient stirring but the continued in-feed of fresh wort may inhibit or repress some processes. For example, in the same way that free glucose represses maltose uptake, the presence of some amino acids inhibits the uptake of others. With stirred immobilised yeast reactors, there is the added complication that diffusion gradients will exist between the fluid and matrix phases. The severity of these effects will depend upon the type of support, the degree of agitation and the flow rate amongst other factors. Bearing in mind these complexities it is unsurprising that it is difficult to predict how any combination of wort, yeast and reactor type will interact and produce beer of a given quality. However, it would be suspected that for primary fermentation, relatively long residence times with reactors that allow considerable re-cycling, offer the best chance of production of balanced beers.

Others have chosen to overcome the problems of restricted mass transfer by using multi-stage systems consisting of a combination of stirred vessels and freely suspended yeast and bioreactors containing immobilised yeast. Yamauchi *et al.* (1994, 1995a, b) studied primary fermentation using a 'continuous stirred tank reactor' (CSTR) the outflow of which was linked to the in-feed of an immobilised yeast 'packed bed reactor' (PBR). The CSTR was supplied continuously with wort. The yeast loading was reduced in the process flow issuing from the CSTR by continuous centrifugation. Conditions in the CSTR were aerobic to encourage yeast growth and anaerobic in the PBR to promote ethanol formation. Using this process, beer quality and composition was comparable with those made by conventional batch fermentation.

It was observed that biomass yield in the CSTR was approximately ten-fold that of the PBR. Higher alcohol formation occurred largely in the CSTR, whereas ester levels were much higher in the PBR. This pattern was ascribed to the relatively high levels of growth in the CSTR allowing the formation of higher alcohols. These, together with ethanol were then available for esterification in the PBR where growth was restricted. The significant influence was considered to be the ratio of assimilation of fermentable extract to amino nitrogen in each vessel. Virtually no sulphite was generated in the CSTR but comparatively high concentrations were formed in the PBR. Levels in the latter continued to increase over a period of 30 days' continuous use. The observations were ascribed to the differential effects of growth on uptake of amino acids in each vessel. Thus, methionine inhibits sulphite production and this was present in the CSTR because of the continual in-feed of fresh wort. However, in the PBR, the concentration of methionine gradually fell to a non-inhibitory level, thereby allowing reduction of wort sulphate to sulphite.

The production of organic acids differed in each vessel and from a conventional batch fermentation. Overall, higher concentrations of succinate and lower levels of acetate were formed in the multi-vessel system compared to batch fermentation. Virtually no acetate was formed in the CSTR but succinate was produced in both vessels. Elevated succinate in beers produced using immobilised reactors has also been reported by Shindo et al. (1993a). These authors concluded that this was produced through the methylcitric acid pathway due to an increased assimilation of isoleucine compared to freely suspended yeast systems.

Contrary to multi-stage continuous reactors, it has been suggested that primary fermentation can best be achieved by an immobilised yeast reactor followed by a stirred fermenter containing free suspended yeast (Masschelein & Andries, 1995; Andries et al., 1997b). In this scenario, the immobilised reactor provides a constant inoculum of young yeast cells for the stirred reactor. There is no need for intermediary removal of yeast and growth in the stirred reactor is limited by the degree of wort attenuation achieved in the first vessel. Theoretically, this approach should allow better control over ester production since higher alcohol formation may proceed in the immobilised reactor via the biosynthetic route but with limited amino acid uptake. In the subsequent stirred vessel, amino acid assimilation may occur to provide substrates for esterification using the already present higher alcohols and ethanol.

Viewed as a whole these results suggest that a two-stage bioreactor used for primary fermentation offers a practical means of maintaining the volumetric productivity gains of immobilisation and producing standard beer. However, some anomalies remain, suggesting that further development is required.

With regard to continuous flavour maturation and low-alcohol or alcohol-free beer, the physiological effects of yeast immobilisation are to some extent less problematic. The simplest arrangement is that in which primary fermentation is performed in a standard batch fermenter. Green beer is then passed through an immobilised yeast reactor, which has the task of simply reducing diacetyl to a sub-flavour threshold concentration. In practice, there is an added complication in that it is necessary to ensure that all α-acetolactate in the green beer is converted to diacetyl prior to application onto the immobilised bioreactor. Otherwise diacetyl may be formed in finished beer when no yeast is present to remove it. The conversion is carried out by subjecting beer to a controlled heat treatment (5–10 minutes at 90°C). Strictly anaerobic conditions should be observed to remove the possibility of beer oxidation and the free cell concentration should be reduced to as low a level as possible, by efficient centrifugation, to avoid the formation of yeast thermal decomposition products.

A more complex system is where both primary and secondary fermentation are carried out in separate immobilised bioreactors. Ryder *et al.* (1995) discuss the observation that green beer issuing from immobilised yeast bioreactors used for primary fermentation has elevated levels of α-acetolactate. This may be related to the relatively low rates of assimilation of amino nitrogen compared to sugar utilisation or specific alterations in the pattern of uptake of valine and isoleucine might be implicated. The former would seem the most plausible explanation. Thus, where glycolytic flux rates are high and uptake of amino nitrogen is restricted it would be predicted that the yeast would seek to synthesise amino acids *de novo*. In this case, rates of synthesis of α-acetolactate from pyruvate may also be enhanced. Where primary fermentation only is carried using an immobilised yeast bioreactor care must obviously be exercised to ensure that the relatively high levels of α-acetolactate are removed in subsequent warm conditioning. If a second immobilised yeast reactor is used for maturation, no problems should arise as long as the intermediate heat step is performed correctly, since yeast has a very high capacity for reduction of the resultant diacetyl.

An elegant approach was described by Kronlof and Linko (1992) in which primary fermentation was performed using a single immobilised cell reactor with a genetically modified yeast strain containing α-acetolactate decarboxylase (see Section 4.3.4). This modification allows the conversion of α-acetolactate directly to acetoin without the formation of diacetyl. Pilot scale trials with this yeast allowed the production of beer with standard analysis in 2–6 days. No maturation period was needed. Of course, the question of public acceptance of beer made with genetically modified yeast remains a commercial bar to such approaches.

For alcohol-free or reduced-alcohol beer production by fermentation, there are three requirements. First, limited alcohol production; second, formation of a normal spectrum of beer flavour compounds other than ethanol; and, third, reduction of carbonyls which contribute to undesirable 'worty' characters. Debourg *et al.* (1994) have demonstrated that the reduction of wort carbonyls by yeast is complex and involves several enzymes. However, immobilisation had no effect on the ability of yeast to perform this task. Yeast with a repressed physiology, as would be the case in limited wort fermentation, was the most efficient at reduction of wort carbonyls.

Conversion of sugar to alcohol may be limited by use of a combination of low temperature and anaerobiosis. Several reports have indicated that these conditions may be readily achieved using immobilised yeast bioreactors (Breitenbucher & Mistler, 1995; van de Winkel *et al.*, 1995; van Dieren, 1995; Mieth, 1995). The first of these reports compared the sugar spectrum of the in-feed wort and exiting beer. This showed that, during contact with the yeast, there was a reduction in the concentration of sucrose, roughly a doubling of fructose, a slight increase in glucose and no change in levels of maltose and maltotriose. As only sucrose hydrolysis and some uptake of glucose had occurred, limited fermentation was assured since the continued presence of glucose would repress assimilation of maltose.

Production of other flavour metabolites in low- or zero-alcohol beer fermentations, particularly esters, is subject to the same restrictions as discussed already with regard to applications of immobilised yeast to primary fermentation. Indeed, because of limited ethanol and higher alcohol formation it would be predicted that it would be an even more serious problem. Using a reactor with a high mass transfer rate and limited aerobiosis, ester levels were found to be low compared to alcohol-free beers produced from a fully fermented beer with subsequent de-alcoholisation (van de Winkel, 1995). A possible compromise is to use a combination of slightly increased fermentation in the immobilised yeast bioreactor to generate flavour metabolites and reduce wort carbonyls followed by de-alcoholisation. Such an approach has been tested successfully at pilot scale as part of a fully continuous process for alcohol-free beer production (Dziondziak & Seiffert, 1995). However, this does lack the elegance and simplicity of the single pass immobilised yeast reactor approach. In general, it would appear that processes for alcohol-free beer by limited fermentation using immobilised yeast are likely to be most successful where beers with low ester levels are acceptable. For more flavoursome beers, this approach clearly has limitations.

5.7.1.3 *Reactor types.* There is no reason why immobilised yeast cannot be used in conventional batch fermentation (Atkinson & Taidi, 1995). The advantages afforded would be rapid process times due to the high biomass concentration and ease of separation of green beer. However, undoubtedly the most gains are made using a continuous process. A plethora of fermenter designs have been suggested for use with immobilised yeast for fuel alcohol production. However, for commercial beverage processes the two most common are packed and stirred bed types. Some typical configurations are shown in Fig. 5.42(a) and (b).

In a packed bed reactor, the liquid flow passes through an unstirred bed of the support matrix together with immobilised yeast cells. Therefore, it is a fully closed two-phase system. This method has the advantage of simplicity and the plug flow arrangement may ensure that substrates do not fall to limiting levels and avoids cells being exposed to inhibitory concentrations of metabolites (Masschelein & Andries, 1995). Theoretically, maintenance of ideal plug flow conditions in a fixed bed reactor would allow the various stages of a batch fermentation to be mimicked, and therefore this approach should be the most suitable for carrying out primary fermentation.

In practice, these ideal conditions are difficult to achieve. Fixed bed reactors are prone to channelling, difficulties are encountered in allowing escape of carbon dioxide and some support media, particularly the soft-bead types, are liable to compression, thereby restricting flow and causing high cross-bed pressures to develop. The efficiency of transfer of metabolites and nutrients between cells and their surrounding environment is low in fixed bed reactors since fluid flow is linear and of low velocity. This effect is exacerbated with soft-bead type supports where the

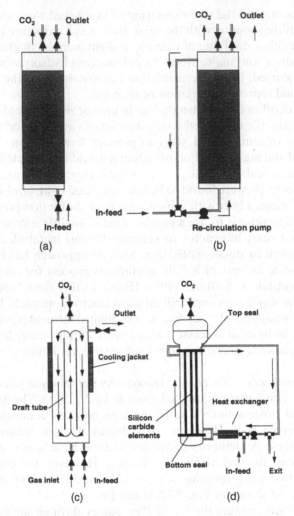

**Fig. 5.42**  Configurations of immobilised yeast bioreactors. (a) Fixed bed. (b) Fluidised bed. (c) Gas lift. (d) Loop reactor. (Re-drawn from Andries *et al.*, 1997b.)

relatively slow rate of fluid flow may be insufficient to generate a satisfactory diffusional gradient through the bead. The generally low mass transfer also makes dissipation of heat difficult and this further mitigates against the use of packed bed reactors for primary fermentation where exothermy is considerable and attemperation is critical to product and yeast quality. Some of these disadvantages can be minimised by using a combination of upward flow and a solid support medium, which is not readily compressed. Axelsson (1988) described a novel horizontal packed bed reactor with baffles, which significantly reduced carbon dioxide hold-up and much increased ethanol productivity. However, this approach has not been applied to beer production and fixed bed reactors are most suitable for either partial primary fermentation as in alcohol-free beer production or rapid maturation processes.

Stirred bed reactors are provided with a means of increasing mass transfer rates by forced agitation. This may simply be achieved by providing the vessel with a motorised impeller; however, care must be taken to ensure that the support is not damaged. More commonly, liquid fluidised bed reactors are used. These vessels have a continuous bottom in-feed and top located exit point. A proportion of the process flow exiting from the top is continuously recirculated back into the bottom of the reactor. This arrangement prevents compaction of the bed and thereby ensures relatively high rates of mass transfer between the fluid and cell support medium and presents little restriction to removal of carbon dioxide. The efficiency of mixing can be regulated by control of the relative rates of recirculation and medium addition. Any support medium can be used with fluidised bed reactors although those with a density slightly greater than the suspending medium are the most appropriate since this avoids the possibility of wash-out at high flow rates. Cho and Choi (1981) compared the efficiency of ethanol production from glucose in similar sized packed bed and fluidised bed laboratory reactors containing yeast entrapped in calcium alginate gel. They concluded that the fluidised bed reactor was twice as efficient as the packed bed type.

Other approaches for increasing yeast metabolic activity using bioreactors with improved mass transfer characteristics have been suggested. Mensour *et al.* (1995) described a gas lift fermenter containing yeast immobilised in κ-carrageenan. This type of fermenter has been widely used with filamentous fungi because it has good mixing characteristics but avoids damage to hyphae, which may occur with conventional stirred reactors. Use of such vessels with comparatively fragile gel beads represents a similar application. It comprises a cylindrical vessel with a central draft tube. Wort is continuously fed from the bottom and product from the top. Mixing is achieved by introducing a stream of air and carbon dioxide into the base of the draft tube. This encourages vertical flow up the draft tube and downward movement in region between draft tube and the outer wall of the vessel. Removal of carbon dioxide was via a flared vessel top. κ-carrageenan beads were chosen since they have a density close to that of wort, thereby facilitating ease of mixing. Overall, productivity of the vessel was controlled by a combination of regulating flow rate, temperature and the proportion of air in the gas stream. Although used only at pilot scale this vessel was used for primary fermentation with a residence time of 20 hours. If used in conjunction with an immobilised yeast reactor for accelerated maturation a total process time of two days was claimed. No data was provided regarding beer analysis.

Andries *et al.* (1997b) described a reactor in which yeast was immobilised in silicon carbide. The support consists of a number of ceramic elements, containing silicon carbide matrix, which is colonised by yeast cells, and a number of open channels for circulation of process fluid. When used for primary fermentation, wort is oxygenated and continuously fed into the bottom of the reactor. Fermenting wort is recycled through the silicon carbide elements and the void between the elements and the vessel wall via an external loop. A proportion is withdrawn from the external loop and fed into a second conical vessel. The contents of this vessel, which are inoculated with yeast issuing from the immobilised yeast reactor, are also mixed by circulation through an external loop. Claimed advantages for this system are high rates of mass transfer, reduced risk of clogging, ease of cleaning, ease of carbon dioxide removal and suitability for use with unfiltered wort. It has been used for the production of both ales and lagers apparently of perfectly acceptable flavour at a semi-industrial scale.

### 5.7.2   Commercial systems

Commercial scale immobilised yeast reactors have been developed for the production of low- or zero-alcohol beers and for continuous maturation. Application at production scale for primary fermentation has yet to be realised; however, a few systems have been developed to semi-industrial scale. For completeness, representative examples of these are described here.

5.7.2.1   *Alcohol-free beer.*   The layout of a typical plant for the production of alcohol-free beer is shown in Fig. 5.43 and is as described by van Dieren (1995) and Mieth (1995). This system uses yeast immobilised on DEAE-cellulose (Spezyme GDC®) and contained within a specifically designed vessel termed the 'Immocon' reactor (Fig. 5.44).

The wort used is of low fermentability and experiences an extensive boiling stage to reduce as fully as possible the concentrations of volatile aldehydes and carbonyls. To avoid oxidation careful handling is required to minimise oxygen pick-up. After cooling, the wort is acidified with lactic acid. This is also produced by a bioreactor, in

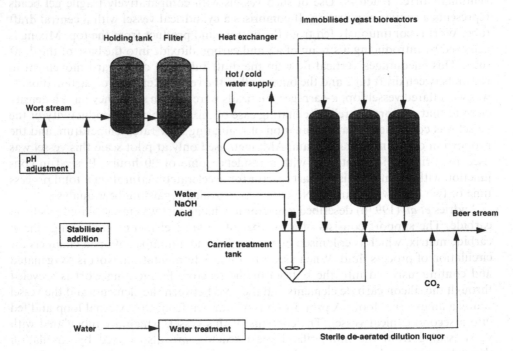

**Fig. 5.43**   Schematic of a plant for alcohol-free beer production using immobilised yeast bioreactors (redrawn from van Dieren, 1995).

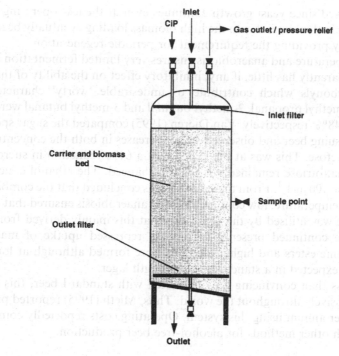

**Fig. 5.44** Immocon immobilised yeast bioreactor (re-drawn from Mieth, 1995).

this case containing immobilised bacteria, *Lactobacillus amylovorus*. The acidified wort is stored in a holding tank. Before delivery to the immobilised yeast bioreactors, the wort is filtered to remove solids (which could cause clogging) and it is cooled by passage through a heat exchanger. The process is described by Back and Pittner (1993) and Pittner *et al.* (1993).

The reactors have a capacity of 1.5 m$^3$ and each contain 1 m$^3$ (400 kg) of carrier. The plant is of a modular design with series of reactors. This arrangement provides operational flexibility to allow for fluctuations in demand and permits regeneration and start-up of individual reactors without disrupting production. Regeneration is performed by treating the used carrier with 2% w/v sodium hydroxide at a temperature of 80°C, followed by neutralisation with sterile carbonated water. Regenerated carrier is returned to the reactor and this is then inoculated with a pure yeast culture. Aerated wort at a temperature of 12°C is then circulated for 24 hours through the through the reactor. This promotes yeast growth and allows attachment to occur. At this stage, the reactors contain approximately 50 × 10$^6$ yeast cells per gram carrier. Wort flow commences at a rate of approximately 5 hl h$^{-1}$ and the temperature is held at 4°C. To minimise wort oxidation, strictly anaerobic conditions are maintained. Gradually the flow-rate is increased to 20 hl h$^{-1}$ and the temperature reduced to between 0 and 1°C. During this stabilisation phase, the yeast concentration doubles to 100 × 10$^6$ cells per gram carrier. Wort flow is from top to bottom and because of the lack of mixing is virtually of an ideal plug type. When working at full capacity, individual reactors have productivity of the order of 20 hl h$^{-1}$. Steady state conditions are never achieved since yeast growth continues even at the low operating tempera-ture. Reduction in flow rates due to the high biomass loading eventually becomes rate limiting, thereby providing the requirement for periodic regeneration.

The low temperature and anaerobiosis ensures very limited fermentation and yeast growth but apparently has little, if any, inhibitory effect on the ability of the yeast to reduce the carbonyls which contribute to undesirable 'worty' character. Thus, reduction of 2-methyl propanal, 2-methyl butanal and 3-methyl butanal were reduced by 100, 61 and 48%, respectively. Van Dieren (1995) compared the sugar spectrum of the wort and issuing beer and observed small increases in both the concentrations of glucose and fructose. This was at the expense of a 64% reduction in sucrose levels. Maltose and maltotriose remained virtually unchanged. The ethanol concentration never rose above $100\,\mu g\,l^{-1}$. From these data it was concluded that the combination of plug flow, low temperature, low contact time and anaerobiosis ensured that only 20% (7 mM) glucose was utilised by the yeast and that this mainly derived from sucrose hydrolysis. The continued presence of glucose repressed uptake of maltose and maltotriose. Some esters and higher alcohols were formed although at lower levels than would be expected in a standard, full strength lager.

Despite a less than convincing flavour match with standard beer, this plant has been used extensively throughout the world. Thus, Mieth (1995) reported production of 700 000 hl per annum using this system. Operating costs reportedly compare very favourably with other methods for alcohol-free beer production.

5.7.2.2 *Continuous maturation.* Production scale plant using immobilised yeast reactors for continuous beer maturation has been in operation at the Sinebrychoff Brewery in Helsinki, Finland, since 1990. A new installation capable of producing 1 million hl of beer per annum came on-stream in 1993 (Pajunen & Gronqvist, 1994). The decision to use this approach was taken at a time when an entirely new brewery was constructed on a green-field site (Pajunen & Jaaskelainen, 1993). This is worthy of note, as there will always be a natural inertia which will tend to resist replacing existing plant and process with a substitute that is novel and relatively unproven. Undoubtedly this factor will continue to be the biggest barrier to the more widespread adoption of techniques such as continuous maturation.

The plant is similar, with some modifications, to that used for the production of alcohol-free beer, as described previously (Section 5.7.2.1) and it also uses the same DEAE-cellulose carrier (Spezyme GDC®). Layout of the plant is illustrated in Fig. 5.45. Green beer is delivered from conventional fermenters via a high-performance continuous centrifuge. This serves two purposes. It reduces the yeast count in the green beer stream to a minimum level ($< 10^5$ cells $ml^{-1}$), which is essential for avoiding the development of yeast autolytic off-flavours in the subsequent heat treatment. In addition, centrifugation reduces the solids loading of the process flow entering the immobilised yeast reactors, thereby reducing the likelihood of clogging. The authors also noted that the clarification step resulted in an absence of tank bottoms down-stream of the centrifuge and that this contributed to overall process efficiency by minimising beer losses.

After centrifugation the beer is subjected to a heat treatment (90°C for 7 min.) during which α-acetolactate is converted to diacetyl. It is essential to maintain

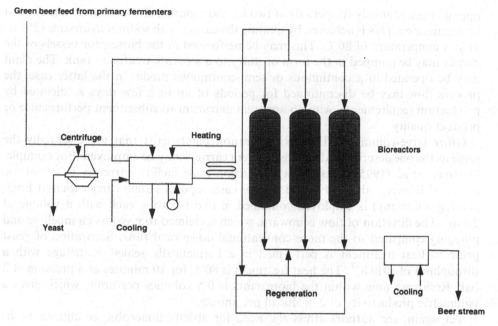

**Green beer feed from primary fermenters**

Centrifuge

Heating

Bioreactors

Yeast

Cooling

Regeneration

Cooling

**Beer stream**

**Fig. 5.45** Continuous maturation using immobilised yeast.

anaerobic conditions during this treatment in order to avoid undesirable oxidation reactions. In addition, in the absence of oxygen, a proportion of the α-acetolactate is converted directly to acetoin, thereby reducing the requirement for diacetyl reduction in the subsequent bioreactor step. The heat treatment has other benefits. It reduces the microbiological loading of the beer which in turn lowers the risk of contamination of the bioreactors. Further, it denatures proteases, which if allowed to persist, can reduce the concentration of foam positive proteins in the finished beer (Muldbjerg *et al.*, 1993). No changes in flavour or colour were observed during the heat treatment apart from increases in the concentrations of the carbonyls, furfural and phenyl ethanal.

Following the heat treatment the beer is cooled to 15°C and then passed through the immobilised yeast bioreactors. The production scale plant has four bioreactors each with a capacity of 7 m³. Contact time within the reactors is 2 hours. Diacetyl is converted to acetoin and 2,3-butanediol, carbonyls are reduced to the same levels, or less, than present in the beer prior to the heat treatment and residual fermentable sugars are assimilated (Gronqvist *et al.*, 1993). A further benefit is that the concentration of some short-chain fatty acids declines during passage through the bioreactors. This was ascribed to non-specific binding to the carrier as opposed to assimilation by the yeast. No changes were observed in the levels of esters and higher alcohols.

The beer issuing from the bioreactor has a yeast count of less than 105 cells ml⁻¹ and is reportedly relatively clear. At this stage, it is cooled to −1.5°C, prior to further stabilisation treatment. As with the alcohol-free beer bioreactor described in Section 5.7.1.1, the modular design provides operational flexibility. Each reactor may be operated continuously for periods of two to four months, after which time they must be regenerated. This is achieved by treating the carrier with sodium hydroxide (2% w/v) at a temperature of 80°C. This may be performed in the bioreactor vessels or the carrier may be pumped in the form of slurry to a separate treatment tank. The plant may be operated in a continuous or semi-continuous mode. In the latter case, the process flow may be discontinued for periods of up to a few days as dictated by production requirements, with no apparent detriment to subsequent performance or product quality.

Other large-volume continuous maturation plants exist that are essentially the same as the one described, although different carriers may be employed. For example, Hyttinen et al. (1995) described a production scale facility currently in use at the Hartwall Brewery, also in Finland. In this case, a porous glass carrier (Schott Engineering, Germany) is employed contained in two reactors, each with a volume of 2.5 m$^3$. The direction of flow is upward, which is claimed to give less channelling and plugging compared to the more conventional downward flow. Separation of yeast prior to heat treatment is performed in a hermetically sealed centrifuge with a throughput of 30 hl h$^{-1}$. The heat treatment is 80°C for 10 minutes at a pressure of 3 bar. Residence time within the bioreactors is 0.5 volumes per hour, which gives a volumetric productivity of 2 × 10$^6$ hl per annum.

Yet again, the authors stress the need for strictly anaerobic conditions to be maintained throughout the process to maintain product quality. Provided this was done, the total VDK (α-acetolactate + diacetyl) decreased to 30% of the initial concentration after the heat treatment and to 17% after passage through the bioreactor. Presumably, the large change after the heat treatment represents the proportion of α-acetolactate converted directly to acetoin. This is in accord with Inoue et al. (1991) who contended that the bulk of α-acetolactate is converted directly to acetoin by heat treatment (70°C for 30 minutes) under strictly anaerobic conditions and in the absence of yeast. These authors suggested that this approach could be used to replace the immobilised yeast step, or at least reduce the requirement for diacetyl reduction by yeast. Nonetheless, unless very careful control could be guaranteed it would be suspected that the majority of brewers would view such a drastic heat treatment with some trepidation.

The glass carrier would seem to have some operational advantages compared to DEAE-cellulose. Thus, Hyttinen et al. (1995) claimed that with careful handling the same column could be kept in constant operation for periods of up to one year, before regeneration became necessary. However, this did include some downtime for periodic re-pitching. Regeneration is accomplished in situ by treatment with hydrogen peroxide, acid washing and rinsing.

Another facility, albeit only at pilot scale, is described by Groneick et al. (1997). This has been installed at the Iserlohn Brewery in Germany and also uses sintered glass as a carrier. The plant is of the same design as described already; however, the authors make the valuable point that the totally inert nature of the carrier means that the process does not contravene the German beer purity laws.

5.7.2.3 *Primary fermentation with immobilised yeast.* There is no production scale immobilised yeast bioreactor system currently in use that is capable of carrying out primary fermentation. The immediate likelihood of success remains a remote possibility; however, a few systems have progressed to pilot scale. These share the common feature of having two or more vessels to allow the separation of the stages of primary fermentation necessary for the development of a balanced spectrum of flavour components (see Section 5.7.1). Two systems are described here which appear to show encouraging results.

Yamauchi *et al.* (1994) reported results obtained with a pilot scale three-stage system shown in diagrammatic form in Fig. 5.46. The first vessel consists of an aerobic stirred reactor of 200 hl capacity (100 hl working volume) maintained at 13°C by cooling through external jackets. A mechanical agitator is provided from the base of which sterile air is introduced. This arrangement ensures good mixing and oxygen solution. Wort is fed in at the top of the vessel close to the wall to avoid undue foaming. The rate of wort addition is controlled automatically, in response to continuous in-tank measurement of the specific gravity within the vessel. The aerobic and highly agitated conditions promote yeast growth at the expense of wort amino nitrogen. Consequently, there is a marked fall in pH and production of considerable higher alcohols.

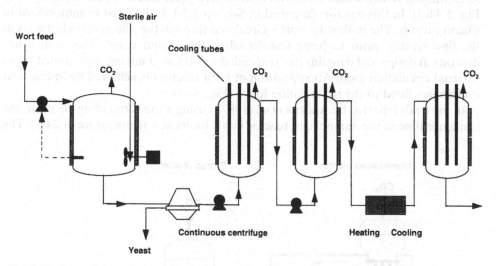

**Fig. 5.46**  System for continuous primary fermentation and maturation using immobilised yeast (adapted from Yamauchi *et al.*, 1994).

The process stream issuing from the stirred reactor is fed through a continuous centrifuge which reduces the yeast count to less than $1 \times 10^6$ cells ml$^{-1}$, prior to infeed into two linked immobilised yeast reactors. These vessels each have bed volumes of 100 hl and use a ceramic bead carrier. The high rates of exothermy associated with primary fermentation and the lack of mixing necessitate high cooling capacity to maintain the set-point temperature of 8°C. To accommodate this, both external wall jackets and internal heat exchanging tubes are provided. In the second stage bioreactor the apparent extract falls from 8°P to 2.5°Plato. It is during this treatment the bulk of the ethanol and esters are formed.

The third stage vessel is of similar design to the second and is used for continuous maturation. Prior to entry, the green beer is passed through an in-line heat exchanger where it is heated (60 to 80°C for 23–60 min) to convert α-acetolactate to diacetyl and acetoin, then cooled to approximately 0°C.

The total residence time for the process is given as 3–4 days. Times for the individual steps are 20–24 hours in the stirred vessel, 24–48 hours in the second stage immobilised yeast reactors and 24 hours in the maturation vessel. This gives a total process time of 72–96 hours. Comparative data for conventionally and continuously produced beers indicated significant differences, apparently related to altered patterns of yeast growth. Thus, the trial beer contained more ethanol, had a higher pII, increased sulphur dioxide, elevated bitterness and lower levels of higher alcohols. There were also differences in the contents of organic acids, notably reduced pyruvate and acetate but increased citrate, succinate and lactate. Apart from the differences in beer analysis, the process was shown to be unstable such that although six months' operation was achieved, it was not possible to attain a steady state and this resulted in continuously changing performance.

Andries *et al.* (1997b) reported experiences with a two-stage continuous primary fermentation system which utilises an immobilised yeast reactor of the type shown in Fig. 5.44(d). In this reactor, described in Section 5.7.1.3, the yeast is immobilised in silicon carbide. The in-flowing wort is circulated through the silicon carbide matrix in the first reactor, prior to being transferred to the second vessel. This is of cylindroconical design and contains free suspended yeast. Good mixing is promoted by an external circulation loop. Attemperation of both reactors is achieved by in-line heat exchangers fitted to the recirculation loops (Fig. 5.47).

In the trials reported by Andries *et al.* (1997b) using a top-fermenting ale yeast, the residence time in the immobilised reactor was 8 hours at a temperature of 24°C. The

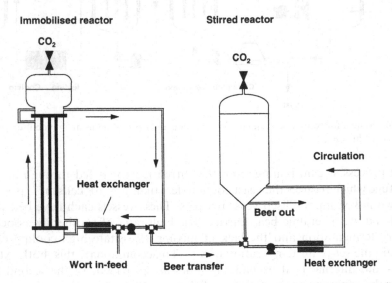

**Fig. 5.47**  Two-stage immobilised yeast reactor for primary fermentation (adapted from Andries *et al.*, 1997b).

free and immobilised yeast concentrations were respectively $55 \times 10^6$ and $187 \times 10^6$ cells $ml^{-1}$. The second stage vessel had a residence time of 16 hours and was also maintained at 24°C. The suspended yeast count in this vessel was $80 \times 10^6$ cells $ml^{-1}$. Apparent attenuation rates in each stage were respectively 35% and 78%. The system was reportedly run for periods of up to 6 months at a 'semi-industrial' scale and during this period steady-state conditions were apparently achieved. Both lager and ales were produced that were claimed to be good matches for conventionally fermented beers. Clearly, this approach would seem to offer much promise.

## 5.8 Pilot scale fermentation systems

Many pilot scale fermentation systems have been built specifically to model, on a semi-industrial scale, experimental designs derived from smaller laboratory scale prototypes. Some of these have been described already in previous sections of this chapter. For example, pilot scale fermentation systems have been much employed in the development of continuous and immobilised fermentation systems.

In addition to these experimental systems, more conventional but relatively small-scale fermenters are used as part of pilot scale breweries. These breweries may be used to investigate new processes at a scale intermediate between laboratory and production plant but are probably used most often to develop new beers. In this sense, the fermenters need to be capable of replicating, on a small scale, the behaviour of production scale fermenters. The requirement to produce, on a small scale, beers of the same quality as those made in large-scale breweries is now even more pertinent, further to the current upsurge in the 'brewpub' or 'microbrewery' (Cottone, 1985; Lewis & Lewis, 1996). It is true that many of these enterprises produce top-fermented ales where small volume and shallow vessels do not present any difficulties.

The design of scaled-down versions of cylindroconical vessels, as used for the production of bottom-fermented beers, in particular the optimum aspect ratio, is perhaps more problematic. Since there is a relationship between degree of agitation, vessel height and extent of yeast growth during fermentation it would be predicted that to achieve a match with production scale fermenters, small vessels should have a relatively higher aspect ratio. Thus, Meisel and Huggins (1989) concluded that an appropriate aspect ratio for a 32 litre capacity cylindroconical fermenter was 10 to 6:1. Fermentation performance and beer quality obtained with this design compared very well with a 660 hl production vessel.

Cooling jacket performance should be rated in accordance with the capacity of the vessel. It is not possible to duplicate accurately the effects due to the high hydrostatic pressures associated with tall vessels although elevated pressure may be applied to the contents of a small fermenter by restricting the escape of carbon dioxide. If possible the mains attached to pilot scale vessels should be of a diameter appropriate to the capacity of the vessel. Where pilot scale fermenters are devoted largely to experimental work, it is useful to provide several ports, at various heights, suitable for the installation of various probes. When not in use, the ports may be sealed with hygienic blanking devices. In addition, it is useful to be able to remove samples from several heights within the vessel. However, it should be appreciated that a multiplicity of over-intrusive probes will disrupt the normal patterns of mixing and agitation to a greater extent in a small vessel than would be the case in a production scale fermenter.

### 5.9 Laboratory fermentation systems

It is necessary to have appropriate laboratory scale fermentation systems in order to study the biochemistry that underpins brewery fermentation, to assess the properties of individual yeast strains, to screen and select new yeast stains and to develop novel processes. Much of this work may be carried out using conventional laboratory apparatus. This may range from the simple Erlenmeyer shake flask to the highly sophisticated fermenter fitted with a plethora of sensors and control devices capable of continuous monitoring and regulation of most process variables. A description of such apparatus may be found in any general text describing the methods used for the cultivation of micro-organisms.

This general purpose apparatus is designed for growing yeast and other micro-organisms principally under aerobic conditions and this usually involves some form of agitation to promote oxygen transfer. In order to study the behaviour of yeast under conditions comparable with a brewery fermentation, namely transient aero-biosis and lack of mechanical agitation, some specialised apparatus has been developed.

### 5.9.1  *Mini-fermenter*

Mini-fermenters are a simple and inexpensive method of carrying out multiple small-scale fermentations under controlled conditions (Quain *et al.*, 1985). This approach allows simultaneous operation of several tens of fermenters. They may be con-veniently used for screening the fermentation performance of many yeast strains under identical conditions. Alternatively, they may be used to study the effects on a single strain of varying conditions such as pitching rate, initial dissolved oxygen concentration and wort composition.

They consist of 120 ml glass 'hypovials' (up to 100 ml operating volume) fitted with gas-tight butyl septa held in place by metal crimped caps. Gaseous exchange is achieved via inlet and outlet tubes, each fitted with sterile filters made from cut down Pasteur pipettes, attached by silicone tubing to hypodermic needles the tips of which pass through the septa (Fig. 5.48). The contents of each hypovial are stirred using magnetic followers. Attemperation may be achieved by placing the fermenters in a suitable temperature controlled room. Alternatively, the vials may be partially immersed in an attemperated water bath. In this case, it is convenient to use multiple place immersible magnetic stirrers.

Vials may be sterilised by autoclaving whilst already filled with medium. In this case, temporary foam bungs are used which are replaced by sterile septa during initial setting up. The gassing tubes and needles are sterilised separately and fitted to the filled vials after attachment of septa and caps. Inoculation is accomplished by injecting an appropriate volume of yeast slurry through the septum using sterile hypodermic syringes and needles. Control of the atmosphere in each vial may be

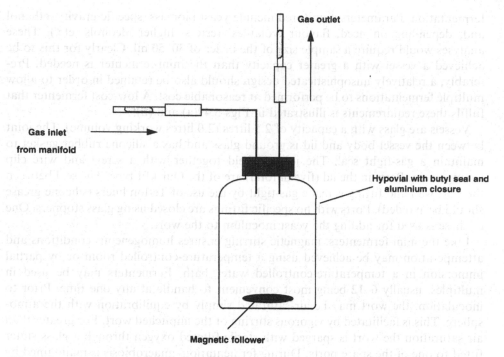

**Fig. 5.48** Hypovial mini-fermenters (from Quain *et al.*, 1985).

achieved by purging the headspace with an appropriate gas (air, oxygen, nitrogen or carbon dioxide). This is facilitated by connecting each inlet tube to the gas supply via a manifold and controlling the flow rate by the use of a rotameter. During operation, the progress of the fermentation may be conveniently monitored by recording weight loss at appropriate intervals. Alternatively, or in addition, multiple identical sets of fermenters may be used from which individual vials may be broached and analysed at intervals, as required.

Mini-fermenters can be usefully applied to studying the effects of varying the initial oxygen concentration on fermentation performance of individual yeast strains (Boulton & Quain, 1987; Bamforth *et al.*, 1988). Here, the hypovials are calibrated with a volume scale. Prior to filling, the air is displaced from each vial by sparging with sterile nitrogen and the yeast inoculum is added. Sterile wort is dispensed from two reservoirs, one made anaerobic by sparging with nitrogen and the other saturated with oxygen. Addition of a proportion of wort from each reservoir, using the volume calibration scale on each vial, allows the establishment of a gradation of initial dissolved oxygen concentrations.

### 5.9.2 Stirred laboratory fermenters

Mini-fermenters are useful where relatively large numbers of individual fermentations are to be performed. They are, however, cumbersome where several samples of reasonable volume require to be removed for off-line analysis during the course of the fermentation. Parameters monitored include yeast biomass, specific gravity, ethanol and, depending on need, flavour 'volatiles' (esters, higher alcohols, etc.). These analyses would require a sample size of the order of 30–50 ml. Clearly for this to be achieved a vessel with a greater capacity than the mini-fermenter is needed. Preferably, a relatively unsophisticated design should also be retained in order to allow multiple fermentations to be performed at reasonable cost. A low-cost fermenter that fulfils these requirements is illustrated in Figs 5.49(a) and (b).

Vessels are glass with a capacity of 2.5 litres (2.0 litres working volume). The joint between the vessel body and lid is ground glass and has a silicone rubber gasket to maintain a gas-tight seal. The joint is held together with a screw and wire clip arrangement. Ports in the lid (five in total) are of the Quickfit type. The seal between the port wall and fitting is made gas-tight by the use of Teflon liners (silicone grease should be avoided). Ports with no specific fittings are closed using glass stoppers. One of these is used for adding the yeast inoculum to the wort.

Like the mini-fermenters, magnetic stirring ensures homogeneous conditions and attemperation may be achieved using a temperature-controlled room or by partial immersion in a temperature-controlled water bath. Fermenters may be used in multiples, usually 6–12 being most convenient to handle at any one time. Prior to inoculation, the wort may be air-saturated simply by equilibration with the atmosphere. This is facilitated by vigorous stirring of the unpitched wort. For greater than air saturation the wort is sparged with sterile-filtered oxygen through a glass sinter fitted to one of the spare ports. During fermentation, anaerobiosis is maintained by flushing oxygen-free nitrogen through the headspace. As with the mini-fermenters, a manifold system and rotameter arrangement are used to deliver gas at an appropriate rate to each fermenter. Appropriate sterile filters, located in the gassing and exhaust lines, provide a microbiological barrier.

### 5.9.3 Tall tubes

Small scale stirred laboratory fermenters are useful where it is important to maintain homogeneous conditions within the vessel. Such an approach offers more precise control over the conditions within the fermenter. However, it does not mirror what happens in a production scale vessel and it is recognised that efficient stirring tends to promote yeast growth. Consequently, crop sizes are usually significantly bigger than those obtained from unstirred fermenters using identical combinations of wort and yeast. Altered patterns of yeast growth brought about by stirring will also influence the production of metabolites related to growth. When designing experiments, these caveats must be borne in mind in choosing a fermenter most suited to the particular avenue of investigation.

In order to study some aspects of yeast growth and fermentation it is necessary to use fermenters where, as in a production scale vessel, the activity of the yeast provides the only source of agitation. For example, when studying the flocculence characteristics of yeast strains, in particular, whether they are top or bottom fermenting. The industry standard fermenter used for such investigations is the 'E.B.C. tall tube' (Institute of Brewing, 1997).

The fermenter consists of a cylindrical glass tube (150 cm × 5 cm diameter) with an

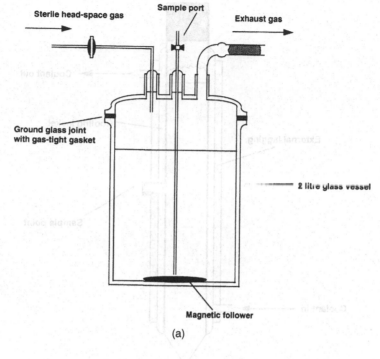

Sterile head-space gas

Sample port

Exhaust gas

Ground glass joint
with gas-tight gasket

2 litre glass vessel

Magnetic follower

(a)

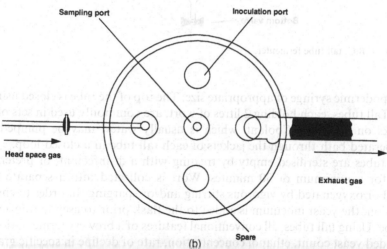

Sampling port

Inoculation port

Head space gas

Exhaust gas

Spare

(b)

**Fig. 5.49** (a) Simple stirred laboratory fermenter. (b) Top view.

outer jacket through which attemperated coolant is circulated. The base of the tube terminates in a tap to facilitate removal of yeast and beer at the end of the fermentation (Fig. 5.50). A sample point is located roughly a third of the way down the vertical side of the cylinder. This consists of a short glass side-arm fitted with a silicone rubber seal. Samples are removed by piercing the seal with a sterile needle fitted

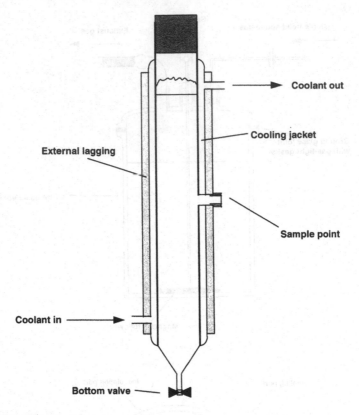

**Fig. 5.50**   E.B.C. tall tube fermenter.

to a hypodermic syringe of appropriate size. The top of the tube is closed using a foam bung. Tall tubes, each holding 2 litres of wort, are commonly used in sets of 12 fitted to racks on a trolley. Coolant, which is usually water, may be pumped from an attemperated bath through the jackets of each tall tube in a closed loop.

The tubes are sterilised empty by treating with a disinfectant, or preferably with steam for a minimum of 30 minutes. Wort is collected into a separate flask and aerated or oxygenated by vigorous stirring and/or sparging. In order to ensure good dispersion, the yeast inoculum is added to the flask prior to aseptic transfer into the tall tube. Using tall tubes, all conventional features of a brewery fermentation, such as suspended yeast count, ethanol concentration, rate of decline in specific gravity, etc., can be monitored by analysis of appropriate samples. In addition, the composition and flavour of the green beer can be assessed at the end of fermentation. Perhaps most importantly the technological behaviour of individual yeast strains can be compared by direct observation.

More sophisticated versions of the simple E.B.C. tall tube have been developed. For example, Sigsgaard and Rasmussen (1985) working at the Carlsberg Research Centre in Copenhagen, Denmark, described a system of 60 × 2 litre tall tubes. Termed the 'Multiferm', it was originally introduced to evaluate the fermentation performance of genetically modified yeast strains. The system was entirely automatic, all operations being controlled by a microcomputer. The latter also functioned as a data-logger.

Individual tubes were constructed from stainless steel and attemperated by circulating coolant through external jackets. At the top of each tube, a valve arrangement allowed a single aperture to function as an exit for evolved carbon dioxide during fermentation and as an entry point for CiP. The base of each tube terminated in a stainless steel container, which could be lowered, automatically, to allow access for yeast inoculation, prior to wort addition and removal of the crop after fermentation. A tube passing through the wall of the vessel, close to and directed towards the base, was used for wort addition and removal of beer at the end of fermentation. During wort addition, the downward flow ensured that the yeast inoculum was resuspended and dispersed. Each tall tube was fitted with a sample point, roughly half way down the vertical wall, consisting of an aperture closed by a silicone rubber membrane.

Two banks of 30 tall tubes were each mounted on circular motorised carousels. The role of these was to rotate each tall tube in turn into a station where automated procedures could take place. Cooled water for attemperation was supplied to individual jackets from a reservoir located in the centre of each carousel. After adding the pitching yeast slurry and raising the bottom containers, fermentations were initiated by adding the wort. This was metered in automatically using a dosing pump. During collection, the wort was aerated in-line by sparging with compressed air. When the fermentation was in progress samples were removed automatically, at selected time intervals, using a needle and tube arrangement driven by a two-step motor. Samples were delivered into evacuated 10 ml tubes held in racks. Preservation of samples of partially fermented wort in the interval between removal from the tall tubes and subsequent analysis was accomplished by a combination of storage at low temperature and addition of $1000\,\mathrm{mg\,l^{-1}}$ copper.

At the end of fermentation, the tall tubes were cooled to 5°C to encourage yeast sedimentation and green beer was removed from the bottom-located drain tube. The yeast crop was recovered from the bottom stainless steel container and could be transferred directly into a centrifuge, after which treatment the size of the yeast crop was measured by weighing. Empty tall tubes were cleaned by successive treatments with detergent and hot demineralised water, introduced into the top of each tube via the carbon dioxide exhaust line. During this treatment, the bottom container was replaced temporarily with a cup containing a bottom drain.

A further incarnation of the automated E.B.C. tall tube has now been developed, also at the Carlsberg Research Centre in Copenhagen (Skands, 1997). This comprises eight pairs of 2 litre stainless steel tubes with automatic sampling and logging facilities. In this case, full and individual control of temperature profile, wort oxygenation and pressure is provided.

A less sophisticated tall tube-type fermenter was described by Briggs et al. (1985). These were used as part of a six-line microbrewery designed to produce a brew-length of approximately 3.5 litres of hopped wort. The fermenters consisted of stainless steel cylinders, 87 cm in length and with an internal diameter of 7.7 cm, giving a total capacity of approximately 4 litres. The cylinder ends were sealed with detachable stainless steel discs. The top disc was perforated by a number of hygienic fittings, which allowed aseptic removal of samples, gas rousing and escape of carbon dioxide. The exhaust gas escape valve could be sealed to allow pressurisation up to 2 bar g. There were no facilities for attemperation of vessels, other than location in a temperature-controlled room.

# 6 Fermentation management

## 6.1 Wort collection

Wort collection describes that part of the brewing process in which wort is delivered from the brewhouse to the fermenting vessels. This must be done in such a way that after the process is complete appropriate initial conditions are established within the filled fermenting vessels. At first sight this may appear a relatively straightforward operation; however, it involves several distinct steps which if performed incorrectly will have adverse effects on both fermentation performance and the quality of the resultant beer.

The wort parameters that have to be controlled during the collection process are: volume, specific gravity, temperature, sterility and clarity. In addition, at some point in the process oxygen and yeast must be added. The total time taken to fill the fermenter and the sequence and timing of oxygen and yeast addition has to be considered. Finally, it is usually necessary to make some additions to the wort in fermenter, to achieve the desired composition. The technical complexity of the methods used to control the parameters listed above tends to reflect the sophistication of the brewery.

### 6.1.1 *Wort cooling and clarification*

The final stage of wort preparation is the copper boil. Before the wort can be added to the fermenting vessel it must be cooled. Both the boiling and cooling stages are accompanied by precipitation of solid materials. These are referred to as 'hot' and 'cold break', respectively, or 'trub', the German for sediment or clouding. Trub is a heterogeneous complex formed from a coagulation reaction between wort proteins and other nitrogenous components with polyphenols. In addition, insoluble minerals may be present, together with a lipid fraction derived from malt and hops and some hop resins. Specific process plant is provided for the separation of hot break. This may be of several types depending on the brewery in question but in modern plants the whirlpool is the most commonly used method (see Section 2.3.6).

In traditional breweries, the hot wort was cooled in a shallow open vessel often referred to as a 'coolship'. Cooling was achieved simply by exposing the wort to the ambient temperature of the room. Occasionally ceiling mounted fans were used to accelerate the process, or more efficient cooling could be obtained by allowing ingress of refrigerated air into the room. Open wort coolers had the advantages that aeration was achieved naturally by exposure to air and the shallow design of the vessel facilitated sedimentation of cold break such that relatively clear wort could be decanted, leaving the sludge behind. However, these advantages were far outweighed by poor hygienic design. In addition, the manner of cooling made it impossible to avoid some oxidation of wort components since inevitably exposure to air commenced when the wort was still hot. The risks of wort contamination could be minimised by controlling the atmosphere of the room in which the coolships were placed. Thus, the rooms were designed to avoid the possibility of condensation forming on walls, ceilings and other internal structures such that it could drip back into the vessels. Furthermore, the supply of cool air was filtered to minimise levels of airborne contamination. Nevertheless, this operation was fraught with risk and could only be managed by restricting the period that the wort spent in the coolships to no longer than a couple of hours.

An alternative to the open coolships was the use of closed wort receiving vessels. These took several forms but were generally deeper than coolships (1–2 metres as opposed to less than 20 cm). Cooling could be accelerated by the provision of internal coils through which cold water or brine was circulated. The design allowed for settlement of solid material and subsequent separation from the clarified wort during run-off. The enclosed design presented a barrier to contamination with air-borne micro-organisms. However, there was less opportunity to ensure that the wort became saturated with air.

In the interests of efficiency, the static cooling vessels have given way to in-line methods of temperature control. Originally, especially in the United Kingdom, coolships were replaced by horizontal refrigerators. These consisted of slightly inclined shallow vessels containing coils fed with a coolant. The hot wort was allowed gently to pass over the coils, thereby lowering the temperature and allowing cold break to form and settle out. Such coolers were extravagant in usage of floor space, and subsequently they were replaced with vertically oriented refrigerators. In these, the wort was allowed to trickle in a thin stream down a series of plates through which a suitable coolant was circulated. Such coolers could be quite efficient and allowed natural aeration of worts to occur. However, no means of cold break separation was provided. If the latter were required then additional plant downstream of the cooler was needed. Of course, any open cooling system required careful control of the atmosphere of the surrounding room to minimise the risk of microbial contamination.

To maintain the efficiency benefits of in-line cooling and provide adequate microbiological control most modern breweries now use enclosed heat exchangers for wort cooling. Typically these are plate heat exchangers or 'paraflows', as originally introduced by Seligman of A.P.V. Limited (Dummett, 1982). They are made up of arrays of elements each of which consist of a stainless steel plate which has holes in each corner and a series of grooves (Figs 6.1 and 6.2). Individual plates are joined together within an enclosing frame. The cavities between each plate are connected by the four corner holes, thereby providing channels for liquid flow. The grooves in the plates are compressed together to form serpentine channels for liquid flow. A series of gaskets between the plates provide watertight seals and control the direction of liquid flow such that coolant and wort circulate through alternate cavities in a counter-current fashion.

In order to maximise rates of heat transfer the plates are constructed from thin stainless steel, typically 0.5 mm, and the grooves are designed to maximise turbulent flow. Paraflows are sized with respect to the desired wort flow rate and the degree of

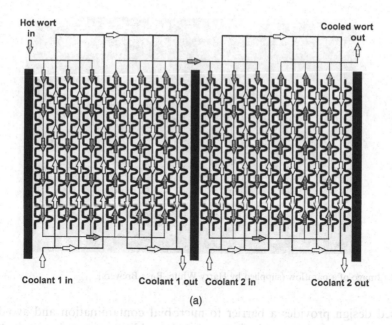

(a)

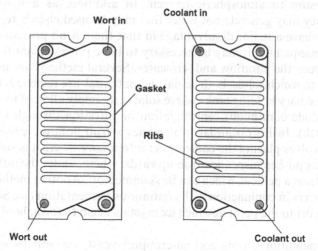

**Fig. 6.1** (a) Side view schematic of a two-stage paraflow. (b) End view of two adjoining plates of wort paraflow.

cooling required. Usually the temperature of the wort issuing from the paraflow is controlled automatically by varying the coolant flow rate. The temperature of the coolant depends on the required cooling duty. In the case of relatively warm ale fermentations a single-stage water-cooled system is adequate. For lager worts it is usual to supplement the water-cooling stage with additional plates through which a refrigerant such as brine, ethanol or glycol is circulated.

Paraflow wort coolers are highly efficient and provide very accurate control of the temperature of wort delivered to the fermenter. They are easily cleanable and totally

**Fig. 6.2**    Image of a paraflow (supplied by Harry White, Bass Brewers).

enclosed design provides a barrier to microbial contamination and avoids uncontrolled exposure to atmospheric oxygen. In addition, as a by-product of their
operation they may generate hot water that may be used elsewhere.

Paraflows have a major disadvantage in that there is no provision for wort clarification. Consequently, it may be necessary to have process plant for solids removal
located between the paraflow and fermenter. Several methods are used, which differ
in the extent to which solids, both actual and potential, are removed. For example, the
use of intermediary holding tanks where solids are simply allowed to settle out. In-line
methods include continuous centrifugation or filtration through kieselguhr (diatomaceous earth). In European lager breweries a trub flotation process is sometimes
used that involves placing the cooled wort into a tank where it is subjected to forced
aeration. This pushes suspended trub upwards, where, under the influence of surface
tension, it forms a pellicle, which can be skimmed off. Another method, used by New
Zealand brewers in conjunction with continuous fermentation (see Section 5.6.2.2), is
to chill the wort to the point at which ice crystals begin to form, hold for a period and
then filter.

Lager fermentations, using bottom-cropping yeast, are perhaps less tolerant of the
presence of trub. A characteristic of these beers is their excellent foaming properties
and the presence of any head-negative materials would be most undesirable. In
addition, wort solids in fermenter will tend to settle and contaminate the yeast crop,
possibly necessitating a subsequent cleansing step. Compared to ales, lager beers
require a high degree of carbonation and excessive loss of carbon dioxide during
fermentation due to the presence of nucleating materials would require re-adjustment
of gas levels at the conditioning stage.

The requirement to clarify wort, the stage at which it is performed and the manner
in which it is accomplished is dependent on the type of fermentation. It is also a
somewhat contentious issue. The presence in fermenter of appreciable wort solids has
potentially both negative and positive influences. It has been suggested that

unsaturated fatty acids in the lipid fraction of cold break may contribute to yeast nutrition and reduce or supplement the requirement for wort oxygenation (see Section 3.5). Siebert *et al.* (1986) noted that turbid worts produced more rapid fermentations than clear worts. However, these authors concluded that this was due to the particles providing nucleation sites for formation of carbon dioxide bubbles. In the opinion of the authors this alleviated inhibition of yeast growth by high concentrations of dissolved carbon dioxide. This view was supported by the observation that the stimulatory effects of trub particles could be replicated by the addition of activated carbon or kieselguhr. Lentini *et al.* (1994), also concluded that acceleration of fermentation rate due to trub particles was a physical effect. However, they also reported influences of trub on beer flavour. Thus, levels of linolenic acid in the lipid fraction of trub were shown to vary inversely with acetate ester concentrations in beer. It was also shown that trub had the ability to bind added zinc, such that it may not be available to the yeast. The presence of excessive trub may have other undesirable effects. The lipids may have deleterious effects on beer foam, and other components may produce undesirable flavours and decreased flavour stability (Olsen, 1981; Schisler *et al.*, 1982; Carpentier *et al.*, 1991).

Traditional ale fermentations performed in square fermenters are probably the most tolerant of the presence of cold break. These fermentations are managed in such a way that there is an opportunity to remove trub. Thus, much solid material rises to the surface with the first yeast head, at which point it may be skimmed off before the formation of the second clean yeast head. The latter is retained for subsequent repitching. Other unwanted solid material forms a sediment which is separated when the vessel is emptied. In the dropping system (see Section 5.3.3) a similar solids separation is accomplished by starting fermentation in one vessel, then, after an interval to allow sedimentation of trub, transferring the clarified fermenting wort to another vessel. Ales have relatively low carbon dioxide contents, and therefore efficient removal of this gas during fermentation is an advantage. In addition, ale yeasts tend to have quite low oxygen requirements, typically air saturation being sufficient to produce profuse yeast growth. This may be partially a result of some unsaturated fatty acid being provided by the wort.

Continuous free or immobilised yeast fermentation systems are the least tolerant of worts containing high solids contents. Clogging of bioreactors can restrict process flow, necessitating frequent repacking and consequent loss of overall productivity due to excessive downtime. With such systems (see Sections 5.6 and 5.7), production of bright wort is essential. Hence, the practice of the New Zealand brewers of chilling wort and filtering prior to delivery to the continuous fermenting vessels.

In the case of conventional batch fermentations, the advantages and disadvantages of wort clarification have to be weighed based on considerations of cost versus process performance. Whatever the decision, it is certainly true that wort solids have to be removed at some stage in the brewing process. The formation of wort solids continues, albeit at a lower rate, after the initial cooling step. Therefore, a case may be made for delaying clarification to conditioning where removal of trub, residual yeast and chill haze materials can be combined in a single process. However, in order to maintain good fermentation control and ensure superior beer quality, the weight of evidence suggests that it is preferable to remove the bulk of the trub before fermentation commences. However, the disadvantage is that if a wort paraflow cooler is employed then there will be an on-cost due to the need for subsequent wort clarification plant. In addition, there will be a decrease in overall process efficiency because of wort losses entrained with the solids.

### 6.1.2 Wort oxygenation

Traditional fermentation systems rely on wort becoming saturated with air during the processes of cooling and transfer into fermenting vessel, as discussed already. Modern installations, which are largely enclosed and essentially anaerobic, require provision for forced addition of oxygen. This is accomplished by the addition of oxygen, in-line, either as a pure gas or in the form of air, to the wort as it is delivered to the fermenting vessel. Regulation of the concentration of wort dissolved oxygen is a crucial part of the strategy by which yeast growth is controlled. Therefore, it is essential that dosing of oxygen is precise and repeatable. In general, the most sophisticated wort oxygenation systems are required for large capacity fermentations, particularly those performed at high gravity. In the case of small-volume fermentations using sales gravity worts, the requirements of the oxygenation system are perhaps less demanding.

Oxygen may be added to the wort on the hot or cold side of the paraflow. Adding oxygen to hot wort has the advantage that the risk of introducing microbial contaminants with the gas is small. Furthermore, efficient solution is favoured because of the good mixing characteristics of the paraflow. However, a large proportion of the added oxygen may never become available to the yeast since it may be wasted in wort oxidation reactions. Furthermore, it is more difficult to attain high oxygen concentrations since the solubility of oxygen in wort decreases with increase in temperature. For these reasons, oxygenation of wort on the hot side of the paraflow is only suitable for those fermentations whose oxygen requirement can be satisfied by air saturation or less.

Addition of oxygen to cooled worts has the advantage that solubility is increased; however, the sterilising effect of the hot wort is lost and a suitable steam sterilisable microbiological filter must be used. In order to make use of the dissolving capabilities of the paraflow it is common to add the gas between the two cooling stages. If oxygen or air must be added after the paraflow, it is essential to improve the gas transfer characteristics of the system. This may be accomplished by introducing the gas in the form of fine bubbles via a stainless steel sinter or candle, preferably followed by an in-line mixer. Plant specifically designed to promote efficient solution of oxygen in wort was described by Ringholt (1997). This consisted of a cylindrical vessel through which the wort passed during collection. The core of the cylinder contained an infusion chamber into which air was introduced. The wort was injected into the infusion chamber via perforations in the wall. The infusion chamber was divided into four sections by a series of three plates located on the long axis of the cylinder. These plates served to close some perforations in response to fluctuations in the wort flow rate such that a constant flow velocity was maintained in the aerator. In this way, a constant aeration rate was achieved. The design of the infusion chamber facilitated good turbulent mixing of wort and air. An in-line mixer down-stream of the aeration chamber further encouraged gas solution.

It is possible to use air or oxygen as the source gas although if the former is used this does, of course, limit the oxygen concentration that can be achieved. The solubility of oxygen is influenced by temperature and the concentration of the wort, as shown in Figs 6.3(a) and (b). A desired dissolved oxygen concentration may be achieved by adding air or oxygen at a given rate which takes into account the temperature and gravity of the wort, the wort flow rate and the pressure within the wort main. This may be calculated from the following relationship (assuming perfect solution of the gas):

$$FO_2 = \frac{RT}{PO_2} \times \frac{CFw}{32000} \ m^3 \ s^{-1}$$

Where:

$FO_2$ is the gas flow rate;
R is the gas constant ($8.3143 \times 10^{-5} \ m^3bar/^\circ Kmol$);

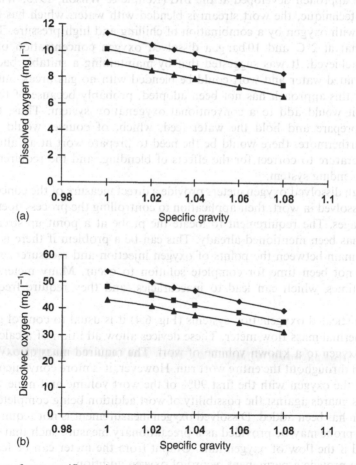

(a)

(b)

Fig. 6.3 Effect of wort specific gravity and temperature on the solubility of oxygen (♦, 10°C; ■, 15°C; ▲ 20°C) with (a) air-saturated wort at NTP and (b) oxygen-saturated wort at NTP.

T is the temperature at NTP;
PO$_2$ is the partial pressure of oxygen in the gas stream;
C is the desired oxygen concentration (mg kg$^{-1}$);
Fw is the wort flow rate (kg s$^{-1}$).

Where pure oxygen is used, PO$_2$ equals the hydrostatic pressure within the wort main. Where air is used the value of PO2 reduces to the value of the hydrostatic wort main pressure multiplied by the factor 0.2094.

In practice, it may be difficult to ensure perfect solution of the inflowing gas. Better control may be achieved by locating a dissolved oxygen meter up-stream from the gas addition point. Using a suitable controller, the rate of gas addition may be regulated automatically to give a desired dissolved oxygen concentration. It should be noted that the dissolved oxygen probe must be located at a point before yeast addition in order to avoid errors of underestimation due to yeast oxygen uptake.

The problems of inefficient oxygen solution in wort mains have been addressed by a novel approach developed at the BRi (Rennie & Wilson, 1975; Wilson, 1975). With this technique, the wort stream is blended with water, which has been super-saturated with oxygen by a combination of chilling and high pressure. Thus, it was claimed that at 2°C and 10 bar g a dissolved oxygen concentration of 700 mg l$^{-1}$ could be achieved. It was suggested that by maintaining a suitable back-pressure, the oxygenated water and wort could be blended with no gas break-out. Although ingenious, this approach has not been adopted, probably because of the complexities that it would add to a conventional oxygenation system. Thus, the requirement to prepare and hold the water feed, which, of course, would have to be sterile. Furthermore, there would be the need to prepare wort at an altered gravity and temperature to correct for the effects of blending, and the requirement for an accurate blending system.

Although dissolved oxygen meters provide a direct measure of the concentration of oxygen dissolved in wort, their application to controlling the process does have some disadvantages. The requirement to locate the probe at a point up-stream of yeast addition has been mentioned already. This can be a problem if there is insufficient length of main between the points of oxygen injection and measurement such that there has not been time for complete solution to occur. Many meters have slow response times, which can lead to inaccuracies, and they require frequent maintenance.

In sophisticated oxygenation systems (Fig. 6.4) it is usual to control gas addition using a thermal mass flow meter. These devices allow addition of a calculated total mass of oxygen to a known volume of wort. The required mass of oxygen may be metered in throughout the entire wort run. However, it is more convenient to arrange to add all the oxygen with the first 90% of the wort volume and none with the last 10%. This guards against the possibility of wort addition being completed before all the oxygen has been added. Dissolved oxygen measurement is not required although an in-line probe may be provided as a precautionary measure such that wort flow is suspended if the flow of oxygen fails. Output from the meter can be fed to a data-logger and provide a permanent record of oxygen addition.

Thermal mass flow meters allow precise control of addition of oxygen; however, it

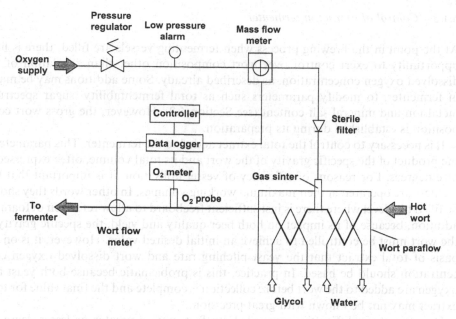

**Fig. 6.4** System for wort oxygenation.

is still necessary to ensure complete solution. Stainless steel sinters with a large surface area and pore size of approximately 20 μm, used in combination with an in-line mixer and pressurisation of the wort main, provide a suitable method. Although pressurisation of the wort main aids oxygen solution, some gas break-out in the fermenter, which will be at atmospheric pressure, is inevitable. Furthermore, it is not possible to measure the dissolved oxygen concentration in fermenter since by the time wort collection is completed, the yeast will also be present and assimilation of oxygen will have commenced.

In practice, it is usual to arrive at an optimal oxygen dosage rate based on empirical observation of fermentation performance. It follows that the most important consideration is to use a system that provides a consistent wort dissolved oxygen concentration. It is important to ensure that all potential sources of wort oxygenation are eliminated other than that added deliberately. If dilution water is added after wort, it should be either deaerated or included in the calculation of oxygen added. Post-pitching air rousing in fermenter, which is sometimes used as a method for ensuring proper dispersal of yeast, is not recommended where precise control of wort oxygen concentration is considered to be critical.

The practice of adding oxygen over the whole period of wort collection has been questioned. Lodolo *et al.* (1999) reported that the optimum time for oxygen addition was four hours after pitching. The authors claimed that this allowed the dissolved oxygen tension to be reduced from $16 \, \text{mg} \, l^{-1}$ to $8 \, \text{mg} \, l^{-1}$ with no alteration in fermentation performance. For brewing at production scale, it was recommended that all the yeast was pitched with the first batch of anaerobic wort. Oxygen was added with subsequent batches of wort.

### 6.1.3 *Control of extract in fermenter*

At the point in the brewing process when fermenting vessels are filled, there is little opportunity to exert control over wort composition other than regulation of the dissolved oxygen concentration, as described already. Some additions may be made, in fermenter, to modify parameters such as total fermentability, sugar spectrum, metal ion and mineral salt content (see Section 6.2). However, the gross wort composition is established during its preparation.

It is necessary to control the total extract added to the fermenter. This parameter is the product of the specific gravity of the wort and its total volume, often expressed in litre degrees. For reasons of efficiency of vessel utilisation, it is important that fermenters are operated at their maximum working volumes. In other words they should be filled to a level where there is just sufficient freeboard to allow retention of foam. In addition, because of its impact on both beer quality and yield, the specific gravity of the wort must be controlled to achieve an initial desired value. However, it is on the basis of total extract that the yeast pitching rate and wort dissolved oxygen concentration should be based. In practice, this is problematic because both yeast and oxygen are added to the wort before collection is complete and the final value for total extract may not be known with great precision.

How are these difficulties reconciled? In fact, with a greater or lesser degree of success depending on how carefully the process is managed. The total extract delivered to the fermenter is controlled by operations within the brewhouse, notably, the quantities of raw materials used and the manner of their treatment during conversion to clarified and cooled wort. In a typical modern brewery, the value for total extract may be measured at the whirlpool stage. Based on this measurement the total yeast to be pitched and mass of oxygen to be added should be calculated. This is rarely done and it is assumed that operations in the brewhouse are sufficiently controlled to produce a consistent batch of wort. Yeast pitching and oxygenation rates are usually then based on this nominal value.

In some breweries, the wort may be held in a receiving vessel located between brewhouse and fermenter. At this stage, adjustments to gravity and clarity may be made and the total extract measured, thereby allowing an accurate calculation of the quantity of yeast to be pitched and oxygen to be added. Unfortunately, this method is extravagant in the use of expensive vessels, slows process times and presents potential risks for microbial contamination. It is of most use where fermenters may be filled with a single batch of wort, which is not the case in many modern breweries. For these reasons, it is more common to fill fermenters directly from the whirlpool.

The quantity of extract transferred to fermenter is dictated by the efficiency of whirlpool operation. Very large capacity fermenting vessels require several individual batches of wort to be discharged from the whirlpool and this magnifies the effects of losses of extract. Wort delivered from the brewhouse is concentrated and needs to be diluted to achieve a desired gravity. This may be performed in fermenter by off-line measurement of gravity and subsequent dilution. If this is approach is used, it is essential to make sure that the contents of the vessel are mixed to homogeneity. This may be achieved by gas rousing, pumping wort through an external loop to promote mixing or use of a mechanical stirrer. The latter option is preferable since it is efficient and carries only a small risk of disturbing the wort oxygen concentration. External loop systems do not provide good mixing if dilution water has been added in such a way that there has been appreciable layering. Gas rousing is the least satisfactory method of mixing the contents of fermenters. If an inert gas is used such as nitrogen or carbon dioxide, some dissolved oxygen will be lost due to gas stripping. Conversely, if air or oxygen is used, control of wort oxygen concentration is lost. In deep vessels, dilution water should be added from the bottom since the density differential will promote mixing. The dilution water should be deaerated to avoid the inadvertent and uncontrolled addition of oxygen.

Wort dilution may be performed in-line during collection using an automatic blending system. This approach probably offers the best chance of maintaining rigorous control of the collection process. Degrees of automation are possible. The simplest system is that in which the volume and specific gravity of the wort are measured in the whirlpool. With knowledge of the desired final gravity, the quantity of dilution water required may be calculated and blended, in-line with the wort, as it is delivered to the fermenter. The blending system should preferably be a proportioning system, which adds the water at a constant ratio throughout the entire run. This provides wort of a consistent gravity and temperature and therefore, constant oxygen solubility. Automatic in-line or in-tank measurement of specific gravity is possible (see Section 6.3.2.3). Such sensors are suitable for automatic control of wort dilution. Systems such as this have been developed; however, they are used for dilution of high gravity beers after fermentation, or possibly monitoring of fermentation progress and not for control of wort collection.

The final stage of wort collection is to provide a record of the total extract in the fermenter. This requires a measure of the specific gravity and the volume. In some countries such as the United Kingdom and Belgium, which used to levy excise duties based on wort sugar content, there was a statutory requirement to maintain an accurate record of wort volume and concentration. In consequence, all fermenting vessels were gauged to allow volume assessment via dip measurements. Since by the time wort collection was completed, the yeast would have been added and fermentation commenced, it was not possible to measure the initial specific gravity. Instead, this was inferred from a value termed the original gravity (OG) of the wort. This used a correction factor for the quantity of fermentable sugar used in yeast metabolism but not converted to ethanol (see Section 1.4). Although excise payment is now based on ethanol concentration in beer, records of OG and wort volume are still maintained to allow assessment of fermentation efficiency.

The change in excise law to end-product duty allows greater latitude in the methods used to assess total extract. For example, vessel volumes may be measured using in-tank level sensors. Daoud (1991) describes the application of ultrasound to level measurement. Here a transducer in the base of the vessel senses liquid depth by the time taken to detect the return of a pulsed signal reflected back from the gas-liquid interface. Depth measurement must take into account possible interfering factors. For example, differential volume changes in liquid and vessel due to variations in temperature, and the need for the liquid to be absolutely still. This avoids errors due to swirl and the effects on volume of entrained gas. An alternative method of wort volume measurement is to use in-line flow meters to record the total volume of liquid added to fermenter. Such devices may be needed as part of the control systems for yeast pitching and wort oxygenation.

### 6.1.4  *Control of yeast pitching rate*

The pitching rate is defined as the concentration of yeast suspended in wort at the start of fermentation. The process of inoculating wort with yeast is described as 'pitching' and the yeast reserved for this purpose 'pitching yeast' (for a review see O'Connor-Cox, 1998b). The term, as the word suggests, derives from the physical act of throwing yeast into the fermenting vessel. For example, in Scandinavian countries it was custom to prepare a circular structure made from many wooden laths woven together and called a pitching wreath. This was recovered from a previous fermentation. The large number of crevices and surfaces in the wreath provided places for yeast cells to lodge and survive the period of storage between fermentations. These provided the inoculum, when the wreath was pitched into a subsequent fermentation. In the United States the term 'brink' yeast is synonymous with pitching, deriving in this case more obviously from that which is present at the start of the fermentation.

The aim of the process is to obtain a defined suspended yeast count in wort at the start of fermentation. In fact, this parameter is rarely measured since by the time wort collection is completed the yeast may have started to multiply, thereby negating the value of measurements. Instead, it is usual to rely on a related parameter such as addition of a known wet weight of yeast, or volume of yeast slurry. Control of pitching rate has two components. First, an analysis of the yeast reserved for pitching, and, second, a method for accurately transferring a desired quantity of this yeast into the wort. Several approaches have been developed to accomplish these two tasks. These are associated with varying degrees of precision and each has advantages and disadvantages.

### 6.1.4.1  *Direct weight of yeast cake.*

In some traditional fermentations, particularly using top-cropping ale types, it is common practice to store the skimmed yeast in the form of pressed cake. In subsequent fermentations, the pitching rate is controlled simply by addition of a defined weight of yeast cake. The pressed yeast may be added directly to the fermenter although it is more usual to re-suspend it in cold water in a separate pitching tank to facilitate subsequent dispersion in the wort.

The quantity of yeast used is based on an assumption that there is a defined correlation between yeast wet weight and cell count. Thus, the UK brewers' 'rule of thumb' that 1 lb pressed yeast per UK barrel is equivalent to $10 \times 10^6$ cells per ml in the pitched wort. In fact, this relationship is obviously an approximation, there being some variation depending on yeast cell size. However, this method provides reasonably precise control of pitching rate since yeast skimmed from top-cropping fermenters tends to be relatively free from contaminating non-yeast wort solids, the major potential source of error. It does have the disadvantage that the yeast is inevitably exposed to considerable atmospheric oxygen during handling. This is especially the case where slurried cake yeast is pitched into empty open square fermenters before the wort is pumped in.

From a fermentation control standpoint, exposure of pitching yeast to oxygen is undesirable since some limited sterol synthesis may occur. This promotes excessive yeast growth in the subsequent fermentation if the wort oxygen concentration is not reduced (see Section 6.4.1). This is a small disadvantage in the case of typically small-scale ale fermentations, which use this pitching rate control system. Thus, maintenance of high process efficiency through controlled yeast growth is less important than providing conditions that ensure a rapid onset of fermentation and a profuse crop. Exposure of pitching yeast to air and oxygenation of wort promotes both of these.

### 6.1.4.2 *Metered addition of yeast slurry.*

Pitching yeast is most commonly stored in the form of a slurry in which the cells are suspended in beer derived from the previous fermentation. The pitching rate of a subsequent fermentation is controlled by the metered addition of a known weight or volume of slurry, either directly to the empty fermenting vessel or more usually in-line with the wort during collection. Analysis of the pitching yeast slurry is required in order to make the calculation of the quantity of yeast to be added. In the majority of breweries, the analysis is performed in the laboratory on samples removed from yeast storage vessels.

Two approaches to the analysis are possible. First, determination of the proportion of suspended solids, expressed as percent weight to weight or weight to volume. The relative proportions of yeast and beer in a sample of slurry are determined by centrifugation and direct weighing of pellet and barm ale. Occasionally, the slurry may be analysed using a graduated centrifuge tube and the solids content defined as percent suspended solids volume to volume. In the interests of precision, this latter approach cannot be recommended. On a separate portion of the sample, the yeast viability may be determined using a microscopic counting and dye staining method, such as the methylene blue test (see Section 7.4.1). This allows a correction factor to be introduced so that pitching rates are controlled in terms of viable spun solids. Such tests are valuable aids in that they also allow assessment of pitching yeast quality or condition (see Sections 7.4.1 *et seq.*). Thus, it would be usual to reject pitching yeast with viability less than a minimum quality limit.

The second method of yeast slurry analysis is to determine the cell count per unit mass or volume of slurry. This has the benefit that unlike the simpler centrifugation procedure there is an opportunity to correct for the presence of non-yeast solids. Furthermore, It allows pitching rates to be controlled directly in terms of yeast count. Two methods of cell counting are commonly used.

First, using an electronic particle counter, which gives a rapid and automatic measure of the suspended yeast cell count. Second, direct microscopic enumeration using a counting chamber such as the haemocytometer (*EBC Analytica Microbiologica II*, Methods 3.1.1.1 and 3.1.1.2, respectively).

Both counting methods have advantages and disadvantages. Electronic counting is rapid and simple but can usually cope only with cells which are borne singly, and errors may accrue due to the presence of non-yeast solids. Some of the more sophisticated devices are capable of discriminating between particles of varying sizes and indeed will produce a size distribution curve. However, yeast cells in flocs or chains may still be scored as single cells and the presence of non-yeast solids with a particle size similar to that of yeast will introduce further errors of under-estimation.

Prior to counting highly flocculent yeast it is necessary to disrupt cell clumps with an ultrasonic treatment or use of a chemical deflocculent such as maltose. These treatments tend to be of limited efficacy with some yeast strains. Electronic counters give no information regarding yeast viability and a separate analysis is required to make correction for this parameter.

Direct counting with a haemocytometer requires no sophisticated apparatus other than a suitable microscope, and used in conjunction with a vital stain it allows simultaneous assessment of total and viable yeast count. It is possible to discriminate between yeast cells and non-yeast particles and chain-forming strains may be counted without problem. As with the electronic approach, large yeast flocs must be disrupted before enumeration is possible. The major drawback with direct microscopic counting is that it requires trained personnel. Even in the hands of skilled operators, large errors in repeatability and precision are the norm. Siebert and Wisk (1984) compared both microscopic and electronic yeast counting and concluded that there was a nine-fold difference in precision in favour of the latter method. Furthermore, electronic counting was easier, less tiring to operators and more rapid. However, correction for yeast viability was not considered in this study.

Miller *et al.* (1978) suggested that the total viable fraction of pitching yeast slurries could be determined accurately and speedily by measurement of ATP concentration using firefly bioluminescence (see Section 8.3.4.1). The authors reported that the ATP content of yeast cells, in any given physiological state, is relatively constant. Since it is absent from dead cells and non-yeast solids, the concentration of this metabolite in pitching yeast slurries showed a positive correlation with viable yeast concentration. A method was developed in which yeast slurry was sampled from storage vessels and analysed photometrically. The results allowed computation of an optimum pitching rate, which in production trials was shown to be 2.0 µg ATP per ml wort. When used at production scale, this method was shown to produce consistent fermentation performance. The method has obvious attractions, but unfortunately yeast slurries require to be diluted some 100-fold to obtain a result. This implies that careful and skilled handling of samples is required in order to avoid what could be potentially very large experimental errors.

With knowledge of the yeast concentration, it is possible to calculate the quantity of slurry to be added to fermenter to achieve a desired pitching rate. This may be a manual operation, in which the required quantity of slurry is pumped from storage vessel to fermenter. Alternatively, automatic systems use a controller which receives output from load cells fitted to the pitching yeast storage vessels or an in-line flow meter. When the appropriate quantity of slurry has been transferred directly to fermenter, or injected into the wort main during collection, the controller terminates the process by switching off the pitching pump and closing the pathway between storage vessel and fermenter.

6.1.4.3 *Cone to cone pitching.* In some breweries that use cylindroconical fermenting vessels it is practice to pitch with yeast taken from the cone crop of one fermentation and transferred directly into the empty cone of a new fermentation. This procedure avoids the need for intermediate storage tanks. The yeast should be in good condition since there has been no opportunity for deterioration due to prolonged storage. Similarly, there should be no exposure to atmospheric oxygen which could influence the requirement for subsequent wort oxygenation. On the other hand, it is an inflexible system since dedicated storage tanks free fermenting vessels to be filled or emptied as dictated by the requirements of production.

As with conventional pitching systems, this method also requires removal of a yeast sample from fermenter for off-line analysis so that the quantity of slurry to be transferred can be assessed. This is a highly inaccurate procedure. Yeast crops in the cones of cylindroconical fermenters are heterogeneous due to layering of yeast cells and non-yeast solids (Boulton & Clutterbuck, 1993). Furthermore, stratification of the yeast crop in terms of cell size/age is also possible (see Section 6.7.2). In consequence, analysis of small off-line samples does not provide a representative assessment of the total yeast crop. Furthermore, removal of yeast from fermenter cones may be associated with complex mixing effects due to vessel geometry. This can cause a central plug of yeast to exit first, followed by the portion of the crop nearest to the walls. It is, therefore, difficult to control accurately the transfer of a known quantity of viable yeast from one fermenter to another.

A method has been described which reportedly surmounts some of these problems (Anonymous, 1992). This approach employs in-line sensing devices, which, by monitoring flow rate and density, provide a continuous measure of the proportion of solids suspended in the yeast crop as it is transferred from cone to cone. Used in conjunction with a batching device, the system claimed to provide accurate control of pitching rate when in routine use at production scale.

#### 6.1.4.4 *Use of near infra-red turbidometry.* Quantification of suspended cell counts, or other particles, indirectly by turbidometry is a well tested method. It relies on there being a correlation between cell concentration and light scattering, usually measured at a particular wavelength. Turbidometric sensors have been designed for use in the quantification of yeast cell concentration for the control of yeast pitching rate (Riess, 1986). This approach has the major advantage that in-line measurements can be made and, therefore, a fully automated pitching rate control system is possible (Fig. 6.5). Yeast concentration is quantified by measurement of turbidity in response to incident radiation in the near infra-red range, which is claimed to give the best linear relationship. Two NIR sensors are used. The first is located in the wort main, up-stream of the yeast injection point and provides a measure of the turbidity due to the unpitched wort. The second sensor is placed down-stream of the yeast injection point and an in-line mixer, but before the oxygen injection point, provides a measure of the turbidity due to the pitched wort. A back-pressure orifice valve located between the control and motoring apparatus prevents gas breakout which would interfere with readings. In operation the difference in turbidity reading between the two sensors is maintained at a set-point, by modulation of the yeast slurry injection control valve, thereby maintaining a constant yeast count within the pitched wort.

The value of the set-point is derived empirically for individual combinations of yeast strain and wort quality. The operating range of the detection system is $1 \times 10^6$ to $50 \times 10^6$ cells ml$^{-1}$. This has implications regarding the manner in which the pitching system must be used. Where yeast is pitched in-line with the wort it is best practice to add the slurry as quickly as possible. This ensures that the fermentation

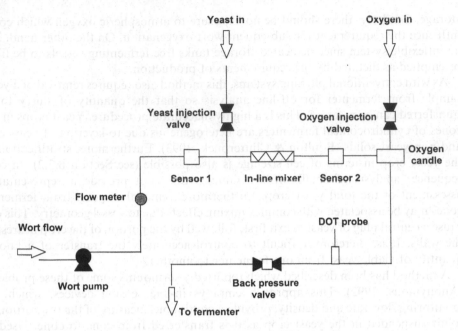

**Fig. 6.5**  Automatic pitching rate control system using NIR turbidometry.

has a rapid start and all of the yeast is exposed, more or less simultaneously, to the transition from storage conditions to suspension in oxygenated wort. Arguably, this offers the best chance of ensuring good control of fermentation, as discussed subsequently (see Section 6.4). The operating range of the turbidometric pitching control system requires the yeast to be dosed in throughout all or most of the period in which wort is pumped into the fermenter. This is probably not a serious problem if the collection time is short and it promotes efficient dispersion of yeast throughout the wort. However, with very large capacity fermenters, taking several hours to fill, continuous pitching is undesirable. Thus, it may be a cause of lack of control since yeast pitched early, and therefore exposed to oxygenated wort for considerable time, will be in a different physiological state to that pitched later.

The turbidometric approach provides demonstrable improvements in pitching rate control compared to a conventional system based on metered addition of a known wet weight of yeast. The data in Table 6.1 shows the range of values of suspended yeast count, measured with a haemocytometer, of samples removed from production scale fermenters at the completion of wort collection. The data is for six different yeast strains, where pitching rate was controlled by conventional methods and using the NIR turbidometric control system. It may be seen that with most beer qualities the range of yeast counts associated with the automatic system were significantly reduced compared to those which used conventional pitching rate control. However, in two instances (ale 1 and ale 2), the automatic system was of no advantage. Coincidentally, these two fermentations used flocculent yeast strains and this highlights one of the disadvantages of turbidometry. Thus, the correlation between particle number and turbidity is highest where the particles are of

**Table 6.1**  Yeast counts measured in fermenter at the end of wort collection for various beer qualities using conventional and NIR turbidometric pitching rate control (Boulton & Besford, unpublished data).

| Beer quality | Target pitching rate (cells ml$^{-1}$ × 10$^6$) | Measured cell counts (cells ml$^{-1}$ × 10$^6$) | | | |
| | | Conventional* | | NIR turbidometric | |
| | | Range | Mean | Range | Mean |
|---|---|---|---|---|---|
| Lager 1 | 10–14 | 9.7–24.0 | 13.4 | 9.2–16.1 | 12.0 |
| Lager 2 | 10–12 | 10.1–16.7 | 12.1 | 9.7–11.7 | 10.8 |
| Lager 3 | 10–13 | 8.8–14.1 | 11.5 | 8.8–11.4 | 10.1 |
| Ale 1 | 10–14 | 9.8–13.5 | 11.1 | 10.4–15.7 | 12.6 |
| Ale 2 | 10–14 | 10.3–13.2 | 11.4 | 10.6–15.6 | 12.8 |
| Ale 3 | 10–14 | 11.6–17.4 | 14.2 | 11.1–12.5 | 11.7 |

* Pitching rate controlled on the basis of addition of known weight of slurry of pre-determined centrifuged wet solids content and corrected for viability as measured by the methylene blue procedure.

a uniform size. This is not the case with flocculent yeast and significant errors may arise.

The data in Table 6.1 also shows that though greater consistency of pitching rate was achieved, on average the turbidometric system tended to pitch fewer yeast cells than the conventional technique. This potential for underpitching can arise in two ways. First, the automatic system has no correction for viability, and, second, although the double sensor arrangement allows for non-yeast solids in worts it does not correct for such material, which may be entrained in the injected yeast slurry. In the case of the fermentations discussed here, the effects of underpitching would seem to be significant because on average vessel residence times were longer where the turbidometric system was employed (Table 6.2). Of course, it is possible that other factors were involved in these discrepancies, for example, the effects of continuous yeast dosing versus one-shot pitching, as discussed earlier. Nevertheless, the turbidometric system has utility and any tendency to underpitch could be corrected for by simply increasing the set-point.

Another yeast sensor using an NIR detection system was described by Byrnes and

**Table 6.2**  Vessel residence times, measured as time to application of crash cooling, for fermentations using conventional and turbidometric pitching rate control. Each datum point is the mean for 6 individual fermentations (Boulton & Besford, unpublished data).

| Beer quality | Time to crash cool (h) | |
| | Conventional pitching rate control | NIR turbidometric pitching rate control |
|---|---|---|
| Lager 1 | 185 | 202 |
| Lager 2 | 223 | 233 |
| Lager 3 | 235 | 262 |
| Ale 1 | 139 | 132 |
| Ale 2 | 120 | 122 |
| Ale 3 | 74 | 96 |

Valentine (1996), which apparently overcomes some of the disadvantages of the system described already. The sensor utilises what was described as a selective narrow band in the near infra-red, which allowed good discrimination between yeast cells, other particulates and gas bubbles. The linear operating range extended to $2 \times 10^9$ cells $ml^{-1}$ and therefore would be suitable for measurement of yeast concentrations in pitching slurries. This would allow the setting up of an automatic pitching system in which the concentrated slurry could be added quickly, as in conventional pitching, and overcome the problem of having to dose in yeast throughout wort collection. It would still be necessary to correct for viability by separate analysis.

More recently, the application to the enumeration of suspended yeast particles of another optical system has been described (Holmes & Teass, 1999; Holmes, 2000). This system was applied to the control of yeast pitching rate in an automatic system. In comparison to an older manual system based on laboratory analysis, improvements in fermentation consistency were reported.

6.1.4.5  *Use of radiofrequency permittivity.*  A biomass sensor has been developed which detects cells by virtue of their dielectric properties (Harris & Kell, 1986; Harris *et al.*, 1987). Suspensions of cells possess several recognisable frequency-dependent dielectric dispersions (Schwan, 1957). One of these, is called the β-dispersion. It is caused by a combination of build up of charge at the surface of the relatively non-conducting plasma membranes and partially restricted lateral motions of the charged lipids and protein components within the membrane. These are generated in response to an applied field (Woodward & Kell, 1990, 1991; Ferris *et al.*, 1990). When the applied signal is of the frequency of radiowaves, the value of the β-dispersion of a suspension of yeast cells is proportional to the volume fraction of biomass bound by a biological membrane. Since yeast cells of a given strain and in the same physiological

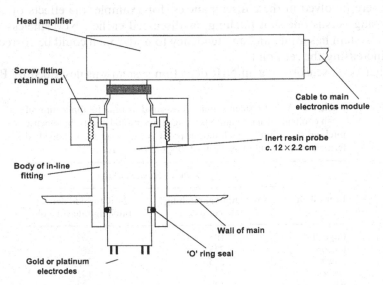

**Fig. 6.6**  Biomass meter, head amplifier and probe assembly for in-line measurement of yeast concentration (from Boulton & Clutterbuck, 1993).

condition are approximately the same size, the build-up of charge, measured as capacitance, is proportional to the number of cells in the operating field.

Yeast cells, which would be classed as non-viable by vital staining methods such as methylene blue, have disrupted plasma membranes. Cells with disrupted membranes have no dielectric response and do not contribute to the measured capacitance. Similarly, non-yeast particulates are incapable of building a charge. Therefore, the biomass meter is responsive only to the viable fraction of a yeast suspension and the reading is unaffected by trub.

The device available commercially is designed specifically for in-line measurement of yeast concentration in brewing applications such as the control of pitching rate. The sensor consists of a probe made from an inert resin in which are embedded four electrodes made from gold or platinum. The probe is resistant to brewery cleaning and sterilisation regimes. Attached to the probe is a small electronics module, which contains amplification circuitry and generates the radiofrequency field. This module is connected by cabling to an electronics module. This is provided with controls for calibration and a visual display of yeast concentration. A signal of 4–20 mA, proportional to the yeast concentration reading, is output from the principal electronics module for connection to an external controller. An RS232 interface can be used for the same purpose. A multiplexer is available which allows connection of banks of four separate probe and amplifier assemblies.

The meter must be calibrated for each individual yeast strain. This entails identifying and entering strain-specific calibration parameters into the memory of the meter. This places a limit on the number of strains that can be handled, usually a maximum of ten. When calibrated, the meter reading corresponds to viable yeast concentration. This is usually expressed as spun solids, or suspended cell count; however, any desired unit may be used. It does highlight one drawback of the meter. In order to set up the calibration it is necessary to use slurries whose yeast concentration has been measured by conventional means. As discussed already, these conventional methods do lack precision and inevitably there will be a one-off error in the initial calibration. Obviously, the meter cannot have a precision greater than the precision of the method used to analyse the slurries employed in the calibration procedure.

In practice, the measure of concentration used is somewhat irrelevant since in an automatic pitching rate control system it would never be necessary to refer to it. The pertinent parameter in this case is that of repeatability. Thus, the control system uses a derived measure of yeast concentration and is only of significance in the initial setting up to achieve a desired pitching rate in fermenter. Once this is done, all that is required is that the yeast concentration is measured in a repeatable manner. The calibration for any particular strain may be selected manually, although in automatic systems this would be accomplished using the external controller. The principal features of the biomass meter are shown in Figs 6.7 to 6.9.

In operation, the biomass meter provides an instantaneous reading of yeast concentration. Output is linear within the range $1 \times 10^8$ to $1.5 \times 10^9$ cells ml$^{-1}$ (approximately 5–60% wet weight to volume). Therefore, unlike the turbidometric sensor, it is not suitable for measurement of yeast concentration in pitched wort. Readings may be taken in-line or in-tank; however, the operating field of the probe extends to only a few centimetres in front of the electrodes. Hence, if in-tank

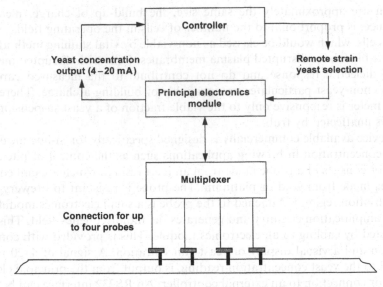

**Fig. 6.7**  Schematic layout of in-line yeast concentration measurement using the biomass meter.

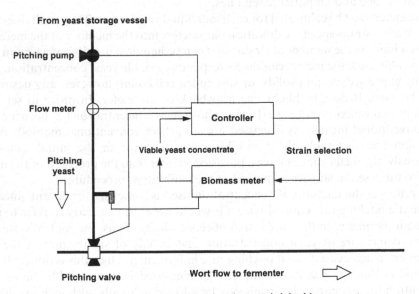

**Fig. 6.8**  Automatic pitching rate control system using the permittivity biomass meter.

measurements are made, the tanks must be stirred to homogeneity. In the case of in-line measurements (Fig. 6.7), periodic fluctuations in yeast concentration as slurry passes the probe will be taken into account. However, it is assumed that there is not a gradient in yeast concentration across the section of the pipe. Provided that homogeneous conditions are maintained, readings are unaffected by the morphology of the yeast. Thus, there is no effect due to the use of chain forming or flocculent strains.

**Fig. 6.9** Biomass meter (courtesy of Aber Instruments).

A major advantage of the biomass probe is that it is responsive only to viable yeast cells – at least those cells that would be classed as viable using the methylene blue staining test (see Section 7.4.1). The data in Table 6.3 compares biomass meter predicted yeast concentrations, measured as dry weight, to actual measured dry weight for a number of pitching yeast slurries removed from brewery storage vessels. In addition, the viability of the yeast in each slurry was determined by methylene blue staining, allowing the viable yeast dry weights to be computed. As may be seen, there was a good correlation between measured viable dry weights and those predicted by the biomass meter, although the former were on average slightly higher. This may be explained in that the measured dry weights contained varying proportions of trub and this was not taken into account with the biomass meter derived values. This illustrates a further benefit of the meter. The correlation between biomass meter output and viable yeast mass has been confirmed by others, for example, Kronloff (1991). This author also reported that the device had utility for biomass measurement in immo-

**Table 6.3** Comparison of biomass concentration for pitching yeast slurries measured directly and using a permittivity biomass meter (from Boulton et al., 1989).

| Sample | Actual dry weight (mg ml⁻¹) | Viability (%)* | Viable dry weight (mg ml⁻¹) | Biomass meter predicted dry weight (mg ml⁻¹)† |
|---|---|---|---|---|
| 1 | 80.2 | 73.1 | 59.4 | 64.2 |
| 2 | 74.2 | 92.2 | 68.4 | 59.0 |
| 3 | 64.9 | 82.8 | 53.7 | 58.4 |
| 4 | 82.3 | 88.9 | 73.2 | 68.4 |
| 5 | 85.0 | 76.9 | 65.4 | 51.0 |
| 6 | 87.2 | 73.6 | 64.2 | 62.6 |
| 7 | 36.8 | 78.2 | 28.8 | 26.0 |

* Viability determined by methylene blue staining.
† Values obtained from calibration curves of biomass meter output versus yeast dry weight.

bilised yeast bioreactors since readings were unaffected by the presence of inert carrier.

A schematic diagram of an automatic pitching rate control system using the biomass meter is shown in Fig. 6.8. The system operates as follows. The volume of wort to be pitched and the desired pitching rate are selected and the controller computes the total viable yeast mass or cell number required. The sequence commences when the controller selects the calibration for the appropriate yeast strain, opens the valve path from the yeast storage vessel to the wort main and switches on the yeast pitching pump. As the yeast slurry is injected into the flowing wort, output from the biomass meter and a flow cell allow calculation of the total viable yeast mass, or total yeast cell number, pitched. When this figure matches the calculated required value, the controller terminates the sequence by closing the valve path and switching off the yeast pitching pump. It is possible to substitute the in-line flow meter for output from a load cell located on the yeast storage vessel and obtain acceptable results. However, flow cells offer the most precise control.

The automatic control system described here produces demonstrable improvements in the consistency of fermentation performance, compared to conventional methods. This is illustrated in the data presented in Fig. 6.10. Here is shown the standard deviation in vessel residence time for 2000 hl lager fermentations for a period of months before and after commission of an automatic pitching rate control system of the type depicted in Fig. 6.8.

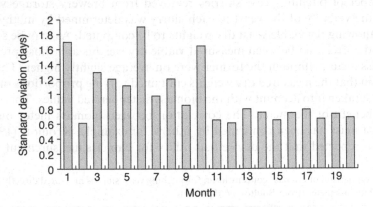

**Fig. 6.10** Consistency of performance, shown as standard deviation in vessel residence time, for 2000 hl lager fermentations over a period of 20 months. Values for each month were calculated from data taken from at least 30 individual fermentations. Pitching rate control was changed at the end of month 11 from a system based on metered addition of yeast slurry of pre-determined viable spun solids to an automatic system using a capacitance biomass meter (Boulton & Box, unpublished data).

Cahill *et al.* (1999b) observed a decrease in mean cell volume during storage of pitching yeast. They suggested that this could lead to over-pitching errors when using total cell weight or volume as the controlling parameter. They concluded that an image analyser could be used to measure cell volume and correct for this deviation. Control of pitching rate using the biomass meter would produce this error since it measures the total enclosed biovolume within the operating field. Therefore, it corrects automatically for any variation in yeast cell volume.

### 6.1.5  *Timing of wort collection*

Proper management of wort collection requires the establishment at the start of fermentation of controlled values for temperature, yeast pitching rate, dissolved oxygen concentration and wort specific gravity. Precise regulation of these parameters is a vital prerequisite to good fermentation control. However, the timing and order in which these events occur also has an important influence on fermentation performance and beer composition. In the case of small volume fermentations, collection times are short, and therefore all yeast cells are exposed to oxygenated wort at roughly the same time. With large capacity fermentations the time taken to complete wort collection may be considerable and a decision has to be taken as to the timing and sequence of addition of the various components which contribute to establishing initial conditions.

To ensure a defined starting point to fermentation it may be considered preferable to fill the fermenter with oxygenated wort at a desired temperature and then initiate the process by pitching in the yeast. This is an undesirable course of action in two respects. First, the risk of microbiological contamination of the unpitched wort would be unacceptable, and, second, delaying pitching to the end of wort collection postpones the start of fermentation and is therefore wasteful of process time.

An alternative option is to pitch yeast continuously throughout wort collection. This has the advantage that the yeast is well dispersed when collection is complete. A variation to this approach is that where several batches of wort are required to fill a fermenter, discrete portions of yeast slurry may be dosed into some or all batches of wort. Continuous dosing and discrete multiple pitchings are not to be recommended since it is very difficult to predict the effects on fermentation performance.

The consequences of discontinuous pitching throughout collection are illustrated in the following extreme but nonetheless real example. In this instance, there was a mismatch between the capabilities of the brewhouse and the capacity of fermenters. Cylindroconical ale fermentations of 1950 hl and original gravity of 15°Plato required five individual batches of wort to fill each vessel. All batches of wort were saturated with air and collection took some 16 hours to complete. The wort was pitched with yeast at a rate of approximately 1 g per litre wet weight. In total, 195 kg of pressed yeast was added in the form of a slurry containing 50% wet weight to volume. Of this, 40% of the total yeast was pitched with the first batch of wort and the remaining 60% with the fourth batch. During fermentation, the temperature was maintained at 22°C. The attenuation gravity of 3.5°Plato was achieved in 50–60 hours.

The rationale behind the pitching regime was that the comparatively small proportion of yeast added with the first worts provided microbiological protection during the prolonged collection period. The larger portion of yeast added near the end of collection was presumed to be responsible for the bulk of the fermentation. In fact, samples removed during the collection period and analysed for yeast concentration revealed that this was not the case (Fig. 6.11). Here is shown the actual measured yeast counts together with predicted yeast counts, assuming no growth had occurred

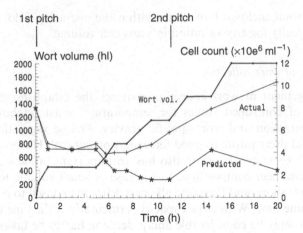

**Fig. 6.11** Wort collection of 2000 hl cylindroconical ale fermentation filled with 5 × 400 hl batches of wort. Yeast slurry was pitched with the first and fourth batches of wort at the times indicated. The wort volume is shown together with the measured yeast count and predicted yeast count, assuming no growth (Boulton & Clutterbuck, unpublished data).

during wort collection and taking into account the dilution effect of the increasing wort volume.

As may be seen during the first 5 hours there was no increase in cell count and the actual and predicted yeast counts could be superimposed. Some 6 hours into collection the actual and predicted cell counts began to gradually diverge indicating that cell proliferation had commenced. At the end of addition of the third batch of wort, immediately prior to injection of the second tranche of pitching yeast, the measured cell count was $5.7 \times 10^6\,\mathrm{ml}^{-1}$. This compared to a predicted count of $1.7 \times 10^6\,\mathrm{ml}^{-1}$, assuming no growth had occurred. After 20 hours, when collection was complete, the actual cell count was $10.3 \times 10^6$, compared with a predicted count of $2.5 \times 10^6\,\mathrm{ml}^{-1}$. At this time, the yeast pitched with the fourth batch of wort would have been exposed to oxygenated wort for 8 hours. The first pitched yeast, after 8 hours, had increased in number by a factor of 1.5. Assuming the same behaviour for the later pitched yeast and subtracting this figure from the total count at 20 hours, this indicated that at this time the first pitched yeast had increased 8-fold, or 3 budding cycles. In fact, microscopic examination of the yeast cells at 20 hours revealed a more complex picture. Thus, prior to the addition of the second batch of yeast the majority of the cells were in the form of short chains and clearly in the process of active proliferation. At 20 hours, the majority of cells had a similar appearance; however, there were also significant numbers of larger single cells showing no obvious signs of budding. This suggested that the late-pitched yeast (which at the time of addition would still have been in the stationary phase) was unable to compete for wort nutrients and oxygen with the actively growing yeast cells derived from the first pitch.

When stationary phase yeast is exposed to oxygen, whether suspended in wort or not, it has a relatively low affinity for assimilation of oxygen. With prolonged exposure to oxygen, there is a marked increase in the ability of the yeast to take up this nutrient (see Section 6.4.2.2). At a temperature of 20°C, the specific oxygen uptake rate (% decrease in oxygen saturation per minute per gram dry weight of yeast) increases roughly five-fold during 3–5 hours' exposure to oxygen (Boulton *et al.*, 1991). Under carbon catabolite repressed conditions, such as would be the case in aerated wort, this relatively higher affinity for uptake of oxygen is maintained. It is likely, therefore, that in the case of this fermentation, the later pitched yeast would have a much lower affinity for oxygen than that pitched earlier and possibly would be at a competitive disadvantage. If yeast is pitched continuously throughout prolonged wort collection, the situation is even more complex. In this case, it would be predicted that when collection was complete there would be a mixed population of yeast with varying affinity for oxygen, and therefore inconsistent physiological condition. Clearly, this cannot be conducive to good fermentation control.

This example highlights the fact that yeast pitching and wort oxygenation must be co-ordinated processes. Thus, the length of time that yeast is exposed to oxygen is as significant to subsequent fermentation performance as the total mass of oxygen supplied. The data in Fig. 6.12 shows a recording of the output from a dissolved oxygen probe suspended in fermenter during wort collection of the fermentation described in Fig. 6.11. It may be seen that as each new batch of wort was added to the vessel it was accompanied by an additional charge of oxygen. This dissolved oxygen decreased due to consumption by the yeast during the periods when wort pumping stopped. Therefore, the rate of decline of dissolved oxygen concentration during these periods provided an indication of the oxygen uptake rate due to the yeast. These rates increased with each consecutive addition of batch of wort. The increased affinity of the yeast for oxygen was further demonstrated by the fact that rates of increase and peak values of DOT associated with fresh wort additions decreased with each consecutive batch of wort. In part, this must have been due to the gradual increase in cell count. However, this would not explain the increase in oxygen uptake rates between

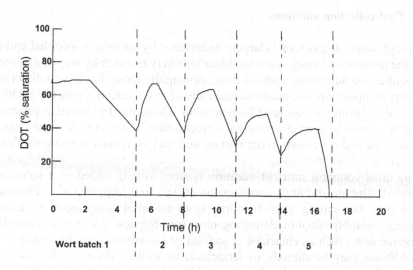

**Fig. 6.12** Trace of dissolved oxygen concentration of wort in fermenter during collection of a 2000 hl ale fermentation filled with 5 × 400 hl batches of air saturated wort (Boulton & Clutterbuck, unpublished data).

the first and second batches of wort since during this same time period there was little change in yeast concentration (Fig. 6.11). In fact, during the first 8 hours of collection, the yeast count declined slightly due to dilution by the in-flowing wort. The oxygen uptake rate did not increase markedly with collection of the fourth batch of wort. This coincided with addition of the remainder of the pitching yeast, perhaps also indicating that the later pitched yeast made only a small contribution to total oxygen consumption and by inference to the subsequent fermentation.

O'Connor-Cox and Ingledew (1990) studied the effects on fermentation performance of the timing of oxygen addition. This has been discussed in Section 6.1.2. However, these authors concluded that it might be advantageous to pitch yeast with the first brew-length under anaerobic conditions. Oxygen should only be added with subsequent brew-lengths. The utility of this approach has as yet been explored at laboratory scale only. Undoubtedly it merits further investigation.

Several general lessons may be drawn from these examples. In the first instance, the introduction of large capacity vessels without concomitant improvements in the capabilities of the brewhouse should perhaps be discouraged. If this is not possible, and fermenters require several hours to fill with multiple batches of wort, the manner of oxygenation and pitching should be given careful consideration. In order to establish a rapid and consistent start to fermentation it is preferable to pitch all the yeast over as short a time period as possible and at the beginning of collection. The oxygenation regime should be chosen on the basis of that which provides the desired fermenter residence time, extent of yeast growth relative to the pitching rate and beer analysis. Whatever regime this happens to be, it should be adhered to with the greatest possible rigour in order to maximise the chances of obtaining consistent fermentation performance.

## 6.2   Post-collection additions

Although wort composition is largely determined by the time it is cooled and pumped into the fermenting vessel, some modifications may be made by way of a miscellany of post-collection additions. Some of these, principally those that have a direct influence on wort composition, are made immediately after collection is complete. Others may be dosed in during the course of fermentation. These may be viewed as process aids or additions made to correct abnormal fermentation behaviour. A third class may be added at the end of primary fermentation and may be viewed as being the first part of post-fermentation processing.

The most common nutrient addition is zinc, usually added as a solution of the hydrated sulphate and at a concentration of 0.05 to 0.15 ppm $Zn^{2+}$. This metal ion, which may be limiting in malt worts, is an essential component of several yeast enzymes, notably alcohol dehydrogenase (see Section 2.4.6). Occasionally other inorganic salts, such as chlorides of sodium or potassium, may be added.

Additions may be directly to fermenter, or in-line, via a small dissolving tank located after the wort cooler. Inorganic chloride supplements are made because they are flavour active, contributing to beer fullness. Salts are added to fermenter rather than during wort preparation because their presence during wort boiling may contribute to corrosion of stainless steel vessels. Flavour considerations apart, they will change the mineral composition of the wort. They will contribute to yeast nutrition of yeast and may influence the expression of yeast flocculence.

Enzymes may be added to fermenter to modify the sugar spectrum of the wort. For example, wort with a high dextrin content may be treated with amyloglucosidases to increase the proportion of fermentable sugars. This increases the potential yield of ethanol. This approach has been used to produce so-called 'low carbohydrate' beers. Thus, the reduction in dextrin concentrations and concomitant increased wort fermentability allows the production of more fully attenuated beers. Commercial preparations of these enzymes tend to be of relatively low activity and are suited to addition to fermenter because of the opportunity for providing a long exposure time.

Antifoam may be added to wort at the end of collection in a single dose to control fobbing. Alternatively, it may be added throughout fermentation as required. To avoid the possibility of microbial contamination the antifoam should be sterilised prior to application. The total dosage rate should be limited to avoid the possibility of carry-over into beer. Failure to do this may result in beers with poor foaming properties.

Post-collection additions may be made where fermentation performance is abnormal. This may take the form of a very slow attenuation rate or failure to achieve the desired final gravity – a 'sticking or hung fermentation'. The causes and treatments of abnormal fermentation performance are discussed in Section 6.5.4. Typically, more oxygen may be added, preferably via a stainless steel candle or some other means of improving gas transfer rates, located at the bottom of the fermenter. In deep fermenting vessels, such treatments should be approached with caution since the possibility of sudden and catastrophic breakout of carbon dioxide from the lower super-saturated regions is a very real risk. Treatment with oxygen or air has the additional benefit of agitating the contents of the fermenter thereby re-suspending yeast, which may have settled out. Should application of oxygen fail to re-start the fermentation it may be necessary to add further yeast. The yeast must be accompanied by more oxygen since conditions within the fermenter will be anaerobic and addition of yeast alone will have little effect.

If addition of further oxygen and/or yeast fails to restart the fermentation, a supplement of vitamins and metals may be made. Such preparations are available commercially and sold under the generic name of 'yeast foods'. They usually consist of mixtures of vitamins, trace metals, lipids and phosphate. Cruder preparations are simply dried powdered extracts of yeast. Dosage rates are given by the manufacturer. However, caution should be exercised since over-dosage, particularly in a stuck fermentation where the yeast may be very stressed, can alter the nutritional balance of the wort such that adventitious microbial contamination becomes a risk.

In traditional ale fermentations the process of clarifying the green beer is sometimes started in fermenter when the attenuation gravity has been reached and the yeast top crop removed. Thus, silica-based auxiliary finings may be added directly to fermenter after cooling and cropping. These contain approximately 2% silica in the form of silicic acid. In beer, which has a pH of approximately 4.0, the silicic acid forms an insoluble precipitate and co-flocculates with suspended proteinaceous matter

(Vickers & Ballard, 1974). This pre-treatment improves the fining behaviour of the beer down-stream of the fermenter.

### 6.3 Monitoring fermentation progress

The progress of fermentation must be monitored throughout. This is necessary to confirm that performance is as expected, to provide a means of early identification of non-standard behaviour and to verify as rapidly as possible that the end-point has been reached. Historically, these tasks were the prerogative of the skilled brewer whose experience with particular fermentations was sufficient to gauge progress based on simple visual observation. As measuring devices became available, visual observation was augmented by physical monitoring. In all but the most primitive of breweries, such empirical approaches have now been entirely superseded by real measurements.

This change was a consequence of the widespread adoption of closed fermenting vessels. Thus, where open fermenters are used, the progress of the process can be seen, for example, by the formation of a yeast head. The timing of formation, the size and appearance of such yeast heads gives much valuable information regarding the vigour and progress of fermentation. In closed vessels, alternative indirect monitoring methods are required. Temperature may be measured using a thermometer located inside the vessel. Other measurements are most commonly made, off-line, on samples of fermenting wort removed from vessels. In recent years, several in-line sensors have been introduced which allow automatic monitoring of fermentation progress. These devices provide continuous output such that any deviation from the norm is quickly identified allowing rapid remedial action to be taken. In addition, they provide an output, which is suitable for regulating the activity of some other device, thereby providing the possibility of introducing systems for the automatic control of fermentation.

Provision of monitoring devices offers an opportunity to maintain records of individual fermentation performance. These data can be used to set up a permanent archive, which can be an invaluable aid to fermentation management. With access to such data, long-term drift in performance can be identified. The effects on fermentation performance of changes in process or plant can be judged.

There are three aspects to monitoring fermentation progress. First, the rate of fermentation is controlled by application of cooling and it is therefore necessary to monitor temperature. Second, the progress of fermentation is gauged, usually by periodic or continuous measurement of wort specific gravity, or a related parameter. Third, in lager fermentations it may be necessary to define the end-point by measurement of the concentration of vicinal diketones (see Sections 3.7.4.1 and 6.3.9).

### 6.3.1   *Monitoring temperature*

Although temperature may be measured off-line it is most usually and conveniently monitored using thermometers located in pockets placed in the walls of fermenting vessels. The readings can simply be used to control temperature via manual application of cooling but in most cases output is linked directly to a controller which provides automatic attemperation. In large capacity cylindroconical vessels, particularly those used for both fermentation and conditioning, the location and number of temperature transmitting probes is of crucial importance to the maintenance of proper attemperation. This subject is discussed in Section 5.4.2.

Platinum resistance probes, conforming to BS 1904 class A or B, are most commonly used for temperature measurement because of their good hygienic design, accuracy and linearity of output. Care must be taken to ensure that the chosen instrument is suitable for the application. Thus, Davies (1992) reported that the BS 1904 standard does not stipulate a response time and that testing of 12 commercially available models revealed a wide range of response times to changes in temperature, one being as slow as 22 seconds.

Where large numbers of fermenting vessels are in simultaneous use, it is necessary to have some type of microprocessor-driven supervisory system. This monitors and controls the temperature of the entire tank farm – for example, the ACCOS system described by Wogan (1992). Typically, the tank farm comprises a few tens of vessels in which all stages of the fermentation cycle will be represented at any one time. Thus, vessels will be empty, being filled, cleaned, in the middle of primary fermentation, on crash cool, or on hold after cooling. The supervisory system receives and records temperature data from each vessel. It confirms that the values are appropriate and within range for the process stage that each vessel is undergoing and applies control, as necessary. During primary fermentation temperature control would typically be $\pm 0.05°C$ of the set-point. Alarms are raised if temperatures are outside pre-set limits for any particular process stage. Should an alarm condition be detected, automatic processes are put into a hold situation during which the process is held under conditions that cause the least possible compromise to the integrity of the product.

### 6.3.2 Monitoring wort gravity

Measurement of the reduction in wort specific gravity, or a derived unit, as sugar is utilised by yeast, is the most commonly used method of gauging fermentation progress. Several different and arbitrary scales have been used throughout the world. Some of these are given here, although at present only degrees Plato and saccharin (present gravity) are in common usage.

(1) **Specific gravity (also termed relative density):** this is the ratio of the density of liquid at a specified temperature (often 20°C) and the density of water at the temperature of its maximum density 4°C (39.2°F). Specific gravity is without dimension.

(2) **Present gravity (degrees saccharin):** the specific gravity, multiplied by 1000 and minus 1000, measured at 20°C and expressed in degrees.

(3) **Degrees Balling:** a scale devised by Carl Joseph Napoleon Balling in 1843, from tables relating density of wort to sucrose concentration measured at 17.5°C (63.5°F). A Balling saccharometer is a hydrometer calibrated to give a measure of wort sugar as percent weight. Thus, 1°Balling is equal to 1 g sugar per 100 g wort. One degree Balling is equal to 3.8°saccharin.

(4) **Degrees Plato:** Balling's tables were slightly erroneous and were later corrected by the German chemist Plato. The tables bearing his name have an identical basis to that of Balling. The difference brought about by the correction is illustrated in Fig. 6.13.

(5) **Degrees Brix:** a specific gravity scale also based on sugar concentration expressed as percentage weight to weight but in this case measured at 15°C (59°F).

(6) **Régie:** a measure of wort density used in France. 'Legal density' is defined as the ratio of the mass of 50 cm$^3$ liquid at 15°C and the mass of an equal volume of pure water at 4°C. One degree Régie (R°) is equal to: (Legal density − 1000) × 100. One degree Régie is equal to 2.6°Balling.

(7) **Belgian degree:** a value derived from the specific gravity by subtracting 1 and multiplying by 100.

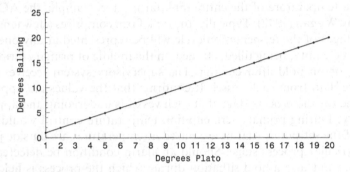

**Fig. 6.13**   Relationship between wort density measured by the Balling and Plato scales.

6.3.2.1 *Sampling from fermenters.*   Before off-line analyses can be made it is necessary to remove samples from the fermenter. In traditional open fermenters samples may be obtained using a dropping can. This device allows samples to be removed from any desired location within a vessel (Fig. 6.14). It consists of a stainless steel container with a capacity of approximately 500 ml. The base is fitted to the body via a screw thread and gasket and is made of heavy gauge metal to ensure that the can sinks. There is a handle to which is attached a length of chain. The can aperture is closed with a rubber bung, which is also attached to a chain. The latter runs through a ring in the handle, which retains the bung when it is removed from the neck of the bottle. Samples are obtained by lowering the can into the fermenter with the bung in place. When the can is submerged to a desired depth the bung is pulled out by a sharp tug on the appropriate chain. The can and sample may then be retrieved. Dropping cans may be used to sample from deep vessels although in the experience of this author with difficulty.

Modern closed fermenting vessels are fitted with sample valves to facilitate off-line analyses (see Section 8.3.2.1). It is essential that these are of good hygienic design and are properly maintained. This important aspect of brewery management is frequently neglected. Thus, poorly maintained or badly designed sample cocks can provide a source of contamination to both the sample and the fermenter contents. Perryman

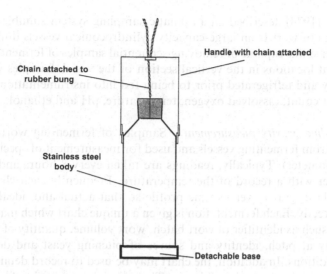

Handle with chain attached

Chain attached to rubber bung

Stainless steel body

Detachable base

**Fig. 6.14** A dropping can.

(1991) lists a number of attributes that sample cocks should possess. In the case of fermenting vessels they should be suitable for removal of both bulk samples for analyses such as present gravity and smaller aseptic samples for microbiological assessment. This requires that the sample valve can be cleaned in place with the vessel and there should be no dead spaces to harbour sources of contamination. If necessary, it should be possible to sterilise the valve prior to sample removal, either by application of steam and/or flooding with alcohol. The design should facilitate easy and rapid operation. They should be relatively maintenance free and have a long operating life.

The location of sample valves is important. A fundamental assumption of off-line analysis for monitoring fermentation progress is that the sample is representative of the whole of the contents of the fermenter. This requires that the fermenting wort is homogeneous, a condition that is probably achieved only during active primary fermentation. Some stratification is likely towards the end of fermentation when mixing due to natural convection currents is limited. Thus, at the end of fermentation some errors are an inevitable consequence of off-line analysis, wherever sample points are located. This may be of small significance with regard to monitoring specific gravity. However, measurement of vicinal diketone concentration is performed when primary fermentation is nearing completion. This requires accuracy since a certain minimum concentration must be achieved before the fermentation is judged to be finished (see Section 6.3.9). It is better, therefore, to perform analyses on samples which may overestimate the true concentration. This is best done by avoiding locating sample points close to areas where the yeast concentration may be very high, for example, in the cone of a cylindroconical fermenter. Here the high yeast density is capable of reducing vicinal diketone concentrations at a more rapid rate than that in other parts of the vessel where the yeast count is more sparse. This may lead to significant underestimates of analysis.

Moll *et al.* (1978) described an automatic sampling system suitable for removing small volumes of wort from large-capacity cylindroconical vessels during fermentation. The device was capable of removing sequential samples of fermenting wort from three different locations in the vertical section of the vessel. Samples were degassed automatically and refrigerated prior to being fed into instrumentation for measurement of yeast count, dissolved oxygen, temperature, pH and ethanol.

6.3.2.2 *Off-line gravity measurement.* Samples of fermenting wort are removed periodically from fermenting vessels and used for measurement of specific gravity (or a related parameter). Typically, readings are taken every 8 hours and plotted on a chart together with a record of the temperature. Frequently each chart has a pre-drawn 'standard' gravity versus time profile so that actual and 'ideal' data can be compared directly. Each fermentation is given a unique chart which may also contain information such as identifier of wort batch, wort volume, quantity of yeast pitched, yeast viability at pitch, identity and source of pitching yeast and details of post-collection additions. In addition, the chart may be used to record details of any non-standard process changes or breakdowns. The charts are retained to form the basis of an archive, which allows tracking of individual batches of beer. The charts may be maintained as 'hard copy' or in electronic form.

Measurement of the specific gravity of samples removed from fermenting vessels is commonly made using a hydrometer (or 'saccharometer') calibrated in one of the units described earlier. The sample must be de-gassed before readings are made to avoid errors due to carbon dioxide break-out. It is essential to record the temperature of the sample as measurements are made to allow appropriate compensation to be made. Specific apparatus has been designed for making measurements with a saccharometer. This consists of a cylinder, usually made from copper or stainless steel, which has an outer jacket through which water is circulated to maintain a constant temperature. The cylinder is filled to the brim and after allowing a period for attemperation and gas dispersal the saccharometer is suspended in the liquid and the reading made. Saccharometers are designed for measurement of gravity over a limited range of sugar concentrations. Where worts of widely different strengths are used, a complementary range of saccharometers is required.

The specific gravity of off-line samples may be determined automatically. For example, the apparatus based on the oscillating U-tube method (Jiggens, 1987). Here the sample is pumped into a glass U-tube that is clamped at its open end. The chamber in which the U-tube is located is attemperated to 20°C. The specific density of the sample is derived from the degree of damping to the oscillation (the time period of the oscillation) caused by the liquid contained within it. Measurement is based on the natural frequency (f) of oscillation of the U-tube which my be related to density in the following relationship:

$$f = \frac{1}{2\pi} \sqrt{\frac{c}{m + \rho v}}$$

where f is the natural frequency of the system; c is the elasticity constant; m is the mass of the tube, v the volume of the tube and $\rho$ the density of the liquid in the tube.

The time period (T) of the oscillation is equal to the reciprocal of the natural frequency. Therefore, from the relationship above:

$$T = 2\pi\sqrt{\frac{m + \rho v}{c}}$$

Design of the instrument allows insertion of two constants, A and B, which are related to the parameters discussed already as shown:

$$A = \frac{4\pi^2 v}{c}$$

$$B = \frac{4\pi^2 m}{c}$$

Therefore:

$$\rho = \frac{T^2 - B}{A}$$

During calibration, the instrument constants are determined from measurements made when the U-tube is filled with either air or water. The numerical values for A and B are retained in the memory of the instrument allowing the microprocessor to calculate the density of unknown samples. In operation, the tube is oscillated to its natural frequency. The time period of the frequency is determined by comparison of the number of oscillations with an internal quartz clock. From this value, the density of the liquid in the U-tube may be calculated.

Very accurate measurements may be obtained with this instrument; however, some sample preparation is necessary. Yeast cells and trub must be removed, either by centrifugation or by filtration. In addition, where necessary the sample must be degassed. This is conveniently achieved by brief immersion in an ultrasonic bath.

6.3.2.3 *Automatic measurement of gravity.* Several methods for measuring wort gravity automatically during fermentation have been developed. These have the advantage that off-line sampling is not required and gravity is measured continuously. This provides early warning of non-standard behaviour and rapid identification of the achievement of racking gravity. Output from automatic gravity measuring systems can be used in control regimes. For example, regulation of attemperation in response to changes in wort gravity. On the other hand, automatic gravity monitoring systems are expensive and obviously must be fitted to each fermenter. Bearing costs in mind, they are appropriate for use only with high capacity vessels.

Gravity sensing devices may be sited within the fermenting vessel. Alternatively, they may be located within a loop of main through which the wort is circulated, out of and back into, the vessel. Sensors must be located in such a way that a representative reading of the whole of the contents of the fermenter is obtained. As with manual sampling this can be a problem where a single point sensor is used, particularly towards the end of fermentation when layering may occur in large vessels. Locating sensors in a circulation loop is an advantage in this respect since continuous pumping is an aid to good mixing. However, some gravity sensors are prone to errors due to the presence of gas bubbles. Gas breakout can be a problem in external loops and this must be controlled by fitting appropriate valves to maintain sufficient back-pressure to keep carbon dioxide in solution. Invasive devices such as probes introduce a potential source of contamination and a cleaning problem. Perhaps in this regard, external loop systems present a greater problem than *in situ* systems. Whichever method is adopted, hygiene is a critical design parameter.

The oscillating U-tube method of gravity determination, as described for analysis of off-line samples (Section 6.3.2.2) can also be used to make in-line measurements (Jiggens, 1987). In this case, the measuring U-tube is located in an external pressurised loop. For improved robustness the U-tube is made from stainless steel as opposed to glass in the laboratory instrument and is of a wider bore to accommodate high flow rates. Since it is not possible to make measurements at 20°C a temperature sensor is fitted to the U-tube to allow appropriate compensation to be made. Output from the instrument can be fed to a recorder, visual display of gravity or as a voltage or current for input to a control device.

A few devices have been developed in which the sensors are located within fermenting vessels. In the Gravibeam system (Dutton, 1990), a displacer of known volume is immersed in the fermenting wort. The displacer experiences an upthrust, which is detected by a highly accurate load cell. The value of the upthrust is proportional to the density of the wort. After calibration, output from the load cell is expressed directly as specific gravity. The device is intrusive. However, it is constructed entirely from stainless steel and is designed to be cleaned with the vessel CiP. The system can be retrofitted to existing fermenters but it is necessary to break into the wall of the vessels to do this. It is reportedly unaffected by gas breakout, the presence of yeast or trub and pressure variations. Temperature compensation is provided by a platinum resistance thermometer fitted to the displacer. Output from several sensors, each located in a separate fermenter, is fed, via a serial data link, to a microprocessor for visual display and/or use in a control system.

Density may be computed from differences in pressure measured simultaneously at different depths within vessels. Thus, at a constant vertical distance between two points the differential in pressure is a direct function of the density of the liquid phase. The Platometer (Fig. 6.15) described by Moller (1975) uses two stainless steel diaphragms separated by 45 cm and located close to the inner wall of the fermenting vessel. Pressure is sensed by the diaphragms and transmitted to a transducer, which converts the pressure differential into an electrical signal. The pressure sensors and transducer are linked by water-filled columns. This arrangement means that the diaphragms and transducers are balanced by equal sized columns of both wort and water, at the same temperature. Therefore, the electrical output from the transducer provides a temperature-compensated indication of specific gravity.

The transducer consists of a housing made from iron, which is divided into three compartments by two transmission diaphragms. The two outer compartments terminate the water columns and the inner compartment is filled with an incompressible

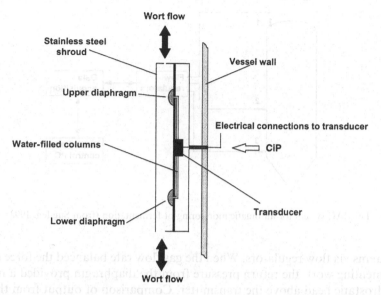

**Fig. 6.15** Platometer automatic gravity sensor.

fluid. The centres of the transmission diaphragms are linked by a rigid connecting arm. The latter is also attached to a leaf spring. Movement of this arm provides a measure of the pressure differential of the system. An electrical signal is generated by allowing movement of the spring to actuate moving plate condensers. The resultant capacitance is then converted to a signal that is directly proportional to specific gravity.

The pressure sensing diaphragms are shrouded in a stainless steel casing provided with apertures at top and bottom to allow access to wort. The casing gives protection to the device and more importantly minimises errors due to short-term variations in pressure due to random convection currents. An armature serves as the method of attaching the casing to the vessel wall. This also provides a duct for electrical wiring and an entry point for dedicated cleaning in place. The system reportedly gave an accuracy of ±0.1°Plato.

Gravity determination by differential pressure measurement has been described by others, for example, Cumberland *et al.* (1984). This report described a device that used pressure cells separated by 3 m of vertical tank wall. Other sensors measured temperature and the tank pressure (difference between top pressure and that of the atmosphere) and the difference between the tank bottom and atmospheric pressure. The vessel level contents could be deduced from the difference between the tank and bottom pressures. Output from the gravity meter was in the form of a 4–20 mA current suitable for input to the vessel temperature controller.

Sugden (1993) reported use of an automated fermentation monitoring and control system termed FerMAC, which has been installed in several UK breweries. This used three pneumatic pressure balance transmitters located in the vessel as shown in Fig. 6.16. The transmitters comprised stainless steel diaphragms attached to the inner wall of the vessel. Gas, either nitrogen or carbon dioxide, is fed to the back of the

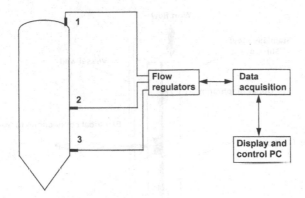

**Fig. 6.16**  FerMAC system for automatic monitoring of fermentation (from Sugden, 1993).

diaphragms via flow regulators. When the gas flow rate balanced the force exerted by the fermenting wort, the return pressure from the diaphragm provided a measure of the hydrostatic head above the transmitter. Comparison of output from the top and bottom sensors (1 and 3) provided a level reading. Similarly, the difference in output from the two lower sensors (2 and 3) could be related to wort gravity. It was reported that careful positioning of the sensors was required. The top one should be located such that CiP sprays did not cause damage. The lower transmitter was best placed at the top of the cone of the vessel to avoid false readings due to the diaphragm becoming covered with yeast.

The data acquisition module received output from sensors fitted to up to 15 fermenters. It converted the transmitted pressure data into digital readings of level, gravity, volume and top pressure. Together with output from a thermometer, the data was transferred to a display and control PC. This provided a graphical representation of change in gravity and temperature with time for each fermentation, and tabulated data describing tank volumes, yeast pitching details and records of CiP.

The system was shown capable of measuring gravity to an accuracy of 1°saccharin or 0.5°Plato. It was proposed that the system could be used to control fermentation rate by relating the gravity output to the attemperation system. In addition, the processor could be interfaced with the CiP and tank filling and emptying systems to provide a complete fermentation management system.

Hees and Amlung (1997) described yet another differential pressure system for monitoring gravity in large cylindroconical fermenters. This system uses three pressure sensors, one located in the cone, the second just below the surface of the wort and the third in the headspace. Output from these is used to calculate wort specific gravity, tank volume and headspace pressure. Output from the top transmitter allows compensation for changes in headspace pressure. The authors claimed that this could affect the accuracy of gravity measurement. This system provided continuous gravity output, which correlated closely with that obtained from off-line measurements and with a precision of ± 0.2%. As with other pressure systems, the output could be used for automatic control. In addition, the headspace sensor was used to provide a signal for automatic control of a carbon dioxide recovery system.

On-line gravity measurement can be made using an ultrasonic sensor, as described by Forrest and Cuthbertson (1986). The sensor measures the time taken for an ultrasonic signal to pass between two fixed points where the intervening space is filled with the liquid which is to be analysed. The time, or sound velocity, is related to the density of the liquid. Gas bubbles interfere and it is necessary to locate the sensors within an external loop. The loop, through which the wort is circulated, is operated under back pressure to prevent gas break-out. An ultrasonic transmitter and receiver are located at one end of the pipe such that the signal passes through the liquid, is reflected off the other end of the pipe and then detected on its return. The method of measurement is affected by temperature and a compensating thermometer is located close to the sensors. Calibration is empirically derived from comparison of output with off-line measurements.

The ultrasonic sensor is relatively inexpensive. It was shown to have an accuracy of $\pm 0.5°$saccharin and has the advantage of being totally non-invasive. However, this has to be balanced against the requirement to use an external loop. Probably this sensor is more suited for application to measurement of the gravity of bright beer. For example, it is particularly suited to on-line dilution of high-gravity beers prior to packaging (Forrest, 1987; Forrest et al., 1989). These authors described an in-line sensor which used a combination of ultrasonic and refractometric measurement to measure beer original gravity.

### 6.3.3 Monitoring $CO_2$ evolution rate

Formation of carbon dioxide during fermentation is stoichiometric with ethanol formation and sugar utilisation. Therefore, the profile of carbon dioxide evolution can be used to monitor fermentation progress. Daoud and Searle (1990) studied patterns of carbon dioxide evolution in trial fermentations from laboratory (1.5 litre) to pilot scale (100 hl). At laboratory scale, the authors demonstrated correlation coefficients of 0.9944 between $CO_2$ evolved and ethanol production and 0.99 between $CO_2$ evolved and carbohydrate utilisation. In 100 hl fermentations, no gas evolution was observed until the wort became saturated. After this time, rates of approximately 1.0 g $CO_2$ per litre per degree gravity drop were measured.

Stassi et al. (1987, 1991) used thermal mass flow meters to measure $CO_2$ evolution rates at both laboratory and production scale. They also noted a high correlation between $CO_2$ formation and decline in gravity. In addition, the former also correlated with ethanol formation, the extent of yeast growth, decline in wort pH and the concentration of dissolved sulphur dioxide. At laboratory scale, an automatic control system was established in which a set-point rate of $CO_2$ evolution was maintained in a feedback loop in which temperature was variable.

Eyben (1989) described a totally automated fermentation monitoring and control system used at the Sebastien Artois Brewery in France. This was also based on the correlation between $CO_2$ evolution and gravity decrease. A single tank fermentation and conditioning process was controlled by measuring total carbon dioxide evolution using a vortex flow-meter. The fermentation, which lasted for a total of 12 days, had three distinct phases. In the first period of active primary fermentation the temperature was held at 11°C and the top pressure 20 g cm$^{-2}$. In the second phase of warm conditioning, the temperature was allowed to increase to 16°C and the top pressure to 700 g cm$^{-2}$. During the second phase, wort attenuation and $CO_2$ evolution were completed. The third phase commenced when the vessel contents were cooled to 0°C whilst maintaining the same high top pressure. This allowed an appropriate level of carbonation to be achieved. After 12 days, the yeast was harvested and the vessel was racked. Whilst the vessel was emptied the $CO_2$ top pressure was maintained to ensure anaerobiosis during transfer of beer to filtration.

During primary fermentation, the evolved $CO_2$ was directed through the flow cell then collected via a triple exhaust-gas network, operating at atmospheric, 20 and 700 g cm$^{-2}$, respectively. The triple arrangement allowed maintenance of the same pressure at the inlet and outlet of the flow meter. Entry into individual legs of the $CO_2$ network was via automatic valves. In the flow cell, vortices proportional to gas flow rates were detected by a piezoelectric sensor and converter. This was mounted on the outside of the flow cell and provided a 4–20 mA output. The flow meter had an operating range of 0–30 m$^3$ h$^{-1}$ with an accuracy of $\pm 1\%$. The flow meter was not affected by the presence of moisture in the gas phase or by variations in temperature.

The system was used in the following manner. During early primary fermentation, $CO_2$ was vented to atmosphere. When a pre-determined volume of gas had passed through the flow meter (c. 12 hours after collection), the atmospheric valve was closed and flow directed to the $CO_2$ recovery system working at a top pressure of 20 g cm$^{-2}$. The fermentation was allowed to proceed at a temperature of 11°C until a second pre-determined total volume of $CO_2$ was recovered. At this time the exit gas flow was switched, again automatically, to the $CO_2$ recovery system operating at a top pressure of 700 g cm$^{-2}$ and the temperature allowed to increase to 16°C. After a further five days, when the wort was completely attenuated and $CO_2$ evolution had ceased, cooling was applied and the fermenter contents were chilled to 0°C.

Determination of the volumes of $CO_2$ required to trigger the changes described above were based on theoretical calculations. These took into account the capacity of the vessel, the volume and fermentability of the wort, the beer type and the growth characteristics of the yeast strain. Measurements from flow cells were made every 20 minutes, the total evolved gas being calculated by a microprocessor based on extrapolation. This system allowed a single flow meter to monitor gas evolution simultaneously from eight vessels such that 24 measurements were made per hour per flow meter. A single supervisory computer collected and recorded data from subordinate microprocessors as well as calculating the set-point gas volumes on which process steps were based.

Monitoring fermentation progress via profiles of $CO_2$ evolution is an attractive option. The sensors are non-invasive, relatively inexpensive, and, as in the example described above, several vessels can be serviced by a single device. An output is provided for use in feedback control loop systems. It has the major advantage that the status of the entire body of fermenting wort is assessed as opposed to single point measurements such as may be made with an automatic gravity sensor. Thus, problems associated with wort heterogeneity are obviated. The major potential problem is that there is little or no opportunity to gather data during early fermentation. In the first few hours of fermentation, the critical processes of oxygen assimilation and yeast sterol synthesis take place. In this phase, little or no $CO_2$ formation occurs and, as discussed earlier, even when gas evolution begins there is the period of inertia due to saturation of the wort. In the event of inadequate control of wort oxygenation or pitching rate, non-standard performance must be identified as quickly as possible. Thus, there is but a short period in which appropriate corrective action can be taken (see Section 6.5.4). By the time $CO_2$ evolution is detectable, this window of opportunity is probably closed.

Annemüller and Manger (1997) addressed this problem, suggesting that it would be more appropriate to monitor the dissolved carbon dioxide concentration in very early fermentation in the period when the wort has not had a chance to become saturated. Undoubtedly this is true but it adds a further level of complexity to the control system.

### 6.3.4 Monitoring exothermy

Glycolysis is an exothermic process, and therefore the profile of heat generation during fermentation can be used to monitor progress (Mou & Cooney, 1976). Ruocco et al. (1980) developed such a system in which exothermy was determined by periodically measuring the increase in temperature of wort when the coolant supply was switched off. The numerical value of the exotherm measured at any particular time was based on calculations that took into account the observed increase in temperature as a function of time and the total quantity of wort present. Exotherm measurements made throughout fermentation were put together to form a profile, which showed good correlation with the profiles of gravity attenuation and yeast growth. These profiles were specific for individual fermenters and wort qualities. Averaging many sets of data allowed an 'ideal' exotherm profile to be constructed which could be used as a standard for comparison with new data.

Exothermy measured at any time during fermentation provides an instantaneous indication of fermentation rate (attenuation rate). Parameters which influence fermentation rate have a concomitant effect on exothermy. For example, increasing the temperature set-point increases the value of exothermy. The total exotherm, which may be determined by calculating the area encompassed by the profile of exothermy versus time, is a function of the total quantity of fermentable sugar present in the wort. It follows, therefore, that measurement of exothermy can be used to control fermentation by comparing the measured profile with a standard. In the event of under-achievement, the set-point temperature may be allowed to rise until correspondence is achieved. Second, through automatic integration of the exotherm/time profile, achievement of the desired attenuation gravity can be inferred.

As with measurement of rates of $CO_2$ evolution, monitoring fermentation through determination of exothermy has the advantage of being non-invasive and provides an output suitable for use in automated control systems. It does have some disadvantages. If temperature readings are made at single specific locations within the vessel there is an assumption that the wort is perfectly mixed. This is not so at all times and clearly it would be necessary to take an averaged output from several thermometers. The nature of the coolant is also critical. The trials reported by Ruocco et al. (1980) used vessels cooled by direct ammonia expansion. This system has little residual effect when the coolant valves are closed. With refrigerants such as glycol there is a considerable residual effect which would tend to underestimate the true value for exothermy.

Like on-line measurement of gravity or rate of $CO_2$ evolution, there is apparently little exothermy during the critical early phase of fermentation, which perhaps again mitigates against this approach to monitoring for use in interactive control strategies. However, this is not necessarily strictly true. Ruocco et al. (1980) reported that in some trials an early and transient peak in exothermy was sometimes but not always observed. They speculated that this burst of exothermy could be associated with sterol synthesis during the aerobic phase of fermentation. If this observation was verified it could perhaps be used as a probe to provide confirmation that the conditions established at the completion of wort collection were appropriate. If not, corrective action could be initiated during early fermentation and the exothermy profile obtained thereafter could be used simply to monitor and confirm satisfactory progress. It is certainly true that when pitching yeast is exposed to oxygen under non-growing conditions, sterol synthesis and glycogen dissimilation proceed. This is an exothermic process (see Section 6.4.2.2 and Fig. 6.31).

### 6.3.5   Monitoring pH

The transformation from wort to beer is accompanied by a decline in pH, typically from just over pH 5.0 to around pH 4.0. This change is a consequence of yeast metabolism, involving excretion of several organic acids and proton extrusion in response to assimilation of sugars (see Section 3.7.1). The pattern of pH change is characteristic for a given fermentation. Therefore, on-line measurement could be used to monitor fermentation progress. This is particularly so since pH electrodes capable of withstanding the rigours of the production environment are now readily available.

The most dramatic changes in pH occur during early fermentation and the minimum value is achieved before wort attenuation is complete (see Section 3.2). Often, there is a modest increase in pH from the mid-point onwards. In this regard, therefore, pH is not a particularly useful monitor of overall fermentation performance, and certainly it is of no value in identifying the end-point. However, the rapid decrease, which occurs in the first few hours after pitching, presents early identification of non-ideal performance.

Leedham (1983) described a control regime in which initial rates of pH decline were measured immediately after the completion of collection and compared with stored profiles pre-determined for particular fermentation types. The system was designed to optimise wort oxygenation. During collection, the wort was oxygenated to a less than optimum concentration. At the end of collection, the pH of the wort was measured at intervals of 15 minutes. After each measurement was made, aliquots of oxygen were added to the vessel if the rate of pH decline was less than the stored control profile. No action was taken if the pH value was at, or below the stored figure. If, after 10 cycles, the pH reading was still too high, the temperature set-point was raised progressively, up to a maximum of 2°C above the normal value.

The method has utility; however, it is not capable of responding to a situation where the pH decline is greater than the standard. This could occur in the event of accidental over-pitching/wort oxygenation, or more likely where yeast had a less than usual requirement for wort oxygenation because of exposure of pitching yeast to air during storage and handling (see Section 6.3.6).

### 6.3.6 *Monitoring rate of oxygen assimilation*

Oxygen added with wort at the beginning of fermentation is used by yeast to synthesise lipids, principally sterols and unsaturated fatty acids, which are essential to proper membrane function (see Section 3.5.1). The quantity of sterol synthesised by individual yeast cells is governed by availability of oxygen and the pitching rate. The quantity of sterol synthesised per yeast cell controls the extent of yeast growth during fermentation.

Oxygen falls to an undetectable concentration a few hours after the completion of wort collection. Since conditions are anaerobic during most of fermentation, monitoring oxygen concentration provides little useful information. Thus, as with monitoring of pH, it is not possible to correlate oxygen concentration with wort attenuation. However, also in common with pH decline, measurement of oxygen consumption by yeast affords a method for early prediction of non-ideal behaviour and an opportunity for remedial action to be taken.

It has been demonstrated that the rate at which yeast assimilates exogenous oxygen gives an indication of its physiological condition. In particular, it can be related to the size of the sterol pool (Kara *et al.*, 1987; Boulton *et al.*, 1991). Measurement of the rate at which oxygen is taken up during the aerobic phase of fermentation can be used to assess the sterol content of the yeast at pitch (providing the yeast concentration and temperature are defined). The appropriate quantity of oxygen may then be added to ensure that additional sterol synthesis brings the total concentration up to an optimum value for that fermentation.

In practice, this is a difficult undertaking. When wort collection is finished, the yeast concentration may be unknown since cell proliferation may have already commenced. Dosing oxygen into fermenter is inefficient and difficult to quantify. The requirement to adjust conditions post-collection prevents making a rapid start to the fermentation. A potential approach is to assess the yeast oxygen uptake rate in-line during wort collection, as shown in Fig. 6.17. This system consists of a loop of coiled stainless steel tubing attached to the wort main at a point after addition of oxygen and yeast. At each terminal of the loop an in-line dissolved oxygen meter is provided. The coil, which is attemperated, must be long enough to provide sufficient distance between the two dissolved oxygen meters to allow significant uptake of oxygen by the yeast.

In operation, wort flow commences at a known rate and oxygenation and yeast pitching are initiated. Yeast is dosed accurately into the wort at a rate controlled by a biomass meter (see Section 6.1.4.5). When steady flow rates have been established, the flow through the coil is initiated by automatically opening valves $V_1$ and $V_2$. Measurement of the decrease in dissolved oxygen concentration between the meters A and B allows a microprocessor to compute the oxygen uptake rate. With input from the wort flow meter and the biomass meter, this may be converted into a specific oxygen uptake rate. The specific rate of oxygen uptake may be compared with reference data for that particular combination of yeast strain and wort type. With

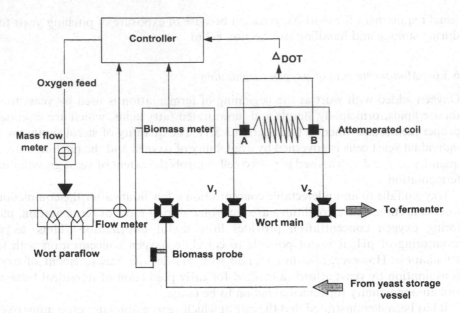

**Fig. 6.17**  System for automatic monitoring and control of wort oxygenation (from Boulton & Quain, 1987).

knowledge of the total volume of wort to be transferred, the controller directs a mass flow meter to dose in the appropriate quantity of oxygen.

This approach has not yet been implemented other than in laboratory simulation trials. It offers several advantages. A single unit may service several vessels, no modifications to fermenters are needed and control is exerted during collection such that there is no delay in starting the fermentation.

### 6.3.7  Monitoring yeast growth

Monitoring yeast growth is an obvious 'direct' method of assessing fermentation progress and overall performance. In laboratory or small pilot scale fermenters which have mechanically driven agitators, off-line analysis of samples for yeast count or by measurement of dry or wet weight are suitable methods which supply reliable data. In production scale fermenters, heterogeneity of vessel contents makes analysis of off-line samples taken from a single point of dubious value. Therefore, it is not usual to monitor yeast concentration during commercial fermentation. It is however, normal practice to assess the magnitude and quality of the yeast crop.

The quantity of yeast transferred to storage vessels gives a rough indication of the extent of growth during the fermentation from which it was taken. With any given fermentation, this should be relatively constant. Abrupt changes, particularly if carried forward into subsequent fermentations, should be taken as evidence of process or raw material changes, which require investigation (see Section 6.5.4). Yeast quality, usually assessed via a viability measurement, provides information of the preceding fermentation only. Thus, if the yeast crop is small and the viability is low it is likely that the fermentation performance would have been less than ideal. Unfortunately, this is unlikely to be a surprising correlation.

Automatic monitoring of yeast concentration during fermentation has been proposed but hampered by a lack of suitable sensors. An automatic sampling system linked to an electronic particle counter (Moll et al., 1978; Lehoel & Moll, 1987). To overcome the problem of heterogeneity in vessels, samples were removed from three locations at various heights in the vessel and analysed in sequence. Output from the counter provided cell number, size distribution and cell size. At present, this information is perhaps of academic interest only since it is unclear how it could be used.

There are no methods currently in use for automatic measurement of yeast concentration in vessel during fermentation. Turbidometric sensing would not be possible and in any case would not distinguish between viable and dead cells. Ultrasonic sensing of suspended solids has been proposed as a method of detecting yeast cells in-line (Behrman & Larson, 1987). Unfortunately, gas bubbles interfere and therefore this approach would be unsuitable.

The biomass meter described in Section 6.1.4.5 would perhaps be of use in this regard. It gives an instantaneous measure of viable yeast concentration and it is relatively unaffected by gas bubbles and non-yeast solids. Individual instruments can be multiplexed with several probes, and therefore it would be possible to locate these at various heights within a fermenting vessel and obtain a real-time profile of the distribution of yeast cells throughout the vessel. This could be of utility in the case of large cylindroconical fermenters. It would be possible to identify the point in fermentation when the bulk of the viable yeast was in the cone and cropping could proceed. Early removal of yeast crops is beneficial to beer quality, by minimising the adverse effects due to yeast autolysis. In addition, it benefits crop quality, by removing yeast from the stressful conditions experienced in the depths of fermenters. However, the biomass meters are relatively expensive, more so when multiplexed, and installation into every fermenting vessel is not likely to be a financially viable proposition for the near future.

### 6.3.8 Monitoring ethanol formation

Measurement of ethanol formation during fermentation provides a direct indication of process efficiency. Monitoring changes in the concentration of this metabolite should be a prime candidate for assessing progress of fermentation as a whole. Ethanol concentration may be measured on line using refractometry (Forrest et al., 1989). Automatic on-line sampling followed by gas chromatographic analysis for ethanol determination has also been proposed (Klopper et al., 1987; Pfisterer et al., 1988). In both of these cases, the intended application was automatic control of ethanol concentration during dilution of high-gravity beers.

Knol et al. (1988) investigated, at pilot scale, monitoring ethanol concentration during primary fermentation using on-line sampling and analysis by either HPLC or gas chromatography. Two sampling devices were tested. First, a membrane device located in a by-pass loop through which volatile components of the wort, including ethanol, diffused into a flow of nitrogen carrier gas and thence to a gas chromatograph for analysis. Second, a membrane sampling device in which liquid was separated from solids. A sample of the clarified liquid injected automatically onto a HPLC analyser. The HPLC approach showed a good correlation with off-line analysis and the liquid sampling and filtration device provided satisfactory performance. The diffusion sampling approach was less satisfactory. Diffusion is temperature dependent and it was necessary to include accurate compensation for this. The gas chromatographic analysis of ethanol showed significant deviation from off-line results. It was concluded that this was due to baseline drift and correction required frequent re-calibration of the instrument.

Anderson (1990) described results of preliminary trials in which ethanol was monitored by automated sampling of the fermenter head space and subsequent gas chromatographic analysis. Using stream-switching 16 fermenters could be monitored simultaneously, each being sampled hourly. The results were fed to a supervisory microprocessor. This provided data logging facilities and the possibility of using the ethanol input in a feed-back control loop linked to the attemperation system. During the first 10 hours of fermentation, a good correlation was demonstrated between ethanol formation and decrease in wort gravity. Over a longer time-scale the correlation became increasingly poor.

Anderson (1990) concluded that, at least for control purposes, a parameter that showed significant change during very early fermentation would be more appropriate and that ethanol did not fit into this category. In the real commercial world, it would be difficult to justify to accountants the cost of on-line chromatographs dedicated to every fermenter. In this respect, use of a single instrument to monitor several fermenters via sampling head-space gases is a much more attractive proposition. In addition, this approach obviates the need for intrusive sensors or the use of external loops.

In addition to measurement of ethanol, having a facility for on-line chromatography offers further interesting possibilities. These techniques are used already for laboratory determination of essential flavour metabolites such as esters, higher alcohols and vicinal diketones. The concentrations of these compounds in beer are much influenced by fermentation performance (see Section 3.7). At present, it is assumed that if the rate and extent of wort attenuation are within specification then so will be the resultant beer. This assumption has to be lived with. However, a control regime based on monitoring the formation of essential flavour components of beer, as well as some aspect of wort attenuation, may well offer new opportunities. For example, it would offer the promise of fermenting at ultrahigh gravity without the unbalanced formation flavour compound. Nevertheless, until more robust and economic sensors, capable of measuring the concentrations of flavour compounds, are developed, such applications at commercial scale remain unlikely.

### 6.3.9   Monitoring vicinal diketone concentration

For many lager-type fermentations, the process is judged complete when the concentration of the vicinal diketones, diacetyl and its precursor α-acetolactate, have fallen below a pre-determined threshold level. Thus, diacetyl has a strong butterscotch flavour, considered undesirable in some beers. The precursor, α-acetolactate, is derived from pyruvate and is excreted by yeast during primary fermentation. Subsequently it undergoes spontaneous decomposition to form diacetyl. The diacetyl is then reduced by yeast to the much less flavour-active acetoin and 2,3-butanediol (see Section 3.7.4.1).

Vicinal diketone (VDK) concentrations are monitored during the latter part of primary fermentation. Chilling cannot be applied until a desired maximum threshold VDK concentration. Analyses are usually performed on off-line samples removed from fermenters at regular intervals, typically every 8 hours. VDK concentration is determined using either spectrophotometric or gas chromatographic procedures (*EBC Analytica*, 1987; American Society of Brewing Chemists, 1992; IOB Recommended Methods of Analysis, 1991). Recently a differential-pulse polarographic method has been proposed as a more accurate substitute for the colorometric approaches (Rodrigues *et al.*, 1997).

Whichever method is used, it is essential that samples are subjected to a heat treatment prior to analysis to ensure that all the precursor α-acetolactate is converted to diacetyl. Thus, the spontaneous decomposition of α-acetolactate to diacetyl is the rate-determining step in the sequence of reactions. Subsequent reduction of diacetyl by yeast is rapid, and therefore in the presence of yeast the concentration of free diacetyl is always low. Should the yeast be removed from the beer prematurely the pool of α-acetolactate may subsequently decompose to diacetyl and persist because further metabolism is not possible. Hence, the requirement to monitor total VDK, as opposed to free diacetyl.

Off-line analysis of vicinal diketones is time-consuming and requires skilled personnel. In most breweries, several hours might elapse between sampling and obtaining a result. Clearly, it would be advantageous if on-line analysis was available. No such methods are currently in use although automatic sampling coupled with a heat treatment and gas chromatography is feasible although complex. An alternative approach has been described by Denk (1997). This author described a method in which the pattern of VDK formation and reduction could be predicted using a software modelling approach based on an artificial neural network (see Section 6.4.2.3). The model builds a simulation of fermentation held in the software of a computer. The underlying premise is that by inputting several sets of real data into a model system the program is capable of undergoing a learning process. Subsequent input of a partial set of data allows output of realistic profiles of missing data. Thus, provided some easily measurable parameters are monitored in early fermentation, the real change in concentration of another parameter, which might be much more difficult to measure, may be predicted with accuracy. The greater the number of sets of data used to develop the model, the greater the precision of the model.

In initial experiments performed at pilot scale (150 litres) input parameters were temperature, pressure, gravity, turbidity, pH and ethanol. Of these, temperature and pressure were both monitored and controlled at set-points; the remainder were simply monitored. In-line data were monitored during the first 15 hours of fermentation after which time the profiles of each parameter, including VDK, could be predicted with accuracy. In later trials, the ethanol input was dispensed with because of the difficulties and cost of off-line measurement of this metabolite. Ultimately it was intended to apply the system to 2500 hl fermenters, with one system servicing several vessels. At the time of writing there have been no reports of this application in brewing at commercial scale.

### 6.3.10  *Miscellaneous*

A few methods have been developed which utilise sophisticated optical devices for monitoring closed fermentation vessels. Hosokawa *et al.* (1999) described a portable camera system which could be used to examine the inner surfaces of large vessels. It consisted of a video camera attached to a controllable boom which, together with a monitor and dedicated lighting system, allowed direct visual examination of all parts of the empty vessel. This was considered essential as a means of checking the efficacy of CiP.

Wasmuht and Weinzart (1999) also recommended the use of a camera located inside the fermenting vessel, in this case to monitor fermentation progress. The system was called TopScan. The camera, together with suitable lighting, was mounted in the top plate of the vessel and directed vertically downwards. A visual image of the surface of the fermenting wort was channelled to a monitor in the control room. This allowed direct viewing of the stages in traditional lager fermentation, such as low and high krausen. In this way, the advantages of closed vessels and ability to judge fermentation progress by direct observation could be combined. As in the previous example, the system also allowed vessel CiP to be assessed.

Excessive foaming in production-scale fermenters can be a problem, which if uncontrolled can result in product loss. Ogane *et al.* (1999) described an in-tank system which allowed automatic measurement of the foam depth in fermenter. This used a laser-based sensor mounted in the top-plate of the vessel. Output from the sensor was directed to a computer via an RS232 interface, hence providing a detailed record of foaming throughout fermentation. The claimed advantages were that improved control of foaming by early detection of problems allowed vessels to be operated with a minimum free-board.

## 6.4  Fermentation control

The aims of fermentation control may be summarised as:

(1) production of beer of consistent flavour and analytical composition;
(2) maximisation of rates of conversion of sugar to alcohol, within the constraints of maintaining high product quality;
(3) ensuring that vessel residence times are consistent and as short as possible, both to maximise fermenter productivity and to allow brewing to be performed to a predictable time-table.

In the majority of fermentations, control is exerted through regulation of initial conditions and thereafter via attemperation. The methods used to control the initial parameters, namely, wort composition, temperature, yeast pitching rate and wort dissolved oxygen concentration, are discussed in Section 6.1. In practice, wort composition and the total extract delivered to the fermenter are controlled up-stream of the fermenter. Therefore, the variables available for manipulation are restricted to temperature, pitching rate, pressure and dissolved oxygen concentration. The first requirement of fermentation control is to choose appropriate values for these parameters and put systems in place to ensure that the desired values are achieved. In the vast majority of breweries, this – that may be described as a 'passive' system – is the extent of fermentation control. A second possibility is to use an interactive control strategy. This recognises the probability that there will be some variability that a passive control system is unable to correct. In particular, variations in wort composition and yeast physiological condition are likely.

Three levels of sophistication of active control strategy are possible. First, the use of preliminary treatments to remove sources of inconsistency. This approach obviates the necessity to modulate the values of key user-variables. For example, variability in the physiological condition of pitching yeast requires compensation by selecting an appropriate pitching rate and/or wort dissolved oxygen concentration. A pre-treatment which eliminates the physiological variability and produces pitching yeast of consistent condition would allow fixed values of pitching rate and dissolved oxygen concentration to be used. A treatment such as this would be oxygenation of yeast before pitching to promote sterol synthesis (see Section 6.4.2.2).

The second option is to make off-line assessments immediately prior to pitching which allow appropriate values for user variables to be chosen for that particular fermentation. For example, the so-called 'yeast vitality' tests whose aim is to produce a rapid assessment of yeast physiological condition, and thereby allow an optimum pitching rate and/or wort oxygenation regime to be chosen (see Section 7.4.2). An automated example of this type of approach is to measure the specific rate of oxygen uptake, by yeast, during wort collection. From this, the optimum oxygen concentration for that particular fermentation may be computed and added to the inflowing wort (see Section 6.3.6).

The third option is the use of interactive automatic control systems. This requires one or more parameters to be monitored, in real-time. In response, automatic adjustments are made to appropriate control elements. In theory, this ensures that the fermentation proceeds according to a desired profile. Parameters that may be measured on line and that produce outputs suitable for use in feedback loop systems are discussed in Section 6.3. They include exothermy and profiles of $CO_2$ evolution and ethanol production. The use of these and other parameters in automatic control systems in brewery fermentations is discussed more fully in Section 6.4.2.

### 6.4.1  *Effect of process variables on fermentation performance*

Interactions between yeast pitching rate, temperature and dissolved oxygen concentration in fermentation are complex. They influence fermentation efficiency, measured both as vessel residence time and the balance between ethanol yield and yeast growth. In addition, although wort composition and the yeast strain are important determinants of beer flavour, they influence the formation of flavour-active metabolites. Choice of appropriate values for temperature, pressure, pitching rate and wort oxygen concentration involves some compromise. Thus, inevitably there will be a play-off between fermentation efficiency and production of beer with a balanced spectrum of flavour components. Although the combined effects of process variables on fermentation performance are complex, it is possible to make some generalisations regarding their individual contributions. These are discussed in this section. An attempt has been made to distinguish between effects on fermentation efficiency and those that influence product quality. Where possible, the effects of simultaneously varying more than one parameter are described.

6.4.1.1 *Temperature.* Temperature has an effect primarily on fermentation rate through its effect on yeast growth and metabolic rate. The data in Fig. 6.18 shows attenuation profiles for pilot-scale high-gravity lager fermentations performed at a range of temperatures between 11°C and 25°C. All other process variables were, as near possible, identical. As may be seen, between the two extremes of temperature, there was a difference of nearly 100 hours in the time taken to achieve final gravity. Based on this result it may be supposed that it must be advantageous to perform fermentations at as high a temperature as possible. Most brewing yeast strains have a maximum growth temperature within the range of 30 to 35°C, suggesting that very rapid fermentations could be achieved. In fact, several factors preclude the use of very high temperatures. Some loss of volatile flavour components and ethanol by gas stripping is inevitable in all fermentations. The severity of this effect is obviously correlated with the vigour of fermentation. At very high temperatures, the rate of loss becomes unacceptable. Indeed, at 30°C, it would be necessary to control the loss of water due to evaporation and it would be difficult to retain the fermenting wort within the vessel!

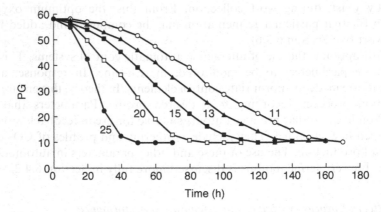

**Fig. 6.18** Pilot scale (800 hl) lager fermentations performed at the temperatures shown (°C). In all fermentations the pitching rate was $12 \times 10^6$ cells ml$^{-1}$ and the initial wort dissolved oxygen concentration was 25 ppm (Boulton & Box, unpublished data).

Of greater significance is the effect of temperature on the formation of yeast-derived flavour components. In general, ales are fermented at higher temperatures (typically 18 to 25°C) compared to lagers (6 to 15°C). Ales tend to be more highly flavoured than lagers, whereas lagers contain greater concentrations of more subtle flavour components. The greater flavour intensity of ales is in part due to the malts and hops used in their preparation. In addition, the beers also usually contain greater concentrations of higher alcohols and esters, which are produced during fermentation.

Some of these differences are a consequence of the yeast strains used. Generally, ale strains produce more esters and higher alcohols than lager strains. However, increased temperature also produces elevated levels of higher alcohols and possibly esters (Stevens, 1960; Mandl *et al.*, 1975; Kumada *et al.*, 1975; Posada et al, 1977; Miedener, 1978). It is probable that it is only in the case of higher alcohols that the increase can be ascribed directly to increase in temperature. In the case of esters, effects due to oxygen may also be implicated. Thus, elevation in temperature reduces the solubility of oxygen. There is a negative correlation between oxygen availability and ester synthesis. Decreased oxygen availability reduces yeast growth extent because of sterol limitation. Reduced utilisation of wort amino nitrogen and sugars for biomass formation allows a greater proportion of the pools of oxo-acids and acetyl-CoA to be used for the synthesis of esters. In fact, it may be shown that yeast growth extent and beer ester concentrations are unchanged at different temperatures when oxygen concentration is maintained at a constant value.

The temperature at which fermentation is conducted may influence oxygen requirement. As the temperature at which yeast is grown is reduced, the cells incorporate a progressively greater proportion of unsaturated fatty acid residues into the plasma membrane in order to maintain fluidity (Boulton & Ratledge, 1985). It follows that *de novo* unsaturated fatty acids synthesis during fermentation will require more oxygen at lower fermentation temperatures.

Fermentation temperature influences the profile of formation and reduction of VDK. Thus, increased fermentation temperature increases the peak value of VDK. In addition, at elevated temperatures the VDK peak occurs earlier in the fermentation and the rates of formation and decline are more rapid (Fig. 6.19). Thus, elevated temperature favours rapid primary and secondary fermentation.

The choice of temperature has practical implications for fermentation management. The use of high temperatures results in increased fermentation rate but some of the potential advantage is lost because of the concomitant increase in time taken to

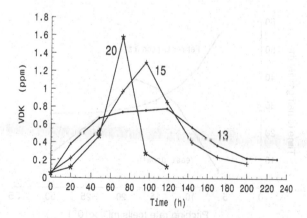

**Fig. 6.19** Profile of vicinal diketone formation and reduction during high-gravity (1060) lager fermentations performed at the temperatures shown (°C). In all cases pitching rates were $15 \times 10^6$ cells ml$^{-1}$ and the initial oxygen concentration was 25 ppm. Vicinal diketones were determined by gas chromatography (*EBC Analytica*, 1987) (Boulton & Box, unpublished data).

cool at the end of fermentation. Vigorous fermentations require more careful handling with respect to control of fobbing. Care must be taken to protect yeast in large
closed vessels when fermentations are performed at high temperatures. In traditional
top-cropping ale fermentations performed in open square vessels, the yeast is
removed comparatively early in the process. Consequently, it is not subject to heat
stress. Conversely, similar fermentations performed in large-capacity closed vessels
provide an environment in which it is much more difficult to control the temperature
of the yeast. Thus, when yeast cells are packed into the cone of a vessel, attemperation
is inefficient and difficult. Clearly, this problem will be exacerbated where a high
fermentation temperature is used. In such cases, the yeast should be removed from the
vessel as soon as is possible and vessels should be fitted with cone cooling jackets (see
Section 5.4.1).

6.4.1.2 *Yeast pitching rate.* Fermentations are pitched within the range of $5 \times 10^6$
cells ml$^{-1}$ for a sales-gravity ale up to $20 \times 10^6$ cells ml$^{-1}$ for a high-gravity lager. In
weight terms these rates equate to *c.* 1.5 g l$^{-1}$ to 6.0 g l$^{-1}$ wet weight (*c.* 0.3–1.2 g l$^{-1}$ dry
weight). The pitching rate must be based on the gravity of the wort and should
produce a fermentation that is as rapid as possible without compromising beer quality
or the size of the yeast crop.

The pitching rate influences both fermentation rate and the extent of yeast growth.
Predictably, within certain limits, there is a positive correlation between yeast pitching
rate and fermentation rate. At low values, there is also a positive correlation between
pitching rate and the extent of new yeast growth during fermentation. However, at
progressively higher pitching rates new yeast growth reaches a peak and thereafter
declines (Fig. 6.20). The pattern of yeast growth may be explained in terms of the ratio
of the initial yeast cell concentration and the concentration of available nutrients. In
most microbial batch cultures, the number of cells in the inoculum is small compared
to the terminal cell concentration when growth is limited by exhaustion of a nutrient.

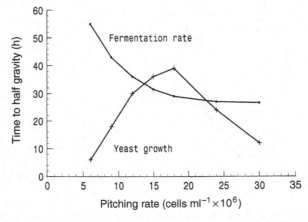

**Fig. 6.20**  Effect of pitching rate on fermentation rate (shown as time to half gravity) and yeast growth
extent (weight of total crop minus weight of pitched yeast) for pilot scale (8 hl) high-gravity lager fermentations (Boulton & Box, unpublished data).

In a brewery fermentation, the inoculum is large, typically a quarter to a fifth of the total crop. In this situation, the proportion of nutrients assimilated by each cell in the inoculum becomes significant. If the size of the inoculum (pitching rate) is increased, this ratio decreases and the effect becomes more significant such that total growth extent is restricted by nutrient availability.

It follows from the data shown in Fig. 6.20 that selection of the optimum pitching rate involves a trade-off between fermentation rapidity and loss of efficiency due to the proportion of wort sugars used for new biomass formation. Whilst it is true that very high pitching rates would give both short fermentation times and low growth, this option is precluded because of unacceptable shifts in flavour due to the limited yeast growth. Thus, low yeast growth extent results in elevated levels of esters. From a practical standpoint, if yeast growth is very restricted, insufficient yeast may be available for re-pitching. In addition, the quality of the crop may be reduced because of the relatively high proportion of cells derived from the pitching yeast.

The patterns of yeast growth rate and extent are further modulated by the oxygen concentration supplied to the wort (Fig. 6.21). Here is shown the effect on attenuation rate of varying both pitching rate and initial dissolved oxygen concentration in a high-gravity lager fermentation. Increase in both parameters resulted in faster fermentation rates such that the most rapid rates were obtained with high pitching rates and high dissolved oxygen concentration.

The effects on yeast growth of simultaneously varying pitching rate and dissolved

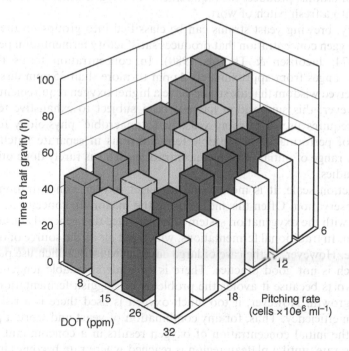

**Fig. 6.21** Effect of varying pitching rate and initial dissolved oxygen concentration on fermentation rate (shown as time to half gravity) for pilot scale (8 hl) high-gravity lager fermentations (Boulton & Box, unpublished data).

oxygen are complex because, depending on the values, different nutrient limitations may result. Where oxygen is limiting, the pattern of growth versus pitching rate is as shown in Fig. 6.20. At any given pitching rate, as the oxygen concentration is increased, more growth is permitted and the peak is moved towards the right until growth becomes limited by some other component of the wort. At different pitching rates the balance between limitation by oxygen or another nutrient is altered and different patterns of growth result.

The effects of varying pitching rate and oxygen concentration on the formation of volatile flavour components are also complex. The concentrations of esters formed during fermentation are influenced by the nature of the nutrient, which is limiting growth. Where growth is limited by oxygen, it would be predicted that ester synthesis would be promoted since less acetyl-CoA would be used for biomass formation, and therefore more would be available for esterifcation. However, where another nutrient restricts growth, such as amino nitrogen, lack of oxo-acid skeletons may reduce ester formation.

### 6.4.1.3  *Wort dissolved oxygen concentration.*

Oxygen is required by yeast during fermentation for the synthesis of sterols and unsaturated fatty acids, both of which are essential for membrane function (see Section 3.5). In controlled fermentations, the initial concentration limits the extent of yeast growth. Dilution of the sterol and unsaturated fatty acid pools between pitched cells and progeny, during growth under anaerobic conditions, produces the requirement for oxygen when the cropped yeast is re-pitched into a fresh batch of wort.

Reportedly, brewing yeast strains can be classified into groups on the basis of a minimum oxygen concentration that produces satisfactory fermentation performance (Kirsop, 1974; Jacobsen & Thorne, 1980). In concentration terms the oxygen requirement ranges from approximately 4 ppm to more than 35 ppm dissolved oxygen. It is received wisdom that ale strains have a higher oxygen requirement than lager strains; however, this supposition has not been subject to exhaustive testing. The evaluation requires use of pitching yeast with 'anaerobic' physiology followed by assessment of performance of multiple fermentations in separate batches of wort containing a range of initial oxygen concentrations. This is rarely done other than in academic studies.

At production scale, it is more usual to select oxygen concentration based on empirical observation. Often this means using the maximum concentration that can be achieved with the oxygenation system at hand, since this generally ensures a rapid fermentation. In traditional fermentations, which use air as the source of oxygen, this is acceptable. However, in the case of large-capacity vessels, which use pure oxygen, this approach is not good practice. There is an understandable tendency to over-oxygenate worts because it avoids the problems of sluggish fermentation rates and poor yeast growth. However, if too much oxygen is used there is a risk of loss of fermentation efficiency. Thus, for any combination of yeast and wort, a progressive increase in the initial concentration of oxygen results in a concomitant increase in fermentation rate, until a plateau region is reached where rate becomes independent of oxygen concentration. However, the increase in oxygen concentration is also accompanied by a progressive increase in yeast growth as the additional available oxygen promotes biomass formation. This shift in the pattern of carbon assimilation is balanced by a reduction in the yield of ethanol (Fig. 6.22).

The inflection point where fermentation rate and oxygen concentration become independent of each other, varies with yeast strain. It is related to the oxygen requirement group to which the strain belongs. If the yeast is a member of the highest oxygen-requiring group, the inflection point may never be reached.

Since oxygen availability and yeast growth are related, it follows that the formation of yeast metabolites related to yeast growth would also be affected. In particular, beer ester levels decrease in inverse proportion to the increase in yeast growth extent.

As discussed in Section 6.4.1.2, the effects of oxygen on fermentation rate, yeast growth, ethanol yield and formation of beer flavour components related to yeast growth are also modulated by the pitching rate. The patterns shown in Fig. 6.22 would all have been subtly shifted if a different pitching rate had been used. The key parameter is the ratio of oxygen supplied per yeast cell since it is this that controls the quantity of sterol formed per cell. At commercial scale, the time taken to fill vessels and in consequence the time during which yeast cells are exposed to oxygen is also influential, as discussed in Section 6.1.5. There is also the possibility of variable

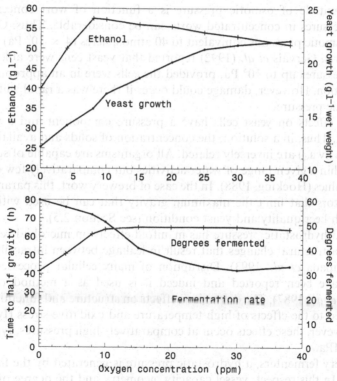

**Fig. 6.22** Effect of varying initial worty oxygen concentration on yeast growth, ethanol and fermentation rate in stirred laboratory fermentations using high-gravity (1060) wort. The pitching rate was $15 \times 10^6$ viable cells $ml^{-1}$ and the temperature was maintained at 11°C throughout the fermentations. Varying oxygen concentrations were obtained as described in Bamforth et al., 1988).

pitching yeast physiology, which produces an altered requirement for wort oxygenation. Exposure of pitching yeast to oxygen during storage can permit limited sterol synthesis. A further possibility is that should the fermentation conditions be such that oxygen is not the limiting factor, it follows that the cropped yeast will not be entirely depleted in sterol. In both of these scenarios, there will be a reduced requirement for oxygen in the next fermentation to produce optimum performance. To maintain control of fermentation this should be corrected for, as discussed in Section 6.4.2.

### 6.4.1.4 *Pressure.*

During fermentation, pressure effects may be manifest in three ways. All have the potential to influence performance. First, there is a hydrostatic pressure due to the height of the fermenting vessel. Second, the yeast cells are subject to an osmotic pressure, or variable water activity, which is related to the composition of the wort. Third, in closed vessels it is possible to restrict the outflow of carbon dioxide and allow the fermenter to pressurise. Extremes of pressure have the potential to exert deleterious effects on yeast cells. However, fermentations may be conducted under moderate top pressure with no effect to yeast. Indeed, application of pressure during fermentation is used as a strategy for manipulating ester formation (see Section 3.7.3).

The magnitude of osmotic pressure is a function of wort concentration. The osmotic pressures in concentrated worts can be considerable. Thus, Owades (1981) reported osmotic pressures equivalent to 40 atmospheres ($4 \times 10^6$ Pa) in some high-gravity worts. Gervais *et al.* (1992) reported that yeast cells were able to withstand osmotic pressures up to $10^8$ Pa, provided the cells were in an appropriate physiological condition. However, damage could occur if there was a rapid shift from low to high osmotic pressure.

Osmotic effects on yeast cells have a pressure component and a water activity component. Thus, in a solution, the concentration of solids and availability of water (water activity $a_w$) are inversely related. All organisms are capable of survival and/or growth within a given range of water activities but comparatively few can withstand very low values (Hocking, 1988). In the case of brewery wort, this parameter is one of several factors that limit the maximum gravity that can be used without compromising both beer quality and yeast condition (see Section 2.5).

Increased hydrostatic pressure has manifold effects on microbial cells. In yeast it causes ultrastructural changes that result in leakage between intracellular compartments (Shimada *et al.*, 1993). Disruption of many cellular processes due to high pressure have been reported and indeed it is used as a method of sterilisation (Kamihira *et al.*, 1987). The deleterious effects on structure and function of cells have been likened to the effects of high temperature and oxidative stress (Iwahashi *et al.*, 1995). However, these effects occur at comparatively high pressures, typically greater than 100 MPa.

In brewery fermenters, a hydrostatic pressure is generated by the liquid height in the vessel. In this respect, vessel capacity, geometry and the degree of agitation are important. Clearly, in tall vessels, the effect will be magnified. During primary fermentation, agitation is efficient, and yeast cells will circulate continuously throughout the vessel. Consequently, they will be subjected to a constantly changing pressure

environment. Undoubtedly this has effects on yeast metabolism and perhaps contributes to the differences in beer flavour when similar worts are fermented in vessels
of different aspect ratio.

Top-pressurisation of vessels involves two inseparable components. First, the
effects of elevated pressure itself, and, second, those due to the accompanying
increased concentration of dissolved carbon dioxide. The effects of both of these
components on ethanol fermentation have been studied in the context of fuel alcohol
production (Thibault et al., 1987; L'Italien et al., 1989). Here the impetus was the
possibility of using super critical carbon dioxide for in situ recovery of ethanol. These
authors reported that hyperbaric conditions (7 MPa) inhibited ethanol formation.
The inhibition was reversible by reducing the pressure. Nevertheless, even at these
high pressures, it was possible to maintain an ethanol productivity of $10.9\,\mathrm{g\,l^{-1}\,h^{-1}}$,
more than would be possible in brewery fermentation. Inhibition by carbon dioxide
appears to be more significant than effects simply due to pressure.

Top pressurisation of brewery fermentations is used as a means of modulating yeast
growth and the concentrations of beer flavour components where alteration of other
control parameters has produced an imbalance. Smaller pressures are used than those
described above. In general, the treatment is used to reduce the extent of yeast growth
and the concentration of volatile beer components. However, conflicting reports have
appeared in the literature as to the precise effects. Rice et al. (1976) described the
effects of carbon dioxide top pressure of 22.7 psig (0.16 MPa) on 100 litre lager fermentations. The extent of yeast growth and concentrations of beer volatiles produced
at 22°C were the same as those observed at 15°C with no top pressure. The increased
pressure had no effect on fermentation rate. Other small-scale trials (Arcay-Ledezma
& Slaughter, 1984) reported that pressurisation at 2 atmospheres ($2 \times 10^5$ Pa)
reduced fermentation rate, yeast growth extent and higher alcohols. In addition, at
the end of fermentation, levels of vicinal diketones were elevated. In this case a top-
fermenting ale yeast was used.

At production scale, pressure fermentation has been used successfully to compensate for the effects on beer flavour of elevated temperature (Nielsen et al.,
1986,1987). In this case, a pressure of 1.2 atmospheres (0.12 MPa) was used. The
pressure was applied gradually during fermentation. This reduced yeast growth and
volatile beer components compared to unpressurised fermentations. Both esters and
higher alcohols were influenced but to varying degrees and with no clear pattern. An
alternative treatment reported by the same authors and claimed to be as effective as
pressurisation in producing the effects described above, was to carbonate the wort. In
this case wort was collected with 8 ppm oxygen and carbonated to 0.3%.

Miedener (1978) described fermentations in which increased pressure was used to
reduce the concentrations of higher alcohols performed at elevated temperatures. The
fermentation was allowed to proceed at atmospheric pressure until approximately
50% attenuation was achieved. At this point the release of $CO_2$ was restricted and a
top-pressure of roughly 1.8 bar ($1.8 \times 10^5$ Pa) was allowed to build up. This author
suggested that a suitable value for the top pressure was equal to a tenth of the fermentation temperature in degrees Celsius.

Miedener (1978) reported a number of analytical differences between beers from
normal and pressurised fermentations. The pH of the latter was higher, possibly a

consequence of increased yeast shock excretion of nucleotides and amino acids in response to the higher pressure (see Section 3.7.6). Head retention values were also lower. This was ascribed to an increased content of short-chain ($C_6$–$C_{10}$) fatty acids. Possibly, this suggests that the pressure caused some yeast autolysis. In flavour terms, there was no significant alteration in the concentrations of higher alcohols, except for 2-phenylethanol. In the pressure fermented beer this increased by a remarkable two-fold. Ethyl acetate and iso-amyl acetate concentrations were not significantly different but the ethyl esters of hexanoic, octanoic and decanoic acids were approximately doubled in the pressure beers. Both dimethyl sulphide and total vicinal diketones were reduced in the pressure-fermented beers.

Posada (1978) described the effects of pressurisation in spheroconical fermenting vessels (see Section 5.4.3.3). It was confirmed that pressure caused a reduction of yeast growth extent. However, this report made the point that the modulating effects of pressure on the formation of some beer flavour components is yeast strain specific. For example, with two different yeast strains, elevated pressure produced opposite changes in the concentration of 2-phenylethanol. Similarly, the concentrations of other flavour components could be shifted upwards or downwards, by pressure, dependent on the yeast strain. Presumably, these strain dependent variations are the source of the sometimes conflicting reports of the effects of pressure fermentation on beer flavour components.

### 6.4.2 Automatic control regimes

It may be appreciated that the combination of the effects of the variables which contribute to the initial conditions established at the start of fermentation are complex and interdependent. It is necessary to choose appropriate values for each user variable, and most often this is achieved by a combination of trial and error tempered by previous experience. Unfortunately, even when the most stringent control is applied to each variable, not all inconsistencies in fermentation performance will be eliminated. This is a consequence of variability in the composition of the wort and the physiological condition of the pitching yeast.

By the time the wort has been collected into fermenter there is little opportunity for modifications to be made other than those discussed in Sections 6.1.3 and 6.2. At present, therefore, there is no mechanism for correcting for variability of wort composition. Several approaches for the rapid assessment of the physiological condition of pitching yeast have been proposed (see Section 7.4.2). These so-called 'yeast vitality' tests may be viewed as an extension to the conventional viability test. The aim is to produce information regarding the viable fraction of yeast that is predictive of subsequent performance in fermenter. Results from such tests may be used to form a judgement of fitness to pitch for a particular batch of yeast. Alternatively, the results of the test may be used to calculate the pitching rate that will produce the desired fermentation performance.

The logical development of vitality testing is that assessment of yeast condition should be carried out in fermenter as part of an automatic interactive feed-back control system. Such an approach opens the possibility of correcting for variability in both yeast condition and wort composition.

6.4.2.1 *Economics.* The greater the sophistication of the fermentation control system, the more the potential cost of installation and operation. Therefore, before installing such systems it is essential to assess the potential economic benefit and compare this with expenditure. This is a difficult calculation since the real cost penalty of variable product quality is not easy to quantify. Certainly, the costs of sophisticated on-line monitoring and control devices can be justified only where several large capacity fermenting vessels are used. In breweries that are more traditional, it should be sufficient to ensure adequate fermentation control by careful manual regulation of the pertinent parameters.

The potential savings to be made by ensuring rigorous fermentation control may be illustrated in the following example. Assume a brewery with 20 × 1500 hl vessels in which fermentations with optimal control take 10 days to complete. Allowing 3 days for filling, cooling, emptying and cleaning, the potential annual productivity is:

$$20 \times 1500 \times \frac{365}{13} = 842\,308 \text{ hl}$$

Assume that, if poorly controlled, the fermentation may take up to a further 3 days to complete. In the worst case the annual productivity would fall to:

$$20 \times 1500 \times \frac{365}{16} = 684\,375 \text{ hl}$$

Assuming the same worst-case vessel turn-round time of 16 days, the annual productivity for each vessel would be 34 219 hl. Therefore, to make up the potential annual shortfall of 157 933 hl, there would be a requirement for an additional five fermenters. Assuming a capital cost of £350 000 per fermenter, the potential savings due to an efficient control regime would equate to a saving in new plant costs of £1.75 million. This figure does not include the revenue costs of operating the additional vessels. Alternatively, the same figures could be expressed in terms achieving the same volume output using the same fermenter capacity. In this case, the control system would provide a potential increase in fermenter productivity of more than 20%.

Set against these potential savings is the cost of installation and operation of the control system. Obviously these may be considerable, particularly the initial capital costs of fitting sensors to every vessel. In this regard, control systems that use existing measuring devices, such as thermometers, are particularly attractive. Where new sensors have to be fitted, those that are non-invasive and preferably able to be shared by a number of fermenters attract a significant cost advantage.

6.4.2.2 *Fermentation control by yeast oxygenation.* Variable pitching yeast physiology can arise in several ways. Inconsistencies in performance of the fermentation from which the yeast was cropped and less than ideal conditions of handling and storage in the intervening period between cropping and re-pitching are probably common causes. An intermediate stage between a passive control system and an automatic interactive regime is to take positive steps to eliminate sources of inconsistency. With respect to yeast, this can be partially achieved by stringent control of storage conditions. Another option is to subject pitching yeast to a pre-treatment that ensures that all the cells are in a desired physiological condition.

A prime cause of inconsistency of pitching yeast condition is variable exposure to oxygen during handling and storage. In this case, yeast may couple dissimilation of stored glycogen to limited sterol synthesis. Failure to correct for this pre-formed sterol by reduction of the concentration of oxygen supplied with the wort will result in a rapid but inefficient fermentation due to excessive yeast growth (Boulton & Quain, 1987). Since the amount of pre-formed sterol is not quantified and is variable, inconsistent fermentation performance and beer analysis will result. A logical extension of this effect is to take positive steps to expose pitching yeast to oxygen in a controlled treatment and encourage sterol synthesis. Historically such a treatment was common practice, particularly in German breweries as a means of improving the vigour of pitching yeast. Proper regulation of the process ensures that all yeast cells achieve a consistent level of sterol synthesis and this source of variability in physiological condition is eliminated. After the treatment, in theory, the yeast will have no requirement for wort oxygenation and good control of fermentation can be achieved by precise regulation of pitching rate.

During the oxygenation process the synthesis of sterol and concomitant utilisation of the carbon and energy released by degradation of glycogen is accompanied by an increase in the rate at which yeast assimilates oxygen. The maximum observed yeast oxygen uptake rate ($\Delta_{DOT}$) coincides with the achievement of steady state maximum and minimum values of sterol and glycogen, respectively, and therefore this parameter can be used as a monitor of the oxygenation process (Fig. 6.23).

The data shown in Figs 6.24(a) and (b) depict the results of a laboratory trial in which a sample of pitching yeast slurry was oxygenated as described in the legend to Fig. 6.23. Samples were removed from the slurry at the times indicated by arrows (Fig. 6.24(a)) and pitched into aliquots of anaerobic wort. The resultant attenuation profiles were as shown in Fig. 6.24(b). As may be seen, samples of yeast removed before the point of maximum $\Delta_{DOT}$ produced fermentations with slow but progressively increasing rates. Yeast removed at, or after, the point at which the maximum $\Delta_{DOT}$ was observed produced essentially identical fermentation profiles.

The total quantity of sterol synthesised during the oxygenation process was of the same magnitude as that seen in a conventional wort-oxygenated fermentation. Thus, the initial sterol content of approximately 0.2% cell dry weight increased five-fold to around 1% cell dry weight. Sterol synthesis was accompanied by dissimilation of squalene. The decrease in the size of the squalene pool was sufficient to account for the quantity of sterol synthesised (Fig. 6.25(c)). This suggests that carbon released by glycogen dissimilation was not utilised directly for sterol synthesis *de novo*.

Oxygenation was accompanied by the formation of ethanol, possibly indicating the likely fate of the carbon derived from glycogen breakdown (Fig. 6.25(b)). In the data shown, the high initial value of ethanol reflected that present in the barm ale entrained in the yeast slurry. The combination of simultaneous exposure to oxygen and ethanol had no effect on yeast viability. This remained high and essentially unchanged throughout the process. The yeast dry weight decreased during the treatment, presumably also reflecting glycogen breakdown (Fig. 6.25(a)). In addition, the pH fell from an initial value of 4.2 to approximately 3.75. Probably this effect is analogous to

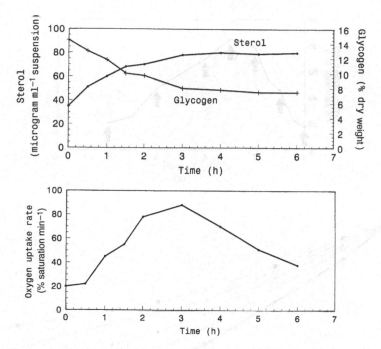

**Fig. 6.23** Changes in the intracellular concentrations of sterol and glycogen compared with the rate of uptake of oxygen of a slurry of pitching yeast during exposure to oxygen. Yeast (35% wet weight to volume) was removed from a brewery storage vessel and oxygenated without dilution at a temperature of 20°C (redrawn from Boulton & Quain, 1987).

that seen in the spontaneous first part of the acidification power test, which has also been related to glycogen breakdown (see Section 7.4.2).

The spectrum of individual sterols formed during oxygenation is shown in Fig. 6.26(a). Predictably, ergosterol showed the greatest increase during oxygenation. However, comparatively large amounts of zymosterol were also synthesised. Interestingly, in relative terms, zymosterol was the only sterol to increase during the oxygenation process, whereas, as a proportion of the whole, ergosterol decreased (Fig. 6.26(b)). This was perhaps surprising since zymosterol is reportedly not inserted into membranes but is esterified and stored in lipid particles (Leber *et al.*, 1992).

The changes observed during prolonged oxygenation are shown in Fig. 6.27. Following the initial peak of oxygen uptake rate, which occurred after about 4 hours, this parameter increased again, to an even higher value. The second peak of oxygen uptake rate occurred after some 16 hours' continuous oxygenation. This second burst of high oxygen consumption was accompanied by a concomitant fall in the concentration of exogenous ethanol. Therefore, this probably indicated derepression and the attainment of respiratory competence. Sterol increased in concert with the first peak of oxygen uptake rate; however, there was no further increase thereafter. Thus, the relatively high sterol concentration associated with derepressed yeast, typically 5% of the cell dry weight (Quain & Haslam, 1979), was not seen here.

The storage properties of oxygenated and non-oxygenated yeast are illustrated in

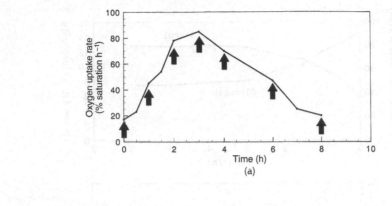

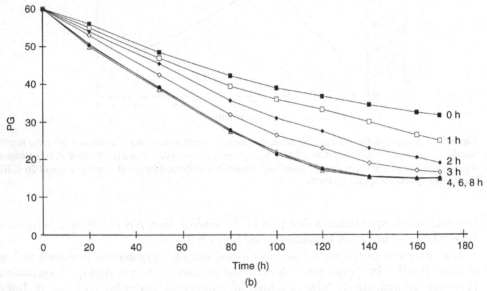

**Fig. 6.24** Profile of change in oxygen uptake rate of yeast slurry exposed to oxygen. Samples were removed at the times indicated by the arrows and (a) oxygen uptake rates were measured off-line using a polargraphic method and (b) evaluated by assessment of their fermentation performance in anaerobic wort in EBC tall tubes (re-drawn from Boulton *et al.*, 1991).

Fig. 6.28. Within the time-scale of production-scale brewing and at normal storage temperatures (2°C) there were no significant differences. However, at elevated temperatures (18°C) oxygenated yeast lost viability more rapidly than the non-oxygenated control. It must be assumed that this reflected the low glycogen content, and, hence, reduced ability to withstand prolonged starvation, of the oxygenated yeast. Synthesis of trehalose which has been reported to occur during yeast oxygenation (Callaerts *et al.*, 1993) and which has been also observed to occur in the studies described here, did not apparently confer an increased ability to withstand prolonged storage.

The yeast oxygenation method of fermentation control has been applied at

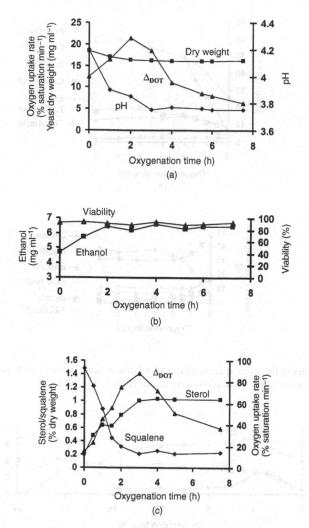

**Fig. 6.25** Biochemical changes associated with oxygenation of a slurry of lager yeast (c. 20% wet weight to volume) suspended in beer. During treatment the slurry was maintained at 25°C.

production scale. In this patented process (Quain & Boulton, 1987c), sufficient yeast is oxygenated to pitch 1600 hl of high-gravity lager wort. The process is performed in a tank designed to deliver high rates of oxygen transfer and maintain the slurry at a constant temperature (Figs 6.29 and 6.30). During the treatment, the dissolved oxygen tension is maintained at a constant value by a feedback control loop system. Oxygen uptake rates are inferred from the rate of oxygen addition that is required to maintain a constant concentration in the slurry. When the maximum $\Delta_{DOT}$ is achieved, the process is judged complete. At this stage the yeast may be pitched immediately into anaerobic wort or chilled and stored until required. After oxygenation, the yeast has a more stable physiology than conventional cropped yeast of 'anaerobic' physiology and therefore it is advantageous to oxygenate as soon as

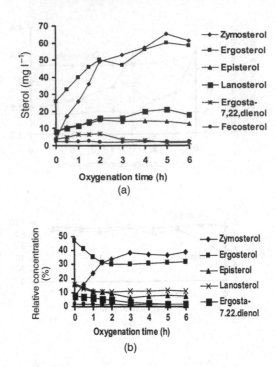

**Fig. 6.26**  Changes in the sterol spectrum (a), and relative concentration of each sterol (b) during oxygenation of a lager yeast slurry. Conditions were as described in Fig. 6.23.

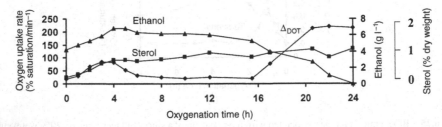

**Fig. 6.27**  Effect of prolonged oxygenation of a lager yeast slurry on the rate of oxygen uptake, total intracellular sterol concentration and exogenous ethanol concentration. The conditions were as described in Fig. 6.23.

possible after cropping. During the oxygenation process the slurry may be supplemented with glucose or maltose. This compensates for depletion of the endogenous glycogen pool, which can occur if the yeast is stored under inappropriate conditions or held for too long in fermenter before cropping.

Some of the changes seen in production-scale oxygenation of pitching yeast are shown in Figs 6.31(a), (b) and (c). In the trial shown, the slurry was supplemented with maltose (3% w/v) added after 30 minutes' oxygenation. The oxygen uptake maximum, inferred as oxygen demand to maintain a constant DOT, occurred after 1 to 2 hours. The peak $\Delta_{DOT}$ coincided with maximum sterol synthesis, exothermy and

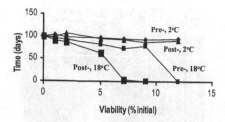

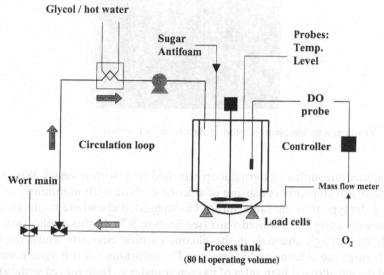

**Fig. 6.28** Effect on the viability of yeast during storage, at the temperatures indicated, before and after oxygenation. Yeast (35% wet weight to volume) suspended in beer was oxygenated for 4 hours while being maintained at a temperature of 25°C.

**Fig. 6.29** Schematic diagram of the equipment used for oxygenating yeast at production scale.

minimum pH. As with the laboratory trials discussed previously, the process was accompanied by ethanol formation and glycogen dissimilation. The latter was associated with a concomitant fall in cell dry weight. Yeast viability did not decrease to any significant degree during 6 hours' oxygenation.

At production scale, it was demonstrated that the use of oxygenated yeast produced more consistent attenuation profiles compared with conventional wort oxygenated fermentations using similar yeast and wort (Figs 6.32(a) and (b)). The improved consistency of attenuation profile observed for yeast oxygenated fermentations was accompanied by more consistent patterns of yeast growth extent and concentrations of beer esters and higher alcohols. In addition, there was an overall improvement in fermentation efficiency. Thus, yeast growth was reduced by approximately 20% compared to average values for wort oxygenated control fermentations. Surprisingly, concentrations of trial beer esters were not significantly different to those of the controls (Figs 6.33 and 6.34).

Masschelein *et al.* (1995) described a yeast oxygenation system in which yeast slurry

**Fig. 6.30**   Yeast oxygenation plant (courtesy of Bass Brewers Limited).

was circulated through an external loop attached to a storage vessel. Within the loop were cylindrical elements consisting of silicone carbide with aluminium oxide membranes of the type that the same author has suggested elsewhere might be applied to fermentations using immobilised yeast (see Section 5.7). In this application, the yeast is circulated through channels in the silicone carbide elements whilst the oxygen is forced through the silicone carbide matrix. The advantage of this system was that the enclosed design allowed high rates of oxygen transfer to be achieved without fobbing. Results were presented showing similar relationships between sterol synthesis, glycogen dissimilation and oxygen consumption rates as shown in Fig. 6.23.

6.4.2.3   *Interactive control regimes.*   Interactive control regimes involve a combination of automatic monitoring of suitable parameter(s) and in response, corrective action(s) to ensure that the process proceeds according to a pre-determined path. A requirement of such systems, therefore, is availability of suitable sensors, knowledge of how changes in measured parameters relate to fermentation progress and beer quality and a means of applying corrective actions should deviations occur.

The means by which fermentation progress may be monitored in-line are described in Section 6.3. It is possible to monitor parameters such as oxygen uptake rate, carbon dioxide evolution, ethanol production, decline in gravity and rate of exothermy. The profiles obtained may be compared with a pre-determined ideal and corrections made by automatic adjustment of temperature. Undoubtedly such methods would lead to improvements in consistency of fermentation performance. However, such simple approaches are likely to be at best only partial solutions. Thus, there remains insufficient understanding of how changes in parameters that can be measured during

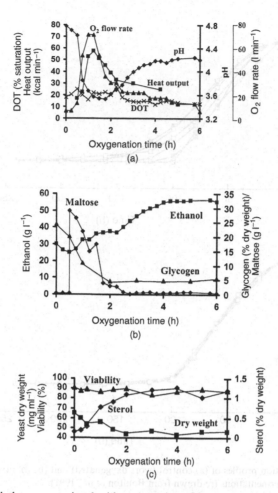

**Fig. 6.31** Biochemical changes associated with oxygenation of lager yeast slurry (45% wet weight to volume) suspended in beer, using the equipment shown in Fig. 6.27. During oxygenation the yeast was maintained at a temperature of 25°C. Maltose (5% g l⁻¹ final concentration) was added 30 minutes after the commencement of oxygenation.

early fermentation, relate to the timing of the end-point of fermentation and beer quality.

A development which perhaps offers a way forward is the use of software models of brewery fermentations. Thus, the 'number-crunching' ability of modern computers offers the potential of identifying hitherto obscured patterns and relationships between control parameters and fermentation performance. Two general approaches have been proposed: those based on self-teaching systems such as neural networks and those that use simulation models.

The concept of artificial neural networks was developed by McCulloch and Pitts (1943) to simulate the function of human neurons. The premise is that multiple data inputs are fed into a computational system that uses the received information to generate a predictive model. The software does not contain a mathematical model of

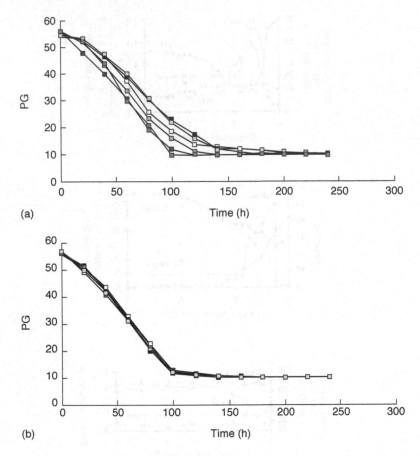

**Fig. 6.32** Attenuation profiles of (a) control (wort oxygenated) and (b) trial (yeast oxygenated) 1600 hl high-gravity lager fermentations (re-drawn from Boulton *et al.*, 1991).

the system under consideration in which real data are used as the basis of calculation. Instead, the input data are used in a 'learning by experience' process in a way analogous to the development of neuron interconnections in the human brain. All data inputs are assigned relative discrete weighting. These weighted data are combined in the network and used to form the basis of decisions.

For application to the study of biological systems, a back-propagation neural network system is often used. This is an example of a neural network with supervised learning, in which a desired output is compared with actual output and the weights between the individual elements of the network updated accordingly. Each network consists of a number of layers, each made up of a number of individual processing elements. In each case, there is an input layer, an output layer and any number of intervening hidden layers. Thus, in terms of a brewing fermentation the input layer could receive information such as pitching rate, aspects of wort composition, dissolved oxygen tension, etc. The output layer could be a time-based parameter such as the rate of formation of ethanol, $CO_2$ or yeast biomass. Each processing element

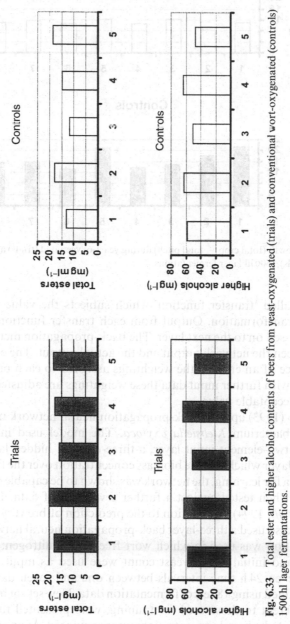

**Fig. 6.33** Total ester and higher alcohol contents of beers from yeast-oxygenated (trials) and conventional wort-oxygenated (controls) 1500 hl lager fermentations.

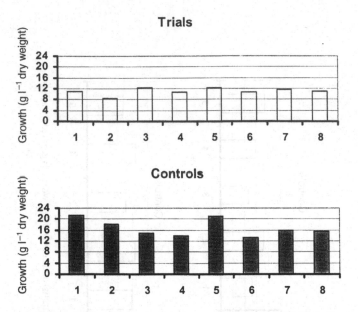

**Fig. 6.34** Yeast growth (total crop − total pitch) during yeast-oxygenated (trials) and conventional wort-oxygenated (controls) 1500 hl lager fermentations.

contains a so-called 'transfer function' which subjects the value of the input to a mathematical transformation. Output from each transfer function within a layer is summed and passed on to the next layer. The back-propagation method compares the difference between the network output and the actual output. The error is assumed to be a consequence of an error in the weightings assigned to each processing element. By comparison with further input data these weightings are adjusted until the error is reduced to an acceptable value.

Syu and Tsao (1993) applied back-propagation neural network modelling to batch growth of the bacterium *Klebsiella oxytoca*. The model used initial glucose concentration in a two-element input layer, a three-element hidden layer and an eight-element output layer which was the biomass concentration over the first 8 hours. Using seven sets of data for learning, the network was shown to be capable of both simulation and prediction when tested against a further seven sets of data. In a later communication (Syu *et al.*, 1994), application to the prediction of brewery fermentation was described. This also used a three-layer back-propagation neural network. In this case a 3-3-7 configuration was used in which wort free amino-nitrogen, dissolved oxygen concentration and initial viable yeast count were used as input and ethanol concentration at seven 24 h time intervals between one and seven days as output. The network was trained using 18 sets of fermentation data. New sets of initial fermentation conditions, different to those used for training, were presented to the network and actual and predicted ethanol concentrations were compared. At production scale (1500 hl), a good correlation between actual and predicted ethanol concentrations was obtained up to 96 h. After this time significant deviation occurred. The discrepancy was ascribed to the fact that temperature was not included as input in the network.

The requirement to consider temperature has been addressed by Gvazdaitis *et al.* (1994). In this case, it was reported that in large fermentation vessels temperature fluctuations limited the efficacy of predictive models. In order to improve attemperation it was necessary to have an accurate model of exothermy due to the fermentation and precise knowledge of the cooling behaviour of the vessel. The progress of fermentation was predicted using a neural net, which had a derivative of present gravity and diacetyl concentration as output. Nine input elements were used and a single hidden layer containing ten elements. Vessel cooling behaviour was modelled using a separate difference equation. Cooling behaviour during both primary fermentation and the diacetyl stand were encompassed. The combination of the models produced a dynamic temperature controller, which minimised coolant costs with no loss of fermentation control.

Software-based fermentation control systems with a different philosophy to neural networks are those based on real models. Several mathematical models of the biochemical reactions which underpin brewery fermentation have been proposed (Engasser *et al.*, 1981; Bezenger & Navarro, 1988; Gee & Ramirez, 1988; Jarzebski *et al.*, 1989; Pascal *et al.*, 1995). Using relationships given in these models, bilinear simulations of brewery fermentation have been developed (Johnson & Burnham, 1996; Johnson *et al.*, 1996a, b). These mathematically based systems allow prediction of the effects of all major fermentation variables on times to achieve attenuation gravity and diacetyl specification. This development is still in its infancy but it is anticipated that the simulations will be used to develop software control models for application in automatic on-line regimes.

### 6.5 Fermentation management

Beer quality, yeast type, fermenter design and operation are all intimately related. In particular, the design of many traditional fermenters arose because they were suited to the quality of beer produced and the properties of the yeast used (see Chapter 5). Modern vessels tend to be used for any beer quality and yeast type although perhaps with some compromises being made. All fermentations proceed through a number of stages. Several variations are possible and the terminology can be somewhat confusing.

(1)  **Primary fermentation:** This is the true fermentation, which commences as soon as yeast, oxygen and wort come together. This stage ends when all fermentable sugars have been utilised, in which case the wort is termed 'fully attenuated'. Alternatively, primary fermentation may be arrested by treatments that encourage the yeast to separate out from the wort. In this instance, the wort would be termed 'partially attenuated'. The present gravity of the wort at the end of primary fermentation is the 'attenuation' or 'racking gravity'.

(2)  **Secondary fermentation:** Secondary fermentation refers to any part of the fermentation process that occurs after primary fermentation and involves reactions requiring the presence of yeast cells. These reactions may take place in a brewery vessel or within a cask or bottle.

(3) **Conditioning:** Conditioning is the stage in the process that occurs after primary fermentation and prior to packaging, in which the immature green beer is converted into mature beer. An obligatory part of conditioning is adjusting the carbonation level of the beer and thus the presence of $CO_2$ in beer, or its development, is referred to as 'condition' or 'coming into condition', respectively. If the $CO_2$ is derived from the action of yeast as in a cask ale this is termed 'natural' or 'cask conditioning' and is an example of secondary fermentation. The process of conditioning traditional lager beers by holding for a period of cold storage in the presence of yeast and fermentable sugar is termed 'lagering', or 'cold conditioning', and is another example of secondary fermentation.

In some cases, beer in fermenter is held at a relatively high temperature after the completion of primary fermentation. The object of this treatment is to ensure that all available α-acetolactate is decarboxylated to diacetyl and the yeast has an opportunity to reduce diacetyl to acetoin and butanediol (see Section 3.7.4.1). Therefore, this phase may be referred to as a 'diacetyl' or 'VDK stand/rest'. It may also, however, be termed warm conditioning or in German, *Ruh* storage. It is another example of a secondary fermentation.

Apart from adjustment of carbonation, the conditioning process also involves clarification by promoting precipitation of materials, which if not removed may cause hazes in packaged beers. This is also accomplished by storage at low temperature and the processes involved are purely physicochemical in nature. In addition, flavour maturation may occur by venting undesirable volatile components of beer to atmosphere and carbonation may be controlled by gas sparging. None of these adjustments requires the presence of yeast cells. This stage may also be referred to as cold conditioning, although, particularly in America, it may be called 'ageing'.

### 6.5.1  Traditional top-cropping systems

Within this category, especially in the case of UK ale fermentations, a variety of systems are used. Some of these, such as the Burton Union system or the Yorkshire square, are unique and their peculiarities are described in Sections 5.3.4 and 5.3.5. In the case of closed or open square vessels, the fermentation process is comparatively simple. Essentially the process consists of just a single stage, since there may be no requirement for a separate conditioning phase. Thus, conditioning and secondary fermentation may be performed in the cask. Primary fermentation is rapid, typically 72–120 h, because temperatures tend to be relatively high (15 to 22°C). Occasionally the dropping system may be used where the fermenting wort is transferred from one vessel to another, usually 24 h after pitching (see Section 5.3.3).

Management is restricted to application of appropriate attemperation, removal of the yeast crop (see Section 6.7.1) and ensuring that the desired degree of wort attenuation is achieved. Differing qualities of beer require specific degrees of attenuation. Thus, if a secondary fermentation process is used, as in a cask-conditioned ale, it is usually necessary to ensure that some fermentable residue remains at the end of primary fermentation. Conversely, if a secondary fermentation is not required, or where additional sugar is added to the cask, a fully attenuated primary fermentation may be needed. The degree of attenuation at the end of primary fermentation is regulated by the controlled application of cooling, to both arrest activity and promote separation of yeast. Where a reasonably high residual yeast count is required, little or no cooling is applied and the process stage is ultimately terminated by emptying the vessel.

### 6.5.2 *Traditional lager fermentations*

This type of fermentation is associated particularly with the production of traditional European lager beers. Fermentations are performed in shallow open or closed vessels at low temperatures. Typically, the initial temperature is in the region of 6°C. The combination of low temperature and long processing times produces beers with subtle floral flavour notes, which are not as evident in similar beers produced at higher temperatures in a more rapid process (see Section 6.5.3). Pitching rates are comparatively high (*c.* 20–25 × $10^6$ cells ml$^{-1}$) to ensure a rapid start and reduce the risk of contamination. Satisfactory fermentation progress is judged by the appearance of the surface of the fermenting wort. Thus, the head goes through a number of characteristic stages during the fermentation. The appearance and disappearance of these types of head is used to gauge the degree of cooling to be applied.

Within 8–16 h of starting, a fine white head appears, which is the first visible manifestation of the onset of gaseous $CO_2$ evolution. During this early phase a sediment forms in the base of the vessel, consisting of cold break precipitated from the wort. This sediment has the potential to contaminate the yeast crop and this may be avoided by using a dropping system analogous to that described for top-cropping ale fermentations (see Section 5.3.3). Alternatively, a single vessel system may be used and in this case it is usual to attempt to leave this behind when the overlay of yeast crop is removed (see Section 6.7.2).

Another variation is the system described in German as *Darauflassen*, in which the fermenting vessel is initially part-filled and then after 24 h a further aliquot of unpitched oxygenated wort is added. The sequential addition of wort promotes good yeast growth though a kind of primitive fed-batch approach. After 48 h a rocky or cauliflower head of foam appears on the surface of the vessel, in German termed *krausen*, ('frill' or 'ruffle'). During this phase, the temperature is allowed to increase gradually by a further 2 or 3°C. After approximately 72 h the fermentation reaches its most active phase and the head grows and assumes a deeply separated appearance. This phase is termed 'high krausen' and during this period of maximum exothermy the temperature is held at a maximum of 8 to 9°C. As the degree of attenuation increases, the vigour of the fermentation declines and the head collapses to form a thin brown and bitter-tasting pellicle containing tannins and other materials. This may be removed by skimming, or tanks may be specifically designed so that it is left behind when the vessel is emptied. As with top-cropping ale fermentations, it is necessary to control the end-point of traditional lager fermentations by careful application of cooling. This is to ensure that sufficient suspended yeast cells and fermentable extract remain to fuel the prolonged cold secondary fermentation.

The secondary fermentation is performed in closed chilled 'lagering' tanks. This process may last from a few weeks up to several months. Typically, the beer is run into lagering tanks at a temperature of roughly 5°C and over a period allowed to decrease to 0°C. Occasionally a more rapid treatment lasting just a few weeks at −1°C to 0°C may be practised. If the secondary fermentation is too sluggish, additional fermentable extract and yeast cells may be added in the form of actively fermenting wort, a treatment known as *krausening*. The conditioning process has three functions: adjustment of carbonation, full development of a filterable chill haze and final flavour development. The flavour changes encompass a multitude of complex reactions. These range from simple $CO_2$ gas stripping of volatile components such as acetaldehyde and $H_2S$, through to the production and removal of desirable and undesirable flavour-active by-products of yeast metabolism. Of vital importance during this stage, remaining acetohydroxy acids spontaneously decarboxylate to form undesirable vicinal diketones, in particular diacetyl. Yeast cells must be present to reduce the vicinal diketones to much less flavour-active diols (see Section 3.7.4).

### 6.5.3   *Modern closed fermentations*

Fermentations performed in modern closed vessels such as cylindroconicals may be of either ale or lager types. In the case of ale fermentations, the process is managed as described already for traditional vessels. However, at the end of primary fermentation the bulk of the yeast is induced to settle out in the base of the vessel by the application of rapid chilling. This happens even though the same top-cropping strains of yeast are used. The altered flocculation is probably a consequence of the combination of smaller surface area, more efficient cooling and much better mixing during primary fermentation. Probably these factors prevent the formation of aggregates of yeast cells and entrapped $CO_2$ bubbles, which characterise the formation of yeast heads in top-cropping fermentations.

Typically, ale fermentations performed in cylindroconicals are fully attenuated and after chilling and cropping are subjected to a short low-temperature conditioning process in a separate tank. The conditioning phase is purely to adjust carbonation, precipitate chill haze, and allow some loss of undesirable flavour volatiles through gas purging. On completion of conditioning the beers are filtered and packaged.

Lager fermentations performed in cylindroconical vessels tend not to be of the traditional variety. Instead, there is an increasingly common trend throughout the world to produce lager beers using a rapid and high-temperature fermentation. Primary fermentation may be performed at temperatures anywhere between 10 and 20°C, although usually within the lower quartile of this range. When primary fermentation is completed, the comparatively high temperature is maintained for a period of warm conditioning. This is to allow the reduction of diacetyl to acetoin and 2,3-butanediol. For this reason, this phase is also referred to as the diacetyl rest or stand. When the concentration of diacetyl and its precursor, α-acetolactate, have fallen to a pre-determined concentration, the diacetyl rest is terminated by the application of chilling. This causes the bulk of the remaining suspended yeast to settle into the cone of the vessel from which it is cropped.

The green beer may then be transferred into a separate conditioning tank via a continuous centrifuge, which reduces further the suspended yeast count. Conditioning is carried out at a low temperature for no longer than a few days. As with chilled and filtered ales this part of the process serves simply to adjust carbonation, develop chill haze protection and clarify the beer. Yeast cells play no part in the process, hence the need for prior elimination of diacetyl precursor. In some cases, the period of cold conditioning may be performed in the same vessel as used for primary fermentation (see Section 5.4.3).

### 6.5.4 Troubleshooting abnormal fermentations

Inevitably, all breweries experience occasional abnormal fermentation performance. This should be a rare problem; however, when it occurs it is necessary to have procedures for identifying the causes and putting into action an appropriate plan of countermeasures. The procedures should encompass an analysis of the observed symptoms, investigative tests to obtain further information, identifying causes and taking remedial actions. The action plan needs to consider how best to deal with the individual problem fermentation, and thereby rescue that batch of beer. It should also establish if the individual abnormality is an isolated incident or part of a bigger brewery-wide problem.

Early detection of abnormal fermentation behaviour is dependent on the methods of monitoring performance. Therefore, atypical behaviour is usually identified as deviations from the normal patterns of profiles of physical parameters such as specific gravity and temperature. Nevertheless, the personal experience of the brewer is also particularly useful. The appearance of the fermenting wort can be an invaluable aid for early identification of non-standard behaviour at a time when there may have been little or no change in measurable parameters. In this regard, the use of open fermenting vessels is an obvious advantage. Where closed vessels are employed, the practitioner is more reliant on physical measurement. The discussion here concentrates mainly on large-scale closed fermentations performed in cylindroconical vessels.

The symptoms of abnormal performance are typically fermentation rates that are too slow (or too rapid) and those that indicate altered patterns of yeast growth or physicochemical behaviour (Table 6.4). Clearly, the symptoms are not mutually exclusive. Inevitably, slow and/or sticking fermentations will be viewed as being the most troublesome as they impact directly on the efficiency of the whole brewery, and therefore most of the following discussion relates to dealing with this situation. Should these circumstances arise, it will be necessary to take steps to try to correct the problem and thereby bring the fermentation back on track and minimise effects on the beer. Where abnormalities relate to features such as over-attenuation or the effects of contamination (either microbiological or other), the problem may not be manifest until after the fermentation is complete. In this case, the investigation can only be of historical value.

The causes of fermentation abnormalities are obviously related to abnormalities in those factors that influence performance, i.e., wort composition, control of the major fermentation variables, yeast phenotype/genotype and microbiological integrity (Table 6.5). To establish the precise cause of a problem an ordered hierarchy of investigations should be followed. The majority of slow fermentations probably arise

**Table 6.4** Symptoms of abnormal fermentation performance.

| Stage | Primary symptom | Secondary symptom |
|---|---|---|
| Primary fermentation | Too rapid. | High coolant demand. Excessive fobbing. |
| | Too slow. | Low coolant demand/fails to achieve set-point temperature. Little fobbing. |
| | Sticking (fails to achieve predicted attenuation gravity) but normal rate of attentuation. | |
| | Slow and sticking. | Low coolant demand. Little fobbing. |
| | Over-attenuation at racking gravity. | Normal or excessive primary coolant demand. |
| Secondary fermentation | Slow or fails to achieve diacetyl specification. Very high diacetyl peak | |
| Yeast crop | Smaller than normal. | Low yeast count in green beer. |
| | Smaller than normal. | High yeast count in green beer. |
| | Larger than normal. | Low yeast count in green beer. |
| | Crop forms sooner than expected. | Low or normal yeast count in green beer. |

**Table 6.5** Factors influencing abnormal fermentation performance.

| Parameter | Factor |
|---|---|
| Wort composition | Starting gravity Sugar spectrum FAN/TSN Metal composition (zinc addition) Presence of inhibitors |
| Fermentation variables | Yeast pitching rate Dissolved oxygen tension Temperature |
| Yeast | Correct strain Viability Physiological condition/vitality Genotype |
| Microbiology | Contamination with: other production yeast wild yeast bacteria |

because of a failure to adequately control pitching rate and/or wort oxygenation, or using pitching yeast that was in a stressed condition. Records of the collection of individual fermentations should be sufficiently detailed and accurate to allow confirmation, or otherwise, that correct conditions were established at the outset and that the condition of the pitching yeast was satisfactory. Of course, in-line control and logging systems are particularly suitable for this type of checking procedure (or ensuring that fewer problems arise in the first place).

If control of the initial parameters is in doubt, a sample of the fermenting wort should be removed from the vessel and examined. Direct microscopic examination of fermenting wort is an invaluable diagnostic aid. If the yeast count is lower than predicted and viability is high, an error in pitching rate is likely. If the yeast count is as expected and viability is acceptable but there is little evidence of budding, an error in dissolved oxygen control is probably indicated. In both of these cases, the corrective action should be to add more oxygen, via direct injection of gas, and, if available, mechanical agitation. The window of time for such remedial action is narrow. In addition, if the cell count is very low, more yeast may be added prior to the application of oxygen. If the diagnosis is correct this should restart the fermentation, or increase the rate of attenuation. Additions and/or agitation result in the release of dissolved carbon dioxide which during active fermentation can be sufficiently 'explosive' to result in major losses of fermenting wort from the vessel (see Fig. 6.35).

**Fig. 6.35** Rousing an active fermentation – flooding of Station Street, Burton-upon-Trent (with permission from the *Burton Daily Mail*).

It would not be expected that the normal attenuation rate would be fully re-established since sticking fermentations are notoriously difficult to recover. In extreme cases, it may be necessary to transfer the fermenting wort to another vessel, which may re-start fermentation simply by ensuring that the yeast is re-suspended. Another approach is to divide the problem fermentation between two vessels and top each up with fresh oxygenated wort. The latter measure should be treated with caution since it can result in doubling the problem! These extreme remedial actions produce a strong possibility that the resultant beer may be out of specification and blending at a suitable rate is advised.

Microscopic examination of fermenting wort provides an opportunity to identify heavy contamination with bacteria, although probably not with another yeast.

Microbial contamination should be a very rare occurrence and the consequences would depend on the nature of the foreign organisms. In the case of a gross bacterial contamination, the situation may be irretrievable and the beer would have to be destroyed. The effects of contamination with another yeast may not be evident at all during active fermentation although alterations in cropping behaviour and subsequent beer processing may occur. In addition, over-attenuation of worts may be observed. Of course, changes in beer flavour and aroma may result from contamination with other yeast and a decision regarding appropriate action would need to be based on the severity of the beer abnormality. In any case, it would be advisable to destroy the yeast crop.

Some fermentation abnormalities relate directly to the yeast. Pitching yeast in a stressed physiological condition will probably produce sub-standard fermentation performance, as described already. In this regard, the utility of the so-called vitality tests is discussed in Section 7.4.2. Occasionally, changes in yeast genotype (see Section 4.3.2.6) may produce atypical fermentation performance. Such changes may produce effects such as sudden shifts in yeast flocculence. Accordingly, effects on fermentation may be higher than normal yeast counts in green beer or increased solids content of sedimented yeast crops. The latter situation may also be associated with failure to reach expected attenuation gravity and/or prolonged diacetyl stand-times. If changes in genotype are suspected the yeast should be discarded and a fresh culture introduced.

Sticking fermentations which contain low cell counts and non-budding yeast and which do not respond to oxygenation and/or repitching may indicate a wort nutrient deficiency or the presence of an inhibitor. Further evidence of the latter could be abnormally low yeast viability. A suitable test for these eventualities is to remove a sample of the fermenting wort and incubate it overnight, preferably using an orbital incubator to ensure aerobiosis. Good yeast growth should be observed and the expected attenuation gravity achieved. If not, a wort problem is indicated. Providing there is no reason to suspect that the batch of wort is atypical severe nutrient deficiencies should be very rare. However, inadvertent failure to add a zinc supplement is possible. Proprietary nutritional supplements, so-called 'yeast foods', are available and the effect of dosing these into the wort, at the recommended rate, may be efficacious.

The presence of growth-inhibitory substances in wort is more difficult to detect and deal with. The effects of contamination may become evident in two ways. First, there is the situation where there has been gross contamination of a single batch of wort, which leads to obvious atypical fermentation behaviour. An example of this would be accidental contamination of wort with a cleaning agent such as sodium hydroxide. Here a simple and obvious test, if this type of accident is suspected, is to measure pH. The remedial action would depend on the severity of the contamination. In extreme cases, it might be necessary to destroy the wort; less serious contamination might be solved by judicious blending.

The second type of contamination is that where a low level of a yeast toxin is introduced at some stage of wort production. Contaminants may be introduced with any raw material. For example, in two recent communications (Boeira et al., 1999a, b) inhibition of the growth of brewing yeast by Fusarium mycotoxins was described.

These toxins are possible contaminants of barley, wheat, rice and maize and they may persist into wort and beer. Another obvious source of contamination is via the water supply. Procedures such as carbon treatment of brewing liquor and appropriate checking procedures should preclude this type of problem. Nevertheless, the possibility of introducing toxins in brewing liquor via pollution of groundwater must be guarded against. Since pollution may be a growing threat in many parts of the world great vigilance is required on the part of the brewer. Contamination problems of this type may have multiple effects on yeast metabolism, depending on the nature of the toxin and the concentration. Typically, the initial effects may be slight but possibly cumulative. It is therefore essential to monitor fermentation performance to check for long-term drift.

The ability to distinguish between isolated and more general problems is much dependent on the quality of the records maintained within the brewery. It is usual to retain records, either as hard copy or in electronic form, of profiles of present gravity, temperature and VDK (if appropriate) and analysis of the yeast at pitch and crop. It should be possible to compare these results for any given set of fermentations grouped on the basis of common parameters of the sort detailed in Table 6.6. In addition, similar data should be available to allow comparison of the analysis and organoleptic quality of the beer devolving from each fermentation. Variations in these data as a function of time (or another variable) can be monitored using statistical methods such as the cumulative sum (CUSUM) procedure.

Table 6.6 Identification of common factors possibly relating to atypical fermentation performance.

| Parameter | Potential common variable |
| --- | --- |
| Fermenter | Individual fermenter |
| | Type of fermenter |
| Raw materials | Wort raw materials |
| | Liquor supply |
| | Process aids |
| | CIP agents |
| Yeast | Line |
| | Generation number |
| | Slurry analysis: |
| | viability |
| | solids content |
| | microbiological analysis |

## 6.6 Recovery of carbon dioxide

All breweries require a supply of carbon dioxide for use as a motor gas for transporting liquids from one location to another, for de-aerating water, other process liquids and empty tanks and for adjusting carbonation levels in beers. This requirement may be satisfied by recovering the carbon dioxide evolved during fermentation. Whether or not this is actually done is dependent on economic considerations. Thus, the cost of purchase of carbon dioxide from an external source must be weighed against the capital expenditure of fitting recovery plant and the revenue costs attracted by its operation. However, current concerns with respect to the generation of greenhouse gases suggest that it should not be simply an economic question. Thus, it is now important to consider the ecological price of uncontrolled emission of gases into the atmosphere.

Theoretically, the yield of carbon dioxide in weight terms is roughly equal to half the weight of sugar fermented. Not all of this gas can be collected since some moves forward in solution in the green beer and another portion is lost in early fermentation whilst air is purged out of the fermenter head space. Williams (1990) described installation and operation of carbon dioxide recovery plant at the Charles Wells Brewery in the UK. With an estimated brewing volume of 650 000 hl per annum it was estimated that 2400 tonnes of carbon dioxide could be recovered (3.7 kg hl$^{-1}$).

Recovery of carbon dioxide from fermenters involves a combination of purification, drying, compression, cooling and liquefaction. The purification steps must remove any potential unwholesome flavour taint and most importantly ensure that oxygen levels are reduced to a minimum. This is essential to avoid the possibility of introducing oxygen to beer when recovered carbon dioxide is used as a process gas. Haffmans (1996) provided a specification for recovered carbon dioxide (Table 6.7).

**Table 6.7** Specification for recovered carbon dioxide (from Haffmans, 1996).

| Parameter | Specification |
|---|---|
| Carbon dioxide | > 99.998% (vol./vol.) |
| Hydrogen sulphide | < 50 ppb |
| Dimethyl sulphide | < 50 ppb |
| Oxygen | < 50 ppb (preferably < 5 ppb) |
| Dew point (atmospheric) | −60°C |

A schematic of a typical carbon dioxide recovery plant is shown in Fig. 6.36. The initial gas flow through the bubble pot is vented to atmosphere to ensure that no air is allowed to enter the plant. This cut-off may be automated using an in-line sensor, which measures levels of oxygen in the gas stream. Carbon dioxide is recovered from fermenter via a fob trap and fob detector. Both of these ensure that no liquid foam goes forward into the recovery plant with the gas stream. The latter is collected into a balloon which acts as a buffer tank to compensate for irregular flow rates. When the balloon is full, forward flow is started automatically and the pressure is increased using a preliminary booster compressor. In the gas scrubber, water-soluble impurities such as ethanol and acetaldehyde are removed by a counter-current flow of water. The pressure is further increased by a two-stage oil-free compressor and the gas stream is directed towards columns containing activated charcoal. These also remove ethanol but are particularly aimed at eliminating volatile components such as beer esters, dimethyl sulphide and hydrogen sulphide. The charcoal columns are used in pairs so that one may be used whilst the other is regenerated. This must be performed periodically by treating the contaminated carbon with a flow of steam followed by drying.

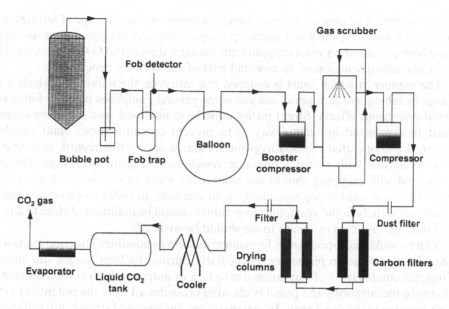

**Fig. 6.36** Carbon dioxide recovery, liquefaction and gas supply plant.

The gas stream is then dried by passage through another column containing moisture retaining materials such as activated alumina or proprietary molecular sieving preparations. Parallel drying columns are also used to allow simultaneous use and re-generation. Moisture must be removed at this stage to prevent ice formation in the liquefaction plant. Filters of various types are placed in the gas stream to remove particulate materials. Liquefaction is achieved using a combination of low temperature and pressure, typically 17.5 bar g and $-24°C$. The liquefaction process assists with purification in that non-condensable oxygen and nitrogen are removed. However, the oxygen concentration at the point of delivery must be less than 0.2% v/v or liquefaction is not possible. Bulk liquid carbon dioxide is stored in an insulated tank. Supply of the gas to the brewery is via an evaporator in which the liquid carbon dioxide is vaporised by heating with steam. At times of high demand, the liquefaction step may be by-passed and the purified carbon dioxide gas used directly.

## 6.7 Yeast recovery

Recovery of yeast during and at the end of fermentation is a three-part process. First, when the fermenter is emptied the yeast concentration within the green beer must be reduced to a level appropriate for the type of beer being produced. Second, a fraction of the recovered yeast must be retained for re-pitching. Third, beer entrained in separated yeast may be recovered. The separation of yeast from green beer during vessel run-down is described in the next section of this chapter. Methods used for removal of the bulk of the yeast crop, especially that destined for re-pitching, are discussed here. The procedures used are much influenced by the type of fermenter and yeast used; however, the same general principles apply. Although numerous reviews have been published on yeast cropping, the reader is directed to O'Connor-Cox (1997) for a readable discussion of an essential part of the brewing process.

The manner in which yeast is cropped can influence the manner in which it performs in subsequent fermentations and some general guidelines must be followed to avoid deleterious effects. Strain purity must be maintained, and therefore cropping must be conducted in such a way as to prevent contamination. Plant should be designed and operated to good hygienic standards and in this regard, cropping systems associated with closed fermenting vessels have a clear advantage. The plant associated with cropping should not subject the yeast to stresses such as excessive shear forces or sudden pressure changes. In addition, in order to avoid unnecessary metabolic activity in the yeast, the temperature should be maintained close to 2 to 4°C and in closed systems exposure to air should be avoided.

Yeast should be cropped from fermenters as soon as possible. The conditions which yeast is exposed to in fermenter, particularly during the later stages, are stressful. Thus, the combination of starvation due to lack of nutrients, high carbon dioxide and ethanol concentrations and possibly elevated pressures all have the potential to cause deterioration in the yeast crop. In extreme cases, the imposed stresses are sufficient to cause cell death and autolysis with consequent adverse effects on beer quality. In this regard, tall bottom-cropping fermenters impose the greatest stresses on yeast. It should be remembered that by the time a crop has formed, the yeast within it plays no further positive role in the fermentation, and therefore it should be removed as soon as is convenient.

Yeast to be used for re-pitching should be as free as possible from contaminating wort solids in order to minimise errors of pitching rate control in the next fermentation. Some entrapment of solids with yeast is inevitable, particularly with bottom-cropping fermenters. However, it is possible to manage the process such that the best possible separation of yeast and solids is achieved.

### 6.7.1  Top-cropping systems

Traditional fermentation vessels used with top-cropping yeast strains are provided with specific systems for skimming the yeast head. This includes a manual operation in which the yeast is pushed into a drain. Frequently a so-called 'parachute' arrangement is employed. This consists of a drain through which the yeast crop is discharged. The inlet is extended into a broad conical collector, the height of which is adjustable to ensure efficient and controllable separation of the yeast head from the beer. Alternatively, the yeast head may be removed with a suction pump.

The timing of crop removal is dependent on the type of fermentation and the properties of the yeast. Management of the process is dependent on visual observation and the experience of the brewer. In particular, in ensuring that the yeast to be retained for re-pitching is relatively free from trub and is the fraction that will produce the required performance in the next fermentation. Thus, during the course of the fermentation, several yeast heads may be formed and these are all cropped but not necessarily retained. The experienced brewer will be aware of how the appearance and timing of formation of the head relates to overall fermentation performance and which yeast cuts should be kept or discarded.

Selection of the appropriate yeast head and the design and operation of top-cropping fermentation vessels assists with separation of yeast from trub. Thus, the first head is heavily contaminated with trub and is discarded. In addition, much of the larger particles of trub form a sediment and in this respect these systems are self-cleansing. In the dropping system this is further refined by starting fermentation in one vessel, then, after 24 h or so, transferring the fermenting wort into a second vessel (see Section 5.3.3). This serves the dual purpose of ensuring that the actively growing yeast is uniformly suspended in the transferred wort and allows much of the sedimented trub to be left behind in the first vessel. It is, of course, profligate in its use of large numbers of vessels.

After removal from the fermenter the cropped yeast is collected into a separate open vessel, a wheeled trolley or more frequently a closed tank termed a 'yeast back'. Yeast to be used for re-pitching may be stored in this form without further treatment. More often the entrained beer (barm ale) is separated from the yeast by pumping the contents of the yeast back, through a cooled plate and frame filter. The yeast cake is recovered from the filter plates and discarded or stored on metal trays or bins in refrigerated rooms. The filtrate can be pumped back into the fermenting vessel from which it originated, or collected into separate recovered beer tanks for blending back at a later stage and at a rate thought appropriate by the brewer. De Clerck (1954) commented that recovered barm ale has a different composition to the mother beer from which it was obtained. It may have very high bitterness levels, due to the propensity of these substances to bind to yeast cell walls (see Section 4.4.3) and it usually contains high concentrations of yeast metabolites derived from shock excretion (see Section 3.7.6). De Clerck (1954) suggests blending back at a low rate to avoid adverse flavour effects, and then only in certain beer qualities. However, it seems likely that these deleterious effects can be ameliorated by careful handling of the slurry and appropriate attemperation.

### 6.7.2 Bottom-cropping systems

In traditional open fermenters used with bottom-cropping yeast, the beer is removed first, at the end of primary fermentation, leaving a yeast sediment in the base of the vessel. This is then removed manually. Attempts are made to ensure that the fraction retained for re-pitching is taken from the middle of the sediment. Thus, the lowest layer is enriched in trub, whereas that which settles last is typically in poor condition.

In closed fermenting vessels, using bottom-cropping yeast strains, it is not possible to see the crop develop, as it is with a top-cropping system. Therefore, it is necessary to gauge the optimum time for yeast removal from other measures of fermentation progress. Usually the fermentation is judged complete when the wort is fully attenuated and the period of warm conditioning, if practised, is over. At this time, the vessel contents are chilled to between 2 and 4°C to encourage the bulk of the remaining suspended yeast to settle into the base of the vessel. In cylindroconical fermenters, this is facilitated by provision of cone cooling jackets. The latter are also useful for maintaining high yeast viability where vessel residence times are long since they prevent localised warming due to the exothermy of the packed yeast. Although chilling encourages yeast sedimentation, much of the crop is formed during the warm phase of active fermentation – especially if the yeast is flocculent. It is a misconception that this yeast contributes to diacetyl reduction during warm conditioning, and for quality reasons, both of beer and yeast, it is better to remove a crop as soon as possible after the achievement of attenuation gravity.

Cahill *et al.* (1999a) studied the efficacy of cooling in sedimented yeast using a specially designed apparatus. This allowed the effects of attemperation of solid yeast plugs to be assessed using a wall jacket arrangement, as is the case in the cones of cylindroconical fermenters. These authors demonstrated that thermal gradients developed rapidly such that a differential of 11°C was measured between the cooling surface and a point 1.2 m into the packed yeast mass.

Loveridge *et al.* (1999) suggested that much of this problem could be avoided by removing an initial 'warm crop' of yeast 24 h after the attainment of attenuation gravity. This was followed by a second conventional crop taken when cooling had been applied after the attainment of VDK specification. The first crop was retained for re-pitching purposes and the second cold crop was disposed of (Figs 6.37(a), (b)). The results were based on comparison of a substantial number of trials (n = 180) and control (n = c. 420) lager fermentations of 15°Plato wort performed in both 2000 and 4000 hl cylindroconicals. The viability of the early-cropped yeast was always greater than that of the conventional cold crop. Furthermore, the suspended solids content of the warm crop was greater than that of the cold crop. A similar differential in mean

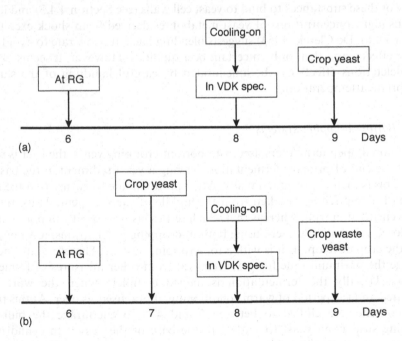

**Fig. 6.37** Comparison of conventional 'cold' yeast cropping (a), and (b) 'warm' yeast cropping fermentation regimes (from Loveridge *et al.*, 1999).

cell size was also noted (Table 6.8). Significant process improvements were noted using this approach (Fig. 6.38). Total residence time for more than 70% of trial fermentations pitched with the warm cropped yeast was less than 10 days, compared with fewer than 30% pitched with the conventional late crop.

**Table 6.8** Analysis of early 'warm' and conventional 'cold' yeast crops (from Loveridge *et al.*, 1999).

| | First crop | | Second crop | |
| --- | --- | --- | --- | --- |
| | Mean | Range | Mean | Range |
| Viability (%) | 92.3 | −1.8/+1.6 | 90.1 | −8.1/+4.5 |
| Cell size (μm) | 7.23 | −0.15/+0.19 | 6.54 | −0.68/+0.58 |
| Slurry wet weight (% weight to volume) | 53.0 | −10.0/+9.0 | 25.0 | −16.0/+13.0 |

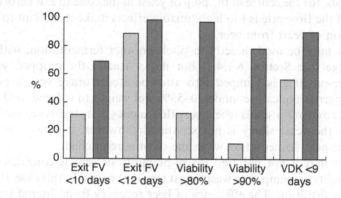

**Fig. 6.38** Comparison of fermentation and yeast performance measures for yeast cropping via conventional (□) and early warm (■) regimes (from Loveridge *et al.*, 1999).

In another recent study, Hodgson *et al.* (1999) investigated yeast cell age distribution in cone crops. Thus, as the crop was removed from the cone of the fermenter, successive samples were removed that could be equated to the distribution from top to bottom of the sedimented yeast. The average age of the cells within each sample was determined by bud scar staining using the fluorochrome, calcofluor. In addition, the predicted future performance of yeast from each sample fraction was assessed in EBC tall tube fermentations. The results indicated that the mean age of cells decreased from bottom to top of the crop. Yeast taken from the middle to top of the crop, judged as being middle to young in age, produced the best fermentation performance. It was recommended that this fraction should be retained for re-pitching purposes.

This recommendation is somewhat contrary to that of the 'warm cropping' procedure described previously. Thus, it would be predicted that the latter approach would select the older and, therefore, lower quality yeast. The possible disadvantages of this need to be weighed against the benefits of removing yeast, as soon as possible, from the hostile environment of the base of a large fermenter. Of course, the difficulties of reconciling these recommendations demonstrate an inherent shortcoming of cylindroconical fermenters as compared with selective top-cropping vessels. A combination of the two approaches would be possible by simply retaining a late cut from an early crop.

Removal of the yeast crop may be a manual process. A sight-glass is fitted to the exit main to allow visual identification of the interface between yeast slurry and green beer. This is used as the signal to open and close relevant valve paths to divert the process flow towards an appropriate destination. Solids which settle into the base of fermenting vessels are not of homogeneous composition. The lowest layer is enriched with trub since this forms the first sediment in early fermentation before appreciable yeast growth has occurred. In order to separate this from the bulk yeast crop the first cut may be diverted to waste tanks. This is judged on a volume basis gained by experience of working with the particular vessels and fermentation in question. Care must be exercised when removing the crop. There is a tendency, particularly in cylindroconicals, for the centre of the plug of yeast in the cone to exit before that close to the walls. If the flow rate is too high, mixing effects make it difficult to engineer a clean separation of yeast from beer.

Yeast crops may be used directly to pitch another fermentation, without intermediate storage (see Section 6.1.4.3) but more usually the cropped yeast slurry retained for re-pitching is pumped into attemperated storage vessels (see Section 7.3.2). Yeast slurries typically contain 30–55% wet weight to volume solids, although this may rise to over 60% solids where very flocculent yeast strains are used. Barm ale collected with the yeast slurry is not separated; however, it is not lost since it is returned to the next fermentation when the yeast is re-pitched.

Yeast slurry that is not required for re-pitching, or that is considered to be of insufficient quality, is pumped to waste yeast storage vessels. In this case, the barm ale is recovered by filtration. The efficiency of beer recovery from filtered yeast may be enhanced by incorporating a washing step, using chilled liquor. Thus, the slurry contains two pools of beer. There is an extracellular fraction located in the interstices between the yeast cells. There is also an intracellular fraction within the yeast cells. Some of the latter can be recovered by resuspension of yeast in cold liquor after filtration and re-treating, or by flushing yeast cake with liquor whilst still in the press. This encourages excretion of ethanol from yeast cells because of the altered equilibrium. The pooled recovered beer is stored in chilled vessels and blended back into beer at a rate that does not impair product quality. In a typical fermentation of 1500 hl, some 5–7 tonnes wet weight of yeast may be recovered. This equates to 12.5–17.5 tonnes of slurry at 40% solids.

The cropping process may be automated using output from a sensor that detects the yeast/beer interface. Optical sensors can be used for this task. A more sophisticated system uses the yeast biomass meter described in Section 6.1.4.5 and shown in schematic representation in Fig. 6.39. In operation, the process is initiated by the controller selecting the calibration on the biomass meter for the yeast strain to be cropped. The valve path is opened and the cropping pump activated. During cropping, the biomass meter measures the viable yeast concentration in the process stream issuing from the fermenter. The first portion of the crop, which is rich in trub, is directed towards a waste tank. When the viable yeast concentration reaches a pre-

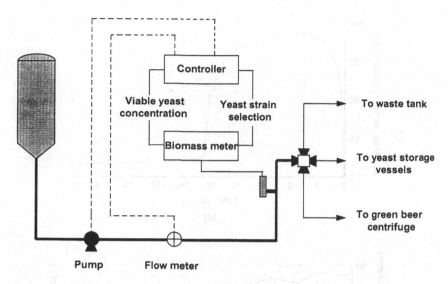

**Fig. 6.39** Automatic control of yeast cropping from a bottom-cropping fermenter (from Boulton & Clutterbuck, 1993).

determined threshold value the valve path is automatically changed and flow is diverted to yeast storage vessels. This continues until the interface between the yeast slurry and green beer is detected by a sudden drop in measured viable yeast concentration. When a second pre-determined threshold value is reached the process flow is either stopped or diverted to the green beer centrifuge. A flow meter located in the fermenter exit main, together with the viable yeast cell concentration output from the biomass meter, allows calculation and loggng of the total mass of viable yeast diverted to storage vessels. This can be fed into the supervisory information management system, which, together with a log of the quantity of yeast pitched, provides a continuously updated record of yeast stocks, their history and fate.

A profile of yeast cropping from an 8 hl pilot scale cylindroconical fermentation is shown in Figs 6.40(a) and (b). A biomass meter probe was located in the main a short distance from the exit valve on the bottom of the cone of the fermenter. At the end of fermentation, the green beer was chilled to 2°C then emptied in the normal manner. During run-down, readings were taken from the biomass meter at the times indicated. At the same time, analyses were performed for total suspended solids and viable yeast count on samples removed from the process stream at a point adjacent to the biomass meter probe. The results confirmed that the first portion of the yeast crop contained high levels of entrained trub. This was not detected by the biomass meter. Instead there was a close correlation between output from the meter and measured viable yeast concentration (Figs 6.40(a) and (b)).

## 6.8 Fermenter run-down

Fermentation is completed when the beer is emptied from the fermenting vessel, releasing it for cleaning and re-use. The process of emptying is termed 'run-down' or

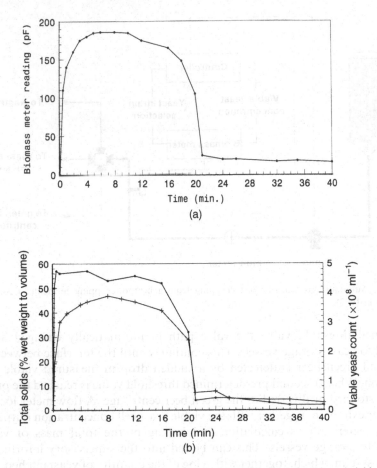

**Fig. 6.40** (a) Readout from a permittivity biomass meter. (b) Profile of total solids and viable yeast count measured in-line during initial run-down of an 8 hl pilot scale cylindroconical fermenter (redrawn from Boulton & Clutterbuck, 1993).

'racking'. Management of this process is dependent on the type of fermentation and the beer quality. The essential features of this stage are control of yeast count and temperature. In addition, it is important to prevent increase in oxygen concentration in the beer as it passes onto the next phase of processing.

In traditional fermentation systems the style of vessel, the yeast strain used and the way the process is managed ensure that the yeast count at the end of fermentation is that which is required. In particular, there has usually been an element of 'natural selection' such that a yeast strain with appropriate properties has been chosen. The essential flexibility of modern fermentation systems, which allow the production of many beer types, is countered by the need to provide a means of controlling the yeast count during racking.

In the case of traditional top-fermented beers, moderate cooling is applied at the end of primary fermentation, usually to between 10 and 15°C. During vessel run-down, the aim is to achieve a suspended yeast count in the region of $1 \times 10^6$ cells ml$^{-1}$

and a residual fermentable extract equivalent to roughly 0.75°Plato. Occasionally silica-based auxiliary finings may be added to fermenter, after cooling, to precipitate proteinaceous materials and promote better fining activity in cask. In some cases, the beer may be fully attenuated, in which case priming sugars are added later in the process to provide fermentable carbohydrate to fuel secondary fermentation. Some breweries reduce the yeast count to a minimum value during racking by the use of a continuous centrifuge, the operation of which is described subsequently. In this case, a further inoculum is made to provide yeast for secondary fermentation. This approach has the advantage that another yeast strain may be selected for use in the secondary fermentation, different to that used in primary fermentation.

For traditional lager fermentations, the required residual extract is 0.5–1°Plato, similar to that for top-fermented ales, however, a yeast count of $2$–$4 \times 10^6$ cells ml$^{-1}$ is required. Racking temperatures are lower than for ales, typically 4–8°C. During transfer to the lagering tanks, it is essential to ensure that the process is anaerobic. Thus, the connecting main and destination tank must be purged with carbon dioxide to remove air before transferring the beer. Where modern vessels are used for the production of chilled and filtered beers, where there is no secondary fermentation, it is necessary to reduce the residual yeast count as much as possible. This is achieved by passing the beer through a continuous centrifuge of a similar type to that shown in Fig. 6.41. This is a cylindrical device constructed of hygienic stainless steel. The cylinder contains a vertically arranged assembly of rapidly rotating discs. Green beer is introduced into the base of the centrifuge where it is thrown onto the discs. Yeast cells and other solids migrate across the surface of the discs from the centre to the outside, under the influence of the centrifugal forces, whilst the clarified beer passes through and exits from the top of the centrifuge. At the outer edge of the discs, the solids are thrown against the inner wall of the centrifuge bowl. At regular intervals,

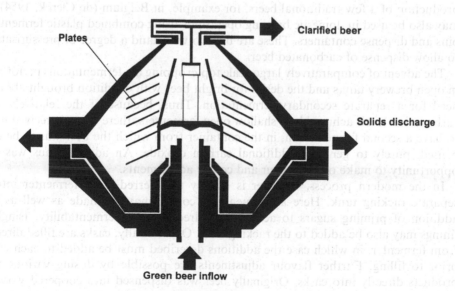

**Fig. 6.41** Cross-sectional view of a continuous intermittent discharge centrifuge.

the two halves of the centrifuge bowl open automatically and the accumulated solids are discharged into a separate collection main.

Centrifuges operate at a constant speed. The efficiency of separation of solids from beer is controlled by adjustment of the number of discs and regulation of the flow rate. The solids discharge is directed towards storage tanks, prior to being sent to a press, where the beer is recovered and the yeast disposed of. Rarely, the yeast may be recovered and utilised for re-pitching (Donnelly & Hurley, 1996).

## 6.9 Secondary fermentations

The changes which occur in the secondary fermentation associated with traditional lagering and warm conditioning processes have been described previously (see Sections 6.5.2 and 6.5.3). The application of immobilised yeast reactors for secondary fermentation is described in Section 5.7.2. Two other secondary fermentation processes are worthy of mention, those used for the production of cask- and bottle-conditioned beers.

### 6.9.1  Cask conditioning

As with so many aspects of brewing, history has produced many variations to the basic process. Undoubtedly beers were originally fermented in the wooden casks from which they were also dispensed, and therefore no distinction was made between primary and secondary fermentation. In this case it would have been difficult to produce a bright product and losses would have been considerable. However, by restricting the escape of carbon dioxide towards the end of fermentation, it would be possible to effect a crude control of carbonation. This approach is still in use for the production of a few traditional beers, for example, in Belgium (de Clerck, 1954). It may also be used in domestic brewing operations using combined plastic fermenting bins and dispense containers. These are built to withstand a degree of pressurisation to allow dispense of carbonated beer.

The advent of comparatively large-scale top-cropping ale fermentations performed in open brewery tanks and the desire for bright beer with condition brought about a need for a separate secondary fermentation. Thus, because of the relatively low carbonation levels achievable in shallow open fermenters, there was a perceived need to have a second fermentation, in the container from which the beer was to be dispensed, purely to generate additional carbon dioxide. An added bonus was the opportunity to make other flavour and colour adjustments.

In the modern process, the beer is usually transferred from fermenter into a separate racking tank. Here adjustments to colour may be made as well as the addition of priming sugars to achieve a desired level of fermentability. Isinglass finings may also be added to the racking tank. Occasionally, casks are filled directly from fermenter, in which case the additions described must be added to each cask, prior to filling. Further flavour adjustments are possible by dosing various hop products directly into casks. Originally beer was dispensed into coopered wooden casks but most breweries now use aluminium replacements, of varying sizes. These have a longer life span, are more hygienic and require less maintenance. The features of a modern aluminium cask are shown in Fig. 6.42.

Filling casks (or 'cask racking') is performed using a multi-head 'racking back'. This consists of a stainless steel main attached to the racking tank. A number of flexible hoses carry beer from this main to the casks. Prior to filling, casks are cleaned and sterilised by steaming and the keystone sealed with a wooden bung. With the cask on its side, the filling head is placed into the bunghole and the valve opened. The beer in the racking main is held under a carbon dioxide top pressure which forces beer into the casks. The volume dispensed may be controlled automatically. When the cask is full, application of carbon dioxide gas pressure forces beer fob out of the cask and back up a separate return pipe. This passes into a collection main from which it is added back to the bulk beer. The opening in the cask is then sealed by driving in a wooden bung known as a 'shive'.

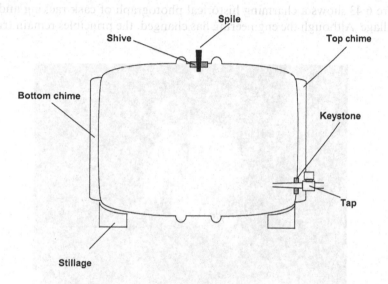

**Fig. 6.42** Cross-section through an aluminium beer cask.

Handling of casks after filling is much dependent on the particular beer type. In addition, physical factors that influence the rate of secondary fermentation are important. These include temperature and the quantity of air introduced into the cask during filling. Generally, casks are held in the brewery for up to seven days, preferably at a temperature within the range 13 to 16°C. During this time sugar is metabolised by the yeast to generate carbon dioxide and there is a concomitant modest increase in ethanol concentration. The process is completed when the cask is taken to the point of dispense. A brass or plastic tap is driven though the wooden bung in the keystone, taking care not to introduce any air and the cask is placed on a framework known as a 'stillage'. Before dispensing the beer, a peg, termed a 'spile', is driven through the shive. This allows a small ingress of air to into the top of the cask so that beer can be dispensed from the tap. Usually about 2 days are allowed to elapse between 'stillaging' and dispense to allow for final development of condition and yeast to form a compact sediment under the influence of the finings. Spiles are made from wood and are available with different degrees of porosity, depending on the cut of the grain (soft or hard), which gives a crude regulation of the quantity of air allowed to enter the cask. Hard pegs may now be made from impermeable plastics.

The beer may be dispensed directly from cask to glass, although more usually a separate cellar is used. Beer is pushed from here to a separate tap under carbon dioxide top pressure or via a pump. Cellarage of cask beers is preferable since a constant cool temperature is more easily achieved. As the beer level falls, the casks are raised using wedges known as 'thralls'. The fining behaviour of each batch of beer is assessed in the brewery by retaining samples, which are dispensed into glass ended casks. These are held at an optimum temperature in the brewery sample cellar and the clarity of the beer judged by direct observation when the cask in placed in front of a bright light source.

Figure 6.43 shows a charming historical photograph of cask racking and tapping post stillage. Although the engineering has changed, the principles remain true today.

**Fig. 6.43** 'Tapping cask beer' (courtesy of the Bass Museum, Burton-upon-Trent, England).

### 6.9.2 *Bottle-conditioned beers*

Rarely, beers may be subjected to a secondary fermentation after packaging into bottles. Essentially the process is similar to cask conditioning although careful control is required. With many traditional beers, a cloudy product is entirely acceptable and in these cases, the only requirement of secondary conditioning is to ensure that sufficient carbon dioxide is generated to suit the style of beer. The presence of suspended yeast cells is of no particular significance; indeed they will enhance the nutritional value of the product.

With some bottle-conditioned beers there is a requirement for both carbonation and clarity and this presents a more demanding technical challenge. In this case a base beer is used which has been brewery cold conditioned and filtered to remove sources of haze. Base beers are as nearly as possible completely attenuated. This allows accurate control of addition of priming sugars. This is important since too much sugar would generate excessive carbon dioxide volumes with potentially disastrous consequences for the bottles.

In a typical process, beer is packaged containing sufficient primings to generate 2–3 volumes of carbon dioxide and yeast at a concentration of $1–2 \times 10^3$ cells ml$^{-1}$. During these operations, great care must be taken to reduce as much as possible the risks of microbiological contamination. After bottling, beers are held in the brewery for a period of about two weeks at 15 to 20°C for condition to develop and allow microbiological checks to be made, prior to release for sale. During this time, the yeast proliferates and the cells gradually settle out. Any yeast strain that is capable of generating condition may be used. However, during the secondary fermentation alterations to flavour also occur as a result of products of yeast metabolism being excreted into the beer. In this respect, a strain providing the desired flavour is important. Frequently yeast strains are used which have a chain-forming morphology. This is useful for formation of a mat of yeast growth, which forms a stable sediment in the bottle. This may more easily be left behind allowing bright beer to be dispensed. Use of bottles with reasonably deep punts is also useful in this regard.

A modern manifestation of bottle-conditioned beers is the practice of sterile packaging beers that contain a very low yeast count into bottle or can. In this instance, the yeast is claimed to give protection against the development of stale flavours associated with oxidation and ageing in small-pack products. Yeast cells are freely suspended throughout the beer but at a concentration low enough to render them invisible to the naked eye. Apart from the need for strict control of hygiene during handling and packaging, it is essential to ensure that there is no opportunity for the yeast cells to proliferate. Therefore, precise control of residual fermentable extract and in-pack oxygen is obligatory.

# 7 Yeast management

## 7.1 Laboratory yeast storage and supply

### 7.1.1 *Maintenance of stock cultures*

The storage and assured supply of brewing yeast strains is an important step in the propagation cycle. Although arguably obvious, it is perhaps not generally appreciated that storage and supply are critical to the ongoing success and consistency of yeast management in the brewery. Indeed as noted by Guldfeldt and Piper (1999), 'despite the fundamental importance of a reliable supply of pure and stable cultures, culture maintenance is often afforded low priority'. This attitude is unfortunate and is inappropriate. Correct and robust procedures for culture maintenance (and supply) are not a 'nice to have' but a 'must have'! To this end, long-term yeast storage should meet four basic criteria (Quain, 1995). Ideally, the method should be genuinely long term in that yeast can be (i) stored for many years without (ii) compromising viability or (iii) genetic stability. Moreover, the method should (iv) be accessible and as technically straightforward as possible. This section identifies the popular options for yeast storage and critically assesses their success in terms of yeast survival (short and long term), genetic stability and simplicity. The strengths and weaknesses of these approaches are summarised diagrammatically in Table 7.1 and reviewed in practical detail by Kirsop (1991).

**Table 7.1** Critical comparison of methods for long-term storage of yeast (after Quain, 1995).

| Method | Survival | Shelflife | Genetic stability | Simplicity |
|---|---|---|---|---|
| Subculture on agar | ☹ | ☹ | ☹ | ☺ |
| Subculture in broth | ☹ | ☹ | ☹ | ☺ |
| Freeze-drying | ☹ | ☺ | 😐 | 😐 |
| Freezing in liquid nitrogen | ☺ | ☺ | ☺ | 😐 |

Performance is graded as: ☺ superior, 😐 intermediate and ☹ poor.

#### 7.1.1.1 *Third-party storage.*
For simplicity, third-party storage cannot be beaten! The responsibility for secure maintenance and storage is handed over to a culture collection that periodically supplies the strains back to the depositor. This option is clearly attractive to small breweries who wish to periodically propagate their strains but do not wish to establish the necessary laboratory facilities and expertise for internal maintenance and storage. Such facilities are offered by commercial culture collections such as the Alfred Jorgensen Laboratory (www.ajl.dk) in Denmark who

supply yeast cultures to more than 100 breweries worldwide. Another approach is take advantage of the secure 'confidential safe deposit' service offered by culture collections such as the National Collection of Yeast Cultures (NCYC) in the UK and the BCCM in Belgium (for details see Section 4.2.3.1). In the case of the NCYC, although marketed as a back-up in the event of catastrophe, the service satisfies the needs of 'third-party' storage of brewing yeasts for periodic propagation. Here, at reasonable cost, strains are preserved by freeze-drying (Section 7.1.1.3) and in liquid nitrogen (Section 7.1.1.4). Two cultures of the strain are supplied annually to the depositor with the option of further cultures charged at the collection's prevailing rates. Although such an approach lacks true flexibility it meets the objectives of simplicity and secure, assured storage. For breweries wishing to store more than one or two strains, it is likely that such an approach would become, over time, prohibitively expensive.

7.1.1.2 *Storage by subculturing.* Subculturing micro-organisms by streaking out on solid agar plates (or slopes) or by inoculation in a liquid broth are fundamental practical tools of microbiology and, by definition, require technical resource and facilities. However, as noted by Kirsop (1991), subculturing is 'simple to do, quick to carry out, and relatively inexpensive'. Unfortunately, the simplicity of subculturing is its undoing! Although universally used across microbiology, it is generally recognised that repeated subculturing of micro-organisms (and subsequent storage at $\leq 4°C$) is a less than ideal protocol for long-term storage. Further, in the case of brewing yeast, viability cannot be assured by storage (at $\leq 4°C$) for longer than 4–6 months. Consequently, the need for periodic subculturing can be labour intensive, particularly with large collections of yeast strains. These considerations are further exacerbated by concerns that these approaches result in 'strain drift' through the formation and enrichment of genetic and phenotypic variants (Wellman & Stewart, 1973). These have been reported for flocculation (Russell & Stewart, 1981; Kirsop, 1991; see Section 4.4.4), respiratory deficient petites (Russell & Stewart, 1981; see Section 4.3.2.7) and, generally, for morphological and physiological properties (Kirsop, 1974). A further practical concern that can be levelled at serial subculturing is that poor technique and cross-contamination can compromise strain identity and purity. Given these many concerns, it is no surprise that subculturing cannot be recommended as an approach to long-term storage of brewing yeasts.

7.1.1.3 *Storage by drying.* Although as described by Kirsop (1991), drying protocols include niche methods such as the 'paper replica method' and the 'silica gel method', the approach described here is restricted to the more popular 'freeze-drying' method.

Lyophilisation or freeze-drying of brewing yeasts has long been practised (Kirsop, 1955) and today remains a popular approach to the long-term storage of yeast and other micro-organisms. Freeze-drying involves water removal by sublimation from the frozen sample using a centrifugal dryer. It is very much a small-scale procedure with typically 100 µl of yeast suspension (at $10^6\,ml^{-1}$) and equivalent volume of suspending 'protectant' being dried and sealed under vacuum in a glass ampoule. Its success is built around the method's relative simplicity (for details see Kirsop, 1974,

1991), low running costs (although start-up can be appreciable) and, perhaps most significantly, the ease of handling, storage and transport when dried. However, from experience, the end user can occasionally experience frustration and difficulty in scoring and breaking glass ampoules.

Although it continues to be used by culture collections such as the NCYC and National Collection for Marine and Industrial Bacteria (NCIMB), freeze-drying has – in recent years – had a bad press. Most damning of all is the general acceptance that post freeze-drying the viability of the culture is usually low. Early observations (Kirsop, 1955) at the then Brewing Industry Research Foundation (now Brewing Research International) reported the average viability for 29 strains of *S. cerevisiae* to be 4.6% two days after drying. Nine months later, the average viability was effectively unchanged at 4.3%. Similar conclusions were made by the same worker (Kirsop, 1991) some 36 years later which showed that at the NCYC the average viability for freeze-dried cultures of the genus *Saccharomyces* was 5%. This, if anything, is higher than other yeast genera in the NCYC! Such poor performance appears to be due to the formation of intracellular ice during the freezing stage of the process. In one report (Berny & Hennebert, 1991), the viability post-freeze-drying of a strain of *S. cerevisiae* increased from 30 to 98% by controlled slow freezing ($3°C$ $min^{-1}$) in the presence of skimmed milk and other protectants. However work in the 'related world' of active dry bakers' yeast suggests that dehydration of the yeast below 15% moisture is the trigger for loss in viability (Bayrock & Ingledew, 1997). A report by Russell and Stewart (1981) found freeze-drying to be the least successful protocol for storage of brewing yeasts, being outperformed by traditional subculturing methods and, most dramatically, by storage in liquid nitrogen. In addition to a catastrophic loss of viability (five of eleven brewing strains were totally dead!), changes in flocculation and increased frequency of petite mutation were seen in some strains. By way of explanation, freeze-drying has been reported to impact at the level of the yeast genome by causing chromosomal breaks (Barros Lopes *et al.*, 1996; see Section 4.3.2.6).

Guldfeldt and Piper (1999) have described an alternative, simpler but lengthy approach to preparing dried yeast used by the Alfred Jorgensen Laboratory to supply pure yeast strains. Yeast cells are suspended in sorbitol (20% v/v), incubated for six hours, filtered under vacuum and then dried over calcium chloride for up to 12 days at 10°C. Although technically much simpler than the standard freeze-drying method, the Guldfeldt and Piper approach suffers similarly when post-storage viability is compared to storage in liquid nitrogen.

Ironically, outside of the rarefied world of culture collections, dried yeast has long found commercial application in the wine and baking industries. Until recently, compared to dried wine yeasts, dried brewing yeasts have fared poorly because of low and inconsistent viability (O'Connor-Cox & Ingledew, 1990). However, lager yeast strains have now been successfully dried using a combination of fed batch propagation and fluidised bed drying – a process similar to that used in the production of active dry bakers' yeast. Although viability post drying is 50–60%, it has been shown with two lager strains to be maintained at this level for up to 12 months at 4°C (Muller *et al.*, 1997) or 10°C (Fels *et al.*, 1999). Although originally marketed for use in primary fermentation (Muller *et al.*, 1997), production scale trials in Israel and

Tanzania (Fels *et al.*, 1999) and in Belgium (Debourg & van Nedervelde, 1999) have focused on the role of dried yeast in yeast propagation. Although fermentation performance is typical of the 'control' wet yeast, there is a need for a more exhaustive analysis of genetic and physiological stability. It remains to be seen whether this approach can go some way to resolving the technical issues surrounding freeze-drying or even be scaled down to achieve similar flexibility.

7.1.1.4   *Storage by freezing in liquid nitrogen.*   Storage in liquid nitrogen is the *crème de la crème* of cell storage methods. It is used to 'successfully store fungi, bacteriophage, viruses, algae, protozoa, bacteria, yeasts, animal and plant cells and tissue cultures' (Snell, 1991). For yeast, cryogenic storage in liquid nitrogen at its boiling point ($-196°C$) has long been recognised as the best approach to maintaining viability and integrity. Its application to brewing yeast has been described at Labatts (Wellman & Stewart, 1973; Russell & Stewart, 1981), the UK's NCYC (Kirsop, 1991), Bass (Quain, 1995) and Scottish Courage (Jones, 1997). If performed correctly, storage in liquid nitrogen has no impact on cell viability over periods of seven months (Guldfeldt & Piper, 1999), two years (Russell & Stewart, 1981), 26 months (Kirsop & Henry, 1984), three years (Wellman & Stewart, 1973) or eight years (Kirsop, 1991). Similarly, the general view is that storage at $-196°C$ does not trigger any genetic changes. Despite this, Wellman and Stuart (1973) reported changes for one lager strain in flocculation and an increase in respiratory mutants. However, subsequent comparative work by the group (Russell & Stewart, 1981) gave liquid nitrogen storage a clean bill of health.

To be successful, storage in liquid nitrogen requires adherence to basic protocols (Kirsop & Henry, 1984; Kirsop, 1991). Cells cannot simply be plunged into liquid nitrogen! Prior to freezing, yeast cells are grown oxidatively (Quain, 1995), an approach which has been shown to best afford protection against the potentially lethal stresses of freezing and thawing (Lewis *et al.*, 1993). Cells are then suspended (at *c.* $100 \times 10^6 \, ml^{-1}$) in fresh media containing a cryoprotecterant (typically glycerol at 2.5–5% v/v). Subsequent freezing is a two step process: (i) the temperature is progressively reduced over two hours from room temperature to $-30°C$ followed (ii) by immersion in liquid nitrogen at $-196°C$. The rate of cooling in the initial step is critically important to the success of cryogenic storage, as it is during this phase that the cells are dehydrated. Over time the fraction of water (in the suspending media) which is frozen increases whilst, in turn, the concentration of salts in the unfrozen fraction rises. This progressive increase in osmotic pressure results in the yeast cells losing water and shrinking with a decrease in surface area and increase in the thickness of the cell wall. It is considered that the rate of shrinkage is a vital element in preventing cell damage through intracellular ice formation (Kirsop, 1991). Indeed a strong correlation has been shown between the incidence of intracellular ice and the loss of yeast viability (Morris *et al.*, 1998). Conversely, revival following storage at $-196°C$ is performed rapidly by immersion in water at 25 to 37°C.

The use of liquid nitrogen lends itself to the storage of collections of microorganisms. In our experience (Quain, 1995), a substantial collection of current production strains (*c.* 25) is maintained alongside (in a separate vessel) a collection (*c.* 50) of 'historical' brewing strains and other yeasts of interest. Typically (Kirsop & Henry, 1984), yeast suspensions are stored in ampoules (0.5 ml) or, preferably in short coloured straws (0.2 ml) derived from children's drinking straws (Fig. 7.1). This later small-scale approach is more flexible, enabling large numbers of straws to be prepared and stored. Perhaps the one disadvantage of such permanent storage is recovery. As noted in Section 7.1.2, this is not quick. Indeed, the ideal and flexible solution is a hybrid of long-term storage in liquid nitrogen coupled with short-term storage (< 6 months) on agar slopes or plates at 4°C.

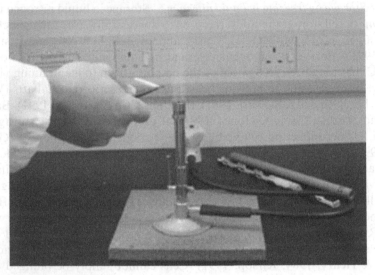

**Fig. 7.1**  Straws and ampoules used for storage of yeast slurries in liquid nitrogen (Kindly supplied by Steph Valenti of Bass Brewers).

In keeping with the scale of cryopreservation, the capital cost of cryovessels (Fig. 7.2) and ancillary equipment can be significant. Further, liquid nitrogen must be safely contained and 'handled' with appropriate safety procedures and precautions. However, tellingly, the need for maintenance is low and is limited to periodic (but regular) topping-up with liquid nitrogen from a 'buffer' tank. This is important as storage in the nitrogen vapour phase (−139°C) reportedly (Kirsop, 1991) allows some biochemical and biophysical processes to continue. Accordingly, similar reservations apply to storage of yeasts at −70°C which, in our experience, also results in significant losses in viability (Wendy Box, unpublished observations).

### 7.1.2  *Deposit, recovery and validation of identity*

As described in Section 7.1.1, long-term storage of yeast strains via serial culture on agar plates or slopes cannot be recommended. This is unfortunate, as this mode of storage is convenient and readily available for sub-culturing at the outset of propagation. Comparatively, storage regimes such as freeze-drying and freezing in liquid nitrogen are inflexible inasmuch that the culture must be recovered, resuscitated and 'grown-up' prior to use. Undoubtedly such constraints have limited the widespread use of liquid nitrogen and, to a lesser extent freeze-drying.

**Fig. 7.2**   Liquid nitrogen storage cryovessels (Kindly supplied by Steph Valenti of Bass Brewers).

Whatever the method, the process significance together with the demands of good laboratory practice, demand some degree of checking of yeast quality both pre- and post-storage. Although perhaps taken as 'a given', it is disappointing that the literature is, with one exception (Quain, 1995), bereft of detailed comment on how such checking may be performed.

Bass Brewers' approach to yeast deposit and supply was described in detail at the EBC in Brussels (Quain, 1995). Here, storage of production yeast in liquid nitrogen is a central service consisting of conventional and 'new' microbiological tests. The package is controlled and documented to meet international quality standards (ISO 9002). The driver for such complexity and control is simple! Yeast is critically important to product quality, consistency and diversity so assurance of the right yeast strain of the right physiological and microbiological quality is paramount. Consequently as noted by Quain (1995), 'there is no room for error in yeast supply'. Risks of contamination or strain mix-up must either be removed or minimised and controlled by applying best operating practices together with appropriate monitoring.

Risks are minimised by a number of basic rules. Irrespective of the number of strains being handled, all activities within 'yeast supply' require two people. This is deemed necessary to negate the risk of operator fatigue, to provide technical support and to observe/confirm actions. Further, only one strain is handled at a time. Total assurance of yeast strain identity and purity are achieved – both going in and coming out of liquid nitrogen – by the use of a variety of microbiological plate tests (Table 7.2, see Section 8.3.3) and RFLP-based DNA fingerprinting (see Section 4.2.6.1). To verify performance – and to instil confidence in the results – each medium is challenged with control micro-organisms that will or will not grow (Table 7.2). Similarly, on deposit into liquid nitrogen strains are identified 'blind' by DNA fingerprinting.

Although originally three times a year (Quain, 1995), yeast supply to Bass Brewers

**Table 7.2**   Microbiological QA of yeast supply (after Quain, 1995).

| Selective media | Solid media | Aerobic incubation (days at 27°C) | + ve control | − ve control |
|---|---|---|---|---|
| – | WL nutrient (WLN) | 3 | Lager yeast | – |
| Wild yeast[a] | WLN + cycloheximide (15 mgl$^{-1}$) | 7 | *Acetobacter aceti* | Ale yeast |
| Wild yeast[a] | YM + copper sulphate (200 mgl$^{-1}$) | 7 | Wild yeast | Ale yeast |
| Wild yeast[a] | YM + Ferulic acid[b] | 2 | Phenolic yeast | Ale yeast |
| Ale/lager yeast | X-α-Gal | 7 | Lager yeast | Ale yeast |
| Bacteria[a] | Raka Ray No. 3 + cycloheximide (15 mgl$^{-1}$) | 7[c] | *Lactobacillus brevis* | Lager yeast |
| Bacteria[a] | MacConkey No. 3 | 3 | *Escherichia coli* | *Micrococcus luteus* |

**Notes**
[a], selective for 'wild yeast' but will support the growth of bacteria (and *vice versa*).
[b], broth.
[c], anaerobic incubation.
All microbiological analyses were in duplicate.
Where appropriate copper sulphate and cycloheximide were added to media at 45 to 47°C.

occurs every six months. Although stored in liquid nitrogen, production yeasts are supplied 'in advance, in excess' to breweries on agar slopes. Strains are recovered from liquid nitrogen by rapid thawing and a 'master culture' prepared that is subject to testing as described above. The number of slopes of each strain that are prepared is determined by the requirements of the breweries together with a back-up of six slopes retained centrally in case of emergencies. To minimise complexity, more slopes than are necessarily required are supplied to breweries to meet unexpected demands of propagation. To enable traceability all slopes are labelled by strain, strain code, best-before date and individually, by number. To assure safe transport the slopes are labelled with a time temperature indicator that changes colour on exposure to temperatures $\geq 37°C$. Finally, to minimise risk, as well as meeting best practice, the entire slope is used in the first stage of laboratory propagation in the brewery.

## 7.2   Yeast propagation

### 7.2.1   *Theoretical*

Maintenance of stocks of pure yeast strains in the laboratory is discussed in Section 7.1. Here are described the methods in which these stocks are used to introduce new yeast cultures into the brewery.

All fermentations generate yeast sufficient to re-pitch two or more further fermentations. In some breweries, particularly traditional ale top-cropping types, this cycle proceeds *ad infinitum* and single yeast cultures have been in use for many years. Frequently, these may be mixed cultures; however, fermentation performance is considered to meet the needs of the process without the need for propagation of fresh yeast. More commonly, however, and without exception in modern breweries, new cultures of yeast are introduced periodically to replace existing stocks.

This is prudent for a number of reasons. Most importantly, it provides an opportunity to introduce a culture of known and guaranteed identity. Continued serial fermentation and cropping carries with it the risk that variants may be selected for within the yeast population. Consequently, over a number of generations of fermentations a gradual drift may occur in the properties of the yeast (see Section 4.3.2.6). This is particularly the case where the cropping regime may inadvertently select for variants. For example, early cropping from the cones of cylindroconical fermenters may select for flocculent sub-populations within the yeast. It is also reported that the formation of petite mutants (see Section 4.3.2.7), which are known to perform poorly in fermentation, is the most common spontaneous mutation in brewing yeast strains (Ernandes *et al.*, 1993). Periodic re-introduction of a new yeast line taken from a master culture limits the opportunities for the occurrence of this source of variation.

Even in the best-managed brewery, the production environment provides opportunities for the introduction of contaminants to bulk yeast. These may be in the form of bacteria (Section 8.1.2) or wild yeast (Section 8.1.3). The consequences of contamination with wild yeast can be significant, resulting in process changes (flocculation, fining) or flavour changes (phenolic, medicinal character). With respect to bacteria, contamination with *Obesumbacterium* spp. (Kara *et al.*, 1987b) can result in the formation of carcinogenic nitrosamines from nitrite (see Section 8.1.5.1). From another standpoint, cross-contamination of yeast strains may occur where several are in use in a single brewery.

Continued serial re-pitching of yeast may be associated with gradual deterioration in yeast condition, which can result in a decline in fermentation performance. This is unlikely in the case of rapid top-cropped fermentations where there is an opportunity to ensure that the fraction of yeast retained is that which is produced when fermentation is at its most vigorous. In addition, such cropping regimes are to some extent self-purifying (see Section 6.7.1). In the case of bottom-cropped fermenters, particularly large-volume cylindroconicals, there is opportunity for high levels of contamination of yeast with trub. The possibility of selecting for non-standard yeast variants has been alluded to already. The use of very large vessels and the tendency towards high-gravity brewing has undoubtedly increased the stresses to which production yeast is subject. It has been demonstrated that senescence of yeast cells (see Section 4.3.3.4) may be associated with declining performance of yeast since some cropping regimes may select for larger and therefore possibly older cells within yeast populations (Barker & Smart, 1996; Smart & Whisker, 1996; Deans *et al.*, 1997).

The frequency of introduction of newly propagated yeast into the brewery is a decision for the individual brewery since there are no immutable rules. A typical regime in a modern brewery built and operated to high standards of hygiene would be to introduce propagated yeast every 15–20 generations. However, some breweries would consider this excessive and only allow 5–10 generations to elapse before introducing new yeast. Two contrary points of view may be considered in this respect. First, a fermentation management decision may have been taken within a particular brewery, which makes it mandatory that new yeast cultures are introduced after a given number of generations. Typically in this scenario a number of different yeast lines, each of varying generational age, would be in use at any given time. Introduction of a new yeast line would be phased to replace a line that has reached the end of its operational lifetime. This method totally disregards the fermentation performance of that particular yeast line.

On the other hand, if a particular yeast line is performing satisfactorily and is microbiologically 'clean', there will be a natural desire to continue brewing with such yeast, even though it may have completed its allotted number of generations. This highlights the need for methods of testing the yeasts' physiological condition, which are predictive of subsequent fermentation performance. In other words the so- called 'vitality tests' described in Section 7.4.2.

Continuing to use yeast that is performing in a satisfactory manner has other advantages. There is an economic cost to propagation. Apart from the capital investment, the revenue costs are obviously proportional to the frequency of use. In addition, it is commonly observed that the first generation fermentation using newly propagated yeast is atypical. In consequence, the first generation beer has to be blended. Occasionally, the non-standard behaviour may be extended over the first few fermentations. Clearly, this would tend to mitigate against frequent propagation.

The reasons for non-ideal behaviour in first fermentations are obscure. Hammond and Wenn (1985) reported the experiences of a brewery where one newly propagated yeast strain produced slow fermentation performance for the first few generations. This was shown to be due to impaired ability of the strain to utilise maltotriose. The causes of the defect were not elucidated and the effect was somewhat transitory, disappearing after 7–10 generations. Since this effect was seen only with a single yeast strain it suggests that the problem was not due to propagation *per se*, but was a feature of the particular yeast. Several reports have described yeast undergoing abrupt changes in flocculence and there is evidence to suggest that this is due to an inherent genetic instability (see Section 4.3.2.6). Altered flocculence and defects in sugar assimilation may be related. Thus yeast that is not capable of utilising maltotriose would stop growing and flocculate before the cells comprising the normal population. Such variant cells could be selected for in cone cropping of fermenters. If such genetic variants can arise in fermenter, it follows that the same phenomenon could occur in propagator. It also follows that should such variants arise it would not be a reflection of propagator performance.

In fact, it can be demonstrated that, provided the design and operation of the propagation plant is adequate, there is no fundamental reason why first-generation propagations should be different from later ones. It may also be surmised that the common contention that newly propagated yeast gives less than ideal performance is a consequence of poor propagator design and operation. In consequence, first-generation fermentation pitching rates are often inadequate and the yeast is in a stressed condition.

The requirements of a propagator are summarised as follows:

(1) Hygiene is of prime importance and the design and operation of the propagation plant must ensure that a pure yeast culture is generated. Since the propagator is to supply yeast for brewing it is essential that it is not a source of contamination. This is an obvious requirement of propagation but one that is not always adhered to.

(2) The terminal cell count must be adequate to achieve the desired pitching rate in the first generation fermentation.

(3) The yeast must be of high viability ($>95\%$).

(4) The physiological condition of the yeast must be consistent and appropriate for subsequent fermentation.

(5) The cycle time of propagation should be as rapid as possible, both for economy and to minimise the risk of contamination, and should use the fewest possible number of vessels.

(6) Terminal cell counts from the final propagation stage should be as high as possible so as to allow high step-up ratios and minimise the effects on the first generation fermentation of the 'barm ale' introduced with the propagated yeast.

### 7.2.2 Propagation systems

The use of pure culture plant in brewing is, of course, not new and the first yeast propagators were introduced by Hansen in 1883 (Curtis, 1971). The process, therefore, has a long history and several distinct systems have been developed. Nevertheless, all propagation regimes basically consist of a sequence of yeast cultures of progressively increasing volume, starting in the laboratory and culminating in a terminal stage which contains sufficient yeast to pitch the first production scale fermentation. Variations on this theme are possible, such as semi-continuous systems, which maintain cultures at the small brewery scale and thereby reduce the requirement for repeated laboratory propagation.

7.2.2.1 *Laboratory propagation.* The aim of the laboratory phase of fermentation is to generate a pure yeast culture of sufficient size to provide an adequate pitching rate for the first stages of brewery propagation. The terminal laboratory culture must be held within a container, which will allow transfer to the brewery under conditions of asepsis, and there be transferred into the brewery propagation vessel under aseptic conditions.

Laboratory propagation uses standard microbiological apparatus. It must be performed to the highest possible standards using skilled personnel. Initial stages may use artificial media such as yeast extract, peptone, glucose. Wort may be used for the terminal laboratory stage; however, it must be sterilised by autoclaving prior to use. A typical laboratory propagation regime is shown in Fig. 7.3. The scheme shown is a suggestion only and several variations are possible. It is sensible to limit as far as possible the number of aseptic transfers, since these represent the points of greatest risk of contamination. In general, a volume scale-up factor of about 1:10 is satisfactory.

The terminal laboratory stage requires a purpose built piece of apparatus, such as

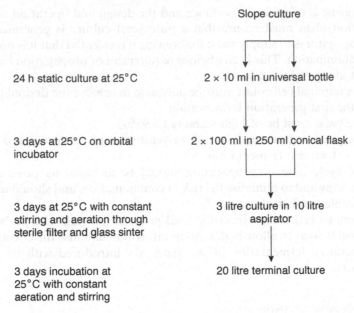

**Fig. 7.3** Schematic flow diagram for laboratory propagation of brewing yeast.

that shown in Fig. 7.4 and described by Boulton and Quain (1999). This consists of a heavy gauge stainless steel flask with a capacity of approximately 25 litres fitted with a number of ports passing through the top-plate assembly. The latter is removable for cleaning purposes. Before use the flask is filled with brewery wort and the flask and contents sterilised by autoclaving. If experience shows it to be necessary, antifoam may be added to the wort before sterilisation. After cooling and prior to inoculation, the wort should be aerated by sparging for at least 30 minutes with air or pure oxygen. Sterility is maintained by passing the gas through a microbiological quality gas filter. High rates of oxygen transfer are facilitated by passing the inflowing gas through a stainless steel candle. Inoculation is via a specific port that terminates in a male fitting, which is wrapped to maintain sterility. Immediately before inoculation the male fitting is joined, using appropriate aseptic precautions, to a matching female fitting, also wrapped, attached to the side-arm of the aspirator used for the 3 litre culture stage (Fig. 7.4). The inoculum culture is transferred by gravity after opening the relevant valves.

After inoculation, the flask is aerated continually with air or pure oxygen. Oxygen transfer rates are further improved by constant agitation using a powerful magnetic stirrer and a follower in the flask. The exhaust gas is vented to atmosphere via another microbiological grade filter. If the gas flow rates are high it is advisable to locate a water condenser between the gas outlet port and the sterile filter. Aseptic sampling is possible via a tube, that extends to near to the bottom of the flask. The sample may be withdrawn by temporarily restricting the outlet gas line.

It is convenient to mount the flask in a purpose-built trolley to facilitate transport. The trolley can be designed to hold the magnetic stirrer and a gas cylinder to provide

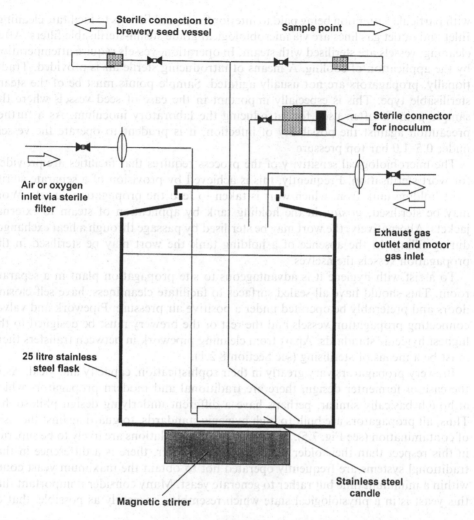

Sterile connection to
brewery seed vessel

Sample point

Sterile connector
for inoculum

Air or oxygen
inlet via sterile
filter

Exhaust gas
outlet and motor
gas inlet

25 litre stainless
steel flask

Stainless steel
candle

Magnetic stirrer

**Fig. 7.4** Laboratory propagation vessel for terminal growth phase (from Boulton & Quain, 1999).

motive power during transfer of the culture from flask to brewery seed vessel. When the culture is ready the flask is disconnected from the gas inlet supply and transferred to the brewery. Connection to the brewery seed vessel is via a dedicated line, which terminates in a sterile wrapped fitting, which is designed to attach securely to the inlet point on the seed vessel. During transfer the culture should be agitated continuously using the magnetic stirrer. Transfer of the culture is achieved by applying top pressure via the gas exhaust line.

### 7.2.2.2 *Brewery propagation.*
The raison d'être of production scale propagation plant is to provide conditions that favour yeast growth. It follows that the same conditions will also favour the growth of contaminants and, therefore, good hygienic design and operation is absolutely essential. Vessels are fabricated from stainless steel

with particular attention being paid to interior finish and fittings to facilitate cleaning. Inlet and outlet gas lines are via microbiological grade steam-sterilisable filters. After cleaning, vessels are sterilised with steam. In operation, vessels require attemperation by the application of cooling. A means of introducing sterile air is provided. Traditionally, propagators are not usually agitated. Sample points must be of the steam sterilisable type. This is especially important in the case of seed vessels where the sample point is often used for introducing the laboratory inoculum. As a further precaution against the possibility of infection, it is prudent to operate the vessels under 0.5–1.0 bar top pressure.

The microbiological sensitivity of the process requires that facilities are provided for wort sterilisation. Frequently, this is achieved by provision of a separate sterile wort holding tank from which wort is taken to feed the propagation vessels. Wort may be sterilised, *in situ*, in the holding tank by application of steam to external jackets. Alternatively, the wort may be sterilised by passage through a heat exchanger during filling. In the absence of a holding tank the wort may be sterilised in the propagation vessels themselves.

To assist with hygiene it is advantageous to site propagation plant in a separate room. This should have all-sealed surfaces to facilitate cleanliness, have self-closing doors and preferably be operated under a positive air pressure. Pipework and valves connecting propagation vessels and the rest of the brewery must be designed to the highest hygienic standards. Apart from cleaning pipework in between transfers there must be a means of sterilising (see Section 8.2.1).

Brewery propagators vary greatly in their sophistication, capacity and yield. As in the case of fermenter design, there are traditional and modern propagators which although basically similar, perhaps have a different underlying design philosophy. Thus, all propagators are built to high hygienic standards, to guard against the risks of contamination (see Fig. 7.5). Certainly, newer installations are likely to be superior in this respect than their older counterparts. However, there is a difference in that traditional systems are frequently operated not to obtain the maximum yeast count within a minimum time but rather to generate yeast. Many consider it important that this yeast is in a physiological state which resembles, as nearly as possible, that of

**Fig. 7.5** Conventional yeast propagator (kindly supplied by Mark Grey and Peter Smith, Bass Brewers).

conventional pitching yeast. This is supposedly achieved by careful control of temperature and oxygenation. Conversely, more modern installations are generally designed and operated with thoughts of maximum yield and shortest cycle times uppermost.

In the brewery, yeast is propagated using sterile wort. Therefore, assuming no supplements are made to the wort, the only means of regulating yeast growth and extent is by manipulation of temperature and oxygenation. All brewing yeast strains have optimal growth temperatures of around 30°C (see Section 4.2.2). Propagation at this temperature would obviously be very rapid; however, this potential benefit is rarely taken advantage of. In traditional systems, relatively low temperatures are used, typically up to 20°C for ale yeasts but lower for lager strains. Commonly a relatively high temperature may be used in the first vessel followed by a gradual reduction in temperature at each subsequent stage with the terminal propagation being performed at the same temperature as the first fermentation (Maule, 1979).

An excess of oxygen during fermentation promotes yeast growth and the same is true in the case of propagation. This strategy also is not adopted with traditional (and many modern) propagators. Usually more oxygen is provided than would be the case for fermentation – typically, intermittent aeration throughout the entire propagation. However, aeration rates are typically low and the absence of mechanical agitation ensures that oxygen transfer rates are generally poor. In this sense, many propagation vessels are little more than hygienically designed fermenters!

The reasons for propagating at relatively low temperatures and limiting the availability of oxygen are two-fold. First, there is a assumption that these conditions will produce yeast with physiology similar to pitching yeast. In addition, this ensures that the yeast suffers no thermal shock. It is difficult to reconcile these assumptions on purely scientific grounds. Second, regard must be made to the 'beer' which will be pitched with the culture yeast. High propagation temperatures and excess oxygen favour elevated levels of higher alcohols and acetaldehyde and reduced esters (see Chapter 3). Clearly where the step-up ratio is small, as is frequently the case with traditional propagators, there is a risk of adverse flavour effects in the first generation beer due to carry-over of non-standard beer. Of course, where propagators are designed to achieve only modest cell yields, and therefore only small scale-up factors can be employed, the problems of non-standard barm ale are to some extent a self-fulfilling prophecy.

Traditional propagation systems use modest scale-up factors between each stage, typically 1 to 5. This can produce a very cumbersome process when reasonably large fermenters require to be serviced. For example, Maule (1979) described a system, illustrated in Fig. 7.6, used in a lager brewery to propagate sufficient yeast for 800 hl fermenters. As may be seen, the entire propagation process from laboratory to first fermentation is lengthy, taking a matter of weeks to complete. A short cut, which circumvents the initial laboratory stages, is to recycle part of an intermediate propagation to provide a new inoculum, as shown in Fig. 7.5. This approach, often termed 'intermittent propagation', produces a much shorter cycle time but does not address one of the prime reasons for propagation, namely the need to guard against selection of non-standard variants. In addition, it is not a suitable system for a brewery that requires to propagate several yeast strains.

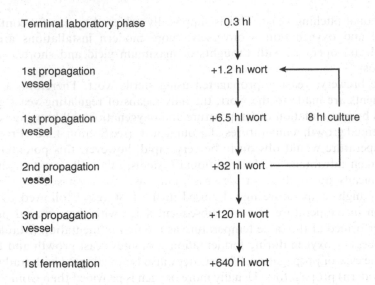

Terminal laboratory phase                             0.3 hl

1st propagation                                      +1.2 hl wort
vessel

1st propagation                                      +6.5 hl wort            8 hl culture
vessel

2nd propagation                                      +32 hl wort
vessel

3rd propagation                                      +120 hl wort
vessel

1st fermentation                                     +640 hl wort

**Fig. 7.6** Scheme for brewery yeast propagation (from Maule, 1979).

The trend towards very large capacity fermentation vessels places great demands on propagation plant. The system described in Fig. 7.6 is capable of providing yeast for an 800 hl fermenter. However, cylindroconicals of twice this capacity or greater are common. Frequently, these large fermenters have been installed within a brewery with an existing propagation plant built to service smaller vessels. In this case, it is common practice to use smaller fermenters, if available, for the first generation fermentations to generate sufficient yeast to pitch larger vessels. Another variation on this theme is to part-fill large fermenters in order to achieve the correct pitching rate using an inadequate propagation plant.

Clearly, strategies such as part filling fermenters is at best a poor compromise and does not make best use of fermenter capacity. Furthermore, such a system becomes very unwieldy in breweries using several yeast strains. For example, a brewery that uses six yeast strains, for a maximum of ten generations, would require propagation plant to be in operation on a permanent basis to avoid the possibility of having no yeast to pitch. An alternative strategy is to use propagators that are capable of delivering high yields within a short process time. Several systems of this type have been designed, which share in common the use of relatively high growth temperatures and aerobic conditions (Geiger, 1993; Schmidt, 1994, 1995; Brandl, 1996; Ashurst, 1990; Von Nida, 1997; Boulton & Quain, 1999; Westner, 1999).

Typically, brewing yeast growing on high-gravity (15–18°Plato) wort under continuously aerobic conditions and temperature within the range 25 to 28°C would yield a terminal yeast count of 200–300 $\times$ $10^6$ cells $ml^{-1}$ within 24–36 hours. Providing the process is terminated when the maximum cell count is achieved and aeration is discontinued, there is no opportunity for diauxy (the transition between fermentative and oxidative physiology) to occur. Consequently, yeast will be in a catabolite repressed condition, essentially similar to conventional pitching yeast. It is, of course, possible to arrange to pitch the culture at a lower than maximum count but during the exponential phase when the yeast is still very active. This ensures that the onset of fermentation is rapid.

High-yield propagators allow larger step-up ratios. Assuming a terminal count from propagator of $250 \times 10^6$ cells ml$^{-1}$ and a target pitching rate in the first fermentation of $15 \times 10^6$ cells ml$^{-1}$, this would allow a scale-up ratio of 1:16. Thus, a typical cylindroconical fermenter of 1500 hl capacity could be serviced by a 100 hl propagator. Furthermore, a single seed vessel of 8 hl working capacity would be sufficient to provide an inoculum for the final propagation stage. It is perfectly feasible to operate a two-stage propagation system of this type within a total seven day cycle time. Although the barm ale arising from propagation will have a non-standard volatile spectrum, the relatively high dilution factor at pitching makes this less of a problem.

A system of this type is shown in Figs 7.7 and 7.8 (Boulton & Quain, 1999). It consists of two tanks each of similar design. Both are serviced by a dedicated CiP system and can be steam sterilised, together with all associated pipework. Sterile wort is delivered to each vessel via an in-line heat exchanger, which serves to both sterilise wort and cool it to a desired starting temperature. Sterile oxygen is introduced to the vessels via a bottom mounted stainless steel diffuser. High rates of oxygen transfer are promoted by constant mechanical agitation and the presence of internal baffles to increase turbulence. Attemperation is achieved by circulating coolant through external jackets. During operation, vessels are top-pressured using sterile inert gas.

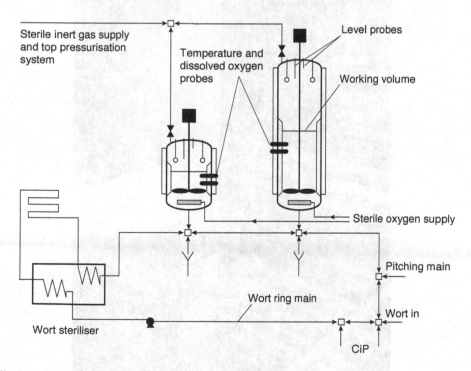

**Fig. 7.7**   Schematic of a two-tank aerobic yeast propagation system (from Boulton & Quain, 1999).

In-tank dissolved oxygen probes can be used to monitor and control oxygenation rates.

Propagation of yeast on wort limits the yield of yeast because metabolism is always catabolite repressed (see Section 3.4.1). Much greater yields could be achieved if the yeast was grown under derepressed conditions such as is practised in bakers' yeast production (Barford, 1987). This approach is capable of generating very high yeast concentrations under appropriate conditions, typically five times greater than those achievable in brewery propagators. In addition, derepressed cells accumulate roughly five times greater concentrations of sterols compared to derepressed cells. Theoretically this yeast should have no requirement for wort oxygenation and produce satisfactory fermentation performance at much lower than usual pitching rates (Quain & Boulton, 1987b).

Derepressed growth of yeast may be achieved by growth on an oxidative carbon source, such as glycerol and ethanol. Alternatively, as is the case in the bakers' yeast industry, a fed-batch approach is used. Here, the principal carbon source is separated from the remainder of the growth medium and added at an exponentially increasing rate during the entire time course of cultivation. This ensures that the carbon source

(a)

(b)

**Fig. 7.8** Two-tank aerobic propagation system at Bass Brewers Limited, Alton Brewery (kindly supplied by Jim Appelbee, Bass Brewers).

remains at low concentrations at all times and repression is not triggered. These approaches have yet to be applied in production scale brewing. However, Quain and Boulton (1987a) reported that brewing yeast strains grown oxidatively on mannitol exhibited excellent storage properties and produced more rapid fermentations than similar yeast grown under repressed conditions. A potential use for these very high yielding systems is that yeast might be used in a 'single trip' or 'pitch and ditch' fermentation regime. There would be an on-cost for the increased requirement for propagation; however, the advantage would be improved fermentation consistency due to use of pitching yeast with consistent physiological condition.

Masschelein *et al.* (1994) described laboratory scale studies in which yeast was propagated in a fed-batch system in which the wort feed was controlled by a computer such that steady state conditions were established. In this instance, the conditions were controlled so as to maintain constant sugar concentration and oxygen tension. Thus, the yeast physiology was not derepressed; however, evidence was provided that it was consistent and highly active.

7.2.2.3  *Use of dried yeast.*  Bakers' yeast propagated under oxidative conditions is frequently dried to render it into a stable form in which it can be more easily stored and transported. Provision of small-scale dried pure cultures for propagation in the brewery laboratory has been discussed (see Section 7.1.1.3). Provision of dried cultures sufficient in quantity for direct use at production scale has been applied to wine yeast and it has been suggested that brewing yeast might be treated similarly. For example, it was proposed as a method of utility where a central facility could supply yeast to a number of satellite breweries, with no propagation facilities of their own (Lawrence, 1986b).

The potential advantages of such approaches are manifold. There would be no requirement for brewery or laboratory propagation plant. If dried yeast were used for all fermentations, there would be no need for, or at least a reduction in, intermediate storage vessels. It would be predicted that the condition of individual batches of the yeast inoculum would be relatively constant, therefore offering the possibility of more consistent fermentation performance. Dried yeast is resistant to storage over periods of several months provided it is kept at cool temperatures and is vacuum packed. This facilitates transport and is particularly suitable for small-scale craft brewers, franchise-brewing operations, or in situations where particular beer qualities may be produced infrequently.

The application to brewing of dried yeast has been described (Muller *et al.*, 1997; Fels *et al.*, 1999). These communications reported the results of fermentation trials at a scale of 300 hl. The dried yeast had a viability of just 65%, compared with a more typical value of 95% for the usual production brewing strain. However, it was claimed that fermenter residence times could be matched, provided that pitching rates were corrected for viability. It was necessary to rehydrate the yeast, prior to pitching, by suspending in wort for 30 minutes at 20°C, before use. The organoleptic properties of beers from fermentations using dried yeast were within the range of conventional controls.

Undoubtedly, this method will be of utility, but at present probably in niche applications. It is unlikely to see widespread adoption in the mainstream large-scale commercial brewing if only because of inertia and cost.

## 7.3  Yeast handling in the brewery

During the interval between cropping from one fermenter and re-pitching into the next, yeast must be stored (see Quain, 1990 and O'Connor-Cox, 1998a for reviews). The yeast must be held under conditions that prevent contamination and minimise any changes in physiological condition, which might compromise the next fermentation into which the yeast is to be pitched. With regard to yeast physiology, the storage phase is a period of starvation and can only be prolonged for a certain length of time. The duration of this period is influenced by the storage conditions and the state of the yeast at the time of cropping. The latter parameter is itself a function of the conditions the yeast was exposed to in the preceding fermentation. It is possible, therefore, to distinguish the 'storage potential' of a particular batch of yeast and the ways in which this potential is modulated by the conditions of storage.

In order to survive periods of starvation the yeast relies on endogenous carbohydrate reserves, which are laid down when exogenous carbon is plentiful. In yeast, two types of storage carbohydrate are recognised, glycogen and trehalose (see Section 3.4.2). Of these two, glycogen appears to be a genuine storage reserve, which is accumulated during mid-fermentation and is utilised during starvation (Quain & Tubb, 1982). Starvation occurs both during storage and, of course, in late fermentation when growth has ceased. It follows, therefore, that the storage potential of the yeast will be compromised if yeast is held in fermenter for a long period, prior to cropping, especially if chilling has not been applied.

Utilisation of glycogen reserves during storage is not a linear event; rather, there is a period of slow dissimilation followed by rapid breakdown phase. Pickerell *et al.* (1991) concluded that the rapidity of glycogen dissimilation was correlated with the temperature and duration of the storage phase. Most importantly, subsequent fermentation performance was impaired (slow attenuation, poor yeast growth, high $SO_2$, acetaldehyde and VDK, high residual $\alpha$-amino nitrogen) if the glycogen content of the pitching yeast fell below a critical value. In the fermentations described, this was 15% of the yeast dry weight, at pitching or 380 mg glycogen per litre of wort (16°Plato).

The pattern of glycogen dissimilation is not linear with time. At first, the rate of breakdown is slow and then there is a second phase, in which the dissimilation rate accelerates. This pattern correlates with changes in yeast viability during storage. The duration of the slow initial phase and the onset of the phase of rapid decline are related to storage temperature. The data given in Fig. 7.9 shows the changes in viability, measured by methylene blue staining, of identical samples of yeast slurries stored at various temperatures. It may be seen that at low temperatures this strain of yeast was remarkably resistant to storage. However, at higher temperatures the onset of the decline in viability occurred after a relatively short period. It is undoubtedly the case that before there was a detectable decline in viability, other physiological changes would have occurred which would certainly have had an adverse impact on fermentation performance, should the yeast have been pitched. Thus, in another report McCaig and Bendiak (1985b) observed that pitching yeast stored at temperatures of 5°C, or below, dissimilated little glycogen and maintained high viability. This yeast

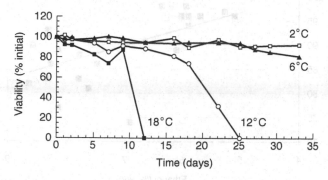

**Fig. 7.9** Effect on viability of storing a slurry of lager yeast (40% wet w/v) at the temperatures shown in the legend. Samples of slurry (1 litre aliquots) were stored under nitrogen gas with gentle continuous stirring (C.A. Boulton, unpublished data).

produced standard fermentation performance and beer quality. However, similar yeast stored at 10, 15, 20 and 25°C for the same period showed a progressive reduction in viability and decline in glycogen reserves. This correlated with increasing impaired fermentation performance when the yeast was pitched into wort.

Others have disputed the importance of maintaining yeast glycogen levels in pitching yeast. Cantrell and Anderson (1983) concluded that no correlation existed between fermentation performance and yeast glycogen content. Sall *et al.* (1988) also reported that the fermentation performance of yeast was independent of glycogen content. In this case, however, the study was performed over just three days, during which time the yeast was stored at 1.6°C. In this time viability remained between 90 and 96% but glycogen levels declined from the relatively low initial concentration of 15.6 to just 9% of the yeast dry weight. These authors suggested some brewing yeast strains were capable of accumulating only low concentrations of glycogen, and, in such strains, this parameter had no relation with fermentation performance. Nevertheless, for most strains the need to conserve glycogen is well established.

Glycogen breakdown in yeast during storage is accompanied by an increase in the extracellular ethanol concentration (see Section 6.4.2.2). Thus, the ethanol concentration in the barm ale of stored yeast correlates with holding time. It has also been demonstrated that the viability of stored yeast correlates with the ethanol concentration in barm ale (Fig. 7.10).

The signal for rapid breakdown of glycogen during fermentation is exposure of yeast to oxygen. In this circumstance, glycogen dissimilation is linked to sterol synthesis (see Section 6.4.2.2). Limited sterol synthesis may also occur if stored pitching yeast is exposed to air. This is of no consequence in traditional fermentations. However, where rigorous control is required as in the case of large volume fermentations, this should be prevented in order to avoid excessive yeast growth. Where control is of importance, pitching yeast should be stored under an inert gas.

Trehalose is accumulated by yeast in response to stress (see Section 3.4.2.2). Starvation is such a stress and indeed during storage some trehalose synthesis occurs at the expense of glycogen dissimilation. For example, the data presented by Sall *et al.*

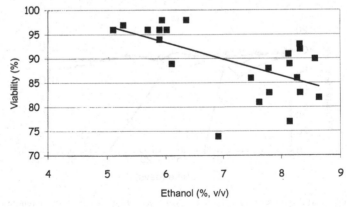

**Fig. 7.10**  Relationship between exogenous ethanol concentration and viability of yeast in storage tank (Loveridge *et al.*, 1999).

(1988) in one case showed an increase during storage in yeast trehalose levels whilst glycogen was declining. However, trehalose levels at pitch do not appear to have a great deal of influence on subsequent fermentation performance. Thus, Guldfeldt and Arneborg (1998) reported that yeast trehalose content in pitching yeast had no effect on growth, attenuation and ethanol production during fermentation, although high trehalose levels did favour maintenance of elevated viability during the initial stages of fermentation.

Trehalose is known to exert protective effects by stabilising membranes. It seems likely, therefore, that during storage of pitching yeast some glycogen may be dissimilated and used to synthesise trehalose in an attempt to prolong survival. This event may not improve the performance of such yeast on pitching since glycogen levels may be reduced. It may be argued that yeast in this form has entered a resting or almost dormant stage in which re-entry into an active growth phase may be a prolonged process. From another point of view it is known that the stresses of very high gravity brewing induce relatively high levels of trehalose accumulation (Majara et al., 1996a, b). This would imply that such yeast when cropped would survive prolonged storage with ease. This possibility appears not to have been tested.

### 7.3.1    Storage as pressed cake

Yeast storage in the form of pressed cake is associated with traditional top-cropped fermenters and in many respects is the least satisfactory method from a quality standpoint, although it has the advantage of economy. Yeast is skimmed, as described in Section 6.7.1, and pressed on a plate and frame filter to recover entrained beer, which is returned to the process stream. Typically, the yeast cake is transferred to clean metal bins and then stored in a refrigerated room at 2 to 4°C until required. Alternatively, it may be removed from the plates of the filter and stored as thin slabs on metal trays.

Yeast storage in this form is prone to infection since the yeast is open to the atmosphere (see Section 8.1.4.2). Obviously, the yeast is also exposed to air with the consequences in terms of limited sterol synthesis, as described in Section 7.3. However, this effect will be limited to the surface of the yeast cake, and therefore some heterogeneity in physiological condition would be predicted. This can be exacerbated where the cold room is of the type that blows cold air over the surface of the yeast.

One of the problems of storing pitching yeast as pressed cake is the difficulty of temperature control. The data in Fig. 7.11 shows the temperature, measured at various depths, in a bin containing 90 kg of yeast cake stored in a cold room maintained at approximately 1°C. The difficulties of achieving adequate attemperation in this situation are highlighted. Thus, yeast cake has poor thermal penetration and this, coupled with exothermy due to the basal metabolism of the yeast, produces the results shown. Prolonged storage in this form may lead to excessive heat generation and autolysis. To avoid this problem it is better to store the yeast in thin sheets.

### 7.3.2    Storage as a slurry

In the majority of breweries, pitching yeast is stored as slurry. This may be pressed yeast removed from a traditional top-cropping fermenter and re-slurried in water. In

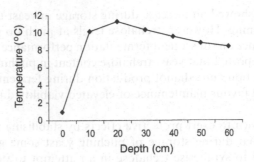

**Fig. 7.11** Temperature profile measured at various depths in a bin containing 90 kg of pressed yeast cake. The bin was located in a cold room attemperated at *c*. 1 to 2°C (A.R. Jones, unpublished data).

some breweries, yeast is stored in the cones of cylindroconical fermenters. This obviates the expense of installing dedicated yeast storage vessels. Although an inexpensive approach, it is not ideal since attemperation of the slurry is impossible. However, it can be a useful temporary measure where the turn-round time between fermenting vessel emptying and refilling is very short.

More commonly, the slurry consists of yeast and entrained beer taken directly from fermenter without further processing. Occasionally the slurry may be stored in open troughs; however, closed vessels are more usual, especially in modern breweries. Storage of yeast in slurry form affords significant advantages. The yeast is normally contained within a vessel, and therefore there is an opportunity to control the risk of infection and provide an inert atmosphere. The yeast is in 'liquid' form, which facilitates transport via pumping. Most importantly, attemperation of yeast slurries is much simpler than with pressed cake. Storage of yeast slurries in closed tanks is particularly suited to breweries that handle several different yeast strains. Thus, it is much simpler to minimise the possibility of cross-contamination. On the debit side, installation of properly designed yeast storage tanks and associated plant is an expensive undertaking. However, when the importance of maintaining yeast stocks within a brewery is reflected upon, it should be considered money well spent.

A typical yeast storage or 'collection' vessel is constructed from stainless steel, with mechanical agitation and dished ends (Figs 7.12 and 7.13). This configuration assists circulation of the slurry and facilitates good draining. Capacities vary from 8 up to about 50 hl depending on the requirements of the brewery. Cooling jackets are usually provided to maintain the yeast slurry at a suitably low temperature. Occasionally, vessels may be unlagged and located in refrigerated rooms, although this is less satisfactory. Attemperation is assisted by provision of a mechanical agitator. As ever, hygiene is of paramount importance. Dedicated CiP is provided and a microbiological quality filter on the gas exhaust main is vital. As with any vessel the sample point must be designed and operated with thoughts of hygiene uppermost and preferably they should be of the steam sterilisable type. Vessels may be mounted on load cells to provide continuous information regarding pitching yeast stocks and usage.

Several yeast storage tanks are typically located in a separate room, for economy each being served by common mains and pumps. It is essential that such common mains are cleaned after every transfer in order to avoid cross-contamination of yeast

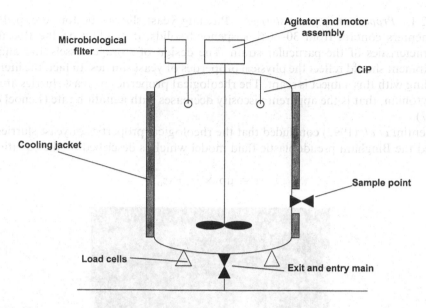

**Fig. 7.12**   Schematic of vessel for storage of pitching yeast slurry.

stocks. After cleaning all mains and valves etc. should be sterilised. Greater flexibility of operation is permitted if separate CiP systems are available for tank and main cleans. These storage tank farms may often have a separate dedicated tank into which an appropriate quantity of yeast is transferred before pitching into fermenter. This separate pitching tank can be used for acid washing (Section 7.3.3). The room in which the tanks are housed must also be built and operated to high hygienic standards. Floors, walls and ceilings should be sealed and easily cleaned with a minimum of fittings. Preferably, the room should be isolated from the rest of the brewery, with access restricted to essential personnel. For a general discussion of measures which need to be taken to control the threat to brewery hygiene, see Section 8.2.

The physical conditions under which pitching yeast is stored, apart from freedom from contamination, relate to time, temperature and exposure to oxygen. The gas in the tank headspace will naturally be rich in carbon dioxide evolved by the yeast. To minimise the risk of contamination, vessels may be top-pressured. If so, an inert gas must be used to avoid pitching yeast being exposed to oxygen. A storage temperature of 2 to 4°C is suitable but care must be taken to ensure that the slurry does not freeze. Mechanical agitation maintains slurry homogeneity and promotes good attemperation. Agitation should not be too vigorous. McCaig and Bendiak (1985a) compared the effects of continuous agitation versus static storage and concluded that the former was associated with loss of viability, excessive glycogen breakdown and impaired fermentation performance. However, totally static storage is unsatisfactory, particularly with very flocculent yeast strains since it is necessary to keep the yeast in suspension to avoid localised heating. This may be achieved by intermittent agitation. Typically yeast is stored safely, under these conditions, for up to three days.

7.3.2.1 *Properties of yeast slurries.*    Pitching yeast slurries bottom cropped from fermenters contain from 30–60% suspended solids, depending on the flocculence characteristics of the particular strain. The design of storage vessels and ancillary equipment should reflect the physical properties of yeast slurries. In fact, the literature dealing with this subject is scant. The rheological properties of yeast slurries are non-Newtonian, that is the apparent viscosity decreases with agitation rate (Lenoel *et al.*, 1987).

Lentini *et al.* (1992) concluded that the rheological properties of yeast slurries best fitted the Bingham pseudoplastic fluid model which is described by the equation:

$$\tau = \mu p \times \gamma + \tau_o$$

(a)

(b)

(c)

**Fig. 7.13** Vessel for storage of pitching yeast slurry (kindly supplied by (a) Steph Valente, Bass Brewers and (b), (c) Ian Dobbs, Bass Brewers).

where:

$\tau$ = Shear stress (Pa)
$\tau_o$ = Yield point (Pa)
$\mu p$ = Pseudoplastic viscosity (Pa × s)
$\gamma$ = Shear rate (s$^{-1}$).

The yield point describes the force required to initiate movement of the slurry. Lentini *et al.* (1992) reported that this parameter was strain-specific and varied between 12 to over 100 Pa. Strains with yield points of less than 30 Pa could be pumped easily, whereas those of 40 Pa or greater were more difficult to handle. By inference, ease of attemperation would also vary with this parameter. The same authors reported that slurry viscosity increased with storage time and that the effect was more pronounced where the temperature was too high. Elevated pH was also associated with increased viscosity. These effects were presumably associated with cell autolysis.

*7.3.2.2 Inter-brewery transport of yeast.* It may be necessary for some brewers to transport yeast in bulk form between sites. This may take the form of a newly propagated culture where this process is performed at a central facility or it may be as pitching yeast cropped from a previous fermentation. In both cases, the yeast will normally be transported as slurry.

Some precautions are needed to ensure that yeast quality does not suffer whilst it is in transit. Depending on the quantity of yeast involved, it is convenient to use small tanks of 8–16 hl capacity specially designed for this purpose (see Fig. 7.14). Tanks should be mounted on skids so that they may be moved by fork-lift truck. They should be made from unlagged stainless steel, of dished-end design with the exit main located flush at the lowest point of the vessel for good draining. All fittings must be of good hygienic design and there should be internal sprayballs for connection to a CiP system. The vessel should be steam sterilised prior to use. A gas exhaust valve mounted on the top of the vessel is required and this must be fitted with a micro-

**Fig. 7.14** Tank used for transport of yeast slurries (kindly supplied by Steph Valente, Bass Brewers).

biological grade filter. It is not practicable to have mechanical stirring and wall cooling in such transportable tanks, and therefore attemperation tends to be poor. Journey times should be given careful consideration, particularly during summer months when tanks may be left in direct sunlight on open lorries. In these circumstances, some deterioration in yeast quality is inevitable.

From a microbiological standpoint, apart from cleaning, the most hazardous processes are emptying and filling the tanks. Usually this requires the use of flexible hose to connect the tank to a convenient brewery outlet main attached to the brewery yeast storage vessels. All such mains and associated fittings must be thoroughly cleaned and sterilised before use. In particular flexible hoses require careful handling since these may easily become foci for infections. The use of dried yeast, which can make easier the job of transport between breweries, is described in Section 7.2.2.3.

### 7.3.3 Acid washing

For many years, it has been recognised that handling of yeast in bulk in the brewery may be associated with low levels of infection with bacteria. Providing fermentation is vigorous and begins with little or no lag this low level of infection can be tolerated since in fermenter most bacteria do not compete successfully with the yeast. However, some bacteria can flourish under fermentation conditions if, for whatever reason, yeast growth is impaired. Typical bacterial contaminants include *Pediococcus* species, *Lactobacillus* species, *Enterobacter agglomerans* and, notably *Obesumbacterium proteus* (see Section 8.1.2). Of these, the latter has been shown to be responsible for the formation of the potential carcinogens ATNCs during fermentation (see Section 8.1.5.1). Therefore, for reasons of product quality and safety, there is a real need to reduce bacterial loading in pitching yeast slurries. This is typically performed by treating yeast with a chemical disinfectant.

Compared to bacteria, yeasts are relatively resistant to low pH, and therefore yeast slurry may be conveniently washed with acid. This is not a new suggestion; for example, Pasteur recommended the use of tartaric acid to reduce the bacterial loading of pitching yeast slurries. Other mineral acids such as hydrochloric, nitric, sulphuric and phosphoric may also be used, with the latter being the most common. Typically, the pH is reduced to a value of pH 2.2–2.5 and the yeast held under these conditions for a few hours, at a temperature of less than 4°C. The oxidising agent, ammonium persulphate (c. 0.75% w/v), either alone or in conjunction with phosphoric acid, is also used (Bruch et al., 1964). Simpson (1987) reported that ammonium persulphate and phosphoric acid used in combination at a pH of 2.8 was more effective than acid alone at the lower pH of 2.2.

Alternatives to acid treatment have been proposed. Comparatively soon after their discovery it was suggested that antibiotics could be used to free yeast from bacteria. Gray and Kazin (1946) made a preliminary study of the use of tyrothricin and Case and Lyon (1956) proposed the use of polymyxin B. Both were shown to be effective and such treatments would have the benefit of persisting and providing residual protection throughout fermentation and beyond. However, these proposals came at a time when the dangers of the profligate use of antibiotics in terms of selection of multiple resistant strains were not recognised. Such use today would be totally unacceptable. More recently, the use of the bacteriocin nisin has been suggested as a safe alternative with the advantages of antibiotics (Ogden, 1987). This polypeptide is accepted for use in the dairy and canning industries and exhibits activity against a wide spectrum of bacteria including Lactobacillus spp. and Gram positive organisms. It has no effect on yeasts, is stable at low pH, and like antibiotics confers residual protection. It has not found great utility in brewing, no doubt because of the reluctance to use additives, which persist into finished beers.

Acid washing supposedly has no effect on brewing yeast. It follows that it is ineffective in removing wild yeast contamination from pitching yeast slurries. More importantly, the suggestion that acid washing has no effect on yeast requires some qualification. Some authors have reported that acid washing improves the performance of yeast in fermentation. Jackson (1988) observed that a pre-treatment 'conditioning' with phosphoric acid at pH 2.2 for 6 hours and 4°C produced more rapid fermentation, shorter diacetyl stand times, reduced beer acetaldehyde concentration and more 'drinkable' beers. The physiological basis of these observations was not explained other than the fact that the acid caused instant de-flocculation and the time of appearance of the first yeast buds was reduced by approximately 8 hours.

More often acid washing has been associated with deleterious effects on yeast. Fernandez et al. (1993) observed a gradual deterioration in pitching yeast condition with time of acid washing, as judged by the acidification power test (see Section 7.4.2.2). The severity of the effects correlated with acidity and was to some extent strain-specific. Simpson and Hammond (1989) studied the effects of acid washing on 16 yeast strains, both ale and lager types. Using acidified ammonium persulphate (0.75% w/v; phosphoric acid at pH 2.1) no effect was observed on viability, flocculation and fermentation performance. However, scanning electron microscopy revealed alterations to the yeast cell surface and leakage of cellular components was evidenced by the appearance of ATP in the external medium. The potential for deleterious effects on yeast due to acid washing was exacerbated by elevated temperature, high ethanol concentrations and over-prolonged exposure to acid. Furthermore, yeast which was already in a stressed condition was more susceptible to damage during acid washing.

Simpson and Hammond (1989) have produced guidelines for acid washing and these are summarised here, together with some observations of the present authors. Yeast should be acid washed in slurry form using food grade phosphoric or citric acid. During the treatment the temperature should be maintained between 2 and 4°C with continuous gentle stirring. The pH of the slurry should be checked with a suitable probe. Care should be taken when adding the acidulant to ensure that mixing efficiency is sufficient to avoid localised high acid concentrations. Automatic systems may be used where acidulant is added via a pump under the control of a pH probe mounted in the tank. Care should be taken to ensure that the point of addition and the pH probe are located such that there is no possibility of over-shooting.

Alternatively, the acid may be dosed in-line as the yeast is transferred from storage to pitching tank. This approach ensures good mixing and in conjunction with an in-line pH probe avoids over addition of acid. The total treatment time should be no longer than two hours. Preferably, the treatment is terminated by immediate pitching into wort. Alternatively, the process may be terminated by addition of food-grade sodium hydroxide to adjust the pH to 4.0–4.5. Although often used, this practice cannot be recommended because of the additional cost, increased complexity and further opportunity for error. Most importantly, the pitching yeast is subjected to an additional and unnecessary stress.

Simpson and Hammond (1989) recommend that yeast recovered from high-gravity fermentations (ethanol > 8% abv) or in a 'distressed condition' should not be acid washed. Distressed yeast was defined as being that which is derived from a previous slow fermentation or that which is heavily contaminated. In our experience, yeast with a viability of less than 80%, as judged by methylene blue staining, should not be acid washed, or better if an alternative is available, not used at all! Similarly, if possible, yeast contaminated with other micro-organisms should be disposed of and not used for brewing. Of course, it is perfectly valid to argue that acid washing represents treatment of a symptom, as opposed to the root cause of hygiene problems (see Section 8.3.1 on the philosophy of QA versus QC). In a modern well-designed and managed brewery the standards of hygiene should be sufficiently high to avoid the need for acid washing. Certainly, the move towards ever-higher-gravity fermentations as practised in many modern breweries would suggest that the opportunity should be taken to eliminate any stresses to which yeast is subject. It is perhaps fair comment that acid washing represents an unnecessary stress.

## 7.4  Assessing yeast condition

Before any batch of pitching yeast is used it is necessary to confirm that it is fit for the purpose. Three types of quality test may be applied. First, the yeast must be free from microbial contamination. Testing the microbiological integrity of yeast is described in Chapter 8. Second, it is usual to determine the viability of the yeast. Most breweries operate a quality reject system in which yeast is discarded if the viability falls below a pre-set value. In addition, viability measurements are used to determine the quantity of slurry required to achieve the desired viable pitching rate. Third, there is a category of quality tests, 'vitality tests', which seek to probe aspects of the physiological condition of yeast. Thus, they provide information regarding the viable fraction of yeast slurries.

### 7.4.1 Assessing yeast viability

The classical microbiological method of assessing viability is to plate out a measured number of cells onto solid media and after a period of incubation, count the resultant colonies (see Section 8.3.2). Each colony is derived from a single live cell. Therefore, the proportion of colonies formed in relation to the total number of cells present in the original sample provides a measure of viability (Institute of Brewing Analysis Committee 1962; ASBC Analysis Committee, 1980). It is possible to accelerate the procedure by counting micro-colonies on a slide culture; however, at best several hours are needed to obtain a result (ASBC Analysis Committee, 1981; Pierce, 1970). This is too long to meet the needs of production brewing where a rapid answer is essential.

Rapid methods for viability assessment rely on the use of vital stains. In the brewing industry, the most commonly used procedure is methylene blue staining, although several others have been used. For example, crystal violet, aniline blue, Rhodamine B and Eosin Y (King et al., 1981; Evans & Cleary, 1985; Koch et al., 1986; Hutcheson et al., 1988). In the methylene blue test, viable cells remain colourless, whereas dead cells are stained blue (see Fig. 7.15). The physiological basis of the test is that viable cells take up the stain at a sufficiently slow rate for it to be

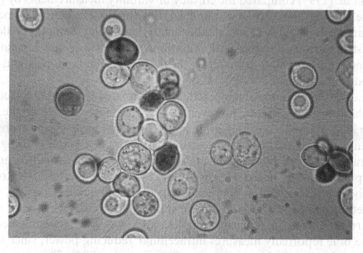

**Fig. 7.15** Viability staining of yeast using methylene blue. (This figure is repeated in the colour section)

oxidised to the colourless 'leuco' form. Conversely dead cells cannot exclude the dye or perform this reaction (Chilver *et al.*, 1978; Jones, 1987). Viability is determined by preparing a suitable dilution of slurry and counting total and stained cells using a haemocytometer and a light microscope.

It is well recognised that the methylene blue method tends to over-estimate viability when compared with plate counting techniques. This effect becomes progressively more pronounced with decrease in viability and it is recommended that the test is used only when viability is greater than 90% (Pierce, 1970; Parkkinen *et al.*, 1976; Chilver *et al.*, 1978; King *et al.*, 1981). The test requires skill on the part of the operator to obtain reproducible results and problems are encountered counting flocculent yeast. Occasionally, particularly in samples containing stressed yeast, cells may stain to a slight extent only, making interpretation somewhat subjective. To overcome some of these problems it has been suggested that the decolourisation reaction could be used as the basis of a spectrophotometric procedure (Bonora & Mares, 1982). However, at high viabilities, where the methylene blue test is valid, the small change in colour makes this test impracticable. Attempts have been made to overcome some of the failings of the methylene blue staining method by use of an alkaline pH and contact with yeast for 15 minutes at 25°C (Sami *et al.*, 1994). The rationale here was that, since entry of the dye is influenced by membrane potential, the lowering of the external $H^+$ concentration would favour entry into very stressed cells, which normally give a false positive viability. Data was provided that supported this contention.

It has been suggested that the disparity between viability determined by methylene blue and plate counting might be overcome by a double staining technique using methylene blue and safranin O (Nishikawa & Nomura, 1974). In this test, a gradation of colour was reported to correlate with physiological condition. Thus, a cell in good condition stained blue/purple, whereas dead cells stained pink/red. In between these two extremes were slightly deteriorated cells, which stained reddish purple and grossly deteriorated cells, which stained brown.

Smart *et al.* (1999) compared the efficacy of viability measurements by plate count, methylene blue/safranin O double staining, citrate methylene blue, alkaline methylene blue, citrate methylene violet and alkaline methylene violet. Samples of ale, lager and cider yeast in various physiological conditions were used in the assessment. These were exponential and stationary phase 'healthy' cells, starved cells and non-viable heat-killed populations.

Predictably, the plate count approach underestimated true viable cell counts in the case of chain-forming strains; however, it was the most reliable of the techniques tested for identifying dead cells. Citrate methylene blue gave a good correlation with healthy and starved yeast but dramatically over-estimated viability using heat-killed cells. Citrate methylene blue and citrate methylene violet were of equal utility where viable populations were tested. However, only the latter was considered to be capable of unequivocal differentiation between viable and non-viable cells.

The methylene blue/safranin O double staining technique was found to give very variable results, even in replicate samples, and was not recommended. Alkaline methylene blue reportedly measures intracellular reducing power, since the high pH removes any barrier to dye penetration (Sami *et al.*, 1994). This method failed to distinguish viable and non-viable cells in a reliable manner, particularly those in the exponential phase of growth. Conversely, alkaline methylene violet could distinguish living, stressed and dead cells. In this sense, the method was of utility in measuring 'vitality' as well as viability, as discussed in Section 7.4.2.

The authors concluded that the methylene blue staining technique produced equivocal staining reactions because of instability of the dye. Thus, they contended that commercial preparations of the dye contain two chromaphores, azure B and Bernthsen methylene violet and some lower azures deriving from oxidative demethylation of methylene blue. The latter reaction was exacerbated by prolonged storage of prepared solutions and this should be avoided.

Tetrazolium salts have been used as indicators of viability of microbial cultures by virtue of their intracellular reduction to coloured formazan deposits (Postgate, 1967). Thom et al. (1993) reported the results of a comparative study of the application of four tetrazolium salts to the measurement of viability. The yeast Candida albicans was included in the study and although results were improved by addition of glucose this organism gave the least satisfactory response. Others have reported good correlation between respiratory activity of Saccharomyces yeasts and staining with the tetra-zolium salt 2-(p-iodophenyl)-3-(p-nitrophenyl)-5-phenyl tetrazolium chloride (INT) (Trevors, 1982; Trevors et al., 1983. It is unlikely that this would be of utility in the case of repressed brewing yeast cells.

The biomass meter described in Section 6.1.4.5 provides an instantaneous measure of yeast concentration. It can be shown that the meter is responsive to that fraction of the population that would be considered viable using the methylene blue test (Boulton et al., 1989). Conversely, it does not respond to the non-viable fraction of the population and it follows that it will not provide a measure of viability per se. However, this can be achieved if the meter reading is used in conjunction with an analysis of total yeast concentration. Conveniently, this removes some of the errors due to the manual aspects of methylene blue counting. Nonetheless, it is still apparently prone to the same over-estimation in the case of very-low-viability samples as the methylene blue method.

Viability may be determined by measurement of the surface electrostatic charge, or zeta potential of yeast cells (see Section 4.4.6.7). It can be shown that a significant difference exists in the magnitude of this parameter for viable (intact membrane) and dead cells (disrupted membrane) (Brown, 1997b). The same author described an apparatus that uses a laser beam arrangement linked to a powerful software package to measure both cell size and zeta potential within a yeast population and thereby calculate the viability. It was further suggested that the magnitude of zeta potential was a general measure of the yeasts' physiological condition.

Several fluorescent dyes have been used in vital staining methods. These rely for their selective staining action on several different facets of cellular metabolism. Some dyes are taken up by all cells and fluorochromes are liberated after modification by enzymes that are present only in viable cells. Other dyes rely on membrane function. Some are excluded by cells with intact membranes, or cells capable of maintaining an electrochemical transmembrane potential. Non-viable cells are not capable of excluding these dyes and these become fluorescent under appropriate conditions. Another class of dyes that also rely on membrane integrity are taken up only by viable cells. Some of the commonly used fluorescent dyes, together with a brief description of their mode of action are given in Table 7.3.

Several authors have reported that viability measurements made with fluorescent dyes provide results that correlate closely with plate counting methods. For example, McCaig (1990) reported that viability measurements made using the fluorochrome Mg-ANS (Table 7.3) were similar to slide culture results, but were significantly different to values obtained by bright field staining with methylene blue or Eosin Y. Similarly, Trevors *et al.* (1983) compared several viabililty methods and reported that Mg-ANS and primulin were both of good utility; however, acridine orange was less accurate.

**Table 7.3** Fluorescent dyes used for viability measurement.

| Dye | Mode of action | Reference |
| --- | --- | --- |
| Fluorescein diacetate | Release of free fluorescein following cleavage by non-specific intracellular esterases in viable cells | Paton and Jones (1975); Chilver *et al.* (1978) |
| Propidium iodide Ethidium homodimer | Excluded by viable cells; bind to nucleic acids in non-viable cells | Bank (1988); Donhauser *et al.* (1993); Hutter (1993) |
| Oxonol dyes | Excluded by viable cells | Dinsdale and Lloyd (1995); Lloyd *et al.* (1996) |
| Rhodamine 123 | Taken up by viable cells with mitochondrial trans-membrane potential | Dinsdale and Lloyd (1995); Lloyd *et al.* (1996) |
| 3,3-Dihexyloxacarbocyanine | Taken up by viable cells with plasmamembrane trans-membrane potential | Lloyd *et al.* (1996) |
| Chemuchrome Y | Uptake followed by enzymic cleavage to release fluorochrome in viable cells | Raynal *et al.* (1994) |
| Mg 1-anilino-8-naphthalene sulphonic acid (Mg-ANS) | Accumulates in viable cells with functional membrane and binds to proteins | King *et al.* (1981); McCaig (1990) |
| Acridine orange | Retained in viable cells | King *et al.* (1981); Trevors (1982); Trevors *et al.* (1982) |

Viability measurements with fluorescent dyes can be performed in the same way as the standard methylene blue haemocyotmeter test although a fluorescent microscope is required (Chilver *et al.*, 1978; McCaig, 1990). Much better results are obtained with automatic measuring devices, which remove the error due to the human operator. For example, Raynal *et al.* (1994) describe the use of fluorescein diacetate and Chemchrome Y to measure viability in conjunction with an electronic image analyser. The most commonly used counting apparatus is the flow cytometer (Petit *et al.* 1993). In this device, cells in a suspension are passed singly through a narrow orifice where each is detected. In the case of fluorescent applications the cells pass through a light beam of appropriate wavelength and and fluorescence is registered by a detector. Other

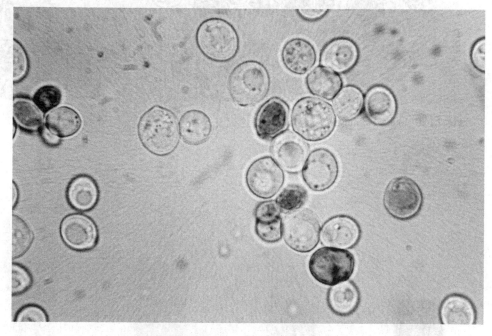

**Fig. 7.15** Viability staining of yeast using methylene blue.

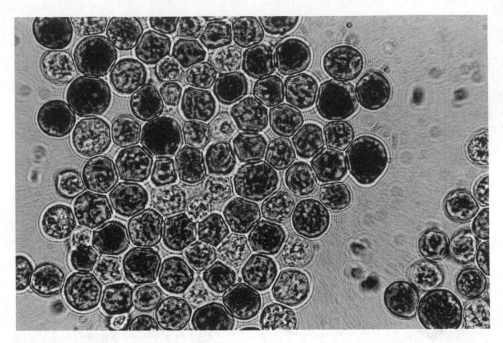

**Fig. 7.16a**  Iodine staining of yeast cells.

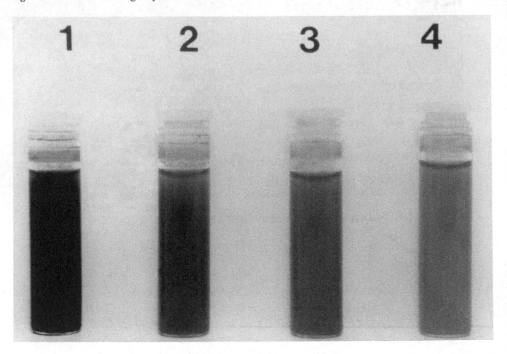

**Fig. 7.16b**  Iodine staining of yeast slurries (Quain & Tubb, 1983, with permission from the Institute of Brewing). Intensity of brown staining is proportional to glycogen content.

detectors measure total cell count by light scattering and possibly other parameters such as cell size distribution. Sophisticated instruments have the facility to separate and collect sub-populations from the stream of cells based on the response from the various detectors (Edwards *et al.*, 1996).

Image analysers and flow cytometers provide rapid and accurate viability measurements compared to manual microscopic counting. However, there is a significant cost penalty. In particular, the more sophisticated flow cytometers cost up to £250 000. Clearly, these will not find use as routine laboratory tools for assessing the quality of production yeast! Less expensive flow cytometers are also unlikely to find application purely for viability measurement; however, they offer the potential for much more delicate probing of yeast physiology, as described in Section 7.4.2.

### 7.4.2 *Yeast vitality tests*

Provided a single method for viability measurement is used and the procedure is performed in a consistent manner, it undoubtedly offers useful comparative information for the individual brewer. In particular, a test such as methylene blue staining is of great use as the basis of a simple decision to use or not use a given batch of pitching yeast. Tests such as this can also be useful, provided the viability is high (> 90%) in establishing a correction factor to ensure that viable pitching rates for all fermentations are the same. Such tests are less useful in identifying variations in pitching yeast physiology, which can produce inconsistencies in fermentation performance and beer quality.

With regard to yeast cells, there is no single definition that encompasses viability. Characteristics of viable cells would include:

(1) capability of cellular proliferation (progression through the cell cycle);
(2) capability of cellular growth (anabolic metabolism)
(3) detectable resting metabolism (oxygen uptake and carbon dioxide evolution);
(4) possession of membrane integrity (controlled selective assimilation of exogenous metabolites and excretion of by-products).

It is obvious that within a population of pitching yeast there may be cells which do not possess all these characteristics but still take an active role in fermentation. For example, they may not be capable, for whatever reason, of multiplication; however, they may still contribute to fermentation by assimilation of wort nutrients and production of beer components. It is unclear which fraction of these populations is detected by standard viability tests. It is also arguable that although the plate or slide test is considered as a standard reference method for viability determination, the test as applied to pitching yeast does not necessarily equate to subsequent growth and fermentation performance.

Simple viability tests are not capable of providing information regarding possible differences in the physiological status of the viable fraction of the yeast population. These differences could span the spectrum from extremely stressed or dying cells through to those which are in a physiological condition that would allow 'super-performance' in fermentation. For example, cells which contain high sterol concentrations would be predicted to undergo more rounds of budding during fermentation than those with basal sterol levels.

Several 'vitality' tests have been proposed which provide information of the physiological condition of the entire population of yeast cells within a slurry. This is an unfortunate choice of terminology since it has no strict scientific meaning other than being a synonym for viability. Of course, it is taken to imply a method that identifies yeast that will produce a vigorous rapid fermentation performance. It does not follow that such a fermentation would be efficient in terms of yeast growth and ethanol yield. It would be better to refer to the methods by the less concise but more accurate and useful definition of 'predictive fermentation tests'. However, vitality testing has become part of the established brewing literature and in order to avoid confusion it is used here.

The requirements of vitality tests are that they should be rapid, simple and preferably use apparatus available in a typical brewery laboratory. The results of such tests may be used to arrive at a simple decision regarding fitness to pitch. However, preferably the result would allow selection of an appropriate pitching rate and wort oxygenation regime, which provides optimum and consistent fermentation performance and beer quality. A plethora of vitality tests have been proposed that may be considered according to which aspect of yeast physiology they seek to probe.

7.4.2.1 *Tests based on cellular composition.* Glycogen is reported to provide carbon and energy for sterol synthesis during the early aerobic phase of fermentation (Quain, 1988). Pitching yeast that has been stored for too long or held under inappropriate conditions may have already expended much of its glycogen reserves, and therefore be unable to efficiently couple oxygen utilisation to sterol synthesis during early fermentation (Section 7.3.2). It would be predicted that this would be manifest as a prolonged lag phase in fermentation. Yeast glycogen content can be determined using the colour reaction with iodine either qualitatively by visual assessment (Fig. 7.16) or quantitatively by measurement of the brown coloration at 660 nm (Quain & Tubb, 1983). Skinner (1996) described another method, which took just over two hours to perform, based on infra-red spectroscopy.

Trehalose levels in yeast have also been proposed as a useful monitor of yeast condition. In particular the relationship between trehalose levels and yeast stress. For example, Majara *et al.* (1996b) observed a positive correlation between trehalose concentration and applied stress. These authors suggested that a sudden increase in trehalose concentration in non-growing cells could be indicative of cells that had been in some way stressed. Probably the carbon for such an increase would have derived from glycogen breakdown. A possible vitality test, therefore, would be to determine the ratio of glycogen to trehalose in pitching yeast. As with glycogen, yeast trehalose content can be determined in a rapid test using near infrared reflectance spectroscopy (Moonsamy *et al.*, 1996). For individual yeast strains used in particular fermentations this ratio should fall within definable limits for yeast in 'good' condition. A significant deviation from this established value would be indicative of abnormal yeast.

A problem with this concept is the situation in which yeast has dissimilated glycogen during storage, in response to inadvertent exposure to oxygen. Such yeast would have a low glycogen content, and therefore appear stressed, but in actuality

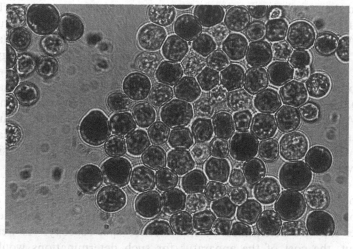

**Fig. 7.16** Iodine staining of (a) yeast cells and (b) yeast slurries (Quain & Rubb, 1983, with permission from the Institute of Brewing). Intensity of brown staining is proportional to glycogen content. (This figure is repeated in the colour section)

would have a reduced requirement for wort oxygenation for standard fermentation performance. This would be detectable by monitoring sterol levels in pitching yeast. A reasonably rapid method for yeast sterol determination has been developed (Rowe *et al.*, 1991). This relies on a change in the absorbance spectrum of the polyene antibiotic filipin, which occurs after it reacts with sterol. Of course, the necessity to monitor sterol, glycogen and possibly trehalose on a routine basis begins to stretch the concept of a rapid inexpensive vitality test!

The intracellular concentration of ATP can be used as a measure of condition. Thus, viable cells must generate sufficient ATP for energy maintenance during periods of starvation. ATP levels in yeast cells (Hysert *et al.*, 1976) are conveniently monitored using bioluminescence (see Section 8.3.3.1). Manson and Slaughter (1986) determined the correlation between ATP content and fermentation performance in samples of yeast that had been stressed by storage at high temperature prior to pitching. A good correlation was obtained, although no better than that seen with methylene blue viability staining. A more accurate reflection of the energy status of a cell is that which takes into account the ratio of concentrations of ATP and ADP to AMP, in other words the concept of 'adenylate energy charge' (Chapman & Atkinson, 1977). Under normal conditions cells maintain a high adenylate energy charge (> 0.75). A low value may be taken as evidence of stress. However, because cells tend to maintain high values for this parameter, a significant decrease is indicative of cells that are close to death, and this would probably be apparent by other simpler tests, such as a viability stain.

The redox state of the cell is of critical influence to many aspects of yeast metabolism, not least glycolysis and ethanol formation. A manifestation of redox is the relative concentrations of NAD/NADH. The intracellular concentrations of these metabolites may be determined by fluorescence spectroscopy. As Lentini (1993) pointed out, the cost of the apparatus for such determinations would probably mitigate against its application for routine testing.

Variable pitching yeast physiology may arise in ways other than in response to environmental stimuli. For example, the mean cell age of the yeast could be a variable factor, perhaps related to the number of generations of fermentation the yeast had passed through, or a particular cropping regime (see Section 4.3.3.4). The age of yeast cells is related to the number of bud scars present on the cell surface. Bud scars consist of the carbohydrate chitin (see Section 4.4.2.3) that does not occur elsewhere in the envelope. Chitin can be detected by specific dyes such as primulin and calcofluor such that the extent of staining reflects cell age.

### 7.4.2.2 Measures of cellular activity.

Another approach to assessing yeast physiology is to measure some aspect of metabolic activity, preferably one that reflects behaviour during fermentation. Several vitality tests of this type have been proposed. The most obvious and direct predictive fermentation test is to perform a laboratory fermentation trial, using the type of apparatus described in Section 5.9. Such tests have the advantage of providing information regarding most of the parameters of interest in a real fermentation and similar wort may be used. Unfortunately, the time required to undertake the test would be too long for routine application. Instead, other aspects of the activity of yeast may be measured which relate to fermentation performance. Possible parameters are rate of oxygen uptake, rate of evolution of carbon dioxide, exothermy and rate of ethanol formation (Boulton & Quain, 1987; Daoud & Searle, 1987).

Measurement of the specific oxygen uptake rate forms the basis of the BRi yeast vitality apparatus, the use of which has been described by Kennedy (1989). It consists of an accurately attempered water bath in which a predetermined quantity of yeast suspension is placed. The specific rate of oxygen uptake is measured using an integral Clark-type electrode with output to a recorder. Results were presented by Kennedy (1989) who indicated that differences in this parameter could be detected in production pitching yeast samples, which were not evident from a simple viability test. Specific rates of oxygen uptake are related to yeast sterol content (see Section 3.5.1).

Another vitality meter has been developed by Muck and Narziss (1988) which measures carbon dioxide formation as increase in pressure in a sealed container. Like the BRi vitality meter, readings are taken whilst a known quantity of yeast is held under controlled conditions. Results are obtained within one hour. Similar measurements may be made with a Warburg manometer or an automated respirometer (Mathieu *et al.*, 1991).

### 7.4.2.3 *Fluorometric vitality tests.*

As discussed in Section 7.4.2.2, several tests of yeast viability have membrane function as their basis. There is the assumption that viable cells will have an intact and functional membrane, whereas non-viable cells will have total loss of membrane function. The disparity between results of tests such as methylene blue staining and slide or plate counting techniques suggest that conditions of membrane competence which lie in between total collapse and fully functional are probable. The fact that sterol concentration in membranes is a known variable, which is influenced by environmental factors, confirms this view. Vitality tests based on limited fermentation rely on the yeast membrane being functional but only in part. Measurement of the specific rate of oxygen uptake more closely probes membrane integrity because of its relation to sterol concentration. Another useful membrane probe is the ability of yeast cells to acidify the external medium, both spontaneously and in response to a supply of exogenous glucose, termed the acidification power test.

Acidification power is defined as the sum of the spontaneous pH change determined after suspending yeast cells in water and the substrate-induced pH change after addition of glucose to the suspension (Sigler *et al.*, 1980, 1981a, b, 1983; Opekarová & Sigler, 1982; Sigler & Hofer, 1991). The observed changes in pH largely reflect the activity of the membrane- bound $H^+$ATPase. Thus, the cell extrudes protons in order to assist with the control of intracellular pH and to maintain a proton electrochemical gradient across the plasma membrane (see Section 4.1.2.2). This gradient supports active transport of nutrients into the cell, and therefore it is essential to the maintenance of viability and is intimately related to metabolic activity during fermentation. Spontaneous acidification is a function of the energy status of the cell and positively correlates with glycogen content. Glucose-induced acidification relates to both membrane state and glycolytic flux. Some counterbalancing activities also occur which have the opposite effect on the external pH. Proton influx occurs as a counter ion in some transport processes and passive proton influx takes place if the membrane becomes de-energised. It has been suggested that this latter effect may be a symptom of ethanol toxicity (Fernanda Rosa & Sa-Correia, 1994). The magnitude of acidification is significantly influenced by univalent cations in sugar-metabolising yeast. Kotyk and Georghiou (1994) reported a 20-fold increase in acidification rate in *S. cerevisiae* in the presence of potassium ions. These authors concluded this was due to an effect of this cation on reactions producing ATP and not the membrane-bound $H^+$ATPase.

An acidification test suitable for application for testing pitching yeast condition was developed by Kara *et al.* (1988). The test is a modification of that of Opekarová and Sigler (1982) in which notably the incubation temperature was changed to 25°C. A known wet weight of yeast is washed in chilled distilled water by repeated suspension and centrifugation. The test commences by suspending the washed yeast pellet in a known volume of distilled water at 25°C. The slurry is incubated with constant stirring for 10 minutes during which time the pH is monitored at intervals of one minute. The change in pH after 10 minutes' incubation is taken as the spontaneous acidification power ($AP_{10}$). After this time, glucose solution was added to the suspension to give a final concentration of 0.1% w/w. The pH was monitored for a further 10 minutes. The change in pH during the second incubation is the glucose-induced acidification power ($AP_{20}$). Using this procedure, the authors demonstrated in laboratory studies an inverse straight-line correlation between acidification power and fermentation performance, measured as time to half gravity. However, they noted that it was unreliable when used with acid washed yeast.

Fernandez et al. (1991) applied the AP test at commercial scale and concluded that it was predictive of fermentation performance. In this study the deleterious effects of storage of pitching yeast under inappropriate conditions were identified and it was observed that in addition to the relation with fermentation rate, a correlation was observed between acidification power and VDK stand-times. Mathieu et al. (1991) measured acidification power on samples of yeast removed from fermenter at intervals after pitching. They observed a sharp drop in $AP_{10}$ during the first few hours of fermentation. This was attributed to the rapid drop in yeast glycogen content associated with the aerobic phase of fermentation. They reported that the correlation between acidification power and fermentation performance (measured as decrease in gravity, °Plato per day) was better if the acidification power was applied to yeast 24 hours after pitching. Accordingly, they modified the test to include a pre-treatment step in which pitching yeast was exposed to wort for 15 minutes prior to acidification measurement. This resulted in a decrease in total acidification power (decrease in $AP_{10}$, slight increase in $AP_{20}$) which was shown to result from an interaction between the yeast cell surface and a trub component, possibly a tannoid. Providing the pre-treatment was performed, these authors also reported a good correlation between acidification power and subsequent fermentation performance.

Others have sought to modify the acidification power test in different ways. Patino et al. (1993) suggested that it would be better to convert each pH reading into the corresponding proton concentration and sum the differences for each of these over the time course of the test. This provided a parameter related to the magnitude of the pH change, which they termed the 'cumulative acidification power'. This corrected for the non-linearity of pH change during the acidification power test. In addition, they proposed that the test was made more reliable by substituting maltose for glucose on the basis that the latter formed the major sugar in worts. Furthermore, since changes in the concentrations of ions other than $H^+$ occur during the test, it would be more meaningful to measure conductance, as opposed to pH.

Iserentant et al. (1996) pointed out that a drawback of the AP test was its relative insensitivity with 'high vitality' yeast. Thus, because of the logarithmic basis of the pH scale, the maximum acidification power is limited to about 2.8. These authors obviated this problem by modifying the procedure such that the pH was maintained at a constant value by titrating with 0.1 M sodium hydroxide. The volume of sodium hydroxide needed to do this was measured and this was referred to as the titrated acidification power. It was demonstrated that this test was useful for yeasts such as those grown under aerobic conditions which would be predicted to have a very high

AP. Conversely, it was not sensitive enough for low vitality yeast, and in this case, it was suggested that conductivity was a more useful measure than pH.

Another of the many effects of yeast metabolism that may be observed during fermentation are changes in extra- and intracellular metal ion concentrations. Mochaba et al. (1996) noted that fluctuations in the concentrations of cations such as potassium, calcium, zinc and magnesium during fermentation were dependent on yeast condition. They developed this into a vitality test based on the release of magnesium ions, which occurs when yeast is pitched into wort (Mochaba et al. 1997). The test involves suspending a small known quantity of washed yeast (0.1 g wet weight) into 10 ml of fresh brewery wort. After one minute contact time a small sample of the suspension is removed and filtered to remove yeast. The magnesium concentration in the filtrate is determined using a commercial spectrophotometric kit. In laboratory and plant trials, it was shown that a positive correlation existed between yeast vitality and magnitude of magnesium release. Yeast judged as being highly vital by this method produced superior fermentation performance in all respects. Thus, more rapid attenuation, greater ethanol yields and shorter VDK stand-times. The differences detected by the magnesium release test were not evident from simple viability testing.

Many of the methods of viability determination that use fluorochromes can also be extended to provide information regarding physiological condition. Imai et al. (1994) developed a procedure using a derivative of fluorescein, 5 (and 6) carboxyfluorescein. This procedure involved liberation by esterases of fluorochrome in viable cells. A vitality test was developed that involved measurement of intracellular pH based on a calibration curve that related fluorescent intensity using a range of buffers of differing pH values. Data was presented that indicated a correlation between intracellular pH of a number of yeast samples and fermentation performance based on rates of sugar consumption. The method was claimed to be superior to the acidification test in that it was capable of distinguishing differences over a greater range of physiological conditions. Thus, a positive correlation between results of the vitality test and fermentation performance was observed for yeast samples with intracellular pH values up to 5.7. However, in a data set of 555 production yeast samples, more than 90% had an intracellular pH greater than 5.7, and therefore would be indistinguishable by the acidification test.

The potential of fluorochromes as monitors of yeast physiological condition can be properly unlocked if used in conjunction with flow cytometry (Donhauser et al., 1993; Edwards, 1996; Hutter, 1997). Flow cytometric tests are rapid and since all cells within the sample population are examined there is no requirement for testing a defined quantity of cells. A wide range of fluorochromes is available which allow many aspects of physiology to be examined. These include viability, membrane competence, intracellular pH, distribution of genealogical age, ploidy, phase in cell cycle and others. Not only is it possible to identify the proportions of cells in a given state within the population but with sophisticated cytometers these sub-populations may be separated and collected for further analysis.

7.4.2.4 *Vitality tests – a summary.* The key element of any vitality test is that it must provide information that can be acted on by the brewer. The potential for influencing the outcome of the brewing process is likely to reflect the effort and cost expended on the test. In this respect, vitality tests may fulfil a number of roles. At their simplest, they will be used to judge batches of pitching yeast on a reject quality limit basis. For the majority of brewers, a viability test will serve for this function. The methylene blue haemocytometer counting method, or other simple microscopic approaches, are equal to this task. The fact that methylene blue staining may significantly over-estimate viability compared to plate counting methods in grossly deteriorated yeast is irrelevant in this case since no brewer given the choice would seriously contemplate use of such yeast. Fluorescent staining techniques more accurately reflect the viability as judged by a colony forming unit method; however, there is a significant additional cost as a fluorescent microscope is needed.

Vitality tests may be used to monitor long-term drift in factors that influence yeast physiological condition, and hence fermentation performance and beer quality. In this case, any drift that resulted in gross deterioration of pitching yeast, would be signalled by the viability methods and again no additional testing would be warranted. However, the change may be of a type that influences yeast physiology without affecting viability. In addition, inherent variability in raw materials contributing to wort composition and inconsistencies in fermentation control and yeast handling can result in subtle variations in yeast physiology which are not detectable by simple viability testing. In these cases additional assessment may be justified.

Of the many tests that have been described, acidification power offers many advantages. It has all the requirements of a routine QA test in that it is rapid, simple, inexpensive and uses readily available laboratory equipment. Most importantly, it is predictive of subsequent fermentation performance such that it allows selection of an appropriate pitching rate. In its original incarnation, it is a less useful test for discriminating yeast samples of very 'high' vitality. The modified acidification tests, such as titrated acidification power, address this problem without the need for much greater outlay in terms of laboratory equipment.

Undoubtedly assessing yeast physiology by flow cytometry is a very powerful and flexible tool, which has the potential to deliver the most detailed analysis. Unfortunately at present this comes at considerable cost. For this reason, this approach is unlikely to find use as a routine method until costs fall by at least an order of magnitude. Until that time, flow cytometers will be destined to be confined to the research laboratory.

## 7.5  Surplus yeast

All breweries generate a surplus of yeast, which must be disposed of, preferably with some financial gain. Usually the yeast will be in the form of pressed cake after processing to recover barm ale. Spent yeast comes in various states of purity and physiological condition depending on which stage of the brewing process it is recovered from. Many breweries have dedicated storage tanks that are used for temporary holding of waste yeast. Associated with the tanks is specific plant for separating the yeast and beer. This may take the form of a large plate and frame filter (Fig. 2.12) or some other similar device.

Much of the yeast that is recovered from conditioning tank bottoms and similar sources is heavily contaminated with other solid materials and in order to facilitate efficient pressing has to be mixed with a filter aid such as kieselguhr. This material has no commercial value and has to be disposed of. In modern breweries, which do not require the presence of yeast during cold conditioning, it would be more appropriate to identify methods for avoiding the formation of yeast-containing tank bottoms. Possibilities would include the use of highly efficient green beer centrifuges capable of removing all yeast from the process stream during fermenter run-down and producing a dry yeast discharge which could be disposed of without further processing.

The highest quality yeast is that destined for use as pitching yeast but surplus to production requirements. Some of this may be sold for use in other fermentation industries, for example, to distillers (Bathgate, 1989). More commonly, such yeast is sold for further processing to produce yeast extracts, which are used in many food products. In addition, brewer's yeast is an excellent source of many vitamins and trace metals, and for this reason, it is used to prepare dietary supplements.

A common goal for many brewers has been to identify methods of increasing the value of spent yeast. Suggestions have included genetic modification to produce heterologous proteins to 'add value' (see Section 4.3.4). Other proposed routes have included the recovery and purification of fine chemicals for use in other chemical and pharmaceutical industries. This avenue has yet to be exploited because of the high initial investment and the uncertainty of finding a market.

# 8 Microbiology

In terms of microbiological threat, yeast propagation, fermentation and yeast handling, are all especially vulnerable. Contamination with bacteria, the wrong production yeast or, indeed, 'wild' yeasts can skew (or worse) product quality, which through yeast recycling can – if not detected – lead to major brewery-wide quality problems. Therefore, given its importance and all-embracing scope, this chapter on 'microbiology' is not restricted to fermentation and yeast but to the entire brewing process. This is necessary, as in the view of the authors a less holistic approach to brewing microbiology would be a missed opportunity! Inevitably only the 'headlines' can be captured here and the interested reader is directed to Rainbow (1981) for a succinct but detailed review or to the 'bible' of brewing microbiologists, the second edition of *Brewing Microbiology* (Priest & Campbell, 1996). This multi-authored text is a veritable treasure-trove of information about what remains the poor relation of brewing science.

## 8.1 Product spoilage

### 8.1.1 *Susceptibility*

In his excellent review on beer spoilage micro-organisms, Rainbow (1981) noted that 'beer is resistant to microbial spoilage because of its relatively low nutritional status, its content of products of yeast metabolism, its adverse values of pH and redox potential and its content of hop bitter substances'. This, in a nutshell, explains the relative robustness of beer to microbial spoilage. Although the 'low nutritional status' is self-evident, it is worth reiterating that yeast removes the vast majority of assimilable nutrients from wort, leaving behind generally high-molecular-weight material of limited appeal to the vast majority of micro-organisms. Whether this is a major factor in providing innate protection is debatable as the nutrient status of beer is boosted by release of nutrients through yeast autolysis and subsequent priming with fermentable sugars. More significant factors in limiting susceptibility to spoilage are 'the products of yeast metabolism', specifically the twin guns of ethanol ($> 3.5\%$ v/v) and pH ($\approx 4$). Telling insights in their contribution to minimising the risk of spoilage comes from experiences with low/no alcohol beers (LAB/NABs) and products with pHs above those of beer. Both are innately more susceptible to spoilage by the usual spectrum of spoilage organisms as well as others that are typically suppressed by these parameters. Additionally, other beer components (hop iso-α-acids, carbon dioxide, etc.) bolster resistance to spoilage. Of course, the robustness of the product to microbial spoilage is best viewed as being an intrinsic 'benefit' and should in no way detract from a commitment to minimising risk through hygienic operations and practices.

The few systematic investigations into the 'spoilability' of beer have focused on the resistance to spoilage by the Gram-positive organisms, *Lactobacillus* and *Pediococcus* (Dolezil & Kirsop, 1980; Fernandez & Simpson, 1995; Hammond *et al.*, 1999). Given the unpredictability of brewing microbiology, it is no surprise to find a somewhat confused picture with little by way of generic insights. As ever, methodology may be at the root of the confusion. The earlier work of Dolezil and Kirsop (1980) was unable to correlate resistance to spoilage with any beer parameters. However, as noted by Fernandez and Simpson (1995), the bacteria used by Dolezil and Kirsop (1980) had not been 'trained' to grow in beer prior to inoculation. Fernandez and Simpson (1995), using 14 strains of hop-resistant lactic acid bacteria and 17 different lagers, were able to correlate resistance to spoilage to a number of beer parameters. These were generally in accord with expectation, such that spoilage was related to beer pH, various nutrients (free amino nitrogen, maltotriose etc.) and the undissociated forms of hop bitter acids and sulphur dioxide. Intriguingly, beer colour correlated strongly with resistance to spoilage. These parameters, together with carbon dioxide (Hammond *et al.*, 1999), may be viewed as providing beer with innate protection to spoilage by lactic acid bacteria. Inevitably though, there are differences between the response of different species of *Lactobacillus*, strains of *Lactobacillus* species and sensitivity of different beers to spoilage. Indeed, it is tempting to conclude that the only predictable thing about microbiology is its lack of predictability!

### 8.1.2 *Spoilage micro-organisms – bacteria*

Despite the ever increasing sophistication of methods to 'fingerprint' bacteria (see Section 8.3.4.2), the typical 'first stab' at identification of an unknown bacterium involves a handful of simple questions and methods. As many brewery bacteria are fastidious in their nutrient requirements and environment, much can be gleaned from simply identifying where the contaminant was isolated. Overlaid on this are considerations such as the growth medium the organism was recovered from and, importantly, the size and shape of the cells under the microscope. Some bacteria are typically spherical (cocci) and others more elongated rods (bacilli). Further diagnostic information can be obtained from whether the cells are single or organised into groups (pairs, tetrads, clusters or chains).

In the majority of cases, depending on need and available resources, the above criteria provide sufficient information for the routine cursory identification of a bacterial contaminant. However, two laboratory methods – the catalase test and the Gram test – are frequently used to describe bacteria and can be used to 'add value' in routine microbiological trouble shooting. Of the two, the catalase test is much the simpler but less revealing diagnostically. The method probes for the presence of catalase, one of a number of enzymes that protect microbial cells from toxic oxygen species. Quite simply, an aliquot of hydrogen peroxide (3% v/v) is added to the microbial colony or culture and the presence or absence of gas (oxygen) evolution scored as, respectively, catalase positive or negative. Typically, anaerobic organisms are catalase negative. Although more convoluted, the Gram test is worthy of greater comment, particularly as it remains the basis of the division of bacteria into two classes, 'Gram positive' and 'Gram negative'.

The Gram test was devised in 1884 by a Danish bacteriologist, Hans Christian Gram, who, at the time was working in the morgue of the city hospital in Berlin (for an historical review see Scherrer, 1984). The method involves the stepwise staining and counterstaining of a thin film of bacteria on a microscope slide and results in Gram positive bacteria being stained purple and Gram negative cells being stained red. Typically, the result is assessed visually without recourse to microscopy. Although long a routine method in bacteriology, it has its limitations, notably being both tedious and time consuming. Further, the outcome is sensitive to factors such as staining technique and reagent stability together with the physiology and age of bacteria under test.

The steps in Gram's staining are detailed in Fig. 8.1. Although subject to debate, it is generally accepted that the two-stage staining with crystal violet and then Gram's iodine (iodine and potassium iodide) leads to the formation of a water insoluble complex. Treatment with ethanol differentiates between bacteria by washing out the purple stain from the Gram-negative cells, which are then counterstained red with safranin. Seemingly, the crystal violet-iodine complex is retained within the Gram-positive cells but lost from the Gram-negative cells. Retention or loss of the complex is explained by the cell wall structure of the two classes of bacteria. Although broadly similar (see Fig. 8.2), the peptidoglycan matrix layer in Gram-positive cells is many times thicker than the Gram-negative cells. Thus, the physical mass and thickness of the cell wall is believed to explain the differing responses of bacteria to the Gram test. It is noteworthy that compared to Gram-negative cells, Gram-positive bacteria are generally more sensitive to growth inhibition by dyes and many antibiotics but are more resistant to enzymic digestion. Unlike other eukaryotic cells, yeast is strongly Gram-positive.

A rapid alternative to the Gram's test has been described (Lin, 1980), which exploits the differing viscosity of pure bacterial colonies when treated with 3% (w/v) potassium hydroxide solution. Gram-negative bacteria form a viscous mass whereas Gram-positive colonies fail to exhibit viscosity. This is thought to reflect the

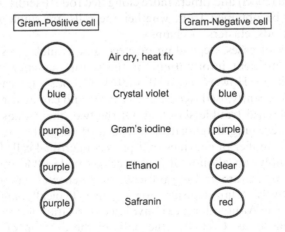

| Gram-Positive cell | | Gram-Negative cell |
|---|---|---|
| ◯ | Air dry, heat fix | ◯ |
| blue | Crystal violet | blue |
| purple | Gram's iodine | purple |
| purple | Ethanol | clear |
| purple | Safranin | red |

**Fig. 8.1**   The Gram stain.

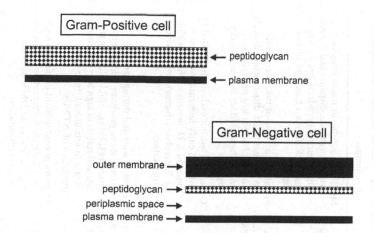

**Fig. 8.2** Differences between the cell walls of Gram-positive and Gram-negative bacteria.

extraction of DNA from the Gram-negative bacteria. The correlation between the 'KOH' method and the Gram test is best described as 'directional'. In a survey of 466 bacteria, 91.4% of the Gram negative and 88% of the Gram positive correlated with KOH lysis (Moaledj, 1986).

Inevitably, the differentiation of bacteria is more complex and extensive than indicated here. Like yeast (Section 4.1.1.1), API test strips have found application in the differentiation and identification of brewery bacteria (see for example Ingledew *et al.*, 1980). However, Priest (1996) has reported reservations in accuracy and repeatability of diagnostic strips. From a taxonomic perspective, the definitive reference work for bacteria is *Bergey's Manual of Determinative Bacteriology* (Holt *et al.*, 1994), which is now in its ninth edition. As with yeast, bacterial taxonomy is increasingly influenced by the application, and consequent insights, of new and evolving molecular methods. A user-friendly overview of the classification of brewing bacteria can be found in Priest (1981).

8.1.2.1 *Gram-negative bacteria.* The major Gram-negative bacteria found in breweries are summarised in Table 8.1. For an authoritative review, see Van Vuuren (1996) in *Brewing Microbiology*. For a wider appreciation of Gram-negative bacteria see *Bergey's Manual of Determinative Bacteriology* (Holt *et al.*, 1994). Specific brewing reviews can be found on *Zymomonas* (Dadds & Martin, 1973), *Enterobacteriaceae* (Priest *et al.*, 1974) and the obligate anaerobes (Chelack & Ingledew, 1987).

Arguably, in comparison to the Gram-positive bacteria, the threat of the various Gram-negative bacteria is under reasonable control. With changes in process, raw materials and market demands some – the acetic acid bacteria (Fig. 8.3) and *Zymomonas* (Fig. 8.4) – have had their day and are now more of a niche concern. *O. proteus* (Fig. 8.5) remains a concern for its role in the formation of ATNCs (Section 8.1.5.1) and ease of recycling via yeast re-pitching. Day-to-day management of this microbiological threat is achieved through regular acid washing (see Section 7.3.3 and

**Table 8.1** Gram-negative bacteria.

| Family | Genus | Major species | Characteristics | Beer spoilage |
|---|---|---|---|---|
| Acetic acid bacteria | *Acetobacter* | *A. aceti*<br>*A. pasteurianus* | Pleomorphic, 0.6–0.8 µm × 1–4 µm, catalase positive, strict aerobes, oxidises ethanol to carbon dioxide and water | Specific to aerobic or microaerophilic environments, i.e. dispense and draught (unpasteurised) products but not keg or smallpack, forms a haze and surface film |
| Acetic acid bacteria | *Gluconobacter* | *G. oxydans* | Pleomorphic, some motile with flagella, 0.6–0.8 µm × 1–4 µm, catalase positive, strict aerobes, oxidises ethanol to acetic acid | Specific to aerobic or microaerophillic environments, i.e. dispense and draught products but not keg or smallpack, forms a haze, surface film and viscous 'ropiness' |
| Enterobacteriaceae ('coliforms') | See below | See below | Typically rods, 0.3–1 µm × 1–6 µm, facultatively anaerobic, diversity of fermentation products (organic acids, butanediol, phenolics), generally sensitive to pH (<4.4) and ethanol (>2%) | Generally (but not exclusively) limited to wort and early fermentation, most common is *O. proteus* – which (within the Enterobacteriacae) best survives fermentation |
| Enterobacteriaceae ('coliforms') | *Obesumbacterium* (syn. *Hafnia*) | *O. proteus* (*H. protea*) | Catalase positive, specific to brewery environments, 'short fat rod' (0.8–1.2 × 1.5–4 µm), ethanol tolerant (<6%), found in yeast heads and slurries and consequent danger of recycling contamination | Grows in wort and during early fermentation, produces sulphur off-products (DMS), inhibits fermentation rate and results in high beer pH, reduces nitrate to nitrite leading to the formation of ATNCs |
| Enterobacteriaceae ('coliforms') | *Citrobacter* | *C. freundii* | Catalase positive, straight rod (1 × 2–6 µm) | Occasional contaminant of pitched wort, can accelerate fermentation, produces organic acids and DMS, does not survive fermentation |
| Enterobacteriaceae ('coliforms') | *Enterobacter* (syn. *Rahnella*) | *E. agglomerans* (*R. aquatilis*)<br>*E. cloacae* | Rods (0.6 × 2–3 µm) | Similar to *O. proteus* in associating with yeast and in surviving fermentation conditions, forms diacetyl and DMS, increases initial fermentation rate, final pH higher |

| | | | |
|---|---|---|---|
| Enterobacteriaceae ('coliforms') | Klebsiella | K. terrigena<br>K. aerogenes<br>K. pneumoniae | Rods (0.3–1 × 0.6–6 µm) | Forms phenolic off-flavours (4-vinylguaiacol) like some wild yeasts and DMS |
| | Zymomonas | Z. mobilis | Rods (1–1.4 × 2–6 µm), anaerobic but tolerate oxygen, catalase positive, ferment glucose efficiently to ethanol, do not ferment maltose, very ethanol tolerant, grows best at 25–30°C | Specific to ale fermentations/breweries (temperature) and glucose primed beers, off-flavours include hydrogen sulphide and acetaldehyde |
| | Pectinatus | P. cerevisiiphilus | Obligately anaerobic (consequently difficult to recover using conventional microbiological methods), curved rods (0.8 × 2–30 µm – elongated in old cultures) | Oxygen sensitivity restricts risk to low oxygen conditions in process or in package, off-flavours include hydrogen sulphide (and other sulphur compounds), acetaldehyde, propionic and other weak acids, grows best at elevated pH (4.5–6) with growth being weaker at pH 3.7–4 |
| | Megasphaera | M. cerevisiae | Obligately anaerobic, cocci (1.0–1.2 µm in diameter), catalase negative, difficult to detect | Oxygen sensitivity generally restricts risk to low oxygen conditions in process or in package, off-flavours (described as 'foul', 'putrid' with a 'faecal' aroma) include hydrogen sulphide, butyric and other short-chain fatty acids, cannot grow at pH <4.1 or ethanol >3.5–5.5% |

**Fig. 8.3** Electron micrograph of *Acetobacter* species (kindly provided by Bill Simpson of Cara Technology).

**Fig. 8.4** Electron micrograph of *Zymomonas mobilis* (kindly provided by Bill Simpson of Cara Technology).

**Fig. 8.5** Electron micrograph of *Obesumbacterium proteus* (kindly provided by Bill Simpson of Cara Technology).

Section 8.2.2) of pitching yeast and scrupulous attention to process hygiene. *Megasphaera* and *Pectinatus* (Fig. 8.6) present a more contemporary threat to product stability. Both are relatively 'new', having been first described as brewing contaminants in the 1970s. However, the threat is growing with contamination by these organisms having been reported in Germany, USA, Scandinavia, Japan and France. *Megasphaera* (Haikara & Lounatmaa, 1987) and *Pectinatus* (Haikara *et al*., 1981) are both 'strict' anaerobes and the by-products of their metabolism are notably noxious. The threat from these organisms is increasing with the worldwide drive to reduce in-process and in-package oxygen concentrations to improve the flavour stability of beer. There is also evidence that *Pectinatus cerevisiiphilus* can survive the early aerobic stages of fermentation and that yeast provides protection to oxygen (Chowdhury *et al*., 1995). This is exacerbated by the difficulties in recovering these organisms via conventional microbiological methods such as membrane filtration. Consequently, less than adequate methods, such as protracted 'forcing' tests, have been used for the routine but retrospective detection of these organisms (Haikara, 1985). There is currently no evidence that *Escherichia coli* and other Gram-negative pathogens can grow in beer (see Section 8.1.5.3). However, it is prudent to maintain a 'watching brief' on developments in the wider food industry, particularly the emergence of acid tolerant strains (Jordan *et al*., 1999).

**Fig. 8.6** Electron micrograph of *Pectinatus cerevisiphilus* (kindly provided by Bill Simpson of Cara Technology).

8.1.2.2 *Gram-positive bacteria*. The major Gram-positive bacteria found in breweries are summarised in Table 8.2. For an authoritative review, see Priest (1996) in *Brewing Microbiology*. For a wider appreciation of Gram-positive bacteria see *Bergey's Manual of Determinative Bacteriology* (Holt *et al*., 1994)

Of the microflora found in the brewery, the Gram-positive lactic acid bacteria are the most feared. In addition to being potent beer spoilers, the lactic acid bacteria have a reputation for being 'difficult' in terms of detection, recovery from spoilt product and typing. These concerns reflect the nutritional fastidiousness of these bacteria and their variable response to the anti-microbial effects of hop iso-$\alpha$-acids. The complex chemistry of the myriad of these hop compounds is outside the scope of this chapter (for a review see Stevens, 1987). The major bittering (and antimicrobial) substances in beer include isohumulone, isocohumulone and isoadhumulone and their *cis* and *trans*

**Table 8.2** Gram-positive bacteria.

| Family | Genus | Major species | Characteristics | Beer spoilage |
|---|---|---|---|---|
| Lactic acid bacteria | Lactobacillus | L. brevis<br>L. damnosus<br>L. casei<br>L. fermentum<br>L. buchneri<br>L. delbrueckii<br>L. lindneri | Aerotolerant anaerobes, catalase negative (but novel protection to reactive oxygen species), 0.6–1.2 × 1–15 μm, metabolism is either homofermentative (main product is lactic acid) or heterofermentative (products include lactic acid, acetic acid, ethanol, and carbon dioxide | Potent beer spoiler growing optimally at pH 4–5, forms 'silky' turbidity, acidity, occasionally viscous 'ropiness' and – notably – diacetyl, impact on beer in process and in package<br>Variable sensitivity to hop iso-α-acids – beer spoilers 10× more resistant than sensitive strains |
| Lactic acid bacteria | Pediococcus | P. damnosus (syn. P. cerevisiae) (almost exclusively in breweries)<br>P. inopinatus | Aerotolerant anaerobes catalase negative (but occasional novel protection to reactive oxygen species), cocci, 0.7 μm in diameter, occur as tetrads, homofermentative | Resistant to hop iso-α-acids and ethanol (<10%, v/v), impact on beer in process and in package<br>Spoilage through diacetyl production ('sarcina sickness') acidity and haze |
| | Bacillus | B. coagulans | Aerobic, catalase positive rods that produce resistant spores, thermophillic | Sensitive to hop iso-α-acids, not beer spoilers, found in hot brewing liquor and sweet wort, form lactic acid in wort at 55–70°C, implicated in the formation of ATNCs in sweet wort |
| | Micrococcus | M. kristinae<br>M. varians | Catalase positive, typically strict aerobes although M. kristinae is a facultative anaerobe, cocci | Widely distributed in breweries, can survive in beer but rarely cause spoilage, M. kristinae reported to spoil products ('fruity' aroma) with high pH, low bitterness |

isomers. Generally, Gram-positive bacteria are sensitive to these isomerised hop acids (see Table 8.2) and, accordingly cannot grow in hopped beers. However as noted in Section 8.1.1, strains of *Lactobacillus* (Fig. 8.7) and *Pediococcus* (Fig. 8.8) able to spoil beer are significantly more resistant to these acids. Work by Simpson and Fernandez (1992) showed great variation in the sensitivity of a selection of Gram negative bacteria to one of the major hop acids, *trans*-isohumulone. The minimum inhibitory concentration (MIC) of *trans*-isohumulone required to inhibit the growth of 42 *Lactobacillus* strains ranged from < 20 μM (19% of strains) through 20–40 μM (45%) to > 80 μM (24%). Those that could resist > 80 μM (120–180 μM) *trans*-isohumulone 'had invariably been isolated from spoilt beer' (Simpson & Fernandez, 1992). Building on this, Simpson and Fernandez (1994) classified beer spoilage lactic acid bacteria as having a MIC of 90–200 μM whereas sensitive, non-spoilage organisms had a MIC of 10–40 μM. To put these results into perspective, the concentration of iso-α-acids in 'normal' UK commercial beers is in the range of 60–70 μM, which is equivalent to 20–25 mg/L of bitterness. However, as would be anticipated for weak acids, the MIC for iso-α-acids varies with pH. Simpson (1993) has reported that the antibacterial activity of *trans*-isohumulone decreases 800-fold as

**Fig. 8.7** Electron micrograph of *Lactobacillus brevis* (kindly provided by Bill Simpson of Cara Technology).

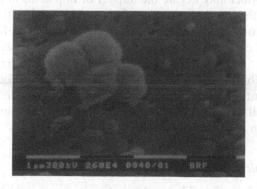

**Fig. 8.8** Electron micrograph of *Pediococcus damnosus* (kindly provided by Bill Simpson of Cara Technology.

the pH is increased from 4 to 7. From a practical perspective a 'change in pH of as little as 0.2 can reduce the protective effect of hop compounds by as much as 50%' (Simpson, 1993).

Although many questions remain to be answered, the work of Simpson and colleagues has provided a solution to the practical issue of whether or not an isolate is a beer spoiler. Typically, lactic acid bacteria isolated from beer will not grow when the colony is transferred to beer. This is unsatisfactory, as the spoilage status of the isolate is not clear. Simpson and Fernandez (1992) resolved this issue by demonstrating that hop-resistant strains could be trained to grow in beer by pregrowth in media containing trans-isohumulone (45 μM). Similarly, it is well known that spoilage lactic acid bacteria can be encouraged to eventually grow in beer by subculture into media containing an increasing ratio of beer. Conversely, hop-sensitive organisms pregrown in the presence of non-inhibitory concentrations (8 μM) could not grow in beer. Although undeniably useful, such approaches are inevitably slow and cannot be applied in routine microbiological testing. A more practical application is in making microbiological media more selective for hop-resistant bacteria. Simpson and Hammond (1991) reported that the inclusion of 20 μM trans-isohumulone in MRS media suppressed the growth of a sensitive Lactobacillus but enabled the growth of a resistant strain (Simpson & Hammond, 1991).

It is noteworthy that hop-sensitive and hop-resistant lactic acid bacteria are indistinguishable from each other in terms of morphology, physiology and metabolism (Fernandez & Simpson, 1993). A promising molecular explanation for hop resistance has been reported by Sami et al. (1997b). The presence or absence of a plasmid gene horA correlates strongly with, respectively, hop resistance or sensitivity. Of 61 strains, which were horA-positive strains, 59 could grow in beer. Conversely, only one of 34 horA-negative strains could grow in beer. Using PCR (see Section 4.2.6), beer spoilage capability (horA-PCR positive) can be assessed in about six hours.

Although not yet watertight, the observations of Sami et al. (1997b) may eventually explain the molecular basis of hop resistance in lactic acid bacteria. Certainly the observation (Sami et al., 1997a) that the horA gene found in hop-resistant strains of L. brevis has homology with multidrug resistance proteins in mammals and Lactococcus adds weight to the argument proposed by Sami et al. (1997b). Although speculative, the horA gene product may negate the impact of hop acids in resistant cells. A physiological spin on the metabolic impact of iso-α-acids on sensitive strains comes from the work of Simpson and his colleagues. In summary (see Fig. 8.9), isohumulones act as ionophores and catalyse the transport of ions across the bacterial membrane. In his Cambridge Prize review, Simpson (1993) has shown the addition of trans-isohumulone to a hop-sensitive strain to result in the import of protons into the cell with a concomitant reduction in intracellular pH. As maintenance of a proton gradient across the cell membrane is necessary for nutrient transport, such circumstances result in starvation and, ultimately, cell death.

### 8.1.3　Spoilage micro-organisms – yeast

A popular definition of 'wild' yeasts in the brewing industry is 'any yeast not deliberately used and under full control' (Gilliland, 1971a). The definition of wild

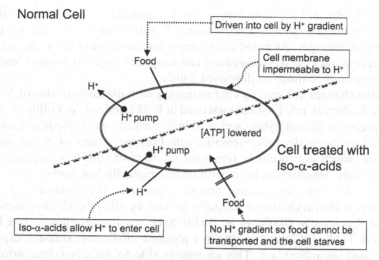

**Fig. 8.9** Effect of iso-α-acids on the physiology of sensitive bacterial cells (after Simpson, 1993).

yeasts is diffuse and, for convenience, is traditionally divided into (i) *Saccharomyces* and (ii) non-*Saccharomyces*. Irrespective of classification, wild yeast contamination of process and product can be a major cause for concern. This reflects their diversity, difficulty of detection and, for *Saccharomyces* wild yeast, similarity to pitching yeast. These issues are compounded by the robustness of wild yeast which is usually equivalent (or exceeds) that of primary yeasts. It is salient to note that the panacea of acid washing is not effective with wild yeasts and control must be achieved via attention to hygienic operations. It is worth noting that 'contamination' need not be overt, characterised by strong off flavours and turbidity. More subtle contamination by other brewing strains can perturb process and product. Excellent reviews on wild yeasts in brewing are to be found in Rainbow (1981) and Campbell (1996) in *Brewing Microbiology*. Other useful but more broad-based reviews are to be found in The Yeasts (Thomas, 1993) and the *Handbook of Food Spoilage Yeasts* (Deak & Beuchat, 1996).

**8.1.3.1  Saccharomyces *wild yeast*.**   These days, the *Saccharomyces* wild yeasts are regarded as more hazardous than the heterogeneous grouping of the non-*Saccharomyces* wild yeasts. As noted elsewhere (Section 4.1.3), of the 75 yeast genera tested by Visser *et al.* (1990), only 23% were able to grow anaerobically and, of these, *S. cerevisiae* stood out as being capable of robust growth in the absence of oxygen. The opportunities during the brewing process and in final package for contamination and growth by facultative anaerobes like the *Saccharomyces* are clearly more significant than for exclusively aerobic non-*Saccharomyces* yeasts. Indeed, the opportunity for the later class is diminishing further with the worldwide focus on tightening dissolved oxygen specifications and decline in more aerobic packaging formats such as cask beer in the UK.

Given the taxonomic consolidation described elsewhere (Section 4.1.1), it is no surprise that typical brewery *Saccharomyces* spoilage yeasts (Wiles, 1953) such as *S.*

*turbidans, S. ellipsoideus, S. willianus* and *S. diastaticus* are now reclassified as *S. cerevisiae*. Spoilage strains of *S. bayanus* and *S. pastorianus* have remained unscathed by taxonomic changes. As noted above, cross contamination of an ale fermentation with a lager strain (i.e. *S. pastorianus*) can trigger changes in process and product, and, consequently, cannot be dismissed lightly.

Dramatic changes are seen by contamination with diastatic strains of *S. cerevisiae* (formerly *S. diastaticus*). First characterised in 1952 (Andrews & Gilliland, 1952), this yeast expresses a glucoamylase and is able to ferment wort oligosaccharides ('dextrins') that are normally unfermentable with brewing strains of *S. cerevisiae*. Contamination, which can occur in fermenter or downstream, has a major impact on product quality, resulting in beers with (i) an atypically low present gravity, a phenomenon called 'superattenuation' and (ii) a phenolic, medicinal aroma. This later observation is characteristic of diastatic as well as other 'wild' (but non-diastatic) strains of *S. cerevisiae* (Ryder *et al.*, 1978). Such yeasts contain an active POF gene (Goodey & Tubb, 1982) that results in a phenolic off-flavour through expression of phenolic acid decarboxylase. This enzyme is able to decarboxylate wort phenolic acids such as ferulic and cinnamic acid to, respectively, 4-vinyl guaiacol (clove/spicy aroma) and styrene (medicinal aroma). It is of interest to note that brewing strains which do not express POF contain a non-functional POF gene (Meaden & Taylor, 1991; Daly *et al.*, 1997). Not surprisingly, the expression of glucoamylase and POF underpins laboratory detection of these yeasts (see Section 8.3.3.2). The genetics and physiology of diastatic strains of *S. cerevisiae* have been subject to widespread study, as the glucoamylase from this yeast was an early candidate for genetic manipulation into brewing strains for the production of 'light' beers (see Section 4.2.4).

Another more threatening class of wild yeasts are the 'killer' yeasts. These pose a 'greater threat than does the presence of other wild yeasts, since they not only compete for substrate but actively kill the indigenous brewing strain' (Hammond & Eckersley, 1984). These yeasts produce exotoxins or 'zymocins' (to which they are immune) that kill susceptible cells belonging to the same species. The *Saccharomyces* killer yeasts are by far the best understood although other distinctly different yeast systems have been described in *Kluyveromyces*, *Pichia* and *Williopsis*. Excellent reviews are to be found in Young (1987), Magliani *et al.* (1997) and Walker (1998).

*Saccharomyces* killer strains have been classified into three main groups – K1, K2 and K28 – according to the toxins they secrete. All are unusual in that the genetic basis is not chromosomal but is coded for by different cytoplasmically inherited double stranded RNA plasmids. Although different, K1 and K2 toxins both bind to β-1,6-glucan receptors in the cell wall and then penetrate the cytoplasmic membrane. These events cause the cell to become 'leaky' such that ions are lost and the cell dies. Conversely, the mode of action of K28 involves binding to the α-1,3-mannose residues in the wall mannoprotein which leads to cell cycle arrest through inhibition of DNA synthesis and the non-separation of mother and daughter cells.

The ecology of the killer yeast phenomenon has been studied primarily in yeast communities such as decaying fruits and in slime growth in trees. The occurrence of killer yeasts in brewing has had a mixed press. In a major survey of 964 yeasts (28 genera) in the National Collection of Yeast Cultures (see Section 4.2.3.1), Philliskirk and Young (1975) found 59 killer strains. The vast majority (38) were of the genus

*Saccharomyces*, of which 27 strains were laboratory haploids, rather than industrial yeasts. No more than four of the killer yeasts were connected with brewing. This is in keeping with the limited number of reports of *Saccharomyces* killer yeasts in brewery fermentations. The first (Maule & Thomas, 1973) described the occasional colonisation of a production-scale two-stage stirred continuous fermenter of the type described by Bishop (1970) (see Section 5.6.2.2). So effective was the killer yeast that on reaching 2–3% of the total yeast mass, the primary yeast viability fell from greater than 95% to less than 20%. From the perspective of *Saccharomyces* wild yeast, it is noteworthy that the beer acquired a 'herbal/phenolic' taint, consistent with the expression of POF as described above. This problem was eliminated, as ever, by effective cleaning and sterilisation. Similarly, the other report of killer yeasts in production brewing (Taylor & Kirsop, 1979) described the contaminated beer as having an unpleasant phenolic aroma. Unlike the earlier report, this example of *Saccharomyces* wild yeast was isolated from a batch fermentation. Not surprisingly, the heavy flocculence of the killer yeast facilitated its transfer between conical fermenters, whereas – in the same brewery – traditional vessels cropped via skimming were not colonised.

The introduction of killer factor into brewing strains was an early candidate in the various 'strain improvement' programmes (see Section 4.2.4) of applied yeast geneticists in the early 1980s (Young, 1981; Hammond & Eckersley, 1984). Simplistically, such developments are of value in managing out the threat of contamination by other primary yeasts or wild *Saccharomyces* strains. The then technical complexity coupled with today's concerns and reticence over genetic modification suggest that contamination is best controlled – as ever – by attention to hygienic practices and CiP.

8.1.3.2  *Non*-Saccharomyces *wild yeast.*   The collection of yeasts grouped together as the 'non-*Saccharomyces* wild yeasts' is notable for its complexity and ambiguity. Indeed, it is perhaps this, together with the frustration of identification, that has hampered the detailed study of the numerous genera of yeasts that are included in the non-*Saccharomyces*. As noted above, this diverse collective is considered to pose less of a general threat to product quality than the *Saccharomyces* wild yeast. Simplistically, the non-*Saccharomyces* wild yeast are less well equipped to spoil beer than the *Saccharomyces* strains. Indeed, 'head to head' there is no contest! The few comparative reports (Visser *et al.*, 1990; Campbell & Msongo, 1991) show, as would be expected, that the *Saccharomyces* yeasts are better adapted to the environment of the brewing process and product. As a class, the non-*Saccharomyces* yeasts grow, at best, poorly under anaerobic conditions and frequently are either unable to ferment sugars or ferment a reduced spectrum of sugars. Despite such a damning comparison, the non-*Saccharomyces* yeasts found in the process or product includes a wide diversity of genera (see Table 8.3). Of those found in breweries, *Candida* and *Pichia* species predominate (Wiles, 1953; Hall, 1971; van der Aa Kuhle & Jespersen, 1998). Although frequently endemic in a brewery or process, in isolation such yeasts are unable to spoil packaged beer in the absence of oxygen. These organisms are typically removed through pasteurisation or sterile filtration. Although they are no threat to product quality, such yeasts are a major irritation inasmuch that they are detected

**Table 8.3** Non- *Saccharomyces* wild yeast.

| Genus | Major species | Characteristics* | Beer spoilage |
|---|---|---|---|
| *Brettanomyces* | *B. anomalus* *B. bruxellensis* *B. lambicus* | Teleomorph (similar morphology, physiology) of *Dekkera*, oxygen stimulates fermentation, ferment glucose, rarely maltose but not sucrose, produce acetic acid, nitrate reducing | Notable for causing off-flavour in bottle conditioned beers, succeed *Saccharomyces* in the spontaneous fermentation of wort (lambic and gueuze) |
| *Candida* | *B. tropicalis* *C. boidinii* *C. vini* (formally *C. mycodema*) | Fermentation typically limited to glucose (*C. tropicalis* can ferment maltose) | Infection limited to the initial 'aerobic' phase of fermentation or unpasteurised draught beers, reports that some strains can grow poorly anaerobically |
| *Cryptococcus* | *C. laurentii* | Unable to ferment but can assimilate a wide range of sugars, some strains produce pigments | Can be found in beer in process or in package, survives but does not spoil |
| *Dekkera* | *D. bruxellensis* *D. hansenii* | Teleomorph of *Brettanomyces* forming ascospores, oxygen stimulates fermentation, ferment glucose, sucrose and maltose (strain variable) | Spoils unpasteurised draught beer |
| *Kluyveromyces* | *K. marxianus* | Glucose is fermented vigorously, other wort sugars are variable as is lactose fermentation, thermotolerant (growth 37-43°C) | Spoils soft drinks, fruit juices and high-sugar products, common contaminant of dairy products |
| *Pichia* (including *Hansenula*) | *P. anomala* *P. fermantans* *P. membranifaciens* | Fermentation usually limited to glucose | Infection limited to the initial 'aerobic' phase of fermentation, can spoil unpasteurised draught beer, forms haze and surface films, *P. membranifaciens* reported to give a sauerkraut flavour |
| *Rhodotorula* | *R. glutinis* *R. mucilginosa* | Some strains are pigmented red, unable to ferment but can assimilate a wide range of sugars | Water borne, found in pitching yeast, can survive but not spoil beer |
| *Torulaspora* | *T. delbrueckii* | Ferments glucose and variably maltose and sucrose, phenotypically close to *Saccharomyces* | Pitching yeast contaminant, can spoil unpasteurised draught beer, capable of poor anaerobic growth |
| *Zygosaccharomyces* | *Z. bailii* *Z. rouxii* | Ferments glucose and variably maltose and sucrose, notably osmophillic | Infamous spoilage organism of soft drinks, fruit juices and high-sugar products |

* The sheer diversity and complexity of cell size and shape together with appearance on plates or in liquid culture undermine the value of any attempt at a generalised description of each genus. The interested reader should consult Kurtzman and Fell (1998) in *The Yeasts, A Taxonomic Study* for specific details.

through routine microbiological monitoring. Irrespective of their limited threat, such yeasts are a valuable indication of the status of the hygiene of the process. Accordingly, steps should be taken to identify the source of such yeasts so that they can be eradicated from the process and the product. After all, it is important to recognise that the presence of non-threatening contaminant yeasts may be the 'tip of the iceberg' that hides the presence of other, potential beer spoilage organisms.

Some non-*Saccharomyces* wild yeasts are typically 'niche' contaminants that come into their own under specific conditions. For example, *Brettanomyces* species which are 'key players' in Belgian beers such as lambic and gueuze (Van Oevelen *et al.*, 1977) are contaminants of bottle conditioned beers (Gilliland, 1961). Here, these slow growing yeasts are described as producing flavours varyingly described as 'harsh', 'mawkish' and 'old beer flavour' (Gilliland, 1961). Unlike the *Saccharomyces*, the *Brettanomyces* are more 'fermentative' under aerobic conditions than anaerobically.

A broader example of niche contaminants is the colonisation of traditional draught 'cask' beers in the UK. Here, it is usual practice for air to be drawn into the container as it is emptied, with the result that any aerobic yeasts that are present can grow and, over time, spoil the beer. The sheer diversity of contaminant wild yeasts found in cask beers was reported in the bad old days of the 1950s when returns 'could be disastrously high on occasions' (Harper *et al.*, 1980). In a survey of various public houses in London and Surrey, Hemmons (1954) analysed 41 samples ex-dispense of 19 different brands. Although *Saccharomyces* wild yeast predominated, aerobic, film-forming yeasts such as *Pichia membranifaciens* and *Candida mycoderma* (now *C. vini*) were also prevalent. A survey some 26 years later (Harper *et al.*, 1980) found a similar spectrum of wild yeasts in dispensed cask beers. This is not surprising, as the origin of contaminant organisms in cask beers is unlikely to have changed in the 1950s, '80s or, indeed, today! Although contaminant yeasts can already be present in the product in low numbers it is more likely that they are introduced via the air, poor cellar hygiene or 'grow back' from dispense lines. Arguably today, cask beer spoilage is less of an issue with improvements in product hygiene, cellar management, the use of inert gas blankets and reduced time on dispense. A further consideration is the commercial decline in cask beer which has resulted in the 'selection' of those 'who look after the beer' continuing to sell these more demanding products.

Similarly, the microbiology of keg beer in the bar and public house can be of concern. Indeed, it is both ironic and disappointing that keg beer leaves the brewery 'commercially sterile' only to become contaminated to a greater or lesser degree on dispense. Remarkably, given its importance in delivering high-quality product, there have been few published reports that focus on the microbiology of keg beers in the trade. Like cask beer, keg beer is subject to varying degrees of contamination by a variety of bacteria and yeasts. In an extensive survey of over 600 samples of keg beers, Harper *et al.* (1980) reported the presence of wild *Saccharomyces* strains together with *Hansenula*, *Pichia*, *Torulaspora* and less commonly *Brettanomyces*, *Debaromyces* and *Kloeckera*. It is noteworthy that Harper *et al.* (1980) reported that 'in general, levels of both yeasts and bacteria were considerably lower than in the case of cask beer'. Given that keg beers, unlike cask conditioned beers, are pasteurised or sterile filtered this observation is both reassuring and no great surprise.

### 8.1.4 *Sources*

Hammond *et al.* (1999) commented that it 'must be realised that the brewing process is not aseptic and the occasional chance contaminant will often be encountered'. Although for some perhaps a rude awakening, this is a helpful, practical and, frankly, realistic comment. The reality is that packaged beer is, at best, 'commercially sterile'. Although this is more than acceptable in terms of product quality and robustness, it does mean that beer can realistically contain relatively low levels of contaminants. However, in this event the loading is such that they are typically undetectable in routine sampling and testing and, most importantly, do not pose a threat during the shelf-life of the product. Consequently, the focus of brewing microbiology has been 'calibrated' to routinely deliver a realistic low loading of micro-organism such that the shelf-life of the packaged product is not compromised. Accordingly, the process is managed to minimise the threat of significant microbial entry and contamination.

Typically this is achieved by minimising the ingress of organisms (closed vessels), cold storage of beer in process and removal from surfaces by robust cleaning (CiP) operations. Operationally, microbial loading is reduced via process steps such as wort boiling, acid washing, filtration, pasteurisation and/or sterile filtration (see Section 8.2).

#### 8.1.4.1 *Biofilms.*

Biofilms have been described as 'the real enemy of process and product hygiene' in brewing (Quain, 1999). Since the 1970s, it has become increasingly obvious that in the 'real world' micro-organisms attach themselves to surfaces and form quasi 'multicellular' mixed communities. These 'biofilms' or 'biosponge' provide both protection and nutrients for micro-organisms in what are typically hostile unwelcoming environments. Biofilms are everywhere! Classic biofilms include dental plaque, slimes in water pipes and drains and biofouling in cooling towers (cf. Legionnaires' disease) and heat exchangers. The growing realisation that biofilms colonise food preparation surfaces and medical implants and catheters has triggered a step change in the profile and financial support for work in this area. Further fuel has been provided by the report from the Centre for Disease Control and Prevention in the USA that 65% of human bacterial infections involve biofilms. A review of the role of biofilms in human infections has been published by Costerton *et al.* (1999). Indeed, it is a sobering thought that it has been estimated that 99% of the planet's bacteria live in biofilms (Coghlan, 1996). By contrast, the free flowing 'planktonic' bacteria so typical of laboratory studies represent only a very small fraction of the microbial community. Seemingly, the decision to enter the biofilm mode of growth is determined by the availability of an external carbon source (Pratt & Kolter, 1999). If the concentration is low, the biofilm option is taken up as, presumably, being better for foraging low levels of nutrients. Conversely, the dispersal of unattached cells and sheared fragments of biofilms act as the 'advance party' to colonise other areas in the environment.

Numerous reviews on biofilms have been published (see Keevil *et al.*, 1995; Coghlan, 1996; Costerton *et al.*, 1995; 1999; Stickler, 1999). This interest in biofilms has spawned some rich descriptions that aid the visualisation of these three-dimensional structures. Of particular note was the article by Coghlan (1996) in the weekly UK popular science magazine *New Scientist*. He reported biofilms as being 'slime cities' resembling 'skyscrapers of ghostly spheres piled one on top of the other' and, most memorably, as looking 'like Manhattan when you fly over it'. Such descriptions infer, quite correctly, that biofilms are organised, arranged structures. Figure 8.10 shows a schematic biofilm consisting of a basal layer 5–10 μm thick covered with pillar and mushroom shaped stacks or towers of microcolonies that rise 100–200 μm above the attached surface. The basal layer consists of a microbially derived extra-polysaccharide (EPS) matrix that aids attachment and provides protection for encapsulated microbes. The chemistry of the EPSs found in biofilms varies greatly – so much so that their stickiness varies from 'velcro' to 'superglue' (Sutherland, 1997). The availability of nutrients, which are transported through channels in the biofilm, determines the density of the 'skyscrapers'. Although an idealised description, biofilms are composed of a complex consortia of microorganisms that provide niche environments for pure and mixed communities. This is key to the success of biofilms, as through pooling metabolic resources, nutrients are ruthlessly utilised. In such an environment, the metabolic by-products of one community are the food of another.

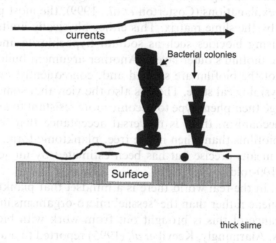

**Fig. 8.10** Schematic diagram of a biofilm (after Keevil *et al.*, 1995 and Coghlan, 1996).

To add further complexity and weight to the view that biofilms are 'rudimentary organs' (Keevil *et al.*, 1995), there is persuasive evidence for cell-to-cell communication within these communities. Signal peptide molecules of the acyl homoserine lactone (AHL) family have been shown to accumulate in growing planktonic cultures of over 30 species of Gram-negative organisms. This 'quorum sensing' (Bassler, 1999) enables bacteria to sense population numbers and to coordinate gene expression of the entire population in direct response to the accumulation of AHLs. Although it is thought that quorum sensing provides a selective advantage in regulating the virulence of pathogenic bacteria, such signalling also plays an important role in biofilm development (Heys *et al.*, 1997). Davies *et al.* (1998) demonstrated that in *Pseudomonas aeruginosa*, 3-oxododecanoyl homoserine lactone ($3OC_{12}$-HSL) is required for biofilm differentiation but not attachment. A mutant unable to secrete this extra-cellular signal produced thin, undifferentiated biofilms, which were sensitive to dispersion by a weak detergent. The addition of $3OC_{12}$-HSL to the mutant restored the formation of normal, biocide resistant biofilms.

Not surprisingly, there has been much interest – particularly in medicine – in prevention of biofilm attachment to surfaces. A popular route has been to incorporate biocides into the substrate so that they are intrinsically resistant to microbial colonisation (Stickler, 1997). Approaches have included coating the surface with silver, copper and incorporation of antibiotics. Indeed, incorporation of the biocide triclosan into materials has been exploited commercially by inclusion into a wide variety of bathroom products, kitchen utensils and household goods (Quain, 1999). Other approaches include physico-chemical modification of surfaces such that they are more hydrophilic (Stickler, 1997) or the use of 'smart' polymers that respond to the environment (Ista et al., 1999). A more recent concept is to interfere with cell-to-cell communication so as to exploit the above observations so that biofilms become more susceptible to removal (Davies et al., 1998).

From our perspective, the real issue is that biofilms are significantly more resistant to removal and death through treatment with detergents and biocides. Although there are a number of explanations (Costerton et al., 1999), the most powerful is the protection afforded by the slime matrix. This either physically limits penetration or, in the case of oxidising biocides such as sodium hypochlorite and peracetic acid, is deactivated in the biofilm's outer layers. Another argument builds on the likelihood that some parts of the biofilm are starved and, consequently, are in a more robust slow growing physiological state. There is also the view that some cells when present in a biofilm change their phenotype to become more resistant to antimicrobial agents. Whatever the mechanism, there is universal acceptance that micro-organisms are more robust in biofilms than when in the free, planktonic form. The degree of protection is by no means precise but has been estimated by numerous authors to be between 10 and 100-fold.

Unfortunately, in the real world there is a mindset that planktonic organisms are the enemy of hygiene rather than the 'sessile' micro-organisms in a biofilm. The real practical significance of this is brought out from work with biofilms in water distribution systems. Alarmingly, Keevil et al. (1995) reported that a free planktonic cell concentration of $10–10^3 \, ml^{-1}$ is equivalent to an attached cell population of $10^5–10^7 \, cm^{-2}$. This observation is of particular interest in the brewing industry where the hygienic status of closed vessels and mains is typically determined by measurement of the microbial loading in the residual rinse liquor. Although undeniably significant in validation of CiP operations, in many respects this insight is unfortunate! The perception that brewing microbiology is less than an exact science and is prone to ambiguity is further undermined by the microbiological loading of rinses effectively representing the tip of the microbial iceberg!

Explicit reports in the brewing press on the importance of biofilms to process and product hygiene have been limited to publications stemming from Diversey (Czechowski & Banner, 1992; Banner, 1994) and from VTT in Finland (Storgårds et al., 1997b; 1999a–c). These groundbreaking studies have clearly shown that archetypal brewery contaminants are adept at forming biofilms on surface materials found in breweries. For example, Czechowski and Banner (1992) reported the ready attachment of *E. agglomerans*, an *Acetobacter* species (Table 8.1) and *L. brevis* (Table 8.2) to stainless steel and other materials. Similarly, Storgårds *et al.* (1997a) demonstrated that 11 out of 20 bacterial species (*Acetobacter* species, *Gluconono-bacter oxydans*, *L. lindneri*, *E. aggolerans*) were capable of forming biofilms on stainless steel. All of the yeasts studied (a diastatic strain of *S. cerevisiae*, *P. membranifaciens* and *B. anomalus*) were found to attach to stainless steel and were classified as 'strong biofilm producers'.

The theme of attachment to some of the different surfaces found in the brewing process was investigated by Storgårds *et al.* (1999a–c), some of whose images are presented in Fig. 8.11. Using a model closed circulating test rig, new and 'aged' coupons of polymeric materials found in gaskets and valves were compared to stainless steel in terms of supporting biofilm formation and 'cleanability' post CiP. With new materials (Storgårds *et al.*, 1999a) both *E. agglomerans* and *P. inopinatus* (Table 8.2) were found to attach readily to EPDM (ethylene propylene diene monomer rubber), PTFE (polytetrafluoroethylene) and Viton (fluoroelastomer). Presumably because of its bacteriostatic properties, new NBR (nitrile butyl rubber or

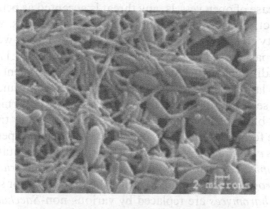

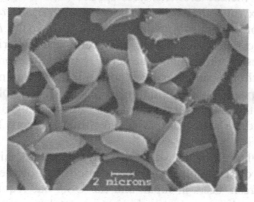

Fig. 8.11 Images of biofilms of yeast/bacteria (kindly provided by Erna Storgårds, VTT, Helsinki, Finland).

Buna-N) was found to support little biofilm growth. In terms of hot CiP, the biofilms were more readily removed from NBR and PTFE than EPDM and Viton.

Outside of the brewery, the ability of micro-organisms to adhere to surfaces within draught beer dispense systems has long been recognised. In work that remains pertinent today, Harper (1981) described the presence of a variety of yeast genera in dispensed beer and noted the attachment of the same yeasts to the lumen surface of dispense pipes. Indeed, without naming them, the characteristics and threats of biofilms were succinctly described in a previous paper by Harper *et al.* (1980). Firstly, they noted that 'many of these organisms produce sticky substances enabling them to adhere readily'. The second point, which reinforces the threat and significance of biofilms, noted that 'if the sanitiser fails to penetrate their accumulations and fails to scour them from the walls, then cleaning is only going to be partially effective'.

The challenge of monitoring and removing attached organisms from surfaces is discussed elsewhere in this chapter.

### 8.1.4.2 *Air and process gases.*

Air hygiene has attracted little attention in the brewing industry. This presumably reflects the view that it is of little importance in brewery hygiene. Either there is no risk as the process is contained within closed vessels or, in the case of open vessels, any threat is accepted as being minor. Although in principle correct, air is a potent source of micro-organisms, which in a brewery, have the capability to spoil or distort product quality. This is well demonstrated by the spoilage of unpasteurised cask beers in cellars (Section 8.1.4.1) and the production of lambic and traditional French ciders by spontaneous fermentation.

The distinctive lambic beers are produced in breweries around the town of Lembeek in Belgium (see Section 2.2). Here hot wort is cooled overnight in shallow open trays whilst air is blown across the surface of the liquid. After transfer into wooden casks, mixed flora fermentation occurs over a two to three year period (van Oevelen *et al.*, 1977; Martens *et al.*, 1991). Although the bulk of the fermentation is performed by *S. cerevisiae, K. apiculata* (Table 8.3) and various *Enterobacteriaceae* (Table 8.1) (*E. cloacae* and *K. aerogenes*) predominate during the first month or so. After about eight months, the *Saccharomyces* are replaced by various non-*Saccharomyces* wild yeasts (Table 8.3). These include *Brettanomyces* species (*B. bruxellensis* and *B. lambicus*) and, to a lesser extent, other wild yeasts (*Candida, Cryptococcus, Torulopsis* and *Pichia*). In terms of bacteria, *Pediococcus* species (principally *P. damnosus* syn. *cerevisiae*) succeed the *Enterobacteriaceae*, and acetic acid bacteria, although less welcome, can be present throughout the latter part of the process. Similarly, the contribution of airborne contaminants is critical in the spontaneous fermentation of apple juice, in the production of traditional French ciders. Laplace *et al.* (1998) have clearly demonstrated the importance of the 'surrounding air' (and biofilms on 'utensils') in providing lactic acid bacteria for the malolactic fermentation in cider making.

The one area of brewing in which the threats of poor air hygiene have been recognised is that of sterile filtration and aseptic filling of beer. The growth of the non-pasteurised 'fresh' beers in Japan and the USA has required a step change in attitude and approach to product and environmental hygiene. Indeed Ryder *et al.* (1994) have described the need for a 'sterile envelope' to enclose the processing of sterile filtered beer and subsequent filling operations. The importance of air hygiene around the bottle or can filler is reinforced by the prescriptive standards of air conditioning in this area. Ryder *et al.* (1994) describe the use of high efficiency particulate air (HEPA) filters capable of removing 99.97% of all particles $\geq 0.2$ μm to deliver air to these areas. In addition to air, the 'sanitary mindset' (Ryder *et al.*, 1994) requires the hygienic management of *all* surfaces in the filler area together with infeeds such as conveyors, bottles and cans.

Although the threat of poor air hygiene in aseptic filling operations is a given, two publications from VTT in Finland (Henriksson & Haikara, 1991; Haikara & Henriksson, 1992) have sought to quantify the microbial risk. Using a portable 'SAS' device that can sample up to 180 litres of air in a minute, the microbial loading in the air was determined (in the autumn and spring) within the bottling halls of ten Finnish breweries. Analysis of anaerobic, mainly lactic acid bacteria in the air (see Table 8.4) identified the bottle filler as being a particular microbial hotspot. Of note was the observation that the airborne loading in breweries experiencing microbiological problems was three times that of other breweries without such concerns. Airborne yeasts were numerically lower (Table 8.4) and the loading was not influenced by season, unlike the anaerobes which were 2–6 times higher in the autumn than in the spring. As would be expected, this survey demonstrated that high airborne counts were associated with environments that typically had higher temperatures and humidity.

**Table 8.4**  Air hygiene in filling area (after Henriksson & Haikara, 1991).

| Sample area | Anaerobic bacteria | | | | Yeast | | | |
|---|---|---|---|---|---|---|---|---|
|  | Autumn mean | Range (cfu/m³) | Spring mean | Range (cfu/m³) | Autumn mean | Range (cfu/m³) | Spring mean | Range (cfu/m³) |
| Incoming air | 50 | < 10–280 | 50 | < 10–410 | 20 | < 10–150 | 30 | < 10–230 |
| Filler | 560 | < 10–3900 | 240 | < 10–1500 | 110 | < 10–780 | 90 | < 10–1300 |
| Outgoing air | 210 | < 10–1100 | 90 | < 10–380 | 40 | < 10–310 | 20 | < 10–100 |

In a subsequent report (Haikara & Henriksson, 1992), the level of anaerobes at the filler was noted as being indicative of the general air hygiene in the bottling hall. Not surprisingly, the filler air hygiene was reduced by improving the cleaning and disinfection regimes in this area. Further, the rich threat from the growth of microorganisms in beer-contaminated conveyor lubricant soaps was dramatically reduced by switching to synthetic lubricants containing biocides. Intriguingly, this study showed that the strictly anaerobic *P. cerevisiiphilus* (see Table 8.1) could be isolated from air around the filler.

Comparable but hitherto unpublished observations (Andrew Jones and David Quain) from a cask (unpasteurised) beer filling hall, suggests a similar airborne loading to those reported in Table 8.4. Here, the results (in July) ranged from 250 (cask washing) through 400 (filling) to 600 cfu m³ (conveyor leaving the hall). As with the Finnish observations, bacterial loading was greater than airborne yeasts. However, an airborne 'phenolic' wild yeast (*P. membranifaciens* – see Table 8.3) was clearly implicated in a previous 'trade problem' in cask beer from this brewery.

Inevitably, the sources of micro-organisms in the air are many and various! The loading of airborne micro-organisms can be assumed to reflect a series of equilibria between the availability of nutrients and appropriate micro-organisms able to exploit these nutrients together with distribution via movement of air. Therefore, not surprisingly, air hygiene reflects the general environment. Consequently, in a brewery, spillages of nutrient-rich soups, such as wort, beer and sugar syrups will act as nuclei for colonisation of settling airborne micro-organisms that can exploit such sources of nutrients. As they dry out these 'puddles of contamination' reseed the air with micro-organisms, which go on to contaminate fresh spillages of nutrient-rich liquids.

Arguably, one of the most underestimated sources of airborne micro-organisms is the delivery of malt in the brewery. In busy breweries, this regular event punctuates the day with plumes of malt dust released into the air. As malt is laden with a variety of micro-organisms (Table 8.5), it would be anticipated that this dust is equally rich in bacteria and yeasts. Surely it is more than a coincidence that the micro-organisms recovered from barley and malt (O'Sullivan *et al.*, 1999) are also typical microflora of beer. For example, the Gram-positive lactic acid bacteria found on malt include *L. plantarum*, *L. fermentum*, *L. brevis* and *L. buchneri* (Table 8.2). Similarly, the Gram-negative bacteria (Table 8.1) included *Enterobacter agglomerans*, *Klebsiella pneumoniae*, *Rahnella aquitilis* and *Citrobacter freundii*. Yeasts found on barley and malt (for a review see Flannigan, 1996) include such familiar genera as *Rhodotorula*, *Hansenula*, *Candida* and *Torulopsis*.

**Table 8.5** Microbiology of kilned and screened malt (cfu g malt$^{-1}$).

| Maltings | Aerobic bacteria | Lactic acid bacteria | Pseudomonads | Coliforms | Yeasts/ moulds | Source |
|---|---|---|---|---|---|---|
| Modern pneumatic (batch size 150 tonne) | $1.7 \times 10^7$ | $1.7 \times 10^5$ | $1.7 \times 10^5$ | $7.5 \times 10^5$ | $2.7 \times 10^4$ | O'Sullivan *et al.* (1999) |
| Traditional floor (batch size 30 tonne) | $5.0 \times 10^5$ | $8.0 \times 10^4$ | $9.0 \times 10^3$ | $2.0 \times 10^4$ | $3.0 \times 10^4$ | O'Sullivan *et al.* (1999) |
| Not reported | $5.5 \times 10^6$ | $5.7 \times 10^4$ | – | – | $1.9 \times 10^4$ | Flannigan (1996) |

From the perspective of product hygiene, the direct role of malt microflora is minimal. O'Sullivan *et al.* (1999) showed the loading of viable organisms to decline dramatically during mashing such that only very low numbers of lactic acid bacteria survive to the copper, where they are obviously killed! Whether malt microflora play a bigger role in brewery hygiene remains to be clarified. However, it is intriguing to speculate that every malt delivery effectively 'inoculates' the wider environment with micro-organisms that can survive alongside yeast, and spoil wort or beer. Although only a hypothesis, it is timely for a study, using molecular fingerprinting methods that can track strains of micro-organisms across the process, to resolve this fascinating concept. If malt microflora are implicated, as would seem likely, as being a threat to brewery hygiene, it will be necessary to identify routes to minimising the generation and distribution of malt dust during delivery and handling.

Although yet to be studied in brewery environment, the movement and distribution of air would be anticipated to play an important role in exacerbating or minimising the threat of poor air hygiene. Certainly, in the wider food industry, mapping airflows around process areas is increasingly recognised as a vehicle for understanding and minimising the microbial threat of airborne micro-organisms. It is noteworthy that airflow patterns can be reconfigured by demolition or the erection of buildings, installation and removal of vessels. In passing, such activities (particularly demolition) can be responsible for the generation and release of micro-organisms. Consequently, it is good practice to reduce any threat to air hygiene by partitioning and damping-down building works.

Although the significance of air hygiene is easily dismissed, particularly in breweries with closed vessels, it should be included in any consideration of 'total hygiene'. The threat is not limited to the high profile, more obvious threat of sterile filling operations but extends brewery-wide. Routes for the ingress and distribution of airborne micro-organisms into the general environment of the brewery should be understood and, where necessary controlled. Points of entry (such as manway doors) into 'closed' vessels should be identified, and managed to control the threat of entry.

Unlike air hygiene which, in the vast majority of breweries, does not feature in the microbiological sampling and testing plan, monitoring the microbiology of process gases is part of the normal QA plan (see Section 8.3.1.2). Process gases, such as nitrogen and carbon dioxide, have long been recognised as posing a threat through the colonisation of gas mains and subsequent transfer into product.

Although such gases are used across the process to minimise and control dissolved oxygen, the major threat is to bright, comparatively 'clean' beer prior to and including packaging. The most common concern is the practice of gas washing in bright beer tanks to reduce dissolved oxygen or to correct carbon dioxide and/or nitrogen concentration. Here, the duration of gas washing coupled with the vulnerability of bright beer make this process particularly susceptible to contamination by 'slugs' of contaminated gas. Similarly, microbiological contamination of process gases is a particular threat to product quality during packaging into bottles, cans and kegs.

Microbiological contamination of process gas is typically caused by beer finding its way into the gas lines. This can result in the colonisation of the gas main by anaerobic organisms, which by definition will be capable of beer spoilage. In terms of HACCP (see Section 8.2), management of process gases to remove or minimise the risk is particularly important. Indeed, process gas quality is a 'critical control point' where the risk is managed by regular cleaning and steam sterilisation. As a second line of defence, process gases are filtered at the point of use to prevent contamination. For this to be effective, the filters themselves must be periodically sterilised and subject to a maintenance schedule of inspection and replacement.

### 8.1.5 *Food safety*

In his Laurence Bishop Memorial Lecture, Long (1999) reviewed food safety issues that have challenged or are challenging the brewing industry. Two drivers are identified that have heightened consumer awareness of food safety. First, the consumer has been 'sensitised' through scares such as BSE in beef products, benzene in carbonated drinks, *E. coli* contamination of inadequately cooked foods and Salmonella in eggs. Second, analytical capability/detection is developing more rapidly than the associated understanding of the toxicological impact of the identified compounds. Indeed, Long (1999) notes that the focus should move to understanding the 'actual balance of toxic and protective components in the food matrix rather than to assess the toxicology of components in isolation'. Thankfully, from a microbiological perspective, there have been relatively few food safety concerns in the brewing industry. The most notable of these is the role of the yeast contaminant *O. proteus* in the formation of non-volatile nitrosamines in beer.

#### 8.1.5.1 *Nitrosamines.*

In the late 1970s, N-nitrosamines were found in cured meat and malt beverages (for a review see Smith, 1994). This was of great concern as nitrosamines are powerful carcinogens in animal systems and there is 'compelling' but indirect evidence of involvement in human cancers (Long, 1999). Consequently, the brewing industry rapidly sought to understand the mechanism of formation of nitrosamines and to eliminate the risk. Control of the *volatile* N-nitrosamines was achieved through changes in the malt kilning process, whereas the *non-volatile* nitrosamines (apparent total N-nitroso compounds – ATNC) necessitated improvements in pitching yeast hygiene.

Work in the late 1980s primarily at the Brewing Research Foundation (now Brewing Research International) unravelled the pathway of ATNC formation during brewery fermentation. In short (Fig. 8.12), nitrate is reduced by *O. proteus* to nitrite which reacts with wort amines to produce ATNC. As nitrate (from liquor or hops) is inevitably present in worts, the only control strategy is to eliminate *O. proteus* from the equation through improved brewery hygiene and/or acid washing (see Sections 7.3.3 and 8.2.2).

*O. proteus* is a Gram-negative member of the Enterobacteriaceae (see Table 8.1)

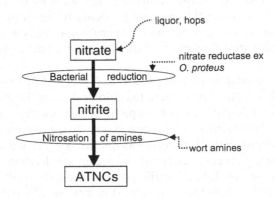

**Fig. 8.12** Formation of ATNCs during fermentation.

which, according to *Bergey's Manual of Determinative Bacteriology* (Holt *et al.*, 1994) is closely related to the *Escherichia* genus. Although found in cooled wort, it is often described as the 'short fat rod of pitching yeast' (van Vuuren, 1996) with which it co-sediments and is recycled. Growth of *O. proteus* is encouraged by the reduction of nitrate to nitrite, which inhibits yeast growth and slows fermentation. Numerically, contamination of pitching yeast with greater than 0.03% *O. proteus* results in ATNC exceeding the Brewers' Society (now Brewers and Licensed Retailers Association – BLRA) recommended limit of 20 $\mu$g l$^{-1}$ (Calderbank & Hammond, 1989).

The quantitative interrelationships between nitrate, nitrite and ATNC were demonstrated by Calderbank and Hammond (1989). Figure 8.13 graphically shows the reduction of nitrate, appearance of nitrite and ATNC during fermentation. Conversion of nitrite to ATNC is by no means quantitative, such that for every 10 mg l$^{-1}$ nitrate reduced about 90 $\mu$g l$^{-1}$ ATNC is formed. This is believed to reflect the loss of nitrite (as nitrous acid) through gas stripping (Simpson *et al.*, 1988) and pH dependency of N-nitrosation of amines (Smith, 1994).

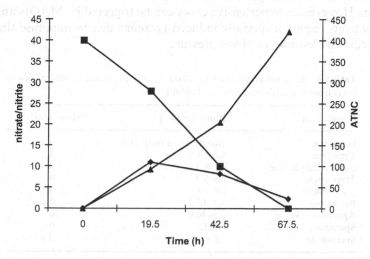

**Fig. 8.13** Formation of nitrite (♦) and ATNC (▲) in nitrate (■) supplemented laboratory fermentations contaminated with *O. proteus*. Concentration of nitrate/nitrite in mg l$^{-1}$ and ATNC in $\mu$g l$^{-1}$. Data from Calderbank and Hammond (1989).

An alternative, less well-recognised route to the formation of ATNC involves Gram-positive *Bacillus* species (Smith, 1994). As they are sensitive to hop acids, these contaminants are typically found in the brewhouse in sweet wort and hot brewing liquor. One, *B. coagulans*, has been shown to produce ATNCs in hot wort (30–70°C) at a rate which is ten times faster than *O. proteus* (Smith *et al.*, 1992). These organisms are notable for their thermotolerance and 'slimy', glutinous appearance which makes them excellent candidates for involvement in surface biofilms (see Section 8.1.4.1).

Smith also investigated the possibility that brewery wild yeasts might provide a further route for ATNC formation (Smith, 1992). Although capable of nitrate reduction in model experiments, ammonia and amino acids were shown to repress this reaction. In brewery practice, contaminating yeasts such as *Hansenula anomala* and *Rhodotorula glutinis* are unable to reduce nitrate and consequently are unlikely to provide a further microbial route to ATNC in beer.

8.1.5.2 *Biogenic amines.* Elevated levels of biogenic amines in cheese, meat and fish products have been linked with a variety of undesirable physiological reactions. Specifically, high levels of histamine are associated with 'histaminic intoxication' and tyramine has been linked with food induced migraines. The major biogenic amines found in beers and their typical concentrations are reported in Table 8.6. Although considered 'too low to produce direct toxicological effects' (Izquierdo-Pulido *et al.*, 1996b), there have been sporadic reports of toxic reactions to beer consumption which have implicated biogenic amines, specifically tyramine. In particular, severe hypertensive events after the consumption of 'modest' amounts of draught beer have been described for two individuals in Canada being treated with monoamine oxidase inhibitors (MAOI) (Tailor *et al.*, 1993; Shulman *et al.*, 1997). Treatment with MAOIs is important in the management of a variety of clinical conditions including depression. However, as hypertensive crises can be triggered by MAOIs and biogenic amines, patients require a specially reduced tyramine diet to minimise the threat of MAOI mediated elevation in blood pressure.

**Table 8.6** Biogenic amines in beer. Data for 195 European canned or bottle beers from Izquierdo-Pulido *et al.* (1996b).

| Biogenic amine | Range (mg l$^{-1}$) | Mean (mg l$^{-1}$) |
| --- | --- | --- |
| Histamine | Not detected (nd)–21.6 | 0.7 |
| Tyramine | 0.6–67.5 | 3.2 |
| β-phenylethylamine | nd–8.3 | 0.4 |
| Tryptamine | nd–5.4 | 0 |
| Cadaverine | nd–39.9 | 0.5 |
| Putrescine | 1.5–15.2 | 4.4 |
| Agmatine | 0.5–40.9 | 10 |
| Spermine | nd–3.9 | 0 |
| Spermidine | nd–6.8 | 0.5 |

Whether beer is best avoided by patients receiving treatment with MOAI has been subject to some debate (Tailor *et al.*, 1993; Izquierdo-Pulido *et al.*, 1996b; Shulman *et al.*, 1997). Although the consensus is to minimise the risk by susceptible individuals not consuming beer, it is clear that this view is driven by the unpredictability of biogenic amine concentration in some beers. For example, in the study of Tailor *et al.* (1994) canned and bottled beers were universally low in tyramine (0–3.2 mg l$^{-1}$), whereas four of 49 draught beers had elevated 'dangerous' levels of tyramine (26–133 mg l$^{-1}$). Similarly, a survey (see Table 8.4) of biogenic amines in 195 European canned or bottled beers showed the widest fluctuations in the 'indicator' amines – tyramine, histamine and cadaverine (Izquierdo-Pulido *et al.*, 1996b). The fermentation style is also important with the more complex, less controlled approaches (lambic, gueze, kriek and Wiessbier) tending toward elevated levels of these amines. Collectively, these observations are in keeping with the view that elevated levels of these amines reflect either unwanted or intentional microbial contamination or complexity. However, it is bacteria not yeast that have been implicated in the formation of biogenic amines through the decarboxylation of amino acids present in wort and beer. The presence of lactic acid bacteria, specifically *Pediococcus* (Izquierdo-Pulido *et al.*, 1996a) and *Lactobacillus* species (Donhauser *et al.*, 1993), has been shown to correlate with elevated levels of tyramine and, to a lesser extent, other biogenic amines. However, it remains unclear whether poor hygiene in the brewery or in trade dispense is the major factor in determining the level of biogenic amines in beer. Although Donhauser *et al.* (1993) implicate wort spoilage in the brewery, it might be argued that the more potent threat originates from poor or infrequent line cleaning and associated unhygienic operations in trade outlets – a view, which is in agreement with the survey of Tailor *et al.* (1993) that elevated levels of tyramine in beer were exclusive to draught products.

8.1.5.3 *Pathogens.* It is generally accepted that pathogenic bacteria cannot grow in beer. Indeed, the whole issue is somewhat academic, as food pathogens have not been associated with either raw materials or the brewing process. Two studies (Bunker, 1955; Sheth *et al.*, 1988) have inoculated beer with low (*c.* 200 ml$^{-1}$ and high (2–20 $\times$ $10^6$ ml$^{-1}$) concentrations of enteric Gram-negative pathogens (*Shigella, Salmonella* and *Escherichia coli*). Irrespective of challenge, all three pathogens were unable to grow in beer, and over time died. Sheth *et al.* (1988), showed that for *E. coli*, an initial loading of 2 $\times$ $10^6$ ml$^{-1}$ was reduced to 5 cfu ml$^{-1}$ over 48 hours. Bunker (1955) reported similar kill kinetics for *E. coli* when four beers were challenged with 200 cfu ml$^{-1}$. However when challenged with 20 $\times$ $10^6$ ml$^{-1}$, *E. coli* could still be detected after 14, and in two cases 21, days.

The consensus is that a combination of low pH and ethanol protect alcoholic beverages from threat by pathogenic bacteria. Indeed, in the experiments reported by Sheth *et al.* (1998), *E. coli* was killed within 4 hours of inoculation into wine. However, the 'success' of pathogenic isolates of *E. coli* (e.g. O157:H7) has been attributed to their acid resistance such that survival at pH 3 has been reported (Jordan *et al.*, 1999). Fortunately, the presence of ethanol (5% v/v) overcomes the resistance of pathogenic *E. coli* to low pH resulting in rapid cell death.

Although not viewed as being a threat to the brewing industry, it is perhaps wise to maintain a 'watching brief' on developments in the wider area of pathogens in food safety. The appearance of more robust pathogens (e.g. acid resistant *E. coli* O157:H7) together with the well documented ability of *E. coli* to survive in beer suggest that it would be foolhardy to ignore a possible future threat to product safety.

## 8.2 Minimising the risk

Although previously noted (Section 8.1.4), the comment of Hammond *et al.* (1999) 'that the brewing process is not aseptic and the occasional chance contaminant will often be encountered' is worth reiterating. The industrial scale of the process coupled with the commercial expectations of product hygiene are such that the risk of spoilage is removed or significantly minimised by tried and tested routes.

Of these, by far the most significant vehicles for minimising risk are 'cleaning-in-place' (CiP) of vessels and mains together with downstream heat treatment (or perhaps sterile filtration) of final product. Typically, these operations are managed and 'made to happen' through quality systems that identify frequency and process validation. Outside the brewery in trade outlets, the threat of product spoilage is managed by good hygienic practices and regular 'line cleaning'.

### 8.2.1 *In the brewery*

The importance of cleaning to product wholesomeness has long been understood. Quite simply, poor hygienic practices result in poor product quality and, consequently poor sales. In other words, there is a commercial imperative to minimise the threat of microbiological contamination and product spoilage. If not a sufficient driver for good hygienic practice, legislative requirements have required breweries to adopt a significantly wider position to assuring product safety and wholesomeness (Mundy, 1997). For example, in Europe the EC Food Hygiene Directive has been implemented in the UK via the Food Safety Act which, in turn has led to an industry code by the Brewers and Licensed Retailers Association. Indeed the scope of such activities (see White, 1994) includes a number of 'big themes' that together minimise (or better still remove!) chemical, physical and microbiological hazards (Table 8.7).

A key platform in meeting the broader based needs of legislation and increasingly customer demand, is the use of HACCP, an acronym for **h**azard **a**nalysis and **c**ritical **c**ontrol **p**oints. The origins of HACCP go back to the 1960s and the 'space race' – the need to ensure safe food for astronauts in space. The philosophy of HACCP – then and now – is that of a preventative strategy based on an analysis of the prevailing

**Table 8.7** Rules of good hygienic practice in the brewing industry (after White, 1994).

| Theme | Focus on | Details include |
|---|---|---|
| Personnel | Motivation, supervision, selection, training and personal hygiene | Training in basic microbiology and hygiene<br>Training in the principles and theory of CiP<br>Housekeeping and pest control |
| Building environment | Design criteria for internal and external areas and surfaces | Drainage, lighting, ventilation<br>Building security, maintenance and cleanliness |
| Plant and equipment | Design criteria and materials of construction | Layout, operation, maintenance and cleanliness |
| Process control | Importance of a quality assurance system | Scope includes raw materials intake through storage to packaged product |
| Product | Importance that the product released to trade meets specification | Sampling and analysis plans, specifications |
| Legislation | Responsibilities of 'awareness' and 'compliance' to legislative requirements | Appropriate references |

conditions. HACCP takes a somewhat pessimistic view in looking at what could go wrong, causes and effect and possible control mechanisms. In essence, HACCP provides a systematic framework for formally identifying potential hazards to food safety. These are assessed in terms of 'risk' and whether they are 'critical' or not. Those that are assessed as being critical are labelled as 'critical control points'. Target specifications are established for each CCP, which are then monitored to ensure the CCPs are controlled. In the event of a CCP moving out of control, corrective or remedial action is taken to bring it back in line.

HACCP analysis of the brewing process enables the chemical, physical and microbiological hazards to be broken down into bite-sized chunks (White, 1994). This enables the identification of the 'key variables', associated control points and a raft of measures for controlling the risk. Although there are pros and cons (Brown, 1997a; Jackson, 1997), HACCP is typically integrated into the brewery quality management system (typically ISO 9000) (Kennedy & Hargreaves, 1997; Mundy, 1997). Indeed, as demonstrated by Kennedy and Hargreaves (1997), there are strong links between the two (Table 8.8).

**Table 8.8**  Relationship between HACCP and ISO 9000 Quality System (after Kennedy and Hargreaves, 1997).

| HACCP | ISO 9000 |
| --- | --- |
| Identify potential hazards, assess risk | Management responsibility; quality system; purchasing; process control; quality planning; design control |
| Determine CCPs | Contract review; document and data control; external legislation; process control; purchasing; design control |
| Establish limits for control | Contract review; inspection and testing; document and data control; legislation; design control |
| Establish system to monitor control | Quality planning; quality system; process control; inspection and testing; inspection and test status; inspection, measuring and test equipment; handling; storage |
| Establish corrective action | Management responsibility; corrective and preventative action, control of non-conforming product |
| Verification procedures, effective operation | Inspection and tasting; internal quality audits |
| Documentation and records | Document and data control; quality records |

As interpreted by the brewing industry, HACCP typically focuses on chemical and physical hazards. Although there are notable exceptions (Kennedy & Hargreaves, 1997), in terms of 'due diligence' and product safety, microbiological hazards are discounted because of the widespread belief that pathogens are not present (Section 8.1.5.3). Whatever, the 'microbiological hazard' is very real when viewed in terms of product spoilage. This commercial 'risk' is controlled through sampling and testing plans (Section 8.3.1.2) and, across the process, through CiP (Section 8.2.1.1) and downstream by pasteurisation and sterile filtration (Section 8.2.2.2).

**8.2.1.1** *CiP and other processes.* Given its importance and application across the brewing process, cleaning in place (CiP) or more generically 'cleaning' has had a bad press! As noted by Ball (1999), 'CiP is one of those basic subjects in our industry, which everyone is supposed to know about; too often people will not admit to only having a small amount of potentially prejudiced knowledge'. Although there are occasional articles in *The Brewer* (Barnett, 1991; Ball, 1999; Pickard, 1999) and *Brewers' Guardian* (Felstead, 1994), reviews on CiP in the brewing industry are few and far between. However, a good process-orientated view can be found in Celis and Dymond (1994), together with an exhaustive review of the chemistry of cleaning in *Brewing Microbiology* (Singh & Fisher, 1996).

Cords and Watson (1998) have identified three main drivers for cleaning and disinfection in the brewery. First, is the removal of unwanted 'soils' that could carryover and contaminate a subsequent batch of beer or wort. Second, and from the perspective of this chapter the most important, is removal of micro-organisms that could spoil or affect product consistency. Third, a factor that cannot be overlooked is the need to remove debris and soil from surfaces that impact on process performance such as wort paraflows and pasteurisers.

The nature and size of the challenge simplifies across the process from the complexity of brewhouse soils to the relative simplicity of beer prior to and during packaging (see Table 8.9). Overlaid on this is the impact of process temperature where, for example in brewhouse operations and wort cooling, soils are baked on to surfaces. Consequently, given the various demands of cleaning, there is no one universal CiP process in breweries. Indeed CiP is all about choice and choosing the best mix of options to meet the demands and constraints of cleaning. Table 8.10 identifies the major options of brewery CiP in terms of (i) choice of detergent (alkali v. acid), (ii) temperature (hot v. cold), (iii) recovery of detergent or single use and (iv) automation (manual v. automatic). Although guidelines exist for best practice, factors such as cost (capital v. revenue), complexity (small v. large brewery) and local preferences impact on choice of CiP operations. Indeed, given the evolution of cleaning practices over the last 25 years, it is common in many breweries for a mix of CiP philosophies to operate side by side.

Whatever the intricacies and choices, effective cleaning is achieved through the synergistic relationship between four parameters (Fig. 8.14) – time, temperature, chemical action and mechanical action. Manipulation of any one of these has a

**Table 8.9** The changing challenge of CiP (after Cords & Watson, 1998).

| Location | Soul composition |
|---|---|
| Brewhouse | Protein, starch, minerals, beerstone, hop materials, fermentable sugars |
| Wort cooler/paraflow | Protein, starch, minerals, beerstone, hop materials, fermentable sugars |
| Fermentation | Protein, non-fermentable sugars, minerals, beerstone, yeast – particularly at the top and bottom of vessels |
| Maturation/conditioning | Protein, non-fermentable sugars, minerals, beerstone, yeast |
| Bright beer tanks and packaging | Beerstone, foam components |

**Table 8.10** CiP – all about choice.

| Options | Guidelines and drivers | Comments |
|---|---|---|
| Alkaline v. acid | **Alkaline** (sodium hydroxide at 2–5% w/v) based detergents are accepted as offering superior cleaning and greater biocidal activity than acid-based approaches. Alkaline detergents are typically used in high-soil process areas (brewhouse, fermenter and conditioning/maturation).<br><br>**Acid** detergents are less favoured but find application in cleaning low-soil areas such as bright beer tanks, packaging lines. A variety of acids are used including phosphoric or, more effectively, a blend of phosphoric with nitric acid.<br><br>Acid detergents are less effective detergents than caustic-based approaches. Although effective against bacteria, acid detergents have little biocidal activity with yeast (cf. acid washing).<br><br>Routinely, acid detergents have been used to periodically remove 'scale' that accumulates over time with alkaline detergents. 'Hard' water scale originates from the precipitation of calcium carbonate or magnesium hydroxide under alkaline (detergent) conditions or at high temperature. Scale protects micro-organisms from detergents and disinfectants.<br><br>Depending on the application, both alkaline and acid detergents can include surface active agents or 'surfactants' to improve the 'wetting power'. Anionic surfactants are high foaming and are used in conveyor lubricants. Non-ionic are most commonly used across the process. Cationic surfactants find application in pasteuriser water treatment. For a discussion see Singh and Fisher (1996). | Czechowski and Banner (1992) have shown caustic-based detergents to be more effective against biofilms than acid-based approaches (which were equivalent to water).<br><br>The major disadvantage of caustic-based detergent is its reaction with carbon dioxide and, to a lesser extent, water hardness salts (calcium bicarbonate, magnesium sulphate and calcium sulphate). This later reaction is minimised by the inclusion of sequestrants (e.g. sodium gluconate) in caustic detergents (Singh & Fisher, 1996).<br><br>Sodium hydroxide is highly effective at mopping-up $CO_2$ to form sodium bicarbonate, which has no significant detergent or biocidal activity. Indeed, caustic cleaning in a high $CO_2$ environment can result in vessel collapse. Depending on the $CO_2$ concentration, losses of sodium hydroxide can be dramatic (Singh & Fisher, 1996). Unfortunately, as control of caustic concentration is typically through measurement of conductivity, the dilution of caustic by sodium bicarbonate is not detected and corrected by the CiP set (Barrett, 1991). Differentiation between them requires an offline titration. Gingell and Bruce (1998) report that 1 $m^3$ (or 10 hl) $CO_2$ at 1 atmosphere at 20°C will neutralise 2 kg sodium hydroxide (= 100 litres at 2% w/v). This reaction can only be minimised by removal of the $CO_2$ from the vessel, which if closed, can take many hours (Singh & Fisher, 1996). Practically, the options are to 'take the hit' and partially vent the vessel which together with the first rinse will reduce the $CO_2$ content or opt for a total dump (single use) approach, or replace with an acid detergent.<br><br>The advantages of acid-based cleaning are significant and include (i) not mopping up $CO_2$ which consequently is not 'diluted', (ii) vessels need not be vented and can be cleaned under a $CO_2$ top pressure, (iii) more easily rinsed and (iv) are used cold with consequent savings in utilities.<br><br>The poor performance of high-soil FVs and DPVs can be overcome by inclusion of an initial 'caustic pre-clean' (pre-water rinse) which removes yeast rings (Gingell & Bruce, 1988) |

*Contd*

**Table 8.10**   *Contd*

| Options | Guidelines and drivers | Comments |
|---|---|---|
| Hot v. cold | 'If the soil goes on hot, it should be cleaned hot' (Platt, 1986). However, high-soil areas such as yeast handling/storage plant are typically cleaned hot. Irrespective of detergent temperature, preclean rinses are invariably cold. Acid detergents are typically used at ambient (cold) temperatures. | Cleaning properties of detergents increases with temperature (see Fig. 8.15). Typical temperatures range from 50–90°C. There is an inevitable play-off between the improved efficacy (reduction in time and/or detergent strength) of high temperature against the cost of heating and maintaining detergent temperature. |
| Recovery v. total dump | **Full recovery** systems offer savings in water, heat, effluent and detergent usage. The downside is the unpredictable rate of detergent deterioration, which reduces cleaning efficacy and can lead to contamination by microbial colonisation of the recovered detergent.<br>**Total dump or single use** systems where rinse liquor and detergents are used once must provide the most effective CiP solution. However, such an approach is significantly more expensive in terms of materials and effluent.<br>Schematic drawings of a single-use, partial-recovery and full recovery CiP systems are presented in Figs 8.16, 8.17 and 8.18. | Single-use systems require to be close to achieve good mixing, with short mains to the tanks they are cleaning. By contrast, with recovery systems the CiP sets are remote from the tanks and the mains are long.<br>Total dump systems are typically used with caustic-based detergents operating in high $CO_2$ environments or in areas of gross soiling. Sodium hypochlorite can be included to enhance causticity (and therefore reduce working concentration) in freshly prepared alkaline detergents. However, care is required in temperature control and avoidance of acid (Ball, 1999). |
| Automated v. manual | **Manual systems** are resource hungry but are cheap and cheerful in terms of capital and maintenance. Process consistency is poor, as is safety. Manual systems are highly flexible.<br>**Automated systems** require little in terms of human resource but are capital hungry. Performance is typically good but they are inflexible with set routes, which can be complex to modify hardware and software. Safety is good. | Although less consistent, manual systems with greater human intervention, are the more 'assured'. Automatic systems, particularly those that are older or less sophisticated, have fewer 'checks and balances'. The temptation to simply press the button and initiate the clean can result in poor performance, which is assumed to have been acceptable. Periodic auditing of automatic systems is an important element of hygiene QA. |

variable impact on the CiP process. Simplistically, to compensate for a reduction in detergent strength, CiP duration (time) can be extended or the temperature increased. Similarly, an increase in detergent strength or temperature can be anticipated to reduce the required cycle time. Arguably, there is less flexibility in the contribution of 'mechanical action' to the cleaning matrix. It is either right or wrong! The velocity with which the detergent and rinses are delivered is critical to the physical removal of soil (and biofilms) from surfaces. To improve shear, flow in mains must be turbulent (rather than laminar) and, in vessels, via sprayballs (fixed low pressure or rotating high pressure) to achieve coverage (Singh & Fisher, 1996). Guidelines for flow rate for mains are 1.5–2.1 m sec$^{-1}$ (Singh & Fisher, 1996; Ball, 1999) and for sprayballs, 1.5–3.5 m$^3$ h$^{-1}$ for each metre of tank circumference. For tanks, which unlike mains

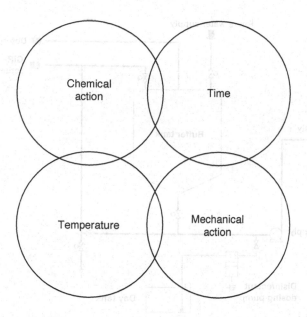

**Fig. 8.14** Key parameters in CiP.

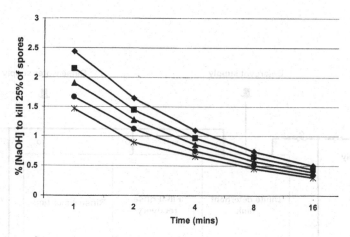

**Fig. 8.15** Impact of temperature on the concentration of sodium hydroxide (w/v) required to destroy 25% of the population of *B. subtilis* spores (from data reported by Singh and Fisher, 1996). Temperatures 49°C (♦), 54.5°C (■), 60°C (▲), 65.5°C (●) and 71°C (✳).

cannot be flooded during cleaning, it is critical that the sprayballs are configured and maintained (to avoid blockages) so that the entire vessel surface is cleaned. Where the desired flow rates cannot be achieved, the best compromise is to extend cycle time.

Inevitably, there is no such thing as a standard CiP cycle! There are, however, general principles that apply to the majority of CiP cycles (see Table 8.11). Fundamentally, CiP cycles invariably include an initial rinse, a detergent clean, and a rinse to remove detergent followed by a sterilising rinse or steam. Two factors that are

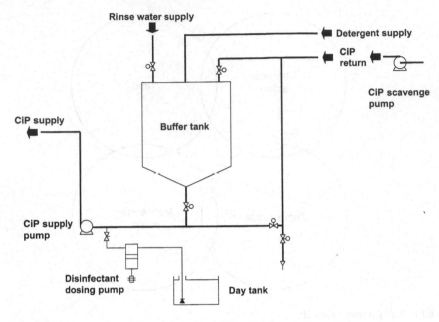

**Fig. 8.16**   Single use/total loss CiP system (redrawn from Singh and Fisher, 1996).

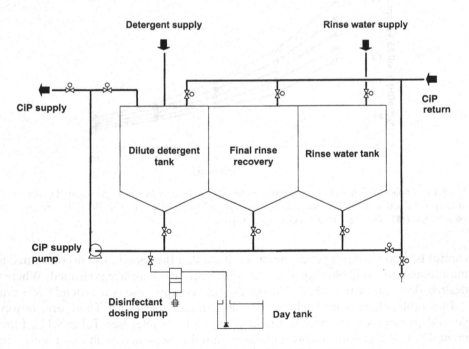

**Fig. 8.17**   Partial recovery CiP system (redrawn from Singh and Fisher, 1996).

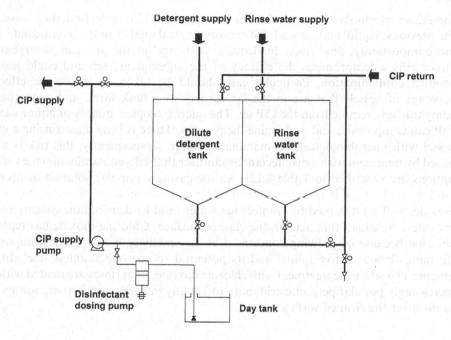

**Fig. 8.18**   Full recovery CiP system (redrawn from Singh and Fisher, 1996).

**Table 8.11**   Typical elements of a CiP cycle.

| CiP cycle step | Comments |
| --- | --- |
| Pre-rinse to drain | Critical that the main/vessel is empty! |
| | Critical to successful CiP and economy of detergent usage |
| | In recovery systems the pre-rinse is recycled from the recovered rinses from the previous CiP |
| | Pulsed or blast rinsing is frequently used in vessels to improve removal of soil |
| Drains/scavenge | Importance can be underestimated |
| | Critical to remove previous step prior to initiating next step |
| Detergent recirculation | Longest part of the cycle (30–60 minutes) – efficacy determined by detergent strength, time, temperature and flow |
| | In alkaline recovery systems detergent strength requires to be assured to account for contact with $CO_2$ |
| Drains/scavenge water rinse | In some alkaline systems, an acid 'neutralising' rinse can be included |
| | Objective is to remove detergent |
| | In recovery systems this rinse is recycled as the pre-rinse of the next CiP |
| Drain/scavenge final rinse | A critical step where the entire CiP operation can be nullified by the microbiological quality of the rinse liquor |
| | Terminal sterilents include peracetic acid, chlorine dioxide or sodium hypochlorite. The latter is being increasingly replaced because of corrosion and the threat of flavour active taints (chlorophenols and chloramines) |
| | Wet, anaerobic steam is used in some areas of the process (packaging lines and associated vessels) |

sometimes overlooked are critical to the success of any CiP cycle; first, the removal of the previous 'liquid' and, second, the microbiological quality of the intermediate and, more importantly, final rinse. Inadequate 'scavenge' of the preclean, detergent, or liquor rinses compromises the efficacy of the subsequent step and could lead to product contamination. Particular care should be taken to ensure the effective scavenge of vessels that are in the far corners of a tank farm, and consequently comparatively remote from the CiP set. The microbiological quality of liquor used in CiP can compromise and undermine the process. There is little sense rinsing a clean vessel with microbiologically contaminated water. Consequently, this risk is minimised by treatment with a disinfectant or sanitiser that kills contaminants (the various options are described in Table 8.12). An increasingly popular solution in brewing (and the wider food industry) is to treat water across a production site with chlorine dioxide. Sufficient is used to disinfect the water (and its distribution system) and to provide a 'residual' that acts on the cleaned surface. Chlorine dioxide has replaced chlorine because of growing concerns over its reactivity with organic compounds (forming flavour active taints) and its potential to corrode stainless steel. In the absence of a site-wide treatment with chlorine dioxide, final rinses are treated with the increasingly popular peracetic acid, both to kill any indigenous micro-organisms and to disinfect the cleaned surface.

**Table 8.12**  Disinfectants and sterilents used in brewing.

| Sterilent | Mode of use | Comments |
|---|---|---|
| Chlorine | Sodium hypochlorite (NaOCl) is typically used in breweries. Heavily influenced by pH. At alkaline pH (9–11) 97–100% is in the form of the hypochlorite ion (OCl$^-$) whereas at pH 4–6 >97% is as 'active chlorine' or hypochlorous acid (HOCl). Compared to the hypochlorite ion, HOCl is 20–100 times more active as a biocide. NaOCl is typically stored at pH 12 and used optimally for disinfection at pH 5. Working concentration between 50–300 mg l$^{-1}$. Oxidising agent. | Inexpensive, non-foaming but can corrode stainless steel at elevated concentrations. Reacts with many organic compounds to form trihalomethanes, which are potential carcinogens. Also forms flavour active compounds such as chlorophenol and chloramines. |
| Chlorine dioxide | Chlorine dioxide is used increasingly throughout the industry both to 'sweeten' liquor across a brewery and in CiP as a terminal disinfectant. As an unstable gas, chlorine dioxide is typically generated at point of use. A favoured approach involves the careful mixing of hydrochloric acid with sodium chlorite solution to produce 2% ClO$_2$ in water. ClO$_2$ does not react with water it behaves as a 'dissolved gas'. Used at 0.1–0.5 mg l$^{-1}$. Oxidising agent with potentially 2.5 × greater effect than chlorine. | The application of ClO$_2$ in the brewing industry has been reviewed by Cadwallader (1992). Less sensitive to pH than chlorine. Does not form trihalomethanes and chlorophenols. Strips biofilms from surfaces. ClO$_2$ must be contained. Although in widespread use there are potentially serious Health and Safety concerns which must be controlled/managed. |

*Contd*

**Table 8.12** *Contd*

| Sterilent | Mode of use | Comments |
|---|---|---|
| Peracetic acid (PAA) | Increasingly popular as a terminal disinfectant. Supplied as a liquid (5 or 15% PAA). In use concentration ranges from 75–300 mg l$^{-1}$ at ambient (or lower) temperature. Strong oxidising agent that decomposes to form acetic acid and free radical oxygen. Wide antimicrobial activity although PAA resistant (but non-spoilage) yeasts can be selected for (e.g. *Cryptococcus laurentii*). | Unpleasant smell, corrosive and highly reactive. Requires careful handling and storage (transitank and bunded). Storage tanks must be vented to atmosphere as oxygen is released over time. Effective against biofilms (Stickler, 1997). Contact with rust should be avoided as this accelerates decomposition. |
| Quarternary ammonium compound ('QACs') | Although effective products, QAC products are rarely used in CiP systems but continue to find application in soak and manual cleaning. Typically used at 200 mg l$^{-1}$. Concerns that QACs are high foaming, difficult to remove by rinsing and taints ('fishy' – Barrett, 1991). Less effective than oxidising disinfectants – resistant organisms can evolve, weak against Gram-negative bacteria. | Expensive. |
| Biguanides | Increasingly niche usage (soak, manual cleaning). Non-oxidising cationic polymers that reportedly disrupt micro-organisms via osmotic shock. Typical in use concentration *c.* 600 mg l$^{-1}$. Below pH 3, biguanides are ineffective and above pH 9 they are precipitated. | Safe. |
| Amphoterics | 'Traditional' user-friendly disinfectant used manually in soak and/or spray applications. Too high foaming for CiP. Depending on pH, ionise to produce cations, anions or zwitterions. Non-oxidising, typically used at *c.* 1000 mg l$^{-1}$. | Safe. |

On commissioning and periodically thereafter, CiP cycles should be reviewed and, where appropriate, optimised. Obviously, the success criteria include cleaning efficacy but also should include financial optimisation. CiP costs money, both directly in terms of utilities, such as CiP liquids, and indirectly in terms of time when the system is out of production. Consequently, cycle times should be considered from the perspective of time, utilities usage, detergent strength and temperature. Against this backdrop is the overriding measure of cleaning efficacy and system hygiene and microbiological performance. Clearly, there is little point in trimming a CiP cycle to reduce costs with an associated reduction in hygienic status.

Routine monitoring of CiP operations is a critical element of product and process hygiene. Despite the apparent logic of this relationship, it is fair to say that until relatively recently CiP processes frequently failed to receive the necessary process monitoring. Thankfully, things have changed! This is due in part to a greater legislative, business and systems emphasis on hygiene together with the development and industry take-up of real-time hygiene testing via ATP bioluminescence. This technology (see Section 8.3.3.1) has overcome one of the great obstacles to microbiology, historical information! Prior to the implementation of ATP bioluminescence, the hygienic status of vessels and mains was routinely assessed via conventional microbiological monitoring of surface swabs or final rinses. Unfortunately, because the nature of microbiological testing (Section 8.3.2), the results of such analyses always arrived two or three days after sampling. Such a lack of timeliness inevitably devalues the result, as it has no real process value other than tracking performance via trend analysis. Given this scenario it is no wonder that routine microbiological monitoring of CiP was viewed by many with little enthusiasm.

Real-time testing of the hygienic status of vessels and mains has facilitated a renewed commitment to validating CiP. As described in Section 8.3.3.1, ATP bioluminescence enables 'go/no go' decision making on the use of plant and enables the option of recleaning and/or further checks on CiP set-up. Although fundamentally a QC test, such real-time testing together with process checks provides a powerful vehicle for assuring CiP operations. Other checks fall into two camps – routine preclean and periodic audits. These are summarised in Table 8.13 and are reviewed by Hammond (1996). Routine preclean off-line tests include detergent strength (e.g. 'causticity' and 'carbonate') together with pH and, where appropriate, temperature. Visual checks on levels, leaks and pumps should also be included and driven by signed-off checklists. A further valuable step in the routine validation of CiP is the visual assessment of surface cleanliness post cleaning. This approach can often reveal the presence of cleaning 'shadows' that are not cleaned or missed through sprayball failure or blockage. A more sophisticated spin on this approach is to use a camera located in the top cover of the (cylindrical) vessel, which enables process events and the assessment of surface cleanliness post CiP. It is claimed that this 'TopScan' approach enables useful insight into CiP process optimisation (Wasmuht & Weinzart, 1999; see Section 6.3). Periodically the performance of CiP sets should subject to a more detailed audit where the focus is on flow rates, efficiency (microbiological and utilities) and capability. The frequency should be gauged by past history and the risk/sensitivity of the units being cleaned. Equally, depending on performance, the

**Table 8.13** Options for the monitoring of CiP operations.

| CiP cycle | Outputs |
| --- | --- |
| Sequence times | Final rinse microbial loading |
| Temperatures | Surface (swab) microbial loading |
| Flow rates | ATP bioluminescence of final rinse sample |
| Detergent – causticity and carbonate v. CiP set conductivity | ATP bioluminescence of surface (swab) sample |
| pH | Product microbial load |
| Terminal sterilent concentration | |

measure of hygiene (ATP bioluminescence) should be extended to include conventional microbiological tests to establish the nature of the microflora and associated product risk.

8.2.1.2 *Pasteurisation.* Described as a 'necessary evil' (Rader, 1979), heat treatment or pasteurisation is by far the most favoured process to minimise the microbiological risk to packaged product. It should be appreciated, however, that cask and bottle conditioned products are not pasteurised and some products ('draft' beers) are made microbiologically stable through a non-thermal process, sterile filtration (see Section 8.2.1.3). With these reservations in mind, pasteurisation is globally a critically important part of the brewing process. Accordingly for a fuller view of pasteurisation than can be presented here, the reader is directed to general reviews on thermal death of brewing organisms (O'Connor-Cox *et al.*, 1991a) and process (Huige *et al.*, 1989; O'Connor-Cox *et al.*, 1991b). The EBC *Manual of Good Practice on Beer Pasteurisation*, (European Brewery Convention, 1995) provides perhaps the definitive tome on pasteurisation with particular emphasis on engineering matters.

Ironically, although pasteurisation of beer is inevitably associated with Pasteur, the process was patented by him for the treatment of wine. Indeed, as reported by Anderson (1995), Pasteur, in his book *Etudes sur la Bière*, 'counsels against the pasteurisation of beer'! Although described in the EBC publication *Beer Pasteurisation, Manual of Good Practice* (European Brewery Convention, 1995) as a 'rather gentle heat treatment', pasteurisation does not aim for 'real' sterility. Rather pasteurisation aims to reduce microbial loading to such an extent that the product is 'commercially' sterile and, as such, is microbiologically stable. As ever, this process is a compromise between, in this case, time and temperature. The mix should achieve adequate microbial kill to assure the product's biological shelf-life but without thermal damage to product quality. To achieve both aims, the longheld view that pasteurisation provides a 'backstop' for poor upstream hygiene can no longer be acceptable. Increasingly, the commercial imperative is to reduce pasteurisation regimes and to tighten dissolved oxygen specifications to limit heat damage and to assure product freshness throughout the shelf-life.

Pasteurisation is practised in two formats, 'tunnel' and 'flash' (the latter occasionally known as 'plate'). The two processes meet different needs and differ fundamentally in duration and maximum temperature. Worldwide, tunnel pasteurisation is the most widely used to assure the commercial sterility of products packaged in can and bottle. Flash (or plate) pasteurisation is a bulk pasteurisation process used to treat beer prior to packaging in kegs or other containers that cannot be tunnel pasteurised.

Tunnel pasteurisation is a comparatively slow process through various 'zones' of fixed temperature (Fig. 8.19). The packaged product is heated in situ by spraying with increasingly hot water, held at a top temperature (e.g. 60°C) for a period of time (10–20 minutes) and then cooled progressively to 10–15°C. Like CiP, the process consumes large volumes of hot water and accordingly, tunnel pasteurisers are designed to recover heat by exploiting every opportunity for heat exchange. Residence times of bottles and cans (which being more conductive require less time) in tunnel pasteurisers is lengthy, in the order of 45 minutes (Dymond, 1992). Conversely, flash

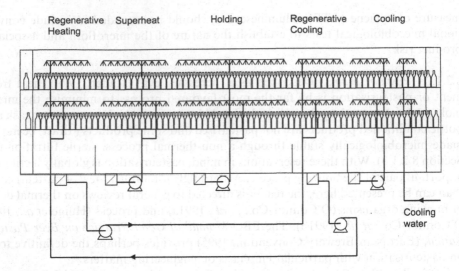

**Fig. 8.19** Tunnel pasteuriser (redrawn from *Beer Pasteurisation, Manual of Good Practice*, EBC Technology and Engineering Forum, 1995).

pasteurisation is a rapid, almost real-time process. Here, using a plate heat exchanger (Fig. 8.20) beer is raised from 2–4°C to *c.* 70°C for 20–30 seconds and then rapidly cooled back to process temperature. Depending on flow rate, total residence time is typically no more than 120 seconds (Dymond, 1992).

Typically, the shelf-life of products emanating from the two processes are distinctly different. Usually homesale largepack (keg) products have a shelf-life of six to eight weeks whereas typically smallpack (cans and bottles) products have 36–52 weeks. Arguably, such differences can be related to the degree of microbiological risk associated with tunnel and plate pasteurisation which, in turn, are benchmarked against sterile filtration in Fig. 8.21. Clearly, *in situ* tunnel pasteurisation is potentially

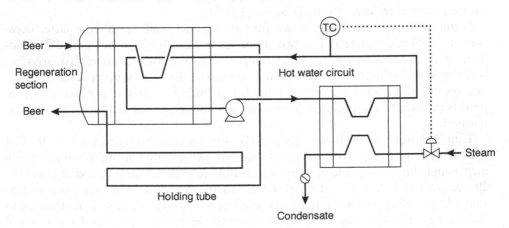

**Fig. 8.20** Flash or plate pasteuriser (redrawn from *Beer Pasteurisation, Manual of Good Practice*, EBC Technology and Engineering Forum, 1995).

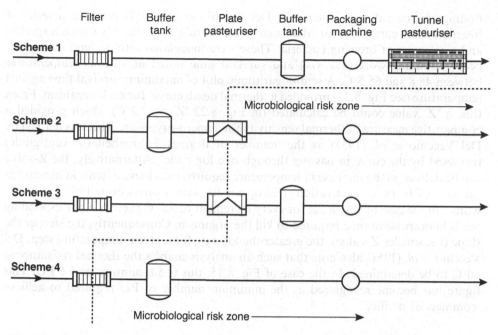

**Fig. 8.21** Comparative microbiological risks of tunnel pasteurisation (scheme 1), flash/plate pasteurisation immediately before prior to packaging (scheme 2), flash/plate pasteurisation into sterile tank prior to packaging (scheme 3) and sterile filtration (scheme 4) (redrawn from *Beer Pasteurisation, Manual of Good Practice*, EBC Technology and Engineering Forum, 1995).

the most effective route to achieving 'commercial sterility' as reinfection under normal process conditions is difficult to conceive. The risk is markedly greater with flash pasteurisation as heat-treated beer is transferred to a 'sterile' buffer tank and then to a packaging line and, typically, keg. In passing, it is noteworthy that Dymond (1992) has described the economic benefits of coupling flash pasteurisation with aseptic packaging into bottle. Whatever the final package, care must be taken to guard against the threat of plate cracking or seal degradation. These events can lead to contamination of pasteurised beer with unpasteurised beer or, more alarmingly from a public relations perspective, secondary refrigerant from the cooling section of the pasteuriser. However, both tunnel and flash pasteurisation are significantly more microbiologically robust than the alternative non-thermal process, sterile filtration. To be successful, this process (see Section 8.2.1.3) requires operating to extremely high hygienic standards to prevent the contamination by micro-organisms post-sterile filtration.

*Measurement of pasteurisation.*   Despite its universal use in the brewing industry, the pasteurisation unit (PU) – one minute of heating at 60°C – was proposed but never published by H.A. Benjamin of the American Can Company. Indeed, the paper in which this relationship was first reported (Del Vecchio *et al.*, 1951) remains, despite subsequent critical comment, one of the seminal papers of brewing microbiology. Accordingly, it is appropriate to consider the paper of Del Vecchio *et al.* (1951) and its findings in some detail. Essentially, Del Vecchio *et al.* (1951) reported a series of 'thermal death curves' for an 'abnormal' yeast, 'torula' (presumably *Candida* species) and a selection of brewing bacteria. These were inoculated into an 'end fermented' beer containing wort (5% v/v) and survival time noted at various temperatures between 48.8 and 65.5°C. A semi-logarithmic plot of maximum survival time against temperature (see Fig. 8.22) provided a 'thermal death curve' for each organism. From this, a 'Z' value could be calculated (in Fig. 8.22 'Z' = 7.2°C) which provided a comparative measure of thermal sensitivity for a micro-organism. This was defined by Del Vecchio *et al.* (1951) as the 'number of degrees Fahrenheit (or centigrade) traversed by the curve in passing through one log cycle'. Alternatively, the Z-value can be defined as the increase in temperature required to achieve a 'tenfold increase in the rate of thermal inactivation' (European Brewery Convention, 1995). In other words, in the case of Fig. 8.22, for every increment of 7.2°C there is a corresponding ten-fold decrease in time required to kill the organism. Consequently, the steeper the slope (i.e. smaller Z-value), the greater the kill rate for a given temperature step. Del Vecchio *et al.* (1951) also note that such an analysis enables the thermal resistance at 60°C to be determined. In the case of Fig. 8.15, this is 5.6 minutes or 5.6 PU. This figure has become recognised as the minimum number of PU required to achieve 'commercial sterility'.

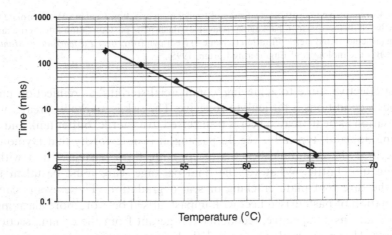

**Fig. 8.22** Example of a thermal death curve for a micro-organism at various temperatures (°C).

The continuing importance of the observations of Del Vecchio *et al.* (1951) is that the familiar pasteurisation tables are derived from the thermal death curve reported for the 'abnormal yeast'. These tables enable the calculation of PUs (or lethal rates) for other temperatures relative to 60°C. Presented graphically (Fig. 8.23), it can be seen that this a logarithmic relationship and, consequently, relatively small increases in temperature result in large increases in PUs. As noted in *Beer Pasteurisation Manual of Good Practice* (European Brewery Convention, 1995), an increase of 2 and 7°C increase the PUs by factors of approximately 2 and 10. These relationships (Fricker, 1984) are explored further in Fig. 8.24. PUs (or lethal rate) can be calculated

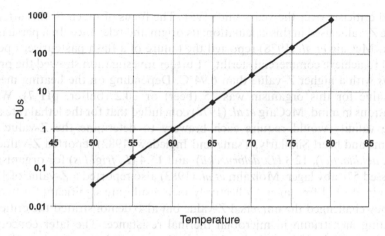

Fig. 8.23  Influence of temperature (°C) on lethal rate as pasteurisation units.

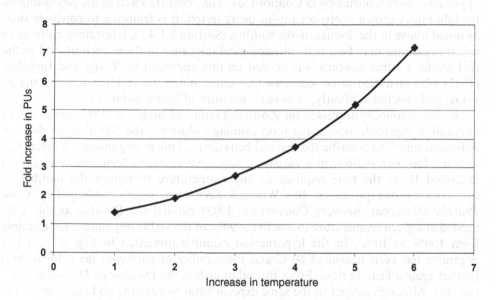

Fig. 8.24  Relationship between increase in temperature (°C) and increase in PUs (from Fricker, 1984).

manually from the equation $L_T = 1.393^{(T-60)}$ where 'T' represents the required temperature. This relationship is derived from the original equation reported by Del Vecchio et al. (1951) where:

$$L_T = \frac{1}{\log^{-1}\dfrac{(60-T)}{Z}}$$

and the value for 'Z' is taken as 6.94°C, the value reported for the 'abnormal yeast'.

The Z-value concept and its role in calculating lethal rates (PUs) have come in for repeated criticism over the years since 1951. The focus of much of this attention has been the Z-value used in this calculation, its origin and relevance. In a persuasive piece of work, McCaig *et al.* (1978) reported the failure of a flash pasteuriser operating at 26.5 PU to achieve commercial sterility. Further investigation showed the presence of *L. brevis* with a higher Z-value than 6.94°C. Depending on the heating menstruum, the Z-value for this organism was 15 (beer) or 20.2 (buffer, pH 7). With these observations in mind, McCaig *et al.* (1978) concluded that for the lethal rate approach to have validity would require each brewery to determine the Z-value for *their* organisms and beer! Similarly, Tsang and Ingledew (1982) reported Z-values (°C) of 11.2 (*P. acidilactici*), 12.3 (*L. delbrueckii*), and 15.4 (*L. frigidus*) for organisms tested in degassed 5% abv lager. Molzahn *et al.* (1983) also reported a Z-value of 12.4°C for *L. brevis* in alcohol-free lager. Collectively such results are significant for two reasons. First, they challenged the universal Z-value but also demonstrated the critical role of the heating menstruum in microbial thermal resistance. The later concern can be extended to include the experimental approach to assess heat resistance (four distinct approaches were reported by O'Connor-Cox *et al.*, 1991a) as well as the physiology of the laboratory-grown micro-organisms being tested. It is tempting to conclude that, as noted above in the discussion on biofilms (Section 8.1.4.1), laboratory cultures of micro-organisms may bear little physiological relevance to those encountered in the real world. Further concern was heaped on this approach by Tsang and Ingledew (1982) who identified three scenarios that undermined the validity of using the Z-value, and concluded bluntly, 'Z is *not* a measure of heat resistance'.

Having so roundly dismissed the Z-value, Tsang and Ingledew (1982) proposed an alternative approach used in the food canning industry – the 'D-value' or decimal reduction time – to describe the thermal behaviour of micro-organisms. As ever, this measure has been defined in a variety of ways. Accordingly, Molzahn *et al.* (1983) described D 'as the time required at any temperature to reduce the number of survivors by one power of 10'. Whereas *Beer Pasteurisation Manual of Good Practice* (European Brewery Convention, 1995) defined the D-value as the 'time needed at a given temperature to inactivate 90% of the viable population for example from 100% to 10%'. In the hypothetical example presented in Fig. 8.25, a test organism has been heated at 60°C and the number of survivors (as a logarithm) plotted against heating time. From this relationship, the D-value or $D_{60}$ value is 1.7 minutes. Although subject to the same experimental constraints and concerns as the Z-value, this approach enables the efficacy of a time/temperature combination to be predicted. If, for the example presented in Fig. 8.25, the loading of the test organism was $20 \times 10^3$ cells $l^{-1}$ (i.e. 20 cells $ml^{-1}$) for every 1.7 minutes at 60°C, the viable loading will reduce by 99% ($20 \times 10^2$ cells $l^{-1}$), 99.9% (20 cells $l^{-1}$), 99.99% (2 cells $l^{-1}$) and so on.

D-values determined at various fixed temperatures can be plotted against temperature to give a 'phantom thermal death curve' (Tsang & Ingledew, 1982). Examples of such curves are presented in Fig. 8.26. It is apparent from this data that at a fixed temperature there is a wide variation in D-value (e.g. $D_{45}$). Further, the gradients the phantom death curves vary widely between micro-organisms. Consequently, the yeast *P. membranifaciens* ($D_{45} = 62.5$ minutes) is more resistant at 45°C than the bacterium *L. frigidus* ($D_{45} = 2.9$ minutes). However, the tables are turned at

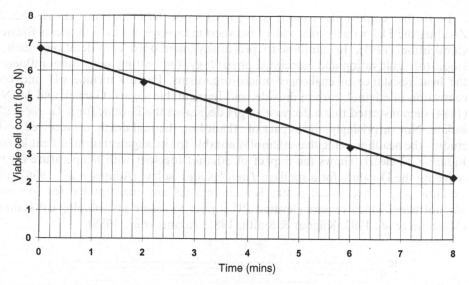

**Fig. 8.25**  Example of a survival curve – semilogarithmic plot of viable cell count against time of incubation at a fixed temperature (°C).

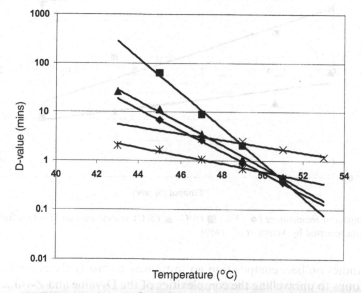

**Fig. 8.26**  Phantom thermal death curves for (■) *P. membranifaciens*, (♦) *P. anomala*, (▲) lager strain of *S. cerevisiae*, (×) *L. frigidus* and (✳) *L. delbrueckii* (from data reported by Tsang and Ingledew, 1982).

51°C where *L. frigidus* ($D_{51}$ = 1.7 minutes) is more resistant than *P. membranifaciens* ($D_{51}$ = 0.4 minutes). Correspondingly the $D_{60}$ values are 0.44 (*L. frigidus*) and 0.00025 minutes (*P. membranifaciens*). The Z-value for each organism can be calculated from the relationship between D-value and temperature from the equation $Z = (T_2 - T_1)/(\log D_1 - \log D_2)$.

To further confuse the issue it is apparent from the literature that, as with the Z-value, the $D_{60}$ value varies widely between micro-organisms (European Brewery Convention, 1995). In addition to previously mentioned considerations such as laboratory cultivation and associated physiology, the heating menstruum influences the heat resistance of challenged micro-organisms. Factors include pH and concentration of ethanol, carbon dioxide, and fermentable sugars. Garrick and McNeil (1984) demonstrated that a decrease in pH decreased the D value for a *Lactobacillus* isolate. Not surprisingly, the impact of ethanol on thermal death has received wider study – the headline being that ethanol is notable for its augmentation of heat killing. Figure 8.27 clearly shows the impact of ethanol together with temperature on the reducing the D-values for a spoilage *Lactobacillus* isolate (Adams *et al.*, 1989). This explains why conventional pasteurisation regimes are unacceptable for low/non-alcohol beers which require elevated PU inputs to assure microbiological stability (Molzahn *et al.*, 1983; Kilgour & Smith, 1985; Adams *et al.*, 1989).

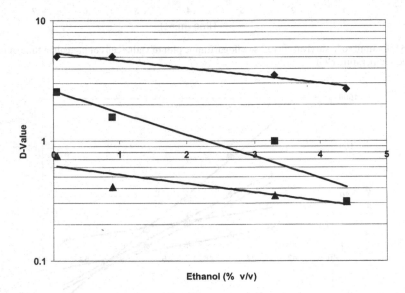

**Fig. 8.27** Impact of temperature (♦ 55°C, ■ 60°C, ▲ 65°C) and ethanol on D-values for *Lactobacillus* E93 (from data reported by Adams *et al.*, 1989).

These studies on beer composition go some way to justify the commitment of the various groups to unravelling the complexities of the D-value and Z-value. However, this insight has prompted Garrick and McNeil (1984) to suggest that pasteurisation regimes may require to be established for each beer according to its composition. The downside is that such experiments are at best 'difficult', a concern that is exacerbated by the lack of a 'standard' method. So much so that O'Connor-Cox *et al.* (1991a) were moved to conclude that the data on 'the relative heat resistance of bacterial and yeast species is confusing and somewhat incomplete'. This, together with the concern that designing pasteurisation regimes around one atypically thermoresistant organism (e.g. the 'abnormal yeast' of Del Vecchio *et al.*, 1951) is generally inappropriate, has to an extent undermined such studies. Under the circumstances, it is no surprise that many of these hard-won publications recommend that further work is undertaken to explore the relationship between time, temperature and microbiology (Tsang & Ingledew, 1982: Garrick & McNeil, 1984; Kilgour & Smith, 1985). The absence of such subsequent studies is perhaps a reflection of a change of attitude to pasteurisation. As noted earlier, it is no longer acceptable for pasteurisation to act as a 'backstop' to mend poor product hygiene. Initiatives such as HACCP and ISO quality standards have focused on prevention of (microbiological) problems. Accordingly, the focus has been on reducing the microbial load to pasteuriser. In terms of process control, this has been recognised as the most effective step in assuring the success of pasteurisation together with reducing the number PUs to achieve 'commercial sterility'. This overcomes many of the complexities and reservations (see for example Kilgour & Smith, 1985) of establishing pasteurisation regimes from D- and Z-values.

*Typical pasteurisation regimes.* Despite the foregoing debate, PUs remain the accepted currency of thermal processing of beer. The number of PUs required to achieve 'commercial sterility' remains a moot point! Laboratory studies over the years have reported that only 1–5 PUs are necessary (Del Vecchio *et al.*, 1951; Tsang & Ingledew, 1982; Garrick & McNeil, 1984). In practice, as a rule of thumb, 15 PUs (equivalent to 15 minutes at 60°C) have been applied 'universally' across the brewing industry worldwide (Del Vecchio *et al.*, 1951; Tsang & Ingledew, 1982; O'Connor-Cox *et al.*, 1991a, b). However, as noted by Dallyn and Falloon (1976) such regimes in tunnel pasteurisers correspond to 30 PU when heating-up and cooling times are included. For many, this is perhaps an idealised, optimistic view of routine pasteurisation regimes used to assure commercial sterility. Some older tunnel pasteurisers do not have the process capability to achieve relatively low PUs. For others, there is a natural tendency to operate at higher temperatures and times to meet the 'just in case' need to satisfy the unexpected microbial loading (the so-called 'safety margin'). Further, a common and unfortunate practice is to inexorably increase pasteurisation regimes (particularly flash pasteurisation) in response to a downturn in microbiological results, real or anticipated trade problems. Unfortunately, the 'after the event' reduction in elevated pasteurisation temperature/time is not pursued with the same vigour or commitment. This invariably results in pasteurisation specifications being higher than strictly required by the microbiological loading pre-pasteuriser.

Although clearly a local issue and one that should be 'tailor-designed to suit the brewery in question' (O'Connor-Cox *et al.*, 1991b), typical PU values have been reported. For example the EBC (European Brewery Convention, 1995) has reported PUs to vary between beer styles, ranging from 15–25 (lager), 20–35 (ales and stouts), 40–60 (low-alcohol beers) and 80–120 (non-alcoholic beers). As noted above, such variability in pasteurisation requirements reflects the impact of product composition on thermal killing of micro-organisms. However, according to O'Connor-Cox *et al.* (1991b), most North American breweries (with presumably conventional products) operate between 5 and 15 PU which, they suggest, may provide an 'excessive safety margin'. Similarly in the UK, as long ago as 1976, there was anecdotal evidence that breweries operate routinely between 10 and 12 PU (Dallyn & Falloon, 1976; O'Connor-Cox *et al.* (1991b). Whatever the reality of 'how many PUs', there is little doubt that PUs are declining worldwide. This is a consequence of drivers such as reducing energy utilisation and the need to minimise thermal damage to product, ambitions which are made possible by improvements in process capability (hardware and product hygiene).

*QA of pasteurisation.* The ultimate measure of the success of pasteurisation is the achievement of 'commercial sterility'. In other words, there is no suggestion of microbial growth over the shelf-life of the product. In reality, pasteurisation regimes invariably achieve this objective, presumably through the inclusion of comfortable 'safety margins'! Process assurance is typically achieved through setting tight quality standards and subsequent monitoring. Pre-pasteuriser microbiological monitoring is usually via a continuous 'drip' sampler of the feed beer. Although inevitably retrospective, this approach enables trend analysis and the identification of deteriorating (or improving) performance. Microbiological monitoring post-pasteuriser ex-final package is increasingly viewed as unnecessary or is a minimal 'comfort' analysis. Again, the results of any analysis are historical and invariably of little statistical relevance. This later point is exemplified by the microbiological analysis of two cans/bottles every hour from a line operating at 2000 cans every minute!

A preferable route is to monitor the process such that its effectiveness is assured. In the case of tunnel pasteurisers, thermographs have long found application in monitoring time and temperature. Monitors such as the 'Redpost' enable various quality parameters (transit time, time above 60°C, maximum temperature, and total PUs) to be downloaded and printed (European Brewery Convention, 1995). Flash pasteurisers are increasingly sophisticated, and accordingly key parameters such as flow rate, pasteurisation temperature, outlet beer temperature and downstream beer pressure are computer controlled to achieve the desired PUs (European Brewery Convention, 1995). In passing *Beer Pasteurisation Manual of Good Practice* (European Brewery Convention, 1995) provides worked 'how to' calculations to determine the PUs from tunnel and flash/plate pasteurisers.

An alternative approach, that has found application post-pasteurisation, is to assay for heat labile yeast derived enzymes that can be found in beer (European Brewery Convention, 1995). The yeast cell wall enzyme invertase (Sections 4.1.2.2, 4.4.2.4), which cleaves sucrose to glucose and fructose, was first introduced in 1902 (Enevoldsen, 1985) as an 'all or nothing' test for pasteurisation. Although considered to be inactivated by 5 PU, there is a view (Rader, 1979; McNeil *et al.*, 1984) that at pHs of 4 and above the invertase method should be used to quantify beers exposed to 10–15 PUs. Building on this approach, Enevoldsen (1981, 1985) proposed a similar approach based on the detection of melibiase, an enzyme found exclusively in the cell wall of lager strains of *S. cerevisiae* (Section 4.1.2.2). Depending on the pH of the beer, Enevoldsen (1985) reported being able to estimate the heat treatment of lager beers between 30 and 125 PUs. Work by Reeves *et al.* (1991) showed melibiase to have a Z-value of 3.5°C and a $D_{60}$ of 185 minutes in a beer of 4.2% (v/v) ethanol and a pH of 4.2. These workers, whilst noting the method's potential application for assurance of thermal killing of yeast, add the caveat that its usefulness is specific to pasteur-isation regimes at 60°C and its use should be modelled for each potential beer.

*Flavour impact of pasteurisation.* The drive to reduce pasteurisation so as to improve product quality is firmly rooted in the negative flavour characteristics attributed to thermal processing. For an excellent, general review of the complex field of flavour stability, the reader is directed to the review of Bamforth (1999). O'Connor-Cox *et al.* (1991a) described over-pasteurised beers as being 'oxidized, bread-crust-like, or possessing a cooked quality'. Chemically, Bamforth (1986) was able to demonstrate the formation of carbonyl compounds on pasteurisation and Back *et al.* (1992) ascribed 2-furfural, heptanal and other compounds as indicators of 'thermal tainting'. It is generally accepted that 'cooked' characters can be avoided, or at least managed, by reducing pasteurisation regimes and, particularly with tunnel pasteurisers, putting an upper 'reject quality limit' on over-pasteurised bottles and cans. A further critical factor in flavour stability is to minimise the dissolved oxygen content of beer during processing and packaging (Bamforth, 1999).

8.2.1.3 *Sterile filtration.* The development of sterile filtration as an alternative process to pasteurisation has been driven by benefits to product quality together with the consequent marketing opportunity to differentiate such products to the con-sumer. Although appropriate to any packaging format, sterile filtration has found especial application for can and bottle (including PET, Reid *et al.*, 1990) products. Replacing pasteurisation with a cold filtration process has supported marketing claims that such products are 'natural', 'fresh' or 'clean'. Indeed, it is a legitimate claim that such beers will not suffer from 'cooked', 'bread-like' or 'pasteurised' flavours described above. Although seemingly new technology, sterile filtration of beer dates back to 1931 with a production unit being used in England (Ryder *et al.*, 1994). Since then sterile filtration has taken off, albeit selectively, representing 63% of the Japanese market in 1990 (Ichikawa & Takinami, 1992) and some 25% of the American market (Ryder *et al.*, 1994). European breweries have been more con-servative in applying the technology with the demands of franchise brewing and the need to match international brands, often driving investment in sterile filtration. Arguably, the wider implementation of sterile filtration has been minimised by the lack of consumer 'pull' and hampered by the more stringent demands of process and product hygiene.

Like pasteurisation, sterile filtration achieves 'commercial sterility' with a typical specification of $< 1$ yeast cell $l^{-1}$ and $< 1$ bacterial cell $l^{-1}$ (European Brewery Con-vention, 1999). Sterile filtration cannot simply be 'bolted in' nor can it be a 'solution to poor hygienic practices upstream' (Dunn *et al.*, 1996). Users of the technology recognise that it is 'higher risk' than pasteurisation (see Fig. 8.21) and repeatedly emphasise the need for a step change in hygienic practice and standards throughout the brewery. Working hand-in-hand with this philosophy is the attitude of the people in the brewery who must have a 'sanitary mindset' (Ryder *et al.* (1994). The take home message is simple! Sterile filtration cannot operate in isolation. Steps must be taken to control and limit the incoming microbial load both to achieve 'clean' beer but also to satisfy process flow rates. The hygiene of subsequent downstream packaging operations is of critical importance and must be controlled to assure microbiological stability. Users of the technology have recognised the need for total hygiene management and, accordingly, have imposed aseptic clean rooms or covering shrouds (Ichikawa & Takinami, 1992; Ryder et al., 1994) around the filler and associated environments. The very real risk of airborne contamination (Section 8.1.4.2) is typically met by the use of clean air conditioning, positive pressure and occasionally air locks. Inevitably, CiP and sanitation is of paramount importance to the routine success of sterile filtration. The regimes used by Sapporo in Japan (Ichikawa & Takinami, 1992) and Miller in the USA (Ryder et al., 1994) both major on the 'before and after' use of hot water (80–85°C) together with regular caustic CiP. Equally the threat of conveyor-mediated contamination (Section 8.1.4.2), is recognised and, consequently, is controlled (Ichikawa & Takinami, 1992; Ryder et al., 1994). These and other considerations are described by Schmidt (1999) in a useful practical overview of the lengths his brewery in Brazil has gone to in assuring the packaging of 'draft beer' in bottle.

As a process, 'filtration' is characterised by variety, be it filtration 'media' or configuration. Accordingly, it is inevitable that sterile filtration regimes vary widely between practitioners and, indeed, within individual companies (see the description of the various systems at Kirin in Japan – Takahashi et al., 1990). Although by no means definitive, sterile filtration operations typically consist of two to three steps (Dunn et al., 1996). First, a conventional kieselguhr filtration of beer from maturation or conditioning tank to a bright beer or holding tank. This step achieves clarification and colloidal stabilisation. In terms of microbial load (Dunn et al., 1996), this filtration step reduces yeast counts from $5 \times 10^6$ cells ml$^{-1}$ to as little as $< 500$ cells ml$^{-1}$ and bacterial counts from as high as $10^4$ cells ml$^{-1}$ to 10 cells ml$^{-1}$. The second step essentially achieves sterile filtration. Here, the microbial load is reduced to a specification of typically $< 1$ yeast cell l$^{-1}$ and $< 1$ bacterial cell l$^{-1}$. As a further guarantee, some breweries include a further membrane filtration stage to assure 'sterility'.

For further details of sterile filtration, particularly to the options for filtration media and plant configuration, the interested reader is directed to *Beer Filtration, Stabilisation and Sterilisation, Manual of Good Practice* (European Brewery Convention, 1999).

### 8.2.2 *In the trade*

The ultimate objective of good process practice, HACCP, QA/QC systems, and real commitment in breweries is to produce beer of excellent microbiological quality. On leaving the brewery, the product in keg is 'commercially sterile' but will contain very low levels of yeasts and bacteria (perhaps $< 1$ cfu in a litre). On dispense, the product will inevitably pick up low levels of 'contaminants' which do no harm whatsoever to the product or the consumer. For the vast majority of restaurants, bars or pubs this is the normal state of affairs for retailing a foodstuff. This is achieved through good hygienic practices and regular cleaning of the dispense lines. This is important as it is universally accepted that 'dispense' is of critical importance to product quality and presentation. As is often said, 'people drink with their eyes'!

Unfortunately, there can be occasions and/or outlets where products are dispensed which are not 'bar bright' or which have obvious flavour defects. The reasons are numerous but include poor working practices, inadequate hygiene education, concern about 'losses', or simply a flagrant disregard for the importance of hygiene in general or line cleaning in particular. Unfortunately, far too many consumers have experienced poor-quality products 'in the trade' caused by microbiological contamination at dispense. Typical problems include flavour defects (acetic acid, diacetyl, phenolics etc.) and a haze or 'cast'. Although some consumers reject or complain about such products, the vast majority will 'vote with their feet' and leave the outlet or change products. The impact of poor product quality on brand and outlet loyalty is inestimable but can be assumed to be considerable.

There are no excuses for this situation! Reports on the impact of poor microbiological hygiene on product quality in the trade are not new. One of the first (Hemmons, 1954) described the wild yeast microflora found in cask beers at dispense (see Section 8.1.3.2). Much of the fundamental work on the microbiology of beer dispense originated from work by the late Jim Hough's team at the British School of Malting and Brewing at the University of Birmingham in England. Initially Hough *et al.* (1976) tracked both the microbial loading and microflora from keg filling to dispense. With hindsight it is not surprising that the results from this work 'pointed the finger' at the hygiene of dispense lines and taps.

Pointedly Hough *et al.* (1976) remarked on the 'dirty and unnecessary custom of bar staff immersing dispense taps in the beer whilst drawing into glasses' and added 'it is particularly objectionable when a dirty glass is refilled'. The reaction within the industry to these hard-hitting observations is not recorded but with general exception of glass hygiene (see Stillman, 1996) may be anticipated as being muted!

Subsequent work from the group at Birmingham focused in greater detail on the microbiology of dispense together with the materials used to make and clean beer lines (Harper *et al.*, 1980; Harper, 1981; Casson, 1982). The microbial diversity of dispensed keg and cask beers was reported in Section 8.1.3.2 (Harper *et al.*, 1980). Again the microbiology and 'handling' of the dispense tap received particularly withering comment. To wit, 'large numbers of yeast and bacteria were encountered, probably due to the aerobic conditions, infrequent cleaning and the practice of immersing the nozzle into the beer during dispense'. These 'field observations' are compelling both in their honesty and in insight. The report of Harper *et al.* (1980) is notable for the realisation that 'infection' is not static but mobile, moving from the tap into the line and, at the other end, from the line into the container. As noted elsewhere (Section 8.1.4.1), this paper was notable in the observations about the adhesion of micro-organisms to surfaces and, consequently, the difficulty in the removal or penetration of the 'biofilm' with line cleaning fluids. The subsequent paper (Harper, 1981) was similarly insightful in clearly demonstrating the difference in performance between yeasts maintained in the laboratory and those isolated from the real-world environment of the public house. The 'natural' isolates had clearly adapted to their environment growing more rapidly in beer and being significantly better at forming biofilms.

Harper's successor Duncan Casson focused his attention on the role of the line and, in particular, what polymer it was made of. Although at the time highly pertinent to the industry, the work described by Casson (1982) has been superseded by developments in beer line technology which, in part, were triggered by these earlier observations. Work at the University of Sunderland sponsored by one of the manufacturers (Premier Python Products) has extended the earlier work from Birmingham. Thomas and Whitham (1996) demonstrated major differences in the attachment of yeasts and bacteria to dispense tubing. Of the three then materials of construction, it is perhaps fortunate that by far the worst in terms of attachment – PVC – is no longer in use! Of the others, nylon was superior (in minimising attachment) over the more commonly used, MDP (medium density polythene). Although minimising microbial attachment to dispense line surfaces is an attractive concept, factors such as cost and gas porosity are bigger determinants of choice. At the end of the day the most effective route to minimising the detrimental effects of colonisation by yeasts and bacteria is the adherence to rules of basic hygiene coupled with regular and effective line cleaning.

In many respects the cleaning of dispense lines mirrors that of CiP in the brewery (Section 8.2.1.1). Although there is a wide range of proprietary line cleaning solutions, they typically revolve around a caustic detergent (sodium hydroxide, Table 8.10), a biocide (sodium hypochlorite, Table 8.12) and chemicals to soften water and ensure 'wetting'. Grant (1986) and Treacher (1995) have published useful and practical reviews on beer line cleaning. As a process there is little debate about line cleaning and the vast majority of practitioners understand what is 'best practice'. In summary after flushing the lines with water, the diluted detergent (as recommended by the supplier) is pulled through the lines and left to stand for about 30 minutes. Occasionally the detergent is 'moved' halfway through this process. Whatever, after the stand, the detergent is chased out by water, to drain. The detergent being 'soapy' provides a useful and simple measure of cleaning status, particularly in the assessment of line rinsing prior to bring beer back on line. Of course, these days line cleaning is far more complex than this 'potted version'! The excellent review of Treacher (1995) should be consulted for guidance on best practice. If performed correctly, it is assumed that the line is clean and accordingly validation is unnecessary. This view is subject to change as ATP bioluminescence provides realtime validation and is increasingly finding application as an audit tool (see Section 8.3.3.1).

There are a few variables in beer line cleaning. Like CiP, the detergent temperature can trigger debate. Some prefer to use cold water whereas others make up the detergent to be 'hand hot'. As to automation (which is increasingly available) most line cleaning is a manual operation and is performed by bar or restaurant staff. Inevitably the biggest and, undoubtedly most important debate is about the frequency of line cleaning. Treacher (1995) recommended that lines are cleaned 'to a minimum of every seven days' with the proviso that some products (low ABV, high sugar) may require cleaning more frequently. Further, Treacher (1995) noted that extending the cycle by two to three days 'can reduce the effectiveness of cleaning by up to 25%'! Although many would agree with these views, the reality is that the frequency of line cleaning in many establishments has been relaxed to every two weeks or more. There are many reasons for this. The most obvious is that line cleaning inevitably results in the loss of beer and, accordingly, has a financial cost. Further, line cleaning cannot be carried during 'opening hours' and has to be added-on to the already long working day. Overlain on top is the view (real or anecdotal) that reducing the frequency of cleaning from weekly to every two weeks has no perceived impact on product quality. Although perfectly reasonable, these views reflect a poor understanding of the role of line hygiene in assuring product quality. It can be argued that through better line hygiene, product quality would be improved, and that more consistent product quality would result in additional sales that would trivialise the significance of beer losses and inconvenience.

Although the focus of this section is (quite rightly) on line cleaning, it is also appropriate to consider the observations from the early work from the University of Birmingham and more recently VTT Biotechnology (Storgårds & Haikara, 1996). Specifically with the emphasis on line cleaning, it is easy to lose sight of the importance of the beginning (keg coupler) or the tap-end of dispense. The cellar end of dispense is important but manageable inasmuch that the keg coupler and keg head can be easily sanitised. Further, the microbiological risk is minimised by the typically low temperature of the cellar together with the 'one touch' connection that is made between line and container. Conversely, from the perspective of risk management, the tap-end of dispense is a greater concern for at least three reasons. First, the tap is inevitably handled by bar staff; second, the environment is warm and aerobic; and, most damning of all, spouts and other removable bits of dispense 'plastic' are rarely cleaned properly. Unfortunately, the preferred option of soaking dispense 'plastics' in line cleaning detergent is rarely taken-up because the line cleaning fluid is typically kept in the cellar and not in the bar dispense area. Consequently, it is common practice to remove spouts and other dispense 'plastic' at the end of the night (or on line cleaning) and transfer to a glass of what is progressively 'beery' water. Although seemingly a good idea, such treatment exacerbates the microbiological contamination of removable spouts, sparklers and other dispense 'furniture'. The harsh reality is that steeping spouts, etc. in 'beery' water is akin to incubating these bits of dispense 'plastic' in a nutrient medium. Accordingly, contamination of dispense 'bits and pieces' is likely to get significantly worse rather than better! The risk requires to be managed, by simply soaking the 'plastics' in line cleaner on a regular (preferably daily) basis.

Over the years a number of publications have sought to quantify the size of microbiological contamination both across the dispense line and in the dispensed product. These reports have also demonstrated the impact of line cleaning on hygiene and, equally, the rapidity with which the system becomes recontaminated (see Hough et al., 1976; Harper et al., 1980; Storgårds & Haikara, 1996). This work has without doubt raised the profile of dispense hygiene and contributed to the debate about best practice in controlling the risk. As noted by Storgårds (1996), in the worst cases, these various reports describe alarmingly contamination as high as $10^5$–$10^7$ cfu per ml of beer. However, these results are typically the 'first runnings' of the day (i.e. beer that has sat overnight in line) or after 250 ml (or so) has been removed. Either way whilst eloquently making the point, these results exaggerate the microbial loading, which, during normal dispense, would be anticipated to be much lower.

A hitherto unpublished study over a two-month period in 1998 monitored the loading of aerobic and anaerobic micro-organisms in lager dispensed from four taps in a busy, well-managed and successful bar. To overcome the 'first runnings' argument, 4 litres of product was pulled through each tap prior to sampling. Further, the sample was taken after the removal of the spout and sparkler, to ensure the recovered micro-organisms were exclusively associated with the beer line. Line cleaning was on a two-week cycle, typically on the Thursday of the second week. To gauge the impact of line cleaning on product microbiology, routine sampling was performed on the Thursday of the first week and the Wednesday and Friday of the second week. The raw microbiological data is presented in Fig. 8.28 and can be seen to be broadly cyclical in tandem with the cleaning cycle. It is noteworthy that the microbial loading was generally lowest from tap 1 and highest from tap 4 with occasional 'blips' in performance. Quantitatively, within a sample, there was little difference between the aerobic and anaerobic counts. The contribution of line cleaning to the microbiology of the product is readily apparent from the results reported in Table 8.14. These fascinating data clearly show that line cleaning reduces the microbial load of the dispensed product. However, these results suggest that line cleaning is of variable efficiency. For example, line cleaning reduced aerobes from $2 \times 10^3$ to $8 \times 10^1$ on one occasion whereas a month later the loading was reduced by only 30% ($9 \times 10^2$ to $6 \times 10^2$). More detailed analysis of the 114 discrete results (anaerobes and aerobes on a tap-by-tap basis) (see Fig. 8.29) shows that the microbial loading was broadly in a band between $10^1$ and $10^3$ per ml of beer. The average of each data set showed aerobes and anaerobes to both be about $10 \times 10^3 \, \mathrm{ml}^{-1}$ (pre-cleaning) and $2.6–3.1 \times 10^2 \, \mathrm{ml}^{-1}$ (post-cleaning). As would be expected from such low microbial loads, there is no suggestion whatsoever that product quality was in anyway compromised. It is anticipated that these results will provide a useful benchmark for what is 'normal' in terms of dispensed product in the trade.

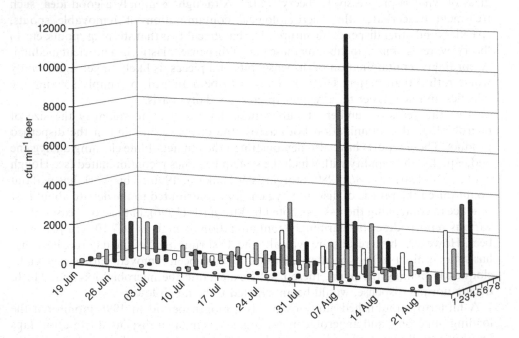

**Fig. 8.28** Trade monitoring of the microbial loading ex-dispense – same brand dispensed from four taps ('1' = tap 1, aerobes, '2' = tap 1, anaerobes and so on); (unpublished results of Alisdair Hamilton, Wendy Box and David Quain).

**Table 8.14** Microbiological loading of beer in the trade pre- and post-line cleaning (unpublished results of Alisdair Hamilton, Wendy Box and David Quain).

| Before/after cleaning (date) | Aerobes (cfu ml$^{-1}$) $\pm$ sem | Anaerobes (cfu ml$^{-1}$) $\pm$ sem |
|---|---|---|
| Before (24/6) | 2023 $\pm$ 989 | 603 $\pm$ 795 |
| After (26/6) | 77 $\pm$ 44 | 25 $\pm$ 13 |
| Before (8/7) | 300 $\pm$ 245 | 184 $\pm$ 123 |
| After (10/7) | 161 $\pm$ 316 | 106 $\pm$ 34 |
| Before (22/7) | 942 $\pm$ 541 | 1350 $\pm$ 979 |
| After (24/7) | 555 $\pm$ 533 | 928 $\pm$ 893 |
| Before (5/8) | 1193 $\pm$ 409 | 1791 $\pm$ 711 |
| After (7/8) | 83 $\pm$ 38 | 255 $\pm$ 208 |

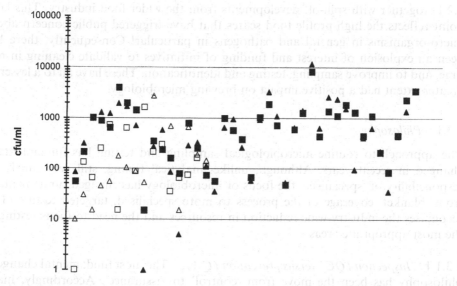

**Fig. 8.29** Raw data from Fig. 8.28 segregated by pre- and post-line cleaning (aerobes, pre ■, post □; anaerobes, pre ▲, post △).

## 8.3 Sampling and testing

For many years, microbiology in the brewery has suffered a bad press. Of its many perceived failings, Quain (1999) focused on its slowness and the equivocal nature of microbiological results. The former consideration is the most damning. Methods that require between three and seven days to deliver a result can offer little to monitoring and process control. Such a lack of real time delivery has undermined the importance of microbiology and reduced it to a 'comfort' analysis that is tracked via charts of compressed and averaged data. The second point ('What do the results mean?') has triggered countless furrowed brows and sleepless nights. Microbiologists – who are frequently seen as bringers of bad tidings – are left to explain to harassed team leaders or departmental managers that they are unable to predict the significance of the (bad!)

results they bring. This is invariably exacerbated by the 'historical' nature of the results ('The beer has gone to trade.') and the uncertainty around the details of the result ('Is it a beer spoiler? 'Is there enough to spoil?'). If this is not enough, lurking beneath such a debate are concerns about the general robustness of brewing microbiology. In particular, the 'quality' of the sample and, particularly, just how representative it is of the vessel it was taken from. For those with the 'knowledge' (often ex-microbiologists), further uncertainty can be introduced by questioning the ability of the media to support the exclusive growth of 'beer spoilers'. Failing these arguments, there is always the indefensible review of the occasions when routine testing failed to flag a spoilage problem and, conversely, the time when the results anticipated a problem which failed to materialise!

Ironically, despite the above pessimism, brewing microbiology has been re-energised through the enhanced profile and importance of 'hygiene' (see Section 8.2.1) together with spin-off developments from the wider food industry. This later point reflects the high-profile food scares that have triggered public concern about micro-organisms in general and pathogens in particular! Consequently, there has been an explosion of interest and funding of initiatives to validate cleaning in real time, and to improve sampling, testing and identification. These have all to a lesser or greater extent had a positive impact on brewing microbiology.

### 8.3.1  *Philosophy*

The approach to routine microbiological sampling and testing has fundamentally changed in recent years. Although, unlike analytical testing, still very much the responsibility of 'specialists', the focus of microbiology has changed fundamentally from 'blanket' coverage of the process to more specialised, targeted testing. This recognises the industry-wide reduction in resources and the need to focus testing in the most appropriate areas.

*8.3.1.1  Inspection (QC) versus prevention (QA).*  The most fundamental change in philosophy has been the move from 'control' to 'assurance'. Accordingly, many breweries have evolved their quality control departments (QC) into quality assurance (QA) departments. This is much more than a fashionable, cosmetic change but a change in emphasis from 'inspection' to 'prevention'. Driven by the implementation of 'systems' (HACCP, ISO 9000), microbiological sampling and testing is now focused on the process rather than the product. The basic premise is that microbiological effort should be directed at the 'critical control points' of the process which together determine product quality.

*8.3.1.2  Sampling plans and specifications.*  Sampling plans and the associated specifications make brewing microbiology happen! Typically, a sampling plan will be highly prescriptive and detail the inevitable (sample, frequency) as well the method and sample volume. The size of the sample is important and, crucially is often too small to realistically achieve occasional counts of micro-organisms. It is all too easy to be deluded into believing that a CCP is in control because of routine zero counts. Accordingly, the sample volume should be stretched so that microbial counts are occasionally observed. This confirms that the sample 'window' is appropriate and gives the assurance that changes in performance can be tracked. Further, it facilitates useful comparison, or benchmarking, of performance against other breweries. Sampling plans and specifications require periodic, at least annual, review to ensure they remain appropriate, focused and responsive to improvement initiatives.

Given the inevitable historical nature of the results, the 'specifications' are invariably a 'target' and, unlike measurements of colour, ABV or $CO_2$ etc., have little authority or impact on the process. Results are tracked to varying degrees of statistical sophistication and regularly reviewed, thereby enabling changes in performance to be readily identified and responded to.

Despite the importance of microbiological sampling plans and specifications, surprisingly little has been published. In a relatively short report for such a big subject, Avis (1990) described the application of HACCP to microbiological process control. Usefully he described CCPs and associated principal sample points (PSPs) which are used to monitor the process. This paper concluded with an example of the then sampling plan for a keg line (see Table 8.15). More recently, Alan Kennedy (Anonymous, 1999) described the minimum sampling plan used by the Scottish Courage Group. (In passing, sampling plans are invariably described as being the 'minimum', a rider that is often forgotten in the translation into practical reality!) Returning to 'specifications', Kennedy usefully described the use of red/amber/green descriptors to 'weight' microbiological results according to micro-organism and process location. In another example, Ichikawa and Takinami (1992) described their sampling plan to meet the more exacting demands of sterile filtration and aseptic packaging.

### 8.3.2 Methods – 'traditional'

Despite the very real developments in 'rapid' methods (see Section 8.3.3), traditional methods continue to be used for the vast bulk of microbiological activities in breweries large and small. These methods have remained relatively unchanged since the days of Louis Pasteur (Ogden, 1993). Typically, these methods revolve around the 'plating out' of a sample, either directly (or after centrifugation) as 'spread plates' ($\leq 0.2$ ml), 'pour' plates (5 ml) or after membrane filtration ($\leq 500$ ml). These approaches differ fundamentally. With the spread plate and membrane filtration approaches, the micro-organisms grow on the surface of the media. With the pour plate technique the sample is added directly to the hot molten media, this then cools and solidifies. Accordingly, the micro-organisms are distributed throughout the media which, together with concerns about the effects of heat shock, has led many microbiologists to stop using pour plates. Irrespective of the approach, after three to seven days' incubation at typically 27°C any microbial colonies that are present are counted and, where deemed necessary, identified. Identification is often 'intuitive' from the selective nature of the media or limited to microscopic examination as 'bacterial rods or cocci' or 'yeast'. More detailed traditional analysis includes the Gram test and catalase test (Section 8.1.2) for bacteria and, as a last resort, diagnostic strip tests (e.g. API, Biolog etc.) for yeast (Section 4.1.1.1) and bacteria (Section 8.1.2).

**Table 8.15** Minimum sampling plan for a keg line (after Avis, 1990).

| Sample | Aerobic media | Anaerobic media | Minimum volume sampled | Volume analysed | Reported as | Frequency | Comments |
|---|---|---|---|---|---|---|---|
| Pre-pasteuriser buffer tank or bright beer tank | WLN | Raka Ray | 10 ml | 1–10 ml | cfu per ml | 1 per tank | Volume analysed depends on anticipated level of contamination |
| Post-pasteuriser and end of beer main | WLN | – | 2 l | 2–10 l | cfu per 250 ml | 2 per shift | Flow rate = 21 h$^{-1}$ |
| Filled kegs | WLN | Raka Ray | 500 ml | 2 × 500 ml | cfu per 250 ml | 1 per tank packaged | Direct sampling |
| Trace kegs | WLN | Raka Ray | 500 ml | 2 × 500 ml | cfu per keg | 1 per quality per shift | Direct sampling |
| Washed, steamed kegs | WLN | – | 500 ml | 250 ml | cfu per keg | 1 per lane per week | 500 ml saline rinse |
| Process gases | WLN | – | 20 seconds full bore | 20 seconds full bore | cfu per 20 secs | 1 per shift | Every point of use |
| CiP rinse liquor | WLN | – | 300 ml | 250 ml | cfu 150 ml | 1 per CiP leg every CiP run | 1 per CiP leg every CiP run |

8.3.2.1   *Hardware.*   Other than the replacement of glass by disposable plastic, relatively little has changed in the last 30 years in terms of the hardware required to perform microbiological testing in the brewery. Simpson (1996) has provided a useful summary of the fundamental items required for the microbiological testing of brewery samples.

Major developments however have occurred in the automation of microbiological 'routines'. Although at a cost, automatic plate readers and plate pourers have facilitated the removal of some of the drudgery of routine microbiology. Equally, the increasing sophistication of anaerobic incubators has provided an attractive alternative to the frustration and low capacity of gas jars. More mundane but equally liberating is the development of disposable sterile membrane filtration funnels that remove, at a stroke, one of the more frustrating elements of microbiological testing. This later development has also facilitated the important move away from pour and spread plating, enabling the wider use of membrane filtration in routine testing.

8.3.2.2   *Media.*   One of the more emotive areas of brewing microbiology is the choice of growth media used to isolate, recover and quantify bacteria and yeast. Despite the efforts of many, there is no one microbiological medium that satisfies the needs of the brewery microbiologist (Smith *et al.*, 1987). Consequently, over the years a raft of different media have been developed that reportedly are the 'best thing' for the selective recovery of wild yeasts, Gram-negative bacteria and so on. Invariably, the subsequent debate then hinges around the 'capability' of the various media to select for the group of interest. This leads to 'make it better' improvements and, consequently, tweaked variants of the original media appear that claim to offer superior performance. As succinctly noted by Casey and Ingledew (1981b):

> 'since 1974, there has been a new medium created almost every year for the lactic acid bacteria. In parallel with this has been a series of papers comparing the new media with the older media and, as different authors have reached different conclusions, it has created a great deal of confusion.'

It is beyond the scope of this chapter to do more than skim the surface of the 'media story'. To obtain a fuller insight into the evolution and composition of microbiological media the interested reader is referred to a series of reviews from Casey and Ingledew from the University of Saskatchewan in Canada. The articles cover general purpose media (Casey & Ingledew, 1981a), and media for lactic acid bacteria (Casey & Ingledew, 1981b), wild yeast and moulds (Ingledew & Casey, 1982b). The series was updated by Smith *et al.* (1987). More recently, Jesperson and co-workers have updated the story for brewery spoilage organisms in general (Jespersen & Jakobsen, 1998) and wild media in particular (van der Aa Kuhle & Jespersen, 1998). Without wishing to rehearse the various debates, some media formulations have found favour and are used by many brewing laboratories in their original form or, inevitably, with the occasional customisation! Where the composition of the media is reported (Tables 8.16–8.20) the details are invariably derived from the appropriate paper by Casey and Ingledew (or *vice versa*). With the exception of the lactic acid 'NBB' medium, the basal media are available commercially in formulations which, on occasion, differ marginally from those described here. The *Oxoid Manual* (Bridson, 1998) is a fund of

**Table 8.16** Composition of WLN medium.

| Component | |
|---|---|
| Glucose | 50 g |
| Monopotassium phosphate | 0.55 g |
| Potassium chloride | 0.425 g |
| Calcium chloride | 0.125 g |
| Magnesium sulphate | 0.125 g |
| Ferric chloride | 2.5 mg |
| Manganese sulphate | 2.5 mg |
| Casein hydolysate | 5 g |
| Yeast extract | 4 g |
| Agar | 20 g |
| Bromocresol green | 2.2 mg |
| pH | 5.5 |
| Distilled water to | 1000 ml |

**Table 8.17** Composition of MYGP + copper.

| Component | |
|---|---|
| Malt extract | 3 g |
| Yeast extract | 3 g |
| Glucose | 10 g |
| Bacto peptone | 5 g |
| Agar | 20 g |
| Copper sulphate (5H$_2$O) | 312 mg |
| Distilled water to | 1000 ml |
| pH | 6.2 |

**Table 8.18** Composition of MRS medium.

| Component | |
|---|---|
| Peptone | 10 g |
| 'Lab-lemco' powder | 10 g |
| Yeast extract | 5 g |
| Glucose | 20 g |
| Tween 80 | 1 ml |
| Dipotassium hydrogen phosphate | 2 g |
| Sodium acetate 3H$_2$O | 5 g |
| Magnesium sulphate 7H$_2$O | 200 mg |
| Manganese sulphate 4H$_2$O | 50 mg |
| Agar | 20 g |
| Final pH | 6–6.5 |
| Distilled water to | 1000 ml |

information on these and other microbiological media, and provides useful guidance on good practice for media preparation and storage.

- *General purpose media* – although a dangerous conclusion to draw, the Wallerstein Laboratories Nutrient (WLN) agar is used almost universally across the brewing industry. The history of the development of this medium is

**Table 8.19** Composition of Raka Ray No. 3 medium.

| Component | |
|---|---|
| Yeast extract | 5 g |
| Trypticase | 20 g |
| Liver concentrate 202–3 | 1 g |
| Maltose | 10 g |
| Fructose | 10 g |
| Betaine hydrochloride | 2 g |
| Diammonium citrate | 2 g |
| Potassium aspartate | 2.5 g |
| Potassium glutamate | 2.5 g |
| Magnesium sulphate (7H$_2$O) | 2 g |
| Manganese sulphate (H$_2$O) | 0.5 g |
| Dipotassium hydrogen phosphate | 2 g |
| N-acetyl-glucosamine | 0.5 g |
| Tween 80 | 10 ml |
| Agar | 20 g |
| Final pH | 5.4 |
| Distilled water to | 1000 ml |

**Table 8.20** Composition of modified NBB medium.

| Component | |
|---|---|
| Casein peptone | 5 g |
| Yeast extract | 5 g |
| Meat extract | 2 g |
| Tween 80 | 0.5 ml |
| Potassium acetate | 6 g |
| Sodium phosphate, dibasic | 2 g |
| L-cysteine monohydrochloride | 0.2 g |
| Chlorophenol red | 70 mg |
| Glucose | 15 g |
| Maltose | 15 g |
| L-malic acid | 0.5 g |
| Agar | 15 g |
| Final pH | 5.8 |
| Beer/distilled water (1:1) to | 1000 ml |

detailed in Casey and Ingledew (1981a). As with all successful media, WLN is purchased commercially from one of the many suppliers of microbiological media. Its composition (see Table 8.16) is notable for containing only glucose (5% w/v) as the sole carbon and for the inclusion of the pH indicator, bromocresol green. In passing, modifying WLN by doubling the concentration of bromocresol green has found application in the simple differentiation (colony colour and size) of closely related brewing yeasts (see Section 4.2.5.1). However, WLN is not a 'selective' medium as it supports the growth of both brewery bacteria and yeast. The media is made selective for bacteria by the inclusion of 15 mg l$^{-1}$ cycloheximide (see Section 4.2.5.1) to inhibit the growth of brewing and some (but not all) 'wild' yeasts and is then known as WLD (Wallerstein

Laboratories Differential) agar. The inclusion of isomerised hop extract (e.g. isohopCO$_2$n at 400 mg l$^{-1}$) in WLN and WLD is useful in suppressing the growth of spore-forming bacillus.

Despite the claims of the original authors (Casey & Ingledew, 1981a), WLD is comparatively poor in terms of supporting the growth of the more fastidious lactic acid bacteria. Despite this, WLD has found application as a 'routine' medium for the detection of aerobic bacteria (e.g. acetic acid bacteria) and some cycloheximide resistant wild yeast in brewing (Quain, 1995; Hammond, 1996; Simpson, 1996; Anonymous, 1999).

• *Wild yeasts* – at the time of writing it seems unlikely that a single method for the unified detection of wild yeasts will be forthcoming in the near future. This is not surprising given the sheer diversity of the *Saccharomyces* and non-*Saccharomyces* wild yeasts (see Section 8.1.3). For a wider review of the pros and cons of many wild yeast media, the interested reader should consult Ingledew and Casey (1982a) together with the update of Smith *et al.* (1987). Ingledew and Casey (1982b) should be consulted for a consideration of the more specialist media for moulds.

Arguably, the nearest thing to a single medium for wild yeasts (both *Saccharomyces* and non-*Saccharomyces*) is the MYGP + copper medium of Taylor and Marsh (1984). Although described in detail in Table 8.17, this medium is normally prepared from commercial MYGP media (YM or 'yeast mould agar') with the addition of sterile filtered copper sulphate (200 mg l$^{-1}$) to tempered medium. To assure consistent results, this medium should be prepared freshly to ensure the 'availability' of the copper. Although Smith *et al.* (1987) were ambivalent about the application of this method, van der Aa Kuhle and Jespersen (1998) reported that this medium detected wild yeast (*Saccharomyces*, *Candida* and *Pichia*) in brewery samples with a success rate of 80%. In this study, other older approaches such as lysine, crystal violet etc. (see Ingledew & Casey, 1982a) were shown to be less successful in detecting wild yeasts in 46–56% of the contaminated samples. Others who have reported the use of copper supplemented media to detect wild yeast include Quain (1995), Hammond (1996), Simpson (1996) and Anonymous (1999).

Inevitably, this medium is open to 'tweaking' the concentration of copper in an attempt to improve its capability. Rather than replacing the existing media, there is an opportunity for an additional medium with a lower concentration of copper to catch the more copper-sensitive strains. For example MYGP + Cu (100 mg l$^{-1}$) may have some utility although it is necessary to test the resistance of primary brewing yeasts as a handful have been found to be capable of growth. Presumably, these strains have acquired enhanced resistance to the toxic effects of copper, through inadvertent exposure to the metal during processing in copper vessels. Resistance is down to sequestration via a metallothionein coded by the gene *CUP1* which, in resistant strains, is tandem repeated up to 15 times (Macreadie *et al.*, 1994).

• *Lactic acid bacteria* – in their review, Casey and Ingledew (1981b) detailed the composition of no less than 18 different microbiological media for the detection of lactic acid bacteria. This is a reflection of the importance of lactic acid bacteria in brewing microbiology and difficulty in cultivation as 'different strains require different growth factors, use different sugars and have different requirements for oxygen' (Smith *et al.*, 1987). Overlain on these considerations is the complex relationship that lactic acid bacteria have with hop acids (see Section 8.1.2.2).

The market leaders for the detection of lactic acid bacteria are MRS (deMan *et al.*, 1960 – Table 8.18), Raka Ray (Saha *et al.*, 1974 – Table 8.19) and the modified NBB medium (Kindraka, 1987; Takemura *et al.*, 1992 – Table 8.20). All three media differ significantly in composition and all have their adherents. NBB ('Nachweismedium fuer Bierschadliche Bacteriem') was first developed as a general purpose medium by Back (1980) before being modified and rebadged by Kindraka (1987) for anaerobic lactic acid bacteria. NBB includes beer and a pH indicator (chlorophenol red). Both MRS and Raka Ray have long been available as dehydrated media whereas NBB has yet to be available commercially, a factor which has doubtless hampered its usage.

The above selection is by no means definitive and many microbiologists may disagree with this choice and, indeed, strongly advocate alternatives such as UBA, VLB-S7 and KOT. However as 'no single medium appears to be capable of detecting all strains of lactic acid bacteria' (Jesperson & Jakobsen, 1996), the debate will doubtless rumble on. Further developments can be anticipated as the 'optimal medium for detection of lactic acid bacteria has yet to be identified' (Jesperson & Jakobsen, 1996).

Whatever the choice of media, lactic acid bacteria are routinely detected after 5–7 days of anaerobic incubation. This is for two reasons: first to eliminate the growth of aerobes and second because the growth rate of *Lactobacillus* and *Pediococcus* species is enhanced anaerobically. Although not requiring oxygen, lactic acid bacteria tolerate oxygen. Recent work (Marty-Teysset *et al.*, 2000) has shown aerobically grown *L. delbrueckii* subsp. *bulgaricus* to reduce oxygen and to accumulate hydrogen peroxide, an oxidative stress which triggers early entry into stationary phase. This results in a concomitant reduction in biomass yield and, from the perspective of growth on agar plates, smaller colonies. In addition, the media are made more selective by the inclusion of inhibitors. These include cycloheximide (to inhibit yeast), 2-phenylethanol (to inhibit Gram-negative bacteria – Casey & Ingledew, 1981b) and vancomycin (to inhibit non-beer spoilage Gram-positive bacteria – Simpson & Hammond, 1987; Simpson *et al.*, 1988b). As noted in Section 8.1.2.2, Simpson and Hammond (1991) have advocated the inclusion of the 20 µM trans-isohumulone in MRS media to better select for potential beer spoilage *Lactobacillus*.

Compared to the developments in the detection of wild yeast and lactic acid bacteria, the Gram-negative bacteria (see Section 8.1.2.1) have, rightly or wrongly, received relatively little attention! However, Casey and Ingledew (1981c) were still able to contribute a review on the subject which was updated by Smith *et al.* (1987) and Jespersen and Jakobsen (1996).

MacConkey's agar (Table 8.21) remains the favourite medium in brewing microbiology to detect specialist Gram-negative bacteria, particularly coliforms. WLD continues to find favour for the enumeration of acetic bacteria and *O. proteus*

**Table 8.21** Composition of MacConkey's agar.

| Component | |
| --- | --- |
| Bactopeptone | 17 g |
| Proteose peptone | 3 g |
| Bacto-lactose | 10 g |
| Bacto-bile salts No. 3 | 1.5 g |
| Sodium chloride | 5 g |
| Bacto-agar | 13.5 g |
| Bacto-neutral red | 30 mg |
| Bacto-crystal violet | 1 mg |
| Distilled water to | 1000 ml |
| pH | 7.1 |

(Fernandez *et al.*, 1993). The more fashionable, strictly anaerobic bacteria *Megasphaera* and *Pectinatus* are detected in a variety of pre-reduced media such as NBB or MRS (Smith *et al.*, 1987; Jesperson & Jakobsen, 1996). Detection of these micro-organisms is complicated by the need to achieve anaerobiosis during sample processing and incubation.

8.3.2.3 *Validation.* The use of controls to validate analytical measurements has long been accepted laboratory practice. However, it is only relatively recently that controls and concepts such as proficiency testing have found application in brewing microbiology. Although by no means 'rocket science', both disciplines add value to routine microbiological testing by adding certainty to the results be they good or bad!

Although far from new, the use of media controls was reported by Quain (1995) in support of the microbiological analyses supporting 'yeast supply'. Here (Table 7.2), the different media are challenged with micro-organisms that should and should not grow on the various microbiological media. Such controls are powerful in confirming that the media 'work'. This approach is especially useful to validate the performance of batches of media that need to be uniquely identified to allow traceability.

Proficiency testing has long been used to assess capability and competency in brewing analysis. Its use in brewing microbiology is relatively recent, owing much to the development of proficiency testing in the water industry. BAPS Microbiology, a popular scheme in the UK, is operated by the Laboratory of the Government Chemist together with BRi. In essence, blind samples are supplied to participating laboratories for qualitative and quantitative analysis using routine microbiological methods. This approach enables the measurement of the capability of the laboratory to recover, quantify and identify what should be familiar brewery micro-organisms. In addition, proficiency testing is useful in validating the competence of test media and the microbiologist. Used appropriately, proficiency testing is a valuable tool for improvement and for assuring performance via an independent route.

In passing, it is worth noting some of the common pitfalls in microbiological testing that undermine performance. Perhaps one of the most common is the failure post-preparation to adjust the pH of the medium. Kindraka (1987) is fulsome in the importance of this adjustment when considering the efficacy of NBB medium. Temperature abuse features frequently in the inconsistent performance of microbiological media. This theme includes the addition of heat sensitive inhibitors (cycloheximide, vancomycin, phenylethanol) prior to autoclaving or directly to very hot media ex-autoclave and 'storage' of media prior to pouring plates at too high a temperature and for too long. The guidelines – which are all too easily ignored – require microbiological media to be stored or 'tempered' at $46 \pm 1°C$ for no longer than three hours. It is good laboratory practice to add heat sensitive materials to tempered media. Similarly, tempered media should be used where larger sample volumes (e.g. 5 ml) are being processed as 'pour plates'. However, in terms of avoiding heat stress and achieving consistency, it is preferable to avoid pour plates and process such volumes via filtration and incubation of the membrane on the top of the plate.

### 8.3.3 *Methods – 'real time' and 'rapid'*

Arguably the nirvana for brewing microbiology is that QA systems and in-line monitoring is so sophisticated and interactive that there is no requirement for conventional sampling and testing! Today's reality is such that the focus has been on accelerating testing methodologies, with the ultimate objective of achieving results in 'real time' on a par with the likes of beer colour, $CO_2$, and ABV. Similarly, much effort has gone into exploiting new technologies for the quicker and less equivocal identification of brewing micro-organisms.

In the last 20 years or so, brewing microbiology has sought to exploit a variety of new technologies that offer improvements over traditional approaches. With the honourable exception of ATP bioluminescence testing in CiP, the 'real-time' objective has been revised down to being quicker or more 'rapid' over and above traditional methods. Although not as dramatic, these developments should not be decried as they frequently reduce testing times by many days. In a world where 'time is money', these developments have value, and frequently potential. For a fuller review than can be given here the interested reader is directed toward Dowhanick (1995), Russell and Dowhanick (1996) in *Brewing Microbiology* and Storgårds *et al.* (1997).

#### 8.3.3.1 *ATP bioluminescence.*

There is no doubt that ATP (adenosine triphosphate) bioluminescence is the great success story of brewing microbiology in the late twentieth century. Indeed, the implementation of this technology has truly achieved a step change in our approach to hygiene testing in particular and microbiology in general. It is now accepted practice to validate CiP performance in real time and, where required, to reclean. With the technology evolving by leaps and bounds, testing is now easily performed in process areas with all-in-one dipstick tests and small portable testing units. Accordingly, hygiene testing is no longer the preserve of the QA laboratory but is performed and responded to by the appropriate people, the process owners. In addition to these tangible changes, ATP bioluminescence has played its part in raising the profile of hygiene and microbiology in the brewing industry. For many, this technology has made microbiology 'real'!

Bioluminescence is the phenomenon in which visible light is emitted by an organism. The history and evolution of ATP bioluminescence (Simpson,1991b) dates back to 1884 when DuBois demonstrated that a crude extract of fireflies (*Photinus pyralis*) emitted light ('bioluminescence') that eventually faded and disappeared. He also coined the terms 'luciferin' and 'luciferase' for, respectively, the reaction's substrate (a 6-hydroxybenzothiazole) and enzyme. Intriguingly, in the wild, bioluminescence acts as a signal to arrange sexual encounters between consenting fireflies! In 1947, McElroy identified the missing piece of the jigsaw with the demonstration that ATP triggered light formation from the firefly extract. The reaction involves the oxidation of luciferin by luciferase with the oxidant being molecular oxygen:

$$ATP + Luciferin + O_2 \xrightarrow{Luciferase} AMP + Oxyluciferin + CO_2 + PPi + Light \ (\lambda_{max} = 532 \ nm)$$

These early observations have spawned a technology built around the release of 'living light' in proportion to very small amounts of ATP — the 'universal' energy molecule found in *all* living cells. The quantum yield from the luciferin–luciferase reaction with ATP approaches unity, with a photon of light being emitted for every molecule of ATP utilised. In passing, bioluminescence in the jellyfish (*Aequorea aequorea*) is triggered by calcium ions. This has been exploited in the laboratory for tracking the movements of $Ca^{2+}$ in biological systems.

Inevitably, today's familiar technology has taken time to deliver with numerous innovations along the way. Initially, in the 1960s, ATP bioluminescence was considered for glamorous applications such as the detection of extraterrestrial life forms in NASA's 'Life on Mars' programme (Simpson, 1991). Since then, although the ambition remained the same, the technology has been brought down to earth for the detection of micro-organisms. However, up and until the late 1980s progress in developing ATP technology was 'exasperatingly slow' (Stanley *et al.*, 1997). Since then, fuelled by innovation, change has been dramatic with the development of simple, user-friendly analyses and portable, robust 'bioluminometers'. The scope of bioluminescence is impressive and can be gauged by presentations at an international conference on bioluminescence (ATP 96) (Stanley *et al.*, 1997). These included personal care products, milk hygiene, animal and poultry carcasses, effluent, pharmaceutical products and brewing.

As described in Section 8.2.1.1, bioluminescence is used in brewing to validate CiP operations in real time, which today is within seconds of sampling! At its simplest, a dirty tank has a high bioluminescence. The source of the ATP can be microbial cells or 'soil'. For this reason it is foolhardy to seek to correlate ATP (RLU) against conventional plate counts. Indeed, failure to achieve a correlation should be expected but it should be appreciated that this does not imply a failure in the capability of either approach! In a landmark paper (see below), Hysert *et al.* (1976) estimated yeast cells to have about 100 times more ATP than bacterial cells. They found 0.24 fmol ATP (or $0.24 \times 10^{-15}$ moles) in an 'average' yeast cell compared to 0.0025 fmol ATP in an 'average' bacterial cell). In terms of 'soil', Simpson *et al.* (1989) reported the concentration of ATP in beer to range widely from a low of 0.01 nM to a high of 100 nM, with a mean of 5 nM. In this context, the report of Boyum and Guidotti (1997) is of particular interest as they showed the glucose dependent efflux of ATP from *S. cerevisiae*. There is no apparent reason why these laboratory observations cannot extrapolate to brewery fermentations. Further work in this intriguing area would be most welcome!

Although differentiation is possible ('total' v. 'free' ATP), the source of the ATP is generally immaterial. From the perspective of poor CiP, it is usually sufficient to recognise that the vessel is not clean and to be unconcerned whether the ATP originates from predominately micro-organisms or beer residues. What is more important is real-time action in response to a real-time result. This enables appropriate corrective action to be taken at the time – an opportunity that was hitherto unavailable through traditional microbiological sampling and testing.

The first reported application of ATP bioluminescence in brewing (Hysert *et al.*, 1976) anticipates much of what has subsequently happened and makes fascinating reading. The technology however is very different with DMSO extraction of ATP and light measurement using a liquid scintillation counter! Tellingly in the final paragraph of this paper, Hysert *et al.* (1976) note the applications as 'monitoring biocide effectiveness in recirculating cleaning systems, evaluating the general microbiological state of brewing facilities and basic biochemical brewing research'. This predication has proved to be remarkably accurate with ATP bioluminescence now being used routinely for CiP validation, but also finding increasing application in reducing incubation times for product testing.

*Hygiene testing.* Since the early work and proposals of Kilgour and Day (1983) and Simpson *et al.* (1989), the application of ATP bioluminescence in hygiene testing has been slowly threatening to come of age. The seminal paper, 'Practical experiences of hygiene control using ATP-bioluminescence' (Ogden, 1993) took the technology out of the laboratory into routine application in the brewery. This two-year study applied bioluminescence in the real time testing of hygiene swabs. 'Go/no go' specifications were established which had to be passed before the plant could be used. As is shown in the paper, this initiative clearly demonstrated a 'significant improvement in the hygiene and cleanliness of the plant, which could only have a beneficial effect on the quality of the product' (Ogden, 1993). Despite this positive endorsement, take-up of the technology immediately thereafter remained patchy.

With hindsight, the most likely explanation for the sluggish response was that the technology was still focused on 'swabs' and was still in the domain of the 'specialist', who was comfortable with user-hostile laboratory-based testing. Indeed, it is instructive to revisit the assay as used at the time by Ogden (1993). Here, swabs were wetted with a nutrient medium and, after use, agitated for 10 seconds in a cuvette containing 200 µl of a 'nucleotide releasing reagent'. After adding the enzyme (100 µl) and mixing, the cuvette was 'loaded' into the bioluminometer and read. Up until the late 1990s developments were primarily embellishments of this basic approach with the inclusion of a rinse method, replacement of pipetting with dropper bottles and general improvements in detection sensitivity.

The step change that did more for the transfer of technology was the introduction of all-in-one 'single shot' testing. This together with the genuinely small, robust bioluminometers with increasingly sophisticated data capture and manipulation, has provided the framework to implement real-time testing in breweries by, as noted above, the right people, the process owners. Popular single-shot systems include the Celsis SwabMate and, from Biotrace, the Aqua-Trace™ and Clean-Trace™. The design of these single shot devices is quite different and accordingly there is no universal bioluminometer. Accordingly, Celsis use the SystemSURE™ and Biotrace the

Uni-Lite® and Uni-Lite® Xcel. The principle behind the single-shot systems can be appreciated from the simplicity of the Biotrace approach. Here dipsticks are available in two formats, with a swab (Clean-Trace™) or with a series of small plastic rings that capture by capillary action a small volume of a final rinse (Aqua-Trace™). The bioluminescence reagents are stored in a separate compartment at the bottom of the dipstick holder. Although robust enough to prevent leakage, the partition is breached by plunging the dipstick through into contact with the reagents. This activates the reaction and, after briefly mixing, the ensuing bioluminescence is quantified in a bioluminometer. Promotional images that capture 'single shot' testing are presented in Fig. 8.30.

Bizarrely, despite the obviousness of 'ATP', there is no universal standard unit for expressing ATP concentration. The output from bioluminometers is invariably RLU or relative light units. Unfortunately, for a known amount of ATP the RLU from one manufacturer's bioluminometer will be different to RLU from another manufacturer's machine. For the consumer's point of view this is unsatisfactory. Consequently, it would be a welcome development and would aid comparison, if the bioluminescence industry could agree to standardise their results in terms of ATP concentration.

Within Bass Brewers Limited, bioluminescence has been implemented with great gusto! Using Biotrace technology, CiP validation is captured within the sampling plan (Section 8.3.1.2) with global specifications for a *pass* ( < 150 RLU), *caution* (150–299 RLU) and *fail* ( > 300 RLU). As described in Section 8.2.1.1, should a rinse fail, the vessel or main is retested, CiP set visually checked and, if necessary recleaned (where commercial pressures allow). Ironically, because of its simplicity and timeliness, there is a 'that's all I need to do' attitude to the use of bioluminescence in the validation of cleaning. This is not the fault of the method but of the user! There remains the need (as described in Section 8.3.1.2) to validate the inputs (causticity, carbonate etc.), the outputs (visual assessment) of a CiP cycle and periodic microbiological checks to characterise the microflora found in the system.

Perhaps the most popular criticism of bioluminescence is that of *sensitivity*. The typical argument is that traditional microbiology has the capability to detect a single micro-organism in up to 1l of sample. Conversely, bioluminescence systems are far less sensitive. Comparison of the different manufacturers' packages has shown varying detection sensitivities for ATP (Table 8.22). Although obviously an important criterion, detection sensitivity has to be considered alongside factors such as cost, sample type, application, ease of use and, where required, data handling. So is

**Table 8.22** Comparison of commercial bioluminometers (unpublished results of Andrew Price, Brewing Research International).

| Bioluminescence 'package' | Minimum detection of ATP (fmol) | Capable of detecting $x$ yeast cells per assay |
|---|---|---|
| Hughes Whitlock Bioprobe | 0.1 | < 5 |
| Celsis SwabMate | 2.7 | 5–50 |
| Biotrace Aqua-Trace™ | 17.5 | 25–250 |
| Biotrace Clean Trace™ | 3.5 | 5–50 |

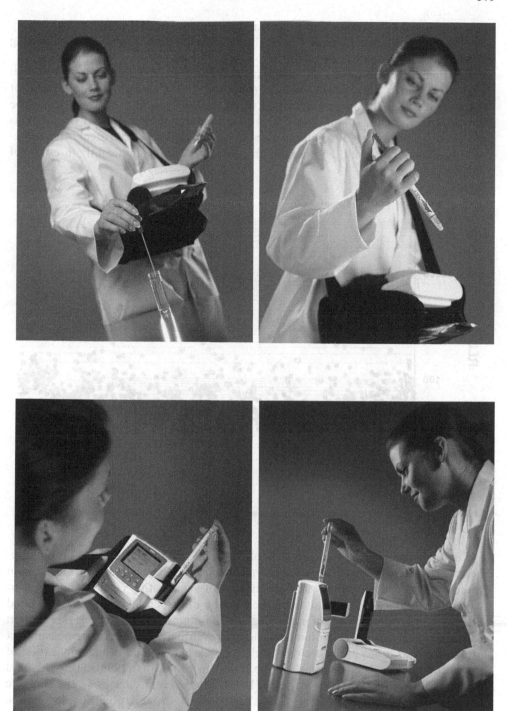

**Fig. 8.30** Images that capture the steps of a 'single shot' bioluminescence test (kindly provided by Gareth Lang, Biotrace Limited).

detection sensitivity important; is the real time advantage of bioluminescence really worth the compromise in sensitivity? Is the use of bioluminescence lowering cleaning standards as opposed to raising them?

Routine CiP rinse results (c. 1000 data points) for a low soiling area (keg line, Fig. 8.31) and a high soiling process (FV, Fig. 8.32) indicate the sensitivity of Aqua-Trace[TM] to be in accord with CiP capability (Quain, 1999). As would be anticipated, the data for the keg line is predominately below 100 RLUs, whereas the more challenging CiP of fermenters is higher and less consistent. It is noteworthy that a 'fail result' is generally unambiguous being significantly in excess of 300 RLUs.

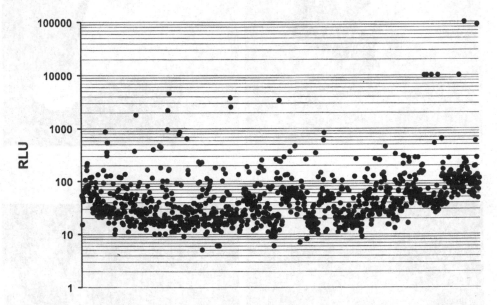

**Fig. 8.31**   Routine bioluminescence data from rinses post-CiP of keg lines (redrawn from Quain, 1999).

For the same reasons that the technology 'took off' in breweries, ATP bioluminescence is increasingly being used in validating the hygienic status of dispense in trade outlets. The immediacy and practicality of such testing clearly lends itself to validating line cleaning and auditing the hygiene of dispense spouts and associated fittings. Given the challenge of improving and maintaining trade hygiene (see Section 8.2.2), it is likely that ATP bioluminescence will eventually have a comparable impact in this arena as it has in the brewery. Relatively early examples of the application of this technology in beer dispense can be found in Storgårds (1996), Storgårds and Haikara (1996) and Orive i Camprubi (1996). More recent data using the Biotrace Clean-Trace[TM] technology (Fig. 8.33), has assessed the microbial loading of the internal surface of 50 dispense lines pre- and post-cleaning. Comparison of this RLU data and that reported above for keg line and FV (Figs 8.31, 8.32), suggests that the standards for what is clean are similar.

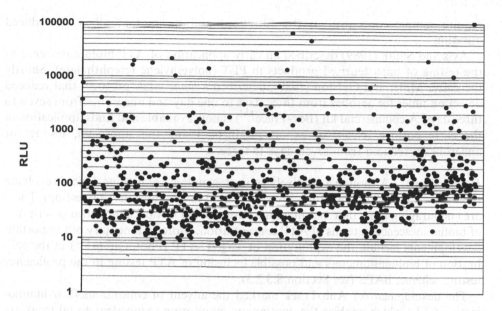

**Fig. 8.32**   Routine bioluminescence data from rinses post CiP of FVs lines (redrawn from Quain, 1999).

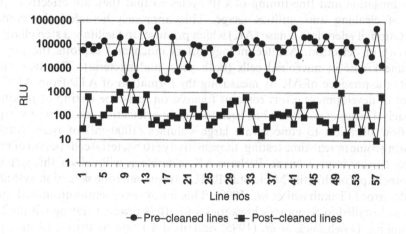

**Fig. 8.33**   Surface bioluminescence of dispense lines pre- and post-cleaning (unpublished results Gareth Lang, Biotrace Limited).

*Product testing.*  The lower sensitivity and lack of selectivity have hampered the application of bioluminescence in final product testing. This is of little general significance as final product testing is of declining significance as prevention (QA) supersedes inspection (QC) in brewing microbiology. However, the more stringent shelf-life demands of export products, sterile filtered products and occasional need for 'positive release', have driven the development of specific bioluminescence protocols that accelerate conventional microbiological testing (for a review see Thompson & Jones, 1997). To improve sensitivity of detection, these hybrid protocols invariably require sample concentration (by filtration) and incubation, albeit of reduced duration.

Avis and Smith (1989) described an early application of ATP bioluminescence to the testing of unpasteurised products in PET (polyethylene terephthalate). Shortly thereafter Miller and Galston (1989), described a *home made* protocol that reduced detection times for aerobes from three days to one day and anaerobes from seven to three days. A commercial kit (Bev-Trace$^{TM}$) is now available and finds application in the early positive release of export products. Using this approach, testing for anaerobes is reduced from seven days to three.

*Developments.* Although a relatively young technology, bioluminescence is evolving rapidly (see Simpson (1999) for a review of recent innovations in ATP testing). There are encouraging signs that manufacturers are moving toward a universal description of bioluminescence in terms of ATP concentration. More rudimentary but important developments include the exploitation of 'caged' ATP (PhotoQuant$^{TM}$) in the calibration of bioluminometers and possible inclusion of ATP testing in the proficiency testing scheme, BAPS (see Section 8.3.2.3).

The development of AutoTrack marked the advent of continuous-flow luminometry (CFL) which enables the continuous monitoring (equivalent to 60 measurements per minute) of ATP during a CiP cycle (Brady, 1999). AutoTrack will enable the optimisation and fine-tuning of CiP cycles so that they are effective – both in terms of cleaning and utilities usage. This approach heralded the commercial exploitation of adenylate kinase (AK) which promises to facilitate a step change in the sensitivity of detection. AK technology offers much and exploits the presence of adenylate kinase in microbial cells (both yeast and bacteria). This new approach exploits the presence of AK by measuring the formation of ATP from ADP.

A lot of development effort remains focused on real-time testing of product. One approach (Davies *et al.*, 1995) has been to improve detection sensitivity by exploiting cross-flow filtration to concentrate large volumes (1000 ml) of beer. Although a promising route to real-time testing, the sensitivity (10 bacterial cells per ml of raw beer) requires further improvement. Perhaps AK technology will deliver this step change.

Another approach – the MicroStar-RMDS (rapid microbe detection system) – has been described (Takahashi *et al.*, 1999). This innovative, semi-automated approach has been installed in Sapporo's breweries and offers product testing within 24 hours of sampling. Dowhanick *et al.* (1995) described a route to truly real-time product testing using a prototype Millipore RMD (rapid micro detection) system. This mix of technologies is reported to detect one to 200 yeast cells in less than 10 minutes although detection of bacteria may be less rapid. This innovative technology couples real-time imaging (CCD camera) of micro-organisms trapped in novel 'hydrophobically gridded' membranes with fibre optics image intensification and (inevitably) computer data analysis. Although clearly highly promising, Dowhanick *et al.* (1995) hint that further development of the RMD approach may hinge around an 'economic cost-benefit evaluation'.

8.3.3.2 *Other rapid methods.* The focus of the 'new microbiology' here and elsewhere, has been on bioluminescence. However, in the early 1980s bioluminescence was one of many putative 'real-time' or 'rapid' technologies that came to the attention of microbiologists. Although bioluminescence won this particular race it is interesting to speculate which of the new crop of methods described below (Section 8.3.4) will come out as the next 'bioluminescence' and change brewing microbiology once more.

With hindsight, the paper by Kilgour and Day at the 1983 EBC recognised the need for change in brewing microbiology. In addition to bioluminescence, they evaluated two other fledgling technologies, DEFT (direct epifluorescent filter technique) and an electrometric method based on growth-related changes in media conductance detected using a 'Malthus' instrument. In passing, it is interesting to note that unlike bioluminescence, these two approaches enabled some degree of identification of the offending micro-organisms. Together these three approaches and to a lesser extent methods such as ELISA (enzyme linked immunosorbent assays) provided the agenda for microbiological research for much of the 1980s and early 1990s.

The 'microcolony' and DEFT methods accelerate detection of micro-organisms by staining them with fluorescent dyes, which can be more readily visualised under a fluorescent microscope. Although frequently described separately, the two methods have much in common. Both involve sample concentration on a membrane, staining and counting the colonies. However DEFT is much quicker in requiring no incubation step prior to analysis whereas the microcolony approach require incubation overnight (aerobes) or for up to three days (anaerobes). In the case of DEFT the membrane is made of polycarbonate and the fluorescent dye is the DNA stain acridine orange. In passing, DEFT and its simpler relative DEM (direct epifluorescent microscopy) are routinely used in visualising and quantifying attached micro-organisms in biofilms (Holah et al., 1988). With the microcolony approach, black membranes are used to improve the visualisation of the optical brighteners (as used in domestic washing powders) which are used to stain the microcolonies.

The implementation of both methods has suffered from their dependency on microscopy. As noted by Storgårds et al. (1997) 'microscopic counting is time consuming and tedious, leads to operator fatigue and limits the capacity' of both approaches. Accordingly, both approaches have been automated (see Parker (1989) for an early example) with varying degrees of success and cost! For details, the interested reader should consult Simpson (1991), Russell and Dowhanick (1996) and Storgårds et al. (1997).

The other approaches that promised but never quite delivered in brewing microbiology were the impedimetric methods. This electrometric method exploited the changes in the electrical behaviour of the growth medium that occur in response to microbial metabolism. Without going into detail (see Russell & Dowhanick, 1996), microbial growth could be detected by an increase in impedance or decrease in conductivity and capacitance. Although the sensitivity was relatively poor (detection of 10 bacteria ml$^{-1}$ required 18 hours' incubation), Evans (1982) described how this approach reduced the time required by a 'forcing test' from weeks to days! Kyriakides and Thurston (1989) enthusiastically 'threw the book' at this technology with particular emphasis on the opportunity to replace conventional plate counts. Unfortunately, the conclusions from this work were lukewarm, with the growth medium lacking the required selectivity to support the growth of the desired contaminants.

Although ELISA is described elsewhere (see Section 4.2.5.5), it is worth touching on the application of immunoassay described by Ziola and colleagues. Using monoclonal antibodies, membrane-filter-based immunofluorescent antibody tests have been worked up for *Pediococcus* species (Whiting *et al.*, 1992), *Pectinatus cerevisiiphilus* (Gares *et al.*, 1993) and *Lactobacillus* species (Whiting *et al.*, 1999). Although the sensitivity is not quite good enough, this technology may yet be fine-tuned to offer a rapid microbiological solution.

### 8.3.4   *Methods – current developments.*

The thrust of much of today's microbiological innovation and development mirrors that described elsewhere (Section 4.2.6) for the differentiation and identification of brewing yeasts. Genetic approaches have been grabbed enthusiastically in the hunt for truly rapid detection and identification of beer spoilage organisms. Methods include PCR (see Section 4.2.6.2) and, more recently, ribotyping, which is restricted to bacteria. Although less to the fore, there are signs that phenotypic methods are now entering the fray.

#### 8.3.4.1   *Phenotypic methods.* 
Pyrolysis mass spectroscopy (PyMS) and Fourier transform infrared spectroscopy (FT-IR) provide a different spin to rapid micro-biology. Both exploit differences in cell composition and both require sophisticated data handling to draw out what are frequently revealing results. Both approaches are genuinely simple and rapid (two minutes for PyMS and 10 seconds for FT-IR) and potentially offer a definitive solution to identification of colonies on plates. Both approaches have found a niche in the differentiation of brewing yeast and accordingly are described elsewhere: Section 4.2.6.5 (PyMS) and Section 4.2.6.6 (FT-IR).

Of the two, PyMS has a higher profile finding particular application in the differentiation of clinically important bacteria (Taylor *et al.*, 1998; Barshick *et al.*, 1999). Preliminary work (Quain, 1999) has used both PyMS and FT-IR to good effect to track the location of environmental isolates of *Lactobacillus*. As discussed below this work has provided insight into the complexity of the different strains of bacterial species. It is anticipated that the use of these approaches in brewing microbiology will grow.

#### 8.3.4.2   *Genotypic methods.* 
The PCR (polymerase chain reaction) approach is now a staple method of molecular biology (for background see Section 4.2.6). For reviews of the application of PCR in brewing microbiology, the interested reader should consult Dowhanick (1995), Russell and Dowhanick (1996) and Storgårds *et al.* (1997). Although PCR has been applied with a vengeance to needs of brewing microbiology (see Table 8.23 for examples), there is a suggestion that the 'bubble has burst' for this technology. Although now more kit-based (and simpler!), the use of PCR in brewing has been frustrated by contamination issues (resulting in the amplification of the wrong bit of DNA) and a variety of the inhibition effects. Most damning of all are the contradictory objectives of simplicity and sensitivity. As succinctly put by Storgårds *et al.* (1997) 'if a simple protocol for sample treatment is used it is not possible, at present, to achieve the required sensitivity'. Although not down and out, PCR requires further effort and commitment to make the transition from the research bench to the brewery QA department.

**Table 8.23** Some examples of the use of PCR in the detection and identification of beer spoilage organisms.

| Year | Target(s) | Reference | Location |
|------|-----------|-----------|----------|
| 1992 | L. brevis, S. cerevisiae | Tsuchiya et al. (1992) | Japan |
| 1993 | L. brevis, L. casei, L. plantarum | DiMichele and Lewis (1993) | USA |
| 1996 | Lactobacilli, Pediococci, Leuconostoc and Saccharomyces species | Tompkins et al. (1996) | Canada |
| 1996 | Lactobacilli, Pediococci, Leuconostoc species | Stewart and Dowhanick (1996) | Canada |
| 1997 | L. brevis, L. casei, L. coryniformis, L. planarum, Pectinatus cereviiphilus, P. frisingensis, Megasphaera cerevisiae | Sakamoto et al. (1997) | Japan |
| 1998 | M. cerevisiae, P. frisingensis | Satokari et al. (1998) | Finland |
| 1998 | S. diastaticus | Yamauchi et al. (1998) | Japan |

Another molecular method, ribotyping, first appeared on the scene in 1988. As yet this is not a rapid method but one that aids identification of bacteria. Fundamentally, ribotyping revolves around the genetic fingerprinting of ribosomal RNA using the RFLP (restriction fragment length polymorphism) approach (see Section 4.2.6.1). Although ribotyping can be performed manually (Simpson & Priest, 1999), its application has been aided by its automation through Qualicon's RiboPrinter. This technology removes much of the human interface and, importantly, enables comparison and identification of the bacteria against a large and growing database. Although the capital cost of this equipment is substantial, ribotyping is increasingly available as a contract analysis. With the exception of a report on the characterisation of *Pectinatus* species (Motoyama *et al.*, 1998), ribotyping in brewing microbiology has focused on the *Lactobacillus* (Funahashi *et al.*, 1998; Storgårds *et al.*, 1998; Quain, 1999; Simpson & Priest, 1999).

Together with some of the newer phenotype-based approaches (see above), the use of ribotyping has been an important force in opening the eyes of microbiologists to the complexity of brewery microflora. As with brewing yeasts there is a vast armada of different strains of, say, *L. brevis* populating brewery environments. As noted by Quain (1999), 'the isolate of *L. brevis* found in the brewhouse was different to that found in stored water and both were distinct from that isolated from yeast'. This sobering and simple fact has often been lost in the race to detect and identify microorganisms.

# References

van der Aa Kuhle, A. & Jespersen, L. (1998) Detection and identification of wild yeasts in lager breweries. *International Journal of Food Microbiology*, **43**, 205–13.

van der Aar, P. (1996) Consequences of yeast population dynamics with regard to flocculence. *Ferment*, **9**, 39 42.

Aastrup, S. & Erdal, K. (1987) A mass balance study of beta-glucan in malt, spent grains and wort using the calcofluor method. *Proceedings of the 21st Congress of the European Brewery Convention*, Madrid, 353–60.

Achsetter, T. & Wolf, D.H. (1985) Proteinases, proteolysis and biological control in the yeast *Saccharomyces cerevisiae*. *Yeast*, **1**, 139–57.

Adams, J., Puskas-Rozsa, S., Simlar, J. & Wilkie, C.M. (1992) Adaptation and major chromosomal changes in populations of *Saccharomyces cerevisiae*. *Current Genetics*, **22**, 13–19.

Adams, M.R., O'Brian, P.J. & Taylor, G.T. (1989) Effect of the ethanol content of beer on the heat resistance of a spoilage *Lactobacillus*. *Journal of Applied Bacteriology*, **66**, 491–5.

Aguilera, A. & Benitez, T. (1985) Role of mitochondria in ethanol tolerance of *Saccharomyces cerevisiae*. *Archives of Microbiology*, **142**, 389–92.

Aigle, M., Erbs, D. & Moll, M. (1983) Determination of brewing yeast ploidy by DNA measurement. *Journal of the Institute of Brewing*, **89**, 72–4.

Aires Barros, M.R., Barros, M.R., Cabral, J.M.S. & Novais, J.M. (1987) Production of ethanol by immobilised *Saccharomyces bayanus* in an extractive fermentation system. *Biotechnology & Bioengineering*, **24**, 1097–1104.

Aires, V., Kirsop, B.H. & Taylor, G.T. (1977) Yeast lipids. *Proceedings of the 16th Congress of the European Brewery Convention*, Amsterdam 255–66.

Akiyama-Jibiki, M., Ishibiki, T., Yamashita, H. & Eto, M. (1997) A rapid and simple assay to measure flocculation in brewer's yeast. *Technical Quarterly of the Master Brewers Association of the Americas*, **34**, 278–81.

Alexandre, H., Rousseaux, I. & Charpentier, C. (1994) Relationship between ethanol tolerance, lipid composition and plasma membrane fluidity in *Saccharomyces cerevisiae* and *Kloekera apiculata*. *FEMS Microbiology Letters*, **124**, 17–22.

Alic, M. (1999) Baker's yeast in Crohn's disease – can it kill you? *American Journal of Gastroenterology*, **94**, 1711.

Alterhum, F., Dombek, K.M. & Ingram, L.O. (1989) Regulation of glycolytic flux and ethanol production in *Saccharomyces cerevisiae*: effects of intracellular adenine nucleotide concentrations on the *in vivo* activities of hexokinase, phosphofructokinase, phosphoglycerate kinase and pyruvate kinase. *Applied and Environmental Microbiology*, **55**, 1312–14.

Amaha, M., Nakakoji, S. & Komiya, Y. (1977) Some experiences in the one-tank process with Asahi large capacity tanks. *Proceedings of the 16th Congress of the European Brewery Convention*, Amsterdam, 545–59.

American Society of Brewing Chemists (1992) Diacetyl. *Methods of Analysis*, 8th edition, published by the American Society of Brewing Chemists, Minnesota, USA, **25**, 1–6.

Amory, D.E., Rouxhet, P.G. & Dufour, J.P. (1988) Flocculence of brewery yeasts and their surface properties: chemical composition, electrostatic charge and hydrophobicity. *Journal of the Institute of Brewing*, **94**, 79–84.

Amri, M.A., Bonaly, R. Duteurtre, B. & Moll, M. (1982) Yeast flocculation: influence of nutritional factors on cell wall composition. *Journal of General Microbiology*, **128**, 2001–8.

Anderson, G.J., Lesuisse, E., Dancis, A., Roman, D.G., Labbe, P. & Klausner, R.D. (1992) Ferric iron reduction and iron assimilation in *Saccharomyces cerevisiae*. *Journal of Inorganic Biochemistry*, **47**, 249–55.

Anderson, R.G. (1989) Yeast and the Victorian brewers: incidents and personalities in the search for the true ferment. *Journal of the Institute of Brewing*, **95**, 337–45.

Anderson, R.G. (1990) Aspects of fementation control. *Ferment*, **3**, 242–9.

Anderson, R.G. (1991) Echoes of journals past. *Ferment*, **4**, 288–97.

Anderson, R.G. (1993) Highlights in the history of international brewing science. *Ferment*, 6, 191–8.

Anderson, R.G. (1995) Louis Pasteur (1822–1895): an assessment of his impact on the brewing industry. *Proceedings of the 25th Congress of the European Brewery Convention*, Brussels, 13–23.

Andersson, L.E. & Norman, H. (1997) Water deaeration, blending and carbonation in high gravity brewing. *Brauwelt International*, 15, 59–61.

Andreason, A.A. & Stier, T.J.B. (1953) Anaerobic nutrition of *Saccharomyces cerevisiae* 1. Ergosterol requirement for growth in a defined medium. *Journal of Cellular and Comparative Physiology*, 41, 23–6.

Andreason, A.A. & Stier, T.J.B. (1954) Anaerobic nutrition of *Saccharomyces cerevisiae* 2. Unsaturated fatty acid requirements for growth in a defined medium. *Journal of Cellular and Comparative Physiology*, 43, 271–8.

Andrews, J. & Gilliland, R.B. (1952) Super-attenuation of beer: a study of three organisms capable of causing abnormal attenuations. *Journal of the Institute of Brewing*, 58, 189–96.

Andrews, J.M.H. (1988) Whirlpool design and manufacture. *Ferment*, 1, 47–8.

Andries, M., Derdelinckx, G., Ione, K.G., Delvaux, F., van Beveren, P.C. & Masschelein, C.A. (1997) Zeolites as catalysts for the cold and direct conversion of acetolactate into acetoin. *Proceedings of the 26th Congress of the European Brewery Convention*, Maastricht, 477–84.

Andries, M., van Beveran, P.C., Goffin, O., Rajotte, P. & Masschelein, C.A. (1997) First results on semi-industrial continuous top fermentation with the Meura-Delta immobilised yeast reactor. *Technical Quarterly of the Master Brewers Association of the Americas*, 34, 119–22.

Annëmuller, G. & Manger, H.-J. (1997) Pitching and starting phase in cylindroconical fermenter – the black box of the fermentation and maturation process? *Brauwelt*, 4, 338–41.

Anness, B.J. (1981) The role of dimethyl sulphide in beer flavour. *Proceedings of the European Brewery Convention Symposium*, Monograph VII, Copenhagen, 135–42.

Anness, B.J. (1984) Lipids of barley, malt and adjuncts. *Journal of the Institute of Brewing*, 90, 315–18.

Anness, B.J. & Bamforth, C.W. (1982) Dimethyl sulphide – a review. *Journal of the Institute of Brewing*, 88, 244–52.

Anness, B.J., Bamforth, C.W. & Wainwright, T. (1979) The measurement of dimethyl sulphoxide in barley and malt and its reduction to dimethyl sulphide by yeast. *Journal of the Institute of Brewing*, 85, 346–9.

Anness, B.J. & Reed, R.J.R. (1985) Lipids in wort. *Journal of the Institute of Brewing*, 91, 313–17.

Anonymous (1991) Woodforde's – 'a scaled up version of a home-brew kit'. *Brewers' Guardian*, January, 26–7.

Anonymous (1992) Monitoring the flow. *Brewers' Guardian*, August, 27.

Anonymous (1995) New interest in older yeast strains keeps NCYC busy. *Brewers' Guardian*, September, 14–15.

Anonymous (1996a) In *Cell Separation and Protein Purification*, Dynal, Technical Handbook, 2nd edition, Norway.

Anonymous (1996b) Search on 'fungal species'. In *Encarta 96 Encyclopaedia CD-ROM* World English edition, Microsoft Home.

Anonymous (1997) Compact high gravity brewing plants. *Brauwelt International*, 15, 260–62.

Anonymous (1999) Microbiological Millennia. *Ferment*, April/May, 45–9.

Anraku, Y., Umemoto, N., Hirata, R. & Wada, Y. (1989) Structure and function of the yeast vacuolar membrane proton ATPase. *Journal of Bioenergetics and Biomembranes*, 21, 589–603.

Anselme, M.J. & Tedder, D.W. (1987) Characteristics of immobilised yeast reactors producing ethanol from glucose. *Biotechnology & Bioengineering*, 30, 736–45.

App, H. & Holzer, H. (1989) Purification and characterisation of neutral trehalase from the yeast ABYS1 mutant. *Journal of Biological Chemistry*, 264, 17583–7.

Arcay-Ledzema, G.J. & Slaughter, J.C. (1984) The response of *Saccharomyes cerevisiae* to fermentation under carbon dioxide pressure. *Journal of the Institute of Brewing*, 90, 81–4.

Armitt, J.D. & Healy, P. (1974) The half-way stress in the fermentation of an Australian beer. *Proceedings of the 13th Congress of the Institute of Brewing (Australia and New Zealand Section)*, Queensland, 91–103.

Arnold, W.N. (1991) Periplasmic space. In *The Yeasts*, 2nd edition (eds A.H. Rose and J.S. Harrison), vol. 4, pp. 279–95. Academic Press, London.

ASBC Analysis Committee (1980) Measurement of yeast viability. *Journal of the American Society of Brewing Chemists* 38, 109–10.

ASBC Analysis Committee (1981) Measurement of yeast viability. *Journal of the American Society of Brewing Chemists* 39, 86–9.

Aschengreen, N.H. & Jepsen, S. (1992) Use of acetolactate decarboxylase in brewing fermentations. *Proceedings of the 22nd Institute of Brewing Convention, Australia and New Zealand Section*, Melbourne, 80–83.

Ashurst, K. (1990) Methods of propagating pitching yeast. *Brewing and Distilling International*, 21, 28–9.

Aswathanarayana, N.V. (1958) Behaviour of the vacuole on stimulation of yeast cells with fresh medium. *Proceedings of the Indian Academy of Sciences*, 47, 225–32.

Atkinson, B. & Taidi, B. (1995) Technical and technological requirements for immobilised systems. *Proceedings of the European Brewery Convention*, Monograph, **XXIV**, Espoo, 17–22.

Atomi, H., Ueda, M., Suzuki, J., Kamada, Y. & Tanaka, A. (1993) Presence of carnitine acetyltransferase in peroxisomes and in mitochondria. *FEMS Microbiology Letters*, **112**, 31–4.

Attfield, P.V., Raman, A. & Northcott, C.J. (1992) Construction of *Saccharomyces* strains that accumulate relatively low concentrations of trehalose and their application in testing the contribution of the disaccharide to stress tolerance. *FEMS Microbiology Letters*, **94**, 271–6.

Augustyn, O.P.H. & Kock, J.L.F. (1989) Differentiation of yeast species, and strains within a species, by cellular fatty acid analysis. 1. Application of an adapted technique to differentiate between strains of *Saccharomyces cerevisiae*. *Journal of Microbiological Methods*, **10**, 9–23.

Augustyn, O.P.H., Ferreira, D. & Kock, J.L.F. (1991) Differentiation between yeast species, and strains within a species, by cellular fatty acid analysis. *Systematic and Applied Microbiology*, **14**, 324–34.

Ault, R.G., Hampton, A.N., Newton, R. & Roberts, R.H. (1969) Biological and biochemical aspects of tower fermentation. *Journal of the Institute of Brewing*, **75**, 260–76.

Austriaco, N.R. Jr (1996) Review: to bud until death: the genetics of ageing in the yeast *Saccharomyces*. *Yeast*, **12**, 623–30.

Avery, S.V. & Tobin, J.M. (1992) Mechanisms of strontium uptake by laboratory and brewing strains of *Saccharomyces cerevisiae*. *Applied and Environmental Microbiology*, **58**, 3883–9.

Avis, J.W. (1990) Microbiological quality hazard analysis. *Proceedings of the 21st Congress of the Institute of Brewing (Australia & New Zealand Section)*. Auckland, 164–7.

Avis, J.W. & Smith, P. (1989) The use of ATP bioluminescence for the analysis of beer in polyethylene terephthalate (PET) bottles and associated plant. In *Rapid Microbiological Methods for Foods, Beverages and Phamaceuticals*, (eds C.J. Stannard, S.B. Petitt and F.A. Skinner), pp. 1–12. Blackwell Scientific Publications, Oxford.

Axelsson, A. (1988) Experimental studies of immobilised yeast packed bed reactors with reduced carbon dioxide entrapment. *Applied Biochemistry & Biotechnology*, **18**, 91–109.

Ayrapaa, T. (1965) The formation of phenylethanol from [14]C-labelled phenylalnine. *Journal of the Institute of Brewing*, **71**, 341–7.

Ayrapaa, T. (1967a) Formation of higher alcohols from [14]C-labelled valine and leucine. *Journal of the Institute of Brewing*, **73**, 17–30.

Ayrapaa, T. (1967b) Formation of higher alcohols from amino acids derived from yeast proteins. *Journal of the Institute of Brewing*, **73**, 30–33.

Ayrapaa, T. (1968) Formation of higher alcohols by various yeasts. *Journal of the Institute of Brewing*, **74**, 169–78.

Azeredo, J., Ramos, I., Rodrigues, L., Oliveira, R. Teixeira, J. (1997) Yeast flocculation: a new method for characterising cell surface interactions. *Journal of the Institute of Brewing*, **103**, 359–61.

Bacallao, R. & Stelzer, E.H.K. (1989) Presentation of biological specimens for observation in a fluorescent confocal microscope and operational principles of confocal fluorescence microscopy. *Methods in Cell Biology*, **31a**, 454–62.

Back, W., Leibhard, M. & Bohak, I. (1992) Flash pasteurisation – membrane filtration. Comparative biological safety. *Brauwelt International*, **1**, 42–9.

Back, W. & Pittner, H. (1993) Kontinuierliche herstellung gesauerter wurze mit half ommibilisierter milschsaurebakterien. *Monatsschrift fur Brauwissenschaft*, **10**, 364–71.

Baldwin, W.W. & Kubitschek, H.E. (1984) Buoyant density variation during the cell cycle of Saccharomyces cerevisiae. *Journal of Bacteriology*, **158**, 701.

Ball, A. (1999) Some thoughts on CiP in the brewing industry. *The Brewer*, March, 120–24.

Ball, W.J. Jr & Atkinson, D.E. (1975) Adenylate energy charge in *Saccharomyces cerevisiae* during starvation. *Journal of Bacteriology*, **121**, 975–82.

Ballou, C.E. (1982) Yeast cell wall and cell surface. In *The Molecular Biology of the Yeast Saccharomyces, Metabolism and Gene Expression*, (eds J.F. Strathern, E.W. Jones and J.R. Broach), pp. 335–60. Cold Spring Harbor Laboratory, USA.

Bamforth, C.W. (1986) Beer flavour stability. *The Brewer*, February, 48–51.

Bamforth, C.W. (1998) Grain to glass – the basics of malting and brewing. In *Beer. Tap into the Art and Science of Brewing*, pp. 35–52. Insight Books, Plenum Press, New York.

Bamforth, C.W. (1999) The science and understanding of the flavour stability of beer: a critical assessment. *Brauwelt International*, **17**, 98–110.

Bamforth, C.W., Boulton, C.A., Clarkson, S.P. & Large, P.J. (1988) The effect of oxygen on brewery process performance. *Proceedings of the 20th Convention of the Institute of Brewing (Australian & New Zealand Section)*, Brisbane, 209–18.

Bandas, E.L. and Zakharov, M. (1980) Induction of rho minus mutations in yeast *Saccharomyces cerevisiae* by ethanol. *Mutation Research*, **71**, 193–9.

Bandlow, W., Strobel, G., Zoglowek, C., Oechsner, U. & Magdolen, V. (1988) Yeast adenylate kinase is active simultaneously in mitochondria and cytoplasm and is required for non-fermentative growth. *European Journal of Biochemistry*, **178**, 451–7.

Bank, H.L. (1988) Rapid assessment of islet viability with acridine orange and propidium iodide. *In vitro Cellular and Developmental Biology*, **24**, 266–73.

Banner, M.J. (1994) Perspectives on conveyor track treatment. *Technical Quarterly of the Master Brewers Association of the Americas*, **31**, 142–8.

Barford, J.P. (1987) The technology of aerobic yeast growth. In *Yeast Biotechnology*, (eds D.R. Berry, I. Russell and G.G. Stewart), pp. 200–230. Allen and Unwin, Hemel Hempstead, UK.

Barker, M.G. & Smart, K.A. (1996) Morphological changes associated with cellular ageing of a brewing yeast strain. *Journal of the American Society of Brewing Chemists*, **54**, 121–6.

Barker, R.L., Irwin, A.J. & Murray, C.R. (1992) The relationship bewteen fermentation variables and flavour volatiles by direct gas chromatographic injection of beer. *Master Brewers Association of the Americas Technical Quarterly*, **29**, 11–17.

Barnett, J.A. (1981) The utilisation of disaccharides and some other sugars by yeasts. *Advances in Carbohydrate Chemistry and Biochemistry*, **39**, 347–404.

Barnett, J.A. (1992a) The taxonomy of the genus *Saccharomyces* Meyen ex Rees: a short review for nontaxonomists. *Yeast*, **8**, 1–23.

Barnett, J.A. (1992b) Some controls on oligosaccharide utilisation by yeasts: the physiological basis of the Kluyver effect. *FEMS Microbiology Letters*, **100**, 371–8.

Barnett, J.A., Payne, R.W. & Yarrow, D. (1990) *Yeasts, Characteristics and Identification*, 2nd edition, Cambridge University Press, Cambridge.

Barrett, M.A. (1991) Detergents and Sterilents. *The Brewer*, May, 199–205.

Barros Lopes, M de., Soden, A., Henschke, P.A. & Langridge, P. (1996) PCR differentiation of commercial yeast strains using intron splice site primers. *Applied and Environmental Microbiology*, **62**, 4514–20.

Barshick, S.A., Wolf, D.A. & Vass, A.A. (1999) Differentiation of microorganisms based on pyrolysis ion trap mass spectrometry using chemical ionization. *Analytical Chemistry*, **71**, 633–41.

Barton, S. & Slaughter, J.C. (1992) Amino acids and vicinal diketone concentration during fermentation. *Technical Quarterly of the Master Brewers Association of the Americas*, **29**, 60–63.

Bassett, D.E. Jr., Boguski, M.S. & Hieter, P. (1996) Yeast genes and human disease. *Nature*, **379**, 589–90.

Bassett, D.E. Jr., Boguski, M.S., Spencer, F. *et al.* (1997) Genome cross-referencing and XREFdb: implications for the identification and analysis of genes mutated in human disease. *Nature Genetics*, **15**, 339–44.

Bassler, B.L. (1999) How bacteria talk to each other: regulation of gene expression by quorum sensing. *Current Opinion in Microbiology*, **2**, 582–7.

Bathgate, G.N. (1989) Cereals in scotch whisky production. In *Cereal Science and Technology* (ed G.H. Palmer), pp. 243–78. Aberdeen University Press, Aberdeen.

Bayrock, D. & Ingledew, W.M. (1997) Mechanism of viability loss during fluidized drying of baker's yeast. *Food Research International*, **30**, 417–25.

Beavan, M.J., Charpentier, C. & Rose, A.H. (1982) Production and tolerance of ethanol in relation to phospholipid fatty-acyl composition of *Saccharomyces cerevisiae* NCYC 431. *Journal of Industrial Microbiology*, **128**, 1445–7.

Becker, J.M. & Naider, F. (1977) Peptide transport in yeast: uptake of radioactive trimethionine in *Saccharomyces cerevisiae*. *Archives for Biochemistry and Biophysics*, **291**, 245–55.

Becker, J.M., Naider, F. & Katchalski, E. (1973) Peptide utilisation in yeast: studies on the methionine and lysine auxotrophs of *Saccharomyces cerevisiae*. *Biochimica et Biophysica Acta*, **291**, 388–97.

Behalova, B. & Vorisek, J. (1988) Increased sterol formation in *Saccharomyces cerevisiae*. Analysis of cell components and ultrastructure of vacuoles. *Folia Microbiologica*, **33**, 292–7.

Behalova, B., Blahova, M. & Behal, V. (1994) Regulation of sterol biosynthesis in *Saccharomyces cerevisiae*. *Folia Microbiologica*, **39**, 287–90.

Behrman, C.E. & Larson, J.W. (1987) On-line ultrasonic particle monitoring of brewing operations. *Master Brewers Association of the Americas Technical Quarterly*, **24**, 72–6.

Bell, D.J., Blake, J.D., Prazak, M., Rowell, D. & Wilson, P.N. (1991a) Studies on yeast differentiation using organic acid metabolites part 1. Development of methodology using high performance liquid chromatography. *Journal of the Institute of Brewing*, **97**, 297–305.

Bell, D.J., Blake, J.D. Prazak, M. Rowell, D. & Wilson, P.N. (1991b) Studies on yeast differentiation using organic acid metabolites part 2. Studies on the organic acid metabolites produced by yeasts grown on glucose. *Journal of the Institute of Brewing*, **97**, 307–15.

Bell, D.J., Blake, J.D., Prazak, M. & Wilson, P.N. (1991c) Studies on yeast differentiation using organic acid metabolites part 3. Studies on the metabolites produced by yeasts grown on a selection of single carbon substrates. *Journal of the Institute of Brewing*, **97** 317–22.

Bell, W. (1995) Blessed is the brewing industry. *Brewers' Guardian*, June, 18.

Bendiak, D. (1994) Quantification of the Helm's flocculation test. *Journal of the American Society of Brewing Chemists*, **52**, 120–22.

Bendiak, D. (1996) Yeast flocculation by absorbance method. *Journal of the American Society of Brewing Chemists*, **54**, 245–8.

Benito, B. & Lagunas, R. (1992) The low affinity component of *Saccharomyces cerevisiae* maltose transport is an artefact. *Journal of Bacteriology*, **174**, 3065–9.

Berminham-MaDonogh, O., Gralla, E.B. & Valentine, J.S. (1988) The copper-zinc dismutase gene of *Saccharomyces cerevisiae*: cloning, sequencing and biological activity. *Proceedings of the National Academy of Sciences USA*, **85**, 4789–93.

Berndt, J., Boll, M., Lowel, M. & Gaumert, R. (1973) Regulation of sterol biosynthesis in yeast: induction of 3-hydroxymethyl-3-glutaryl-coenzyme A to squalene. *Biochemical and Biophysical Research Communications*, **51**, 843–51.

Berny, J.-F. & Hennebert, G.L. (1991) Viability and stability of yeast cells and filamentous fungus spores during freeze-drying: effects of protectants and cooling rates. *Mycologia*, **83**, 805–15.

Berry, D.R. & Watson, D.C. (1987) Production of organoleptic compounds. *In Yeast Biotechnology* (eds D.R. Berry, I. Russell and G.G. Stewart), pp. 345–68. Allen and Unwin, London.

Bezenger, M.C. & Navarro, J.M. (1988) Alcoholic fermentation model accounting for initial nitrogen influence. *Biotechnology and Bioengineering*, **31**, 747–9.

Bishop, L.R. (1938) Experiments on top fermentation. *Journal of the Institute of Brewing*, **44**, 69–73.

Bishop, L.R. (1970) A system of continuous fermentation. *Journal of the Institute of Brewing*, **76**, 172–81.

Bisson, L.F. & Fraenkel, D.G. (1983a) Involvement of kinases in glucose and fructose uptake by *Saccharomyces cerevisiae*. *Proceedings of the National Academy of Sciences of the USA*, **80**, 1730–34.

Bisson, L.F. & Fraenkel, D.G. (1983b) Transport of 6-deoxyglucose in *Saccharomyces cerevisiae*. *Journal of Bacteriology*, **155**, 995–1000.

Bisson, L.F. & Fraenkel, D.G. (1984) Expression of kinase dependent glucose uptake in *Saccharomyces cerevisiae*. *Journal of Bacteriology*, **159**, 1013–17.

Blackwell, K.J., Singleton, I. & Tobin, J.M. (1995) Metal cation uptake by yeast: a review. *Applied and Microbial Biotechnology*, **43**, 579–84.

Blazquez, M.A., Lagunas, R., Gancedo, C. & Gancedo, J.M. (1993) Trehalose 6-phosphate, regulator of glycolysis by inhibition of hexokinases. *FEBS Letters*, **329**, 51–4.

Blomberg, A. & Adler, L. (1989) Roles of glycerol and glycerol 3-phosphate dehydrogenase (NAD) in acquired osmotolerance of *Saccharomyces cerevisiae*. *Journal of Bacteriology*, **171**, 1087–92.

Blomqvist, K., Suihko, M.-L., Knowles, J. and Penttila, M. (1991) Chromosomal integration and expression of two bacterial alpha-acetolactate decarboxylase genes in brewer's yeast. *Applied and Environmental Microbiology*, **57**, 2796–2803.

Bode, H.-P., Dumschat, M., Garoti, S and Fuhrmann, G.F. (1995) Iron sequestration by the yeast vacuole. *European Journal of Biochemistry*, **228**, 337–42.

Boeira, L.S., Bryce, J.H., Stewart, G.G. & Flannigan, B. (1999a) Inhibitory effects of *Fusarium* mycotoxins on growth of brewing yeasts. 1 zearaenone and fumonisin B1. *Journal of the Institute of Brewing*, **105**, 366–74.

Boeira, L.S., Bryce, J.H., Stewart, G.G. & Flannigan, B. (1999b) Inhibitory effects of *Fusarium* mycotoxins on growth of brewing yeasts. 2-deoxynivalenol and nivalenol. *Journal of the Institute of Brewing*, **105**, 376–81.

Bonora, A. & Mares, D. (1982) A simple colorimetric method for detecting cell viability in cultures of eukaryotic microorganisms. *Current Microbiology*, **7**, 217–22.

Boonyarat, D. & Doonan, S. (1988) Purification and structural comparisons of the cytosolic and mitochondrial fumarases from baker's yeast. *International Journal of Biochemistry*, **20**, 1125–32.

Borst-Pauwels, G.W.F.H. (1981) Ion transport in yeast. *Biochemica et Biophysica Acta*, **650**, 88–127.

Bottema, C.D.K., Rodriguez, R.J. and Parks, L.W. (1985) Influence of sterol structure on yeast plasma membrane properties. *Biochimica et Biophysica Acta*, **813**, 313–20.

Boulton, C.A. (1991) Yeast management and the control of brewery fermentations. *Brewers' Guardian*, April, 25–9.

Boulton, C.A. & Clutterbuck, V.J. (1993) Application of a radiofrequency permittivity biomass probe to the control of yeast cone cropping. *Proceedings of the 24th Congress of the European Brewery Convention*, Oslo, 509–16.

Boulton, C.A. & Quain, D.E. (1987) Yeast, oxygen and the control of brewery fermentation. *Proceedings of the 21st Congress of the European Brewery Convention*, Madrid, 401–8.

Boulton, C.A. & Quain, D.E. (1999) A novel system for propagation of brewing yeast. *Proceedings of the 27th Congress of the European Brewery Convention*, Cannes, 647–34.

Boulton, C.A. & Ratledge, C. (1985) Biosynthesis of fatty acids and lipids. In *Comprehensive Biotechnology*, (ed. M. Moo-Young), vol. 1, pp. 459–82. Pergamon Press, Oxford.

Boulton, C.A. Jones, A.R. & Hinchliffe, E. (1991) Yeast physiological condition and fermentation performance. *Proceedings of the 23rd Congress of the European Brewery Convention*, Lisbon, 385–92.

Boulton, C.A., Maryan, P.S., Loveridge, D. & Kell, D.B. (1989) The application of a novel biomass sensor to the control of yeast pitching rate. *Proceedings of the 22nd Congress of the European Brewery Convention*, Zurich, 653–61.

Bourot, S. & Karst, F. (1995) Isolation and characterisation of the *Saccharomyces cerevisiae* SUT1 gene involved in sterol uptake. *Gene*, **165**, 97–102.

Bowen, W.R. & Ventham, T.J. (1994) Aspects of yeast flocculation. Size distribution and zeta-potential. *Journal of the Institute of Brewing*, **100**, 167–72.

Boyum, R. & Guidotti, G. (1997) Glucose-dependent, cAMP-mediated ATP efflux from *Saccharomyces cerevisiae*. *Microbiology*, **143**, 1901–8.

Brada, D. & Schekman, R. (1988) Coincident localisation of secretary and plasma proteins in organelles of the yeast secretary pathway. *Journal of Bacteriology*, **170**, 2775–83.

Bradley, D.E. (1956) Carbon replica technique for microbiological specimens applied to the study of *Saccharomyces cerevisiae* with the electron microscope. *Journal of the Royal Microscopic Society*, **75**, 254–61.

Bradley, L.L. (1997) Uses of iso-alpha acids and chemically modified products. *Ferment*, **10**, 48–50.

Brady, P. (1999) Autotrack. The next generation in monitoring. *The Brewer*, March, 130–33.

Brandl, J. (1996) Yeast propagation, yeast cycles, yeast storage. *Brauwelt International*, **14**, 32–4.

Brandriss, M.C. (1983) Proline utilisation in *Saccharomyces cerevisiae*: analysis of the cloned PUT2 gene. *Molecular and Cellular Biology*, **3**, 1846–56.

Braun, E.L., Fuge, E.K., Padilla, P.A. & Werner-Washbume, M. (1996) A stationary-phase gene in *Saccharomyces cerevisiae* is a member of a novel, highly conserved gene family. *Journal of Bacteriology*, **178**, 6865–72.

Breitenbach-Schmitt, I., Heinisch, J., Schmitt, H.D. & Zimmerman, F.K. (1984) Yeast mutants without phosphofructokinase activity can still perform glycolysis and alcoholic fermentation. *Molecular and General Genetics*, **195**, 530–35.

Breitenbucher, K. & Mistler, M. (1995) Fluidised bed fermenters for the continuous production of non-alcoholic beer with open-pore sintered glass carriers. *Proceedings of the European Brewery Convention*, Monograph **XXIV**, Espoo, 77–88.

Brennan, J.G., Butters, J.R., Cowell, N.D. & Lilly, A.E.V. (1976) Plant hygiene – hygienic design, cleaning and sterilising. In *Food Engineering Operations*, 2nd edition, pp. 421–441. Applied Science Publishers, London, UK.

Brenner, R.R. (1984) Effect of unsaturated fatty acids on membrane structure and enzyme kinetics. *Progress in Lipid Research* **23**, 69–96.

Bridson, E.Y. (1998) *Oxoid Manual*, 8th edition. Oxoid Limited, Hampshire.

Briggs, D.E. (1987) Endosperm breakdown and its regulation in germinating barley. In *Brewing Science*, (ed. J.R.A. Pollock), **3**, *Academic Press, London*. **3**, 441–532.

Briggs, D.E., Hough, J.S., Stevens, R & Young, T.W. (1981) Outline of malting and brewing. In *Malting and Brewing Science*, pp. 1–14. Chapman and Hall, London,

Briggs, D.E., Young, T.W. & Harcourt, (1985) A simple experimental six-line brewery. *Journal of the Institute of Brewing*, **91**, 257–63.

Briones, A.I., Ubeda, J. & Grando, M.S. (1996) Differentiation of *Saccharomyces cerevisiae* strains isolated from fermenting musts according to their karyotype patterns. *International Journal of Food Microbiology*, **28**, 369–77.

British Standards Institution (1975) Boilers and Pressure Vessels Code. *British Standards Institution, Hemel Hempstead*.

Bromberg, S.K., Bower, P.A., Duncombe, G.R. *et al.* (1997) Requirements for zinc, manganese, calcium and magnesium in wort. *Journal of the American Society of Brewing Chemists*, **55**, 123–8.

Brown, A.J.P. (1998) Appendix III: useful world wide web addresses for yeast researchers. In *Yeast Gene Analysis*, Methods in Microbiology, (eds A.J.P. Brown and M.F. Tuite), vol. 26, pp. Academic Press, London. 479–84.

Brown, D.I. (1997a) HACCP – stand alone or in your quality system. *Proceedings of the European Brewery Convention Symposium*, Monograph XXVI, Stockholm, 132–40.

Brown, P. (1997b) Keeping an eye on your yeast. *Brewers' Guardian*, November, 39–40.

Brown, S.W., Sugden, D.A. & Oliver, S.G. (1984) Ethanol production and tolerance in grande and petite yeasts. *Journal of Chemistry, Technology and Biotechnology*, **34B**, 116–20.

Bruch, C.W., Hoffman, A., Gosine, R.M. & Brenner, M.W. (1964) Disinfection of brewing yeast with acidified ammonium persulphate. *Journal of the Institute of Brewing*, **70**, 242–8.

Bruinenberg, P.M., van Dijken, J.P. & Scheffers, W.A. (1983) A theoretical analysis of NADPH production and consumption in yeasts. *Journal of General Microbiology*, **129**, 953–64.

Bryant, T.N. & Cowan, W.D. (1979) Classification of brewing yeasts by principal coordinates analysis of their brewing properties. *Journal of the Institute of Brewing*, **85**, 89–91.

Buhler, T. (1995) Economic methods for wort separation. *Ferment*, **8**, 116–18.

Bunker, H.J. (1955) The survival of pathogenic bacteria in beer. *Proceedings of the 5th Congress of the European Brewery Convention*, Baden Baden, 330–39.

Burns, J.A. (1941) A biological test for the examination of brewery yeast. *Journal of the Institute of Brewing*, **47**, 10–14.

Busturia, A. & Lagunas, R. (1986) Catabolite inactivation of the glucose transport system in *Saccharomyces cerevisiae*. *Journal of General Microbiology*, **132**, 379–85.

Busturia, J.R. & Lagunas, R. (1985) Identification of two forms of the maltose transport system in *Saccharomyces cerevisiae* and their regulation by catabolite inactivation. *Biochimica Biophysica Acta*, **820**, 324–36.

Butke, T.M., Brint, S.L. & Lowe, M.R. (1988) Regulation of squalene epoxidase activity by membrane fatty acid composition in yeast. *Lipids*, **23**, 68–71.

Button, A.H. & Wren, J.J. (1978) Anti-foam agents in brewing. *Journal of the Institute of Brewing*, **78**, 443.

Byers, B. (1981) Cytology of the yeast life cycle. In *The Molecular Biology of the Yeast Saccharomyces, Life Cycle and Inheritance*, (eds J.F. Strathern, E.W. Jones and J.R. Broach), pp. 59–96. Cold Spring Harbor Laboratory, USA.

Byrnes, J. & Valentine, A. (1996) Across the pipe cell count of yeast. *Brauwelt International*, **14**, 44–6.

*C. elegans* Sequencing Consortium (1998) Genome sequence of the nematode *C. elegans*: a platform for investigating biology. *Science*, **282**, 2012–18.

Cabib, E., Roberts, R. & Bowers, B. (1982) Synthesis of the yeast cell wall and its regulation. *Annual Review of Biochemistry*, **51**, 763–93.

Cadwallader, S.D. (1992) The use of chlorine dioxide in water sterilisation. *Ferment*, **4**, 380–87.

Cahill, G., Walsh, P.K. & Donnelly, D. (1999a) A study of thermal gradient development in yeast crops. *Proceedings of the 27th Congress of the European Brewery Convention*, Cannes, 695–702.

Cahill, G., Walsh, P.K. & Donnelly, D. (1999b) Improved control of brewery yeast pitching using image analysis. *Journal of the American Society of Brewing Chemists*, **57**, 72–8.

Calderbank, J. & Hammond, J.R.M. (1989) Influence of nitrate and bacterial contamination on the formation of apparent total N-nitroso compounds (ATNC) during fermentation. *Journal of the Institute of Brewing*, **95**, 277–81.

Calderbank, J., Rose, A.H. & Tubb, R.S. (1985) Peptide removal from all malt wort and adjunct worts by *Saccharomyces cerevisiae* NCYC 240. *Journal of the Institute of Brewing*, **91**, 321–4.

Callaerts, G., Iserentant, D. & Verachtert, H. (1993) Relationship between trehalose and sterol accumulation during oxygenation of cropped yeast. *Journal of the American Society of Brewing Chemists*, **51**, 75–7.

Cameron, D.R., Cooper, D.G. & Neufeld, R.J. (1988) The mannoprotein of *Saccharomyces cerevisiae* is an effective bioemulsifier. *Applied and Environmental Microbiology*, **54**, 1420–25.

Campbell, I. (1971) Detection and identification of yeasts. *Technical Quarterly of the Master Brewers Association of the Americas*, **8**, 129–33.

Campbell, I. (1996) Systematics of yeasts. In *Brewing Microbiology*, 2nd edition, (eds F.G. Priest and I. Campbell), pp. 1–11. Chapman and Hall, London.

Campbell, I. & Msongo, H.S. (1991) Growth of aerobic wild yeasts. *Journal of the Institute of Brewing*, **97**, 279–82.

Candrian, U. (1995) Polymerase chain reaction in food microbiology. *Journal of Microbiological Methods*, **23**, 89–103.

Cantrell, I.C. & Anderson, R.G. (1983) Yeast performance in production fermentations. *Proceedings of the 19th Congress of the European Brewery Convention*, London, 481–8.

Carle, G.F. & Olson, M.V. (1984) Separation of chromosomal DNA molecules from yeast by orthogonal field alternation gel electrophoresis. *Nucleic Acids Research*, **12**, 5647–64.

Carlson, M. & Botstein, D. (1982) Two differentially regulated mRNAs with different 5 prime ends encode secreted and intracellular forms of yeast invertase. *Cell*, **28**, 145–54.

Carpentier, B., van Haecht, J.L. & Dufour, J.P. (1991) Influence of the trub content of the pitching yeast. *Proceedings 3rd Scientific and Technical Symposium, Institute of Brewing Central & South African Section*. Victoria Falls, Zimbabwe, 144–9.

Case, A.A. & Lyon, A.I.L. (1956) Action of polymyxin on some common brewery bacteria. *Journal of the Institute of Brewing*, **62**, 477–85.

Casey, G.P. (1990) Yeast selection in brewing. In *Yeast Strain Selection*, Chapter 4, (ed. C.J. Panchal), Marcel Dekker, New York, 65–111.

Casey, G.P. (1996) Practical application of pulsed field electrophoresis and yeast chromosome fingerprinting in brewing QA and R&D. *Technical Quarterly of the Master Brewers of Association of the Americas*, **33**, 1–10.

Casey, G.P. & Ingledew, W.M. (1981a) The use and understanding of media used in brewing bacteriology 1. Early history and development of general purpose media. *Brewers Digest*, February, 26–32.

Casey, G.P. & Ingledew, W.M. (1981b) The use and understanding of media used in brewing bacteriology II. Selective media for the isolation of lactic acid bacteria. *Brewers Digest*, March, 38–45.

Casey, G.P. & Ingledew, W.M. (1981c) The use and understanding of media used in brewing bacteriology III. Selective media used for gram-negative bacteria. *Brewers Digest*, April, 24–35.

Casey, G.P. & Ingledew, W.M. (1986) Ethanol tolerance in yeasts. *CRC Critical Reviews in Microbiology*, **13**, 219–80.

Casey, G.P., Pringle, A.T. & Erdmann, P.A. (1990) Evaluation of recent techniques used to identify individual strains of *Saccharomyces* yeasts. *Journal of the American Society of Brewing Chemists*, **48**, 100–106.

Casey, G.P., Chen, E.C.-H. & Ingledew, W.M. (1985) High gravity brewing: production of high levels of ethanol without excessive concentrations of esters and fusel alcohols. *Journal of the American Society of Brewing Chemists*, **43**, 179–82.

Casey, G.P., Magnus, C.A. & Ingledew, W.M. (1984) High gravity brewing: effects of nutrition on yeast composition. Fermentative ability and alcohol production. *Applied and Environmental Microbiology*, **48**, 639–46.

Casey, G.P., Xiao, W. & Rank, G.H. (1988) Application of pulsed field chromosome electrophoresis in the study of chromosome XIII and the electrophoretic karyotype of industrial strains of *Saccharomyces* yeasts. *Journal of the Institute of Brewing*, **94**, 239–43.

Casey, W.M., Burgess, J.P. & Parks, L.W. (1991) Effect of sterol side-chain structure on the feed back control of sterol biosynthesis in yeast. *Biochimica et Biophysica Acta*, **1081**, 279–84.

Casey, W.M., Keesler, G.A. & Parks, L.W. (1992) Regulation of partitioned sterol biosynthesis in *Saccharomyces cerevisiae*. *Journal of Bacteriology*, **174**, 7283–8.

Casson, D. (1982) Beer dispense lines. *The Brewer*, November, 447–53.

Cawley, T.N., Harrington, M.G. & Letters, R. (1972) A study of phosphate linkages in phosphomannan in cell walls of *Saccharomyces cerevisiae*. *Biochemical Journal*, **129**, 711–20.

Celis, G. & Dymond, G. (1994) CIP – a key to profitable brewing. *Proceedings of the European Brewery Convention Symposium*, Monograph **XXI**, Nutfield, 92–115.

Chant, J. (1994) Cell polarity in yeast. *Trends in Genetics*, **10**, 328–33.

Chapman, A.C. (1931) The yeast cell: what did Leeuewenhoek see? *Journal of the Institute of Brewing*, **37**, 433–7.

Chapman, A.G. & Atkinson, D.E. (1977) Adenine nucleotide concentrations and turnover rates. Their correlation with biological activity in bacteria and yeasts. *Advances in Microbial Physiology*, **15**, 253–306.

Chapman, C. & Bartley, W. (1968) The kinetics of enzyme changes in yeast under conditions that cause the loss of mitochondria. *Biochemical Journal*, **107**, 455–65.

Chelack, B.J. & Ingledew, W.M. (1987) Anaerobic gram-negative bacteria in brewing – a review. *Journal of the American Society of Brewing Chemists*, **54**, 123–7.

Chen, C.-S., Chan, E., Wang, S.L., Gong, C.S. & Chen, L.F. (1994) Ethanol fermentation in a tower fermenter using self-aggregating *Saccharomyces uvarum*. *Applied Biochemistry & Biotechnology*, **45/46**, 531–44.

Chen, E.C.-E. (1980) Utilisation of fatty acids by yeast during fermentation. *Journal of the American Society of Brewing Chemists*, **38**, 148–53.

Chen, E.C.-H., van Gheluwe, G. & Buday, A. (1973) Effect of mashing temperature on the nitrogenous constituents of wort and beer. *Proceedings of the American Society of Brewing Chemists*, 6–10.

Chen, E.C.-H. (1978) Relative contribution of Ehrlich and biosynthetic pathways to the formation of fusel alcohols. *Journal of the American Society of Brewing Chemists*, **38**, 39–43.

Chen, E.C.-H., Jamieson, A.M. & van Gheluwe, G. (1980) The release of fatty acids as a consequence of yeast autolysis. *Journal of the American Society of Brewing Chemists*, **38**, 13–17.

Cheng, Q. & Michels, C.A. (1991) MAL11 and MAL61 encode the inducible high affinity maltose transporter of *Saccharomyces cerevisiae*. *Journal of Bacteriology*, **173**, 1817 20.

Cherry, J.M., Adler, C., Ball, C. *et al.* (1998) SGD: *Saccharomyces* genome database. *Nucleic Acids Research*, **26**, 82–8.

Cherry, J.M., Ball, C., Weng, S. *et al.* (1997) Genetic and physical maps of *Saccharomyces cerevisiae*. *Nature*, **387**, 67–73.

Chervitz, S.A., Aravind, L., Sherloc, G. *et al.* (1998) Comparison of the complete protein sets of worm and yeast: orthology and divergence. *Science*, **282**, 2022–8.

Cheryan, M. & Mehaia, M.A. (1984) Ethanol production in a membrane recycle bioreactor. *Process Biochemistry*, December, 204–8.

Chester, W.A. (1963) A study of changes in flocculence in a single cell culture of a strain of *Saccharomyces cerevisiae*. *Proceedings of the Royal Society, Series B*, **157**, 223–33.

Chi, Z. & Arneborg, N. (1999) Relationship between lipid composition, frequency of ethanol-induced respiratory deficient mutants, and ethanol tolerance in *Saccharomyces cerevisiae*. *Journal of Applied Microbiology*, **86**, 1047–52.

Childress, A.M., Franklin, D.S., Pinswasdi, C., Kale, S. & Jazwinski, S.M. (1996) *LAG2*, a gene that determines yeast longevity. *Microbiology*, **142**, 2289–97.

Chilver, M.J., Harrison, J. & Webb, T.J.B. (1978) Use of immunofluorescent and viability stains in quality control. *Journal of the American Society of Brewing Chemists*, **36**, 13–18.

Chindampom, A., Iwaguchi, S.I., Nakagawa, Y., Homma, M. & Tanaka, K. (1993) Clonal size-variation of rDNA cluster region on chromosome XII of *Saccharomyces cerevisiae*. *Journal of General Microbiology*, **139**, 1409–15.

Cho, G.H. & Choi, C.Y. (1981) Continuous ethanol production by immobilised yeast in a fluidised bed reactor. *Biotechnology Letters*, **3**, 667–71.

Chowdhury, I., Watier, D. & Hornez, J.-P. (1995) Variability in survival of *Pectinatus cerevisiiphilus*, strictly anaerobic bacteria, under different oxygen conditions. *Anaerobe*, **1**, 151–6.

Christensen, R., Gong, C.S., Tang, R.-T., Chen, L.-F. & Rimedo, N. (1990) A multichamber tower fermenter for continuous ethanol fermentation with a self-aggregating yeast mutant. *Applied Biochemistry & Biotechnology*, **24/25**, 603–11.

Chuang, L.F. & Collins, E.B. (1968) Biosynthesis of diacetyl in bacteria and yeast. *Journal of Bacteriology*, **95**, 2083–9.

Chuang, L.F. & Collins, E.B. (1972) Inhibition of diacetyl synthesis by valine and the roles of alpha-ketoisovalerate acid in the synthesis of diacetyl by *Saccharomyces cerevisiae*. *Journal of General Microbiology*, **72**, 201–10.

Cid, V.J., Duran, A., Rey, F. del., Snyder, M.P., Nombela, C. & Sanchez, M. (1995) Molecular basis of cell integrity and morphogenesis in *Saccharomyces cerevisiae*. *Microbiological Reviews*, **59**, 345–86.

Ciesarova, Z., Smogrovicova, D. & Domeny, Z. (1996) Enhancement of yeast ethanol tolerance by calcium and magnesium. *Folia Microbiologica*, **41**, 485–8.

Clapperton, J.F. (1971) Simple peptides of wort and beer. *Journal of the Institute of Brewing*, **77**, 177–80.

Clarke, B.J., Burmeister, M.S., Krynicki, L. *et al.* (1991) Sulphur compounds in brewing. *Proceedings of the 23rd Congress of the European Brewery Convention*, Lisbon, 217–24.

Clarkson, S.P., Large, P.J., Boulton, C.A. & Bamforth, C.W. (1991) Synthesis of superoxide dismutase, catalase and other enzymes and oxygen and superoxide toxicity during changes in oxygen concentration in cultures of brewing yeast. *Yeast*, **7**, 91–103.

Clemons, K.V., Park, P., McCusker, J.H. *et al.* (1997) Application of DNA typing methods and genetic analysis to epidemiology and taxonomy of *Saccharomyces* isolates. *Journal of Clinical Microbiology*, **35**, 1822–8.

Codon, A.C., Benitez, T. & Korhola, M. (1998) Chromosome polymorphism and adaptation to specific industrial environments of *Saccharomyces* strains. *Applied Microbiology and Biotechnology*, **49**, 154–63.

Coghlan, A. (1996) Slime city. *New Scientist*, 31 Aug., 32–6.

Cohen, J.D., Goldenthal, M.J., Chow, T., Buchferer, B. & Marmur, J. (1985) Organisation of the *MAL* loci of *Saccharomyces cerevisiae*. Physical identification and functional characteristics of the three genes of the MAL6 locus. *Molecular and General Genetics*, **200**, 1–8.

Colaco, C., Sen, S., Thangavelu, M., Pinder, S. & Roser, R. (1992) Extraordinary stability of enzymes dried in trehalose: simplified molecular biology. *Biotechnology*, **10**, 1007–11.

Colin, S., Montesinos, M., Meersman, E., Swinkels, W. & Dufor, J.-P. (1991) Yeast dehydrogenase activities in relation to carbonyl compounds removal from wort and beer. *Proceedings of the 23rd Congress of the European Brewery Convention*, Lisbon, 409–16.

Coolbear, T. & Threfall, D.R. (1989) Biosynthesis of terpenoid lipids. In *Microbial Lipids* (eds C. Ratledge and S.G. Wilkinson), vol. 2, pp. 115–254. Academic Press, London.

Coote, N. & Kirsop, B.H. (1973) The concentration and significance of pyruvate in beer. *Journal of the Institute of Brewing*, **79**, 298–304.

Coote, N. & Kirsop, B.H. (1974) The content of some organic acids in beer and other fermented media. *Journal of the Institute of Brewing*, **80**, 474–83.

Cords, B. & Watson, J. (1998) Single phase cleaning from brewhouse to bright beer tanks. *Proceedings of the 23rd Institute of Brewing Convention, Asia Pacific Section*, Perth, 139–42.

Corran, H.S. (1975) *A history of brewing*. David & Charles, London, UK.

Costa, V., Amorim, M.A., Reis, E., Quintanilha, A. & Moradas-Ferreira, P. (1997) Mitochondrial superoxide dismutase is essential for ethanol tolerance in the post-diauxic phase. *Microbiology*, **143**, 1649–56.

Costerton, J.W., Lewandowski, Z., Caldwell, D.E., Korber, D.R. & Lappin-Scott, H.M. (1995) Microbial biofilms. *Annual Review of Microbiology*, **49**, 711–45.

Costerton, J.W., Stewart, P.S. & Greenberg, E.P. (1999) Bacterial biofilms: a common cause of persistent infections. *Science*, **284**, 1318–22.

Cottone, V. (1985) The microbrewery phenomenon. *Modern Brewery Age*, December, 14–17.

Coutts, M. (1956) Continuous beer fermentation. British Patent 872391.

Coutts, M.W. (1966) The many facets of continuous fermentation. *Proceedings of the 9th Convention of the Institute of Brewing (Australia & New Zealand Section)*, Auckland, 1–7.

Cowan, W.D. & Bryant, T.N. (1981) Antigenic structure of brewing yeasts and its relationship to non-serological methods of classification. *Journal of the Institute of Brewing*, **87**, 45–8.

Cowan, W.D., Hoggan, J. & Smith, J.E. (1975) Introduction of respiratory-deficient mutants in brewery yeast. *Technical Quarterly of the Master Brewers Association of the Americas*, **12**, 15–22.

Cowland, T.W. (1968) Variation in the serological behaviour of yeasts of the genus *Saccharomyces* towards fluorescent antibodies. *Journal of the Institute of Brewing*, **74**, 457–64.

Cox, B.S. (1995) Genetic analysis in *Saccharomyces cerevisiae* and *Schizosaccharomyces pombe*. In *The Yeasts*, 2nd edition, (eds A.H. Rose, A.E. Wheals and J.S. Harrison), vol. 6, pp. 7–67. Academic Press, London.

Crabb, D. & Maule, D.R. (1978) Temperature control and yeast sedimentation characteristics in large storage vessels. *Proceedings of the European Brewery Convention Symposium Monograph* V, Zoeterwoude, 165–80.

Criddle, R.S. and Schatz, G. (1969) Promitochondria of anaerobically-grown yeast. I. isolation and biochemical properties. *Biochemistry*, **8**, 322–34.

Crowe, J.H., Crowe, L.M. & Chapman, D. (1984) Preservation of membranes in anhydrobiotic organisms: the role of trehalose. *Science*, **223**, 701–3.

Crumplen, R., D'Amore, T., Slaughter, C. & Stewart, G.G. (1993) Novel differences between ale and lager brewing yeasts. *Proceedings of the 24th Congress of the European Brewery Convention*, Oslo, 267–74.

Cumberland, W.G., MacDonald, D.M. & Skinner, E.D. (1984) Automated fermenter control at Moosehead Breweries Ltd. *Master Brewers Association of the Americas Technical Quarterly*, **21**, 39–44.

Curtis, N.S. (1971) A century of yeast culture. *Brewers' Guardian*, September, 95–100.

Czechowski, M.H. & Banner, M. (1992) Control of biofilms in breweries through cleaning and sanitizing. *Technical Quarterly of the Master Brewers Association of the Americas*, **29**, 86–8.

Dadds, M.J.S. & Martin, P.A. (1973) The genus *Zymomonas* – a review. *Journal of the Institute of Brewing*, **79**, 386–91.

Dallyn, H. & Falloon, W.C. (1976) The heat resistance of yeasts and the pasteurisation of beer. *The Brewer*, November, 354–6.

Daly, B., Collins, E., Madigan, D., Donnelly, D., Coakley, M. Ross, P. (1997) An investigation into styrene in beer. *Proceedings of the 26th Congress of the European Brewery Convention*, Maastricht, 623–30.

D'Amore, T., Panchal, C.J. & Stewart, G.G. (1988) Intracellular ethanol accumulation in *Saccharomyces cerevisiae* during fermentation. *Applied and Environmental Microbiology*, **54**, 110–14.

Damsky, C.H. (1976) Environmentally induced changes in mitochondria and endoplasmic reticulum of *Saccharomyces carlsbergensis* yeast. *Journal of Cell Biology*, **71**, 123–8.

Dancis, A., Haille, D., Yuan, D.S. & Klausener, R.D. (1994) The *Saccharomyces cerevisiae* copper transport protein (Ctr1p). *Journal of Biological Chemistry*, **269**, 25660–67.

Daoud, I.S. & Searle, B.A. (1990) On-line monitoring of brewery fermentation by measurement of $CO_2$ evolution rate. *Journal of the Institute of Brewing*, **96**, 297–302.

Daoud, I.S. (1991) New techniques for in-line measurement. *Ferment*, **4**, 40–46.

Daoud, I.S. & Searle, B.A. (1987) Yeast vitality and fermentation performance. *Proceedings of the European Brewery Convention Symposium*, Monograph **XII**, Vuoranta, 108–15.

Daveloose, M. (1987) An investigation of zinc concentrations in brewhouse worts. *Technical Quarterly of the Master Brewers Association of the Americas*, **24**, 109–12.

Davies, A.W. (1988) Continuous fermentation – 30 years on. *Proceedings of the 20th Convention of the Institute of Brewing (Australia & New Zealand Section)*, Brisbane, 159–65.

Davies, D.G., Parsek, M.R., Pearson, J.P., Iglewski, B.H., Costerton, J.W. & Greenberg, E.P. (1998) The involvement of cell-to-cell signals in the development of a bacterial biofilm. *Science*, **280**, 295–8.

Davies, M.R. (1990) Aspects of Customs and Excise legislation influencing manufacture of lower alcohol products. *Ferment*, **3**, 230–34.

Davies, S.A. (1992) A supplier's view. The correct application and use of sensors. *Ferment*, **5**, 218–21.

Davies, A.M., Clements, G.J., Derwent, L & Brennan, M. (1995) A new rapid method for the detection of low-level contamination from brewing products. *Proceedings of the 25th Congress of the European Brewery Convention*, Brussels, 637–44.

de Barros Lopes, M., Rainieri, S., Henschke, P.A. & Langridge, P. (1999) AFLP fingerprinting for analysis of yeast genetic variation. *International Journal of Systematic Bacteriology*, **49**, 915–24.

de Barros Lopes, M., Soden, A., Martens, A.L., Henschke, P.A. & Langridge, P. (1998) Differentiation and species identification of yeasts using PCR. *International Journal of Systematic Bacteriology*, **48**, 279–86.

de Clerck, J. (1954) General survey of the brewing process. In *A Textbook of Brewing*. Translated by Kathleen Barton-Wright, pp. 1–3. Chapman & Hall Ltd, London.

De Keersmaecker, J. (1996) The mystery of lambic beer. *Scientific American*, August, 56–62.

De Man, J.C., Rogasa, M. & Sharpe, M.E. (1960) A medium for the cultivation of lactobacilli. *Journal of Applied Bacteriology*, **23**, 130–35.

De Vries, S. & Marres, C.A.M. (1987) The mitochondrial respiratory chain of yeast. Structure and biosynthesis and the role in cellular metabolism. *Biochimica et Biophysica Acta*, **895**, 205–39.

De Vries, S., van Witzenburg, R., Grivell, L. & Marres, C.A. (1992) Primary structure and import pathway of the rotenone-insensitive NADH-ubiquinone oxidoreductase of mitochondria from *Saccharomyces cerevisiae*. *European Journal of Biochemistry*, **203**, 587–92.

Deak, T. & Beuchat, L.R. (1996) *Handbook of Food Spoilage Yeasts*. CRC Press, London.

Deans, K., Pinder, A., Catley, B.J. & Hodgson, J.A. (1997) Effects of cone cropping and serial re-pitch on the distribution of cell ages in brewery yeast. *Proceedings of the 26th Congress of the European Brewery Convention, Maastricht*, 469–76.

Debourg, A. & van Nedervelde, L. (1999) The use of dried yeast in the brewing industry. *Proceedings of the 27th Congress of the European Brewery Convention, Cannes*, 751–60.

Debourg, A., Goossens, E., Villaneuba, K.D., Masschelein, C.A. & Pierard, A. (1991) The role of the mitochondrial ILV enzymes in the over-production of vicinal diketones by petite mutants. *Proceedings of the 23rd Congress of the European Brewery Convention, Lisbon*, 265–72.

Debourg, A., Laurent, M., Dupire, S. & Masschelein, C.A. (1993) The specific role of yeast enzymatic systems in the removal of flavour potent wort carbonyls during fermentation. *Proceedings of the 24th Congress of the European Brewery Convention, Oslo*, 437–44.

Debourg, A., Laurent, M., Goossens, E., Borremans, E., van de Winkel, L. & Masschelein, C.A. (1994) Wort aldehyde reduction potential in free and immobilised yeast systems. *Journal of the American Society of Brewing Chemists*, **52**, 100–106.

Del Vecchio, H.W., Dayharsh, C.A. & Baselt, F.C. (1951) Thermal death time studies on beer spoilage organisms – Part I. *Proceedings of the American Society of Brewing Chemists*, 45–50.

Delente, J., Akin, C., Krabbe, E. & Ladenburg, K. (1968) Carbon dioxide in fermenting beer – Part II. *Technical Quarterly of the Master Brewers Association of the America*, **5**, 228–34.

Dengis, P.B. & Rouxhet, P.G. (1997) Flocculation mechanisms of top and bottom fermenting brewing yeast. *Journal of the Institute of Brewing*, **103**, 257–61.

Dengis, P.B., Nelissen, L.R. & Rouxhet, P.G. (1995) Mechanisms of yeast flocculation: comparison of top- and bottom-fermenting strains. *Applied and Environmental Microbiology*, **61**, 718–28.

Denk, V. (1997) New method of on-line determination of diacetyl in real time by means of a software sensor. *Cerevisia – Belgian Journal of Brewing and Biotechnology*, **22**, 30–35.

Derdelinckx, G. & Neven, H. (1996) Current and recent knowledge of top fermentation and conditioning by refermentation. *Cerevisia – Belgian Journal of Brewing and Biotechnology*, **21**, 41–58.

Deters, J.H., Muller, I. & Hamberger, H. (1978) Bulk isolation of yeast mitochondria. *Methods in Cell Biology*, **20**, 107–12.

Deutch, C.E. & Parry, J.M. (1974) Sphaeroplast formation in yeast during the transition from exponential phase to stationary phase. *Journal of General Microbiology*, **80**, 259–68.

D'Hautcourt, O. & Smart, K.A. (1999) Measurement of brewing yeast flocculation. *Journal of the American Society of Brewing Chemists*, **57**, 123–8.

Dickenson, C.J. (1983) Dimethyl sulphide – its origin and control in brewing. *Journal of the Institute of Brewing*, **89**, 41–6.

Dickinson, F.M. (1996) Purification and some properties of the magnesium activated cytosolic aldehyde dehydrogenase of *Saccharomyces cerevisiae*. *Biochemical Journal*, **315**, 393–9.

van Dieren, B. (1995) Yeast metabolism and the production of alcohol-free beer. *Proceedings of the European Brewery Convention*, Monograph **XXIV**, Espoo, 66–75.

Dihanich, M., Schmid, A., Opplinger, W. & Benz, R. (1989) Identification of a new pore in the mitochondrial outer membrane of a porin-deficient mutant. *European Journal of Biochemistry*, **181**, 703–8.

DiMichele, L.J. & Lewis, M.J. (1993) Rapid, species-specific detection of lactic acid bacteria from beer using the polymerase chain reaction. *Journal of the American Society of Brewing Chemists*, **51**, 63–6.

Dinsdale, M.G. & Lloyd, D. (1995) Yeast vitality during cider fermentation: two approaches to the measurement of membrane potential, *Journal of the Institute of Brewing*, **101**, 453–8.

Dixon, I.J. & Leach, A.A. (1968) The adsorption of hop substances on the yeast cell wall. *Journal of the Institute of Brewing*, **74**, 63–7.

D'mello, N.P., Childress, A.M., Franklin, D.S., Kale, S.P., Pinswasdi, C. & Jazwinski, S.M. (1994) Cloning and characterization of *LAG1*, a longevity-assurance gene in yeast. *Journal of Biological Chemistry*, **269**, 15451–9.

Does, A.L. & Bisson, L.F. (1989) Comparison of glucose uptake kinetics in different yeast. *Journal of Bacteriology*, **171**, 1303–8.

Dolezil, L. & Kirsop, B.H. (1980) Variations amongst beers and lactic acid bacteria relating to beer spoilage. *Journal of the Institute of Brewing*, **86**, 122–4.

Dombek, K.M. & Ingram, L.O. (1986a) Determination of the intracellular concentration of ethanol in *Saccharomyces cerevisiae* during fermentation. *Applied and Environmental Microbiology*, **51**, 197–200.

Dombek, K.M. & Ingram, L.O. (1986b) Magnesium limitation and its role in apparent toxicity of ethanol during yeast fermentation. *Applied and Environmental Microbiology*, **52**, 975–81.

Dombek, K.M. & Ingram, L.O. (1987) Ethanol production during batch fermentation with *Saccharomyces cerevisiae*: changes in glycolytic enzymes and internal pH. *Applied and Environmental Microbiology*, **53**, 1286–91.

Dombek, K.M. & Ingram, L.O. (1988) Intracellular accumulation of AMP as a cause for the decline in the rate of ethanol production by *Saccharomyces cerevisiae* during batch fermentation. *Applied and Environmental Microbiology*, **54**, 98–104.

Donhauser, S. (1997) Characterisation of yeast species and strains. *Brauwelt International*, **15**, 52–8.

Donhauser, S., Eger, C., Hubl, T., Schmidt, U. & Winnewisser, W. (1993) Tests to determine the vitality of yeasts using flow cytometry. *Brauwelt International*, **3**, 221–4.

Donhauser, S., Wagner, D. & Geigher, E. (1993) Biogenic amines. *Brauwelt International*, **11**, 100–107.

Donnachie, I. (1979) *A History of the Brewing Industry in Scotland*. John Donald Publishers, Edinburgh, Scotland.

Donnelly, D. & Hurley, J. (1996) Yeast monitoring: the Guinness experience. *Ferment*, **9**, 283–6.

Doran, P.M. & Bailey, J.E. (1986) Effects of immobilisation on growth, fermentation properties and macromolecular composition of *Saccharomyces cerevisiae* attached to gelatin. *Biotechnology & Bioengineering*, **28**, 73–87.

Dowhanick, T.M. (1995) Advances in yeast and contaminant determination: the future of the so called 'rapid' methods. *Cerevisia – Belgian Journal of Brewing and Biotechnology*, **20**, 40–50.

Dowhanick, T.M., Sobczak, J., Presente, E. & Russell, I. (1995) Trial studies on the rapid quantitative detection of brewery microorganisms using a prototype ATP bioluminescence system. *Proceedings of the 25th Congress of the European Brewery Convention*, Brussels, 645–51.

Drawert, F. & Tressl (1972) Beer aromatics and their origin. *Technical Quarterly of the Master Brewer Association of the Americas*, **9**, 72–6.

Driedonks, R.A., Toschka, H.Y., van Almkerk, J.W., Schaffers, I.M. & Verbakel, J.M.A. (1995) Expression and secretion of antifreeze peptides in the yeast *Saccharomyces cerevisiae*. *Yeast*, **11**, 849–64.

Dror, Y. Cohen, O. & Freeman, A. (1988) Stabilisation effects on immobilised yeast: effect of gel composition in tolerance to water miscible solvents. *Enzyme & Microbial Technology*, **10**, 273–9.

Duarte, F.L., Pais, C., Spencer-Martins, I. & Leao, C. (1999) Distinctive electrophoretic isoenzyme profiles in *Saccharomyces sensu stricto*. *International Journal of Systematic Bacteriology*, **49**, 1907–13.

Duffus, J.H. (1971) The cell cycle in yeast – a review. *Journal of the Institute of Brewing*, **77**, 500–508.

Duffus, J.H. (1976) Isolation of yeast nuclei and methods to study their properties. *Methods in Cell Biology*, **12**, 77–98.

Dufour, J.-P. (1991) Influence of industrial brewing and fermentation working conditions on beer sulphur dioxide level and flavour stability. *Proceedings of the 23rd Congress of the European Brewery Convention*, Lisbon, 209–16.

Dufour, J.-P. & Malcorps, P. (1994) Ester synthesis during fermentation: enzyme characterisation and modulation mechanism. *Proceedings of the 4th Aviemore Conference on Malting, Brewing and Distilling* (ed. I. Campbell), pp. 137–45. Institute Brewing, London.

Dufour, J.-P., Carpentier, B., Kulakumba, M., van Haecht, J.-L. & Devreux, A. (1989) Alteration of $SO_2$ production during fermentation. *Proceedings of the 22nd Congress of the European Brewery Convention*, Zurich, 331–8.

Dujon, B. (1981) Mitochondrial genetics and functions. In *The Molecular Biology of the Yeast Saccharomyces, Life Cycle and Inheritance* (eds J.F. Strathern, E.W. Jones and J.R. Broach), pp. 505–635. Cold Spring Harbor Laboratory, USA.

Dulieu, C., Boivin, P., Malanda, M., Dautzenberg, H & Poncelet, D. (1997) Encapsulation of alpha acetolactate decarboxylase to avoid diacetyl formation. *Proceedings of the 26th Congress of the European Brewery Convention*, Maastricht, 454–60.

Dummett, G.A. (1982) *A History of the A.P.V. Company Limited*. Hutchinson Benham, London.

Dunbar, J., Campbell, S.L., Banks, D.J. & Duncan, N.S. (1990) Diacetyl removal through use of a continuous maturation vessel. *Proceedings of the 21st Convention of the Institute of Brewing (Australia & New Zealand Section)*, Auckland, 131–5.

Dunbar, J., Campbell, S.L., Banks, D.J. & Warren, D.R. (1988) Metabolic aspects of a commercial continuous fermentation system. *Proceedings of the 20th Convention of the Institute of Brewing (Australia & New Zealand Section)*, Brisbane, 151–7.

Dunham, I., Shimizu, N., Roe, B.A. *et al.* (1999) The DNA sequence of human chromosome 22. *Nature*, **402**, 489–95.

Dunn, A.E., Leeder, G.I., Molloy, F. & Wall, R. (1996) Sterile beer filtration. *Ferment*, **9**, 155–61.

Durr, M., Urech, K., Boller, T., Wiemken, A., Schwenke, J. & Nagy, M. (1979) Sequestration of arginine by polyphosphates in vacuoles of yeast (*Saccharomyces cerevisiae*). *Archives for Microbiology*, **121**, 169–75.

Dutton, J. (1990) FV control with real time SG monitoring. *Brewing and Distilling International*, May, 20–21.

Dymond, G. (1992) Pasteurization of beer in plate heat exchangers: lower costs and higher quality. *Proceedings of the 22nd Institute of Brewing Convention (Australia & New Zealand Section)*, Melbourne, 164–74.

Dziondziak, K. & Seiffert, T. (1995) Process for the continuous production of alcohol free beer. *Proceedings of 25th Congress of the European Brewery Congress*, Brussels, 301–8.

Eddy, A.A. (1955) Flocculation characteristics of yeasts. Sugars as dispersing agents. *Journal of the Institute of Brewing*, **61**, 313–17.

Eddy, A.A. (1958) Aspects of the chemical composition of yeast. In *The Chemistry and Biology of Yeasts* (ed. A.II. Cook), pp. 157–249. Academic Press, New York.

Eddy, A.A. & Rudin, A.D. (1958) Part of the yeast surface apparently involved in flocculation. *Journal of the Institute of Brewing*, **64**, 19–21.

Edgely, M. & Brown, A.D. (1983) Yeast water relations: physiological changes induced by solute stress in *Saccharomyces cerevisiae* and *Saccharomyces rouxii*. *Journal of General Microbiology*, **129**, 3453–63.

Edwards, C., Porter, J. & West, M. (1996) Fluorescent probes for measuring physiological fitness of yeast. *Ferment*, **9**, 288–93.

Eger, C., Donhauser, S. & Winnewisser, W. (1995) Rapid immunoassay of beer spoilage. *Brauwelt International*, **13**, 164–71.

Egilmez, N.K. & Jazwinski, S.M. (1989) Evidence for the involvement of a cytoplasmic factor in the ageing of the yeast *Saccharomyces cerevisiae*. *Journal of Bacteriology*, **171**, 37–42.

Elbien, A.D. (1974) The metabolism of trehalose. *Advances in Carbohydrate Chemistry and Biochemistry*, **30**, 227–56.

Eleutherio, E.C.A., de Araujo, P.S. & Panek, A.D. (1993) Role of the trehalose carrier in dehydration resistance of *Saccharomyces cerevisiae*. *Biochimica et Biophysica Acta*, **1156**, 263–6.

Elliott, B., Haltwanger, R.S. & Futcher, B. (1996) Synergy between trehalose and Hsp 104 for thermo-tolerance in *Saccharomyces cerevisiae*. *Genetics*, **144**, 923–33.

Enari, T.-M. (1974) Amino acids, peptides and proteins. *Proceedings of the European Brewery Convention Symposium Monograph*, **1**, Zeist, 73–89.

Enevoldsen, B.E. (1981) Demonstration of melibiase in non-pasteurized lager beers and studies on the heat stability of the enzyme. *Carlsberg Research Communications*, **46**, 37–42.

Enevoldsen, B.E. (1985) Determining pasteurization units from residual melibiase activity in lager beer. *Journal of the American Society of Brewing Chemists*, **43**, 183–9.

Enevoldsen, B.S. (1974) Dextrins in brewing – a review. *Proceedings of the European Brewery Convention Symposium, Monograph* I, Zeist, 158–88.

Enevoldsen, B.S. & Schmidt, F. (1974) Dextrins in brewing. *Journal of the Institute of Brewing*, **80**, 520–33.

Engan, S. (1978) Formation of volatile flavour compounds: alcohols, esters, carbonyls, acids. *Proceedings of the European Brewery Convention Symposium*, Monograph V, Zoeterwoude, 28–39.

Engan, S. (1981) Beer composition: volatile substances. In *Brewing Science*, (ed.) J.R.A. Pollock), vol. 2, pp. 98–105. Academic Press, London.

Engasser, J.M., Marc, I., Moll, M. Duteurte, B. (1981) Kinetic modelling of beer fermentation. *Proceedings of the 18th Congress of the European Brewery Convention*, Copenhagen, 579–86.

Engelberg, D., Mimran, A., Martinetto, H. *et al.* (1998) Multicellular stalk-like structures in *Saccharomyces cerevisiae*. *Journal of Bacteriology*, **180**, 3992–6.

Entian, K.D. & Zimmermann, F.K. (1982) New genes involved in carbon catabolite repression and derepression in the yeast *Saccharomyces cerevisiae*. *Journal of Bacteriology*, **151**, 1123–8.

Ernandes, J.R., Williams, J.W., Russell, I. & Stewart, G.G. (1993) Respiratory deficiency in brewing yeasts – effects on fermentation, flocculation and beer flavour components. *Journal of the American Society of Brewing Chemists*, **51**, 16–20.

Espinel-Ingroff, A., Stockman, L., Roberts, G., Pincus, D., Pollack, J. & Marler, J. (1998) Comparison of the RapID yeast plus system with the API 20C system for identification of common, new and emerging yeast pathogens. *Journal of Clinical Microbiology*, **36**, 883–6.

European Brewery Convention (1995) *Beer Pasteurisation, Manual of Good Practice*, EBC Technology and Engineering Forum, Getranke-Fachverlag Hans Carl, Nurnberg.

European Brewery Convention (1987) Diacetyl and other vicinal diketones in beer. *Analytica*, 4th edition, published by Brauerei-und Getranke-Rundschau, Zurich, Switzerland. **9.11**, E187–E190.

European Brewery Convention (1999) *Beer Filtration, Stabilisation and Sterilisation, Manual of Good Practice*, EBC Technology and Engineering Forum, Fachverlag Hans Carl, Nurnberg.

Evans, H.A.V. (1982) A note on two uses for impedimetry in brewing microbiology. *Journal of Applied Bacteriology*, **53**, 423–6.

Evans, H.A.V. & Cleary, P. (1985) Direct measurement of yeast and bacterial viability. *Journal of the Institute of Brewing*, **91**, 73.

Eyben, D. (1989) An automated method for fermentation process control. *Technical Quarterly of the Master Brewers Association of the Americas*, **26**, 51–5.

Falco, S.C. & Dumas, K.S. (1985) Genetic analysis of mutants of *Saccharomyces cerevisiae* resistant to the herbicide sulfometuron methyl. *Genetics*, **109**, 21–35.

Fels, S., Reckelbus, B. & Gosselin (1999) Dried yeast as an alternative to fresh yeast propagation. *Proceedings of the 7th Institute of Brewing Convention, (Africa Section)*, Nairobi, 147–51.

Felstead, J. (1994) Materials, equipment and installation of pipework systems. Part 2: Designing for hygiene and ease of cleaning. *Brewers' Guardian*, July, 31–4.

Ferea, T.L., Botstein, D., Brown, P.O. & Rosenzweig, R.F. (1999) Systematic changes in gene expression patterns following adaptive evolution in yeast. *Proceedings of the National Academy of Science, USA*, **96**, 9721–6.

Fernanda Rosa, M. & Sa-Correia, I. (1994) Limitations to the use of extra-cellular acidification for the assessment of plasma membrane H + -ATPase activity and ethanol tolerance in yeasts. *Enzyme and Microbial Technology*, **16**, 808–12.

Fernandez, J.L. & Simpson, W.J. (1993) Aspects of the resistance of lactic acid bacteria to hop bitter acids. *Journal of Applied Bacteriology*, **75**, 315–19.

Fernandez, J.L. & Simpson, W.J. (1995) Measurement and prediction of the susceptibility of lager beer to spoilage by lactic acid bacteria. *Journal of Applied Bacteriology*, **78**, 419–25.

Fernandez, J.L., Simpson, W.J. & Dowhanick, T.M. (1993) Enumeration of *Obesumbacterium proteus* in brewery yeasts and characterisation of isolated strains using Biolog GN microplates and protein fingerprinting. *Letters in Applied Microbiology*, **17**, 292–6.

Fernandez, S.S., Gonzalez, G. & Sierra, J.A. (1991) The acidification power test and the behaviour of yeast in brewery fermentations. *Technical Quarterly of the Master Brewers Association of the Americas*, **28**, 89–95.

Fernandez, S.S., Gonzalez, M.G. & Sierra, J.A. (1993) Evaluation of the effect of acid washing on the fermentative and respiratory behaviour of yeasts by the acidification power test. *Technical Quarterly of the Master Brewers Association of the Americas*, **30**, 1–8.

Fernandez, S.S., Machucha, N., Gonzalez, M.G. & Sierra, J.A. (1985) Accelerated fermentation of high gravity worts and its effect on yeast performance. *Journal of the American Society of Brewing Chemists*, **43**, 109–14.

Ferris, L.E., Davey, C.L. & Kell, D.B. (1990) Evidence from its temperature dependence that the beta-dielectric dispersion of cell suspensions is not due solely to the charging of a static membrane capacitance. *European Biophysics Journal*, **18**, 267–76.

Fiechter, A., Fuhrmann, G.F. & Kappeli, O. (1981) Regulation of glucose metabolism in growing yeast cells. *Advances in Microbial Physiology*, **22**, 123–85.

Field, J., Nikawa, J., Brock, D. *et al.* (1988) Purification of a *RAS*-responsive adenyl cyclase complex from *Saccharomyces cerevisiae* by use of an epitope addition method. *Molecular and Cellular Biology*, **8**, 2159–65.

Finnerty, W.R. (1989) Microbial lipid metabolism. In *Microbial Lipids*, (eds C. Ratledge and S.G. Wilkinson), vol. 2, pp. 525–66. Academic Press, London.

Flannigan, B. (1996) The microflora of barley and malt. In *Brewing Microbiology*, 2nd edition, pp. 83–125. (eds F.G. Priest and I. Campbell), Chapman and Hall, London.

Fleet, G.H. (1991) Cell Walls. In *The Yeasts*, 2nd edition, (eds A.H. Rose and J.S. Harrison), vol. 4, pp. 199–277. Academic Press, London.

Fleet, G.H. & Manners, D.J. (1976) Isolation and composition of an alkali-soluble glucan from the cell walls of *Saccharomyces cerevisiae*. *Journal of General Microbiology*, **94**, 180–92.

Flikweert, M.T., van der Zanden, L., Janssen, W.M.T.M., Steensma, H.Y., van Dijken, J.P. & Pronk, J.T. (1996) Pyruvate decarboxylase: an indispensable enzyme for growth of *Saccharomyces cerevisiae* on glucose. *Yeast*, **12**, 241–57.

Florkin, M. (1972) A history of biochemistry. *Comprehensive Biochemistry*, **30**, 129–44.

Forget, C. (1988) *Dictionary of beer and brewing*. Brewers Publication, Boulder, Colorado, USA.

Forrest, I.S. (1987) Methods of in-line OG measurement. *Brewers' Guardian*, September, 8–28.

Forrest, I.S. & Cuthbertson, R.C. (1986) Novel ultrasonic device for in-line measurement of sugar, alcohol and original gravity. *Proceedings of the 2nd Aviemore Conference on Malting, Brewing and Distilling*, Institute of Brewing, 324–7.

Forrest, I.S., Cuthbertson, R.C., Dickson, J.E., Gilchrist, F.H.L. & Skrgatic, D. (1989) In-line measurement of original gravity by sound velocity and refractive index. *Proceedings of the 22nd Congress of the European Brewery Convention*, Zurich, 725–32.

Forsberg, S.L. & Guarente, L. (1989) Communication between mitochondria and the nucleus in regulation of cytochrome genes in the yeast *Saccharomyces cerevisiae*. *Annual Review of Cellular Biology*, **5**, 153–80.

Foury, F. (1997) Human genetic diseases: a cross-talk between man and yeast *Gene*, **195**, 1–10.

Foury, F., Roganti, T., Lecrenier, N. & Purnelle, B. (1998) The complete sequence of the mitochondrial genome of *Saccharomyces cerevisiae*. *FEBS Letters*, **440**, 325–31.

Francke Johannesen, P., Nyborg, M. & Hansen, J. (1999) Construction of *S. carlsbergensis* brewer's yeast without production of sulfite. *Proceedings of the 27th Congress of the European Brewery Convention*, Cannes, 655–62.

Freeman, G.J., Reid, A.I., Martinez, C.V., Lynch, F.J. & Juarez, J.A.G. (1997) The use of ultrasonic radiation to suppress foaming in fermenters. *Proceedings of 26th Congress of the European Brewery Convention*, Maastricht, 405–12.

Fricker, R. (1978) The design of large tanks. *Brewers' Guardian*, May, 28–37.

Fricker, R. (1984) The pasteurisation of beer. *Journal of the Institute of Brewing*, **90**, 146–52.

Fridovich, I. (1986) Biological effects of the superoxide radical. *Archives for Biochemistry and Biophysics*, **247**, 1–11.

Fuge, E.K., Braun, E.L. & Wemer-Washburne, M. (1994) Protein synthesis in long-term stationary-phase cultures of *Saccharomyces cerevisiae*. *Journal of Bacteriology*, **176**, 5802–13.

Fuji, T., Kobayashi, O., Yoshimoto, H., Furukawa, S. Tamai, Y. (1997) Effect of aeration and unsaturated fatty acids on expression of the *Saccharomyces cerevisiae* alcohol acetyl transferase gene. *Applied and Environmental Microbiology*, **63**, 910–15.

Fuji, T., Yoshimoto, H., Nagasawa, N. & Tamai, T. (1996) Acetate ester production by *Saccharomyces cerevisiae* lacking the ATF1 gene encoding alcohol acetyl transferase. *Journal of Fermentation and Bioengineering*, **81**, 538–42.

Fuji, T., Nagasawa, N., Iwamatsu, A., Bogaki, T., Tamai, Y. & Hamachi, M. (1994) Molecular cloning, sequence analysis, and expression of the yeast alcohol acetyltransferase gene. *Applied and Environmental Microbiology*, **60**, 2786–92.

Fujita, K., Iwahashi, H., Kodama, O. & Komatsu, Y. (1995) Induction of heat-shock proteins and accumulation of trehalose by TPN in *Saccharomyces cerevisiae*. *Biochemical and Biophysical Research Communications*, **216**, 1041–7.

Funahashi, W., Suzuki, K., Ohtake, Y. & Yamashita, H. (1998) Two novel beer-spoilage *Lactobacillus* species isolated from breweries. *Journal of the American Society of Brewing Chemists*, **56**, 64–9.

Gadgil, C.J., Bhat, P.J. & Venkatesh, K.V. (1996) Cybemetic model for the growth of *Saccharomyces cerevisiae* on melibiose. *Biotechnology Progress*, **12**, 744–50.

Galazzo J.L. & Bailey, J.E. (1989) *In vivo* nuclear magnetic resonance analysis of immobilisation effects on glucose metabolism of yeast, *Saccharomyces cerevisiae*. *Biotechnology & Bioengineering*, **33**, 1283–9.

Galazzo, J.L. & Bailey, J.E. (1990) Fermentation pathway kinetics and metabolic flux control in suspended and immobilised *Saccharomyces cerevisiae*. *Enzyme & Microbial Technology*, **12**, 162–72.

Galitski, T., Saldanha, A.L., Styles, C.A., Lander, E.S. & Fink, G.R. (1999) Ploidy regulation of gene expression. *Science*, **285**, 251–4.

Gamo, F.J., Moreno, E. & Lagunas, R. (1995) The low affinity component of the glucose transport system in *Saccharomyces cerevisiae* is not due to passive diffusion. *Yeast*, **11**, 1393–8.

Gancedo, C. & Serrano, R. (1989) Energy yielding metabolism. In *The Yeasts*, (eds A.H. Rose and J.S. Harrison), vol. 3, pp. 205–60. Academic Press, London. 205–60.

Gancedo, J.M. (1992) Carbon catabolite repression. *European Journal of Biochemistry* **206**, 297–313.

Garcia, J.C. & Kotyk, A. (1988) Uptake of L-lysine by a double mutant of *Saccharomyces cerevisiae*. *Folia Microbiologica*, **33**, 285–91.

Gares, S.L., Whiting, M.S., Ingledew, W.M. & Ziola, B. (1993) Detection and identification of *Pectinatus cerevisiiphilus* using surface-reactive monoclonal antibodies in a membrane filter-based fluoroimmuno-assay. *Journal of the American Society of Brewing Chemists*, **51**, 158–63.

Garrick, C.C. & McNeil, K.E. (1984) Influence of product composition on pasteurisation efficiency. *Proceedings of the 18th Institute of Brewing Convention (Australia & New Zealand Section)*, Adelaide, 244–51.

Garsoux, G., Haubursin, H., Bilbault, S. & Dufour, J.P. (1993) Yeast flocculation: biochemical characterisation of yeast cell wall components. *Proceedings of the 24th Congress of the European Brewery Convention*, Oslo, 275–82.

Gbelska, Y., Subik, J., Svoboda, A., Goffeau, A. & Kovac, L. (1983) Intramitochondrial ATP and cell functions: yeast cells depleted of intramitochondrial ATP lose the ability to grow and multiply. *European Journal of Biochemistry*, **130**, 281–6.

Gee, D.A. & Ramirez, W.F. (1988) Optimal temperature control for batch beer fermentation. *Biotechnology and Bioengineering*, **31**, 224–34.

Geering, P. (1996) The mash separation process – theory and practice. *Brewers' Guardian*, October, 32–6.

Geiger, E. (1993) Continuous yeast propagation. *Brauwelt International*, **11**, 430–4.

Geiger, E. & Piendl, A. (1976) Technological factors in the formation of acetaldehyde during fermentation. *Technical Quarterly of the Master Brewers Association of the Americas*, **13**, 51–63.

Geiger, K.H. (1961) Advanced brewing methods – fermentation phase. *Technical Quarterly of the Master Brewers Association of the Americas*, 8–14.

Geiger, K.H. & Compton, J. (1961) Continuous fermentation process. US Patent 2967107.

Gelinas, P., Fiset, G., LeDuy, A. & Goulet, J. (1989) Effect of growth conditions on and trehalose content on cryotolerance of baker's yeast in frozen doughs. *Applied and Environmental Microbiology*, **55**, 2453–9.

Gerlach, H. (1995) Cooling of cylindroconical tanks. *Brauindustrie*, **80**, 358–62.

Gervais, P., Marchal, P.A. & Molin, P. (1992) Effects of kinetics of osmotic pressure variation on yeast viability. *Biotechnology and Bioengineering*, **40**, 1435–9.

Gibson, R.M., Large, P.J. & Bamforth, C.W. (1985) The use of radioactive labelling to demonstrate the production of dimethyl sulphide from dimethyl sulphoxide during fermentation of wort. *Journal of the Institute of Brewing*, **91**, 397–400.

Gilliland, R.B. (1951) The flocculation characteristics of brewing yeasts during fermentation *Proceedings of the 3rd Congress of the European Brewery Convention*, Brighton, 35–58.

Gilliland, R.B. (1961) *Brettanomyces*. I. Occurrence, characteristics, and effects on beer flavour. *Journal of the Institute of Brewing*, **67**, 257–61.

Gilliland, R.B. (1971a) Yeast classification. *Journal of the Institute of Brewing*, **77**, 276–84.

Gilliland, R.B. (1971b) Stability and variation in brewing yeast. *Brewers' Guardian*, October, 29–31.

Gingell, K. & Bruce, P. (1998) Latest developments in acid cleaning and related sanitation. *Proceedings of the 23rd Institute of Brewing Convention (Asia Pacific Section)*, Perth, 134–8.

Godia, F., Casas, C. & Sola, C. (1987) A survey of continuous ethanol fermentation systems using immobilised cells. *Process Biochemistry*, April, 43–8.

Goffeau, A., Barrell, B.G. Bussey, H. *et al.* (1996) Life with 6000 genes. *Science*, **274**, 546–67.

Good, L., Dowhanick, T.M., Ernandes, J.E., Russell, I. & Stewart, G.G. (1993) *Rho⁻* mitochondrial genomes and their influence on adaptation to nutrient stress in lager yeast fermentations. *Journal of the American Society of Brewing Chemists*, **51**, 36–9.

Goodacre, R. (1994) Characterisation and quantification of microbial systems using pyrolysis mass spectrometry: introducing neural networks to analytical pyrolysis. *Microbiology Europe*, **2**, 16–22.

Goodacre, R. & Kell, D.B. (1996) Pyrolysis mass spectrometry and its applications in biotechnology. *Current Opinion in Biotechnology*, **7**, 20–28.

Goodey, A.R. & Tubb, R.S. (1982) Genetic and biochemical analysis of the ability of *Saccharomyces cerevisiae* to decarboxylate cinnamic acids. *Journal of General Microbiology*, **128**, 2615–20.

Goossens, E., Debourg, A., Villaneuba, K.D. & Masschelein, C.A. (1991) Decreased diacetyl production by site directed integration of the *ILV5* gene into chromosome XIII of *Saccharomyces cerevisiae*. *Proceedings of the 23rd Congress of the European Brewery Convention*, Lisbon, 289–96.

Goossens, E., Debourg, A., Villaneuba, K.D. & Masschelein, C.A. (1993) Decreased diacetyl production in lager brewing yeast by integration of the *ILV5* gene. *Proceedings of the 24th Congress of the European Brewery Convention*, Oslo, 251–8.

Goossens, E., Dillemans, M., Debourg, A. & Masschelein, C.A. (1987) Control of diacetyl formation by an intensification of the anabolic flux of acetohydroxy acid intermediates. *Proceedings of the 21st Congress of the European Brewery Convention*, Madrid, 553–60.

Götz, R., Schlüter, E., Shoham, G. & Zimmermann, F.K. (1999) A potential role of the cytoskeleton of *Saccharomyces cerevisiae* in a functional organisation of glycolytic enzymes. *Yeast*, **15**, 1619–29.

Gourvish, T.R. & Wilson, R.G. (1994) *The British Brewing Industry 1830–1980*. Cambridge University Press, Cambridge, UK.

Graham, R.K., Skurray, G.R. & Caiger, P. (1970) Nutritional studies on yeasts during batch and continuous fermentation. 1: Changes in vitamin concentrations. *Journal of the Institute of Brewing*, **76**, 366–71.

Grant, L.G. (1986) A review of dispense system cleaning. *The Brewer*, March, 89–92.

Gray, P.P. & Kazin, A.D. (1946) Antibiotics and the treatment of brewer's yeast. *Wallerstein Laboratory Communications*, **9**, 115–27.

Green, S.R. (1962) Past and current aspects of continuous beer fermentation processes. *Communications of the Wallerstein Laboratories*, **25**, 337–47.

Gregory, W.P. (1967) Stainless Steels. *Technical Quarterly of the Master Brewers Association of the Americas*, **4**, 205–8.

Grenson, M. (1992) Amino acid transporters in yeast: structure, function and regulation. In *Molecular Aspects of Transport Proteins*, (ed. J. De Pont), pp. 219–45. Elsevier Science Publishers, London.

Griffin, S. (1996) New Yorkshire squares at Joshua Tetley. *Brewers' Guardian*, August, 12–14, 17.

Grivell, L.A. (1989) Nucleo-mitochondrial interactions in yeast mitochondrial biogenesis. *European Journal of Biochemistry*, **182**, 477–93.

Groneick, E., Groppe, H., Dillenhofer, W. & Ronn, D. (1997) Secondary fermentation of beer with immobilised yeast. *Brauwelt International*, **15**, 264–7.

Gronqvist, A., Siirila, J., Virtanen, H., Home, S. & Pajunen, E. (1993) Carbonyl compounds during beer production and in beer. *Proceedings of 24th Congress of the European Brewery Congress*, Oslo, 421–8.

Grosz, R. & Stephanopoulis, G. (1990) Physiological, biochemical and mathematical studies of micro-aerobic continuous ethanol fermentation by *Saccharomyces cerevisiae*. *Biotechnology and Bioengineering*, **36**, 1008–20.

Groth, C., Hansen, J. & Piskur, J. (1999) A natural chimeric yeast containing genetic material from three species. *International Journal of Systematic Bacteriology*, **49**, 1933–8.

Guarante, L. (1996) Do changes in chromosomes cause ageing. *Cell*, **86**, 9–12.

Guérin, B. (1991) Mitochondria. in *The Yeasts*, 2nd edition (eds A.H. Rose and J.S. Harrison), vol. 4, pp. 541 600. Academic Press, London.

Guidici, P., Caggia, C., Pulvirenti & Rainieri, S. (1998) Karyotyping of *Saccharomyces* strains with different temperature profiles. *Journal of Applied Microbiology*, **84**, 811–19.

Guijo, S., Mauricio, J.C., Salmon, J.M. & Ortega, J.M. (1997) Determination of the relative ploidy in different *Saccharomyces cerevisiae* strains used for fermentation and 'flor' film ageing of dry sherry-type wines. *Yeast*, **13**, 101–17.

Guldfeldt, L.U. & Arneborg, N. (1998) The effect on yeast trehalose content at pitching on fermentation performance during brewing fermentations. *Journal of the Institute of Brewing*, **104**, 37–9.

Guldfeldt, L.U. & Piper, J.U. (1999) Yeast typing and propagation of dry brewing yeast cultures. *Technical Quarterly of the Master Brewers Association of the Americas*, **36**, 1–6.

Gutteridge, C.S. & Priest, F.G. (1996) Methods for the rapid identification of microorganisms. In *Brewing Microbiology*, 2nd edition, (eds F.G. Priest and 1. Campbell), pp. 237–70. Chapman and Hall, London.

Gvazlaitis, G., Beil, S., Kreibaum, U., Simutic, R., Haulik, I., Dorb, M., Schneider, F. & Lubbert, A. (1994) Temperature control in fermenters' application of neural nets and feedback control in breweries. *Journal of the Institute of Brewing*, **100**, 99–104.

Gyllang, H., Winge, M. & Korch, C. (1989) Regulation of sulphur dioxide formation during formation. *Proceedings of the 22nd Congress of the European Brewery Convention*, Zurich, 347–54.

Haboucha, J., Masschelein, C.A. & Devreux, A. (1967) Accelerated discontinuous fermentation and its influence on yeast metabolism. *Proceedings of 21st Congress of the European Brewery Congress*, Madrid, 197–211.

Hackstaff, B.W. (1978) Various aspects of high gravity brewing. *Technical Quarterly of the Master Brewers Association of the Americas*, **15**, 1–7.

Hackwood, W. (1985) *Inns, Ales and Drinking Customs of Old England*. Bracken Books, London.

Hadfield, C., Harikrishna, J.A. & Wilson, J.A. (1995) Determination of chromosome copy numbers in *Saccharomyces cerevisiae* strains via integrative probe and blot hybridization techniques. *Current Genetics*, **27**, 217–28.

Haffmans, B. (1996) Carbon dioxide recovery in breweries. *Brauwelt*, **14**, 86–7.

Haikara, A. (1985) Detection of anaerobic, gram negative bacteria in beer. *Monatsschrift für Brauwissenschaft*, **6**, 239–43.

Haikara, A. & Henriksson, E. (1992) Detection of airborne microorganisms as a means of hygiene control in the brewery filling area. *Proceedings of the 22nd Institute of Brewing Convention (Australia & New Zealand Section)*, Melbourne, 159–63.

Haikara, A. & Lounatmaa, K. (1987) Characterization of *Megasphaera* sp., a new anaerobic beer spoilage coccus. *Proceedings of the 21st Congress of the European Brewery Convention*, Madrid, 473–80.

Haikara, A., Enari, T.-M. & Lounatmaa, K. (1981) The genus *Pectinatus*, a new group of anaerobic beer spoilage bacteria. *Proceedings of the 18th Congress of the European Brewery Convention*, Copenhagen, 229–39.

Hall, J.F. (1954) Survey of British top fermentation yeasts 1. Introduction. *Journal of the Institute of Brewing*, **60**, 482–5.

Hall, J.F. (1971) Detection of wild yeasts in the brewery. *Journal of the Institute of Brewing*, **77**, 513–16.

van Hamersveld, E.H., van Loosdrecht, M.C.M., van der Lans, R.G.J.M. & Luyben, K. Ch.A.M. (1996) On the measurement of the flocculation characteristics of brewers' yeast. *Journal of the Institute of Brewing*, **102**, 333–42.

Hammond, J. (1996) Microbiological techniques to confirm CIP effectiveness. *The Brewer*, August, 332–8.

Hammond, J. (1998) Brewing with genetically modified amylolytic yeast. In *Genetic Modification in the Food Industry. A Strategy for Food Quality Improvement*, (eds S. Roller and S. Harlander), pp. 129–57. Chapman & Hall, London.

Hammond, J., Brennan, M. & Price, A. (1999) The control of microbial spoilage of beer. *Journal of the Institute of Brewing*, **105**, 113–20.

Hammond, J.R.M. (1993) Brewing yeast. In *The Yeasts*, 2nd edition, (eds A.H. Rose and J.S. Harrison), vol. 5, pp. 8–67. Academic Press, London.

Hammond, J.R.M. (1995) Genetically-modified brewing yeasts for the 21st Century. Progress to date. *Yeast*, **11**, 1613–27.

Hammond, J.R.M. (1996) Yeast Genetics. In *Brewing Microbiology*, 2nd edition, (eds F.G. Priest and I. Campbell), pp. 43–82. Chapman and Hall, London.

Hammond, J.R.M. & Eckersley, K.W. (1984) Fermentation problems of brewing yeast with killer character. *Journal of the Institute of Brewing*, **90**, 167–77.

Hammond, J.R.M. & Wenn, R.V. (1985) Atypical carbohydrate utilisation and fermentation performance by newly propagated yeast cultures. *Proceedings of the 20th Congress of the European Brewery Convention*, Helsinki 315–22.

Hampsey, M. (1997) A review of phenotypes in *Saccharomyces cerevisiae*. *Yeast*, **13**, 1099–133.

Han, Y., Wilson, D.B. & Lei, X.G. (1999) Expression of an *Aspergillus niger* phytase gene (*phyA*) in *Saccharomyces cerevisiae*. *Applied and Environmental Microbiology*, **65**, 1915–18.

Hansen, J. & Kielland–Brandt, M.C. (1996) Inactivation of *MET10* in brewer's yeast specifically increases $SO_2$ formation during beer production. *Nature Biotechnology*, **14**, 1587–91.

Hansen, J. & Kielland-Brandt, M.C. (1994) *Saccharomyces carlsbergensis* contains two functional *MET2* alleles similar to homologues from *S. cerevisiae* and *S. monacensis*. *Gene*, **140**, 33–40.

Hansen, J. & Kielland-Brandt, M.C. (1995) Genetic control of sulphite production in brewer's yeast. *Proceedings of the 25th Congress of the European Brewery Convention*, Brussels 319–28.

Hansen, M., Rocken, W. & Emeis, C.-C. (1990) Construction of yeast strains for the production of low carbohydrate beer. *Journal of the Institute of Brewing*, **96**, 125–9.

Hapala, I., Butko, P. & Schroeder, F. (1990) Role of acidic phospholipids in intermembrane sterol transfer. *Chemistry and Physics of Lipids*, **56**, 37–47.

Hardwick, B.C., Donley, J.R. & Bishop Jr., G. (1976) The interconverson of vicinal diketones and related compounds by brewer's yeast enzymes. *Journal of the American Society of Brewing Chemists*, **34**, 65–7.

Hardwick, W.A. (1995) *Handbook of Brewing*, (ed. W.A. Hardwick), Marcel Dekker, New York.

Harold, F.M. (1995) From morphogenes to morphogenesis. *Microbiology*, **141**, 2765–78.

Harper, D.R. (1981) A microbiologist looks at beer dispense. *Brewers' Guardian*, August 23–31.

Harper, D.R., Hough, J.S. & Young, T.W. (1980) Microbiology of beer dispensing systems. *Brewers' Guardian*, January 24–8.

Harris, C.M. & Kell, D.B. (1986) The estimation of microbial biomass. *Biosensor*, **1**, 17–84.

Harris, C.M., Todd, R.W., Bungard, S.J., Lovitt, S.J., Morris, J.G. & Kell, D.B. (1987) Dielectric permittivity of microbial suspensions at radiofrequencies: a novel method for the real time estimation of microbial biomass. *Enzyme and Microbial Technology*, **9**, 181–6.

Harris, G. & Merritt, R. (1962) Effects of aeration on continuous fermentation operating at high yeast concentrations. *Journal of the Institute of Brewing*, **68**, 241–4.

Harris, J.O. (1980) Single tank operation for fermentation and maturation. *Journal of the Institute of Brewing*, **86**, 230–33.

Harris, J.O. & Watson, W. (1968) The use of controlled levels of actidione for brewing and non-brewing yeast strain differentiation. *Journal of the Institute of Brewing*, **74**, 286–90.

Harrison, G.A.F. & Collins, E. (1968) Determination of the taste threshold for a wide range of volatile and non-volatile compounds in brewing. *Proceedings of the American Society of Brewing Chemists*, 83–7.

Hartwell, L.H. & Unger, M.W. (1977) Unequal division in *Saccharomyces cerevisiae* and its implications for the control of cell division. *Journal of Cell Biology*, **75**, 422–35.

Harvey, W. (1992) Marston's maintains a tradition with new Burton Unions. *Brewing & Distilling International*, June, 16–20.

Hatch, F.A. (1936) The microscope in the brewery. *Journal of the Society of Chemistry in Industry*, **55**, 373–6.

Haworth, C.D. (1983) Polypropylene – in brewing today. *The Brewer*, November, 451–3.

Hazel, J.R. & Williams, E.E. (1990) The role of alterations in membrane lipid composition in enabling physiological adaptation of organisms to their physical environment. *Progress in Lipid Research*, **29**, 167–227.

Heelan, J.S., Sotomayor, E., Coon, K. & D'arezzo, J.B. (1998) Comparison of the rapid yeast plus panel with the API 20C yeast system for identification of clinically significant isolates of *Candida* species. *Journal of Clinical Microbiology*, **36**, 1443–5.

van Heerden, I.V. (1989) Sorghum beer – a decade of nutritional research. *Proceedings of the 2nd Institute of Brewing Scientific and Technical Convention (Central & South African Section)*, Johannesburg, 293–304.

Hees, M. & Amlung, A. (1997) Automation of fermentation using continuous extract content measurement in the cylindroconical fermenter. *Brauwelt*, **37**, 1549–51.

Heidlas, J. & Tressl, R. (1990) Purification and properties of two oxidoreductases catalysing the enantioselective reduction of diacetyl and other ketones from baker's yeast. *European Journal of Biochemistry*, **188**, 165–74.

Heinisch, J. & Zimmerman, F.K. (1985) Is the phosphofructokinase reaction obligatory for glucose fermentation by *Saccharomyces cerevisiae? Yeast*, **1**, 173–5.

Helm, E., Nøhr, H. & Thorne, R.S.W. (1953) *Measurement of yeast flocculation and its significance in brewing*, Wallerstein Laboratory Communications, **16**, 315–26.

Hemmons, L.M. (1954) Wild yeasts in draught beer I. An exploratory survey. *Journal of the Institute of Brewing*, **60**, 288–91.

Hennaut, C., Hilger, L.C. & Grenson, M. (1970) Space limitations for permease insertion in the cytoplasmic membrane of *Saccharomyces cerevisiae. Biochemistry and Biophysics Research Communications*, **39**, 666–71.

Henriksson, E. & Haikara, A (1991) Airborne microorganisms in the brewery filling area and their effect on microbiological stability of beer. *Monatsschrift für Brauwissenschaft*, **44**, 4–8.

Henschke, P.A. & Rose, A.H. (1991) Plasma membranes. In *The Yeasts*, 2nd edition (eds A.H. Rose and J.S. Harrison), vol. 4, pp. 297–346. Academic Press, London.

Her Majesty's Stationery Office (1979) Alcoholic Liquor Duties Act. HMSO, London.

Herbert, D. (1961) *Continuous Culture of Microorganisms*. Society of Chemistry in Industry Monograph 12, London.

Hermia, J. & Rahier, G. (1992) The 2001 filter: a new technology for sweet wort production. *Ferment*, **5**, 280–86.

Herrera, V.E. & Axcell, B.C. (1991a) Induction of premature flocculation by a polysaccharide fraction isolated from malt husk. *Journal of the Institute of Brewing*, **97**, 359–66.

Herrera, V.E. & Axcell, B.C. (1991b) Studies on the binding between yeast and a malt polysaccharide that induces heavy yeast flocculation. *Journal of the Institute of Brewing*, **97**, 367–73.

Heumann, K., Harris, C. & Mewes, H.W. (1996) A top-down approach to whole genome visualisation. Intelligent Systems for Molecular Biology, **4**, 98–108.

Heys, S.J.D., Gilbert, P., Eberhard, A. & Allison, D.G. (1997) Homoserine lactones and bacterial biofilms. In *Biofilms: Community Interactions and Control* (eds J. Wimpenny, P. Handley, P. Gilbert, H. Lappin-Scott and M. Jones), pp. 103–12. Bioline, Cardiff.

Hilge-Rotmann, B. & Rehm, H.-J. (1991) Relationship between fermentation capability and fatty acid composition of free and immobilised *Saccharomyces cerevisiae. Applied Biochemistry & Biotechnology*, **34**, 502–8.

Hinchliffe, E., Box, W.G., Walton., E.F. & Appleby, M. (1985) The influences of cell wall hydrophobicity on the top fermenting properties of brewing yeast. *Proceedings of the 20th Congress of the European Brewery Convention*, Helsinki, 323–40.

Hinnebusch, A.G. (1987) The general control of amino acid biosynthetic genes in the yeast *Saccharomyces cerevisiae. CRC Critical Reviews in Biochemistry*, **21**, 277–317.

Hlavacek, I.I. (1977) Pilsner Urquell – the light lager prototype. *Technical Quarterly of the Master Brewers Association of the Americas*, **14**, 94–9.

Hocking, A.D. (1988) Strategies for microbial growth at reduced water activities. *Microbiological Science*, **5**, 280–84.

Hodgson, J., Pinder, A., Catley, B.J. & Deans, K. (1999) Effect of cone cropping and serial re-pitch on the distribution of cell ages in brewery yeast. *Technical Quarterly of the Master Brewers Association of the Americas*, **36**, 175–7.

Hoekstra, S.F. (1974) Fermentable sugars in wort, brewed with adjuncts and fermented with bottom yeast. *Proceedings of the European Brewery Convention Symposium Monograph* 1, Ziest, 189–97.

Hoggan, J. (1978) The design and operation of cylindroconical tanks. *Proceedings of the European Brewery Convention Symposium*, Monograph V, Zoeterwoude, 194–206.

Holah, J.T., Betts, R.P. & Thorpe, R.H. (1988) The use of epifluorescent microscopy (DEM) and the direct epifluorescent filter technique (DEFT) to assess microbial populations on food contact surfaces. *Journal of Applied Bacteriology*, **65**, 215–21.

Holcberg, I.B. & Margalith, P. (1981) Alcoholic fermentation by immobilised yeast at high sugar concentrations. *European Journal of Applied Microbiology & Biotechnology*, **13**, 133–40.

Hollis, R., Price, I.G. & Kirkman, J. D'A. (1989) The use of stainless steel for fermentation and storage tanks at the South African Breweries Ltd. *Proceedings of the 2nd Science & Technical Convention*, Institute of Brewing, Johannesburg, 440–46.

Holmes, C. (2000) Yeast pitching with relatively high variability in yeast slurry concentration. *Proceedings of the 2nd Brewing Yeast Fermentation Performance Congress* (in press).

Holmes, C. & Teass, H.A. (1999) Application of in-pipe yeast counting analyser in medium to small breweries. *Technical Quarterly of the Master Brewers Association of the Americas*, **36**, 335–8.

Holt, J.G., Krieg, N.R., Sneath, P.H.A., Staley, J.T. & Williams, S.T. (eds) (1994) *Bergey's Manual of Determinative Bacteriology*, 9th edition. Williams and Wilkins, Baltimore.

Hook, A. (1994) An effervescent market for German wheat beer. *Brewers' Guardian*, February, 29–35.

Hopkinson, J.H., Newbury, J.E., Spencer, D.M. & Spencer, J.E. (1988) Differentiation between some industrially significant yeasts through the use of Fourier Transform-Infrared spectroscopy. *Biotechnology and Applied Biochemistry*, **10**, 118–23.

Horak, J. (1986) Amino acid transport in eukaryotic microorganisms. *Biochimica Biophysica Acta*, **864**, 223–56.

Hosokawa, M., Mochida, I., Ishiwata, Y., Shimomura, K., Hiraishi, K. & Okada, T. (1999) A portable camera system for observing the inner surface of big cylindroconical tanks. *Technical Quarterly of the Master Brewers Association of the Americas*, **36**, 179–81.

Hottiger, T., Boller, T. & Wiemken, A. (1989) Correlation of trehalose content and heat resistance in yeast mutants altered in the *RAS*/adenylate cyclase pathway: is trehalose a thermoprotectant? *FEBS Letters*, **255**, 431–4.

Hough, J.S. (1957) Characterising the principal components of pitching yeasts. *Journal of the Institute of Brewing*, **63**, 483–7.

Hough, J.S. & Hudson, J.R. (1961) Influence of yeast strain on loss of bittering material during fermentation. *Journal of the Institute of Brewing*, **67**, 241–3.

Hough, J.S. & Ricketts, M.A. (1960) New method for producing beer by continuous fermentation. *Journal of the Institute of Brewing*, **66**, 301–4.

Hough, J.S. & Rudin, A.D. (1958) Experimental production of beer by continuous fermentation. *Journal of the Institute of Brewing*, **64**, 404–10.

Hough, J.S., Briggs, D.E., Stevens, R. & Young, T.W. (1982) Chemistry of wort boiling and hop extraction. In *Malting and Brewing Science*, vol. 2 *Hopped Wort and Beer*, 2nd edition, pp. 456–98. Chapman and Hall, London and New York.

Hough, J.S., Gough, P.E. & Davis, A.D. (1962) Continuous brewing IV. Continuous fermentation on the pilot scale. *Journal of the Institute of Brewing*, **68**, 478–86.

Hough, J.S., Young, T.W., Braund, A.M., Longstaff, D., Weeks, R.J. & White, M.A. (1976) Keg and cellar tank beer in public houses – a microbiological study. *The Brewer*, June, 179–83.

Hsu, W.-P. & Bernstein, L. (1985) A new type of bioreactor employing immobilised yeast. *Technical Quarterly of the Master Brewers Association of the Americas* **22**, 159–61.

Hudson, J.R. (1970) Institute of Brewing: Analysis Committee measurement of viscosity. *Journal of the Institute of Brewing*, **76**, 341–2.

Hudson, J.R. (1973) Wort, the key to beer quality. *Proceedings of the 14th Congress of the European Brewery Convention*, Salzburg, 157–69.

Hudson, J.R. & Button, A.H. (1968) A novel pilot brewery. *Journal of the Institute of Brewing*, **74**, 300–304.

Hudson, J.R. & Stevens, R. (1960) Beer flavour II. Fusel oil content of some British beers. *Journal of the Institute of Brewing*, **66**, 471–4.

Huige, N., Sanchez, G. & Surfus, J. (1989) Pasteurizer operation and control. *Technical Quarterly of the Master Brewers Association of the Americas*, **26**, 24–9.

Hurley, R., Louvois, J. de & Mulhall, A. (1987) Yeasts as human and animal pathogens. In *The Yeasts*, 2nd edn, (eds A.H. Rose and J.S. Harrison), vol. 1, pp. 207–81. Academic Press, London.

Hutcheson, T.C., McKay, T., Farr, L. & Seddon, B. (1988) Evaluation of the stain viablue for the rapid estimation of viable yeast cells. *Letters in Applied Microbiology*, **6**, 85–8.

Hutter, A. & Oliver, S.G. (1998) Ethanol production using nuclear petite yeast mutants. *Applied Microbiology and Biotechnology*, **49**, 511–16.

Hutter, K.-J. (1993) Rapid test methods for dead-or-alive analysis of yeast cells. *Brauwelt International*, **11**, 300–305.

Hutter, K.-J. (1997) Biomonitoring of yeast in production. *Brauwelt International*, **2**, 116–23.

Hysert, D.W., Kovecses, F. & Morrison, N.M. (1976) A firefly bioluminescence ATP assay method for rapid detection and enumeration of brewery microorganisms. *Journal of the American Society of Brewing Chemists*, **34**, 145–50.

Hyttinen, I., Kronlof, J. & Hartwall, P. (1995) Use of porous glass at Hartwall Brewery in the maturation of beer with immobilised yeast. *Proceedings of the European Brewery Convention Symposium*, Monograph, **XXIV**, Espoo, 55–62.

Ichikawa, S. & Takinami, M. (1992) Microbiological control for sterile filtration and aseptic packaging. *Proceedings of the 22nd Congress of the Institute of Brewing Convention (Australia & New Zealand Section)*, Melbourne, 175–82.

Iglesias, R., Ferreras, J.M., Arias, F.J., Munoz, R. & Girbes, T. (1990) Changes in activity of the general amino acid permease from *Saccharomyces cerevisiae var. ellipsoides* during fermentation. *Biotechnology and Bioengineering*, **36**, 808–10.

Imai, T., Nakajima, I. & Ohno, T. (1994) Development of a new method for evaluation of yeast vitality by measuring intracellular pH. *Journal of the American Society of Brewing Chemists*, **52**, 5–8.

Ingledew, W.M. (1975) Utilisation of wort carbohydrates and nitrogen by *Saccharomyces carlsbergensis*. *Technical Quarterly of the Master Brewers Association of the Americas*, **12**, 146–50.

Ingledew, W.M. & Casey, G.P. (1982a) The use and understanding of media used in brewing mycology I. Media for wild yeast. *Brewers Digest*, March, 18–22.

Ingledew, W.M. & Casey, G.P. (1982b) The use and understanding of media used in brewing mycology II. Media for moulds/rapid methods. *Brewers Digest*, April, 22–6, 50.

Ingledew, W.M. & Patterson, C.A. (1999) Effect of nitrogen sources and concentration on the uptake of peptides by a lager yeast in continuous culture. *Journal of the American Society of Brewing Chemists*, **57**, 9–17.

Ingledew, W.M., Sivaswamy, G. & Burton, J.D. (1980) The API 20E microtube system for rapid identification of Gram negative brewery bacteria. *Journal of the Institute of Brewing*, **86**, 165–8.

Inoue, T. (1975) Mechanism of higher alcohol formation during wort fermentation by brewer's yeast. *Reports of the Research Laboratory of the Kirin Brewery Company*, **18**, 13–16.

Inoue, T. (1992) A review of diacetyl control technology. *Proceedings of the 22nd Institute of Brewing Convention (Australia & New Zealand Section)*, Melbourne, 76–9.

Inoue, T. (1988) Immobilised cell biotechnology – a new possibility for brewing. *Journal of the American Society of Brewing Chemists*, **46**, 64–6.

Inoue, T. & Kashihara, T. (1995) The importance of indices relating to nitrogen metabolism in fermentation control. *Technical Quarterly of the Master Brewers Association of the Americas*, **32**, 109–13.

Inoue, T., Maruyama, H., Kajino, K., Kamiya, T., Mitsui, S. & Mawatari, M. (1991) Direct spontaneous conversion of acetolactate into non-diacetyl substance in fermenting wort. *Proceedings of the 23rd Congress of the European Brewery Convention*, Lisbon, 369–76.

Inoue, T., Masayama, K., Yamamoto, Y. & Okada, K. (1968a) Mechanism of diacetyl formation in beer. Part 1 Presence of material X and its chemistry. *Reports of the Research Laboratory of the Kirin Brewing Company*, **11**, 13–16.

Inoue, T., Masayama, K., Yamamoto, Y. & Okada, K. (1968b) Mechanism of diacetyl formation in beer. Part 2 Identification of material X with alpha-acetolactate. *Reports of the Research Laboratory of the Kirin Brewing Company Limited*, **11**, 9–16.

Inoue, T., Murayama, H., Kajino, K., Kamiya, T., Mitsui, S. & Mawatari, M. (1991) Direct spontaneous conversion of acetolactate into non-diacetyl substance in fermenting wort. *Proceedings of the 23rd Congress of the European Brewery Congress*, Lisbon, 369–76.

Inoue, T., Tanaka, J. & Mitsui, S (1992) Development of saké brewing research and technology. *Recent Advances in Japanese Brewing Technology*, **2**, 13–26.

Institute of Brewing (1997) Yeast characterisation: fermentation in EBC tubes. *Institute of Brewing Methods of Analysis*, **2**, 21–5.

Institute of Brewing (1988) Floor, wall and ceiling finishes. Report issued by the Joint Maker/User Committee of the Institute of Brewing and the Allied Brewery Traders' Association. Project 54. Institute of Brewing, London.

Institute of Brewing (1997) Vicinal diketones (various). *Institute of Brewing Methods of Analysis*, **1**, 9.2–9.23.

Institute of Brewing Analysis Committee (1962) Estimation of yeast viability. *Journal of the Institute of Brewing*, **68**, 14–20.

Iserentant, D (1996) Practical aspects of yeast flocculation. *Cerevisia – Belgian Journal of Brewing and Biotechnology*, **21**, 30–33.

Iserentant, D., Geenens, W. & Verachtert, H. (1996) Titrated acidification power: a simple sensitive method to measure yeast vitality and its relation to other vitality measurements. *Journal of the American Society of Brewing Chemists*, **54**, 110–14.

Island, M.D., Naider, F. & Becker, J.M. (1987) Regulation of dipeptide transport in *Saccharomyces cerevisiae* by micromolar amino acid concentrations. *Journal of Bacteriology*, **169**, 2132–6.

Island, M.D., Perry, J.R., Naider, F. & Becker, J.M. (1991) Isolation and characterisation of *Saccharomyces cerevisiae* mutants deficient in amino acid-inducible peptide transport. *Current Genetics*, **20**, 457–63.

Ison, R.W. (1987) A revised method for the application of API 50 CH carbohydrate kits to yeasts. *Letters in Applied Microbiology*, **4**, 9–11.

Ista, L.K., Perez-Luna, V.H. & Lopez, G.P. (1999) Surface-grafted, environmentally sensitive polymers for biofilm release. *Applied and Environmental Microbiology*, **65**, 1603–9.

Iwahashi, H., Obuchi, K., Fujii, S. & Komatsu, Y. (1995) The correlative evidence suggesting that trehalose stabilises membrane structure in the yeast *Saccharomyces cerevisiae*. *Cellular and Molecular Biology*, **41**, 763–9.

Iwahashi, Y. & Nakamura, T. (1989) Localisation of the NADH kinase in the inner mitochondrial membrane of yeast mitochondria. *Journal of Biochemistry*, **105**, 916–21.

Izquierdo-Pulido, M., Font-Fabregas, J., Carceller-Rosa, J.-M., Marine-Font, A. & Vidal-Carou, C. (1996a) Biogenic amine changes related to lactic acid bacteria during brewing. *Journal of Food Protection*, **59**, 175–80.

Izquierdo-Pulido, M., Hernandez-Jover, T., Marine-Font, A. & Vidal-Carou, C. (1996b) Biogenic amines in European beers. *Journal of Agriculture and Food Chemistry*, **44**, 3159–63.

Jackson, A.C. (1996) Composition and its use. International Application (Patent Cooperation Treaty), PCT/GB96/01808.

Jackson, A.P. (1988) Control of process time and beer quality in high gravity fermentations. *Technical Quarterly of the Master Brewers Association of the Americas*, **25**, 104–7.

Jackson, G. (1997) Practical HACCP in the brewing industry. *Proceedings of the European Brewery Convention Symposium*, Monograph **XXVI**, Stockholm, 50–57.

Jacobsen, M. & Thorne, R.S.W. (1980) Oxygen requirements of brewing strains of *Saccharomyces uvarum (carlsbergensis)* – bottom fermentation yeast. *Journal of the Institute of Brewing*, **86**, 284–79.

Jacobsen, M.K. & Bernofsky, C. (1974) Mitochondrial aldehyde dehydrogenase from *Saccharomyces cerevisiae*. *Biochimica Biophysica Acta*, **350**, 277–91.

Jacobsen, T. & Lie, S. (1977) Chelators and metal buffering in brewing *Journal of the Institute of Brewing*, **83**, 208–12.

Jarzebski, A.B., Malinowski, J.J. & Goma, G. (1989) Modelling of ethanol fermentation at high yeast concentration. *Biotechnology and Bioengineering*, **34**, 1225–30.

Javadekar, V.S., Sivaraman, H., Sainkar, S.R. & Khan, M.I. (2000) A mannose-binding protein from the cell surface of flocculent *Saccharomyces cerevisiae* (NCIM 3528): its role in flocculation. *Yeast*, **16**, 99–110.

Jazwinski, S.M. (1990) An experimental system for the molecular analysis of the ageing process: the budding yeast *Saccharomyces cerevisiae*. *Journal of Gerontology: Biological Sciences*, **45**, B68–74.

Jazwinski, S.M. (1999) Longevity, genes and aging: a view provided by a genetic model system. *Experimental Gerontology*, **34**, 1–6.

Jespersen, L. & Jakobsen, M. (1996) Specific spoilage organisms in breweries and laboratory media for their detection. *International Journal of Food Microbiology*, **33**, 139–55.

Jiang, H., Medintz, I. & Michels, C.A. (1997) Two glucose sensing/signalling pathways stimulate glucose-induced inactivation of maltose permease in *Saccharomyces*. *Molecular Biology of the Cell*, **8**, 1293–304.

Jiggens, P. (1987) Gravity measurement by the oscillating U-tube method. *The Brewer*, September, 403–5.

Jimenez, J. & Benitez, T. (1987) Adaption of yeast membranes to ethanol. *Applied and Environmental Microbiology*, **53**, 1196–8.

Jin, Y.L. & Speers, A. (1998) Flocculation of *Saccharomyces cerevisiae*. *Food Research International*, **31**, 421–40.

Jirko, V. (1989) Energy status of starving yeast cells immobilised by covalent linkage. *Biotechnology Letters*, **11**, 881–4.

Jirko, V. Turkova, J. & Krumphanzi, V. (1980) Immobilisation of yeast cells with retention of cell division and extracellular production of macromolecules. *Biotechnology Letters*, **2**, 509–13.

Johnson, C.R. & Burnham, K.J. (1996) Use of bilinear structures for modelling a brewery fermentation process. *Journal of Measurement and Control*, **29**, 262–5.

Johnson, C.R., Burnham, K.J. & James, D.J.G. (1996a) Modelling and simulation of a batch beer fermentation process. *Journal of Systems Science*, **22**, 93–107.

Johnson, C.R., Burnham, K.J. & James, D.J.G. (1996b) Simulation of oxygen and pitching rate effects in a brewery fermentation process. *Proceedings of the International Conference on Systems Engineering, Coventry University*, **1**, 332–6.

Johnson, J.C. (1977) *Yeasts for Food and Other Purposes*, Noyes Data Corporation, Park Ridge, New Jersey.

Johnston, G.C., Pringle, J.R. & Hartwell, L.H. (1977) Coordination of growth with cell division in the yeast *Saccharomyces cerevisiae*. *Experimental Cell Research*, **105**, 79–98.

Jones, H.L. (1997) Yeast propagation – past, present and future. *Brewers' Guardian*, October, 24–7.

Jones, M. (1974) Amino acid composition of wort. *Proceedings of the European Brewery Convention Symposium Monograph* **1**, Ziest, 90–105.

Jones, R.M., Russell, I. & Stewart, G.G. (1986) The use of catabolite derepression as a means of improving the fermentation rate of brewing yeast strains. *Journal of the American Society of Brewing Chemists*, **44**, 161–6.

Jones, R.P. (1987) Measures of cell death and deactivation and their meaning. *Process Biochemistry*, **22**, 129–34.

Jones, R.P. (1988) Intracellular ethanol – accumulation and exit from yeast and other cells. *FEMS Microbiology Reviews*, **54**, 239–58.

Jones, R.P. (1989) Biological principles for the effects of ethanol. *Enzyme and Microbial Technology*, **11**, 130–52.

Jones, R.P. & Greenfield, P.F. (1987) Specific and non-specific inhibitory effects of ethanol on yeast growth. *Enzyme and Microbial Technology*, **9**, 334–8.

Jones, S.T. (1984) Applications of pyrolysis gas chromatography in an industrial research laboratory. *Analyst*, **109**, 823–8.

Jordan, K.N., Oxford, L. & O'Byne, C.P. (1999) Survival of low-pH stress by *Escherichia coli*, O157:H7: correlation between alterations in the cell envelope and increased acid tolerance. *Applied and Environmental Microbiology*, **65**, 3048–55.

Jordan, S.L., Glover, J., Malcolm, L., Thompson-Carter, F.M., Booth, I.R. & Park, S.F. (1999) Augmentation of killing of *Escherichia coli* O157 by combinations of lactate, ethanol, and low-pH conditions. *Applied and Environmental Microbiology*, **65**, 1308–11.

Joslyn, M.A. (1955) Yeast autolysis. 1. Chemical and cytological changes involved in autolysis. *Wallerstein Laboratory Communications*, **18**, 107–21.

Junger, R. & Kruse, E. (1972) Fissure corrosion in aluminium vessels. *Brauwelt*, **112**, 271–2.

Kakar, S.N. & Wagner, R.P. (1964) Genetic and biochemical analysis of isoleucine-valine mutants of yeast. *Genetics*, **49**, 213–22.

Kamihira, M., Taniguchi, M. & Kobayashi, T. (1987) Sterilisation of microorganisms with supercritical carbon dioxide. *Agricultural and Biological Chemistry*, **51**, 407–12.

Kara, B.V., Daoud, I. & Searle, B. (1987) Assessment of yeast quality. *Proceedings of the 21st Congress of the European Brewery Convention*, Madrid, 409–16.

Kara, B.V., Godber, S. & Hammond, J.R.M. (1987) The reduction of nitrate during fermentation. *Proceedings of the 21st Congress of the European Brewery Convention*, Madrid, 663–70.

Kara, B.V., Simpson, W.J. & Hammond, J.R.M. (1988) Prediction of the fermentation performance of brewing yeast with the acidification test. *Journal of the Institute of Brewing*, **94**, 153–8.

Kawai, S., Murao, S., Mochizuki, M., Shibuya, I., Yano, K. & Takagi, M. (1992) Drastic alteration of cycloheximide sensitivity by substitution of one amino acid in L41 ribosomal protein of yeasts. *Journal of Bacteriology*, **174**, 254–62.

Keevil, C.W., Rogers, J. & Walker, J.T. (1995) Potable-water biofilms. *Microbiology Europe*, **3**, 10–14.

Keller, M., McCormick, M. & Efron, V. (1982) *A Dictionary of Words About Alcohol*. Rutgers Centre of Alcohol Studies, Publishing Division.

Kellog, J.A., Bankert, D.A. & Chaturvedi, V. (1998) Limitations of the current microbial identification system for identification of clinical yeast isolates. *Journal of Clinical Microbiology*, **36**, 1197–1200.

Kember, R.M.J. (1990) Hop products and their usage – a review. *Ferment*, **2**, 38–41.

Kempers, J., van der Aar, P.C. & Krotje, J. (1991) Flocculation of brewers' yeast during fermentation. *Proceedings of the 23rd Congress of the European Brewery Convention*, Lisbon, 249–56.

Kennedy, A.I. & Hargreaves, L. (1997) Is there improved quality in brewing through HACCP. *Proceedings of the European Brewery Convention Symposium*, Monograph **XXVI**, Stockholm, 58–70.

Kennedy, R. (1989) Measuring vitality. *Brewers' Guardian*, September, 57–8.

Kielland-Brandt, M.C., Nilsson-Tillgren, Gjermansen, C., Holmberg, S. & Pedersen, M.B. (1995) Genetics of brewing yeast. In *The Yeasts*, 2nd edition, (eds A.H. Rose, A.E. Wheals and J.S. Harrison), vol. 6, pp. 223–54. Academic Press, London.

Kieninger, H. & Durner, G. (1982) Contents of bitter substances and tannins in harvested yeast. Investigations on repeated repitching. *Journal of the Institute of Brewing*, **88**, 189.

Kieninger, H. & Rottger, W. (1974) Carbohydrates in all malt worts – fermentable sugars. *Proceedings of the European Brewery Convention Symposium*, Monograph I, Ziest, 124–34.

Kilgour W.J. & Day, A. (1983) The application of new techniques for the rapid determination of microbial contamination in brewing. *Proceedings of the 19th Congress of the European Brewery Convention*, London, 177–84.

Kilgour, W.J. & Smith, P. (1985) The determination of pasteurisation regimes for alcoholic and alcohol-free beer. *Proceedings of the 20th Congress of the European Brewery Convention*, Helsinki, 435–42.

Kindraka, J.A. (1987) Evaluation of NBB anaerobic medium for beer spoilage organisms. *Technical Quarterly of the Master Brewers Association of the Americas*, **24**, 146–51.

King, A.T., Hodgson, J.A. & Moir, M. (1990) Control of hydrogen sulphide production by brewing yeasts. *Proceedings of the 3rd Aviemore Conference on Malting, Brewing and Distilling*, (ed I. Campbell), pp. 391–4. Institute of Brewing, London.

King, L.M., Schioler, O.D. & Ruocco, J.J. (1981) Epifluorescent method for detection of non-viable yeast. *Journal of the American Society of Brewing Chemists*, **39**, 52–4.

Kirsop, B. (1955) Maintenance of yeasts by freeze drying. *Journal of the Institute of Brewing*, **61**, 466–71.

Kirsop, B. (1974a) Segregation of strains of brewing yeast by their carbon assimilation patterns. *Journal of the Institute of Brewing*, **80**, 555–7.

Kirsop, B.E. (1991) Maintenance of yeasts. In *Maintenance of Microorganisms and Cultured Cells, A Manual of Good Practice*, 2nd edition (eds B.E. Kirsop and A. Doyle), pp. 161–82. Academic Press, London.

Kirsop, B.H. (1974b) Oxygen in brewery fermentations. *Journal of the Institute of Brewing*, **80**, 252–9.

Kirsop, B.H. (1982) Developments in beer fermentation. In *Topics in Enzyme and Fermentation Biotechnology*, (ed. A. Wiseman), 6, pp. 79–131. Ellis Horwood, Chichester.

Kirsop, B. & Henry, J. (1984) Development of a miniaturised cryopreservation method for the maintenance of a wide range of yeasts. *Cryo-Letters*, **5**, 191–200.

Kleber, W. (1987) Systems of fermentation. In *Brewing Science*, vol. 3. (ed. J.R.A. Pollock), Academic Press, London.

Klionsky, D.J., Nelson, H. & Nelson, N. (1992) Compartment acidification is required for efficient sorting of proteins to the vacuole in *Saccharomyces cerevisiae*. *Journal of Biological Chemistry*, **267**, 3416–22.

Klopper, W.J. (1974) Wort composition – a survey. *Proceedings of the European Brewery Convention Symposium*, Monograph 1, Ziest, 8–24.

Klopper, W.J., Angelino, S.A.G.F., Knol, W. & Minekus, M. (1987) Automatic in-line analyses, including biosensors. *Proceedings of the 21st Congress of the European Brewery Convention*, Madrid, 87–103.

Klopper, W.J., Angelino, S.A.G.F., Tuning, B. & Vermeire, H.-A. (1986) Organic acids and glycerol in beer. *Journal of the Institute of Brewing*, **92**, 311–18.

Klopper, W.J., Roberts, R.H., Royston, M.G. & Ault, R.G. (1965) Continuous fermentation in a tower fermenter. *Proceedings of the 10th Congress of the European Brewery Convention*, Stockholm, 238–59.

Knol, W., Minekus, M., Angelino, S.A.G.F. & Bol, J. (1988) In-line monitoring and process control in beer fermentation and other biotechnological processes. *Monatsschrift für Brauwissenschaft.* **7**, 281–7.

Knudsen, J.G., Bell, K.J., Holt, A.D. *et al.* Heat transmission. In *Perry's Chemical Engineers' Handbook*, 6th edition (eds R.H. Perry, D. Green & J.D. Maloney), pp. 1–66. McGraw Hill, New York, USA.

Knudsen, F.B. & Vacano, N.L. (1972) The design and operation of uni-tanks. *Brewers Digest*, **7**, 68–76.

Knudsen, F.B. (1978) Tank hydraulics. *Technical Quarterly of the Master Brewers Association of the Americas*, **15**, 132–9.

Koch, H.A., Bandler, R. & Gibson, R.R. (1986) Fluorescence microscopy procedure for quantitation of yeast in beverages. *Applied and Environmental Microbiology*, **53**, 599–601.

Kodama, K. (1970) Saké yeast. In *The Yeasts*, 1st edition, (eds A.H. Rose, A.E. and J.S. Harrison), vol. 3, pp. 225–82. Academic Press, London.

Kodama, K. (1993) Saké-brewing yeasts. In *The Yeasts*, 2nd edition, (eds A.H. Rose, A.E. Wheals and J.S. Harrison), vol. 5, pp. 129–68. Academic Press, London.

Kodama, Y., Fukui, N., Ashskara, T. *et al.* (1995) Improvement of maltose fermentation efficiency: constitutive expression of *MAL* genes in brewing yeast. *Journal of the American Society of Brewing Chemists*, **53**, 24–9.

Kohno, Y., Egawa, Y., Itoh, S., Nagaoka, S., Takahashi, M. & Mukai, K. (1995) Kinetic study of quenching reaction of singlet oxygen and scavenging reaction of free radical by squalene in n-butanol. *Biochimica et Biophysica Acta*, **1256**, 52–6.

Kolarov, J., Kolarova, N. & Nelson, N. (1990) A third ADP/ATP translocator gene in yeast. *Journal of Biological Chemistry*, **265**, 12711–16.

Kollar, R., Reinhold, B.B., Petrakova, E. *et al.* (1997) Architecture of the yeast cell wall. *Journal of Biological Chemistry*, **272**, 17762–75.

Kopp, F. (1975) Electron microscopy of yeasts. *Methods in Cell Biology*, **11**, 23–45.

Korch, C., Mountain, H.A., Gyllang, H., Winge, M. & Brehmer, P. (1991) A mechanism for sulphite production in beer and how to increase sulphite levels by recombinant genetics. *Proceedings of the 23rd Congress of the European Brewery Convention*, Lisbon, 201–8.

Koshcheyenko, K.A., Turkina, M.V. & Skryabin, G.K. (1983) Immobilisation of living microbial cells and their application for steroid transformations. *Enzyme & Microbial Technology*, **5**, 14–21.

Kotyk, A. & Michaljanicova, D. (1979) Uptake of trehalose by *Saccharomyces cerevisiae*. *Journal of General Microbiology*, **110**, 323–32.

Kotyk, A. & Georghiou, G. (1994) Univalent-cation-elicited acidification by yeasts. *Biochemistry and Molecular Biology International*, **33**, 1145–9.

Koukol, R. (1997) High gravity brewing. *Brauwelt International*, **15**, 248–51.

Kramer, R., Kopp, F., Niedermeyer, W. & Fuhrmann, G.F. (1978) Comparative studies of the structure and composition of the plasmalemma and the tonoplast in *Saccharomyces cerevisiae*. *Biochimica et Biophysica Acta*, **507**, 369–80.

Kreder, G.C. (1999) Yeast assimilation of trub-bound zinc. *Journal of the American Society of Brewing Chemists*, **57**, 129–32.

Kreger-van Rij, N.J.W. (1984) *The Yeasts, a Taxonomic Study*, 3rd edition, Elsevier/North Holland, Amsterdam.

Krems, B., Charizanis, C. & Entian, K.-D. (1995) Mutants of *Saccharomyces cerevisiae* sensitive to oxidative and osmotic stress. *Current Genetics* **27**, 427–34.

Kreutzfeldt, C. & Witt, W. (1991) Structural biochemistry. In *Saccharomyces* (eds M.F. Tuite and S.G. Oliver), Biotechnology Handbooks, vol. 4, pp. 5–58. Plenum Press, New York and London.

Krikilion, Ph., Andries, M., Goffin, O., van Beveren, P.C. & Masschelein, C.A. (1995) Optimal matrix and reactor design for high gravity fermentation with immobilised yeast. *Proceedings of the 25th Congress of the European Brewery Convention*, Brussels, 419–26.

Kronlof, J. (1991) Evaluation of a capacitance probe for the determination of viable yeast biomass. *Proceedings of the 23rd Congress of the European Brewery Convention*, Lisbon, 233–40.

Kronlof, J. & Linko, M. (1992) Production of beer using immobilised yeast encoding alpha-acetolactate decarboxylase. *Journal of the Institute of Brewing*, **98**, 479–91.

Kruckenberg, A.L. & Bisson, L.F. (1990) The *HXT2* gene of *Saccharomyces cerevisiae* is required for the high affinity glucose transport. *Molecular and Cellular Biology*, **10**, 5903–13.

Kumada, J., Nakajima, S., Takahashi, T. & Narziss, L. (1975) Effect of fermentation temperature and pressure on yeast metabolism and beer quality. *Proceedings of the 15th Congress of the European Brewery Convention*, Nice, 615–23.

Kummerle, M., Scherer, S. & Seiler, H. (1998) Rapid and reliable identification of food-borne yeasts by Fourier-transform infrared spectroscopy. *Applied and Environmental Microbiology*, **64**, 2207–14.

Kunau, W.-H. & Hartig, A. (1992) Peroxisome biogenesis in *Saccharomyces cerevisiae*. *Antonie van Leeuwenhoek*, **62**, 63–78.

Kunst, A., Schedding, D.J.-M. & van Schie, B.J. (1996) Glycoprotein from yeast. *International Application (Patent Cooperation Treaty)*, PCT/EP96/02588.

Kuo, S.-C. & Yamamoto, S. (1975) Preparation and growth of yeast protoplasts. *Methods in Cell Biology*, **11**, 169–84.

Kuriyama, H. & Slaughter, J.C. (1995) Control of cell morphology of the yeast *Saccharomyces cerevisiae* by nutrient limitation in continuous culture. *Letters in Applied Microbiology*, **20**, 37–40.

Kurlanzka, A., Rytka, J., Rozalska, B. & Wysocka, M. (1999) *Saccharomyces cerevisiae IRR1* protein is indirectly involved in colony formation. *Yeast*, **15**, 23–33.

Kurtzman, C.P. (1998) Nuclear DNA hybridization: quantitation of close genetic relationships. In *The Yeasts, A Taxonomic Study*, 4th edition, (eds C.P. Kurtzman & J.W. Fell), Elsevier, Amsterdam. pp. 63–8.

Kurtzman, C.P. & Blanz, P.A. (1998) Ribosomal RNA/DNA sequence comparisons for assessing phylogenetic relationships. In *The Yeasts, A Taxonomic Study*, 4th edition, (eds C.P. Kurtzman & J.W. Fell), pp. 69–74. Elsevier, Amsterdam.

Kurtzman, C.P. & Fell, J.W. (1998) *The Yeasts, A Taxonomic Study*, 4th edition, (eds C.P. Kurtzman and J.W. Fell), Elsevier, Amsterdam.

Kurtzman, C.P. & Phaff, H.J. (1987) Molecular taxonomy. In *The Yeasts*, 2nd edition, (eds A.H. Rose and J.S. Harrison), vol. 1., pp. 63–94. Academic Press, London.

Kyriakides, A.L. & Thurston, P.A. (1989) Conductance techniques for the detection of contaminants in beer. In *Rapid Microbiological Methods for Foods, Beverages and Pharmaceuticals*, (eds C.J. Stannard, S.B. Petitt and F.A. Skinner), pp. 101–17. Blackwell Scientific Publications, Oxford.

Ladenburg, K. (1968) System engineering for beer fermentation. *Technical Quarterly of the Master Brewers Association of the Americas*, **5**, 81–6.

van Laere, A. (1989) Trehalose, reserve and/or stress metabolite? *FEMS Microbiology Reviews*, **63**, 201–10.

Lagunas, R. (1979) Energetic irrelevance of aerobiosis for *Saccharomyces cerevisiae* growing on sugars. *Molecular and Cellular Biochemistry*, **27**, 139–46.

Lagunas, R. (1981) Is *Saccharomyces cerevisiae* a typical facultative anaerobe? *Trends in Biochemical Sciences*, **6**, 201–3.

Lagunas, R. (1986) Misconceptions about the energy metabolism of *Saccharomyces cerevisiae*. *Yeast*, **2**, 221–8.

Lagunas, R. (1993) Sugar transport in *Saccharomyces cerevisiae*. *FEMS Microbiology Reviews*, **104**, 229–42.

Lagunas, R. & Gancedo, C. (1983) Role of phosphate in the regulation of the Pasteur effect in *Saccharomyces cerevisiae*. *European Journal of Biochemistry*, **137**, 479–83.

Lagunas, R. & Moreno, E. (1985) The calculation of cellular parameters from the turbidity of yeast cultures may give rise to important errors. *FEMS Microbiology Letters*, **29**, 335–7.

Lagunas, R., Dominguez, C., Busturia, A. & Saez, M.J. (1982) Mechanisms of appearance of the Pasteur effect in *Saccharomyces cerevisiae*: inactivation of sugar transport systems. *Journal of Bacteriology*, **152**, 19–25.

Laidlaw, L., Tompkins, T.A., Savard, L. & Dowhanick, T.M. (1996) Identification and differentiation of brewing yeasts using specific and RAPD polymerase chain reaction. *Journal of the American Society of Brewing Chemists*, **54**, 97–102.

Lancashire, B. (2000) Application of DNA technology for yeast strain and fermentation process development. *The Brewer*, February, 69–75.

Lancashire, W.E. (1986) Modern genetics and brewing technology. *The Brewer*, **72**, 345–8.

Lancashire, W.E. & Wilde, R.J. (1987) Secretion of foreign proteins by brewing yeasts. *Proceedings of the 21st Congress of the European Brewery Convention*, Madrid, 513–20.

Lancashire, W.E., Carter, A.T., Howard, J.J. & Wilde, R.J. (1989) Superattenuating brewing yeast. *Proceedings of the 22nd Congress of the European Brewery Convention*, Zurich, 491–8.

Laplace, J.-M., Apery, S., Frere, J. & Auffray, Y. (1998) Incidence of indigenous microbial flora from utensils and surrounding air in traditional French cider making. *Journal of the Institute of Brewing*, **104**, 71–4.

Larson, J.W. & Brandon, H.J. (1988) Cooling characteristics of Uni-tanks. *Technical Quarterly of the Master Brewers Association of the Americas*, **25**, 41–6.

Laurent, M., Geldorf, B., van Nedervelde, L., Dupire, S. & Debourg, A. (1995) Characterisation of the aldoketoreductase yeast enzymatic system involved in the removal of wort carbonyls during fermentation. *Proceedings of the 25th Congress of the European Brewery Convention*, Brussels, 337–44.

Lawrence, D.R. (1983) Yeast differentiation and identification. *Proceedings of the 19th Congress of the European Brewery Convention*, London, 449–56.

Lawrence, D.R. (1986a) Brewing yeast – past, present and future. *The Brewer*, September, 340–44.

Lawrence, D.R. (1986b) Dried yeast in brewing. *Proceedings of the 2nd Aviemore Conference on Malting, Brewing and Distilling*, (eds I. Campbell and F.G. Priest), pp. 291–4. Institute of Brewing, London.

Lawrence, D.R., Bowen, W.R., Sharpe, F.R. & Ventham, T.J. (1989) Yeast zeta potential and flocculation. *Proceedings of the 22nd Congress of the European Brewery Convention*, Zurich, 505–12.

Lawrence, M. (1990) *The Encircling Hop, a History of Hops and Brewing*. Media Print, Sittingbourne, Kent, UK.

Laws, D.R.J., McGuinness, J.D. & Rennie, H. (1972) The losses of bitter substances during fermentation. *Journal of the Institute of Brewing*, **78**, 314–21.

Leather, R.V. (1994) The theory and practice of beer clarification. Part 1 Theory. *The Brewer*, **80**, 429–33.

Leather, R.V., Dale, C.J. & Morson, B.T. (1997) Characterisation of beer particle charges and the role of particle charge in beer processing. *Journal of the Institute of Brewing*, **103**, 377–80.

Leber, R., Zinser, E., Hrastnik, C., Paltauf, F. & Daum, G. (1995) Export of steryl esters from lipid particles and release of free sterols in the yeast, *Saccharomyces cerevisiae*. *Biochimica et Biophysica Acta*, **1234**, 119–26.

Leber, R., Zinser, E., Paltauf, F. & Daum, G. (1992) Intracellular transport of sterols in the yeast, *Saccharomyces cerevisiae*. *Proceedings of the 16th International Conference of Yeast Genetics and Molecular Biology*, S509.

Leber, R., Zinser, E., Zellnig, G., Paltauf, F. & Daum, G. (1994) Characterisation of lipid particles of the yeast *Saccharomyces cerevisiae*. *Yeast*, **10**, 1421–8.

Lee, F.J. & Hassan, H.M. (1987) Biosynthesis of superoxide dismutase and catalase in chemostat cultures of *Saccharomyces cerevisiae*. *Applied Microbiology and Biotechnology*, **26**, 531–6.

Lee, T.C. & Lewis, M.J. (1968) Mechanism of release of nucleotidic material by fermenting brewer's yeast. *Journal of Food Science*, **33**, 124–8.

Leedham, P.A. (1983) Control of brewery fermentation via yeast growth. *Proceedings of the 19th Congress of the European Brewery Convention*, London, 153–60.

Leemans, C., Dupire, S. & Macron, J.-Y. (1993) Relation between wort DMSO and DMS concentration in beer. *Proceedings of the 24th Congress of the European Brewery Convention*, Oslo, 709–16.

Lees, N.D., Skaggs, B., Kirsch, D.R. & Bard, M. (1995) Cloning of late genes in the ergosterol biosynthetic pathway of *Saccharomyces cerevisiae* – a review. *Lipids*, **30**, 221–6.

Legeay, O., Ratomahenina, R. & Galzy, P. (1989) Study of diacetyl reductase in a polyploid brewing strain of *Saccharomyces carlsbergensis*. *Agricultural and Biological Chemistry*, **53**, 531–2.

Legmann, R. & Marglith, P. (1986) Ethanol formation by hybrid yeasts. *Applied Microbiology and Biotechnology*, **23**, 198–202.

Lehoel, M. & Moll, M. (1987) Controlling the fermentation process in-line. *Brewing and Distilling International*, January, 22–3.

Lerner, C., Taeymans, D. & Masschelein, C.A. (1991) Application of the gas lift principle to the improvement of deep cooling efficiency in one tank operation. *Proceedings of the 23rd Congress of the European Brewery Convention*, Lisbon, 329–36.

Lenoel, M., Meunier, J.-P., Moll, M. & Midoux, N. (1987) Improved system for stabilising yeast fermenting power during storage. *Proceedings of the 21st Congress of the European Brewery Convention*, Madrid, 425–32.

Lentini, A. (1993) A review of the various methods available for monitoring the physiological status of yeast: yeast viability and yeast vitality. *Ferment*, **6**, 321–7.

Lentini, A., Hawthorne, D.B. & Kavanagh, T.E. (1992) A rheological study of yeast during handling and storage. *Proceedings of the 22nd Institute of Brewing Convention (Australia & New Zealand Section)*, Melbourne, 198–9.

Lentini, A., Jones, R.D., Wheatcroft, R. *et al.* (1990) Metal ion uptake by yeast. *Proceedings of the 21st Congress of the Institute of Brewing (Australia & New Zealand Section)*, Auckland, 158–63.

Lentini, A., Takis, S., Hawthorne, D.B. & Kavanagh, T.E. (1994) The influence of trub on fermentation and flavour development. *Proceedings of the 23rd Convention Institute of Brewing (Asia Pacific Section)*, Sydney, 89–95.

Letters, R. (1992) Lipids in brewing, friend or foe? *Ferment*, **4**, 268–74.

Lewis, J.G., Learmonth, R.P. & Watson, K. (1993) Role of growth phase and ethanol in freeze-thaw stress resistance of *Saccharomyces cerevisiae*. *Applied and Environmental Microbiology*, **59**, 1065–71.

Lewis, M.J. (1974) Maximum temperature of growth of yeasts. *Journal of the Institute of Brewing*, **80**, 423.

Lewis, M.J. & Phaff, H.J. (1964) Release of nitrogeneous substances by brewer's yeast. III. Shock excretion of amino acids. *Journal of Bacteriology*, **87**, 1389–96.

Lewis, M.J. & Phaff, H.J. (1965) Release of nitrogenous substances by brewer's yeast. IV. Energetics in shock excretion of amino acids. *Journal of Bacteriology*, **89** 960–66.

Lewis, M.J. & Poerwantaro, W.M. (1991) Release of haze material from the cell walls of agitated yeast. *Journal of the American Society of Brewing Chemists*, **49**, 43–6.

Lewis, M.J. & Lewis, D.J. (1996) Microbrewing in the USA. Change, challenge and opportunity. *Grist International*, **July/August**, 18–22.

Lewis, T.A., Rodriguez, R.J. & Parks, L.W. (1987) Relationship between intracellular sterol content and sterol esterification. *Biochimica et Biophysica Acta*, **921**, 205–12.

Lewis, T.L., Keesler, G.A., Fenner, G.P. & Parks, L.W. (1988) Pleotrophic mutations in *Saccharomyces cerevisiae* affecting sterol uptake and metabolism. *Yeast*, **4**, 93–106.

Lichko, L.P. & Okorokov, L.A. (1990) Phosphohydrolase activities of vacuoles of *Saccharomyces carlsbergensis* yeast cells. *Biokhimiya*, **55**, 210–17.

Lie, S. & Jacobsen, T. (1983) The absorption of zinc by brewer's yeast. *Proceedings of the 19th Congress of the European Brewery Convention*, London, 145–51.

Lie, S., Haukeli, A.D. & Jacobsen, T. (1974) Nitrogenous components in wort. *Proceedings of the European Brewery Convention Symposium Monograph* 1, Ziest, 25–40.

Lie, S., Haukeli, A.D. & Jacobsen, T. (1975) The effect of chelators in brewery fermentation. *Proceedings of the 15th Congress of the European Brewery Convention*, Nice, 601–14.

Lieckfeldt, E., Meyer, W. & Borner, T (1993) Rapid identification and differentiation of yeasts by DNA and PCR fingerprinting. *Journal of Basic Microbiology*, **33**, 413–26.

Lillie, S.H. & Pringle, J.R. (1980) Reserve carbohydrate metabolism in *Saccharomyces cerevisiae*: responses to nutrient limitation. *Journal of Bacteriology*, **143**, 1384–94.

Lin, Y. (1980) Use of potassium hydroxide technique for the differentiation of Gram-positive and Gram-negative bacteria. *Brewers Digest*, **March**, 36–7.

Lindner, P. (1895) *Microscopic Control of Process in the Fermentation Industries* (in German). Paul Parey, Berlin.

Lindquist, S. (1996) Mad cows meet mad yeast: the prion hypothesis. *Molecular Psychiatry*, **1**, 376–9.

Lindquist, S.L. & Craig, E.A. (1988) The heat shock proteins. *Annual Review of Genetics*, **22**, 631–77.

Lindsay, J.H. & Larson, J.W. (1975) Large volume tanks – a review. *Technical Quarterly of the Master Brewers Association of the Americas*, **12**, 264–72.

Link, A.J., Eng, J., Schieltz, D.M. *et al.* (1999) Direct analysis of protein complexes using mass spectrometry. *Nature Biotechnology*, **17**, 676–82.

Linnane, A.W. & Lukins, H.B. (1976) Isolation of mitochondria and techniques for studying mitochondrial biogenesis in yeast. *Methods in Cell Biology*, **12**, 285–310.

Lipke, P.N. & Ovalle, R. (1998) Cell wall architecture in yeast: new structure and new challenges. *Journal of Bacteriology*, **180**, 3735–40.

L'Italien, Y., Thibault, J. and LeDay, D. (1989) Improvement of ethanolic fermentation under hyperbaric conditions. *Biotechnology and Bioengineering*, **33**, 471–6.

Liu, J.-J. & Lindquist, S. (1999) Oligopeptide repeat expansions modulate 'protein-only' inheritance in yeast. *Nature*, **400**, 573–6.

Lloyd Hind, H. (1940) Brewing processes. In *Brewing Science and Practice*, 2, Chapman & Hall, London. pp. 755–70.

Lloyd, D. (1974) *The Mitochondria of Microorganisms*. Academic Press, London.

Lloyd, D. & Cartledge, T.G. (1991) Separation of yeast organelles. In *The Yeasts*, 2nd edition, (eds A.H. Rose and J.S. Harrison), vol. 4, pp. 121–74. Academic Press, London.

Lloyd, D., Moran, C.A., Suller, M.T.E. & Dinsdale, M.G. (1996) Flow cytometric monitoring of Rhodamine 123 and a cyanine dye uptake by yeast during cider fermentation. *Journal of the Institute of Brewing*, **102**, 251–9.

Lloyd, D., Morrell, S., Carilsen, H.N. *et al.* (1993) Effects of growth with ethanol on fermentation and membrane fluidity of *Saccharomyces cerevisiae*. *Yeast*, **9**, 825–33.

Lodder, J. (1970) *The Yeasts, a Taxonomic Study*, 2nd edition, North Holland, Amsterdam.

Lodolo, E.J., du Plessis, G.A. O'Connor-Cox, E.S.C. & Axcell, B.C. (1995) Pantothenate supplementation in brewery wort as a means of reducing sulphur dioxide and acetaldehyde concentrations. *Proceedings Institute of Brewing 5th Central and South African Section Convention*, Victoria Falls, 134–51.

Lodolo, E.J. O'Connor-Cox, E. & Axcell, B.C. (1999) Optimisation of the dissolved oxygen supply for high gravity brewing. *Technical Quarterly of the Master Brewers Association of the Americas*, **36**, 139–54.

Londesborough, J. & Varimo, K. (1984) Characterisation of two trehalases in bakers' yeast. *Biochemical Journal*, **219**, 511–18.

Long, D.G. (1999) From cobalt to chloropropanol: de tribulationibus aptis cerevisiis imbibendis. *Journal of the Institute of Brewing*, **105**, 79–84.

Longo, E. & Vezinhet, F. (1993) Chromosomal rearrangements during vegetative growth of a wild strain of *Saccharomyces cerevisiae. Applied and Environmental Microbiology*, **59**, 322–6.

Lorenz, R.T. & Parks, L.W. (1987) Regulation of ergosterol biosynthesis and sterol uptake in a sterol-auxotrophic yeast. *Journal of Bacteriology*, **169**, 3707–11.

Lorenz, R.T. & Parks, L.W. (1991) Involvement of heme components in sterol metabolism of *Saccharomyces cerevisiae. Lipids*, **26**, 598–603.

Lorenz, R.T., Casey, W.M. & Parks, L.W. (1989) Structural discrimination in the sparking functions in the yeast *Saccharomyces cerevisiae. Journal of Bacteriology*, **171**, 6169–73.

Lorenz, R.T., Rodriguez, R.J., Lewis, T.A. & Parks, L.W. (1986) Characteristics of sterol uptake in *Saccharomyces cerevisiae. Journal of Bacteriology*, **167**, 981–5.

Louis-Eugene, S., Ratomahenina, R. & Galzy, P. (1988) Enzymatic reduction of diacetyl by a strain of *Saccharomyces uvarum. Folia Microbiological*, **33**, 38–44.

Loureiro, V. & Ferrera, H.G. (1983) On the intracellular accumulation of ethanol in yeast. *Biotechnology and Bioengineering*, **25**, 2263–9.

Loveridge, D., Ruddlesden, J.D., Noble, C.S. & Quain, D.E. (1999) Improvements in brewery fermentation performance by 'early' yeast cropping and reduced yeast storage time. *Proceedings of the 7th Institute of Brewing Convention (Africa Section)*, Nairobi, 95–9.

Lowe, D. (1991) Gales gears up to supply its growing estate. *Brewers' Guardian*, May, 25–7.

Lucero, P., Herweijer, M. & Lagunas, R. (1993) Catabolite inactivation of the yeast maltose transporter is due to proteolysis. *FEBS Letters*, **333**, 165–8.

Luckiewicz, E.T. (1978) Computer modelling of fementation. *Technical Quarterly of the Master Brewers Association of the Americas*, **15**, 190–97.

Lyness, C.A., Steele, G.M. & Stewart, G.G. (1997) Investigating ester metabolism: characterisation of the *ATF1* gene in *Saccharomyces cerevisiae. Journal of the American Society of Brewing Chemists*, **55**, 141–6.

Ma, J. & Lindquist, S. (1999) De novo generation of a PRP$^{Sc}$-like conformation in living cells. *Nature Cell Biology*, **1**, 358–61.

Mackie, A. (1985) The production of cask conditioned ales in cylindroconical fermenters. *The Brewer*, July/August, 256–63.

Macreadie, I.G., Sewell, A.K. & Winge, D.R. (1994) Metal ion resistance and the role of metallothionein in yeast. In *Metal Ions in Fungi*, (eds G. Winkelmann & D.R. Winge), pp. 279–310. Marcel Dekker Inc, New York.

MacWilliam, I.C. (1968) Wort composition – a review. *Journal of the Institute of Brewing*, **74**, 38–54.

MacWilliam, I.C. (1970) The structure, synthesis and functions of the yeast cell wall – a review. *Journal of the Institute of Brewing*, **76**, 524–35.

Madeo, F., Frohlich, E. Ligr, M. *et al.* (1999) Oxygen stress: a regulator of apoptosis in yeast. *Journal of Cell Biology*, **145**, 757–67.

Mager, W.H. & Varela, J.C.S. (1993) Osmostress response of the yeast *Saccharomyces cerevisiae. Molecular Microbiology*, **10**, 253–8.

Magliani, W., Conti, S., Gerloni, M., Bertolotti, D. & Polonelli, L. (1997) Yeast killer systems. *Clinical Microbiological Reviews*, **10**, 369–400.

Majara, M., O'Connor-Cox, E.S.C. & Axcell, B.C. (1996a) Trehalose – an osmoprotectant and stress indicator compound in high and very high gravity brewing. *Journal of the American Society of Brewing Chemists*, **54**, 149–54.

Majara, M., O'Connor-Cox, E.S.C. & Axcell, B.C. (1996b) Trehalose – a stress protectant and stress indicator compound for yeast exposed to adverse conditions. *Journal of the American Society of Brewing Chemists*, **54**, 221–7.

Malcorps, P., Cheval, J.M., Jamil, S. & Dufour, J.P. (1991) A new model for the regulation of ester synthesis by alcohol acetyl transferase in *Saccharomyces cerevisiae* during fermentation. *Journal of the American Society of Brewing Chemists*, **49**, 47–53.

Mandl, B. (1974) Mineral matter, trace elements, organic and inorganic acids in hopped worts. *Proceedings of the European Drewery Convention Symposium Monograph* 1, Ziest, 233–8.

Mandl, B., Geiger, E. & Piendl, A. (1975) Formation of higher aliphatic alcohols in relation to substrate utilisation by yeast. *Proceedings of the 15th Congress of the European Brewery Convention*, Nice, 539–63.

Manger, H.-J. & Annemüller, G. (1996) The cooling of fermenting vessels. *Brauwelt*, **136**, 2160–70.

Mannella, C.A. (1992) The ins and outs of mitochondrial membrane channels. *Trends in Biochemical Sciences*, **17**, 315–20.

Manson, D.H. & Slaughter, J.C. (1986) Methods for predicting yeast fermentation activity. *Proceedings of the 2nd Aviemore Conference on Malting, Brewing and Distilling*, (eds I. Campbell and F.G. Priest), pp. 295–7. Institute of Brewing, London.

Mansure, J.J., Souza, R.C. & Panek, A.D. (1997) Trehalose metabolism in *Saccharomyces cerevisiae* during alcoholic fermentation. *Biotechnology Letters*, **19**, 1201–3.

Marder, R., Becker, J.M. & Naider, F. (1977) Peptide transport in yeast: utilisation of leucine and lysine containing peptides by *Saccharomyces cerevisiae*. *Journal of Bacteriology*, **131**, 906–16.

Marinoni, G., Manuel, M., Peterson, R.F. Hvidtfeldt, J., Sulo, P. & Piskur, J. (1999) Horizontal transfer of genetic material among *Saccharomyces* yeasts. *Journal of Bacteriology*, **181**, 6488–96.

Martens, H., Dawoud, E. & Verachtert, H. (1991) Wort Enterobacteria and other microbial populations involved during the first month of lambic fermentation. *Journal of the Institute of Brewing*, **97**, 435–9.

Martin, S., Bosch, J., Almenar, J. & Posada, J. (1975) Large outdoor sphero-conical tanks: a new realisation of fermentation and lagering processes. *Proceedings of the 15th Congress of the European Brewery Convention*, Nice, 301–10.

Martini, A. (1993) Origin and domestication of the wine yeast *Saccharomyces cerevisiae*. *Journal of Wine Research*, **4**, 165–76.

Marty-Teysset, C., de la Torre, F. & Garel, J.-R. (2000) Increased production of hydrogen peroxide by *Lactobacillus delbrueckii* subsp. *bulgaricus* upon aeration: involvement of an NADH oxidase in oxidative stress. *Applied and Environmental Microbiology*, **66**, 262–7.

Masneuf, I., Aigle, M. & Dubourdieu, D. (1996) Development of a polymerase chain reaction/restriction fragment length polymorphism method for *Saccharomyces cerevisiae* and *Saccharomyces bayanus* identification in enology. *FEMS Microbiology Letters*, **138**, 239–44.

Masneuf, I., Hansen, J., Groth, C., Piskur, J. & Dubourdieu, D. (1998) New hybrids between *Saccharomyces* sensu stricto yeast species found among wine and cider production strains. *Applied and Environmental Microbiology*, **64**, 3887–92.

Masschelein, C.A. (1981) Flavour developments in large-capacity vessels. *Brewing & Distilling International*, May, 37–42.

Masschelein, C.A. (1986a) The biochemistry of maturation. *Journal of the Institute of Brewing*, **92**, 213–219.

Masschelein, C.A. (1986b) Yeast metabolism and beer flavour. *Proceedings of the European Brewery Convention*, Monograph **XII**, Vuoranta, 2–22.

Masschelein, C.A. (1989) Recent and future developments of fermentation technology and fermenter design in brewing. In *Biotechnology Applications in Beverage Production*, (eds, C. Cantarelli and G. Lanzarini), pp. 77–91. Elsevier Science Publishers, London & New York.

Masschelein, C.A. (1994) State of the art and future developments in fermentation. *Journal of the American Society of Brewing Chemists*, **52**, 128–35.

Masschelein, C.A. & Andries, M. (1995) Future scenario of immobilised systems: promises and limitations. *Proceedings of the European Brewery Convention*, Monograph **XXIV**, Espoo, 223–38.

Masschelein, C.A., Andries, M., Franken, F., van der Winkel, L. & Beveron van, P.C. (1995) The membrane loop concept: a new approach for optimal oxygen transfer into high cell density pitching yeast suspensions. *Proceedings of the 25th Congress of the European Brewery Convention*, Brussels, 377–86.

Masschelein, C.A., Borremans, E. & van de Winkel, L. (1994) Application of exponential fed-batch culture to brewing yeast propagation. *Proceedings of the 23rd Institute of Brewing Convention (Asia Pacific Section)*, Sydney, 109–13.

Masschelein, C.A. & van der Meersche. (1976) Flavour maturation of beer. *Technical Quarterly of the Master Brewers Association of the Americas*, **13**, 240–50.

Masschelein, C.A., van der Winkel, L. & Debourg, A. (1993) Optimal ester production by the application of the fed-batch principle to brewery fermentation. *Proceedings of the 24th Congress of the European Brewery Convention*, Oslo, 231–40.

Masters, C. (1992) Microenvironmental factors and the binding of glycolytic enzymes to contractile elements. *International Journal of Biochemistry*, **24**, 405–10.

Masy, C.L., Henquinet, A. & Mestdagh, M.M. (1992) Fluorescence study of lectin-like receptors involved in the flocculation of the yeast *Saccharomyces cerevisiae*. *Canadian Journal of Microbiology*, **38**, 405–9.

Mathieu, C., van den Bergh, L. & Iserentant, D. (1991) Prediction of yeast fermentation performance using the acidification power test. *Proceedings of the 23rd Congress of the European Brewery Convention*, Lisbon, 273–80.

Matile, Ph., Moor, H. & Robinow, C.F. (1969) Yeast cytology. In *The Yeast*, 1st edition (eds A.H. Rose and J.S. Harrison), vol. 1. pp. 220–301. Academic Press, London.

Matsuyama, S., Nouraini, S. & Reed, J.C. (1999) Yeast as a tool for apoptosis research. *Current Opinion in Microbiology*, **2**, 618–23.

Maule, A.P. & Thomas, P.D. (1973) Strains of yeast lethal to brewery yeasts. *Journal of the Institute of Brewing*, **79**, 137–41.

Maule, D.R. (1976) Cylindroconical tanks – a study of cooling jacket performance. *The Brewer*, May, 140–44.

Maule, D.R. (1979) Propagation and handling of pitching yeast. *Brewers' Guardian*, May, 76–80.

Maule, D.R. (1986) A century of fermenter design. *Journal of the Institute of Brewing*, **92**, 137–45.

Maule, D.R.J. (1967) Rapid gas chromatographic examination of beer flavour. *Journal of the Institute of Brewing*, **73**, 351–61.

Maurico, J.C. & Salmon, J.M. (1992) Apparent loss of sugar transport activity in *Saccharomyces cerevisiae* may mainly account for maximum ethanol production during alcoholic fermentation. *Biotechnology Letters*, **14**, 577–82.

McCaig, L., Egan, L., Schisler, D. & Hahn, C.W. (1978) Development of required time temperature relationships for effective flash pasteurisation. *Journal of the American Society of Brewing Chemists*, **36**, 144–9.

McCaig, R. (1990) Evaluation of the fluorescent dye 1-anilino-8-naphthalene sulphonic acid for yeast viability determination. *Journal of the American Society of Brewing Chemists*, **48**, 22–5.

McCaig, R. & Bendiak, D.S. (1985a) Yeast handling studies I. Agitation of stored pitching yeast. *Journal of the American Society of Brewing Chemists*, **43**, 114–19.

McCaig, R. & Bendiak, D.S. (1985b) Yeast handling studies II. Temperature of storage of pitching yeast. *Journal of the American Society of Brewing Chemists*, **43**, 119–22.

McCaig, R., McKee, J., Pfisterer, A. & Hysert, D.W. (1992) Very high gravity brewing – laboratory and pilot plant trials. *Journal of the American Society of Brewing Chemists*, **50**, 18–26.

McCullough, M.J., Clemons, K.V., Farina, C., McCusker, J.H. & Stevens, D.A. (1998) Epidemiological investigation of vaginal *Saccharomyces cerevisiae* isolates by a genotypic method. *Journal of Clinical Microbiology*, **36**, 557–62.

McCulloch, W.S. & Pitts, W. (1943) A logical calculus of the ideas imminent in nervous activity. *Bulletin of Mathematical Biology*, **5**, 115–33.

McCusker, J.H., Clemons, K.V., Stevens, D.A. & Davis, R.W. (1994) Genetic characterisation of pathogenic *Saccharomyces cerevisiae* isolates. *Genetics*, **136**, 1261–9.

McMurrough, I. (1995) Scope and limitations for immobilised cell systems in the brewing industry. *Proceedings of the European Brewery Convention*, Monograph **XXIV**, Espoo, 2–16.

McMurrough, I. & Delcour, J.A. (1994) Wort polyphenols. *Ferment*, **7**, 175–82.

McMurrough, I., Hennigan, G.P. & Loughrey, M.J. (1983) Contents of simple, polymeric and complexed flavanols in worts and beers and their relationship to haze formation. *Journal of the Institute of Brewing*, **89**, 15–23.

McNeil, K.E., Hawthorne, D.B. & Murray, P.J. (1984) A simple qualitative method for the detection of yeast invertase in primed beer. *Brewers Digest*, November, 12–13.

Meaden, P. (1986) Genetic engineering of yeast. *Brewers' Guardian*, July, 23–7.

Meaden, P. (1990) DNA fingerprinting of Brewers' yeast: current perspectives. *Journal of the Institute of Brewing*, **96**, 195–200.

Meaden, P.G. (1996a) Yeast genome now completely sequenced. *Ferment*, **9**, 213–14.

Meaden, P.G. (1996b) DNA fingerprinting of Brewers' yeast. *Ferment*, **9**, 267–72.

Meaden, P.G. & Taylor, N.R. (1991) Cloning of a yeast gene which causes phenolic off-flavours in beer. *Journal of the Institute of Brewing*, **97**, 353–7.

Meaden, P.G., Dickinson, F.M., Mifsud, A. *et al.* (1997) The *ALD6* gene of *Saccharomyces cerevisiae* encodes a cytosolic, $Mg^{2+}$ activated acetaldehyde dehydrogenase. *Yeast*, **13**, 1319–27.

van de Meersche, J., Devreux & Masschelein, C.A. (1979) Formation and role of volatile fatty acids during the maturation of beer. *Proceedings of the 17th Congress of the European Brewery Convention*, Berlin, 787–800.

Meilgaard, M. (1974) Flavour and threshold of beer volatiles. *Technical Quarterly of the Master Brewers Association of the Americas*, **11**, 87–9.

Meilgaard, M. (1975) Flavour chemistry of beer. Part 2 Flavour and threshold of 239 aroma volatiles. *Technical Quarterly of the Master Brewers Association of the Americas*, **12**, 151–68.

Meisel, H.R. & Huggins, G. (1989) The development of a microbrewery fermenter. *Proceedings of the 2nd Science & Technical Convention, Institute of Brewing*, Johannesburg, 187–211.

Mensour, N., Margaritis, A., Briens, C.L., Pilkinton, H. & Russell, I. (1995) Gas lift systems for immobilised cell fermentations. *Proceedings of the European Brewery Convention*, Monograph **XXIV**, Espoo, 125–32.

Mestdagh, M.M., Rouxhet, P.G. & Dufour, J.P. (1990) Surface chemistry and flocculation of brewery yeast. *Ferment*, **3**, 31–7.

Meunier, J.-R. & Choder, M. (1999) *Saccharomyces cerevisiae* colony growth and ageing: biphasic growth accompanied by changes in gene expression. *Yeast*, **15**, 1159–69.

Mewes, H.W., Albermann, K., Bahr, M. *et al.* (1997) Overview of the yeast genome. *Nature*, **387 (Supp)**, 7–65.

Meyer, J.Z. & Whittaker, P.A. (1977) Respiratory repression and the stability of the mitochondrial genome. *Molecular and General Genetics*, **151**, 333–42.

Michaljanicova, D., Hodan, J. & Kotyk, A. (1982) Maltotriose transport and utilisation in baker's and brewer's yeast. *Folia Microbiologica*, **27**, 217–21.

Michels, C.A. & Needleman, R.B. (1984) The dispersed repeated family of *MAL* loci in *Saccharomyces* spp. *Journal of Bacteriology*, **157**, 949–52.

Miedener, H. (1978) Optimisation of fermentation and conditioning in the production of lager. *Proceedings of the European Brewery Convention*, Monograph V, Zoeterwoude, 110–34.

Mieth, H.O. (1995) Immobilised yeast plants for alcohol free beer production and rapid maturation. *Proceedings of the 5th Congress of the Institute of Brewing (Central and South African Section)*, Victoria Falls, 166–72.

Miki, B.L.A., Poon, N.H., James, A.P. & Seligy, V.L. (1982) Possible mechanism for flocculation interactions governed by gene *FLO1* in *Saccharomyces cerevisiae*. *Journal of Bacteriology*, **150**, 878–89.

Miller, D. (1990) *Continental Pilsener*. Brewers Publications, Boulder, Colorado, USA.

Miller, L.F., Mabee, M.S., Gress, H.S. & Jangaard, N.O. (1978) An ATP bioluminescence method for the quantification of viable yeast for fermenter pitching. *Journal of the American Society of Brewing Chemists*, **36**, 59–62.

Miller, R. & Galston, G. (1989) Rapid methods for the detection of yeast and lactobacillus by ATP bioluminescence. *Journal of the Institute of Brewing*, **95**, 317–19.

Minetoki T., Bogaki, T., Iwamatsu, A., Fujii, T. & Hamachi, M. (1993) The purification, properties and internal peptide sequences of alcohol acetyltransferase isolated from *Saccharomyces cerevisiae* Kyokai No. 7. *Biosciences Biotechnology and Biochemistry*, **57**, 2094–8.

MIPS (Munich Information Centre for Protein Sequences – Yeast Genome Project) http://www.mips. biochem.mpg.de/mips/yeast/

Mitchell, P. (1979) Keilin's respiratory chain concept and its chemiosmotic consequences. *Science*, **206**, 1148–59.

Mitchison, J.M., Passano, I.M. & Smith, F.H. (1956) Investigative methods for the interference microscope. *Quarterly Journal of the Microscopy Society*, **97**, 287–302.

Mittenbühler, K. & Holzer, H. (1988) Purification and characterisation of acid trehalase from the yeast *such 2* mutant. *Journal of Biological Chemistry*, **263**, 8537–43.

Mittenbühler, K. & Holzer, H. (1991) Characterisation of different forms of yeast acid trehalase. *Archive for Microbiology*, **155**, 217–21.

Mitts, M.R., Grant, D.B. & Heideman, W. (1990) Adenylate cyclase in *Saccharomyces cerevisiae* is a peripheral membrane protein. *Molecular and Cellular Biology*, **10**, 3873–83.

Moaledj, K. (1986) Comparison of Gram-staining and alternate methods, KOH test and aminopeptidase activity in aquatic bacteria: their application to numerical taxonomy. *Journal of Microbiological Methods*, **5**, 303–10.

Mochaba, F., O'Connor-Cox, E.S.C. & Axcell, B.C. (1996) Metal ion concentration and release by a brewing yeast: characterisation and implications. *Journal of the American Society of Brewing Chemists*, **54**, 155–63.

Mochaba, F., O'Connor-Cox, E.S.C. & Axcell, B.C. (1997) A novel and practical yeast viability method based on magnesium release. *Journal of the Institute of Brewing*, **103**, 99–102.

Mogens, J. & Piper, J.U. (1989) Performance and osmotolerance of different strains of lager yeast in high gravity fermentations. *Technical Quarterly of the Master Brewers Association of the Americas*, **26**, 56–61.

Moll, M., Duteurte, B., Scion, G. & Lehuede, J.-M. (1978) New techniques for control of fermentation in breweries. *Master Brewers Association of the Americas Technical Quarterly*, **15**, 26–9.

Moller, N.C. (1975) Continuous measurement of wort/beer extract in a fermenter. *Master Brewers Association of the Americas Technical Quarterly*, **12**, 41–45.

Molzahn, S.W., Hockney, R.C., Kelsey, P. & Box, W.G. (1983) Factors influencing the flash pasteurisation of beer. *Proceedings of the 19th Congress of the European Brewery Convention*, London, 255–62.

Montrocher, R., Verner, M-C., Briolay, J., Gautier, C. & Marmeisse, R. (1998) Phylogenetic analysis of the *Saccharomyces cerevisiae* group based on polymorphisms of the rDNA spacer sequences. *International Journal of Systematic Bacteriology*, **48**, 295–303.

Moonsamy, N., Mochoba, F., O'Connor-Cox, E.S.C. & Axcell, B.C. (1996) Rapid yeast trehalose measurement using near infrared reflectance spectroscopy. *Journal of the Institute of Brewing*, **101**, 203–6.

Moor, H. & Mühlethaler, K. (1963) Fine structure in frozen etched yeast cells. *Journal of Cell Biology*, **17**, 609–28.

Morris, G.J., Coulson, G.E. & Clarke, K.J. (1988) Freezing injury in *Saccharomyces cerevisiae*: the effect of growth conditions. *Cryobiology*, **25**, 471–82.

Morrison, K.B. & Suggett, A. (1983) Yeast handling, petite mutants, and lager flavour. *Proceedings of the 19th Congress of the European Brewery Convention*, London, 489–96.

Mortimer, R.K. & Johnston, J.R. (1959) Life span of individual yeast cells. *Nature*, **183**, 1751–2.

Mortimer, R.K. & Johnston, J.R. (1986) Genealogy of principal strains of the yeast genetic stock centre. *Genetics*, **113**, 35–43.

Mortimer, R.K. & Schild, D. (1981) Genetic mapping in *Saccharomyces cerevisiae*. In *The Molecular*

*Biology of the Yeast Saccharomyces, Life Cycle and Inheritance*, (eds J.F. Strathern, E.W. Jones & J.R. Broach), pp. 11–26. Cold Spring Harbor Laboratory, USA.

Mortimer, R.K., Contopoulou, C.R. & King, J.S. (1995) Genetic and physical maps of *Saccharomyces cerevisiae*. In *The Yeasts*, 2nd edition, (eds A.H. Rose, A.E. Wheals and J.S. Harrison), vol. 6, pp. 471–98. Academic Press, London.

Motoyama, Y., Ogata, T. & Sakai, K. (1998) Characterization of *Pectinatus cerevisiiphilus* and *P. frisingensis* by ribotyping. *Journal of the American Society of Brewing Chemists* **56** 19–23.

Mou, D.-G. & Cooney, C.L. (1976) Application of dynamic calorimetry for monitoring fermentation processes. *Biotechnology and Bioengineering*, **18**, 1371–92.

Mozes, N., Marchal, F., Hermesse, M.P. *et al.* (1987) Immobilisation of microorganisms by adhesion: interplay of electrostatic and non-electrostatic interactions. *Biotechnology & Bioengineering*, **30** 439–50.

Muck, E. & Narziss, L. (1988) The fermentation activity apparatus – a means for determining yeast vitality. *Brauwelt International*, **1** 61–2.

Muldbjerg, M., Meldal, M., Breddam, K. & Sigsgaard, P. (1993) Protease activity in beer and correlation to foam. *Proceedings of the 24th Congress of the European Brewery Convention*, Oslo, 357–64.

Muller, R. (1990) The production of low-alcohol and alcohol-free beers by limited fermentation. *Ferment*, **3** 224–30.

Muller, R. (1991) Immuno-assay and its relevance to brewing analysis. *Ferment*, **4**, 224–8.

Muller, R.E., Fels, S. & Gosselin, Y. (1997) Brewery fermentations with dried lager yeast. *Proceedings of the 26th Congress of the European Brewery Convention* Maastricht, 431–8.

Mundy, A.P. (1997) HACCP for business benefit. *Proceedings of the European Brewery Convention Symposium*, Monograph **XXVI**, Stockholm, 141–51.

Mundy, A.P. (1997) Food safety. *The Brewer*, December, 517–24.

Murphy, A. & Kavanagh, K. (1999) Emergence of *Saccharomyces cerevisiae* as a human pathogen: Implications for biotechnology. *Enzyme and Microbial Technology*, **25**, 551–7.

Murphy, C., Large, P.J., Wadforth, C., Dack, S.J. & Boulton, C.A. (1996) Stain-dependent variation in the NADH-dependent diacetyl reductase activities of lager and ale brewing yeasts. *Biotechnology and Applied Biochemistry*, **23**, 19–22.

Nagodawithana, T.W. & Steinkraus, K.H. (1976) Influence of the rate of ethanol production and accumulation on the viability of *Saccharomyces cerevisiae* in rapid fermentation. *Applied and Environmental Microbiology*, **31**, 158–62.

Nakamura, T., Chiba, K., Asahara, Y. & Tada, S. (1997) Prediction of barley which causes premature yeast flocculation. *Proceedings of the 26th Congress of the European Brewery Convention*, Maastricht, 53–60.

Nakatani, K., Fukui, N., Nagami, K. & Nishigaki, M. (1991) Kinetic analysis of ester formation during beer fermentation. *Journal of the American Society of Brewing Chemists*, **49**, 152–7.

Nakatani, K., Takahashi, T., Nagami, K. & Kumada, J. (1984a) Kinetic study of vicinal diketones in brewing (1): formation of total vicinal diketones. *Technical Quarterly of the Master Brewers Association of the Americas*, **21**, 73–8.

Nakatani, K., Takahashi, T., Nagami, K. & Kumada, J. (1984b) Kinetic study of vicinal diketones in brewing (2): theoretical aspect for the formation of total vicinal diketones. *Technical Quarterly of the Master Brewers Association of the Americas*, **21**, 175–83.

Narziss, L. (1984) The German Beer Law. *Journal of the Institute of Brewing*, **90**, 351–8.

Narziss, L. & Hellich, P. (1972) Rapid fermentation and maturing of beer by means of the bioreactor. *Brewers Digest*, September, 106–18.

Naser, S.F. & Fournier, R.L. (1988) A numerical evaluation of a hollow fibre extractive fermenter process for the production of ethanol. *Biotechnology & Bioengineering*, **32**, 628–38.

Nathan, L. (1930a) Improvements on the fermentation and maturation of beers. Part I. *Journal of the Institute of Brewing*, **36**, 538–44.

Nathan, L. (1930b) Improvements in the fermentation and maturation of beers. Part II. *Journal of the Institute of Brewing*, **36**, 544–50.

Naumov, G.I. (1996) Genetic identification of biological species in the *Saccharomyces sensu stricto* complex. *Journal of Industrial Microbiology*, **17**, 295–302.

Naumov, G.I., Naumova, E.S. & Korhola, M.P. (1995) Chromosomal polymorphism of *MEL* genes in some populations of *Saccharomyces cerevisiae*. *FEMS Microbiology Letters*, **127**, 41–5.

Naumov, G.I., Naumova, E.S. & Sniegowski, P.D. (1998) *Saccharomyces paradoxus* and *Saccharomyces cerevisiae* are associated with exudates of North American oaks. *Canadian Journal of Microbiology*, **44**, 1045–50.

Neal, U.K., Hoffmann, H.-P. & Price, C.A. (1971) Sedimentation behaviour and ultrastructure of mitochondria from repressed and derepressed yeast in *Saccharomyces cerevisiae*. *Plant and Cellular Physiology*, **12**, 181–92.

Needleman, R.B., Kaback, D.B., Dubin, R.A. *et al.* (1984) MAL6 of *Saccharomyces*. A complex genetic locus containing three genes required for maltose fermentation. *Proceedings of the National Academy of Sciences of the USA*, **81**, 2811–15.

Neigeborn, L. & Carlson, M. (1994) Genes affecting the regulation of *SUC2* gene expression by glucose repression in *Saccharomyces cerevisiae*. *Genetics*, **108**, 845–58.

Neigeborn, L., Schwartzberg, P., Reid, R. & Carlson, M. (1986) Null mutations in the *SNF3* gene of *Saccharomyces cerevisiae* cause a different phenotype than do previously isolated missense mutations. *Molecular and Cellular Biology*, **6**, 3569–74.

Nelson, N. (1987) The vacuolar proton-ATPase of eukaryotic cells. *BioEssays*, **7**, 251–4.

Nes, W.D., Janssen, G.G., Crumley, F.G., Kalinoswki, M. & Akihisi, T. (1993) The structural requirements of sterols for membrane function in *Saccharomyces cerevisiae*. *Archives for Biochemistry and Biophysics*, **300**, 724–35.

Neves, M.-J. & Francois, J. (1992) On the mechanism by which a heat shock induces trehalose accumulation in *Saccharomyces cerevisiae*. *Journal of Biochemistry*, **288**, 859–64.

Nevoigt, E. & Stahl, U. (1997) Osmoregulation and glycerol metabolism in the yeast *Saccharomyces cerevisiae*. *FEMS Microbiology Reviews*, **21**, 231–41.

Nielsen, H., Hoybe-Hansen, I., Ibaek, D. & Kristensen, B.J. (1986) Introduction to pressure fermentation. *Brygrnesteren*, **2**, 7–17.

Nielsen, H., Hoybe-Hansen, I., Ibaek., D Kristensen, B.J. & Synnesvedt, K (1987) Pressure fermentation and wort carbonation. *Technical Quarterly of the Master Brewers Association of the Americas*, **24**, 90–94.

Nishikawa, N. & Nomura, B. (1974) Numerical evaluation of yeast deterioration caused by repeated fermentation. *Proceedings of the 13th Institute of Brewing Convention (Australia & New Zealand Section)*, Surfers' Paradise, 105–12.

Nobel, de J.G. & Barnett, J.A. (1991) Passage of molecules through yeast cells: a brief essay-review. *Yeast*, **7**, 313–23.

Nordstrom, K. (1962) Formation of ethyl acetate in fermentation with brewer's yeast. II. Kinetics of formation from ethanol and influence of acetaldehyde. *Journal of the Institute of Brewing*, **68**, 188–96.

Nordstrom, K. (1963) Formation of ethyl acetate in fermentation with brewers yeast. IV. Metabolism of acetyl-CoA. *Journal of the Institute of Brewing*, **69**, 142–53.

Nordstrom, K. (1964) Formation of esters from acids by brewer's yeast. IV. Effect of higher fatty acids and toxicity of lower fatty acids. *Journal of the Institute of Brewing*, **70**, 233–42.

Norton, S. & D'Amore, T. (1994) Physiological effects of yeast cell immobilisation. *Enzyme & Microbial Technology*, **16**, 365–75.

Nothaft, A. (1995) The start-up of an immobilised yeast system for secondary fermentation. *Proceedings of the European Brewery Convention*, Monograph **XXIV**, Espoo, 41–9.

Novak, M. Strehaiano, P. & Mareno, M. (1981) Alcoholic fermentation: on the inhibitory effect of ethanol. *Biotechnology and Bioengineering*, **23**, 201–11.

Novellie, L. & de Schaepdrijver, P. (1986) Modern developments in traditional African beers. *Progress in Industrial Microbiology*, **23**, 73–157.

Novotny, C., Flieger, M., Panos, J. & Karst, F. (1992) Effect of 5, 7-unsaturated sterols on ethanol tolerance in *Saccharomyces cerevisiae*. *Biotechnology and Applied Biotechnology*, **15**, 314–20.

Nurse, P.M. (1975) Genetic control of cell size at cell division in yeast. *Nature*, **256**, 447–51.

Oakley-Gutowski, K.M., Hawthorne, D.B. & Kavanagh, T.E. (1992) Application of chromosome fingerprinting to the differentiation of brewing yeasts. *Journal of the American Society of Brewing Chemists*, **50**, 48–52.

O'Connor-Cox, E. (1997) Improving yeast handling in the brewery, part 1: yeast cropping. *Brewers' Guardian*, **December**, 26–32.

O'Connor-Cox, E. (1998a) Improving yeast handling in the brewery, part 2: yeast collection. *Brewers' Guardian*, **February**, 22–34.

O'Connor-Cox, E. (1998b) Improving yeast handling in the brewery, part 3: yeast pitching and measurement of yeast quality. *Brewers' Guardian*, **March**, 20–25.

O'Connor-Cox, E.S.C. & Ingledew, W.M. (1989) Wort nitrogenous sources – their use by brewing yeast: a review. *Journal of the American Society of Brewing Chemists*, **47**, 102–8.

O'Connor-Cox, E. & Ingledew, W.M. (1990a) Commercial active dry yeasts demonstrate wide variability in viability and in two common indicators of yeast physiological quality. *Brewers Digest*, **June**, 34–6.

O'Connor-Cox, E.S.C. & Ingledew, W.M. (1990b) Effect of the timing of oxygenation on very high gravity brewing fermentations. *Journal of the American Society of Brewing Chemists*, **48**, 26–32.

O'Connor-Cox, E.S.C., Lodolo, E.J. & Axcell, B.C. (1993) Role of oxygen in high gravity fermentations in the absence of unsaturated lipid biosynthesis. *Journal of the American Society of Brewing Chemists*, **51**, 97–107.

O'Connor-Cox, E.S.C., Lodolo, E.J. & Axcell, B.C. (1996) Mitochondrial relevance to yeast fermentative performance: a review. *Journal of the Institute of Brewing*, **102**, 19–25.

O'Connor-Cox, E.S.C., Yui, P.M. & Ingledew, W.M. (1991a) Pasteurisation: thermal death of microbes in brewing. *Technical Quarterly of the Master Brewers Association of the Americas*, **28**, 67–77.

O'Connor-Cox, E.S.C., Yui, P.M. & Ingledew, W.M. (1991b) Pasteurisation: industrial practice and evaluation. *Technical Quarterly of the Master Brewers Association of the Americas*, **28**, 99–107.

van Oevelen, D., Spaepen, M., Timmermans, P. & Verachtert, H. (1977) Microbiological aspects of spontaneous wort fermentation in the production of lambic and gueuze. *Journal of the Institute of Brewing*, **83**, 356–60.

Ogane, O., Yokoyama, F., Katsumata, T., Tateisi, Y. & Nakajima, I. (1999) Measurement of foam height in cylindroconical fermenters using a laser-based distance sensor. *Technical Quarterly of the Master Brewers Association of the Americas*, **36**, 155–7.

Ogden, K. (1987) Cleansing contaminated pitching yeast with nisin. *Journal of the Institute of Brewing*, **93**, 302–7.

Ogden, K. (1993) Practical experiences of hygiene control using ATP-bioluminescence. *Journal of the Institute of Brewing*, **99**, 389–93.

Ogur, M., Joh, R. St & Nagai, S. (1957) Tetrazolium overlay technique for population studies of respiratory deficiency in yeast. *Science*, **125**, 928–9.

Ogur, M., Minckler, S., Lindegren, G. & Lindegren, C.C. (1952) The nucleic acids in a polyploid series of *Saccharomyces*. *Archives for Biochemistry and Biophysics*, **40**, 175–84.

Okolo, B., Johnston, J.R. & Berry, D.R. (1987) Toxicity of ethanol, n-butanol and iso-amyl alcohol in *Saccharomyces cerevisiae* when supplied separately and in mixtures. *Biotechnology Letters*, **9**, 431–4.

Okorokov, L.A. & Lichko, L.P. (1983) The identification of a proton pump on vacuoles of the yeast *Saccharomyces carlsbergensis*. *FEBS Letters*, **155**, 102–6.

Oliver, S.G. (1996) From DNA sequence to biological function. *Nature*, **379**, 597–600.

Oliver, S.G. (1997) Yeast as navigational aid in genome analysis. *Microbiology*, **143**, 1483–7.

Oliver, S.G., Winson, M.K., Kell, D.B. & Baganz, F. (1998) Systematic functional analysis of the yeast genome. *Trends in Biotechnology*, **16**, 373–8.

Olivera, H., Gonzalez, A. & Pena, A. (1993) Regulation of the amino acid permeases in nitrogen limited continuous cultures of the yeast, *Saccharomyces cerevisiae*. *Yeast*, **9**, 1065–78.

Olsen, A. (1981) The role of wort turbidity in flavour and flavour stability. *Proceedings of the European Brewery Convention*, Monograph VII, Copenhagen, 223–36.

O'Malley, W.P. (1961) Continuous processing of beer. *Technical Quarterly of the Master Brewers Association of the Americas*, 15–26.

Omori, T., Ogawa, K., Umemoto, Y., Yuki, K., Kajihara, Y., Shimoda, M. & Wada, H. (1996) Enhancement of glycerol production by brewing yeast (*Saccharomyces cerevisiae*) with heat shock treatment. *Journal of Fermentation and Bioengineering*, **82**, 187–90.

Omori, T., Umemoto, Y., Ogawa, K., Kajiwara, Y., Shimoda, M. & Wada, H. (1997) A novel method for screening high glycerol and ester producing brewing yeasts (*Saccharomyces cerevisiae*) by heat shock treatment. *Journal of Fermentation and Bioengineering*, **83**, 64–9.

Opekarova, M. & Sigler, K. (1982) Acidification power: indicator of metabolic activity and autolytic changes in *Saccharomyces cerevisiae*. *Folia Microbiologica*, **27**, 395–403.

Orive, I. & Camprubi, M. (1996) Hygiene monitoring in draught beer dispense. *Proceedings of the European Brewery Convention Symposium*, Monograph XXV, Edinburgh, 173–85.

O'Sullivan, T.F., Walsh, Y., O'Mahony, A., Fitzgerald, G.F. & van Sinderen, D. (1999) A comparative study of malthouse and brewhouse microflora. *Journal of the Institute of Brewing*, **105**, 55–61.

Otter, G.E. & Taylor, L. (1976) The determination of amino acids in wort, beer and brewing materials using gas chromatography. *Journal of the Institute of Brewing*, **82**, 264–9.

Oura, E. (1977) Reaction products of yeast fermentations. *Process Biochemistry*, **12**, 19–21.

Oura, E., Haarasilta, S. & Londesborough, J. (1980) Carbon dioxide fixation by baker's yeast under a variety of growth conditions. *Journal of General Microbiology*, **118**, 51–8.

Owades, J.L. (1981) The role of osmotic pressure in high and low gravity fermentations. *Technical Quarterly of the Master Brewers Association of the Americas*, **18**, 163–5.

Owens, C.C. (1987) The history of brewing in Burton on Trent. *Journal of the Institute of Brewing*, **93**, 37–41.

Ozcan, S. & Johnston, M. (1999) Function and regulation of yeast hexose transporters. *Microbiology and Molecular Biology Reviews*, **63**, 554–69.

Ozcan, S., Vallier, L.G., Flick, J.S., Carlson, M. & Johnston, M. (1997) Expression of the *SUC2* gene of *Saccharomyces cerevisiae* is induced by low levels of glucose. *Yeast*, **13**, 127–37.

Padmanaban, G., Venkateswar, V. & Rangarajan, P.N. (1989) Haem as a multi-functional regulator. *Trends in Biochemical Sciences*, **14**, 492–6.

Pajunen, E. (1995) Immobilised yeast lager beer maturation: DEAE-cellulose at Sinebrychoff. *Proceedings of the European Brewery Convention, Monograph XXIV*, Espoo, 24–34.

Pajunen, E. & Gronqvist, A. (1994) Immobilised yeast fermenters for continuous lager beer fermentation. *Proceedings of the 23rd Institute of Brewing Convention (Australia & New Zealand Section)*, Sydney, 101–13.

Pajunen, E. & Jaaskelainen, K. (1993) Sinebrychoff Kerava – a brewery for the 90's. *Proceedings of the 24th Congress of the European Brewery Convention*, Oslo, 559–67.

Palkova, Z., Janderova, B., Gabriel, J., Zikanova, B., Pspisek, M. & Forstova, J. (1997) Ammonia mediates communications between yeast colonies. *Nature*, **390**, 532–6.

Palmieri, F., Bisaccia, F., Capobianco, L., Iacobazzi, V., Indiveri, C. & Zara, V. (1990) Structural and functional properties of mitochondrial anion carriers. *Biochimica et Biophysica Acta*, **1018**, 147–50.

Pampulha, M.E. & Loureiro-Dias, M.C. (1990) Activity of glycolytic enzymes of *Saccharomyces cerevisiae* in the presence of acetic acid. *Applied Microbiology and Biotechnology*, **34**, 611–14.

Pampulha, M.E. & Loureiro, V. (1989) Interaction of the effects of acetic acid and ethanol inhibition of fermentation in *Saccharomyces cerevisiae*. *Biotechnology Letters*, **11**, 269–74.

Pande, S.V. & Mead, J.F. (1968) Inhibition of enzyme activities by free fatty acids *Journal of Biological Chemistry*, **243**, 6180–85.

Panek, A.D. & Panek, A.C. (1990) Metabolism and thermotolerance function of trehalose in *Saccharomyces cerevisiae*: a current perspective. *Journal of Biotechnology*, **14**, 229–38.

Panek, A.D. (1991) Storage carbohydrates. In *The Yeasts*, 2nd edition, (eds. A.H. Rose and J.S. Harrison), vol. 4, Academic Press, London. 655–74.

Paquin, C. & Adams, J. (1983) Frequency of fixation of adaptive mutations is higher in evolving diploid than haploid yeast populations. *Nature*, **302**, 495–500.

Park, T.H. & Kim, I.H. (1985) Hollow fibre fermenter using ultrafiltration. *Applied Microbiology & Biotechnology*, **22**, 190–94.

Parker, M.J. (1989) The application of automated detection and enumeration of microcolonies using optical brighteners and image analysis to brewery microbiological control. *Proceedings of the 22nd Congress of the European Brewery Convention*, Zurich, 545–52.

Parkkinen, E., Oura, E. & Soumalainen, H. (1976) Comparison of the methods for the determination of viability in stored baker's yeast. *Journal of the Institute of Brewing*, **82**, 283–5.

Parks, L.W. (1978) Metabolism of sterols in yeast. *CRC Critical Reviews in Microbiology*, **6**, 301–41.

Parks, L.W., Smith, S.J. & Crowley, J.H. (1995) Biochemical and physiological effects of sterol alterations in yeast – a review. *Lipids*, **30**, 227–30.

Parsons, R.P., Wainwright, T. & White, F.H. (1977) Malt kilning conditions and the control of beer dimethyl sulphide levels. *Proceedings of the 16th Congress of the European Brewery Convention*, Amsterdam, 115–28.

Pascal, F., Dagot, C., Pingaud, H., Corriou, J.P., Pons, M.N. & Engasser, J.M. (1995) Modelling of an industrial alcohol fermentation simulation of the plant by a process simulator. *Biotechnology and Bioengineering*, **46**, 202–17.

Pasteur, L. (1876) *Etudes sur la bière*. Guathier-Villars, Paris.

Patino, H., Edelen, C. & Miller, J. (1993) Alternative measures of yeast vitality: use of cumulative acidification power and conductance. *Journal of the American Society of Brewing Chemists*, **51**, 128–32.

Paton, A.M. & Jones, S.M. (1975) Yeast cell death in continuous culture. *Journal of Applied Bacteriology*, **38**, 199–204.

Patterson, C.A. & Ingledew, W.M. (1999) Utilisation of peptides by a lager brewing yeast. *Journal of the American Society of Brewing Chemists*, **57**, 1–8.

Payne, R.W., Kurtzman, C.P. & Fell, J.W. (1998) Key to species. In *The Yeasts, A Taxonomic Study*, 4th edition, (eds C.P. Kurtzman & J.W. Fell), pp. 891–913. Elsevier, Amsterdam.

Pearce, D.A. & Sherman, F. (1998) A yeast model for the study of Batten's disease. *Proceedings of the National Academy of Sciences, USA*, **95**, 6915–18.

Pearce, D.A., Ferea, T., Nosel, S.A., Das, B. & Sherman, F. (1999) Action of *BTN1*, the yeast orthologue of the gene mutated in Batten's disease. *Nature Genetics*, **22**, 55–8.

Peddie, H.A.B. (1990) Ester formation in brewery fermentations. *Journal of the Institute of Brewing*, **96**, 327–31.

Pedersen, M.B. (1993) Instability of the brewer's yeast genome. *Proceedings of the 24th Congress of the European Brewery Convention*, Oslo, 291–8.

Pedersen, M.B. (1994) Molecular analysis of yeast DNA – tools for pure yeast maintenance in the brewery. *Journal of the American Society of Brewing Chemists*, **52**, 23–7.

Pedersen, M.B. (1995) Recent views and methods for the classification of yeasts. *Cerevisia – Belgian Journal of Brewing and Biotechnology*, **20**, 28–33.

Perry, C. & Meaden, P. (1988) Properties of a genetically engineered dextrin-fermenting strain of brewer's yeast. *Journal of the Institute of Brewing*, **94**, 64–7.

Perry, R.H., Green, D., & Maloney, J.D. (eds.) (1963) In *Perry's Chemical Engineers' Handbook*, Ch. 10, pp. 1–66. McGraw-Hill, New York.

Perryman, M. (1991) Practical brewery sampling. *Ferment*, **4**, 366 8.

Pessa, E. (1971) Variations in the acetaldehyde content of beer. *Proceedings of the 13th Congress of the European Brewery Convention*, Estoril, 333–42.

Petersen, R.F., Nilsson-Tillgren, T. & Piskur, J. (1999) Karyotypes of *Saccharomyces sensu lato* species. *International Journal of Systematic Bacteriology*, **49**, 1925–31.

Petit, J.-M., Denis-Gay, M. & Ratinaud, M.-H. (1993) Assessment of fluorochromes for cellular structure and function studies by flow cytometry. *Biology of the Cell*, **78**, 1–13.

Pfisterer, E. & Stewart, G.G. (1976) High gravity brewing. *Brewers Digest*, June, 34–42.

Pfisterer, E.A., Krynicki, C.L., Steer, J.T. & Hogg, W.T. (1988) On-line control of ethanol and carbon dioxide in high gravity brewing. *Technical Quarterly of the Master Brewers Association of the Americas*, **25**, 1–5.

Phaff, H.J. & Starmer, W.T. (1987) Yeasts associated with plants, insects and soil. In *The Yeasts*, 2nd edition, (eds A.H. Rose and J.S. Harrison), vol. 1, pp. 123–80. Academic Press, London.

Phaweni, M., O'Connor-Cox, E.S.C., Pickerell, A.T.W. & Axcell, B.C. (1992) The impact of yeast physiological status on carbohydrate utilisation patterns in high gravity wort fermentations. *Proceedings of the 22nd Institute of Brewing Convention (Australia & New Zealand Section)*, Melbourne, 69–75.

Phaweni, M., O'Connor-Cox, E.S.C., Pickerell, A.T.W. & Axcell, B.C. (1993) Influence of adjunct carbohydrate spectrum on the fermentative activity of a brewing strain of *Saccharomyces cerevisiae*. *Journal of the American Society of Brewing Chemists*, **51**, 10–15.

Philliskirk, G. & Young, T.W. (1975) The occurrence of killer character in yeasts of various genera. *Antonie Van Leeuwenhoek*, **41**, 147–51.

Phipps, M.E. (1990) Brewery mould control. *Brewers Digest*, July, 28–9.

Phowchinda, O., Delia-Dupuy, M.L. & Strehaiano, P. (1995) Effects of acetic acid on growth and fermentative activity of *Saccharomyces cerevisiae*. *Biotechnology Letters*, **17**, 237–42.

Pickard, J. (1999) Cleaning in place system design: The holistic approach. *The Brewer*, March, 126–8.

Pickerell, A.T.W., Hwang, A. & Axcell, B.C. (1991) Impact of yeast handling procedures on beer flavour development during fermentation. *Journal of the American Society of Brewing Chemists*, **49**, 87–92.

Pierce, J.S. (1970) IOB Analysis Committee: Measurement of yeast viability. *Journal of the Institute of Brewing*, **76**, 442–3.

Pierce, J.S. (1987) The role of nitrogen in brewing. *Journal of the Institute of Brewing*, **93**, 378–81.

Pilkington, H., Margaritis, A. Mensour, N., Sobczak, J., Hancock, I. & Russell, I. (1999) Kappa-carrageenan gel immobilisation of lager brewing yeast. *Journal of the American Society of Brewing Chemists*, **105**, 398–404.

Piper, P.W. (1995) The heat shock and ethanol stress responses of yeast exhibit extensive similarity and functional overlap. *FEMS Microbiology Letters*, **134**, 121–7.

Piper, P.W., Talreja, K., Panaretou, B. *et al.* (1994) Induction of major heat shock proteins of *Saccharomyces cerevisiae* including plasma membrane Hsp by ethanol levels above a critical threshold. *Microbiology*, **140**, 3031–8.

Pirt, S.J. (1975) *Principles of Microbe and Cell Cultivation*. Blackwell Scientific Publications, Oxford, London and Edinburgh.

Piskur, J., Mozina, S.S., Stenderup, J & Pedersen, M.B. (1995) A mitochondrial molecular marker, *ori-rep-tra*, for the differentiation of yeast species. *Applied and Environmental Microbiology*, **61**, 2780–82.

Piskur, J., Smole, S., Groth, C., Petersen, R.F. & Pedersen, M.B. (1998) Structure and genetic stability of mitochondrial genomes vary among yeasts of the genus *Saccharomyces*. *International Journal of Systematic Bacteriology*, **48**, 1015–24.

Pittner, H., Back, W., Swinkels, W., Meersman, E., van Dieren, B. & Lommi, H. (1993) Continuous production of acidified wort for alcohol-free beer using immobilised lactic acid bacteria. *Proceedings of the 24th Congress of the European Brewery Convention*, Oslo, 323–9.

van der Plaat, J.B. & van Solingen, P. (1974) cAMP stimulates trehalose degradation in bakers' yeast. *Biochimica et Biophysica Acta*, **56**, 580–86.

Platt, D. (1986) Keeping dirt out and the bugs at bay. *Brewing and Distilling International*, **16**, 20–21.

Plattner, H., Salpeter, M., Saltzgaber, J., Rouslin, W. & Schatz, G. (1971) Promitochondria of anaerobically-grown yeast: evidence for their conversion into functional mitochondria during respiratory adaption. In *Autonomy and Biogenesis of Mitochondria and Chloroplasts*, (eds N.K. Boardman, A.W. Linnane and R.M. Smillie), pp. 175–84. North Holland Publishing Company, Amsterdam.

Polakis, E.S. & Bartley, W. (1965) Changes in the enzyme activities of *Saccharomyces cerevisiae* during aerobic growth on different carbon sources. *Biochemical Journal*, **97**, 284–95.

Portno, A.D. (1967) New systems of continuous fermentation by yeast. *Journal of the Institute of Brewing*, **73**, 43–50.

Portno, A.D. (1968a) Continuous fementation in relation to yeast fermentation. *Journal of the Institute of Brewing*, **74**, 448–56.

Portno, A.D. (1968b) Fermentation by a progressive continuous system. *Journal of the Institute of Brewing*, **75**, 468–71.

Portno, A.D. (1968c) Continuous fermentation of brewer's wort. *Journal of the Institute of Brewing*, **74**, 55–63.

Portno, A.D. (1969) Fermentation in a continuous progressive system. *Journal of the Institute of Brewing*, **75**, 468–71.

Portno, A.D. (1970) Theoretical and practical aspects of continuous fermentation. *Wallerstein Laboratory Communications*, **33**, 149–61.

Portno, A.D. (1978) Continuous fementation in the brewing industry – the future outlook. *Proceedings of the European Brewery Convention Symposium*, Monograph V Zouterwoude, 145–54.

Posada, J. (1978) Reasons to use spheroconical fermenters. *Proceedings of the European Brewery Convention Symposium*, Monograph V, Zouterwoude, 207–18.

Posada, J., Candela, J., Calero, G., Almenar, J. & Martin, S. (1977) Fermentation conditions and yeast performance. *Proceedings of the 16th Congress of the European Brewery Convention*, Amsterdam, 533–44.

Posteraro, B., Sanguinetti, M., D'Amore, G., Masucci, L., Morace, G. & Fadda, G. (1999) Molecular and epidemiological characterisation of vaginal *Saccharomyces cerevisiae* isolates. *Journal of Clinical Microbiology*, **37**, 2230–5.

Postgate, J.R. (1967) Viability measurements and the survival of microbes under minimum stress. *Advances in Microbial Physiology*, **1**, 1–23.

Postma, E., Verduyn, C., Kuiper, A., Scheffers, W.A. & van Dijken, J.P. (1990) Substrate accelerated death of *Saccharomyces cerevisiae* CBS 8066 under maltose stress. *Yeast*, **6**, 149–58.

Postma, E., Verduyn, C., Scheffers, W.A. & van Dijken, J.P. (1989) Enzymatic analysis of the Crabtree effect in glucose-limited chemostat cultures of *Saccharomyces cerevisiae*. *Applied and Environmental Microbiology*, **55**, 468–77.

Powell, C.A., van Zandycvke, S.M., Quain, D.E. & Smart, K.A. (2000) Replicative ageing and senescence in *Saccharomyces cerevisiae* and the impact on brewing fermentations. *Microbiology*, **146**, 1023–34.

Pratt, L.A. & Kolter, R. (1999) Genetic analysis of bacterial biofilm formation. *Current Opinion in Microbiology*, **2**, 598–603.

Priest, F.G. (1981) The classification and nomenclature of brewing bacteria: a review. *Journal of the Institute of Brewing*, **87**, 279–81.

Priest, F.G. (1996) Gram-positive brewery bacteria. In *Brewing Microbiology*, 2nd edition, (eds F.G. Priest and I. Campbell), pp. 127–61. Chapman and Hall, London.

Priest, F.G. & Campbell, I. (eds) (1996) *Brewing Microbiology*, 2nd edition, Chapman and Hall, London.

Priest, F.G., Cowbourne, M.A. & Hough, J.S. (1974) Wort enterobacteria – a review. *Journal of the Institute of Brewing*, **80**, 342–56.

Pringle, J.R. & Hartwell, L.H. (1981) The *Saccharomyces cerevisiae* cell cycle. In *The Molecular Biology of the Yeast Saccharomyces, Life Cycle and Inheritance*, (eds J.F. Strathern, E.W. Jones and J.R. Broach), pp. 97–42. Cold Spring Harbor Laboratory, USA.

Pringle, J.R., Preston, R.A., Adams, A.E.M. *et al.* (1989) Fluorescence microscopy methods for yeast. *Methods in Cell Biology*, **31**, 358–437.

Pronk, J.T., Wenzel, T.J., Luttik, M.A.H. *et al.* (1994) Energetic aspects of glucose metabolism in a pyruvate dehydrogenase negative mutant of *Saccharomyces cerevisiae*. *Microbiology*, **140**, 601–10.

Pruisner, S.B. (1995) The prion diseases. *Scientific American*, January, 48–57.

Quain, D.E. (1981) The determination of glycogen in yeasts. *Journal of the Institute of Brewing*, **87**, 289–91.

Quain, D.E. (1986) Differentiation of brewing yeast. *Journal of the Institute of Brewing*, **92**, 435–8.

Quain, D.E. (1988) Studies on yeast physiology – impact on fermentation performance and product quality. *Journal of the Institute of Brewing*, **95**, 315–23.

Quain, D.E. (1990) Yeast handling and fermentation management in brewing. *Proceedings of the 3rd Aviemore Conference on Malting, Brewing and Distilling*, (ed. I. Campbell), pp. 74–83. Institute of Brewing, London.

Quain, D.E. (1995) Yeast supply – the challenge of zero defects. *Proceedings of the 25th Congress of the European Brewery Convention*, Brussels, 309–18.

Quain, D.E. (1999) The 'new' microbiology. *Proceedings of the 27th Congress of the European Brewery Convention*, Cannes, 239–48.

Quain, D.E. & Boulton, C.A. (1987a) Growth and metabolism of mannitol by strains of *Saccharomyces cerevisiae*. *Journal of General Microbiology*, **133**, 1675–84.

Quain, D.E. & Boulton, C.A. (1987b) The propagation of yeast on oxidative carbon sources – advantages and implications. *Proceedings of the 21st Congress of the European Brewery Convention*, Madrid, 417–24.

Quain, D.E. & Boulton, C.A. (1987c) Oxygenated yeast. UK patent 2197341B.

Quain, D.E. & Duffield, M.L. (1985) A metabolic function for higher alcohol production by yeast. *Proceedings of the 20th Congress of the European Brewery Convention*, Helsinki, 307–14.

Quain, D.E. & Haslam, J.M. (1979) The effects of catabolite derepression on the accumulation of steryl esters and the activity of 3-hydroxymethylglutaryl-CoA reductase in *Saccharomyces cerevisiae*. *Journal of General Microbiology*, **111**, 343–51.

Quain, D.E. & Tubb, R.S. (1982) The importance of glycogen in brewing yeasts. *Technical Quarterly of the Master Brewers Association of the Americas*, **19**, 29–33.

Quain, D.E. & Tubb, R.S. (1983) A rapid and simple method for the determination of glycogen in yeast. *Journal of the Institute of Brewing*, **89**, 38–40.

Quain, D.E., Box, W.G. & Walton, E.F. (1985) An inexpensive and simple small-scale laboratory fermenter. *Laboratory Practice*, **34**, 84.

Quain, D.E., Thurston, P.A. & Tubb, R.S. (1981) The structural and storage carbohydrates of *Saccharomyces cerevisiae*: changes during fermentation of wort and a role for glycogen catabolism in lipid biosynthesis. *Journal of the Institute of Brewing*, **87**, 108–11.

Rachidi, N., Barre, P. & Blondin, B. (1999) Multiple Ty-mediated chromosomal translocations lead to karyotype changes in a wine strain of *Saccharomyces cerevisiae*. *Molecular and General Genetics*, **261**, 841–50.

Rader, S. (1979) Chemical pasteurization test. *Brewers Digest*, November, 45–6.

Radovich, J.M. (1985) Mass transfer effects in fermentations using immobilised whole cells. *Enzyme & Microbial Technology*, **7**, 2–10.

Rainbow, C. (1981) Beer spoilage organisms. In *Brewing Science* (ed. J.R. Pollock), vol. 2, pp. 491–550. Academic Press, London.

Rank, G.H. & Robertson, A.J. (1983) Protein and lipid composition of the yeast plasma membrane. In *Yeast Genetics: Fundamental and Applied Aspects* (eds J.F.T. Spencer, D.M. Spencer and A.R.W. Smith), pp. 225–41. Springer-Verlag, Berlin.

Ratledge, C.R. & Evans, C.T. (1989) Lipids and their metabolism. In *The Yeasts*, 2nd edition (eds A.H. Rose and J.S. Harrison), vol. 3, pp. 367–56. Academic Press, London.

Rattray, J.B.M. (1989) Yeasts. In *Microbial Lipids*, (eds C. Ratledge & S.G. Wilkinson), vol. 1, pp. 555–697. Academic Press, London.

Raynal, L., Barnwell, P. & Gervais, P. (1994) The use of epi-fluorescence to determine the viability of *Saccharomyces cerevisiae* subjected to osmotic shifts. *Journal of Biotechnology*, **36**, 121–7.

Reed, G. & Nagodarithana, T.W. (1991) *Yeast Technology*, Van Nostrand, USA.

Rees, E.M.R. & Stewart, G.G. (1997) The effects of increased magnesium and calcium concentrations on yeast fermentation in high gravity worts *Journal of the Institute of Brewing*, **103**, 287–91.

Reeves, M.J., Crofskey, G.D. & Dunbar, J. (1991) Applicability of the measurement of residual melibiase activity in packaged beer to quantify pasteurisation. *Proceedings of the 21st Congress of the Institute of Brewing Convention (Australia & New Zealand Section)*, Auckland, 201–5.

Regan, J. (1990) Production of alcohol-free and low-alcohol beers by vacuum distillation. *Ferment*, **3**, 235–7.

Reid, G.C., Hwang, A., Meisel, R.H. & Allcock, E.R. (1990) The sterile filtration and packaging of beer into polyethylene terephthalate containers. *Journal of the American Society of Brewing Chemists*, **48**, 85–91.

Reilly, C. (1972) Zinc, iron and copper contamination in home-produced alcoholic drinks. *Journal of the Science of Food and Agriculture*, **23**, 1143–4.

Reinhart, M.P. (1990) Intracellular sterol trafficking. *Experentia*, **46**, 599–611.

Remize, F., Roustan, J.L., Sablayrolles, J.M., Barre, P. & Dequin, S. (1999) Glycerol overproduction by engineered *Saccharomyces cerevisiae* wine yeast strains leads to substantial changes in by-product formation and to a stimulation of fermentation rate in stationary phase. *Applied and Environmental Microbiology*, **65**, 143–9.

Rendueles, P.S. & Wolf, D.H. (1988) Proteinase function in yeast: biochemical and genetic approaches to a central mechanism of post-translational control in the eukaryote cell. *FEMS Microbiology Reviews*, **54**, 17–46.

Rennie, H. & Wilson, R.J.H. (1975) A method for controlled oxygenation of worts. *Journal of the Institute of Brewing*, **81**, 105–6.

van der Rest, M., Kamminga, A.H., Nakano, A., Anraku, Y., Poolman, W.N. (1995) The plasma membrane of *Saccharomyces cerevisiae*: structure, function and biogenesis. *Microbiological Reviews*, **59**, 304–22.

Reuther, H., Brandon, H., Raasch, J. & Raabe, D. (1995) Simulation of fermenter cooling performance. *Monatsschrift für Brauwissenschaft*, **48**, 310–17.

Rhymes, M.R. & Smart, K.A. (1996) Effects of starvation on the flocculation of ale and lager brewing yeast. *Journal of the American Society of Brewing Chemists*, **54**, 50–56.

Rice, J.F., Chicoye, E., Helbert, J.R. & Garver, J. (1976) Inhibition of beer volatiles formation by carbon dioxide pressure. *Journal of the American Society of Brewing Chemists*, **35**, 35–40.

Richards, M. (1967) The use of giant-colony morphology for the differentiation of brewing yeasts. *Journal of the Institute of Brewing*, **73**, 162–6.

Ricketts, R.W. (1971) Fermentation systems. In *Modern Brewing Technology* (ed. W.P.K. Findlay), pp. 83–108. Macmillan Press, London.

Rieger, M., Kappeli, O. & Fiechter, A. (1983) The role of a limited respiration in the complete oxidation of glucose by *Saccharomyces cerevisiae*. *Journal of General Microbiology*, **129**, 653–61.

Riess, S. (1986) Automatic control of the addition of pitching yeast. *Technical Quarterly of the Master Brewers Association of the Americas*, **23**, 32–5.

van Rijn, L.R. (1906) Fermenting liquids. *British patent 18045*.

Ringholt, M. (1997) Automated wort aeration. *Brauwelt International*, **15**, 50–51.

Robinow, C.F. & Johnson, B.F. (1991) Yeast cytology: an overview. In *The Yeasts*, 2nd edition, (eds A.H. Rose and J.S. Harrison), vol. 4, pp. 8–120. Academic Press, London.

Robinow, C.F. (1975) Preparation of yeasts for light microscopy. *Methods in Cell Biology*, **11**, 2–21.

Robinson, J.B. & Srere, P.A. (1985) Organisation of Krebs tricarboxcylic acid cycle enzymes in mitochondria. *Journal of Biological Chemistry*, **260**, 10800–805.

Rodrigues, J.A., Barros, A.A., Machado Cruz, J.M. & Ferreira, A.A. (1997) Determination of diacetyl in beer using differential-pulse polarography. *Journal of the Institute of Brewing*, **103**, 311–14.

Rodrigues de Sousa, H., Madeira-Lopes, A. & Spencer-Martins, I. (1995) The significance of active fructose transport and maximum temperature for growth in the taxonomy of *Saccharomyces sensu stricto*. *Systematic and Applied Microbiology*, **18**, 44–51.

Rodriguez, R.J. & Parks, L.W. (1983) Structural and physical features of sterols necessary to satisfy bulk membrane and sparking requirements in yeast sterol auxotrophs. *Archives of Biochemistry and Biophysics*, **225**, 861–71.

Rodriguez, R.J., Low, C., Bottema, C.D. & Parks, L.W. (1985) Sterol structure and membrane function in yeast. *Biochimica et Biophysica Acta*, **837**, 336–43.

Rogers, P.J. & Stewart, P.R. (1973) Respiratory development in *Saccharomyces cerevisiae* grown at controlled oxygen tension. *Journal of Bacteriology*, **115**, 88–93.

Romano, P. & Suzzi, G. (1992) Production of hydrogen sulphide by different yeast strains during fermentation. *Proceedings of the 22nd Institute of Brewing Convention (Australia & New Zealand Section)*, Melbourne, 96–8.

Romano, P., Suzzi, G., Comi, G. & Zironi, R. (1992) Higher alcohol and acetic acid production by apiculate wine yeasts. *Journal of Applied Bacteriology*, **73**, 126–30.

Rose, A.H., Wheals, A.E. & Harrison, J.S. (eds) (1995) *The Yeasts*, 2nd edition, vol. 6, Academic Press, London.

Rosen, K. (1989) Preparation of yeast for industrial use in the production of beverages. In *Biotechnological Applications in Beverage Production* (eds C. Cantarelli and G. Lanzarini), pp. 164–223. Elsevier Science Publishers, New York.

Rous, C.V. & Snow, R. (1983) Reduction of higher alcohols by fermentation with a leucine-auxotrophic mutant of wine yeast. *Journal of the Institute of Brewing*, **89**, 274–8.

Rowe, S.M., Simpson, W.J. & Hammond, J.R.M. (1991) Spectrophotometric assay of yeast sterols using a polyene antibiotic. *Letters in Applied Microbiology*, **13**, 182–5.

Roy, D.J. & Dawes, I.W. (1987) Cloning and characterisation of the gene encoding lipoamide dehydrogenase in *Saccharomyces cerevisiae*. *Journal of General Microbiology*, **133**, 925–33.

Royan, S. & Subramaniam, M.K. (1956) The nucleus in living yeast. *Proceedings of the Indian Academy of Sciences*, **43**, 228–32.

Rubio, R.S., Lecuona, J.M., Ruano de la Haza, G. & Alcacer, L. (1987) Automatic control of high gravity beer dilution, *Brauwelt International*, **11**, 144–5.

Rudin, A.D. & Hough, J.S. (1959) Influence of yeast strain on production of beer by continuous fermentation. *Journal of the Institute of Brewing*, **65**, 410–14.

Ruocco, J.L., Coe, R.W. & Hahn, C.W. (1980) Computer assisted exotherm measurement in full-scale brewery fermentations. *Master Brewers Association of the Americas Technical Quarterly*, **17**, 69–76.

Russell, I. & Dowhanick, T.M. (1996) Rapid detection of microbial spoilage. In *Brewing Microbiology*, 2nd edition, (eds F.G. Priest and I. Campbell), pp. 209–35. Chapman and Hall, London.

Russell, I. & Stewart, G.G. (1981) Liquid nitrogen storage of yeast cultures compared to more traditional storage methods. *Journal of the American Society of Brewing Chemists*, **39**, 19–24.

Ryder, D.S., Barney, M.C., Daniels, D.H., Borsh, S.A. & Christiansen, K.J. (1994) The philosophy and practice of process hygiene in sterile filtration and aseptic filling in modern brewery operations. *Proceedings of the European Brewery Convention Symposium*, Monograph **XXI**, Nutfield, 64–80.

Ryder, D.S., Bower, P.A., Bromberg, S.K. *et al.* (1995) Challenges for yeast physiology in immobilised systems. *Proceedings of the European Brewery Convention Symposium*, Monograph **XXIV**, Espoo, 175–92.

Ryder, D.S., Murray, J.P. & Stewart, M. (1978) Phenolic off-flavour problem caused by *Saccharomyces* wild yeast. *Technical Quarterly of the Master Brewers Association of the Americas*, 15, 79–86.

Saha, R.B., Sondag, R.J. & Middlekauf, J.E. (1974) An improved medium for the selective culturing of lactic acid bacteria. *Journal of the American Society of Brewing Chemists*, 5, 9–10.

Sakamoto, K., Funahashi, W., Yamashita, H. Eto, M. (1997) A reliable method for detection and identification of beer-spoilage bacteria with internal positive control PCR (IPC-PCR). *Proceedings of the 26th Congress of the European Brewery Convention*, Maastricht, 631–8.

Salerno, L.F. & Parks, L.W. (1983) Sterol uptake in the yeast *Saccharomyces cerevisiae*. *Biochimica et Biophysica Acta*, 752, 240–43.

Salgueiro, S.P., Sa-Correia, I. & Novais, J.M. (1988) Ethanol-induced leakage in *Saccharomyces cerevisiae*: kinetics and relationship to yeast ethanol tolerance and alcohol fermentation productivity. *Applied and Environmental Microbiology*, 54, 903–9.

Sall, C.J., Seipp, J.F. & Pringle, A.T. (1988) Changes in brewer's yeast during storage and the effect of these changes on subsequent fermentation performance. *Journal of the American Society of Brewing Chemists*, 46, 23–5.

Sambrook, P. (1996) *Country House Brewing in England 1500–1900*. Hambledon Press, London.

Sami, M., Ikeda, M. & Yabuchi, S. (1994) Evaluation of the alkaline methylene blue staining method for yeast activity determination. *Journal of Fermentation and Bioengineering*, 78, 212–16.

Sami, M., Yamashita, H., Hirono, T. *et al.* (1997a) Hop-resistant *Lactobacillus brevis* contains a novel plasmid harboring a multidrug resistance-like gene. *Journal of Fermentation and Bioengineering*, 84, 1–6.

Sami, M., Yamashita, H., Kadokura, H., Kitamoto, K., Yoda, K. & Yamasaki, M. (1997b) A new and rapid method for determination of beer-spoilage ability of lactobacilli. *Journal of the American Society of Brewing Chemists*, 55, 137–40.

Samuel, D. (1996) Archaeology of ancient Egyptian beer. *Journal of the American Society of Brewing Chemists*, 54, 3–12.

Samuel, D. (1997) Fermentation technology 3000 years ago – the archaeology of ancient Egyptian beer. *Society for General Microbiology Quarterly*, 24, 3–5.

Samuel, D. & Bolt, D. (1995) Rediscovering ancient Egyptian beer. *Brewers' Guardian*, December, 27–31.

Sato, M., Watari, J., Sahara, H. & Koshino, S. (1994) Instability in electrophoretic karyoytpe of brewing yeasts. *Journal of the American Society of Brewing Chemists*, 52, 148–51.

Satokari, R., Juvonen, R., Mallison, K., von Wright, A. & Haikara, A. (1998) Detection of beer spoilage bacteria *Megasphaera* and *Pectinatus* by polymerase chain reaction and colorimetric microplate hybridization. *International Journal of Food Microbiology*, 45, 119–27.

Satyanarayana, T., Chervenka, C.H. & Klein, H.P. (1980) Subunit specificity of the two acetyl-CoA synthetases of yeast as revealed by an immunological approach. *Biochimica et Biophysica Acta*, 614, 601–6.

Satyanarayana, T., Mandel, A.D. & Klein, H.P. (1974) Evidence for two immunologically distinct acetyl-coenzyme A synthetases in yeast. *Biochimica et Biophysica Acta*, 341, 396–401.

Savelkoul, P.H.M., Aarts, H.J.M., de Haas, J. *et al.* (1999) Amplified-fragment length polymorphism analysis: the state of the art. *Journal of Clinical Microbiology*, 37, 3083–91.

Schaaf, I., Heinisch, J. & Zimmerman, F.K. (1989) Overproduction of glycolytic enzymes in yeast. *Yeast*, 5, 285–90.

Schaffner, G. & Matile, P. (1981) Structure and composition of baker's yeast lipid globules. *Biochemistry and Biophysiology*, 176, 659–66.

Schalk, H.A. (1906) Brewing. British patent no. 13915.

Schaus, O.O. (1971) Brewing with high gravity worts. *Technical Quarterly of the Master Brewers Association of the Americas*, 8, 7–10.

Schermers, F.H., Duffus, J.H. & Macleod, A.M. (1976) Studies on yeast esterase. *Journal of the Institute of Brewing*, 82, 170–74.

Scherrer, R. (1984) Gram's staining reaction, gram types and cell walls of bacteria. *Trends in the Biochemical Sciences*, May, 242–5.

Scherrer, R., Loudon, L. & Gerhardt, P. (1974) Porosity of the yeast cell wall and membrane. *Journal of Bacteriology*, 118, 534–40.

Schisler, D.O., Ruocco, J.J. & Mabee, M.S. (1982) Wort trub and its effect on fermentation and beer flavour. *Journal of the American Society of Brewing Chemists*, 40, 57–61.

Schlesinger, M.J. (1990) Heat shock proteins. *Journal of Biological Chemistry*, 265, 12111–14.

Schmidt, H.-J. (1994) Accelerated yeast propagation. *Brauwelt International*, 12, 66–72.

Schmidt, H.-J. (1995) Accelerated propagation of pure culture yeast. *Brauwelt International*, 13, 130–49.

Schmidt, H.-J. (1999) Draft beer – maintaining product safety in the process. *Brauwelt International*, 17, 424–8.

Schneible, J. (1902) Manufacture of fermented liquors. US patent 700833.

Schoffel, F. (1970) Corrosion of aluminium fermenters with copper attemperators. *Brauwelt*, 110, 527–32.

Schofield, M.A., Rowe, S.M., Hammond, J.R.M., Molzahn, S.W. & Quain, D.E. (1995) Differentiation of brewery yeast strains by DNA fingerprinting. *Journal of the Institute of Brewing*, **101**, 75–8.

Schuch, C. (1996) Temperature distributions in cylindroconical tanks. *Brauwelt*, **136**, 594–7.

Schuch, C. & Denk, V. (1996) Die visualisierung von stromungen in einem model eines zylindrokonischen tanks. *Monatsschrift für Brauwissenschaft*, 3 Apr, 98–103.

Schuddenat, J., de Boo, R., van Leeuwen, C.C.M., van den Broek, P.J.A. & van Steveninck, J. (1989) Polyphosphate synthesis in yeast. *Biochimica et Biophysica Acta*, **1010**, 191–8.

Schulthess, D. & Ettlinger, L. (1978) Influence of the concentration of branched amino acids on the formation of fusel alcohols. *Journal of the Institute of Brewing*, **84**, 240–43.

Schur, F., Hug, H. & Pfenniger, H. (1974) Dextrins, beta glucans and pentoses in all-malt worts. *Proceedings of the European Brewery Convention Symposium*, Monograph I, Zeist, 150–57.

Schwan, H.P. (1957) Electrical properties of tissue and cell suspensions. *Advances in Biology and Medical Physiology*, **5**, 147–209.

Schwartz, D.C. & Cantor, C.R. (1984) Separation of yeast chromosome-sized DNSs by pulsed field gel electrophoresis. *Cell*, **37**, 67–75.

Schwarz, J.G. & Hang, Y.D. (1994) Purification and characterisation of diacetyl reductase from *Kluyveromyces marxianus*. *Letters in Applied Microbiology*, **18**, 272–6.

Schweizer, E. (1989) Biosynthesis of fatty acids and related compounds. In *Microbial Lipids*, (eds C. Ratledge and S.G. Wilkinson), vol. 2, pp. 13–50. Academic Press, London.

Schwencke, J. (1991) Vacuoles, internal membranous systems and vesicles. In *The Yeasts*, 2nd edition (eds A.H. Rose and J.S. Harrison), vol. 4, pp. 347–432. Academic Press, London.

Scott, C.D. (1987) Immobilised cells: a review of recent literature. *Enzyme & Microbial Technology*, **9**, 66–73.

Scott, J.A. & O.Reilly, A.M. (1995) Use of flexible sponge matrix to immobilise yeast for beer fermentation. *Journal of the American Society of Brewing Chemists*, **53**, 61–71.

Seaton, J.C., Suggett, A. & Moir, M. (1981) The role of sulphur compounds in beer flavour. *Proceedings of the European Brewery Convention Symposium*, Monograph VII, Copenhagen, 143–55.

Seddon, A.W. (1975) Continuous tower fermentation – experiences in establishing large scale commercial production. *Technical Quarterly of the Master Brewers Association of the Americas*, **3**, 130–37.

Seoighe, C. & Wolfe, K.H. (1998) Extent of genomic rearrangement after genome duplication in yeast. *Proceedings of the National Academy of Sciences, USA*, **95**, 4447–52.

Serrano, R. (1977) Energy requirements for maltose transport in yeast. *European Journal of Biochemistry*, **80**, 97–102.

Servouse, M. & Karst, F. (1986) Regulation of early enzymes of ergosterol biosynthesis in *Saccharomyces cerevisiae*. *Biochemical Journal*, **240**, 541–7.

SGD (*Saccharomyces* Genome Database) http://genome-www.stanford.edu/Saccharomyces.

Shankar, C.S. & Umesh-Kumar, S. (1994) A surface lectin associated with flocculation in brewing strains of *Saccharomyces cerevisiae*. *Microbiology*, **140**, 1097–101.

Shardlow, P.J. (1972) The choice and use of cylindro-conical fermentation vessels. *Technical Quarterly of the Master Brewers Association of the Americas*, **9**, 1–5.

Shardlow, P.J. & Thompson, C.C. (1971) The Nathan system conical fermenters. *Brewers Digest*, **46**, 76–81.

Sharma, S.C. (1997) A possible role of trehalose in osmotolerance and ethanol tolerance in *Saccharomyces cerevisiae*. *FEMS Microbiology Letters*, **152**, 11–15.

Sharma, S. & Tauro, P. (1987) Control of ethanol production by yeast: pyruvate accumulation in slow ethanol producing *Saccharomyces cerevisiae*. *Biotechnology Letters*, **9**, 585–6.

Shearwin, K. & Masters, C. (1990) The binding of glycolytic enzymes to the cytoskeleton – influence of pH. *Biochemistry*, **22**, 735–40.

Sheehan, C.A., Weiss, A.S., Newsom, I.A., Flint, V. & O'Donnell, D.C. (1991) Brewing yeast identification and chromosome analysis using high resolution chef gel electrophoresis. *Journal of the Institute of Brewing*, **97**, 163–7.

Sherlock, G. & Rosamond, J. (1993) Starting to cycle: G1 controls regulating cell division in budding yeast. *Journal of General Microbiology*, **139**, 2531–41.

Sheth, N.K., Wisniewski, T.R. & Franson, T.R. (1988) Survival of enteric pathogens in common beverages: an *in vitro* study. *American Journal of Gastroenterology*, **83**, 658–60.

Shimada, S., Andou, M., Naito, N., Yamada, N., Osumi, N. & Hayashi, R. (1993) Effects of hydrostatic pressure on the ultrastructure and leakage of internal substances in the yeast *Saccharomyces cerevisiae*. *Applied and Environmental Microbiology*, **40**, 123–31.

Shimizu, F., Sone, H. & Inoue, T. (1989) Brewing performance of a genetically transformed yeast with acetolactate decarboxylase activity. *Technical Quarterly of the Master Brewers Association of the Americas*, **26**, 47–50.

Shimizu, I., Nagai, J., Hatanaka, H. & Katsuki, H. (1973) Mevalonate synthesis in the mitochondria of yeast. *Biochimica et Biophysica Acta*, **296**, 310–16.

Shimwell, J.C. & Kirkpatrick, A.I.C. (1939) A new light on the Sarcina question. *Journal of the Institute of Brewing*, **45**, 137–45.

Shinabarger, D.L., Keesler, G.A. & Parks, L.W. (1989) Regulation by haem of sterol uptake in *Saccharomyces cerevisiae*. *Steroids*, **53**, 607–23.

Shindo, S., Sahara, H. & Koshino, S. (1993) Relationship of production of succinic acid and methyl citric acid pathway during alcohol fermentation with immobilised yeast. *Biotechnology Letters*, **15**, 51–6.

Shindo, S., Sahara, H., Koshino, S. & Tanaka, H. (1993) Control of diacetyl precursor (alpha-acetolactate) formation during alginate fermentation with yeast cells immobilised in alginate fibres with double gel layers. *Journal of Fermentation and Engineering*, **76**, 199–202.

Shindo, S., Sahara, S., Watanabe, N. & Koshino, S. (1994) Main fermentation with immobilised yeast using a fluidised bed reactor. *Proceedings of the 23rd Institute of Brewing Convention (Australia & New Zealand Section)*, Sydney, 109–13.

Shulman, K.I., Tailor, S.A.N., Walker, S.E. & Gardner, D.M. (1997) Tap (draft) beer and monamine oxidase inhibitor dietary restrictions. *Canadian Journal of Psychiatry*, **42**, 310–12.

Shuttlewood, J.R. (1984) How to engineer cylindroconicals correctly. *Brewing & Distilling International*, August, 22–30.

Siebert, K.J. & Wisk, T.J. (1984) Yeast counting by microscopy or by electronic particle counting. *Journal of the American Society of Brewing Chemists*, **42**, 71–9.

Siebert, K.J., Blum, P.H., Wisk, T.J., Stenroos, L.E. & Anklam, W.J. (1986) The effect of trub on fermentation. *Technical Quarterly of the Master Brewers Association of the Americas*, **23**, 37–43.

Sigler, K. & Hofer, M. (1991) Mechanisms of acid extrusion in yeast. *Biochimica et Biophysica Acta*, **1071**, 375–91.

Sigler, K., Knotkova, A., Paca, J. & Wurst, M. (1980) Extrusion of metabolites from baker's yeast during glucose-induced acidification. *Folia Microbiologica*, **25**, 311–17.

Sigler, K., Knotkova, A. & Kotyk, A. (1981a) Factors governing substrate-induced generation and extrusion of protons in the yeast *Saccharomyces cerevisiae*. *Biochimica et Biophysica Acta*, **643**, 572–82.

Sigler, K., Kotyk, A., Knotkova, A. & Opekarova, M. (1981b) Processes involved in the creation of buffering capacity and in substrate-induced proton extrusion in the yeast *Saccharomyces cerevisiae*. *Biochimica et Biophysica Acta*, **643**, 583–92.

Sigler, K., Pascual, C. & Romay, C. (1983) Intracellular control of proton extrusion in the yeast *Saccharomyces cerevisiae*. *Folia Microbiologica*, **28**, 363–70.

Sigsgaard, P. (1996) Strain selection and characterisation. *Ferment*, **9**, 43–5.

Sigsgaard, P. & Rasmussen, J.N. (1985) Screening of the brewing performance of new yeast strains. *Journal of the American Society of Brewing Chemists*, **43**, 104–8.

Silhankova, L. (1985) Yeast mutants excreting vitamin B and their use in the production of thiamine-rich beers. *Journal of the Institute of Brewing*, **91**, 78–81.

Silhankova, L., Savel, J. & Mostek, J. (1970a) Respiratory deficient mutants of bottom brewer's yeast. I. Frequencies and types of mutant in various strains. *Journal of the Institute of Brewing*, **76**, 280–88.

Silhankova, L., Mostek, J., Savel, J. & Solinova, H. (1970b) Respiratory deficient mutants of bottom brewer's yeast. II. Technological properties of some RD mutants. *Journal of the Institute of Brewing*, **76**, 289–95.

Simpson, K.L. & Priest, F.G. (1998) Characterization of some lactic acid bacteria from Scotch whisky distilleries. *Proceedings of the 5th Aviemore Conference on Malting, Brewing and Distilling*, (ed. I. Campbell), pp. 275–8. Institute of Brewing, London.

Simpson, W.J. (1987) Kinetic studies of the decontamination of yeast slurries with phosphoric acid and acidified ammonium persulphate and a method for the detection of surviving bacteria involving solid medium repair in the presence of catalase. *Journal of the Institute of Brewing*, **93**, 313–18.

Simpson, W.J. (1991a) Rapid microbiological methods in the brewery. *Brewers' Guardian*, June, 30–34.

Simpson, W.J. (1991b) Shedding light on brewery hygiene. *Journal of Biological Education*, **25**, 257–62.

Simpson, W.J. (1993) Studies on the sensitivity of lactic acid bacteria to hop bitter acids. *Journal of the Institute of Brewing*, **99**, 405–11.

Simpson, W.J. (1996) Microbiology for the small brewer. Part 2: Traditional test methods, alternatives and services. *Brewers' Guardian*, March, 21–5.

Simpson, W.J. (1999) Developments in ATP-bioluminescence technology in the brewing industry. *Brewers' Guardian*, May, 24–8.

Simpson, W.J. & Fernandez, J.L. (1992) Selection of beer-spoilage lactic acid bacteria and induction of their ability to grow in beer. *Letters in Applied Microbiology*, **14**, 13–16.

Simpson, W.J. & Fernandez, J.L. (1994) Mechanism of resistance of lactic acid bacteria to trans-isohumulone. *Journal of the American Society of Brewing Chemists*, **52**, 9–11.

Simpson, W.J. & Hammond, J.R.M. (1987) The response of brewery microorganisms to vancomycin. *Journal of the Institute of Brewing*, **93**, 459.

Simpson, W.J. & Hammond, J.R.M. (1989) The response of brewing yeast to acid washing. *Journal of the Institute of Brewing*, **95**, 347–54.

Simpson, W.J. & Hammond, J.R.M. (1991) Antibacterial action of hop resin materials. *Proceedings of the 23rd Congress of the European Brewery Convention*, Lisbon, 185–92.

Simpson, W.J., Fernandez, J.L. & Hammond, J.R.M. (1992) Differentiation of brewery yeasts using a disc-diffusion test. *Journal of the Institute of Brewing*, **98**, 33–6.

Simpson, W.J., Hammond, J.R.M. & Kara, B.V. (1988) Apparent total N-nitrose compounds (ATNCs). Incidence, mechanism of formation, fermentation management and measurement. *Ferment*, **1**, 45–8.

Simpson, W.J., Hammond, J.R.M. & Miller, R.B. (1988) Avoparcin and vancomycin: useful antibiotics for the isolation of brewery lactic acid bacteria. *Journal of Applied Bacteriology*, **64**, 299–309.

Simpson, W.J., Hammond, J.R.M., Thurston, P.A. & Kyriakides, A.L. (1989) Brewery process control and the role of 'instant' microbiological techniques. *Proceedings of the 22nd Congress of the European Brewery Convention*, Zurich, 663–74.

Sims, A.P. & Barnett, J.A. (1991) Levels of activity of enzymes involved in anaerobic utilisation of sugars by six yeast species: observations towards understanding the Kluyver effect. *FEMS Microbiology Letters*, **77**, 295–8.

Sims, A.P., Stalbrand, H. & Barnett, J.A. (1991) The role of pyruvate decarboxylase in the Kluyver effect in the food yeast. *Candida utilis. Yeast*, **7**, 479–87.

Sinclair, D., Mills, K. & Guarente, L. (1998) Ageing in *Saccharomyces cerevisiae. Annual Reviews in Microbiology*, **52**, 533–60.

Sinclair, D.A. & Guarente, L. (1997) Extrachromosomal rDNA circles – a cause of ageing in yeast. *Cell*, **91**, 1033–42.

Sinclair, D.A., Mills, K. & Guarente, L. (1997) Accelerated ageing and nucleolar fragmentation in yeast SGS1 mutants. *Science*, **277**, 1313–16.

Singer, S.J. & Nicolson, G.L. (1972) The fluid mosaic model of the structure of the cell membrane. *Science*, **175**, 720–31.

Singh, M. & Fisher, J. (1996) Cleaning and disinfection in the brewing industry. In *Brewing Microbiology*, 2nd edition, (eds F.G. Priest and I. Campbell), pp. 271–300. Chapman and Hall, London.

Skands, B. (1997) Studies of yeast behaviour in fully automated test plant. *Proceedings of the 26th Congress of the European Brewery Convention*, Maastricht, 413–21.

Skinner, K.E. (1996) Estimation of yeast glycogen content from the mid-infra red spectra of yeast. *Journal of the American Society of Brewing Chemists*, **54**, 71–5.

Slaughter, J.C., Flint, P.W.N. & Kular, K.S. (1987) The effect of carbon dioxide on the absorption of amino acids from a malt extract medium by *Saccharomyces cerevisiae. FEMS Microbiology Letters*, **40**, 239–43.

Smart, K. (1999) Ageing in brewing yeast. *Brewers' Guardian*, February, 19–24.

Smart, K.A. & Whisker (1996) Effect of serial re-pitching on the fermentation properties and condition of brewing yeast. *Journal of the American Society of Brewing Chemists*, **54**, 41–4.

Smart, K.A., Boulton, C.A., Hinchliffe, E. & Molzahn, S.W. (1995) Effect of physiological stress on the surface properties of brewing yeasts. *Journal of the American Society of Brewing Chemists*, **53**, 33–8.

Smart, K.A., Chambers, K.M., Lambert, I. & Jenkins, C. (1999) Use of methylene violet staining procedures to determine yeast viability and vitality. *Journal of the American Society of Brewing Chemists*, **57**, 18–23.

Smit, G., Straver, M.H., Lugtenberg, B.J.J. & Kijne, J.W. (1992) Flocculence of *Saccharomyces cerevisiae* cells is induced by nutrient limitation, with cell surface hydrophobicity as a major determinant. *Applied and Environmental Microbiology*, **58**, 3709–14.

Smith, C.E., Casey, G.P. & Ingledew, W.M. (1987) The use and understanding of media used in brewing microbiology – update 1987. *Brewers Digest*, October, 12–16, 43.

Smith, N.A. (1992) Nitrate reduction and ATNC formation by brewery wild yeasts N-nitrosation in brewing. *Journal of the Institute of Brewing*, **98**, 415–20.

Smith, N.A. (1994) Nitrate reduction and N-nitrosation in brewing. *Journal of the Institute of Brewing*, **100**, 347–55.

Smith, N.A., Smith, P. & Woodruff, C.A. (1992) The role of *Bacillus* spp. In N-nitrosamine formation during wort production. *Journal of the Institute of Brewing*, **98**, 409–14.

Smits, G.J., Kapteyn, J.C., van den Ende, H. & Klis, F.M. (1999) Cell wall dynamics in yeast. *Current Opinion in Microbiology*, **2**, 348–52.

Snell, J.J.S. (1991) General introduction to maintenance methods. In *Maintenance of Microorganisms and Cultured Cells. A Manual of Good Practice*, 2nd edition (eds B.E. Kirsop and A. Doyle), pp. 21–30. Academic Press, London.

Soares, E.V. and Mota, M. (1997) Quantification of yeast flocculation. *Journal of the Institute of Brewing*, **103**, 93–8.

Sols, A. (1981) Multi-modulation of enzyme activity. *Current Topics in Cellular Regulation*, **19**, 77–101.

Soltis, D.E. & Soltis, P.S. (1995) The dynamic nature of polyploid genomes. *Proceedings of the National Academy of Sciences, USA*, **92**, 8089–91.

Soltoft, M. (1988) Flavour active sulphur compounds in beer. *Brygmesteren*, **2**, 18–24.

Sone, H., Fujii, T., Kondo, K. & Tanaka, J. (1988a) Molecular cloning of the gene encoding alpha-acetolactate decarboxylase from *Enterobacter aerogenes*. *Technical Report of the Research Laboratory of the Kirin Brewing Company*, **31**, 1–4.

Sone, H., Kondo, K., Fujii, T., Shimizu, F., Tanaka, J. & Inoue, T. (1988b) Fermentation properties of brewer's yeast having alpha-acetolactate decarboxylase gene. *Technical Report of the Research Laboratory of the Kirin Brewing Company*, **31**, 5–9.

Sone, H., Kondo, K., Fujii, T., Shimizu, F., Tanaka, J. & Inoue, T. (1987) Fermentation properties of brewer's yeast having alpha-acetolactate decarboxylase gene. *Proceedings of the 21st Congress of the European Brewery Convention*, Madrid, 545–52.

Soumalainen, H. (1981) Yeast esterases and aroma esters in alcoholic beverages. *Journal of the Institute of Brewing*, **87**, 296–300.

Speers, R.A. & Ritcey, L.L. (1995) Towards an ideal flocculation assay. *Journal of the American Society of Brewing Chemists*, **53**, 174–7.

Speers, R.A., Tung, M.A., Rurance, T.D. & Stewart, G.G. (1992) Biochemical aspects of yeast flocculation and its measurement: a review. *Journal of the Institute of Brewing*, **98**, 293–300.

Spencer, J.F.T. & Spencer, D.M. (1997) Ecology: where yeasts live. In *Yeasts in Natural and Artificial Habitats*, (eds J.F.T. Spencer and D.M. Spencer), pp. 33–58. Springer-Verlag, Berlin and Heidelberg.

Sperka-Gottlieb, C.D.M., Hermetter, A., Paltauf, F. & Daum, G. (1988) Lipid topology and physical properties of the outer mitochondrial membrane of the yeast *Saccharomyces cerevisiae*. *Biochimica et Biophysica Acta*, **946**, 227–34.

Sprague, G.F., Jr (1995) Mating and mating-type interconversion in *Saccharomyces cerevisiae* and *Schizosaccharomyces pombe*. In *The Yeasts*, 2nd edition, (eds A.H. Rose, A.E. Wheals and J.S. Harrison), vol. 6, pp. 411–59. Academic Press, London.

Stanley, G.A. & Pamment, N.B. (1993) Transport and intracellular accumulation of acetaldehyde in *Saccharomyces cerevisiae*. *Biotechnology and Bioengineering*, **42**, 24–29.

Stanley, P.E., Smither, R. & Simpson, W.J. (1997) Preface. In *A Practical Guide to Industrial Uses of ATP Bioluminescence in Rapid Microbiology*, (eds P.E. Stanley, R. Smither and W.J. Simpson), pp. vii–viii. Cara Technology Limited, Lingfield.

Stassi, P., Goetzke, G.P. & Fehring, J.F. (1991) Evaluation of an insertion thermal mass flowmeter to monitor carbon dioxide evolution rate in large scale fermentations. *Technical Quarterly of the Master Brewers Association of the Americas*, **28**, 84–8.

Stassi, P., Rice, J.F., Munroe, J.H. & Chicoye, E. (1987) Use of carbon dioxide evolution rate for the study and control of fermentation. *Technical Quarterly of the Master Brewers Association of the Americas*, **24**, 44–50.

Steele, D.B. & Stowers, M.D. (1991) Techniques for selection of industrially important microorganisms. *Annual Reviews in Microbiology*, **45**, 89–106

Stephen, D.W.S., Rivers, S.L. & Jamieson, D.J. (1995) The role of the *YAP1* and *YAP2* genes in the regulation of the adaptive oxidative stress responses of *Saccharomyces cerevisiae*. *Molecular Microbiology*, **16**, 415–23.

Stevens, B.J. (1981) Mitochondrial structure. In *The Molecular Biology of the Yeast Saccharomyces, Metabolism and Gene Expression*, (eds J.F. Strathern, E.W. Jones and J.R. Broach), pp. 471–504. Cold Spring Harbor Laboratory, USA.

Stevens, B.J. (1977) Variation in number and volume of the mitochondria in yeast according to growth conditions. A study based on serial sectioning and computer graphics reconstruction. *Biology of the Cell*, **28**, 37–42.

Stevens, R. (1960) Volatile products of fermentation: A review. *Journal of the Institute of Brewing*, **66**, 453–71.

Stevens, R. (1987) The chemistry of hop constituents. In *An Introduction to Brewing Science and Technology*, Series II, vol. 1 (ed. R. Stevens), pp. 23–54. The Institute of Brewing, London.

Stewart, G.G. (1995) Adjuncts. In *Handbook of Brewing*, (ed. W.A. Hardwick), Marcel Dekker, New York. 121–32.

Stewart, G.G. & Russell, I. (1981) Yeast flocculation. In *Brewing Science* (ed J.R.A. Pollock), pp. 61–92. **2**, Academic Press, London.

Stewart, G.G. & Russell, I. (1986) One hundred years of yeast research and development in the brewing industry. *Journal of the Institute of Brewing*, **92**, 537–58.

Stewart, G.G., Zheng, X. & Russell, I. (1995) Wort sugar uptake and metabolism – the influence of genetic and environmental factors. *Proceedings of the 25th Congress of the European Brewery Convention*, Brussels, 403–10.

Stewart, R.J. & Dowhanick, T.M. (1996) Rapid detection of lactic acid bacteria in fermenter samples using a nested polymerase chain reaction. *Journal of the American Society of Brewing Chemists*, **54**, 78–84.

Stickler, D. (1999) Biofilms. *Current Opinion in Microbiology*, **2**, 270–75.

Stickler, D.J. (1997) Chemical and physical methods of biofilm control. In *Biofilms: Community Interactions and Control* (eds J. Wimpenny, P. Handley, P. Gilbert, H. Lappin-Scott & M. Jones) pp. 215–25. Bioline, Cardiff.

Stillman, C.G. (1996) Glass washing. *Proceedings of the European Brewery Convention Symposium*, Monograph **XXV**, Edinburgh, 139–44.

Storgards, E. (1996) Microbiological quality of draught beer – is there reason for concern? *Proceedings of the European Brewery Convention Symposium* Monograph **XXV**, Edinburgh, 92–103.

Storgards, E. & Haikara, A. (1996) ATP bioluminescence in the hygiene control of draught beer dispense systems. *Ferment*, **9**, 352–60.

Storgards, E., Juvonen, R., Vanne, L. & Haikara, A. (1997a) Detection methods in process and hygiene control. *Proceedings of the European Brewery Convention Symposium*, Monograph **XXVI**, Stockholm, 95–107.

Storgards, E., Pihlajamaki, O. & Haikara, A. (1997b) Biofilms in the brewing process – a new approach to hygiene management. *Proceedings of the 26th Congress of the European Brewery Convention*, Maastricht.

Storgards, E., Suihko, M.-L., Pot, B., Vanhonacker, K., Janssens, D., Broomfield, P.L.E. & Banks, J.G. (1998) Detection and identification of *Lactobacillus lindneri* from brewery environments. *Journal of the Institute of Brewing*, **104**, 47–54.

Storgards, E., Simola, H., Sjoberg, A.-M. & Wirtanen, G. (1999a) Hygiene of gasket materials used in food processing equipment part 1: new materials. *Transactions of the Institute of Chemical Engineers, Part C*, **77**, 137–45.

Storgards, E., Simola, H., Sjoberg, A.-M. & Wirtanen, G. (1999b) Hygiene of gasket materials used in food processing equipment part 2: aged materials. *Transactions of the Institute of Chemical Engineers, Part C*, **77**, 146–55.

Storgards, E., Yli-Juuti, P., Salo, S., Wirtanen, G. & Haikara, A. (1999c) Modern methods in process hygiene control – benefits and limitations. *Proceedings of the 27th Congress of the European Brewery Convention* Cannes, 249–58.

Stratford, M. (1989) Yeast flocculation: calcium specificity. *Yeast*, **5**, 487–96.

Stratford, M. (1992a) Yeast flocculation: a new perspective. In *Advances in Microbial Physiology*, (ed. A.H. Rose), vol. 33, pp. 2–71. Academic Press, London.

Stratford, M. (1992b) Yeast flocculation: reconciliation of physiological and genetic viewpoints. *Yeast*, **8**, 25–38.

Stratford, M. (1992c) Yeast flocculation: receptor definition by mnn mutants and concanavalin A. *Yeast*, **8**, 635–45.

Stratford, M. (1993) Yeast flocculation: flocculation onset and receptor availability. *Yeast*, **9**, 85–94.

Stratford, M. (1994) Another brick in the wall? Recent developments concerning the yeast cell envelope. *Yeast*, **10**, 1741–52.

Stratford, M. (1996) Yeast flocculation: restructuring the theories in line with recent research. *Cerevisia – Belgian Journal of Brewing and Biotechnology*, **21**, 38–45.

Stratford, M. & Assinder, S. (1991) Yeast flocculation: Flo1 and NewFlo phenotypes and receptor structure. *Yeast*, **7**, 559–74.

Stratford, M. & Carter, A.T. (1993) Yeast flocculation: lectin synthesis and activation. *Yeast*, **9**, 371–8.

Stratford, M. & Keenan, M.H.J. (1987) Yeast flocculation: kinetics and collision theory. *Yeast*, **3**, 201–6.

Stratton, M.K., Campbell, S.J. & Banks, D.J. (1994) Process developments in the continuous fermentation of beer. *Proceedings of the 23rd Institute of Brewing Convention* (Asia Pacific Section), Sydney, 96–100.

Straver, M.H. & Kijne, J.W. (1996) A rapid and selective assay for measuring cell surface hydrophobicity of brewer's yeast cells. *Yeast*, **12**, 207–13.

Straver, M.H., Aar, P.C.V.D., Smit, G. & Kijne, J.W. (1993) Determinants of flocculence of brewers' yeast during fermentation in wort. *Yeast*, **9**, 527–32.

Straver, M.H., Smit, G. & Kijne, J.W. (1994b) Purification and partial characterization of a flocculin from brewer's yeast. *Applied and Environmental Microbiology*, **60**, 2754–8.

Straver, M.H., Traas, V.M., Smit, G. & Kijne, J.W. (1994a) Isolation and partial purification of mannose-specific agglutinin from brewer's yeast involved in flocculation. *Yeast*, **10**, 1183–93.

Strehaiano, P. & Goma, G. (1983) Effects of initial substrate concentration on two wine yeasts: relation between glucose sensitivity and ethanol inhibition. *American Journal of Enology and Viticulture*, **34**, 1–5.

Streiblova, E. (1988) Cytological methods. In *Yeast, a Practical Approach*, (eds I. Campbell & J.H. Duffus), pp. 9–49. IRL Press, Oxford.

Subden, R.E. (1990) Wine yeast: selection and modification. In *Yeast Strain Selection*, Chapter 5, (ed. C.J. Panchal), pp. 113–37. Marcel Dekker, New York.

Subik, J., Kolarov, J. & Kovac, L. (1972) Obligatory requirement of intramitochondrial ATP for normal functioning of the eukaryotic cell. *Biochemistry and Biophysical Research Communications*, **49**, 192–8.

Sudbery, P.E., Goodey, A.R. & Carter, B.L. (1980) Genes which control cell proliferation in the yeast *Saccharomyces cerevisiae*. *Nature*, **288**, 401–4.

Sugden, S. (1993) In-line monitoring and automated control of the fermentation process. *Brewers' Guardian*, June, 21–32.

Suihko, M.-L., Penttila, M., Sone, H. *et al.* (1989) Pilot brewing with alpha-acetolactate decarboxylase active yeasts. *Proceedings of the 22nd Congress of the European Brewery Convention*, Zurich, 483–90.

Sun, J., Kale, S.P., Childress, A.M., Pinswasdi, C. & Jazwinski, S.M. (1994) Divergent roles of *RAS1* and *RAS2* in yeast longevity. *Journal of Biological Chemistry*, **269**, 18638–45.

Sutherland, I.W. (1997) Microbial biofilm exopolysaccharides – superglues or velcro? In *Biofilms: Community Interactions and Control* (eds J. Wimpenny, P. Handley, P. Gilbert, H. Lappin-Scott and M. Jones), pp. 33–9. Bioline, Cardiff.

Syu, M.-J. & Tsao, G.T. (1993) Neural network modelling of batch cell growth pattern. *Biotechnology and Bioengineering*, **42**, 376–80.

Syu, M.-J., Tsao, G.T., Austin, G.D., Gelotto, G. & D'Amore, T.D. (1994) Neural network modelling for predicting brewery fermentations of batch cell growth pattern. *Journal of the American Society of Brewing Chemists*, **52**, 15–18.

Szabó, I., Báthori, G., Wolff, D., Starc, T., Cola, C. & Zoratti, M. (1995) The high conductance channel of porin-less yeast mitochondria. *Biochimica et Bioiophysica Acta*, **1235**, 115–25.

Szlavko, C.M. (1974) The influence of wort glucose levels on the formation of aromatic higher alcohols. *Journal of the Institute of Brewing*, **80**, 534–9.

Tada, S., Takeuchi, T., Sone, H. *et al.* (1995) Pilot-scale brewing with industrial yeasts which produce the alpha-acetolactate decarboxylase of *Acetobacter aceti* ssp. *xylinum*. *Proceedings of the 25th Congress of the European Brewery Convention*, Brussels, 369–76.

Tailor, S.A.N., Shulman, K.I., Walker, S.E., Moss, J. & Gardner, D. (1993) Hypertensive episode associated with phenelzine and tap beer – a reanalysis of the role of pressor amines in beer. *Journal of Clinical Psychopharmacology*, **14**, 5–14.

Takahashi, S., Okada, A. & Masuoka, S. (1990) Total sterilization system in the brewing process and sterile filtration techniques. *Proceedings of the European Brewery Convention Symposium*, Monograph **XVI**, Leuven, 214–27.

Takahashi, T., Nakakita, Y., Monji, Y., Watari, J. & Shinotsuka, K. (1999) Application of new automatic MicroStar RMDS (ATP bioluminescence system) for rapid quantitative detection of brewery contaminants. *Proceedings of the 27th Congress of the European Brewery Convention*, Cannes, 259–66.

Takayanagi, S. & Harada, T. (1967) Experiences with new types of free standing fermentation and lagering tanks. *Proceedings of the 11th Congress of the European Brewery Convention*, Madrid, 473–87.

Takemura, O., Nakatani, K., Tanigucha, T. & Murakami, M. (1992) Practice on the microbial control for the production of non-pasteurised bottled beer. *Proceedings of the 22nd Congress of the Institute of Brewing Convention (Australia & New Zealand Section)* Melbourne, 148–53.

Takeo, K., Nishimura, K. & Miyaji, M. (1988) Topographical stability of lateral regions of the *Schizosaccharomyces pombé* plasma membrane as revealed by invagination characteristics. *Canadian Journal of Microbiology*, **34**, 1063–8.

Tamai, Y. (1996) Alcohol acetyl transferase genes and ester formation in brewer's yeast. *Biotechnology for Improved Foods and Flavours*, ACS Symposium Service 637, 196–205.

Tamai, Y., Momma, T., Yoshimoto, H. & Kaneko, Y. (1998) Co-existence of two types of chromosome in the bottom fermenting yeast *Saccharomyces pastorianus*. *Yeast*, **14**, 923–33.

Tamanoi, F. (1988) Yeast *RAS* genes. *Biochimica et Biophysica Acta*, **948**, 1–15.

Taylor, D.G. (1990) The importance of pH control during brewing. *Technical Quarterly of the Master Brewers Association of the Americas*, **27**, 131–6.

Taylor, G.T. & Marsh, A.S. (1984) MYGP + copper, a medium that detects both *Saccharomyces* and non-*Saccharomyces* wild yeast in the presence of culture yeast. *Journal of the Institute of Brewing*, **90**, 134–45.

Taylor, G., Goodacre, R., Wade, W.G., Rowland, J.J. & Kell, D.B. (1998) The deconvolution of pyrolysis mass spectra using genetic programming: application to the identification of some *Eubacterium* species. *FEMS Microbiology Letters*, **160**, 137–46.

Taylor, L. (1974) Fermentable sugars and dextrins in top fermentation wort. *Proceedings of the European Brewery Convention Symposium*, Monograph I, Zeist, 208–25.

Taylor, R. & Kirsop, B.H. (1979) Occurrence of a killer strain of *Saccharomyces cerevisiae* in a batch fermentation plant. *Journal of the Institute of Brewing*, **85**, 325.

Teunissen, A.W.R.H. & Steensma, H.Y. (1995) Review: The dominant flocculation genes of *Saccharomyces cerevisiae*. *Yeast*, **11**, 1001–13.

Thatiparnala, R., Rohani, S. & Hill, G.A. (1992) Effects of high product and substrate inhibitions on the kinetics and biomass and product yields during ethanol batch fermentation. *Biotechnology and Bioengineering*, **40**, 289–97.

Thevelein, J.M. (1984) Regulation of trehalose mobilisation in fungi. *Microbial Reviews*, **42**, 48–59.

Thevelein, J.M. (1994) Signal transduction in yeast. *Yeast*, **10**, 1753–90.

Thevelein, J.M. & Hohmann, S. (1995) Trehalose synthase: guard to the gate of glycolysis in yeast? *Trends in Biochemical Sciences*, **20**, 3–10.

Thibault, J., LeDay, D. & Cote, F. (1987) Production of ethanol by *Saccharomyces cerevisiae* under high pressure conditions. *Biotechnology and Bioengineering*, **30**, 74–80.

Thom, S.M., Horobin, R.W., Seidler, E. & Barer, M.R. (1993) Factors affecting the selection and use of tetrazolium salts as cytochemical indicators of microbial viability and activity. *Journal of Applied Bacteriology*, **74**, 433–43.

Thomas, D.S. (1993) Yeasts as spoilage organisms in beverages. In *The Yeasts*, 2nd edition, (eds A.H. Rose and J.S. Harrison), vol. 5, pp. 517–61. Academic Press, London.

Thomas, K. & Whitham, H. (1996) Improvement in beer line technology. *Proceedings of the European Brewery Convention Symposium*, Monograph **XXV**, Edinburgh, 124–61.

Thompson, A.N. & Jones, H.L. (1997) ATP bioluminescence in modern brewing. In *A Practical Guide to Industrial Uses of ATP Bioluminescence in Rapid Microbiology*, (eds P.E. Stanley, R. Smither and W.J. Simpson), pp. 125–9. Cara Technology Limited, Lingfield.

Thompson, C.C. & Cameron, A.D. (1971) The differentiation of primary brewing yeasts by immunofluorescence. *Journal of the Institute of Brewing*, **77**, 24–7.

Thorn, J.A. (1971) Yeast autolysis and its effect on beer. *Brewers Digest*, October, 110–13.

Thorne, R.S.W. (1975) Brewing yeasts considered taxonomically. *Process Biochemistry*, September, 17–28.

Thorpe, E. & Brown, H.T. (1914) Reports on the determination of the original gravity of beers by the distillation process. *Journal of the Institute of Brewing*, **20**, 569–713.

Thurman, C.W. & Witheridge, J.R. (1995) *Statistical Handbook, a compilation of drinks industry statistics*. Brewing Publications Ltd, Portman Square, London.

Thurston, P.A., Quain, D.E. & Tubb, R.S. (1981) The control of volatile ester biosynthesis in *Saccharomyces cerevisiae*. *Proceedings of the 18th Congress of the European Brewery Convention*, Copenhagen, 197–206.

Thurston, P.A., Quain, D.E. & Tubb, R.S. (1982) Lipid metabolism and the regulation of volatile ester synthesis in *Saccharomyces cerevisiae*. *Journal of the Institute of Brewing*, **88**, 90–94.

Timmins, E.M., Quain, D.E. & Goodacre, R. (1998) Differentiation of brewery yeast strains by pyrolysis mass spectrometry and Fourier transform infrared spectroscopy. *Yeast*, **14**, 885–93.

Tompkins, T.A., Stewart, R., Savard, L., Russell, I. & Dowhanick, T.M. (1996) RAPD-PCR characterisation of brewery yeast and beer spoilage bacteria. *Journal of the American Society of Brewing Chemists*, **54**, 91–6.

Treacher, K. (1995) A critical review of beer cleaning systems. *Brewers' Guardian*, August, 19–25.

Tredoux, H.G., Kock, J.L.F., Lategan, P.M. & Muller, H.B. (1987) A rapid identification technique to differentiate between *Saccharomyces cerevisiae* strains and other yeast species in the wine industry. *American Journal of Enology and Viticulture*, **38**, 161–4.

Trevors, J.T. (1982) Tetrazolium reduction in *Saccharomyces cerevisiae*. *European Journal of Applied Microbiology and Biotechnology*, **15**, 172–4.

Trevors, J.T., Merrick, R.L., Russell, I. & Stewart, G.G. (1983) A comparison of methods for assessing yeast viability. *Biotechnology Letters*, **5**, 131–4.

Trivedi, N.B. & Jacobsen, G. (1986) Recent advances in baker's yeast. *Progress in Industrial Microbiology*, **23**, 45–71.

Trocha, P.J. & Sprinson, D.B. (1976) Localisation and regulation of early enzymes of sterol biosynthesis in yeast. *Archives for Biochemistry and Biophysics*, **174**, 45–53.

Trumbly, R.J. (1992) Glucose repression in the yeast *Saccharomyces cerevisiae*. *Molecular Microbiology*, **6**, 15–21.

Tsang, E.W.T. & Ingledew, W.M. (1982) Studies on the heat resistance of wild yeasts and bacteria in beer. *Journal of the American Society of Brewing Chemists*, **40**, 1–8.

Tsuchiya, Y., Kaneda, H., Kano, Y. & Koshino, S. (1992) Detection of beer spoilage organisms by polymerase chain reaction technology. *Journal of the American Society of Brewing Chemists*, **50**, 64–7.

Tubb, R.S. (1981) Developing control over yeast. *Brewers' Guardian*, July, 23–31.

Tubb, R.S. (1984) Genetic development of yeast strains. *Brewers' Guardian*, September, 34–7.

Tubb, R.S. & Liljestrom, P.L. (1986) A colony-colour method which differentiates alpha-galactosidase-positive strains of yeast. *Journal of the Institute of Brewing*, **92**, 588–90.

Tubb, R.S., Searle, B.A., Goodey, A.R. & Brown, A.J.P. (1981) Rare mating and transformation for construction of novel brewing strains. *Proceedings of the 18th Congress of the European Brewery Convention*, Copenhagen, 487–96.

Tullin, S., Gjermansen, C. & Kielland-Brandt, C. (1991) A high affinity uptake system for branched amino acids in *Saccharomyces cerevisiae*. *Yeast*, **7**, 933–41.

Turakainen, H., Kristo, P. & Korhola, M. (1994) Consideration of the evolution of the Saccharomyces cerevisiae *MEL* gene family on the basis of the nucleotide sequences of the genes and their flanking regions. *Yeast*, **10**, 1559–68.

Tyagi, R.D. & Ghose, T.K. (1982) Studies on immobilised *S. cerevisiae* – I. Analysis of continuous rapid ethanol fermentation in immobilised cell reactor. *Biotechnology & Bioengineering*, **24**, 781–95.

Tyson, C.B., Lord, P.G. & Wheals, A.E. (1979) Dependency of size of *Saccharomyces cerevisiae* cells on growth rate. *Journal of Bacteriology*, **138**, 92–8.

Tzagoloff, A. & Dieckmann, C.L. (1990) *PET* genes of *Saccharomyces cerevisiae*. *Microbiological Reviews*, **54**, 211–25.

Ulenberg, G.H., Gerritson, H. & Huisman, J. (1972) Experiences with a giant cylindro-conical tank. *Technical Quarterly of the Master Brewers Association of the Americas*, **9**, 117–22.

Umesh-Kumar, S. & Nagarajan, L. (1991) Cell wall antigenic homology in certain species of *Saccharomyces* and an enzyme-linked immunosorbent assay for rapid identification of *Saccharomyces cerevisiae*. *Folia Microbiologica*, **36**, 305–10.

Uno, I., Matsumoto, K., Adachi, K. & Ishikawa, T. (1983) Genetic and biochemical evidence that trehalose is a substrate of cAMP dependent protein kinase in yeast. *Journal of Biological Chemistry*, **285**, 10867–72.

van Urk, H., Mak, P.R. Scheffers, W.A. & van Dijken, J.P. (1988) Metabolic responses of *Saccharomyces cerevisiae* CBS 8066 and *Candida utilis* CBS 621 upon transition from glucose limitation to glucose excess. *Yeast*, **4**, 283–91.

van Urk, H., Schipper, D., Breedveld, G.J., Mak, R., Scheffers, W.A. & van Dijken, J.P. (1989) Localisation and kinetics of pyruvate-metabolising enzymes in relation to aerobic alcoholic fermentation in *Saccharomyces cerevisiae* CBS 8086 and *Candida utilis* CBS 621. *Biochimica et Biophysica Acta*, **992**, 78–86.

van Urk, H., Voll, W.S.L., Scheffers, W.A. & van Dijken, J.P. (1990) Transient state analysis of metabolic fluxes in Crabtree positive and Crabtree negative yeasts. *Applied and Environmental Microbiology*, **56**, 282–6.

Vágvölgyi, C., Kucsera, J. & Ferenczy, L. (1988) A physical method for separating *Saccharomyces cerevisiae* cells according to their ploidy. *Canadian Journal of Microbiology*, **34**, 1102–4.

Vakeria, D. (1991) The genetic improvement of brewing yeast. *Brewers' Guardian*, May, 29–31.

Vakeria, D. & Hinchliffe, E. (1989) Amylolytic brewing yeasts: their commercial and legislative acceptability. *Proceedings of the 22nd Congress of the European Brewery Convention*, Zurich, 475–82.

Van Loon, A.P.G.M., Pesold-Hurt, B. & Schatz, G. (1986) A yeast mutant lacking manganese superoxide dismutase is hypersensitive to oxygen. *Proceedings of the National Academy of Sciences USA*, **83**, 3820–24.

Vaughan-Martini, A. & Kurtzman, C.P. (1985) Deoxyribonucleic acid relatedness amongst species of the genus. *Saccharomyces stricto sensu*. *International Journal of Systematic Bacteriology*, **35**, 508–11.

Vaughan-Martini, A. & Martini, A. (1993) A taxonomic key for the genus *Saccharomyces*. *Systematics and Applied Microbiology*, **16**, 113–19.

Vaughan-Martini, A. & Martini, A. (1998) *Saccharomyces* Meyen ex Rees. In *The Yeasts, A Taxonomic Study*, 4th edition, (eds C.P. Kurtzman and J.W. Fell), pp. 358–71. Elsevier, Amsterdam.

Vickers, J. & Ballard, G. (1974) Amelioration of colloidal conditions of beer. *The Brewer*, January, 19–24.

Viegas, C.A., Sa-Correia, I. & Novais, J.M. (1985) Synergistic inhibition of the growth of *Saccharomyces cerevisiae* by ethanol and octanoic or decanoic acids. *Biotechnology Letters*, **7**, 611–14.

Vietor, R.J., Angelino, S.A.G.F. & Voragen, A.G.J. (1991) Arabinoxylans in barley, malt and wort, *Proceedings of the 23rd Congress of the European Brewery Convention*, Lisbon, 139–46.

Vietor, R.J., Voragen, A.G.J. & Angelino, S.A.G.F. (1993) Composition of non-starch polysaccharides in wort and spent grain from brewing trials with malt from a good malting quality barley and a feed barley. *Journal of the Institute of Brewing*, **99**, 243–8.

Viljoen, B.C. & Kock, J.L.F. (1991) The value of pyrolysis gas-liquid chromatography in the taxonomy of some ascogenous yeasts. *Systematic and Applied Microbiology*, **14**, 178–82.

Villaneuba, K.D., Goossens, E. & Masschelein, C.A. (1990) Subthreshold vicinal diketone levels in lager brewing yeast fermentations by means of ILV5 gene amplification. *Journal of the American Society of Brewing Chemists*, **48**, 111–14.

Vincent, S.F., Bell, P.J.L., Bissinger, P. & Nevalainen, K.M.H. (1999) Comparison of melibiose utilizing baker's yeast strains produced by genetic engineering and classical breeding. *Letters in Applied Microbiology*, **28**, 148–52.

van der Walt, J.P. (1987) The typological yeast species, and its delimitation. In *The Yeasts*, 2nd edition, (eds A.H. Rose and J.S. Harrison), vol. 1, pp. 95–121. Academic Press, London.

de Virgilio, C., Hottiger, T., Dominguez, J., Boller, T. & Wiemken, A. (1994) The role of trehalose synthesis for the acquisition of thermotolerance in yeast. *European Journal of Biochemistry*, **219**, 179–86.

Visser, W., van Sponsen, E.A., Nanninga, N., Pronk, J.T., Kuenen, J.G. & van Dijken, J.P. (1995) Effects of growth conditions on mitochondrial morphology in *Saccharomyces cerevisiae*. *Antonie van Leeuwenhoek*, **67**, 243–53.

Visser, W., Scheffers, W.A., Batenburg-van der Vegte & van Dijken, J.P. (1990) Oxygen requirements of yeasts. *Applied and Environmental Microbiology*, **56**, 3785–92.

Visser, W., van der Baan, A.A., Batenburg-van der Vegte, W., Scheffers, W.A., Kramer, R. & van Dijken, J.P. (1994) Involvement of mitochondria in the assimilatory metabolism of anaerobic *Saccharomyces cerevisiae* cultures. *Microbiology*, **140**, 3039–46.

Von Nida, L. (1997) Aerobic yeast propagation. *Brauwelt International*, **15**, 147–51.

Vos, P., Hogers, R., Bleeker, M. *et al.* (1995) AFLP: a new technique for DNA fingerprinting. *Nucleic Acids Research*, **23**, 4407–14.

Vrieling, A.M. (1978) Agitated fermentation in high fermenters. *Proceedings of the European Brewery Convention*, Monograph V, Zoeterwoude, 135–44.

van Vuuren, H.J.J. (1996) Gram-negative spoilage bacteria. In *Brewing Microbiology*, 2nd edn, (eds F.G. Priest and I. Campbell), Chapman and Hall, London, 163–91.

van Vuuren, H.J.J. & van der Meer, L. (1987) Fingerprinting of yeasts by protein electrophoresis. *American Journal of Enology and Viticulture*, **38**, 49–53.

Wada, Y., Ohsumi, Y., Tanifuji, M., Kasai, M. & Anraku, Y. (1987) Vacuolar ion channel of the yeast *Saccharomyces cerevisiae*. *Journal of Biological Chemistry*, **262**, 17260–63.

Wainer, S.R., Bouveris, A. & Ramos, E.H. (1988) Control of leucine transport in yeast by periplasmic binding proteins. *Archives for Biochemistry and Biophysics*, **262**, 481–90.

Wainwright, T. (1973) Diacetyl – a review. Part 1, analytical and biochemical considerations. Part 2, brewing experience. *Journal of the Institute of Brewing*, **79**, 451–70.

Wales, D.S., Cartledge, T.G. & Lloyd, D. (1980) Effect of glucose repression and anaerobiosis on the activities and subcellular distribution of tricarboxylic acid cycle and associated enzymes in *Saccharomyces cerevisiae*. *Journal of General Microbiology*, **116**, 92–8.

Walker, G.M. (1998) *Yeast Physiology and Biotechnology*. Wiley, Chichester.

Walker, G.M. & Maynard, A.L. (1997) Accumulation of magnesium ions during fermentative metabolism in *Saccharomyces cerevisiae*. *Journal of Industrial Microbiology and Biotechnology*, **18**, 1–3.

Walker, G.M., Birch, R.M., Chandrasena, G. & Maynard, A.I. (1996) Magnesium, calcium and fermentative metabolism in industrial yeast. *Journal of the American Society of Brewing Chemists*, **54**, 13–18.

Walkey, R.J. & Kirsop, B.H. (1969) Performance of strains of *Saccharomyces cerevisiae* in batch fermentation. *Journal of the Institute of Brewing*, **75**, 393–8.

Walmsley, R.M. (1994) DNA fingerprinting of yeast. *Ferment*, **4**, 231–4.

Walmsley, R.M., Wilkinson, B.M. & Kong, T.H. (1989) Genetic fingerprinting for yeasts. *Bio/Technology*, **7**, 1168–70.

Walsh, R.M. & Martin, P.A. (1977) Growth of *Saccharomyces cerevisiae* and *Saccharomyces uvarum* in a temperature gradient incubator. *Journal of the Institute of Brewing*, **83**, 169–72.

Walther, P., Müller, M. & Schweingruber, M.E. (1984) The ultrastructure of the cell surface and plasma membrane of exponential and stationary phase cells of *Schizosaccharomyces pombé* grown in different media. *Archives for Microbiology*, **137**, 128–34.

Walton, E.F., Carter, B.L.A. & Pringle, J.R. (1979) An enrichment method for temperatuire sensitive mutant and auxotrophic mutants of yeast. *Molecular and General Genetics*, **171**, 111–14.

Walworth, N.C., Goud, B., Ruohola, H. & Novick, T.J. (1989) Fractionation of yeast organelles. *Methods in Cell Biology*, **31**, 335–57.

Wang, S.-S. & Brandriss, M.C. (1987) Proline utilisation in *Saccharomyces cerevisiae*. Sequence regulation and mitochondrial localisation of the PUT1 gene product. *Molecular and Cellular Biochemistry*, **7**, 4431–40.

Wanke, V., Vavassori, M., Thevelein, J.M., Tortora, P. & Vanoni, M. (1997) Regulation of maltose utilisation in *Saccharomyces cerevisiae* by genes of the *RAS*/protein kinase A pathway. *FEBS Letters*, **402**, 251–5.

Wasmuht, K. & Weinzart, M. (1999) Observing fermentation with the help of a new control system referred to as 'TopScan' *Brauwelt International*, **17**, 512–13.

Watari, J., Nomura, M., Sahara, H., Koshino, S. & Keranen, S. (1994) Construction of flocculent brewer's yeast by chromosomal integration of the yeast flocculation gene *FLO 1*. *Journal of the Institute of Brewing*, **100**, 73–7.

Watson, K. & Caviccioli, R. (1983) Acquisition of ethanol tolerance on yeast cells by heat shock. *Biotechnology Letters*, **5**, 683–8.

Weete, J.D. (1989) Structure and function of sterols in fungi. *Advances in Lipid Research*, **23**, 115–67.

Weissler, H.E. (1965) The maximum density temperature of malt beverages. *Journal of the American Society of Brewing Chemists*, **23**, 167–9.

Welhoener, H.J. (1954) Continuous fermentation and maturation process for beer. *Brauwelt*, **104**, 624–6.

Wellman, A.M. & Stewart, G.G. (1973) Storage of brewing yeasts by liquid nitrogen refrigeration. *Applied Microbiology*, **26**, 577–83.

Wenzel, T.J., Luttik, M.A.H., van den Berg, J.A. & Steensma, H.Y. (1993) Regulation of the *PDA1* gene encoding the E1 alpha subunit of the pyruvate dehydrogenase complex from *Saccharomyces cerevisiae*. *European Journal of Biochemistry*, **218**, 405–11.

Wemer-Washburne, M., Braun, E.L. Crawford, M.E. & Peck, V.M. (1996) Stationary phase in *Saccharomyces cerevisiae*. *Molecular Microbiology*, **19**, 1159–66.

Westner, H. (1999) KHS two-tank system for safe yeast propagation. *Brewing and Distilling International*, September, 24.

Wheals, A.E. (1987) Biology of the cell cycle in yeasts. In *The Yeasts*, 2nd edition, (eds A.H. Rose and J.S. Harrison), vol. 1, pp. 283–390. Academic Press, London.

Wheals, A.E. (1995) Introduction. In *The Yeasts*, 2nd edition, (eds A.H. Rose, A.E. Wheals and J.S. Harrison), vol. 6, pp. 1–5. Academic Press, London.

White, F.H. (1994) Good hygiene practice in response to legal requirements. *Proceedings of the European Brewery Convention Symposium*, Monograph **XXI**, Nutfield, 2–10.

White, F.H. & Portno, A.D. (1978) Continuous fermentation by immobilised brewer's yeast. *Journal of the Institute of Brewing*, **84**, 228–30.

Whitear, A.I. & Crabb, D. (1977) High gravity brewing – concepts and economics. *The Brewer*, **63**, 60–63.

Whiting, G.C. (1976) Organic acid metabolism by yeasts during fermentation of alcoholic beverages – a review. *Journal of the Institute of Brewing*, **82**, 84–92.

Whiting, M.S., Chrichlow, M., Ingledew, W.M. & Ziola, B. (1992) Detection of *Pediococcus* spp. In brewing yeast by a rapid immunoassay. *Applied and Environmental Microbiology*, **58**, 713–16.

Whiting, M.S., Gares, S.L., Ingledew, W.M. & Ziola, B. (1999) Brewing spoilage lactobacilli detected using monoclonal antibodies to bacterial surface antigens. *Canadian Journal of Microbiology*, **45**, 51–8.

Whitworth, C. (1978) Technological advances in high gravity fermentation. *Proceedings of the European Brewery Convention Symposium*, Monograph **V**, Zoeterwoude, 155–67.

Wickner, R.B. (1991) Methods in classical genetics. In *Saccharomyces*, (eds M.F. Tuite and S.G. Oliver), Biotechnology Handbooks, vol. 4, pp. 101–47. Plenum Press, New York and London.

Wickner, R.B. (1997) *Prion Diseases of Mammals and Yeast: Molecular Mechanisms and Genetic Features*. Medical Intelligence Unit, Landes Bioscience, Springer Verlag, Heidelberg, Germany.

Wickner, R.B., Taylor, K.L., Edskes, H.K., Maddelein, M-L., Moriyama, H. & Roberts, B.T. (1999) Prions in *Saccharomyces* and *Podospora* spp: protein-based inheritance. *Microbiology and Molecular Biology Reviews*, **63**, 844–61.

Wicksteed, B.L., Collins, I., Dershowitz, A. *et al.* (1994) A physical comparison of chromosome III in six strains of *Saccharomyces cerevisiae*. *Yeast*, **10**, 39–57.

Wiemken, A. (1976) Isolation of vacuoles from yeast. *Methods in Cell Biology*, **12**, 99–110.

Wiemken, A., Matile, P. & Moor, H. (1970) Vacuolar dynamics in a synchronously budding yeast. *Archives for Microbiology*, **70**, 89–103.

Wiesmuller, L. & Wittinghofer, F. (1994) Signal transduction pathways involving RAS. *Cellular signalling*, **6**, 247–67.

Wightman, P., Quain, D.E. & Meaden, P.G. (1996) Analysis of production brewing strains of yeast by DNA fingerprinting. *Letters in Applied Microbiology*, **22**, 85–9.

Wilcocks, K.L. & Smart, K.A. (1995) The importance of surface charge and hydrophobicity for the flocculation of chain-forming brewing yeast strains and resistance of these parameters to acid washing. *FEMS Letters*, **134**, 293–7.

Wiles, A.E. (1953) Identification and significance of yeasts encountered in the brewery. *Journal of the Institute of Brewing*, **59**, 265–84.

Wilkinson, N.R. (1991) Process plant design. *Ferment*, **4**, 388–401.

Willetts, A. (1988) Use of immobilised microbial cells for accelerated maturation of beer. *Biotechnology Letters*, **7**, 473–8.

Williams, J. (1990) $CO_2$ recovery at Charles Wells Brewery. *Brewing and Distilling International*, **21**, 21–5.

Williams, R.P. & Ramsden, R. (1963) Continuous fermentation process and apparatus for beer production. British patent 926 847.

Williamson, A.G. & Brady, J.T. (1965) Development and operation of a commercial continuous brewery. *Technical Quarterly of the Master Brewers Association of the Americas*, **2**, 79–84.

Wills, C. (1990) Regulation of sugar and ethanol metabolism in *Saccharomyces cerevisiae*. *CRC Critical Reviews in Biochemistry and Molecular Biology*, **25**, 245–80.

Wills, C. (1996) Some puzzles about carbon catabolite repression in yeast. *Research in Microbiology*, **147**, 566–72.

Wilson, R.J.H. (1975) Oxygenation and decarbonation. *Brewers' Guardian*, October, 47–50.

Windig, W., Hoog, G.S. de & Haverkamp, J. (1981/1982) Chemical characterisation of yeasts and yeast-like fungi by factor analysis of their pyrolysis-mass spectra. *Journal of Analytical and Applied Pyrolysis*, **3**, 213–20.

van de Winkel, L., McMurrough, I., Evers, G., van Beveren, P.C. & Masschelein, C.A. (1995) Pilot scale evaluation of silicon carbide immobilised yeast systems for continuous alcohol-free beer production. *Proceedings of the European Brewery Convention*, Monograph **XXIV**, Espoo, 90–96.

van de Winkel, L., van Beveran, P.C., Borremans, E., Goossens, E. & Masschelein, C.A. (1993) High performance immobilised yeast reactor design for continuous beer fermentation. *Proceedings of the 24th Congress of the European Brewery Convention*, Oslo, 307–14.

van de Winkel, L., van Beveren, P.C. & Masschelein, C.A. (1991) The application of an immobilised yeast loop reactor to the continuous production of alcohol-free beer. *Proceedings of the 23rd Congress of the European Brewery Convention*, Lisbon, 577–84.

Winkler, K., Kienle, I., Burgert, M., Wagner, J.C. & Holzer, H. (1991) Metabolic regulation of trehalose content of vegetative yeast. *FEBS Letters*, **291**, 269–72.

Winzeler, E.A., Shoemaker, D.D., Astromoff, A. *et al.* (1999) Functional characterisation of the *S. cerevisiae* genome by gene deletion and parallel analysis. *Science*, **285**, 901–6.

de Witt, R.S. & Hewlett, P.J.H. (1974) A composite cellar block for fermentation and cold storage using large vertical tanks. *Proceedings of the 13th Convention of the Institute of Brewing (Australia & New Zealand Section)*, Queensland, 113–20.

Wogan, R. (1992) Process control and MIS – the ACCOS family. *Brewing and Distilling International*, August, 20–21.

Woldringh, C.L., Fluiter, K. & Huls, P.G. (1995) Production of senescent cells of *Saccharomyces cerevisiae* by centrifugal elutriation. *Yeast*, **11**, 361–9.

Wolfe, K. & Shields, D. (1997) Molecular evidence for an ancient duplication of the entire yeast genome. *Nature*, **387**, 708–13.

Wood, K.A., Quain, D.E. & Hinchliffe, E. (1992) The attachment of brewing yeast to glass. *Journal of the Institute of Brewing*, **98**, 325–7.

Woodward, A.M. & Kell, D.B. (1990) On the non-linear dielectric properties of biological systems, *Saccharomyces cerevisiae*. *Bioelectrochemistry and Bioenergetics*, **24**, 83–100.

Woodward, A.M. & Kell, D.B. (1991) Dual-frequency excitation: a novel method for probing the non-linear properties of biological systems, and its application to suspensions of *Saccharomyces cerevisiae*. *Bioelectrochemistry and Bioenergetics*, **25**, 395–413.

Wright, D.M., Thompson, A.N., Hodgson, J.A. & Silk, N.A. (1994) The use of yeast fingerprinting (CHEF) analysis as a Quality Assurance test in the brewing industry. *Proceedings of the 4th Aviemore Conference on Malting, Brewing and Distilling*, (eds I. Campbell and F.G. Priest), pp. 349–52. Institute of Brewing, London.

Wright, R. & Rine, J. (1989) Transmission electron microscopy and immunocytochemical studies of yeast. *Methods in Cell Biology*, **31**, 473–513.

Wright, R.M., Repine, T. & Repine, J.E. (1993) Reversible pseudohyphal growth in haploid *Saccharomyces cerevisiae* is an aerobic process. *Current Genetics*, **23**, 388–91.

Wright, R.M., Simpson, S.L. & Lanoli, B.D. (1995) Identification of a low specificity, oxygen, heme and growth phase regulated binding activity in *Saccharomyces cerevisiae*. *Biochemical and Biophysical Research Communications*, **216**, 458–66.

Xu, J., Boyd, C.M., Livingston, E., Meyer, W., Madden, J.F. & Mitchell, T.G. (1999) Species and genotypic diversities and similarities of pathogenic yeasts colonising women. *Journal of Clinical Microbiology*, **37**, 3835–43.

Yamagishi, H., Otsuta, Y., Funahashi, W., Ogata, T. Sakai, K (1999) Differentiation between brewing and non-brewing yeasts using a combination of PCR and RFLP. *Journal of Applied Microbiology*, **86**, 505–13.

Yamamura, M., Takeo, K. & Kamihara, T. (1991) *Saccharomyces* yeast cells grown at elevated temperatures are susceptible to autolysis. *Agricultural and Biological Chemistry*, **55**, 2861–4.

Yamano, S., Tomizuka, K., Sone, H. *et al.* (1995) Brewing performance of a brewer's yeast having alpha-acetolactate decarboxylase from *Acetobacter aceti* ssp. *xylinum* in brewer's yeast. *Journal of Biotechnology*, **39**, 21–6.

Yamauchi, H., Yamamoto, H., Shibano, Y., Amaya, N. & Saeki, T. (1998) Rapid methods for detecting *Saccharomyces diastaticus*, a beer spoilage yeast, using the polymerase chain reaction. *Journal of the American Society of Brewing Chemists*, **56**, 58–63.

Yamauchi, Y. & Kashihara, T. (1995) Kirin immobilised system. *Proceedings of the European Brewery Convention Symposium*, Monograph **XXIV**, Espoo, 99–177.

Yamauchi, Y., Kashihara, T., Murayama, H. *et al.* (1994) Scale-up of immobilised bioreactor for continuous fermentation of beer. *Technical Quarterly of the Master Brewers Association of the Americas*, **31**, 90–94.

Yamauchi, Y., Okamoto, T., Murayama, H., Nagara, A. Kashihara, T. (1995) Rapid fermentation of beer using an immobilised yeast multistage bioreactor – control of sulphite formation. *Applied Microbiology & Biotechnology*, **53**, 277–83.

Yamauchi, Y., Okamoto, T., Murayama, H. *et al.* (1995) Rapid fermentation of beer using an immobilised yeast multistage bioreactor system – balance control of extract and amino acid uptake – *Applied Microbiology & Biotechnology*, **53**, 261–76.

Yarrow, D. (1998) Methods for the isolation, maintenance and identification of yeasts. In *The Yeasts, A Taxonomic Study*, 4th edition, (eds C.P. Kurtzman and J.W. Fell), pp. 77–100. Elsevier, Amsterdam.

Yoda, A., Kontani, T., Ohshiro, S., Yagishita, N., Hisatomi, T. & Tsuboi, M (1993) DNA-length polymorphisms of chromosome III in the yeast *Saccharomyces cerevisiae*, *Journal of Fermentation and Bioengineering*, **75**, 395–8.

Yoshihisa, T., Ohsumi, Y. & Anraku, Y. (1988) Solubilisation and purification of α-mannosiadase, a marker enzyme of vacuolar membranes in *Saccharomyces cerevisiae*. *Journal of Biological Chemistry*, **263**, 5158–63.

Yoshioka, K. & Hashimoto, N. (1982a) Ester formation by alcohol acetyl transferase from brewer's yeast. I. Properties of alcohol acetyl transferase. *Reports of the Research Laboratory of the Kirin Brewing Company*, **25**, 1–6.

Yoshioka, K. Hashimoto, N. (1982b) Ester formation by alcohol acetyl transferase from brewer's yeast. II. Purification of acetyl alcohol transferase. *Reports of the Research Laboratory of the Kirin Brewing Company*, **25**, 7–12.

Yoshioka, K. & Hashimoto, N. (1984) Ester formation by alcohol acetyl transferase from brewers yeast. IV. Cellular fatty acid and ester formation by brewers yeast. *Reports of the Research Laboratory of the Kirin Brewing Company*, **27**, 23–30.

Young, E.T. & Pilgrim, D. (1985) Isolation and DNA sequence of *ADH3*. A nuclear gene encoding the mitochondrial isozyme of alcohol dehydrogenase in *Saccharomyces cerevisiae*. *Molecular and Cellular Biology*, **5**, 3024–34.

Young, M., Davies, M.J., Bailey, D. *et al.* (1998) Characterisation of oligosaccharides from an antigenic mannan of *Saccharomyces cerevisiae*. *Glycoconjugates Journal*, **15**, 815–22.

Young, T.W. (1981) The genetic manipulation of killer character into brewing yeast. *Journal of the Institute of Brewing*, **87**, 292–5.

Young, T.W. (1987) Killer yeasts. In *The Yeasts*, 2nd edition, (eds A.H. Rose and J.S. Harrison), vol. 2, pp. 131–64. Academic Press, London.

Young, T.W. & Hosford, E.A. (1987) Genetic manipulation of *Saccharomyces cerevisiae* to produce extracellular protease. *Proceedings of the 21st Congress of the European Brewery Convention*, Madrid, 521–8.

Younis, O.S. & Stewart, G.G. (1998) Sugar uptake and a subsequent ester and higher alcohol production by *Saccharomyces cerevisiae*. *Journal of the Institute of Brewing*, **104**, 255–64.

Younis, O.S. & Stewart, G.G. (1999) Effect of malt wort, very high gravity malt wort, and very high gravity adjunct wort on volatile production in *Saccharomyces cerevisiae*. *Journal of the American Society of Brewing Chemists*, **57**, 39–45.

Zamaroczy de, M. & Bernardi, G (1986) The primary structure of the mitochondrial genome of *Saccharomyces cerevisiae* – a review. *Gene*, **47**, 155–77.

Zinser, E., Paltauf, F. & Daum, G. (1993) Sterol composition of yeast organelle membranes and sub-cellular distribution of enzymes involved in sterol metabolism. *Journal of Bacteriology*, **175**, 2853–8.

Zitomer, R.S. & Lowry, C.V. (1992) Regulation of gene expression by oxygen in *Saccharomyces cerevisiae*. *Microbiological Reviews*, **56**, 1–11.

Zolan, M.E. (1995) Chromosome-length polymorphism in fungi. *Microbiological Reviews*, **59**, 686–98.

# Index